ELSEVIER'S DICTIONARY OF THE GENERA OF LIFE

ELSEVIER'S DICTIONARY OF THE GENERA OF LIFE

compiled by

Dr. Samuel A. Tsur (Mansoor), M.D.
Bombay - London - Akko (Acre), Israel

1999
ELSEVIER
Amsterdam – Lausanne – New York – Oxford – Shannon – Singapore – Tokyo

ELSEVIER SCIENCE B.V.
Sara Burgerhartstraat 25
P.O. Box 211, 1000 AE Amsterdam, The Netherlands

First edition 1999

Library of Congress Cataloging in Publication Data
A catalog record from the Library of Congress has been applied for.

ISBN: 0-444-82905-9

♾ The paper used in this publication meets the requirements of ANSI/NISO Z39.48-1992 (Permanence of Paper).
Printed in The Netherlands.

This book is dedicated to my dearest wife **RACHELLE**
and our three loyal children
Mr. Maurice Tsur, Dr. Daniel S. Tsur D.V.M.
and
our charming daughter, **Mrs. Muzie Olshevitzky**

Preface

This dictionary is the alphabetical compilation of most of the recognized Genera of Life on earth, of the Animal and Vegetable kingdoms, including bacteria and other parasitic forms of existence. Besides the medical profession, this book should be of equal interest to students of botany and zoology.

The genera are listed with at least one example of their respective species, each described in detail. Wherever possible, the family to which each genus belongs has also been included. More than 5,280 genera have been tabulated in this project, and the list is far from complete! An index of the more common or lay names of the various genera has been included, to enable the student or reader to more easily identify the scientific or Latin name of the specific genus searched for.

Finally it must be recalled that a reference book of this scope has its specific limitations, in that however exhaustive the information within its covers, there is no limit whatsoever to the addition of new names and fresh discoveries with regard to the Genera of Life.

Dr. Samuel A. Tsur (Mansoor), M.D.
Akko, Israel

A

Abelia A genus of late blooming plants of the Honeysuckle family, *Caprifoliaceae*. e.g. A. chinensis (= A. rupestris) - A small shrub with short-stemmed, oval-lanceolate leaves and masses of small, tubular, white, rose-tinted, fragrant flowers. Used in shrubberies or to form loose edges. Place of origin: Central and eastern China; introduced into Europe in 1844.

Abelpharus A genus of Reptiles of the Skink family, *Scincidae* (order of Lizards and Snakes, *Squamata*). e.g. A. kitaibelii - The Snake-eyed Skink, 10-11 cm long; unlike most skinks it lays eggs. Distribution: Balkans, Aegean islands, and western Asia.

Abies A genus of Trees of the Pine family, *Pinaceae*. e.g. A. balsamae - Balsam of Fir or Canada Balsam. Used in microscopy as a mounting medium. A. canadensis (= Tsuga c.) - The Hemlock Spruce (see **Tsuga**).

Abietinaria A genus of Sea Fir animals, colonial relatives of the Sea Anemones. e.g. A. abietina - size up to 30 cm (1 ft); sturdy colonies; simple stalk with alternate branches. Attaches itself to rocks and mussel shells.

Abra A genus of Lamelli branch Molluscs of the superfamily *Tellinacea*. These deposit-feeding species have long, separate, inhalant and exhalant siphons and are also very mobile, with a large foot. Burrowing and also "netling" in crevices, they bore into rock and wood. e.g. A. nitida - Glossy Furrow Shell; size 1.5 cm; long and oval. Normally lives in mud below the tideline, especially in estuaries or in less salty water. Native in Britain.

Abramis A genus of Fishes of the Minnow and Carp family, *Cyprinidae*. e.g. A. brama - The Common Bream; 30-50 cm long; a freshwater fish found in the lower reaches of large rivers, valley reservoirs and lakes. Native in Britain and a large part of Europe (see **Carassius**).

Abraxas A genus of Insects of the Geometer or Looper Moth family, *Geometridae*. e.g. A. grossulariata - The Magpie Moth; size up to 40 mm; it has white wings with black spots and two yellow stripes on the front pair. The caterpillars live on gooseberries, and are known as loopers or measuring worms, as they move by humping their bodies into a loop then throwing themselves forward. Distribution: the Tropics, as well as the Neoarctic region. Native in Britain.

Abroma A genus of plants of the family *Sterculiaceae*. e.g. A. augusta - Devil's Cotton; Olatkambal. Used in India as an emmenagogue.

Abrostola A genus of Moths of the order *Lepidoptera*. e.g. A. triplasia - it has a greyish-black coloring, a perfect camouflage, resembling a piece of mud or bird's droppings. Its green or pink caterpillar lives on nettles, and is very abundant in Central Europe.

Abrus A genus of plants of the Pea family, *Leguminosae*. e.g. A. precatorius - Abrus Seed; Indian Liquorice; Prayer Beads. Contains the toxic albumin abrin (a phytotoxin). Deaths of children have occurred from eating one or more seeds.

Absidia A genus of pathogenic fungi. e.g. A. corymbifera - a species causing mycosis in man.

Abudefduf A genus of Fishes of the Damsel family. e.g. A. saxatilis vaigiensis - The Sergeant Major fish, so-called because of its stripes; dark-blue to grey; found in various regions. Especially common in the Red Sea.

Abutilon A genus of Trees and Shrubs of the Mallow family, *Malvaceae*. e.g. A. hybridum - A hybrid shrub obtained from crossing the Brazilian species A. striatum with A. darwinii. A tropical plant frequentlygrown in greenhouses in many varieties. The flower is bell-shaped and often yellow.

Acacia A genus of Trees or Bushes of the Pea family, *Leguminosae*. e.g. A. arabica - The Gum Arabic tree; its bark, Acacia Bark, contains stannic and gallic acids. Used as an astringent in gargles and lotion.

Acalypha A genus of Shrubs of the family *Euphorbiaceae*. e.g. A. indica - The Indian Acalypha plant. It contains the alkaloid acalyphine - an expectorant and emetic.

Acanthamoeba A genus of free-living Amoebae, capable of causing meningoencephalitis in humans.

Acantharus A genus of Fishes (subphylum *Vertebrata*). e.g. A. leucosternon - The Surgeon-fish.

Acanthaster A genus of Starfishes of the class *Asteroidea*. e.g. A. planci (= A. plancii) - A species of carnivorous echinoderms, the Poisonous Starfish or Crown of Thorns, the most destructive starfish in the world, found in the Indo-Pacific region and the Red Sea. This giant asteroid feeds on the polyps which make up coral reefs.

Acanthia A genus of insects, the Bedbugs. e.g. A. lecturalia - The Common Bedbug (see **Cimex**).

Acanthinula A genus of Molluscs of the family *Valloniidae*. e.g. A. aculeata - a species found in forests amidst damp leaves, in decaying wood, etc.

Acanthisitta A genus of New Zealand Wrens of the family *Acanthisittidae* = *Xenicidae*. These birds are also known by the generic name, *Xenicus*. e.g. A. chloris - The Rifleman, the smallest and best-known species of the New Zealand wrens. It is a yellowish-green bird feeding on insects and spiders, and lives in open forests up to 4000 feet.

Acanthocardia A genus of Molluscs of the Cockle family, *Cardiidae*. e.g. A. tuberculata - The Rough Cockle, an edible species found along the Atlantic coast of Europe and the Mediterranean Sea.

Acanthocephalus A genus of Acanthocephalans or thorny-headed worms of the phylum *Acanthocephala*. The species is parasitic in amphibians and fishes having no digestive tract, circulatory system or sense-organs.

Acanthochaenus A genus of Pricklefishes of the family *Stephanoberycidae*. These are elongated fishes with large eyes, the head being covered with spines, and the body with spiny scales (see **Stephanoberyx**).

Acanthocheilonema A genus of filarial nematodes. e.g. A. perstans (= *Dipetalonema* p.; *Filaria* p.) - A species causing acanthocheilonemiasis. A. streptocerca - a species causing elephantiasis.

Acanthocinus A genus of Long-haired Beetles of the family *Cerambycidae*. e.g. A. aedilis - one of the first spring Cerambycids, appearing on pine logs in early March. It blends in color completely with the log. Found in Europe and East Asia.

Acanthocybium A genus of fast-swimming Fishes (subphylum *Vertebrata*). e.g. A. solandri - The Wahoo fish, a fast-swimming fish of the tropical Atlantic, travelling in open water at speeds exceeding 70 km per hour.

Acanthodactylus A genus of Birds of the Swallow family, *Hirundinidae*. e.g. A. scutellatus - The Sand Swallow, a small swift-flying migrating bird.

Acantholienon A genus of plants of the Sea-lavender family, *Plumbaginaceae*. e.g. A. venustum - The Prickly Thrift, a perennial species found in the Mediterranean and Near East regions.

Acanthophacetus A genus of small fishes, better known by the generic name, *Lebistes*, of the Guppy family, *Poeciliidae*. e.g. A. reticulatus (= *Lebistes* r.) - The Guppy Fish (see **Lebistes**).

Acanthophis A genus of Reptiles of the Elapid Snake family. Also known by the generic name, *Acanthopis*. e.g. A. antartica (= A. antarticus) - The Death Adder, found in Australia and New Guinea.

Acanthopis See **Acanthophis**.

Acanthopsetta A genus of Fishes of the Spiny Flounder family, *Hippoglossidae*. They have a large mouth and a straight lateral line, and are occasionally caught in the seas off Japan.

Acanthoptilum A genus of Sea Pens of the order *Pennatulacea*. e.g. A. gracile - this coelenterate forms fleshy colonies which originate from one polyp. This is anchored by a long stalk in sandy mud, and is found in shallow bays on the American Pacific coast.

Acanthozostera A genus of Chiton Molluscs of the subclass *Polyplacophora* (= *Loricata*) In these species lunar periodicity occurs in spawning.

Acanthus A genus of plants of the family *Acanthaceae*. e.g. A. spinosus - The Bear's Breech, an herbaceous perennial, with deep-green, pinnatifid, spiny leaves, and rosy-white flowers with flaring lips and purplish-green hoods. This is the plant whose leaf pattern inspired the design of the Greek Corinthian columns in the 5th century B.C. Place of origin: Southern Europe.

Acarapis A genus of Mites of the order *Acari*. e.g. A. woodi - The mite of the honey bee, causing Isle of Wight Disease, with paralysis of the flight muscles of the honey bee.

Acarus A genus of small, often ectoparasitic Mites of the order *Acari* (*Acarina*) of ticks and mites. e.g. A. hordei - The Barley Bug, infecting the skin of man. A. scabei (= A. scabiei; A. siro; *Sarcoptes* s.; S. scabiei) - The Itch Mite, which bores beneath the skin, forming canaculi or burrows, transferred mainly by actual contact, and causing in humans and various domestic animals, the skin disease known as scabies or sarcoptic mange. The female is larger than the male, but both are under 0.05

cm (0.02 in) long. Distribution: Worldwide (see **Sarcoptes**).

Accipeter See **Accipiter**.

Accipiter A genus of Birds of the Old World Eagle, Harrier, Hawk and Vulture family, *Accipitridae* (True Hawk subfamily, *Accipitrinae*). Also known by the generic name, *Accipeter*. e.g. A. gentilis (= *Accipeter* g.) - The Goshawk, the "Gentle Falcon" of olden times, a notorious and much-persecuted bird of prey, up to 60 cm (2 ft) long (the hen being much larger than the cock), and with a wing-span of over 90 cm (3 ft). A resident bird, nestling in densely wooded coniferous forests, where it leads a secretive life. Distribution: Northwest Africa, Europe (except Britain, Iceland and Portugal), Asia, and North America.

Acentropus A genus of Insects of the Moth family, *Pyraustidae*. The larvae and pupae are aquatic, and live in tubes composed of fragments of the leaves of water plants.

Acer A genus of Shrubs and Trees of the Sycamore or Maple family, *Aceraceae*. e.g. A. japonicum - The Fullmoon Maple; Japanese Maple - a slow-growing shrub or small tree, 6-9 m (20-30 ft) tall, with soft, green palmate leaves which color to a rich crimson in autumn. Place of origin: Japan and Central China.

Aceraria A genus of Nematode worms. e.g. A. spiralis - a species found in the esophagus of fowls.

Aceras A genus of plants of the Orchid family, *Orchidaceae*. e.g. A. anthropophorum - The Man Orchid, a perennial plant, 20-30 cm high; greenish-yellow flowers, often edged with reddish brown; lips with bifid, central, and two lateral lobes; grows in scrub, open woods and grassland. Native in Britain.

Acerina A genus of Fishes of the Perch family, *Percidae*. e.g. A. cornua (= A. cernua) - The Ruffe or Pope fish, up to 25 cm (10 in) long; a small perch-like fish; it has a body with dark spots, an

4

undivided or fused dorsal fin with 13-16 hard and 11-15 soft rays and pelvic fins set far forward. Found in deeper, calm, fresh and brackish waters. Distribution: east Britain, central and northern Europe.

Acestrura A genus of Birds of the Hummingbird family, *Trochilidae*. This genus has the highest energy output per unit of weight of any living warm-blooded animal. It has been calculated that a hummingbird would use about 155,000 calories a day in energy, compared (theoretically) weight for weight, with 3,500 calories for an average male human being. e.g. A. bombus - The diminutive Bee Hummingbird, the average weighing about 2.2 g.

Acetobacter A genus of microorganisms of the family *Pseudomonadaceae*. Ellipsoid torod-shaped cells, important for their role in the production of vinegar. e.g. A. aceti (= *Mycoderma* a.).

Achatina A genus of Gastropod Molluscs of a family of large Snails. e.g. A. achatina - The Giant African Snail, the largest known landsnail, originally found only in East Africa and Madagascar, but today worldwide. It reproduces at a staggering rate; it has been estimated that one snail can theoretically produce 11,000,000 descendants in 5 years.

Acherontia A genus of Insects of the Hawk Moth family, *Sphingidae*. e.g. A atropos - The Death's Head Hawk Moth, size 12 cm; named after Atropos, eldest of the three fates in Greek mythology, and characterized by the startling pattern on the thorax, which resembles a human skull. This insect makes a squeaking noise if handled. Distribution: Europe, Africa.

Acheta A genus of Insects of the True Cricket family, *Gryllidae*. e.g. A. domesticus - The European House Cricket, an insect also found in the United States. Lives in meadows, woods and sandy places.

Achillea A genus of plants of the Daisy family, *Compositae*. e.g. A. millefolium - Milfoil; Nosebleed; Yarrow plant. A perrenial plant, 15-30 cm high, with white or pink ray-florets. It contains aconitic acid (= achilleic acid), an alkaloid (achilleine), a bitter principle (ivain), and tannin. Has been used in medicine from early times as a diaphoretic and hemostatic. Grows in meadows and pastureland. Distribution: Europe; native in Britain.

Achimenes A genus of plants of the family *Gesneriaceae*. e.g. A. erecta - an erect plant with opposite, hairy leaves and underground scaly, cone-like tubers. The flowers have tubular corollas with spreading lobes.

Achirus A genus of Fishes of the Sole family, *Soleidae*. e.g. A. achirus - a fish of commerce. It migrates over 700 miles up the Amazon River.

Achnatherum A genus of plants of the Grass family, *Gramineae*. e.g. A. calamagrostis - Silver Rough-grass, a perennial plant, 30-120 cm high; panicle up to 30 cm long; lemma long-hairy on back. Grows on stony mountain pastures, dry turf, scree and rocks. Distribution: South and Southcentral Europe.

Achorion A genus of Fungi. e.g. A. schonleinii (= *Tricopyton* s.) - the favus fungus.

Achras A genus of Tropical Trees and Shrubs of the Sapodilla family, *Sapotaceae*. e.g. A. sapota (= A. zapota) - a tropical American evergreen tree with alternate, simple leaves, durable reddish wood and a milky or sticky, latex. It yields the chicle gum used in chewing-gum, and an edible sweet-fleshed fruit, the Sapodilla Plum (Span: zapotella; Fr: nahuatyl).

Achromobacter A genus of gram-negative coccobacilli, now included in the genus *Acinetobacter*. e.g. A. anitratus (= *Acinetobacter* a.) - see **Acinetobacter**.

Achyranthes A genus of plants of the Amaranth family, *Amaranthaceae*. e.g.

A. aspera - The Atkumah or Chirchira plant, used in India as a decoction diuretic.

Acilius A genus of asects of the True Water Carnivorous Beetle family, *Dytiscidae*. e.g. A. sulcatus - The Ground Acilius, 16-18 mm long; the pronotum has two black bands. A diving beetle found in pools, ponds and calm waters of rivers; also in large puddles. Native in Britain.

Acineta A genus of Suctorian Holotrichs of the order *Suctorida*. The adult form is sessile and without cilia. These protozoans have no mouth, and the prey is captured by long knob-ended tentacles, which paralyze it and suck it dry.

Acinetobacter A genus of gram-negative coccobacilli, of the family *Brucellaceae*, and formerly classified as, or including the genera *Achromobacter*, *Herellea*, *Mima*, *Moraxella* and of some species of *Pseudomonas* (sometimes known as the *Mimaherellea* group). e.g. A. anitratus (= A. calcoaceticus; *Achromobacter* a.; *Bacterium* anitratum; *Herellea* vaginicola; *Moraxella* glucidolytica; *Pseudomonas* calcoacetica) - an opportunistic pathogen, being part of the normal flora of the vagina, skin, respiratory and gastro-intestinal tracts. Frequently mistaken for neisseria species in smears of spinal fluid and pus. Associated with conjunctivitis, meningitis, septicemia, gastro-enteritis, urinary tract infections, vaginitis, etc. A. lwoffi (= *Achromobacter* l.; *Moraxella* l.; *Mima* polymorpha) - part of the normal flora of the skin, mouth and external genitalia, but may be an opportunistic pathogen. Commonly isolated from sputum and urine, but may not be etiologically significant in local infections. May cause septicemia, subacute bacterial endocarditis, and pneumonia in debilitated individuals.

Acinonyx A genus of Mammals of the Cat family, *Felidae*. e.g. A. jubatus - The Cheetah or "Hunting Leopard", the fastest terrestrial animal over short distances, with a probable maximum speed of 96 km/h (60 miles/h) over level ground. Found in East Africa, Iran and Afghanistan.

Acinos A genus of plants of the Thyme family, *Labiatae*. e.g. A. arvensis - Basil-thyme, an annual to perennial plant, 5-60 cm high; aromatic, with bluish-violet flowers up to 10 mm long. Grows in arable fields, short grassland, and on rocks. Native in Britain.

Acipenser A genus of Fishes of the Sturgeon family, *Acipenseridae*. Also known by the generic name, *Huso*. e.g. A. huso (= *Huso* h.) - The Russian Sturgeon or Giant Beluga fish, the largest bony fish in the world, from which is prepared Isinglass or Sturgeon's bladder. From its large number of eggs (roe), black caviar is processed (see **Huso**).

Aciptilia A genus of Insects of the Plume Moth family (order *Lepidoptera*). e.g. A. balliodactyla - The Grape Plume Moth.

Aclypea A genus of Insects of the Carrion Beetle family, *Silphidae*. e.g. A. opaca - this species differs from others in the family in that these are plant-feeders, subsisting chiefly on various grasses, even weeds. Sometimes regarded as a pest, for it feeds on sugar beet. Distribution: Europe, North America.

Acmaea A genus of Gastropod Molluscs of the family *Acmaeidae*. e.g. A. dorsuosa - The Japanese limpet, a marine snail known to have a life-span of 17 years.

Acneta A genus of Insects of the True Cricket family, *Gryllidae*. e.g. A. domestica - The House Cricket; male 19 mm; female 21 mm; brown head with yellow spots; female with ovipositor 11-14 mm long. Native in Britain.

Acocanthera See **Acokanthera**.

Acokanthera A genus of plants of the Periwinkle family, *Apocynaceae*. e.g. A. schimperi (= *Acocanthera* s.) - From the plant wood is obtained Ouabain or

Strophanthin G, a cardiac glycoside, with action similar to that of digitalis (see **Strophanthus**).

Acomys A genus of Mammals of the Rodent Mouse family, *Muridae*. e.g. A. caharinus - The Spiny Mouse; size up to 12.8 cm; coat light brown; bristly pointed hairs down center of the back; long whiskers; brittle tail, 12 cm long; insect-eating. Habitat: in semi-deserts on the limestone rocks of Crete and Cyprus.

Aconitum A genus of plants of the Buttercup family, *Ranunculaceae*. e.g. A. napellus - Common Monkshood or Aconite plant, a leafy, herbaceous, perennial species, 50-150 cm high, with hood-shaped upper perianth-segment, and extremely poisonous root-stocks. The dried root (Aconite Root; Monkshood Root; Wolfsbane Root) contains the alkaloid aconitine, used in the past as a powerful arrow poison. Formerly used extensively in linaments in the treatment of neuralgia, sciatica and rheumatism. Place of origin: widely distributed in Europe and Asia; native in Britain.

Acorus A genus of plants of the Arum family, *Araceae*. e.g. A. calamus - Sweet Flag, a perennial plant, 60-125 cm high, with fragrant sword-shaped leaves and numerous flowers; grows in ponds, ditches and on banks of rivers and lakes. The root, Calamus Rhizome or Sweet Flag Root, is used as an aromatic bitter and carminative. Native in Britain since 1660. Originally a native of southern Asia and central and western North America. Introduced into Europe about the middle of the 16th century.

Acrida A genus of Insects of the Locust order, *Caelifera*. e.g. A. nasuta - a locust or grasshopper species, up to 7 cm long, with a long conical head, slender jumping legs, and very long flattened antennae. Lives on stony fields and among bushes. Distribution: the Mediterranean region.

Acris A genus of Amphibians of the typi-cal Treefrog family, *Hylidae*. These are the Cricket Frogs of the United States.

Acrobates A genus of Mammals of the Marsupial Phalanger family, *Phalangeridae*. e.g. A. pygmaeus - The Feather-tailed Glider or Pigmy Flying Phalanger, a marsupial dwarf found in Australia, barely 15 cm (6 in) long (including its bushy tail). It looks rather like a mouse, and has a membrane stretched out between the extended limbs.

Acrocephalus A genus of Birds of the Old World or True Warbler subfamily, *Sylviinae* (family *Slyviidae* - the great Warbler family, *Muscicapidae*). e.g. A. scirpaceus - The Reed-Warbler, an inconspicuous brownish-green migrant bird, about 12.5 cm (5 in) long, the commonest member of this genus, found amidst reeds bordering ponds, as well as in bushes far away from water. Their call is not pleasant, being rather like a croak. Distribution: Europe, Asia, Northwest Africa; native in Britain.

Acrocinus A genus of Insects of the Longicorn or Long-horned Beetle family, *Cerambycidae*. e.g. A. longimanus - The Harlequin Beetle; it has brown, grey, and reddish markings. The larvae are wood-boring, and sometimes damage trees used for timber. Distribution: the tropical forests of the Amazon basin.

Acronicta A genus of Insects of the Noctuid Moth family, *Noctuidae*. Also known by the generic names, *Acronycta* or *Apatele*. e.g. A. psi (= *Acronycta* p.; *Apatele* p.) - The Grey Dagger Moth; the pattern of the forewings resembles the Greek letter "psi" (ψ), hence the scientific name. Its protective coloring, dark markings on a lightgrey background, makes it resemble the bark of the trees on which it lives.

Acronycta See **Acronicta**.

Acropora A genus of True or Stony Corals of the order *Scleractinia*, also known by the generic name, *Madrepora*. The living tissue in these perforate corals is responsible for the deposition of most of the calcareous material that builds the gigantic coral reefs of all tropical seas. e.g. *Acropora* sp. - The Staghorn Coral, one species quite abundant along the Red Sea coast and the Gulf of Eilat (Akaba). A. cervicornis (= *Madrepora* c.) - a West Indian reef coral.

Acrosiphonia A genus of plants of the Green Algae family. e.g. A. arcta - Acrosiphonia, up to 8 cm high, with freely-branching uniaxial filaments, producing in addition to the ordinary branches, small hooks and coils, which unite filaments in groups. Distribution: the North Sea; also in the Baltic.

Acrotus A genus of Fishes of the Ragfish order, *Malacichthyes* (= *Icosteiformes*). These fishes have limp bodies, rather like a wet rag; their skeleton is very poorly developed and is mainly cartilaginous. One of the degenerate deep-sea forms of these fishes is this genus (i.e. *Acrotus*).

Acryllium A genus of Birds of the Guinea-fowl family, *Numididae*. e.g. A. vulturinum - The Vulturine Guinea-fowl of East Africa, the most ornamental bird species of this family, much of its plumage being cobalt blue, patterned with black and white.

Actaea A genus of plants of the Buttercup family, *Ranunculaceae*. e.g. A. spicata - The Common Baneberry, a perennial plant, 20-60 cm high; fruit, a black ovoid berry. Distribution: Europe; native in Britain.

Actaeon A genus of marine gastropods or Sea Slugs of the order *Cephalaspidae* (= *Bullomorpha*). These species inhabit sandy or muddy buttoms and retain a sizeable shell, in this genus the barrel shell, into which the animal can completely withdraw.

Actias A genus of Insects of the Emperor Moth family, *Saturniidae*. e.g. A. luna - The Luna Moth of North America, a beautiful pale green species equipped with purple-brown markings, eyespots, and long tails on its hind wings.

Actinia A genus of Anthozoan Coelenterates of the order of Sea Anemones, *Actiniaria*. e.g. A. equina - The Beadlet Anemone, height up to 3 cm; slender tentacles with red, green, or yellow columns. Distribution: Baltic and Mediterranean Seas.

Actinobacillus A genus of microorganisms of the family *Brucellaceae*; also called by the generic name, *Pseudomonas*. These are gram-negative coccobacilli. e.g. A. actinoides (= *Actinomyces* a.) - a species isolated from chronic pneumonias in calves and rats (see **Actiomyces**). A. lignieresii - the causative organism of actinobacillosis or "wooden tongue", an actinomycosis-like disease of cattle, sometimes occurring in man. A. mallei (= *Pseudomonas* m.) - a species causing glanders (see **Pseudomonas**).

Actinodiscus A genus of Sea Anemones of the class *Anthozoa*. Many species of this coelenterate are found along the Red Sea coast. They are closely related to the stony corals, having a soft hollow body with a mouth surrounded by tentacles capable of a potent sting. This is due to thousands of stinging cells that erupt when touched, paralysing the prey (usually fish and crustaceans), which is then brought into the mouth to be digested. Other genera are *Cerianthus, Cryptodendrum, Euptasia, Gyrostoma, Triactis*, etc.

Actinomyces A genus of microorganisms of the family *Actinomycetaceae*, formely known by the generic names, *Micromyces* or *Streptothrix*. These are anaerobic, gram-positive bacilli with a structural branching tendency resembling that of a fungus. e.g. A. actinoides (= *Actinobacillus* a.) (see **Actinobacillus**). A. baudetii - a species causing actinomycosis in cats and dogs. A. bovis - the specific etiological agent of actinomycosis or "lumpy jaw" in cattle. Has been isolated from the human oral cavity. A. griseus - like the *Streptomyces* g., it produces the antimicrobial agent, streptomycin. A. israelii - a specific parasite of the mouth, proliferating in necrotic tissues; occasionally causes human actinomycosis. A. madurae - the pale or colorless species of granular fungi seen on white mycetoma or White Maduromycosis (Madura Foot) - see **Madurella** A. naeslundii - has been recovered from the human mouth and saliva. Pathologically uncertain. A. odontolyticus - has been cultured from carious dentine in the mouth. A. subtropicus - a species producing the antibiotic substance albomycin (see **Streptomyces** griseoflavus).

Actinophrys A genus of Heliozoan Protozoa of the order *Heliozoa*. These protozoa are "naked" and do not possess a stiffening skeleton.

Actinoptychus A genus of Diatoms of the group *Bacillariophyta*. e.g. A. adriaticus - microscopic unicellular algae; the surface view is usually circular.

Actinosphaerium A genus of Heliozoan Protozoa of the order *Heliozoa*; the cytoplasm is divided into two concentric spheres and there is no stiffening skeleton. e.g. A. eichorni - a sun animalcule species of this genus.

Actitis A genus of Birds of the family *Scolopacidae* (Woodcocks, Sandpipers and Curlews). e.g. A. macularia - The Spotted Sandpiper, found throughout North America; a species closely allied to the Common Sandpiper (*Tringa* hypoleucos).

Actophilornis A genus of Birds of the family of Jacanas or Lilytrotters, *Jacanidae*. e.g. A. africana (= A. africanus) - The African Jacana or Lilytrotter, 12 in long, brown with a blue forehead; is found in mountain regions walking on moss-covered rocks, or in swamps.

Acuaria A genus of Filarian parasites.

e.g. A. spiralis - a species parasitic in fowls.

Aculeatus A genus of plants of the Lily family, *Liliaceae*. e.g. *Aculeatus* sp. (= *Ruscus* aculeatus) - The Butcher's Broom, a dense, evergreen, monoecious shrub, 90 cm (3 ft) high, with large scarlet berries appearing in the autumn. Its grooved stem bears stiff, dark green, flattened, spiny-pointed branchlets (cladodes) resembling leaves, which have been used to make brooms, and were once employed to sweep butchers' counters and chopping-blocks. Place of origin: Northern Asia and the Mediterranean region.

Adalia A genus of Insects of the Ladybird Beetle family, *Coccinellidae*. Also known by the generic name *Coccinella*. e.g. A. bipunctata (= *Coccinella* b.) - The Two-spotted or Two-spot Ladybird Beetle, 4-5 mm long; it has red elytra (hard anterior wings) with a black spot on each. Widely distributed over Europe and North America.

Adamsia A genus of Sea Anemones of the order *Actiniaria*. A commensal animal, always found in association with a Hermit Crab. Creamy in color, with purplish spots. Now known by the generic name, *Calliactis* (see **Calliactis**).

Adansonia A genus of tropical Trees of the family *Bombacaceae*. e.g. A. digitata - The Baobab tree, 18 m (60 ft) tall, with an extremely thick trunk. It has compound palmate leaves, and the solitary flowers are up to 15 cm (6 in) across. The fruit is edible, and a strong fiber is produced from the bark. A native of West and Central Africa.

Addax A genus of Mammals of the Horned Ungulate family, *Bovidae*, related to the ox and goat, and belonging to the Antelope subfamily, *Antilopinae*. e.g. A. nasomaculatus - The Addax Antelope of the North African deserts. It has long ribbed horns in an open spiral.

Sometimes also called *Oryx* nasomaculatus (see **Oryx**).

Addox See **Addax**.

Adelges A genus of Insects of the family *Adelgidae*. e.g. A. viridis (= A. abietis) - A species producing pineapple-shaped galls on young spruce shoots.

Adenophora A genus of plants of the Bellflower family, *Campanulaceae*. e.g. A. liliifolia - The Gland- bellflower, a perennial plant, 30-100 cm high; serrate leaves; drooping, bluish lilac flowers. A rare species growing in damp woods. Place of origin: Southeast and Central Europe.

Adenostyles A genus of plants of the Daisy family, *Compositae*. e.g. A. glabra - Glabrous Adenostyles, a perennial plant, 30-80 cm high; hairy leaf-veins; grows in mountain forests, alpine grassland, damp scree, and rock-crevices. Place of origin: southcentral Europe.

Adhatoda A genus of plants of the family *Acanthaceae*. e.g. A. vasica - The leaves of this species contain an alkaloid, vasicine (peganine) and an organic acid, adhatodic acid. They are smoked as cigarettes for the relief of asthma.

Adiantum A genus of plants of the Polypody or Fern family, *Polypodiaceae* (= *Aspleniaceae*). e.g. A. capillus-veneris - Maidenhair; Maidenhair Fern; Rock Fern Venus Hair - a perennial plant, up to 50 cm high; with unequally lobed frond-segments. Grows in crevices of basic rocks. The dried fronds of this plant are mucilaginous and expectorant, and are used as an ingredient of cough syrups in many European countries. Place of origin; South and Western Europe; distribution cosmopolitan.

Adioryx A genus of Fishes of the Grouper and Sea bass family, *Serranidae*. e.g. A. diadema - The Soldier Fish, an exclusively night fish, dwelling at the the entrance of small caves in coral reefs, and particularly active at night. A predator fish, often found in small groups, preying only on live fish. Distribution: the Red Sea.

Adleria A genus of Insects of the Gall Wasp family, *Cynipidae*. Also known by the generic name, *Cynips*. e.g. A. gallae tinctoria (= *Cynips* g. t.) - The Gall Wasp, the larvae of which stimulate the twigs of the plant *Quercus* infectoria (the Beech family, *Fagaceae*), to produce gallotannic acid. Used as an astringent in the treatment of hemorrhoids.

Adonis A genus of plants of the Buttercup family, *Ranunculaceae*. e.g. A. vernalis - The Herba Adonidis plant or Spring Pheasant's-eye, a perennial plant, 15-60 cm high; 2- or 3-pinnate, thread-like leaves; solitary, large, yellow flowers, 3-7 cm wide. Grows in heath meadows, on sunny hillocks, and pine woods. Used as a cardiac stimulant. Introduced and naturalized in Britain; widespread in Europe and western Asia.

Adoxa A genus of plants of the Moschatel family. e.g. A. moschatellina - The Moschatel, a perennial plant, 3-5 cm high; head-like inflorescences with 5 to 7 flowers. Native in Britain.

Adoxophyes A genus of Insects of the Bell Moth family, *Tortricidae*. e.g. A. orana - The Summer Fruit Tortrix Moth, a pest of apple and pear trees and cultivated plants, damaging the leaves and the fruit. It lays flat scale-like eggs in an overlapping fashion, like tiles on a roof.

Aechmea A genus of plants of the Pineapple family, *Bromeliaceae*. e.g. A. fasciata - an evergreen epiphyte; has a rosette of curved, oblong leaves marked with green and white bands. The pink flowers are subtended by pointed, rose-pink bracts and occur in a dense terminal head on an erect central stem. Used as an indoor plant. There is much confusion between this plant and the closely related *Billbergia*, and in fact this species is sometimes known as B. rhodocyanea. Distribution: Cosmopolitan. Very common in Brazil (see **Billbergia**).

Aechmophorus A genus of Birds (class *Aves*). e.g. A. occidentalis - The Western Grebe of North America; a large bird having a body temperature approximating that of a human being (38.5° C = 101.3° F).

Aëdes A genus of Mosquitoes of the family *Culicidae*, formerly known by the subgenus name *Stegomyia*. e.g. A. aegypti - a culicine species of mosquitoes transmitting dengue and yellow fever; may also be a vector of filariasis and encephalitis (see **Stegomyia**).

Aegilops A genus of plants of the Grass family, *Graminae*. The species are involved by natural hybridization in the development of bread-wheat (see **Triticum**).

Aegithalos A genus of Long-tailed Tit Birds of the family *Aegithalidae*. e.g. A. caudatus - The Long-tailed Tit or Titmouse; a small resident bird, 14 cm long (including the 7.5 cm tail), found in broad-leaved and mixed woods, gardens, and parks. It has a coat of black and pink, white apron, white head, and black eyebrows. Distribution: Europe, Siberia, and the Middle East. Native in Britain.

Aegle A genus of plants of the Rue family, *Rutaceae*. Some species are classified under the generic name, *Citrus*. e.g. A. marmelos - Bael Fruit; Bengal Quince; Indian Bael plant; is mildly astringent and is used in India in the treatment of diarrhea and dysentery. A. sepiaria (= *Citrus* trifoliata) - The Hardy Orange (see **Citrus**).

Aegolius A genus of Birds of the Typical Owl family, *Strigidae*. e.g. A. funereus - The Boreal or Tengmalm's Owl, length 25 cm; light-colored, sharply defined eye-discs; the legs are thickly covered with feathers which reach as far as the talons. A night bird of prey inhabiting deep coniferous forests at both lowland and mountainous elevations. Distribution: Central and North Europe; Asia; North America.

Aegopodium A genus of plants of the Carrot family, *Umbelliferae*. e.g. A. podagraria - Goutweed or Ground Elder, a perennial plant, 20-120 cm high; lower leaves in three segments, each divided into three. Grows in hedges and open woods, and as a weed on waste ground. Introduced and naturalized in Britain.

Aegosoma A genus of Insects of the Long-horned or Longicorn Beetle family, *Cerambycidae*. e.g. A. scabricorne - a rare European species living on deciduous trees.

Aegypius A genus of Birds of the Vulture family, *Accipitridae*. e.g. A. monachus - The European Black Vulture, Cinereous or Hooded Vulture, one of the heaviest birds of prey, weighing on the average 9.22 kg (20.3 lb); average length 103 cm (41 in), with a wing-span of 255 cm (81/2 ft). The plumage is brown, with a black neck-ruff. A resident bird found in open country, but favoring mountainous elevations, nesting in trees or on rocks. Distribution: southern Europe, the Balkans, Asia, and North Africa.

Aegyptianella A genus of parasites, e.g. A. Pullorum - A species found in the blood of fowls.

Aelia A genus of Insects of the True Bug family, *Pentatomidae*. e.g. A. acuminata - Bishop's Mitre, a common and plentiful species, named after its typical appearance and found on grass blades including cereals, on which the eggs are laid in spring.

Aelosoma A genus of Earthworms of the class *Oligochaeta*. e.g. A. kashyapi - an exceedingly small species, 1 mm long.

Aeonium A genus of plants of the Cactus family, *Crassulaceae*. e.g. A. canariense (= *Sempervivum* c.) - The Aeonium Cactus, an evergreen sub-shrub, up to 45 cm (18 in) high, forming a large bowl-shaped rosette on a short, thick stem. Grows on rockeries, or as an indoor or greenhouse plant. Place of origin: Teneriffe (Canary Islands); introduced into Europe in 1699.

Aepyceros A genus of Mammals of the Antelope subfamily, *Antilopinae* (Cattlefamily, *Bovidae*). e.g. A. melampus - The Impala, 97.5 cm (39 in) high; found mainly in East Africa. The male has lyrate horns, and its leaping powers are extraordinary.

Aepyornis A genus of large, now extinct Birds (class *Aves*). e.g. A. maximus - The huge Elephant Bird of Madagascar (extinct since c. 1660), whose eggs had a capacity of up to 8 liters.

Aequipecten A genus of Molluscs of the Scallop family, *Pectinidae*; also known by the generic name, *Chlamys*. e.g. A. opercularis (= *Chlamys* o.) - The Queen Scallop; size of shell 9 cm; valve with 18-22 ribs; these bivalve molluscs have long tentacles, and the gill plates (ctenidia) are ciliated and enlarged for food collection. Initially attached by sally, the adults become free and swim away, holding the valves horizontally. The Queen Scallop is on firm sandy, muddy or shelly sea bottoms, and is widely gathered as a popular delicacy. Distribution: along the Atlantic coast of Europe and the Mediterranean Sea.

Aerobacter A genus of microorganisms of the family *Enterobacteriaceae*; an atypical variety of *Bacillus* coli. Also known by the generic name, *Enterobacter*. e.g. A. aerogenes (= *Aero* a.; *Bacillus* lactis a.; *Enterobacter* a.) - causes urinary tract and other infections in man (see **Enterobacter**).

Aeromonas A genus of microorganisms of the family *Pseudomonadaceae*. These are gram-negative bacilli; rod-shaped cells isolated from lakes, rivers, wells, soil, or food; primarily pathogenic to fish and amphibians, but has also caused human infection. Some species also classified under the generic name, *Plesiomonas*. e.g. A. hydrophila - occurs naturally in lakes, soil, or food. Occasionally isolated from feces of asymptomatic individuals, but has been reported to cause diarrhea, wound infections, osteomyelitis, urinary tract infections, etc. A. shigelloides (= *Plesiomonas* s.) - has been isolated from the stools of asymptomatic individuals. Has been associated with diarrhea. Originally found in association with Shigella gastro-enteritis and dysentery.

Aeronautes A genus of Birds of the Swallow family, *Hirundinidae*. e.g. A. saxatalis - The White-breasted Swallow, a swift-flying migratory bird.

Aeschna A genus of Hawker Dragonflies of the family *Aeshnidae*. Better known by the generic name, *Aeshna* (see **Aeshna**).

Aeschrion A genus of plants of the family *Simarubaceae*; also known as *Picraena* or *Picrasma*. e.g. A. excelsa (= *Picraena* e.; *Picrasma* e.) - The Jamaica Quassia plant; the dried wood or Quassia Wood (= Bitter Wood; Quassiae Lignum is used as a bitter stomachic and appetizer. It contains the active principle quassin, and infusions of quassia have been used as an enema to expel threadworms, and externally as lotions for pediculosis (see **Quassia**).

Aesculus A genus of plants of the Horse Chestnut family, *Hippocastanaceae*. e.g. A. hippocastanum - The Horse Chestnut, a large, deciduos, shady tree, up to 30 m (100 ft) high; it has a thick trunk covered with a dark scaling bark, and smooth twigs bearing large brown sticky buds. The leaves are palmate, and the flowers ornamental pink or white, growing in erect conical clusters. The fruit consists of a prickly capsule containing a large, inedible seed, the horse-chestnut, which contains the glycoside Esculoside (Esculin). Used as an anticoagulant. Place of origin: a native of southeast Europe, Iran, and the Himalayas. Has been widely planted in North America, Europe, and in Britain (where it is often found semi-wild).

Aeshna A genus of Insects of the Hawker

Dragonfly family, *Aeshnidae*. Also known by the generic name, *Aeschna*. e.g. A. cyanea (= *Aeschna* c.) - The Southern Aeshna or Aeschna; size 7cm; forehead with dark spot; top of thorax dark brown, with two oval spots; color varies with age. A swift flier, with a well defined beat, over the pools and ponds where it hunts insects. One of the commonest members of the family of dragon-flies. Native in Britain; quite common in the United States.

Aethia A genus of deep-diving flying Birds (class *Aves*). e.g. A. cristatella - The Crested Auklet; reported to have been recovered from the stomachs of the Cod fish (*Gadus* callarias), caught at a depth of 60.9 m (200 ft).

Aethionema A genus of plants of the Crucifer or Cress family, *Cruciferae*. e.g. A. saxatile - Candy Mustard, a perrenial plant, 5-20 cm high; with oblong-linear leaves, white or reddish flowers, and broadly winged fruits. Grows on basic limestone rocks and screes at mountainous levels up to 1,850 m. Distribution: South and Central Europe.

Aethopyga A genus of Birds of the Sunbird family, *Nectariniidae*. e.g. A. siparaja - The Yellow-backed Sunbird, an Asiatic species. It feeds on nectar as well as small insects found inside the blossom.

Aethusa A genus of plants of the Carrot family, *Umbelliferae*. e.g. A. cynapium - Fool's Parsley, a perennial plant, 7-125 cm high; ternately 2-pinnate leaves, three bracteoles on outer side of partial, deflexed umbels. A poisonous plant, growing as a weed on cultivated ground. Native in Britain.

Afrixalus A genus of Frogs. e.g. A. fulvovittatus - The Bamboo Frog.

Afrosubulitermes A genus of diminutive termites. The smallest known termites belong to this genus, measuring about 3.5 mm (0.13 in) in length.

Agabus A genus of Insects of the Carnivorous Water Beetle family,

Dytiscidae. e.g. A. bipustulatus - Two-spotted Agabus, a carnivorous water beetle, 12 mm long, with a black body and head often with two black spots; antennae red; and long, flattened, long-haired hind legs. A true waterbeetle. Native in Britain.

Agalma A genus of Siphonophores of the order *Siphonophora*. These coelenterates are very small and transparent, and are very often unnoticed. They are found in all seas, but prefer warmer waters.

Agama A genus of Agamid Lizards of the family *Lacertidae* (= *Agamidae*). e.g. A. agama - The Common Agama of Africa (= Common African Lizard). It is polygamous and can change color rapidly.

Agamodistomum A genus of Trematode parasites. e.g. A. ophthalmobium - has been found in the crystalline lens of the human eye.

Agamofilaria A genus of filarian Nematode worms. e.g. A. georgina - a filarian worm parasite.

Agamomermis A genus of Nematode parasites. e.g. A. culicis - a species parasitic in the mosquito.

Agamonematodum A genus of minute Nematode larval parasites, causing "creeping eruption". e.g. A. migrans.

Agapanthus A genus of indoor flowering plants of the Lily family, *Liliaceae*. e.g. A. africanus (= A. umbellatus) - The African Lily (= Lily of the Nile) - a perennial plant with blue or white tubular flowers carried on large round heads or umbels. Height 75 cm (21/2 ft). Place of origin: Cape Province (South Africa); introduced into Europe in 1629.

Agapornis A genus of Birds of the Parrot family, *Psittacidae*; small short-tailed parrots, commonly known as lovebirds because of the great affection displayed between and female of the species. e.g. A. fischeri - Fischer's Lovebird, a popular pet, with its coral red bill, orange head, green body, reddish-brown wings, and green, blue-topped tail. It has a white ring round the eyes. Habitat: the forests of Northwest Tanganyka.

Agaricus A genus of Mushrooms of the family *Agaricaceae*; also known by the generic name, *Amanita*. e.g. A. campestris - the common edible mushroom. A. muscarius (= *Amanita* muscaria) - The "fly agaric" poisonous mushroom containing the deadly alkaloid muscarine (see **Anita**).

Agathis A genus of plants of the family *Pinaceae*. e.g. A. australis - The Austrial Copal or Kauri Gum plant; from this species is obtained a gum resin used in dentistry for covering cement fillings in tooth cavities and in taking impressions for dentures. Copal is also expressed from other plants in various countries (see **Hymenoea** ... Brazilian copal; **Trachylobium** ... Zanzibar copal; and **Vateria** ... Indian copal).

Agathosma A genus of plants of the family *Rutaceae*. e.g. A. betulina (= *Barosma* b.) - The Buchu plant; from the dried leaves is prepared an infusion used in diuretic mixtures for urinary tract infection.

Agave A genus of plants of the family *Agavaceae*. e.g. A. victoriae-reginae - a succulent plant with rosettes of closely packed, leathery, dark green leaves tapering to a point, with conspicuous white lines along the entire edges. This cactus plant is used for its foliage outdoors or in greenhouses. Place of origin: Northern Mexico; introduced into Europe in the second half of the 19th century.

Agelaius A genus of American Blackbirds of the family *Icteridae*. e.g. A. phoeniceus - The Red-winged Blackbird, the most abundant land bird in America, breeding in reedy marches from Canada to Costa Rica.

Agelastica A genus of Leaf Beetles of the family *Chrysomelidae*. e.g. A. alni - a species found in the Palaearctic region and introduced to North America.

Agenus A genus of Scorpionfishes of the suborder *Scorpaenoidea*. e.g. A. cataphractus - The Pogge or Armed Bullhead, inhabiting the North Atlantic.

Ageratum A genus of border plants of the family *Compositae*. e.g. A. houstonianum - The Floss Flower or Ageratum plant, an annual plant reaching a height of about 1 ft 8 in (50 cm) with tassel-like bright blue, to pink and white flowers appearing in clusters. Leaves ovate to heart-shaped. Used for edging borders, flowerbeds, window boxes, etc. Place of origin: Mexico. Introduced into Europe in 1822.

Agkistrodon A genus of venomous Snakes of the family *Crotalidae*; known also by the generic name, *Ancistrodon*. e.g. A. contortrix mokeson (= *Ancistrodon* c. m.) - The Northern Copperhead snake, probably the oldest venomous snake on record; collected near Boston, Massachusetts in May 1941 and died in Philadelphia Zoological Gardens on 7th April 1971, having been in captivity approximately 29 years and 11 months. A sensitive pit between the eyes and the nostril, acts as a heat receptor and detects warm-blooded prey. Its coloring is copper-reddish and its maximum length is about 120 cm (4 ft). Distribution: the forests of the eastern United States, excepting Florida (see **Ancistrodon**).

Aglaiocercus A genus of fast-flying seabirds (class *Aves*). e.g. A. kingi - The Blue-throated Sylph; this bird has been known to attain, in a straight course, a flight velocity over 85 km/h.

Aglais A genus of Insects, also called by the generic name, *Aglivis*, of the Tortoiseshell Butterfly family, *Nymphalidae*. e.g. A. urticae (= *Aglivis* u.) - The Small Tortoiseshell Butterfly; size 5.5 cm; with fox-colored wings, and three black spots on the front edge of each forewing. Found from lowland to mountain elevations, migrating to great heights. The greatest height reliably reported is 5,791 m (19,000 ft) for this species, seen flying over the Zemu Glacier (Sikkim) in the eastern Himalayas. Distribution: Eurasia; native in Britain.

Aglaophenia A genus of Hydra polypi of the family *Aglaopheniidae*. The species are arranged in clumps up to about 50 mm high and resemble a bird feather. e.g. A. tubulifera - abundant on rock wells at depths of 1-2 m; found on the shores of western Europe.

Aglivis A genus of Insects of the family *Nymphalidae* (see **Aglais**).

Agonum A genus of Insects of the Ground Beetle family, *Carabidae*. e.g. A. sexpunctatum - a species notable for its metallic color. Its scutum is usually green but may be blue, and the elytra golden-red, bronze, blue-violet or entirely black. Found chiefly in foothills and mountains.

Agonus A genus of Fishes of the Bullhead or Scorpion Fish family, *Cottidae*. e.g. A. cataphractus - The Armed Bullhead or Pogge Fish, 20 cm long; lower jaw has numerous barbels. Found by river-mouths and harbours. Native in Britain.

Agrilus A genus of the Metallic Wood-borer Beetle family, *Buprestidae*. e.g. A. pannonicus - a species found in young oak thickets, clearings, and forest margins, or in stumps or felled oak trunks.

Agrimonia A genus of plants of the Rose family, *Rosaceae*. e.g. A. eupatoria - Common Agrimony, a perennial plant, 20-80 cm high; leaves grey-hairy beneath. Widespread and common in Europe. Native in Britain.

Agriocharis A genus of Birds of the Turkey family, *Meleagrididae*. e.g. A. ocellata - The Ocellated Turkey of Mexico and northern Central America.

Agriocnemis A genus of small Insects of the order of Dragonflies, *Odonata*. e.g. A. naia - The smallest dragon-fly in the world. A specimen preserved in the British Museum of Natural History, has a wingspan of 17.6 mm (0.69 in) and an overall length of 18 mm (0.7 in).

Agrion A genus of Insects of the Damselfly family, *Agriidae*. Also known by the generic name, *Calopteryx*. e.g. A. virgo (= *Calopteryx* v.) - The Damselfly or Demoiselle Agrion; length of body 5-6 cm, wingspan 7 cm; wings of the male are dark blue to a glossy blue-green, of the female a dull greyish-green to brown. Like other damselflies, it is capable of folding its wings over the back when at rest. Found by the banks of still and slow-flowing waters, around rivers and streams. Distribution: Europe, Asia. Native in Britain.

Agriotes A genus of Insects of the Click Beetle family, *Elateridae*. e.g. A. lineatus - The Lined Click Beetle, up to 10 mm long; each elytron has eight rows of spots. Native in Britain.

Agriotypus A genus of parasitic Insects (suborder *Parasitica*) of the family *Ichneumonidae*. They have a very unusual habit in that the adults swim beneath water to find their hosts (the larvae of caddis-flies).

Agrobacterium A genus of microorganisms of the family *Rhizobiaceae*. Short flagellated rods found in the soil. e.g. A. tumefaciens.

Agromyza A genus of Flies of the family *Agromyzidae*. The leaf-miners; they usually form long serpentine mines.

Agropyron A genus of plants of the Grass family, *Gramineae*. Also known by the generic names, *Agropyrum* or *Triticum*. The species have been involved by natural hybridization in the development of bread-wheat (see **Triticum**). e.g. A. repens (= *Agropyrum* r.; *Triticum* r.) - Couch-grass or the Common Couch (= Agropyrum; Graminis Rhizoma; Scutch; Triticum; Twitch) - a perennial plant, 20-150 cm high; rhizomatous, with spikelets falling entire at maturity. A troublesome weed of cultivated fields and gardens, growing also by roadsides, on waste land and sandy places near the sea. It contains glucose, mannitol, inositol, and triticin (a carbohydrate resembling inulin). Used as a mild diuretic in the treatment of urinary tract infections. Native in Britain (see **Triticum**).

Agropyrum See **Agropyron**.

Agrostemma A genus of plants of the Pink family, *Caryophyllaceae*. Also known by the generic name, *Lychnis*. e.g. A. githago (= *Lychnis g.*) - The Corn Cockle, an annual plant, up to 70 cm high; leaves with apressed white hairs; large, solitary flowers with purple petals; a poisonous plant, widespread but local; growing in cornfields. Native in Britain.

Agrostis A genus of plants of the Grass family, *Gramineae*. e.g. A. tenuis - The Common Bent, a perennial plant, 30-75 cm high; tufted grass with ligules up to 2 mm long; open loose panicles. Grows in lowland grasslands, on heaths, moorland, pastures and waste ground. Native in Britain.

Agrotis A genus of Insects of the Noctuid Moth family, *Noctuidae*. e.g. A. segetum - The Turnip Moth, found in fields and gardens. The caterpillar is polyphagous, and is found on grasses and vegetables, where it is a pest, especially of turnips and swedes.

Agrumaenia A genus of Insects of the Burnet and Forester Moth family, *Zygaenidae*. e.g. A. carniolica - The Burnet Moth, size 12-15 mm; has a number of large bright-red spots on the wings. After hibernating, the caterpillar pupates in a parchment-like cocoon. Distribution: Eurasia.

Agrypnus A genus of Insects of the Click Beetle family, *Elateridae*. e.g. A. murinus - A species found in forests, gardens, meadows and fields, up to mountain elevations.

Aguila A genus of Birds of the Eagle family, *Accipitridae*. Better known by the generic name, *Aquila*. e.g. A. ver-reauxi - The Verreaux's eagle (see **Aquila**).

Agulla A genus of Insects of the Snakefly family, *Raphiidae*. It has an unusually long prothorax, a long ovipositor, and the larvae live under bark.

Agyroneta A genus of Arachnids of thge Water Spider family, *Agelenidae*, better known by the generic name, *Argyroneta* (see **Argyroneta**).

Ahaetulla A genus of venomous Snakes of the class *Reptilia*. e.g. A. natsuta (= A. nasuta) - The Indian Long-nosed Tree Snake; unlike other species it can see forward along a groove in its nose, which is a great asset in hunting. Its green body is almost indistinguishable from the surrounding lianas and other vegetation. Distribution: the jungles of Malaya and the East Indies.

Ailanthus A genus of Trees of the family *Simarubaceae*. Also known by the generic name, *Ailantus*. e.g. A. glandulosa (= *Ailantus g.*) - its bark is used as a purgative and anthelmintic.

Ailantus See **Ailanthus**.

Ailuropoda A genus of Mammals of the Panda subfamily, *Ailurinae*. e.g. A. melanoleuca - The Giant Panda, the rarest species of bear and the most valuable land animal in the world; about 165 cm (51/2 ft) long; it has a thick cream-and-black fur, with circular black spots around the eyes; the ears are black. Habitat: eastern Tibet and southwest China. It was first discovered in Szechwan, western China towards the end of the 19th century by the French missionary, Père David on the farthest reaches of the eastern Himalayas.

Ailurus A genus of Mammals of the Panda subfamily, *Ailurinae*. e.g. A. fulgens - The Red or Lesser Panda, up to 60 cm (2 ft) long, with a dark stripe from each eye to the corner of the mouth, and its long bushy tail, 45 cm (11/2 ft) long, which is rigid. Found in Asia, from Nepal to Yunnan, living in high altitudes.

Aiptasia A genus of Sea Anemones of the order *Actiniaria*. e.g. A. diaphana - a species 3-5 cm in diameter; tentacles rusty yellow, and unable to contract. Distribution: from the Mediterranean Sea northwards along the Atlantic coast. Native in Britain.

Aira A genus of plants of the Grass family, *Gramineae*. e.g. A. praecox - Early Hair-grass, an annual plant, 3-10 cm high dense pannicles, with branches little longer than spikelets; a common species, growing on cultivated and waste land and in dry places. Distribution: North, Central and Western Europe. Native in Britain.

Aix A genus of Birds of the Duck family, *Anatidae*. e.g. A. galericulata - The Mandarin Duck, size 43 cm; male with an orange "sail" on each wing, female has a white chin. A resident bird introduced from East Asia as an ornamental species, and now widely established in Europe. It nests on moors, marches, and damp meadows; also in trees. Also introduced and naturalized in Britain.

Ajaia A genus of Birds of the Spoonbill family, *Threshkiornithidae*. e.g. A. ajaia (= A. ajaja) - The Roseate Spoonbill of South America and Florida. The beak tip is flat and wide like that of a duck, but more markedly so. The bird sweeps this spoon-shaped bill back and forth, filtering crustaceans from the water.

Ajuga A genus of plants of the Thyme family, *Labiatae*. e.g. A. reptans - The Common Bugle, a perennial plant, 4-50 cm high; stoloniferous (producing creeping stems or stolons), with the lower bracts resembling stem leaves. A widespread and common species, growing in meadows, on dry turf and at the edges of woods. Distribution: Europe; native in Britain.

Akabaria A genus of Octocorals, a group of corals that do not usually deposit calcareous skeletons. Each living individual or polyp has eight feathered tentacles surrounding the mouth. Species of Akabaria octocoral are found hidden in reef cracks and crevices, and are unique to the Gulf of Eilat (Akaba) in the Red Sea region.

Alabes A genus of Swamp-eel Fishes of the order *Synbranchii* (*Synbranchiformes*). These eel-like forms, the Alabes or Glove-eels have no affinity with the true eels, and are found in Australia and Tasmania; they have 75 vertebrae, well-developed dorsal and anal fins, and pelvics on the throat.

Alaria A genus of Trematode Fluke-worms of the order *Digenia*. These digenetic trematodes have very complex life-cycles, many species requiring four hosts: snails, amphibians, rats or mice, and mammals such as cats, dogs, mink or weasels.

Alauda A genus of Birds of the Lark family, *Alaudidae*. e.g. A. arvensis - The Skylark; size 18 cm; long tail with white outer edges; found in fields, meadows and other tree-less places in lowlands and on mountains. A high perching song-bird of the Old World, and a partial migrant. Native in Britain.

Albizia A genus of Trees and Shrubs of the family *Leguminosae*. e.g. A. julibrissin - The Mimosa; Pink Siris; or Silktree - a 40 ft (13 m) high tree, with doubly pinnate leaves made up of 20 to 30 pairs of leaflets. Place of origin: Subtropical Asia.

Albula A genus of Fishes (subphylum *Vertebrata*). e.g. A. vulpes - The Bonefish. Found in the tropical Atlantic and a magnificient sprinter, capable of dashing off at a speed of over 64 km/h (40 miles/h).

Albulina A genus of Insects of the Blue Butterfly family, *Lycaenidae*. e.g. A. orbitulus - size 2.6 cm; in the male the upper side is silvery grey; the female is brown. Found in Europe.

Alburnoides A genus of Fishes of the Minnow and Carp family, *Cyprinidae*. e.g. A. bipunctatus - The Schneider Fish; found in shallow waters in the current of the upper reaches of rivers, together with minnows. Feeds on insects and insect larvae (see **Carassius**).

Alburnus A genus of Bleak Fishes of the Minnow and Carp family, *Cyprinidae*. e.g. A. alburnus - The Bleak Fish, size 10-15 cm; resembles the Schneider (*Alburnoides* bipunctatus), but is more shallow-bodied. Found in deeper parts of slow-flowing waters in the middle and lower reaches of larger rivers. Distribution: Europe (see **Carassius**).

Alca A genus of Birds of the Auk, Guillemot, and Puffin family, *Alcidae*. e.g. A. torda - The Razor-billed Auk or Razorbill, 40 cm (10 in) long; thick bill with white marking; white eye-stripe in summer. A partial migrant bird, found on coastal cliffs and solitary islands feeding on fish and crustaceans, and nesting in colonies. Distribution: all over the North Atlantic and Arctic Ocean. Native in Britain.

Alcaligenes A genus of microorganisms of the family *Achromobacteraceae*; also known by the name *Alkaligenes*. Many species are now placed in the genus *Acinetobacter* or *Pseudomonas*. These are gram-negative coccobacilli or rods, usually saprophytes, and are considered to be part of the normal intestinal flora. e.g. A. faecalis (= *Alkaligenes* f.; *Bacillus* faecalis alkaligenes) - causes urinary tract and other infections. A. odorans (= *Pseudomonas* o.) - occasionally an opportunistic pathogen, that has been isolated from the feces. Less frequently recovered from urine, sputum and wounds.

Alcea A genus of plants of the Mallow family, *Malvaceae*; better known by the generic name, *Althaea*. e.g. A. officinalis (= *Althaea* o.) - The March Mallow (see **Althaea**).

Alcedo A genus of Birds of the Kingfisher family, *Alcedinidae* (subfamily *Alcedininae*. e.g. A. atthis (= A. acis) - The Common Kingfisher, size 16.5 cm. A predatory bird, characterized by its brilliant plumage, large head and long sharp bill. Feeds on small fish and aquatic insects in the vicinity of calm and flowing waters. A partial migrant. Habitat: all of Europe, except the north. Native in Britain.

Alcephalus A genus of Mammals of the Antelope subfamily, *Antilopinae*. e.g. A. caama - The Hartebeest, a large, awkward animal with V-shaped or lyre-shaped horns projecting from a lumpy forehead. Distribution: the African savannas.

Alces A genus of Mammals of the True Deer family, *Cervidae*. e.g. A. alces gigas - The Alaskan Moose or Elk, the largest deer in the world. Adult bulls stand c. 1.83 m (6 ft) at the withers, and scale 499-543 kg (1,100-1,200 lb). The male has palmate antlers with many tines; the female has no antlers. Found in the forested areas of Alaska, U.S.A. and the Yukon, Canada.

Alchemilla A genus of plants of the Rose family, *Rosaceae*. e.g. A. vulgaris - The Common Lady's Mantle, a small perennial herb, 10-50 cm high; with palmate, 7-11 lobed leaves, and minute green flowers with no petals, borne in dense clusters. The fruit is a single achene. Grows commonly on meadows, wet grassland, in woods and on scrubland. Distribution: Europe, Asia, and eastern North America. Native in Britain.

Alcippe A genus of Babbler Birds of the subfamily *Timaliinae*. e.g. A. poiocephala - The Nun-babbler of Malaya; a poor flyer with short, rounded wings and a slender bill, it closely resembles the Olive-backed Jungle Fly Catcher bird, *Rhinomyias* olivacea.

Alcyonidium A genus of Moss Animals or Sea Mats of the phylum *Polyzoa*, also known as *Bryoza* or *Ectoprocta*. These are small colonial animals found mainly in the sea, the species forming erect finger-like colonies with the zooids embedded in a firm gelatinous matrix. e.g. A. gelatinosum - size up to 50 cm; sponge-

18

like, slimy, gelatinous colonies branching like a tree. Native to the Baltic Sea.

Alcyonium A genus of Sea-animals, the Coelenterates of the order *Alcyonacea* (= *Alcyonaria*). e.g. A. digitatum - Dead Men's Fingers; these form colonies of up to 200 mm, composed of polyps, each furnished with 8 branching arms; The polyps are white and retractile, capable of completely concealing themselves inside the calcareous skeleton. They live on the seabottom often attached to the cells of oysters, and are found off the shores of the British Isles.

Aldrovanda A genus of plants of the Sundew family. e.g. A. vesiculosa - Aldrovanda, a perennial plant, 3 cm high; stem submerged, with leaves specially adapted for trapping insects; roots absent. Occurs in South and Central Europe.

Alectis A genus of Fishes (subphylum *Vertebrata*). e.g. A. crinitus - The African Pompano fish.

Alectoria A genus of Lichens, small, slow-growing plants, capable of surviving prolonged desiccation. Also known by the generic name, *Alextoria*. e.g. A. fuscescens (= *Alextoria* f.) - Wiry Lichen, a perennial plant, size 30 cm; consists of a grey thallus of finely branched filaments. Found growing on the barks of trees, particularly the conifers. Distribution: Europe; native in Britain.

Alectoris A genus of Birds of the Partridge family, *Phasianidae*. e.g. A. graeca - The Rock Partridge or Chukor, 35 cm (14 in) in length; strikingly colored yellow, black, grey and rufous; a fowl-like resident gamebird nesting on the ground, and found on dry, warm, and stony hillsides. Distribution: the Balkans and Carpathian mountains; also in the French Alps above 1,500 m (5,000 ft).

Alectra A genus of plants of the Figwort or Snapdragon family, *Scrophulariaceae*. e.g. A. parasitica var.

chitrakutensis - a parasitic plant, the rhizome of which is used in Indian native medicine in the treatment of leprosy.

Alepisaurus A genus of Fishes of the Lancetfish family, *Alepisauridae*. e.g. A. ferox - The Longnose Lancetfish, nearly 2 m long; among the largest of deep-sea predators, having a long, high, dorsal fin and a large mouth furnished with long, sharp, fang-like teeth. Diet consists mainly of other deep-sea fish.

Aletris A genus of plants of the Lily family, *Liliaceae*. e.g. A. farinosa - The Ague Root plant (= Colic Root; Star Grass; Unicorn Root). The root is used as a so-called "uterine tonic".

Alextoria See **Alectoria**.

Alginobacter A genus of microorganisms of the family *Enterobacteriaceae*; short motile rods found in the soil. e.g. A. acidofaciens.

Alginomonas A genus of microorganisms of the family *Pseudomonadaceae*; motile coccoid rods in algae and seawater. e.g. A. alginica.

Alisma A genus of plants of the Water-plantain family. e.g. A. plantago aquatica - The Common Water-plantain, a perennial plant, 20-100 cm high; long-stalked leaves, ovate, rounded or half heart-shaped at base (in water forms, the submerged leaves are linear); straight style. Distribution: widespread in Europe; native in Britain.

Alkalescens dispar A genus of gram-negative bacilli, rod-like microorganisms intermediate in evolution between *Escherichia* and *Shigella*. Sometimes included as a serotype of *Escherichia* coli. May cause dysentery, urinary tract infections, and bacteremia.

Alkaligenes See **Alcaligenes**.

Alkanna A genus of plants of the Borage family, *Boraginaceae*. e.g. A. tinctoria (= *Anchusa* t.) - Alkanet Root; Anchusa - contains a red dye alkannin (see **Anchusa**).

Allactaga A genus of Mammals of the Jerboa Rodent family, *Dipodidae*. e.g.

A. jaculus - The Fine-toothed Jerboa, a nocturnal rodent hiding in underground burrows by day. Found in the steppe and semi-steppe regions in Asia.

Allenopithecus A genus of Mammals of the Old World Monkey family, *Cercopithecidae*. e.g. A. nigroviridis - Allen's Swamp Monkey, a gentle good-natured monkey, the size of a cat. Its fur is greyish-brown with a greenish tinge. It lives in large packs in forests on the banks of rivers in Gabon and northwest Congo. It utters a high-pitched whistle.

Allescheria A genus of moldlike Fungi. e.g. A. boydii - a species found in human lesions.

Alliaria A genus of plants of the Cress or Crucifer family, *Cruciferae*. e.g. A. officinalis (= A. petiolata) - Garlic Mustard; Hedge-garlic; Jack-by-the-hedge; or Sauce-alone; a bienniel to perennial plant, 12 to 110 cm high; leaves kidney- to heart-shaped, and smelling of garlic; the flowers are white. It grows in shady woods. Distribution: most of Europe and Asia; native in Britain. Naturalized in North America.

Alligator A genus of Reptiles of the Crocodile family, *Alligatoridae*. e.g. A. mississipiensis - The American Alligator of the southeast United States, a long-living reptile reaching the age of 50 years or more. It grows to a length of 3.6 m (12 ft) or more. It has abroad flattened snout, and is sometimes bred in "alligator farms".

Allium A genus of plants of the Daffodil family, *Amaryllidaceae*. e.g. A. sativum - The Garlic plant, a perennial, 30-60 cm high; has expectorant, antiseptic, and diuretic properties, and has been used in the treatment of chronic bronchitis; also as a gargle or throatspray. Distribution: cosmopolitan.

Allolobophora A genus of Earthworms of the family *Lumbricidae*. e.g. A. longa - a species of relatively long-lived segmented worms, living as much as 10 years or more.

Alloteuthis A genus of Cephalopod Molluscs. e.g. A. subulata - The North Sea Squid, 13 cm long; transparent. Silvery grey skin with dark spots; arms half the total length; narrow, short fins. Distribution: along the Atlantic coasts and in the North Sea.

Almyracuma A genus of Crustaceans of the order *Cumacea*. e.g. A. proximoculi - A cumacean species which has penetrated farthest from the sea into the brackish shallow waters near the mouth of a river on Cape Cod, Massachusetts.

Alnus A genus of Trees of the Birch family, *Betulaceae*. e.g. A. glutinosa - The Birch, Common Birch, or Alder tree; a north temperate deciduous tree up to 20 m tall; stalked buds; the minute male and female flowers are borne in separate catkins. Abundant by lakes and rivers. Distribution: Europe and Asia; native in Britain, and naturalized in North America.

Alocimua A genus of Snails. e.g. A. longicornis (= *Bythnia* longicornus) (see **Bythnia**).

Aloë A genus of plants of the Lily family, *Liliaceae*, sometimes known by the generic name, *Gasteria*. e.g. A. curacao - The Aloes plant; it contains the glycoside barbaloin, and is used as a purgative. A. verrucosa (= *Gasteria* v.) - The Ox-tongue (see **Gasteria**).

Aloides A genus of Bivalve Molluscs. e.g. A. gibba - Basket Shell; size 1.3 cm; valves very dense and of unequal size. Habitat: the North Sea; native in Britain.

Alonella A genus of Crustaceans of the order of Water Fleas, *Cladocera*. One of the world's smallest known crustaceans, some species measuring less than 0.25 mm (0.0098 in) in length.

Alonopsis A genus of Crustaceans of the order of Water Fleas, *Cladocera*. e.g. A. ambigua - a tiny creature, the size of a flea, inhabiting both marine and fresh waters, and an important part of plankton for the nutrition of fish and other aquatic animals.

Alopecosa A genus of Arachnids of the Wolf Spider family, *Lycosidae*. e.g. A. aculeata - about 5 mm long; very common in forests in the Palaearctic and Nearctic regions.

Alopecurus See **Alopercurus**.

Alopercurus A genus of plants of the Grass family, *Gramineae*, also called *Alopecurus*. e.g. A. pratensis (= *Alopecurus* p.) - The Meadow Foxtail, a perennial plant, 60-120 cm high; a species with spike-like panicles like all the others in this genus of Foxtail grasses. The culm does not root at the nodes, and the glumes and lemma are acute. A widespread plant, common in meadows, pastures, and sandy places. Distribution: Europe; native in Britain.

Alopex A genus of Mammals of the Fox family, *Canidae*. e.g. A. lagopus - The Arctic Fox; size 65 cm, with a tail 33 cm long. Two forms exist: the White Fox, which is a light brown in summer and white in winter, and the Blue Fox, greyish-blue throughout the year. Found around the North Pole and the tundra of northern Asia and North America.

Alopias A genus of Mackerel or Thresher Sharks, better known by the generic name, *Alopius*. e.g. A. vulpinus (= *Alopius* v.) - The Thresher Shark (see **Alopius**).

Alopius A genus of Mackerel or Thresher Sharks (Maneaters), of the family *Lamnidae* (= *Isuridae*), one of the most active and ferocious of this family. Also known by the generic name, *Alopias*. e.g. A. vulpinus (= *Alopias* v.) - The Thresher Shark; its tail is bent upwards from its base, and is as long as the rest of its body. It is said to destroy schools of fish by thrashing with its tail, hence its name.

Alopochen A genus of Geese of the family *Anatidae*. e.g. A. aegyptiacus - The Egyptian Goose, a sheld goose of the Old World, known to migrate at great heights (of 8,046 m; 26,400 ft; 5 miles). Distribution: North and Central Africa; also Syria.

Alosa A genus of Fishes of the Herring family, *Clupeidae*. e.g. A. alosa - The Allis Shad or American Shad fish; up to 71 cm in length; black spot behind gill-covers; it spawns far up rivers, the young returning to the sea where they live for several years. The cycle is complete when they return once again to spawn in freshwaters. Habitat: the Atlantic Ocean; North and Baltic Seas.

Alouatta A genus of New World Monkeys of the family *Cebidae*. e.g. A. caraya - The Black Howling Monkey; only the males are black, the females and young being pale yellow. The largest of the New World monkeys, the male being the size of a fairly large dog. Bearded, with a prehensile tail longer than the body. Distribution: From Ecuador to Paraguay.

Aloysia A genus of aromatic plants of the family *Verbenaceae*, formerly known by the generic name, *Lippia*. e.g. A. citriodora (formerly *Lippia* c.) - The Lemon Verbena plant, a deciduous shrub about 2 m (61/2 ft) high, with lance-shaped, lemon-scented leaves and insignificant, small, pale purple flowers. The leaves yield an aromatic oil, Verbena Oil, used in perfumery and for scenting linen, potpourri, etc. The leaves are also used locally as a digestive tea infusion. Place of origin: southern South America; introduced into Europe in 1784.

Alpinia A genus of plants of the family *Zingiberaceae*. e.g. A. galanga - The Chinese Ginger or East Indian Root plant; used as a carminative and flowering agent.

Alsophila A genus of Insects of the Geometer or Looper Moth family, *Geometridae*. e.g. A. pometaria - The American Fall Cankerworm; the larvae are serious defoliators, and like others of the Geometer family, progress along the true stem with characteristic looping movements; they are variously known as measuring worms, inchworms, loopers or earth-measurers. The adult female moth is wingless.

Alstonia A genus of plants of the Periwinkle family, *Apocynaceae*. e.g. A. scholaris - The Chhatim or Dita Bark; used in India and the Far East in the treatment of diarrhea and malaria.

Alstroemeria A genus of herbaceous plants of the family *Alstroemeriaceae*. e.g. A. pelegrina - The Peruvian Lily; height 60 cm (2 ft), with thick fibrous roots, leafy stems, and terminal umbels of lilac, red-purple spotted, showy flowers. The roots are a local source of starch. Place of origin: Chile and Peru; in Europe since 1754.

Althaea A genus of medicinal and aromatic plants of the Mallow family, *Malvaceae* (Gr. altheo = to cure). Also known by the generic name, *Alcea*. e.g. A. officinalis - The Marsh Mallow or Marshmallow, an erect, hairy, perennial plant, 60-150 cm (2-5 ft) high; stem and leaves velvety, with pale, pink flowers clustered in axils of leaves. Grows in upper margins of marshes and ditch sides, and banks near the sea. The root is used as a demulcent and emollient, for irritation and inflammation in stomatitis and pharyngitis, and externally as a poultice. Distribution: Europe, western Asia, and North Africa. Native in Britain.

Althea A genus of Shrubs of the Mallow family, *Malvaceae*. Also known by the generic name, *Hibiscus*. e.g. A. frutex (= *Hibiscus* syriacus) - The Althea Shrub or Rose of Sharon, a deciduous shrub, 1-3 m (4-10 ft) high; with large pink, lilac, red, violet or white, sometimes blotched flowers. Place of ori-gin: India and China; in Syria as a cultivated plant only. Introduced into Europe in the late 16th century.

Alucita A genus of Insects of the Moth family, *Alucitidae*. e.g. A. desmodactyla - 5 to 6 mm long; each wing is divided into six plumes (24 plumes in all). Distribution: southern Europe.

Alyssum A genus of plants of the Cress or Crucifer family, *Cruciferae*; the species have other generic names. e.g. A. maritimum (= *Lobularia* maritima) - The Sweet Alyssum plant; a spreading, mat-forming, woody-stemmed perennial, up to 22 cm (9 in) high; with small white or pinkish, honey-scented flowers growing in dense terminal clusters. Place of origin: Europe; western Asia (see **Lobularia**). A. saxatile (= *Aurininia* saxatilis) - Basket-of-Gold; Gold Dust; Golden Alison; Golden-Tuft; Rock Alyssum; a shrubby-based perennial plant, 10-30 cm high; with small golden-yellow numerous flowers in close flat-topped corymbs; and grey-hairy, roundish, glabrous fruits. Grows on arable and waste land, and in grassy fields. Place of origin: Central and southeast Europe. Introduced and naturalized in Britain.

Alytes A genus of Amphibians of the Toad family, *Discoglossidae*. e.g. A. obstetricans - The Midwife Toad; size 4.5-5.5 cm; a nocturnal amphibian, found mostly in mountainous regions. A small bluish-grey, burrowing species with warts, and vertical pupils. Movements of the male at spawning, result in ropes of spawn being tangled round its thighs, where they remain (hence the name Midwife), until the tadpoles hatch. Also known as the Bell Toad because of its call note. Distribution: Europe, from Spain to the Black Forest; introduced and naturalized in Britain.

Amanita A genus of Mushrooms of the Gill Fungus family, *Agaricaceae*. e.g. A. mappa - The toadstool containing the hallucinogen, Bufotenine (see **Piptadenia**). A. muscaria (= *Agaricus* muscarius) - Fly Agaric; a poisonous plant containing the deadly alkaloid, muscarine. This mushroom species has a cap 8-20 cm wide, scarlet or yellowish-red, and covered with thick, white or yellowish fragments of volva. Grows in mixed woods, particularly near birch trees and conifers. Found in Europe; native in Britain (see **Agaricus**). A. pantherina - Panther Cap, a very poisonous mushroom species of the Gill Fungi family, containing the two nerve-paralysing poisons, atropine and muscarine. The cap is 5-10 cm wide, brown, with white fragments of volva, and deeply grooved at the margin. Native in Britain. A. phalloides - Death Cap, a deadly species containing the poisonous glycoside, amanitine. Cap 7-12 cm wide, and greenish or greenish-white in color. Native in Britain. A. verna - "Fools" Mushroom, a deadly but uncommon species, found growing scattered in woodlands. Cap up to 8 cm wide, white, convex and finally flat. Native in Britain. A. virosa - Avening Angel, a deadly poisonous mushroom, cap 5-10 cm wide, white, sticky, conical. Native in Britain.

Amaranthus A genus of late blooming plants of the Amaranth family, *Amaranthaceae*. e.g. A. caudatus - The Love-Lies-Bleeding or Foxtail plant; an annual ornamental species, producing flowers in long, pendent racemes 45 cm (18 in) long; crimson colored; also a pale-green variety. Place of origin: the Tropics; introduced into Britain in 1596.

Amaroucium A genus of Sea-squirts of the class *Ascidiaceae*. These chordate animals develop colonial forms in which numerous small individuals are invested by a common tunic (see **Botryllus**).

Amaryllis A monotypic genus of plants of the Daffodil family, *Amaryllidaceae*. e.g. A. belladonna - The Belladonna Lily, a showy late-flowering bulb with large, funnel-shaped, 6-parted, rose-red or pink, sweet-scented flowers on stout stems. Place of origin: South Africa.

Amaurobius A genus of Spiders of the order *Araneae*. e.g. A. socialis - A web-building species of Australia, showing a tendency towards social habits, building communal webs measuring up to 3.65 m (12 ft) in length and 1.21 m (4 ft) in width.

Amazona A genus of Birds, the Amazons of South America, of the Parrot family, *Psittacidae*. e.g. A. dufresniana - A species of rare Puerto Rican parrots, believed to have a total world population of less than 100. A. farinosa - The Common Green Parrot. A. ochrocephala - The Yellow-headed Amazon - the best talker.

Amblyglyphidodon A genus of Fishes (subphylum *Vertebrata*). e.g. A. leucogaster - The Yellow-belly Damsel Fish; a territorial species, in that it lives within definite territorial boundaries and rarely wanders. It is aggressive towards other fish encroaching upon its home. Found along the shores of the Red Sea.

Amblyomma A genus of Ticks. e.g. A. hebraeum - The Bont Tick, transmitter of African "heart-water" disease and South African tick bite fever.

Amblyopsis A genus of Blind Fishes (subphylum *Vertebrata*). e.g. A. spelaea - The Kentucky Blind-fish; this blind species live incaves; their eyes have disappeared completely or are mere vestiges under the skin.

Amblyorhynchus A genus of Lizards of the Iguana family, *Iguanidae*. e.g. A. cristatus - The Marine Iguana, a heavy sea-weed eating lizard found in the Galapagos Islands, eastern Central Pacific. A large lizard about 41/2 ft long and weighing up to 22 lb, and in danger of extinction because of domestic dogs and cats released on the islands.

Amblyosomus A genus of Mammals of the Golden Mole family, *Chrysochloridae*. e.g. A. hottentottus - The Red Golden Mole; a burrowing animal whose fur has a red golden coppery sheen. Found in eastern South Africa.

Amblystoma See **Ambystoma**.

Ambystoma A genus of Tiger Salamanders of the family

Ambystomatidae; also known by the generic name, *Amblystoma*. e.g. A. tigrinum diaboli - The Devil's Lake Tiger Salamander, found in brackish or highly saline lakes in Mexico and the United States. A neotenous species with a tendency to remain in the larval state, although gaining sexual maturity (neoteny).

Ameba See **Amoeba**.

Ameiurus A genus of North American freshwater Catfishes of the family *Ameiuridae* (= *Ictaluridae*). Also known by the generic name, *Ictalurus*). e.g. A. nebulosus (= *Ictalurus* n.) - The Brown Bullhead or Horned Pout, one of the best-known catfishes; length 45 cm, weight 2 kg, with sharp pectoral spines; the long pectoral barbels help the fish to locate its food in the muddy waters it inhabits. Found commonly in the streams, lakes, and ponds of the east and middle United States. First introduced into Europe, including Britain, in 1885 as a pond fish from North America.

Amelanchier A genus of plants of the Rose family, *Rosaceae*. e.g. A. ovalis - The Common Service-berry, a plant growing up to 3 m high; with narrow, wedge-shaped sepals and globose, black, sweet and edible fruits. Distribution: in open woods in South and Central Europe.

Amia A genus of Fishes of the order *Protospondyli* (= *Amiiformes*). e.g. A. calva - The Bowfin or Grindle fish, about 75 cm (2 1/2 ft) long, the only surviving representative species of this ancient primitive genus, now confined to the Mississippi basin and the Great Lakes of North America.

Amidostomum A genus of Roundworms, also known by the generic name, *Strongylus*. e.g. A. anseris (= *Strongylus* nodularis) - The Gizzard Worm, found in the intestines of geese.

Amitermes A genus of Insects of the family *Termitidae*. e.g. A. meridionalis - The celebrated "Magnetic" Termite of Northern Australia, whose huge mound-like nests always point north and south, with broad faces to east and west, apparently for maximum warmth from the sun.

Ammi A genus of plants of the Carrot family, *Umbelliferae*. e.g. A. visnaga - from the fruit is extracted an active principle, khellin, used as a coronary and bronchial dilator.

Ammocrypta A genus of Fishes of the Perch family, *Percidae*. e.g. A. pellucida - The Eastern Sand Darter fish, a very tiny, almost transparent species.

Ammodorcas A genus of Mammals of the Gazelle subfamily, *Antilopinae*. e.g. A. clarkei - The Dibatag or Clarke's Gazelle; it has a long thin neck. Found in Africa.

Ammodytes A genus of Fishes of the Sand-eel family. e.g. A. lanceolatus - The Greater Sand-eel, up to 30 cm long; bark brownish or yellowish-green; underside silvery. These fishes are widespread along the Atlantic, the Baltic coast, and in the North Sea. Native around Britain.

Ammomanes A genus of Birds of the Lark family, *Alaudidae*. e.g. A. deserti - The Desert Lark of North Africa; it has a long, frail bill and sandy-grey plumage harmonizing with the ground.

Ammoperdix A genus of Desert Birds of the Partridge family, *Phasianidae*. e.g. A. heyi - The See-see Partridge.

Ammophila A genus of Hymenopterous Insects of the Digger Wasp family, *Sphecidae*. e.g. A. sabulosa - The Sand Wasp; size 2 cm; abdomen club-shaped, halfred and half blue. The nest is covered by the female in sand; it paralyses caterpillars and drags them into a prepared hole as food for its young. Distribution: the sandy regions of the Palaearctic. Native in Britain.

Ammospermophilus A genus of Burrowing Rodents. e.g. A. leucurus - The desert antelope Ground Squirrel, host to the plague-transmitting flea, *Thrassia* francisi.

Ammotragus A genus of Sheep of the subfamily, *Caprinae*. e.g. A. lervia - The Barbary Sheep (= Aoudad; Arni); up to 1 m (3 ft 4 in) high. It has long hair on its neck and fore-limbs. In appearance and habits it resembles a goat, with horns equally developed in both sexes, sweeping in a wide semicircle away from the head. The Barbary Sheep is a very hardy animal and can withstand great differences inclimate. Its habitat extends from the highest peaks of the Atlas Mountains in North Africa to the Red Sea.

Amoeba See **Entamoeba**. Also called *Ameba*. A name given to various protozoa of the class *Rhizopoda*, many of whom are now assigned to other taxonomic groups. e.g. A. gruberi (= *Dimastigamoeba* g.) - see **Dimastigamoeba**. A. histolytica (= A. dysenteriae; *Entamoeba* h.) - order *Amoebina*, causative organism of amoebiasis. A. limax (= *Endolimax* nana) - order *Rhizomastigina*. A. meleagridis (= *Histomonas* m.) - order *Rhizomastigina* (see **Histomonas**). A. proteus - order *Amoebina*, the most commonly studied organism in schools. Also known as *Chaos* diffluens.

Amoebobacter A genus of microorganisms of the family *Thiorhodaceae* - cells usually occurring without a common capsule. e.g. A. granula.

Amoebotaenia A genus of Cyclophyllidean Tapeworms of the family *Davainiidae*. Parasitic in poultry.

Amoebus See **Amoeba**; **Entamoeba**.

Amomum A genus of plants of the Ginger family, *Zingiberaceae*. e.g. A. aromaticum - The Bengal Cardamom plant; used in India as a substitute for cardamom.

Amorpha A genus of Insects of the Hawk Moth family, *Sphingidae*. e.g. A. populi - The Poplar Hawk Moth; wings colored orche and blackish-brown.

Lives on poplars and willows throughout the temperate and warm regions of Europe.

Amorphophallus A genus of plants of the Arum family, *Araceae*. e.g. A. bulbifer - A large tuberous plant, with dull green, variegated, dissected, compound leaves up to 1.2 m (4 ft) long, and a cup- orbell-shaped green spathe at the base of a pinkish spadix, about 7.5 cm (3 in) long. Place of origin: Tropical regions of Asia, Australia, and the Pacific.

Amphibolurus A genus of Lizards of the family *Agamidae*. e.g. A. barbatus - The Australian Bearded Lizard, about 50 cm long, inhabiting the dry parts of Australia. When it opens its mouth a collar of scales at the sides of its throat bristles out like stiff hairs.

Amphileptus A genus of ciliate parasitic microorganisms. e.g. A. branchiarum - A species found on the gills of frog tadpoles.

Amphilina A genus of Monozoic cestode worms of the order *Amphilinidea*. There is no alimentary canal and the active body may be regarded as a single proglottis. It has a terminal retractile proboscis, and is parasitic in rabbit-fishes.

Amphimallon A genus of Insects of the Chafer and Dung Beetle family, *Scarabaeidae*. Also known by the generic name, *Amphimallus*. e.g. A. solstitialis (= *Amphimallus* s.) - The June Bug or Cockchafer (= Midsummer Chafer); size 1.7 cm; a small, hairy, yellowish-brown Chafer Beetle species; head reddish-yellow, elytra pale yellow; found in large swarms in meadows, grass clearings, forest margins, and around trees in parks and gardens. Feeds on tree foliage. Distribution: Central Europe; native in Britain.

Amphimallus See **Amphimallon**.

Amphimerus A genus of Trematode Worms. e.g. A. noverca - a species infesting dogs in India; also reported humans.

Amphiophiura A genus of Brittlestar Fishes, the Echinoderms, spiny-skinned marine invertebrate animals with a hard calcareous internal skeleton. The species are found in the ocean at depths exceeding 4 miles (6,432 m; 21,120 ft).

Amphioxus A genus name formerly used for the Lancelet (subphylum *Cephalochordata*) a fish-like chordate animal, now known by the generic name, *Branchiostoma* (see **Branchiostoma**).

Amphipholis A genus of Brittle-stars of the class *Ophiuroidea*, an almost cosmopolitan animal, found between tidemarks, capable of making its arms luminescent, and of brooding its young.

Amphipnous A genus of Cuchias or Swamp Eel Fishes of the order *Synbranchii* (= *Synbranchiformes*). e.g. A. cuchia - The Cuchia, found in India; a fish equipped with only three pairs of gills.

Amphiprion A genus of Fishes (subphylum *Vertebrata*). e.g. A. bicinctus - The Clown Fish, a species of Damsel fishes living in symbiosis with the poisonous Sea Anemone in the Red Sea region.

Amphisbaena A genus of Worm Lizards of the family *Amphisbaenidae*. e.g. A. alba - The White Amphisbaena of tropical America; the largest of the worm lizards, about 2 ft long and as thick as a man's finger. Often found in manure heaps, feeding on ants and termites.

Amphistoma A genus of parasitic Trematode Worms. e.g. A. watsoni (= *Cladorchis* w.; *Watsonius* w.) - see **Cladorchis**.

Amphitirte A genus of Polychaete Worms of the family *Terebellidae*. e.g. A. edwardsi - The terebellid worm living in mud burrows, sharing the same home with another polychaete, the scaleworm, *Lepidastheria* argus.

Amphitrite A genus of Polychaete Worms of the family *Terebellidae*. e.g.

A. figulus (= A. brunnea) - A species found on both sides of the Atlantic.

Amphiuma A genus of Amphibians of the Newt and Salamander family, *Amphiumidae* (= *Salamandridae*). e.g. A. means - The Two-toed Amphiuma or Lamper Eel; an eel-like salamander species, also known as the Congo Eel; up to 1 m (3 ft) long, and with a maximum lifespan of some 25 years. Lives in still waters such as swamps. Distribution: the United States.

Amphiura A genus of Echinoderms of the Brittle-star or Brittle-starfish class, *Ophiuroidea*. e.g. A. filiformis - The Burrowing Brittle-star; size up to 12 cm, with a disc diameter of 8 mm, and five arms with spines along the edges, (each 5-7 cm long). A phosphorescent animal, and a scavenger feeding mainly on molluscs. Distribution: the seas of Europe; native in these as around the British Isles.

Amygdalus A genus of plants of the Rose family, *Rosaceae*. e.g. A. communis var. amara (= *Prunus* c. var. a.) - Bitter Almond (= Amygdala Amara); contains the glycoside amygdalin, yielding hydrocyanic acid (HCN).

Anabas A genus of Fishes of the Labyrinth Fish or Climbing Perch family, *Anabanthidae*. e.g. A. testudineus - The Climbing Perch Fish; like other fishes in this family, these species have an extra respiratory organ formed from the first gill arch; this organ has a labyrinthine structure, hence the name.

Anableps A genus of four-eyed Fishes of the family *Anablepidae*; the species is viviparous.

Anabrus A genus of Insects of the Bush-cricket or Katydid family, *Tettigoniidae*. e.g. A. simplex - The Mormon Cricket, a wingless insect of North America and a serious pest of crops in the Great Plains.

Anacamptis A genus of plants of the Orchid family, *Orchidaceae*. e.g. A. pyramidalis - The Pyramidal Orchid, a perennial plant, 20-60 cm. Grows in woods, copses, and in scrub. Distribution: Europe; native in Britain.

Anacardium A genus of Trees and Shrubs of the Cashew family, *Anacardiaceae*. e.g. A. occidentale - The Cashew Tree, a tropical species with opposite compound leaves, and small flowers in dense clusters; the fruit is the succulent Cashew Nut of economic importance.

Anacridium A genus of Insects of the Locust (Grasshopper) family. e.g. A. aegyptium - The Egyptian Locust; size up to 7 cm; long wings; keel of pronotum crenate, body brown and finely spotted. Occurs in swarms in North Africa.

Anacyclus A genus of plants of the Daisy family, *Compositae*. e.g. A. pyrethrum - Pyrethrum or Pellitory plant; The root is used in lozenges and pastilles to promote secretion of saliva in dryness of the mouth and throat.

Anagallis A genus of plants of the Primrose family, *Primulaceae*. e.g. A. arvensis - Scarlet Pimpernel, an annual plant, 5-60 cm high; red or blue corolla lobes, fringed with short hairs. Grows commonly incultivated and waste land, and in pastures. Distribution: Europe; native in Britain.

Anagasta A genus of Insects of the Moth family, *Phycitidae*. e.g. A. kuehniella - The Mediterranean Flour Moth, an insect pest of stored foodstuffs.

Anagyris A genus of plants of the Pea family, *Leguminosae*. e.g. A. foetida - Stinking Wood; a shrub or small tree 1-3 m high; with densely hairy, stinking leaves (of 3 leaflets); yellow flowers, often with a black spot. Grows on dry hillocks in the Mediterranean region; also in hedgerows and in scrub (Maquis).

Anaimos A genus of Birds of the Flowerpecker family, *Dicaeidae*. The species look rather like kinglets or gold crests and are found in southeast Asia and Australasia.

Anajapx A genus of Diplurid Insects of the family *Projapygidae*. e.g. A. harmosus - A rare Californian species, with two short tubular cerci (terminal abdominal processes). Found in Mexico, South America, and the Mediterranean countries.

Analges A genus of Arachnids of the order of Mites, *Acari*; known as the Hair or Feather Mites.

Anamirta A genus of plants of the family *Menispermaceae*. e.g. A. paniculata (= A. cocculus) - The seeds of this plant contain picrotoxin, a powerful stimulant of the central nervous system.

Anampses A genus of Fishes of the Wrasse family, *Labridae*. e.g. A. meleagrides - The Speckled Wrasse, a brilliantly colored fish, typical of this genus and found in the Red Sea.

Ananas A genus of plants of the Pineapple family, *Bromeliaceae*. e.g. A. comosus - The Pineapple Tree, a native of Brazil, now cultivated in numerous forms in other tropical countries. It has a basal rosette of stiff, spiny, lanceolate leaves up to 1.5 m (5 ft) long. The edible head is formed by the fleshy fruits and persistant bracts. From the plant is derived a mixture of proteolytic enzymes, known as the bromelains or bromelins (= Plant Protease Concentrate), which has been proposed for the treatment of soft tissue inflammation and edema associated with trauma and surgery.

Anaplasma A genus of microorganisms of the family *Anaplasmataceae*, order *Rickettsiales*; parasitic in ruminant animals and transmitted by arthropods. e.g. A. centrale (see **Hippobosca**).

Anarhicas A genus of Fishes of the Wolf Fish family. e.g. A. lupus - The Wolf Fish; 175 cm long; grey, with 10-12 in conspicuous transverse bands. A sea and shore fish, inhabiting the northern waters of Europe; native around Britain.

Anarynchus A genus of Birds of the Plover and Lapwing family, *Charadriidae*. e.g. A. frontalis - The Wrybill or Crooked-billed Plover, found in New Zealand; it is a shore bird with a unique feature: a right-hand lateral twist to its beak, supposed to be an adaptation to its habit of seeking out insects and other invertebrate prey from under stones and pebbles.

Anarrhinum A genus of plants of the Figwort or Snapdragon family, *Scrophulariaceae*. e.g. A. bellidifolium - The Daisy-leaved Snapdragon, a biennial to perennial plant, up to 70 cm high; pale violet flowers growing in racemes. Grows in cultivated fields on light soils. Distribution: the Mediterranean region.

Anas A genus of Birds of the Duck, Goose, and Swan family, *Anatidae*. e.g. A. platyrhynchos - The Mallard Duck, one of the broad-billed dabbling ducks, size 58 cm; a partial migrant bird. Also known as the Wild Duck; it is from this species that the Domestic Duck originates, and with which it still interbreeds occasionally. Distribution: Europe Africa, Asia, and North America.

Anasa A genus of Heteropterous Insects of the Squash Bug family, *Coreidae*. e.g. A. tristis - The North American Squash Bug, a predominantly phytophagous insect.

Anaspides A genus of Crustaceans of the order *Anaspidacea*. e.g. A. tasmaniae - The largest species of this order; found in Tasmania, reaching a length of about 5 cm (2 in). The body is elongated with all the segments distinct, except the first which joins with the head. The eyes are set on short movable stalks. It lives in freshwater.

Anastomus A genus of Openbill or Shell-stork Birds of the Stork family, *Ciconiidae*. e.g. A. lamelligerus - The African Open-bill; its plumage is dark and the feathers of the neck, belly and thigh end in a long narrow plaque with a horny appearance. Its beak when closed, has an obvious gap roughly in the middle. It lives in swamps and on sandy river banks in East Africa, and nests in colonies.

Anatis A genus of Insects of the Ladybird Beetle family, *Coccinellidae*. e.g. A. ocellata - The Eyed Ladybird, size 8-9 mm; a beetle with multiple-spotted elytra (up to 10 black light-edged spots). It feeds on plant lice and is found chiefly in coniferous forests. The larvae are spotted red. Distribution: Europe; native in Britain.

Anax A genus of Insects of the Hawker Dragon-fly family, *Aeshnidae* (order *Odonata*). e.g. A. imperator - The Emperor Dragonfly, size 7-8 cm, the largest British dragonfly; eyes large and green; a swift-flying insect hunter species, preying on large insects such as butterflies, and found by brooks and rivers; also by still water. Estimated as travelling at a speed of 8 m/sec (28.57 km/h; 17.86 miles/h). Distribution: Europe; native in Britain.

Anchusa A genus of plants of the Borage family, *Boraginaceae*; also known by the generic name, *Alkanna*. e.g. A. tinctoria (= *Alkanna* t.) - The Alkanet plant; its root (Alkanet Root), contains a red dye alkannin. Used for coloring toilet preparations of an oily or spirituous nature.

Ancistrocactus A genus of plants of the Cactus family, *Cactaceae*. e.g. A. scheeri - This species has tubercles tipped with radiating spines of which the central one is long and hooked. The small flowers are greenish-yellow. Place of origin: Mexico.

Ancistrodon A genus of poisonous Serpents of the Pit Viper family, *Crotalidae*. Also known by the generic name, *Agkistrodon*. e.g. A. contortrix (= *Agkistrodon* c.) - The Copperhead Snake, a poisonous pit viper found in the United States. A sensitive pit between the eyes and the nostril acts as a heat receptor and detects warm-blooded prey (see **Agkistrodon**).

Ancylostoma A genus of Nematode parasites, the "hookworms", of the order *Strongylida*, also known by other generic names such as, *Ankylostoma*, *Dochmius*, *Sclerostoma*, and *Uncinaria*. e.g. A. duodenale (= *Ankylostoma* d.; *Dochmius* duodenalis; *Sclerostoma* d.; *Uncinaria* d.) - The Common Hookworm.

Ancylus A genus of Gastropod Molluscs of the Limpet family, *Ancylidae*. e.g. A. fluviatilis - The Freshwater or River Limpet; the shell is 4.7 to 7.3 mm long, cup-shaped, thin-walled, and marked with radial grooves. It lives on stones in clean flowing streams, feeding on algae and aquatic moss, chiefly in foothills and mountains. Distribution: Europe; native in Britain.

Anda A genus of Euphorbicaceous Trees of Brazil. e.g. A. assu - from this species is expressed a purgative oil.

Andira A genus of plants of the Pea family, *Leguminosae*. e.g. A. araroba - from this plant is prepared a powder, Crude Chrysarobin or Goa Powder, used in the treatment of psoriasis and other chronic skin diseases.

Andrena A genus of Insects of the Solitary Bee family, *Apidae*. e.g. A. fulva (= A. armata) - The Solitary Bee or Tawny Mining Bee, found in fields, city parks and gardens, where in some measure, it supplements or serves as a substitute for the honeybee, which is less often found in large cities (see **Anthophora**; **Bombylius**).

Andricus A genus of Insects of the Gall Wasp family, *Cynipidae*. e.g. A. fecundator - The Artichoke Gall-fly; the asexual females develop inside the artichoke galls on the twigs of the oak tree. Their off-spring, the sexual generation, develop inside tiny rose-like galls in the flowers.

Androctonus A genus of Arachnids of the Scorpion order, *Scorpiones*. e.g. A. australis - The North African Scorpion, the world's most dangerous scorpion; a large scorpion delivering a massive neurotoxic venom which has been known to kill a man in four hours, and a dog in about seven minutes. A specific antitoxin has been produced by the Pasteur Institute in Algiers.

Andromeda A genus of Shrubs and Trees of the Heath family, *Ericaceae*. e.g. A. mariana - contains a poisonous narcotic, andromedotoxin. A. polifolia - The Marsh Andromeda; 10-40 cm high; a dwarf ever-green shrub with long-stalked, pinkish-red flowers in terminal umbellate clusters, growing mainly on mountains, high moorlands, bogs and wet heaths. Distribution: North and Central Europe; native in Britain.

Andropogon A genus of Grasses of the Valerian family, *Valerianaceae*. These include the broom corn, kafir corn and the sorghum plant. e.g. A. nardus - False Spikenard, an aromatic and stimulant East Indian grass (see **Nardostachys**). A. sorghum - The Sorghum plant.

Androsace A genus of plants of the Primrose family, *Primulaceae*. e.g. A. helvetica - Swiss Rock-jasmine, a perennial plant, 2-5 cm high, forming silver-grey cushions; growing in rock-crevices, on scree and in mountain pastureland. Place of origin: the Alps, up to 3,500 m.

Anemone A genus of plants of the Buttercup family, *Ranunculaceae*. e.g. a. pulsatilla (= *Pulsatilla* vulgaris) - The Pasque Flower plant; it contains the active principle, anemonin and has been used in the treatment of amenorrhea and dysmenorrhea. Distribution: northern and central Europe; western Asia.

Anemonia A genus of Sea Anemones of the order *Actiniaria*. e.g. A. sulcata - The Snakelocks Sea Anemone or Opelet Anemone, 4-20 cmhigh; yellowish-brown to bright green in color due to many minute plants living in its tissues in symbiosis. It has a large number

(150-200) stinging tentacles up to 10 cm long. Lives on rocks and feeds on molluscs, crustaceans and small fish, along the coast of the Mediterranean Sea and the Atlantic coast of Europe to depths of about 6 m.

Anethum A genus of plants of the Carrot family, *Umbelliferae*. e.g. A. graveolens (= *Peucedanum* g.) - The Dill plant; Dill Fruit is used as a carminative.

Angelica A genus of plants of the Carrot family, *Umbelliferae*. e.g. A. archangelica - Common Angelica, a perennial plant, 30-200 cm high; with round, hollow leaf-stalks, and greenish-white or green flowers. Grows in still and running water, and in deep meadows. The fruit and rhizome are used as diaphoretics and expectorants. Place of origin: most of Europe; introduced and naturalized in Britain.

Angiococcus A genus of microorganisms of the family *Myxococcaceae*. e.g. A. cellulosum.

Angiopteris A genus of Fern plants of the order *Marattiales*. e.g. A. erecta - A species of large, primitive tropical and subtropical ferns. It has long, pinnately compound leaves which arise from a short, thick, vertical stem.

Angiostrongylus A genus of microscopic parasitic worms. e.g. A. cantonensis - A common parasite of rodents in the Far East and the Pacific. Man may become infected by eating freshwater prawns and snails containing the larvae. These migrate to the brain where they cause an eosinophilic meningitis. The disease is known as Angiostrongyliasis.

Anguilla A genus of freshwater Fishes of the Eel family, *Anguillidae*. e.g. A. anguilla - The Common Eel or European Freshwater Eel, one of the largest-lived fishes; length up to 150 cm, weighing around 4 kg. The dorsal, caudal, and anal fins run together, and the mouth opening is large. Widely found in freshwater with a muddy bottom. Feeds on crustaceans, small fish, and aquatic insects. A migratory fish, spawning in the Atlantic in the Sargasso Sea.

Anguillula A genus of Nematode Worms. e.g. A. stercoralis (= *Strongyloides* s.) - A parasite causing anguilluliasis or anguillulosis (see **Strongyloides**).

Anguillulina A Nematode Worm, normally present in the bulb of onions, sometimes infecting man.

Anguina A genus of Nematode parasites of the order *Tylenchida*, the Eelworms. e.g. A. tritici - A long-living eelworm, recorded as living in laboratory storage in a dried state for 28 years before it came back to life.

Anguis A genus of Lateral-fold Lizards of the Slow-worm family, *Anguidae*. e.g. A. fragilis - The Anguid, Blind-worm, or Slow-worm Lizard, a leg-less reptile, 40-50 cm long; the back is bluish-grey to dark greyish-brown with a fine black line down the center. It lives in moist regions under stones, tree trunks, and fallen leaves. A harmless creature, it is frequently killed because of its resemblance to a snake. It has an authentic age of "more than 54 years". Distribution: almost all of Europe except Ireland.

Anhalonium A genus of plants of the Cactus family, *Cactaceae*. e.g. A. lewinii (= *Lophophora* williamsii; *Mammillaria* l.) - The Cactus plant; contains mescaline (peyote), a drug used by the natives in Mexico to produce a state of intoxication, marked by feelings of ecstasy (see **Lophophora**).

Anhima A genus of Birds of the Screamer family, *Anhimidae*. e.g. A. cornuta - The Horned Screamer, about 75 cm (30 in) long; lives in the rain-forests of tropical South America. As big as a turkey, it is greyish-black above and white below, with white rings around the eyes and neck.

Anhinga A genus of Birds of the Darter family, *Anhingidae*. A close relative of the cormorants, the darters or anhingas, sometimes called the snake-birds, are distinguished by their long and slender necks, which terminate in long, straight, pointed bills. These are fish-eating, web-footed birds of tropical Africa and America. e.g. A. melanogaster - The Indian Darter or Snakebird; its method of fishing is special; it spears its prey as it hunts it through the water, throws the fish into the air, and swallows it as it falls.

Anhydrophryne A genus of Amphibians of the Frog family, *Ranidae*. e.g. A. rattrayi - Rattrayi's Frog; a species living on the forest floor in South Africa, whose eggs develop in a hole in the ground and whose tadpoles are so helpless that they drown if placed in water.

Anisomorpha A genus of Insects. e.g. A. buprestoides - An orthopterous insect, also known as the Walking Stick.

Anisoplia A genus of Chafers and Dung Beetles of the family *Scarabaeidae*.

Anisotremus A genus of Fishes of the Grunt family, *Pomadasyidae*, so called because they can make a grunting noise with their air bladders. e.g. A. virginicus - The Porkfish; an inhabitant of coral reefs.

Ankylostoma A genus of nematode parasites, also called *Ancylostoma*, the Hookworms. e.g. A. duodenale (= *Ancylostoma* d.) - The Common Hookworm.

Annarrhicas A genus of Wolf Fishes of the family *Anarrhichadidae*. e.g. A. lupus - The Atlantic Wolf Fish; length 120 cm; occurs at depths of 100-300 m, in the northern parts of the northeast Atlantic.

Anoa A genus of Cattle of the family *Bovidae*. e.g. A. depressicornis - The Wild Dwarf Buffalo of the Celebes (= Pygmy Buffalo or Celebes Anoa); the smallest of the wild Cattle, only 100 cm (40 in) high at the shoulders. Its horns are straight and sweep back like those of an antelope. Found in the jungle swamps of India, Burma, and some of the islands of Indonesia.

Anobium A genus of Beetles of the family *Anobiidae*. e.g. A. punctatum - The Woodworm or Furniture Beetle, length 3-5 mm; developing in old wood, in areas such as floors, furniture, rafters, etc. They bore tunnels in the wood, which is their sole food, cellulose being digested with the aid of symbiotic organisms that live in special gut diverticula in their bodies.

Anodonta A genus of Freshwater Molluscs of the Swan Mussel family, *Unionidae*. e.g. A. cygnea - The Swan Mussel, 170-220 mm long, with a thin, oval, greenish-yellow shell, found on the muddy bottom of ponds, reservoirs, and slow-flowing rivers. As in other of its species, the eggs hatch into small parasitic larvae called glochidia, which infest fish. Distribution: Europe; native in Britain.

Anomala A genus of Insects of the Cockchafer or Dung Beetle family, *Scarabaeidae*. e.g. A. dubia - The Bronze Anomala, 1.5 cm long; head and pronotum green; elytra yellow or green; found on bushes and cereal crops. Distribution: Europe; native in Britain.

Anomalurus A genus of Mammals of the Scaly-tailed Flying Squirrel family, *Anomaluridae*, found in tropical Africa. These highly specialized rodents are confined to the high forests, where they sleep in hollow trees during the day and emerge at dusk to make long gliding flights from tree to tree in search of food, mainly fruit and nuts. e.g. A. peli - The Scaly-tail of West Africa, as large as a domestic cat, and conspicuously black and white.

Anomia A genus of Bivalve Molluscs of the Saddle Oyster or Jingle Shell family, *Anomiidae*. e.g. A. ephippium - The Saddle Oyster; it has a thin brownish-red to brownish-yellow shell, 50-65 mm long; the lower valve has a hole for the byssus or tuft of horny threads by which the animal anchors itself to surrounding objects. Found at shallow depths in the sea, from low water to about 150 m, attaching itself to wood or other hard objects on the sea bottom. Distribution: the coasts of Europe; native around Britain.

Anomma A genus of large Ants, more than 3.75 cm (11/2 in) long, found in Africa.

Anonymus A genus of Polyclad Tubellarian Flatworms of the suborder *Cotylea*. This genus of large marine forms has a plicate pharynx, a many-branched intestine, and a small pseudo-sucker.

Anopheles A genus of Mosquitoes transmitting malaria, the most dangerous insects in the world, which excluding wars and accidents, have probably been responsible directly or indirectly for 50% of all human deaths since the Stone Age. This genus, about 6 mm in size, is characterized by long slender palpi, nearly as long as the beak, and also by the holding of the body at an angle with the surface on which it rests, while the head and beak are in line with the body. Some species are vectors of *Wuchereria* bancrofti. The chief malaria-carrying (54) *Anopheles* species in the world, include: aconitus, albimanus, albitarsis, amictus, annularis, annulipes annulipes, aquasalis, bancrofti, barbirostris barbirostris, bellator, claviger, culcifacies, darlingi, farauti, fluviatilis, funestus, gambiae, hancocki, hargreavesi, hyrcanus nigerrimus, hyrcanus sinensis, jeyporiensis candidiensis, jeyporiensis jeyporiensis, kochi, labrianchiae atroparvus, labrianchiae labrianchiae, leucosphyrus leucosphrus, lungae, maculatus maculatus, maculipennis freeborni, mangyanus, messeae, minimus, minimus flavirostris, moucheti moucheti, moucheti nigeriensis, multicolor, nili, pattoni, pharoensis, philippinensis, pretoriensis, pseudopunctipennis pseudopunctipennis, punctimacula, punctulatus punctulatus, quadrimaculatus, sacharovi, sergentii, stephensi

stephensi, subpictus subpictus, sundaicus, superpictus, umbrosus, varuna.

Anoplius A genus of Insects of the Spider-hunting Wasp family, *Pompilidae*. e.g. A. viaticus - 7-20 mm long; locally abundant on flowering *Umbelliferae* (Carrot family). The larvae feed on spiders which the female hunts, paralyses, and carries to the underground nest. Distribution: all over Europe.

Anoplocephala A genus of Tapeworms (class *Cestoda*) of the family *Anoplocephalidae*. These belong to the order *Cyclophyllidea*.

Anous A genus of Birds of the Tern family, *Laridae*. e.g. A. stolidus - The Common Noddy or Noddy Tern of America. Like other terns, these tropical birds have pointed bills, very long, narrow pointed wings, and forked tails. They are sometimes known as Sea-swallows.

Anser A genus of Birds of the Duck, Goose, and Swan family, *Anatidae*. e.g. A. anser - The Wild or Grey (= Greylag) Goose, size 76-89 cm, the largest European Wild Goose; bill orange, head light-colored, legs flesh-colored. A partial migrant. From these wild mobile species was derived the ponderous Domestic Goose. Distribution: northern and central Europe; central Asia. Native in Britain. A. anser domesticus - The Domestic Goose, the largest-lived domesticated bird excluding the ostrich, having a usual natural lifespan of about 25 years. Derived originally from the Wild Goose, A. anser.

Antedon A genus of Sea-lilies of the class *Crinoidea*. e.g. A. parviflora - The Feather-star Sea-lily, one of the smallest crinoids, found in Japanese waters and having a span of c. 40 mm (1.57 in).

Antennaria A genus of plants of the Daisy family, *Compositae*; also known by the generic name, *Gnaphalium*. e.g. A. dioica (= *Gnaphalium* dioicum) -

Cat's Foot plant; a perennial plant, 2-25 cm high; the flower heads are used as an expectorant and a poultice for bruises and wounds. Widespread on heaths, dry hillocks, slopes and in open woods. Distribution: Europe; native in Britain.

Antennarius A genus of Fishes (subphylum *Vertebrata*). e.g. A. molluccensis - The Yellow Angler fish; a small string-like filament is attached above its mouth, ending in a fleshy bulb. This is used as bait to attract small fish, to be swallowed by the Angler's lightning-swift gulp.

Antestia A genus of Heteropterous Terrestrial Bugs or Geocorisae, of the Shield-bug family, *Pentatomidae*. e.g. A. variegata - An insect pest, attacking fruit trees and coffeeplants in Africa.

Anthaxia A genus of the Metallic Wood-borer Beetle family, *Buprestidae*. e.g. A. nitidula - A species common in warm regions, such as central and southern Europe and North Africa. Larval development takes place in dying fruit trees.

Anthemis A genus of plants of the Daisy family, *Compositae*; variously known by other generic names such as, *Chamaemelum* and *Matricaria*. e.g. A. nobilis (= *Chamaemelum* nobile; *Matricaria* chamomilla; M. recutita) - The Common or True Chamomile, the Wild Chamomile, or the American Garden Chamomile; a perennial plant, 10-40 cm high, with white female ray florets and many small bisexual yellow disc florets borne on blunt, oblong, coni-cal, receptacle scales. Grows on sandy places, pastures, and by roadsides; also cultivated in gardens. In Britain the Common or True Chamomile is *Chamaemelium* nobile or British Camomille; the Wild Chamomile is *Matricaria* chamomilla (= M. recutita) or German Chamomile (= Camomille Romaine) of Central Europe; and the American Garden Chamomile is *Anthemis* nobilis. The plant is strongly aromatic and from the flower heads,

Chamomile Flowers, is extracted a volatile oil, Oil of Chamomile. Used as an aromatic bitter, carminative and febrifuge. An infusion, "Chamomile Tea" is a domestic remedy for indiges-tion. Distribution: central and western Europe (native in Britain), western Asia to India. Has also been introduced into North America and Australia.

Antheraea A genus of Insects of the Emperor Moth family, *Saturniidae*. This species produces silk in usable quanti-ties.

Anthericum A genus of plants of the Lily family, *Liliaceae*. e.g. A. ramosum - Spiderwort, a perennial plant, 30-100 cm high, with white flowers on branched inflorescences. Grows in dry woods and grasslands. Distribution: Central Europe.

Anthias A genus of Fishes (subphylum *Vertebrata*). e.g. A. squamipinnis - The Sea Goldfish, one of the most common schooling reef fish in the Red Sea region. All male sea goldfish are born females, and the larger individuals undergo a spontaneous sex change.

Anthobothrium A genus of Tapeworms (Merozoic cestodes) of the subclass *Eucestoda* (= *Cestoda*; *Merozoa*). e.g. A. auriculatum - A tapeworm found in the spiral valve of the skatefish (family *Rajoidae*).

Anthocharis A genus of Insects of the Whites and Yellows Butterfly family, *Pieridae*. e.g. A. cardamines - The Orange Tip Butterfly, size 4.5 cm; the male has a large orange patch at the tip of each forewing, the female a smaller black patch. It hibernates in the chrysalis stage. Distribution: Europe; United States.

Anthocoris A genus of Flower Bugs of the family *Anthocoridae*. e.g. A. nemo-rum - The Common Flower Bug, 3-4.5 mm long; very common indeciduous woodlands and gardens. Distribution: most of Europe, northern Asia, and North Africa.

Anthomyia A genus of small black Houseflies. e.g. A. canicularis - Larvae of this species are sometimes present in human feces.

Anthonomus A genus of Beetles of the Weevil family, *Curculionidae*. e.g. A. pomorum - The Apple Blossom Weevil; found in fruit orchards. The larvae which live in the apple blossom buds, are seen more often than the beetles themselves.

Anthophora A genus of Solitary Bees of the family *Apidae*. The species are host tothe Oil beetles (*Meloidae*), in that their nests are used by the latter in their larval development (see **Andrena**).

Anthornis A genus of Birds of the Honey-eater family, *Meliphagidae*. e.g. A. melanura - The New Zealand Bell-bird; it has a highly specialized, extensile, brush-like tongue which can be used as a probe brush, and sucking-tube for drawing nectar, and also swallowing the insects assembled drinking it.

Anthoxanthum A genus of plants of the Grass family, *Gramineae*. e.g. A. odoratum - Sweet Vernal-grass, a perennial plant, 30-50 cm high; grows in dense panicles, with unbranched culms and thinly hairy glumes. Strongly scented with coumarin. Widespread and common in dry meadows and pastures, also in open woodlands. Distribution: Europe; native in Britain.

Anthrenus A genus of Insects of the Carpet, Hide, or Museum Beetle family, *Dermestidae*. e.g. A. scrophulariae - The Carpet Beetle, up to 4 mm long; black body with three indistinct, transverse bands. The adult is harmless, living on flowers. The larvae feed on furs, carpets, fabrics, feathers and natural history collections. A. verbasci - A Museum Beetle species, the most feared pest of collections.

Anthreptes A genus of Birds of the Sunbird family (class *Aves*). e.g. A. mallaecensis - The Brown-throated Sunbird, a perching bird; it builds a characteristic suspended nest with a lateral entrance.

Anthriscus A genus of plants of the Carrot family, *Umbelliferae*. e.g. A. sylvestris - Cow Parsley, a biennial plant, 15-150 cm high; 715-rayed umbels; fruit smooth or with bristly tubercles. Widespread in waste and grassy places, by hedgerows and at edges of woods. Distribution: Europe; native in Britain.

Anthrophora A genus of Insects of the Solitary Bee family. e.g. A. parietina - Size 13-15 mm; it has a compressed body, yellowish-brown thorax, with dense hairs. It builds its nest in clay walls. Distribution: Europe.

Anthropoides A genus of Birds of the Crane family, *Gruidae*; also known by the generic name, *Anthyopoides*. e.g. A. virgo (= *Anthyopoides* v.) - The Demoiselle Crane, almost 90 cm, 3 ft) tall; a graceful bird with dazzling white tufts of feathers on each side of its head; smaller than other cranes. Always nests on dry ground. A partial migrant. Distribution: Central Asia, North Africa and southeast Europe.

Anthurium A genus of plants of the Arum family, *Araceae*. e.g. A. andreanum - The Flamingo Flower or Painter's Palette, an ever-green plant up to 45 cm (18 in) high, with drooping lanceolate leaves and orange-red, palette-shaped flower spathes, each with a protruding white spadix. An indoor flowering plant originally from Columbia.

Anthus A genus of Perching Birds of the Pipit family, *Motacillidae*. e.g. A. trivialis - The Tree Pipit, 15 cm long; plumage has vivid brown streaks; reddish legs. A partial migrant found in open woodlands, heaths, trees and bushes, from lowland to mountainous elevations. Distribution: North and Central Europe; native in Britain.

Anthyllia A genus of plants of the Pea family, *Leguminosae*. e.g. A. vulneraria

34

- Kidney Vetch, a perennial plant, 5-50 cm high; with pinnate stem leaves and woolly, inflated calyces. Widespread in dry places. Distribution: southern Europe.

Anthyllis A genus of plants of the Pea family, *Leguminosae*. The species belong to the Clover subfamily, *Papilionaceae*.

Anthyopoides A genus of Birds of the Crane family, *Gruidae*; also known by the generic name, *Anthropoides*. e.g. A. virgo (= *Anthropoides* v.) - The Demoiselle Crane (see **Anthropoides**).

Antiaris A genus of Artocarpus Trees. e.g. A. toxicaria - The poisonous Upas tree of Java.

Antidorcas A genus of True Antelopes from Africa, of the subfamily *Antilopinae*. e.g. A. marsupialis - The Springbok or Springbuck, once involved in huge migrations ("trekbokkens") across the high open plains of the western parts of southern Africa in the 19th century. Now found in reasonable abundance only in the Kalahari Desert and in parts of S. West Africa, where these gazelle herds no longer threaten to disturb the delicate balance of nature.

Antilocapra A genus of Antelope-like ungulates, the Pronghorns, of the family *Antilocapridae*. e.g. A. americana - The Pronghorn Antelope of the western U.S.A., the fastest of all terrestrial animals over a sustained distance, with an estimated cruising speed of 56 km/h (35 miles/h) for 6.4 km (4 miles). A protected animal, often referred to as the American Antelope; it is not an antelope, but a hollow-horned Ungulate.

Antilope A genus of Antelopes of the subfamily *Antilopinae*. e.g. A. cervicapra - The Blackbuck, inhabiting the plains of India, especially in areas of tall grass. In this species sexual dimorphism is carried very far: the males are nearly black with long spiral horns, while the females are light beige and hornless. A. rupicapra - The Chamois, a horned,

goat-like antelope of the high Alps and European mountains. Its skin makes a very soft, warm leather used for making gloves, polishing glass, etc. (see **Ovibos**).

Antipataria A genus of Black Corals of the octocoral family, species of which are found in the deep waters of the Red Sea coral reefs, and are treasured by jewellers due to their dark colored skeletons.

Antipathes A genus of Black or Thorny Corals of the order *Antipatharia*. The body consists of a brown or black skeletal axis consisting of horny material and covered with thorns, with polyps spread along the axis. These animals live in deep abyssal waters and little is known about their biology.

Antirrhinum A genus of flowering plants of the Figwort or Snapdragon family, *Scrophulariaceae*. e.g. A. majus - The Snapdragon, an erect woody-based perennial plant, 25-30 cm (10-32 in) high, with reddish-purple flowers in dense terminal racemes. Also cultivated in a wide range of colors, from white and yellow to mauve, scarlet, and crimson. Place of origin: Mediterranean region; naturalized in the UK.

Anurida A genus of Springtail Insects (order *Collembola*), of the family *Poduroidea*. e.g. A. maritima - A species found on the seashore in Europe; they have a rough, warty integument with short appendages.

Aoliscus A genus of Fishes (subphylum *Vertebrata*). e.g. A. punctulatus - The Razorfish, a curious little fish with a long compressed body, that swims in a vertical position, with its long tube-like snout pointing upwards.

Aonyx A genus of Otters of the family *Mustelidae*, animals completely adapted to an aquatic life. The short dense fur is impermeable to water, and the broad flattened tail is used for steering. e.g. A. cinerea - The Indian Small-clawed Otter, also found from the East Indies to the Himalayas and China.

Aotes See **Aotus**.

Aotus A genus of Douroucoulis New World Monkeys, also called Night-apes or Owl-faced Monkeys, of the family *Cebidae*. Sometimes, also known by the generic name, *Aotes*. e.g. A. trivirgatus - The Three-banded Douroucouli or Night Ape; greyish-brown, with three black stripes along the head. About 32.5 cm (13 in) long, with a tail measuring nearly 50 cm (20 in). Found in Central and South America, from Guiana south and west to Brazil and Peru.

Apamea A genus of Moths of the family *Noctuidae*. e.g. A. monoglypha - The Dark Arches Moth; below the margin of the forewings is a light pattern with a clearly evident letter "W".

Apanteles A genus of parasitic Insects (suborder *Parasitica*) of the family *Ichneumonidae*. e.g. A. glomeratus - A common parasite of the Cabbage White caterpillars; its numerous sulfur-yellow cocoons can be seen among the larvae they have killed.

Apatele A genus of Insects of the Moth family, *Noctuidae*. e.g. A. psi (= *Acronicta* p.) - The Grey Dagger Moth (see **Acronicta**).

Apatosaurus A genus of extinct animals, the dinosaurs (see **Brontosaurus**).

Apatura A genus of Insects of the Tortoiseshell Butterfly family, *Nymphalidae*. e.g. A. iris - The Purple Emperor Butterfly, size 6.5 cm; the male shave blackish-brown wings with a purplish sheen. Found in damp forests flying around tree tops, especially over oak woods. Distribution: temperate parts of Europe; Japan. Native in Britain.

Apeltes A genus of Fishes of the Stickleback family, *Gasterosteidae*. This is the American form of the Fifteen spine Stickleback, a marine shore fish.

Apera A genus of plants of the Grass family, *Gramineae*. e.g. A. spicu-venti - Loose Silky-bent, an annual grass plant, 40-100 cm high; has a large, open, very delicate panicle and sharp, pointed leaves. Widespread, growing commonly on cultivated and waste land and in dry places. Distribution: Europe; native in Britain.

Aphanes A genus of plants of the Rose family, *Rosaceae*. e.g. A. arvensis - Parsley Piert, an annual plant, 1-30 cm high; flowers grow in dense leaf-opposed clusters. Commonly found on arable land and bare grassland. Distribution: Europe; native in Britain.

Aphanizomenon A genus of Blue-green Algae of the group *Cyanophyta*. e.g. A. flos-aquae - A planktonic species which floats just beneath the surface of lakes, multiplying rapidly and coloring the water green.

Aphelandra A genus of indoor plants of the family *Acanthaceae*. e.g. A. squarrosa - The Zebra plant, a small evergreen shrub with shining green elliptic leaves with slender points, the main veins ivory colored, giving a zebra effect. The yellow leaves are enclosed in deep-yellow, red-edged bracts. Height 45 cm (18 in). Used as a decorative indoor plant. Place of origin: Brazil.

Aphelenchus A genus of Nematode Worms. e.g. A. avenae - An amazing species, which in times of stress (e.g. lack of oxygen), manufactures its own alcohol! Once it has stored enough alcohol, it goes into a state of suspended animation (cryptobiosis), which can preserve its life indefinitely. When the emergency is over the worm surfaces again, drinks the alcohol, and carries on as usual.

Aphelocheirus A genus of Insects, the Amphibisorisae, of the Swimming Bug family, *Naucoridae*. These are broad, flat insects living on the surface of the water. They have adopted plastron respiration, and therefore do not need to visit the surface to obtain oxygen. e.g. A. aestivalis - A "swimming bug", 8.5-10 mm long with compressed body.

36

Aphia A genus of Fishes of the Goby family, *Gobiidae*. e.g. A. minuta - The Transparent Goby, a marine fish, 7 cm long; completely transparent. An active species, living in tidal pools; if left stranded, it is capable of jumping across the sand to find water. Habitat: the seas of Europe.

Aphiochaeta A genus of Insects of the Fly family. e.g. A. ferruginea - A species found in South America and India, causing cutaneous miasis in man.

Aphodius A genus of Insects of the Dung Beetle family, *Scarabaeidae*. e.g. A. fimetarius - Found from early spring in horse and cattle dung, where the eggs are laid.

Aphredoderus A genus of Fishes of the Perch family, *Aphredoderidae*; order *Salmopercae* (= *Percopsiformes*). e.g. A. sayanus - The Pirate Perch fish, a species of archaic fishes that are relics of an earlier fauna; it has no adipose fin. Found in the cooler parts of North America; common in the Great Lakes.

Aphriza A genus of Birds of the Plover family, *Charadriidae*. e.g. A. virgata - The Surfbird; mottled grey and white; it has a blacktail with a white band at the base. Found on wild surf regions from Alaska to Chile.

Aphrocalistes A genus of Marine Sponges, better known by the generic name, *Aphrocallistes* (see **Aphrocallistes**).

Aphrocallistes A genus of Marine Sponges of the suborder *Hexasterophora*; the typical spicule is a hexaster with a six-rayed star shape. e.g. A. beatrix (= *Aphrocalistes* b.) - A Japanese deep-sea sponge, resembling a fragile cockade of Meissen porcelain.

Aphrodite A genus of Segmented Worms (phylum *Annelida*) of the family *Aphroditidae* e.g. A. aculeata - The Sea Mouse, an elongated oval worm with a short and wide body (100-200 mm long and 50 mm wide), covered with numerous bristly hairs. Lives on soft sea bottoms and feeds on minute animals. Distribution: Atlantic coast of Europe and the Mediterranean Sea. Native on British coasts.

Aphrophora A genus of Homopterous Insects of the family *Cercopidae*. e.g. A. alni - The European Spittlebug, a small brown insect. Its larva feeds on various bushes and herbs.

Apis A genus of Insects of the Bee family, *Apidae*. e.g. A. mellifera - The Hive or Honey Bee (= Worker Honey Bee), length 14-18 mm; lives in colonies numbering 40,000-80,000 individuals, each "ruled" by a queen which lays the eggs. From the salivary secretions is prepared Royal Jelly (Queen Bee Jelly).

Apium A genus of plants of the Carrot family, *Umbelliferae*. e.g. A. graveolens - The Celery plant or Wild Celery, a biennial plant, 30-100 cm high; pinnate basal leaves, umbel with 6-12 rays, bracts and bracteoles absent. Grows in damp places near the sea, by rivers and ditches. Used as an antispasmodic and diuretic. Distribution: worldwide. A. petroselinum - The Parsley plant (see **Petroselinum**).

Aplocera A genus of Insects of the Butterfly and Moth family, *Geometridae*. e.g. A. plagiata - The Treble Bar Moth, found on dry and warm hillsides, in steppes and forest margins.

Aplocheilichthys A genus of Fishes of the suborder *Cyprinodontei*. e.g. A. pelagicus - An open-water fish inhabiting the larger lakes of Africa.

Aplodinotus A genus of Fishes (subphylum *Vertebrata*). e.g. A. grunniens - The Freshwater Drum fish.

Aplodontia A genus of Mountain Beavers of the family *Aplodontidae*. Found in North America, it has no close relationship with the True Beavers (family *Castoridae*). It is a primitive rodent, perhaps in the ancestral line to the squirrel. e.g. A. rufa - The Sewellel or Mountain Beaver, the sole species of this family, brown in color, about the size of a rabbit, and almost tailless. Lives in the coastal areas of the western United States up to altitudes of 9,000 ft, making extensive burrows, usually near water.

Aploppus A genus of Weeds or Weed plants, poisonous to cattle and sheep. e.g. A. heterophyllus - The Jimmey Weed, a plant causing Milk Sickness (see **Eupatorium** urticaefolium).

Aplysia A genus of Sea Slug, Gastropod Molluscs of the order *Anaspidea* (= *Aplysiomorpha*). e.g. A. punctata - The Sea Hare, size up to 8 cm; it possesses a small delicate shell which covers the viscera, and is bounded by upwardly extending folds of the foot, the parapodia. It inhabits areas of green seaweed on which it feeds. Distribution: the coasts of Europe; native around Britain.

Apodemus A genus of Mammals of the Old World Rat and Mouse subfamily, *Murinae* (family *Muridae*). e.g. A. sylvaticus - The Long-tailed Field-mouse or Wood Mouse, 9 cm long with an equally long tail; found in fields, copses, woods, cemetries, parks and gardens; in winter it enters buildings. Rusty colored. Mostly nocturnal, it burrows and hoards, and does a great deal of damage. Distribution: throughout Europe, Asia, and North Africa.

Apoderus A genus of Insects of the Weevil (Beetle) family, *Curculionidae*. These are their relatives, the Leaf Roller beetles. e.g. A. coryli - The Nut Apoderus, a Leaf Roller species, 6-8 mm long; head black; thorax, elytra, and upper leg joints red. The female rolls hazel leaves. Distribution: Europe; native in Britain.

Aporia A genus of Insects of the Whites and Yellows Butterfly family, *Pieridae*. e.g. A. crataegi - The Black-veined White Butterfly, size 6.5 cm, with white wings, and black veins running to the outer edge; lays 150-200 eggs on hawthorne and sloes. Formerly the caterpillars were feared pests of fruit trees, but today the species are much rarer and no longer pose a problem. Distribution: Europe; once common in Britain, now extinct there.

Aporocactus A genus of plants of the Cactus family, *Cactaceae*. e.g. A. flagelliformis - The Rat's Tail Cactus, a free-flowering plant, with showy vermilion-colored flowers. Used as a pot plant for decorating windows, balconies, etc. Place of origin: Central America; introduced into Europe in the late 17th century.

Aporrectodea A genus of Earthworms of the family *Lumbricidae*. e.g. A. caliginosa - A common species, 60-170 mm long, with variable color (greyish-blue, grey, pink, and brown). Also has a variable number of segments, ranging from 100 to more than 200. Distribution: Europe.

Aporrhais A genus of Conch Molluscs of the order *Mesogastropoda*. e.g. A. pespelecani - Pelican's Foot, a strombid or conch mollusc, 5 cm in size; outer lip finger-shaped; a herbivore animal, burrowing in muddy gravel offshore or at moderate depths in north temperate seas. Distribution: widespread in European waters; native around Britain.

Aposeris A genus of plants of the Daisy family, *Compositae*. e.g. A. foetida - The Aposeris, a perennial plant, 15-40 cm high; resembles the Common Dandelion (see **Taraxacum**); stinking, milky sap; no pappus, and solitary heads. Grows in woods and scrub, and in meadows in the Alps. Distribution: Europe.

Aprocta A genus of Filarial Worms. e.g. A. semenova - A species infesting the eyes of birds.

Aptenodytes A genus of Birds of the Penguin family, *Spheniscidae*. e.g. A. forsteri - The Emperor Penguin of Antarctica, the deepest-diving bird in the world, with a recorded dive of 265 m (869 ft) below the ice surface, at the diving station where this figure was recorded. It lays the smallest egg laid by any bird in relation to its body weight - exactly 1.4% of its total body weight. It has the lowest egg output of any living bird, producing only one egg annually.

Apterona A genus of Insects of the Bagworm Moth family, *Psychidae*. e.g. A. crenulella - 6 to 7 mm long; occurring in the so-called helix form; the caterpillars make coiled cases resembling small snail-shells.

Apterygiformes A former genus name of Birds, now used to include the order *Apterygiformes*, of the Kiwi family of New Zealand. These birds lay the largest eggs laid by any bird in proportion to their size, nearly one quarter of the body-weight of the hen. The egg of the chicken-sized Kiwi is more than ten times as large as a hen's egg. This however can be disastrous, and not unfrequently, female kiwis containing fully developed eggs have been found dead in their nesting burrows (see **Aptereryx**).

Apteryx A genus of Birds of the Kiwi family, *Apterygidae*, a single family of the order *Apterygiformes*. e.g. A. australis (formerly called *Apterygiformes* a.) - The Common or Brown Kiwi; it lays the largest egg laid by any bird; with a long incubation period of 75-80 days. A nocturnal animal of New Zealand, its name Kiwi is derived from its cries of ki-i-wi at night, when it emerges to feed on earthworms, insects, larvae, soft fruit, and leaves (see **Apterygiformes**).

Apus A genus of Birds of the Swift family, *Apodidae*. e.g. A. apus - The Common Swift, 16.25 cm (61/2 in) long; the most aerial of all land birds, being known to remain aloft for at least nine months at a stretch, and possibly much longer. An all-black migrant bird, with a light-colored patch on the throat; it is unmistakable because of its shrill cry repeated all the time it is on the wing. It has long sickle-shaped wings,

and a short forked tail. Distribution: Europe, Asia, northwest Africa.

Aquarius A genus of Insects of the order of True Bugs, *Heteroptera*. These are the Water-striders, Pond-skaters or freshwater Bugs which run only on the surface of the water. e.g. A. najas - The Great Pond Skater, size of male 13 mm, the female is larger; it runs rapidly on its four hind legs, the two front legs holding the prey. Distribution: Europe; native in Britain.

Aquila A genus of Birds of the Eagle subfamily, *Accipitrinae* (family *Accipitridae*). e.g. A. chrysaetos - The Golden Eagle, length 82 cm; body dark brown, scrown and nape golden-yellow; the tarsi are feathered down to the toe. A resident bird, found in vast lowland forests in Asia; in Europe it has been forced into the mountains. It has extremely acute vision, and has been known to detect a 450 mm (18 in) long hare at a range of 1,965 m (2,150 yd), or even up to 3 km (approx. 2 miles) in good light and against a contrasting background. Distribution: Europe, Asia. It is the only eagle native to Britain, where it is now restricted to the Scottish Highlands.

Aquilegia A genus of climbing and spreading plants of the Buttercup family, *Ranunculaceae*. e.g. A. vulgaris - Columbine, a perennial plant, 30-90 cm high, with bluish-colored flowers, and strongly hooked spurs. Distribution: west, central, and southern Europe; native in Britain.

Ara A genus of Birds of the Macaw Parrot and Parakeet family, *Psittacidae*. These are strikingly colored, long-tailed birds found in the New World. e.g. A. macao - The Scarlet Macaw, 90 cm (3 ft) long; an arboreal tree-bird feeding on fruit, berries, and nuts which it cracks open with its strong hooked beak. Native to the tropical rain forests of Central America, from Mexico to Bolivia.

Arabidopsis A genus of plants of the Cress or Crucifer family, *Cruciferae*. e.g. A. thaliana - The Thale Cress, an annual plant, 6-30 cm high, with stalk-less stem-leaves, and linear fruit. Widespread and growing in waste land, hedge rows and grassy places, and on walls and river-banks. Distribution: Europe; native in Britain.

Arabis A genus of plants of the Cress or Crucifer family, *Cruciferae*. e.g. A. albi-da (= A. caucasia; A. caucasica) - Rock Cress or Garden Arabis, a low, mat-forming, perennial plant, 10-40 cm (4-16 in) high, with white fragrant flowers in loose clusters. Place of origin: Southeast Europe and the Middle East; introduced into Britain in 1798.

Arachis A genus of plants of the Pea family, *Leguminosae*. e.g. A.hypogaea - The Common Peanut or Groundnut; a small herb with shortly stalked yellow flowers. After fertilization the stalk of the ovary elongates and the young fruits are pushed underground, where they ripen. The seeds of the mature fruits yield Arachis Oil (= Groundnut Oil; Peanut Oil), used internally as a nutrient, demulcent, and mild laxative, in commerce as a valuable source of cooking-oil, and externally as an emollient, lubricant, and protective. A native of Brazil, the plant is now widely cultivated in the warmer parts of North America, Africa, and Asia.

Arachnocampa A genus of Dipterous Insects of the family *Mycetophilidae*, also known as the fungus gnats. e.g. A. luminosa - A species found in New Zealand; its larvae can produce light, and the glowworm cave at Waitomo in New Zealand, is celebrated for their luminous activities.

Arachnoidiscus A genus of Diatoms of the class *Bacillariophyta*. e.g. A. ehren-bergii - Microscopic unicellular organisms with a circular surface or valve view.

Arachnospermum A genus of plants of

the Daisy family, *Compositae*. e.g. A. laciniatum - Spider-seed, an annual plant, 20-40 cm high branched stem, deeply pinnate leaves, solitary heads with sulfur-yellow florets. Found in scattered localities, preferring calcareous soils, it grows by waysides, on sandy banks, waste land, heaths and in grassy places. Distribution: Europe; native in Britain.

Aralia A genus of aromatic and diaphoretic plants of the family *Araliaceae*. e.g. A. quiquefolia - Ginseng; used as a sedative in fatigue and neurasthenia. Ginseng is also prepared from the Panax plant (see **Panax**). A. elegantissima (= *Dizygotheca* e.) - False Aralia (see **Dizygotheca**). Other species belong to the Spignet or American aralia plants, which are aromatics and stimulants.

Aramides A genus of Birds of the Rail, Moorhen and Coot family, *Rallidae*. e.g. A. ypecaha - The Ypecaha Wood-rail of South America, from Brazil to Argentina; one of the largest rails, 18 in long, brown with a grey head and throat, blackish tail and chestnut breast.

Aramus A genus of Birds of the Limpkin family, *Aramidae*. e.g. A. guarauna - The Limpkin or Courlan, the sole member of the family, a sedentary wader of tropical America; has the build of a crane. Its melancholy call has earned it a variety of names: Clucking hen, Courlan, Lamenting Bird, and Crazy Widow. It frequents marshes and waterways, hunting large snails, and is found as far north as the states of Georgia and Florida.

Araneaus See **Araneus**.

Araneus A genus of Arachnids of the Orb Web Spider family, *Araneidae*; also known by the generic name, *Araneaus*. e.g. A. diadematus (= *Araneaus* d.) - The Garden Spider, Garden Cross Spider, or Diadem Spider; male 4-8 mm; female 10-15 mm; yellowish-brown, with white cross on the abdomen. Found on shrubs and trees, and in woods. One of the commonest European species. Native in Britain.

Arapaima A genus of Fishes of the sub-order *Osteoglossoidea*. e.g. A. gigas - The large pike-like Arapaima Fish of tropical rivers also called the "Pirarucu". The largest freshwater fish in the world, it is found chiefly in the Amazon drainage regions of Brazil and Peru. It is armed with hard bony scales, and can reach a length of 2 to 2.5 m.

Araschnia A genus of Insects of the Tortoiseshell Butterfly family, *Nymphalidae*. e.g. A. levana - The male is 2.7 cm in size; the female is 3.5 cm, with the upper side reddish-yellow with blackish spots, and white spots at the wing tips. It lays its eggs on nettles. Distribution: Europe.

Araucaria A genus of Trees of the Pine family, *Pinaceae*. e.g. A. araucana - The Chile Pine or Monkey Puzzle tree, 45 m (150 ft) tall, with a slender, cylindrical trunk and whorls of spreading branches. The stiff, pointed, dark-green leaves persist for 10 to 15 years. The globose cone is up to 17.5 cm (7 in) long. Place of origin: a native of Chile, Tierra del Fuego, and northern Patagonia.

Arbutus A genus of Trees and Shrubs of the Heath family, *Ericaceae*. e.g. A. unedo - The Strawberry Tree, up to 10 m (321/2 ft) tall; reddish-brown bark, shining above; white or pink flowers and red edible berries. Can also be a shrub, and is cultivated the world over for its fruit. A. uva-ursi (= *Arctostaphylos* u.) - The Bearberry Tree (see **Arctostaphylos**).

Arca A genus of Molluscs of the Ark Shell family, *Arcidae*. e.g. A. noae - The Noah's Ark Shell, a longish (70 mm), boat-shaped, edible mollusc, found at various depths on stony and sandy bottoms, often in the shells of other molluscs or on rocks. Distribution: the Mediterranean and Atlantic regions.

Arcella A genus of Testacean or Testate Amoebas (class *Rhizopoda*) of the order *Testacea*. e.g. A. dentata - A testate amoeba; these organisms are 25-100 µ in size, and are found on surfaces of decaying plants in freshwater. They build a shell or test in which they live, pale brown in color and shaped like the cap of a young mushroom. The organism lives outside the test and extends blunt pseudopodia through a hole, where in the mushroom, the stalk joins the cap.

Archerpeton A genus of Prehistoric Extinct Reptiles (class *Reptilia*), the species of which lived about 290 million years ago, and whose remains have been discovered in Nova Scotia, Canada.

Archichauliodes A genus of Insects of the family of Dobson-flies, *Corydalidae*. e.g. A. dubitatus - The New Zealand Dobson-fly; it has a wingspan of up to 10 cm (4 in).

Archidiskodon A genus of Prehistoric Mammoths. e.g. A.imperator - The Imperial mammoth, whose tusk measuring 0.6 m (2 ft) in maximum circumference, was discovered near Post, Texas, U.S.A. in 1933.

Archilochus A genus of Birds, also known by the genus name, *Archulochus*, of the Hummingbird family, *Trochilidae*. e.g. A. colubris (= *Archulochus* c.) - The Ruby-throated Hummingbird; credited with amazing bursts of speed, i.e. up to 46.4 km/h (29 miles/h).

Architeuthis A genus of large Squids of the order *Decapoda* (class *Cephalopoda*). e.g. A. dux - The Giant Squid or Cephalopod, the largest invertebrate. It inhabits the North Atlantic and is preyed upon by sperm whales. It may attain a size of 21.64 m (72 ft) or more.

Archulochus A genus of Birds of the Hummingbird family, *Trochilidae*, better known by the generic name, *Archilochus*. e.g. A. colubris (= *Archilochus* c.) - The Ruby-throated Hummingbird (see **Archilochus**).

Arctia A genus of Insects of the Tiger Moth family, *Arctiidae*. e.g. A. caja (= A. caia) - The Common Garden Tiger Moth of Britain size 7 cm; found in several color forms, with forewings yellowish-white to ochre with a dark pattern, and hind wings vermilion or ochre with dark patches. Distribution: the northern part of the northern hemisphere i.e. Europe and North America.

Arctica A genus of Bivalve Molluscs of the order *Veneroida*. e.g. A. islandica - Iceland Cyprina, size of shell up to 11 cm; large, strong, dark-colored valves with large teeth. Distribution: the seacoasts of Europe, especially the Baltic.

Arctictis A genus of Mammals of the Civet, Genet, and Mongoose family, *Viverridae*. e.g. A. binturong - The Binturong of Asia, related to the Palm Civet. One of the very few mammals outside Australia and South America to have a prehensile tail. About 90 cm (3 ft) long, with a tail which is almost as long. The fur is black to blackish-brown.

Arctium A genus of plants of the Daisy family, *Compositae*. e.g. A. lappa (= A. majus) - The Great Burdock, a biennial plant, 50-200 cm high; lower leaves ovate-heart-shaped at base and up to 50 cm long. Burdock Root was formerly used as a diuretic and diaphoretic. Grows in waste places and by waysides. Distribution: Europe; native in Britain.

Arctocebus A genus of Mammals of the Loris or "Slow Lemur" family, *Lorisidae*. e.g. A. calabarensis - The Angwantibo, a slow-moving mammal, having only rudimentary fingers. It lives in the forests of India, Sri Lanka (Ceylon), and southeast Asia, its only protection from predatory animals being its strictly arboreal and nocturnal habits.

Arctocephalus A genus of Fur Seals of the Eared Seal family, *Otariidae*. They are not true seals in that they possess obvious external ears, but there are other differences. The Fur Seals are also sometimes called Sea Bears. e.g. A. philippii philippii - The Juan Fernandez Fur Seal, which was believed to have become extinct in 1917, was rediscovered on the Juan Fernandez Islands, some 640 km (400 miles) west of Chile in November 1968. A. australis - The Southern Fur Seal, the Australian Seal of the Antarctic, found only in the Falkland Islands and around the neighbouring coast of South America.

Arctogadus A genus of Cod Fishes of the subfamily *Gadinae*, the East Siberian cods; have not been studied in detail, nor have they been extensively fished.

Arctomys A genus of large fur-bearing Rodents or Marmots. e.g. A. bobac - The Tarbagan; this marmot species is a natural reservoir for the plague, which is transmitted by a flea, *Ceratophyllus silantiewi*.

Arctonoë A genus of Scaleworms of the family *Polynoidae*. e.g. A. vittata - This segmented worm lives under the mantle of the Keyhole Limpet (*Diodora* aspera) on the American west coast.

Arctonyx A genus of Mammals of the Badger, Otter, Skunk, and Weasel family, *Mustelidae*. e.g. A. collaris - The Hog Badger; has a pig-like snout and is found in the Himalayas to Yunnan.

Arctostaphylos A genus of plants of the Heath family, *Ericaceae*; also known by the generic name, *Arbutus*. e.g. A. uva-ursi (= *Arbutus* u.) - Uva Ursi or Bearberry Bärentrauben-blätter; Busserole; an evergreen shrub up to 30 cm high with 5-toothed flowers. The dried leaves are used as a diuretic and astringent; also as a fresh infusion for their antiseptic effect in urethritis and cystitis. Distribution: Europe; native in Britain.

Arctous A genus of plants of the Heath family, *Ericaceae*. e.g. A. alpina - Black Bearberry, an evergreen shrub up to 30 cm high; with soft leaves, white and greenish flowers, and bluish-black globose drupes or fleshy fruits containing seeds enclosed by a hard stone). Grows on heaths and moors up to 2,400 m. Distribution: North and Central Europe; native in Britain.

Arcyria A genus of Slime Molds of the class *Myxomycetes* (= *Myxomycophyta*). e.g. A. nutans - A species found commonly on decayed wood. The deep yellow sporangia are in groups. Each has a fine network within which the spores are produced.

Ardea A genus of Birds of the Bittern, Egret, and Heron family, *Ardeidae*. e.g. A. purpurea - The Purple Heron, length about 90 cm (3 ft); crown and crest black, neck rust-red with black stripes. A migrant bird, found mostly near water, such as sweet water reservoirs, with thick stands of reeds. It feeds on fish, frogs, small mammals and insects. Distribution: southern Europe; Asia; Africa.

Ardeola A genus of Birds of the Bittern, Egret, and Heron family, *Ardeidae*. e.g. A. ibis (= A. "Bulbulcus" ibis) - The Cattle Egret or Buff-backed Heron, size 51 cm; red bill and legs; white plumage, with a plume of conspicuous long, single feathers of a yellowish-brown to reddish-brown color. A common heron formerly inhabiting Asia and Africa, now worldwide, where it associates with water buffalo and cattle, riding on their backs and feeding on insects that presumably disturb the beasts. Migrant to South Spain and Portugal, living in southern Europe, and nesting on the ground in heronries, almost always near freshwater.

Ardeotis A genus of Birds of the Bustard family, *Otididae*. e.g. A. kori - The Kori or Giant Bustard, the largest of the bustards, 135 cm (41/2 ft) long and a swift runner. Dark brown in color, it is a ground dweller, inhabiting dry, open country in South and East Asia. Bustards are closely related to cranes, but have a shorter neck and a larger body and tail.

Ardisia A genus of Trees and Shrubs of the family *Myrsinaceae*. e.g. A. paniculata - A woody evergreen from Assam, India, which is sometimes grown in warm greenhouses for its white flowers and bright red fruits.

Arduenna A genus of parasitic blood-sucking Worms. e.g. A. strongylina - Found in the pig stomach.

Areca A genus of Asiatic Trees of the Palm family, *Palmae* (= *Arecaceae*). e.g. A. catechu - The fruit, Betel Nut or Betel Seed is used as a veterinary anthelmintic. In the East it is often chewed, wrapped in the leaves of the Betel plant, *Piper* betle.

Aremonia A genus of plants of the Rose family, *Rosaceae*. e.g. A. agrimonoides - Agrimony, a perennial plant, 5-40 cm high, flowers with 6-10-lobed involucre (bract surrounding each flower or inflorescence). Place of origin: southeast Europe. Introduced and naturalized in Britain.

Arenaria A genus of Birds of the Plover family, *Charadriidae*. e.g. A. interpres - The Turnstone, size 23 cm; black and white headblack breastband, and rust-colored back. A migrant bird, common on stony shores and rocky islands. The female lays its eggs in a depression in the ground. Distribution: West European coasts; native in Britain.

Arenicola A genus of Segmented Worms of the Lugworm or Lobworm family, *Arenicolidae*. e.g. A. marina - The Common European Lugworm, 10-20 cm long; greenish or yellowish-black, with 13 pairs of blood-red gill tufts along the middle of the body. Found on soft sea bottoms, where they dig L-shaped or U-shaped burrows or tubes to a depth of 30 cm, in which they live, feeding on sand-containing minute organisms. Used by fishermen as the best bait for sea fish. Distribution: Europe, eastern America. Native in Britain.

Argas A genus of Ticks. e.g. A. persicus -

A poultry parasite; its bite causes local inflammation in humans.

Argema A genus of Insects of the Giant Silkworm Moth family, *Saturniidae*. e.g. A. mittrei - The Comet Moth, a lovely creature, the largest of allthe Saturnids, with wings protracted into long tails, making it the largest moth in the world. It is a vivid yellow with darker markings, mostly the color of fired clay. It is found in the southern parts of Madagascar, and is one of the most coveted specimens in Butterfly collections.

Argiope A genus of True Spiders of the order *Araneae*. e.g. A. bruenichi - One of the most decorative of the True Spiders, building its web in the grass above streams, spinning globular cocoons for its eggs. Distribution: southern and central Europe.

Argonauta A genus of Cephalopod Molluscs of the Octopus family, *Octopodidae*. e.g. A. argo (= A. argus) - The Paper Nautilius or Mediterranean Paper Nautilius; size of male 1 cm, of female 20 cm; the female builds apaperthin, boat-shaped shell, which looks as if it were made of opaque glass, and in which the eggs are incubated. The males are small, with no shell, and resemble Octopuses. Distribution: the Mediterranean Sea.

Argulus A genus of Crustaceans of the Fish Lice family (order, *Ostracoda*). e.g. A. nobilis - The Carp Louse; length 25 mm; a freshwater species parasitic on the Alligator Gar fish in the southern United States.

Argusianus A genus of Birds of the Pheasant family, *Phasianidae*. e.g. A. argus - The Argus Pheasant, the giant of the pheasant group; over 1.8 m (6 ft) long, of which over 1.2 m (4 ft) is the tail; a forest bird of Malaya, Sumatra, and Borneo. The secondary wing feathers and extremely long tail are decorated with eye-spots with brilliantly colored centers, hence the name Argus, a reference to the mythical Argus with a hundred eyes.

Argynnis A genus of Insects of the Tortoiseshell Butterfly family, *Nymphalidae*. e.g. A. paphia - The Silver-washed Fritillary, a tortoise shell butterfly, with a wingspan 6 cm long, and three rows of black spots on the edges of the brownish-yellow wings. It lays its eggs on violets, nettles, raspberries and blackberries. Found in woods and forests. Distribution: Europe; native in Britain.

Argyrodes A genus of Arachnids of the order of True Spiders, *Araneae*; this genus of small silver spiders shows a tendency to social behavior, and members may often be found living commensally in the webs of the much larger Nephila.

Argyroneta A genus of Arachnids of the Water-Spider family, *Agelenidae*; sometimes called by the generic name, *Agyroneta*. e.g. A. aquatica (= *Agyroneta* a.) - The Water Spider; it lives gregariously in calm or slow-flowing, clean waters with abundant vegetation. An air-breather, it is the only species known, to spend its life submerged. It constructs a "diving bell" which it fills with air, and in this the spider lives, moults, and lays its eggs.

Argyropelecus A genus of Fishes of the Marine Hatchetfish family, *Sternoptychidae*. e.g. A. lychnus - The Silvery Hatchet fish, 9-10 cm long; a deep-seafish, deep-bodied, laterally compressed, with large telescopic eyes aimed upward at an angle. Found at depths of 350-700 m in tropical and temperate regions of all oceans, where it is important as food for tuna and other fish that forage at greater depths.

Argyrosomus A genus of Fishes (subphylum *Vertebrata*). e.g. A. regius - The Meagre Fish, up to 2.3 m long; the first dorsal fin has 10-11 spines. A predaceous species, found in the English

Channel, the Atlantic coast of Europe, and the Mediterranean.

Arianta A genus of Gastropod Molluscs of the Garden Snail family, *Helicidae*. e.g. A. arbustorum - The Copse Snail; the shell is 25 mm wide and 20 mm high. It is yellow-spotted in color, with a chestnut navel and a dark lengthwise band; the inside of the lip is white. Found in damp forests, in hedgerows, on walls, stones and meadows, from lowlands to mountain elevations. Feeds on the leaves of herbaceous plants and on fruits. Distribution: Europe; native in Britain.

Arion A genus of Gastropod Molluscs of the family *Arionidae*. e.g. A. ater - The Large Black Slug, 12-15 cm long; the body is black, brick-red or brown, and the respiratory pore is towards the front of the right side of the shield. It is found in moist regions such as broad-leaved forests. An omnivore, it feeds on vegetable matter as well as living and dead animals. Distribution: Europe; native in Britain.

Ariosoma A genus of Fishes of the order of Conger Eels *Apodes* (= *Anguilliformes*) These are elongated, serpentine fishes with a naked skin, living in these as off the Atlantic and Pacific shores of the American continent.

Arisaema A genus of plants of the Arum family, *Araceae*. e.g. A. ringens - A species from Japan; it has a tuber which produces two trilobed leaves about 45 cm (18 in) long. The spadix is completely hidden by the green and white tubular spathe, which has a curved hood.

Arisarum A genus of plants of the Arum family, *Araceae*. e.g. A. vulgare - The Common Arisarum, a perennial plant, 15-30 cm high; with a hood-like, brownish-red spathe (a large sheath-like bracten closing the spadix). Grows mainly in cultivated fields and grassy places. Distribution: the Mediterranean region.

Aristeomorpha A genus of Arthropods of the Crustacean order, *Decapoda*. e.g. A. foliacea - A long-tailed seawater crustacean; the male is 30 cm long with a short, serrated rostrum or snout; the female is 20 cm longwith a long, pointed rostrum. Lives in lower depths, below 250 m. Distribution: the Mediterranean Sea.

Aristolochia A genus of plants of the Birthwort family, *Aristolochiaceae*. e.g. A. serpentaria - The Virginia Snakeroot plant, a species of birth-worts; the dried rhizome and roots are the source of serpentaria, an aromatic and astringent bitter.

Arius A genus of Fishes of the Bagrid Catfish family, *Bagridae*. These are warmwater fishes with an adipose fin and naked skin.

Arixenia A genus of Insects, the Earwigs, of the order *Dermaptera*. The species are wingless, viviparous, rather bristly insects whose cerci are soft, flexible structures, living in bat-inhabited caves in southeast Asia. This genus belongs to the suborder *Arixeniina*.

Arizona A genus of microorganisms of the family *Enterobacteriaceae*. Gram-negative rods, resembling and in the same tribe as *Salmonella* and *Citrobacter*. e.g. A. hinshawii - A species isolated from snakes, fowl, and mammals. In man it may produce gastroenteritis, urinary tract infections, meningitis, otitis media, etc.

Armadillidium A genus of Arthropods of the Crustacean family, *Armadillidae* (= *Armadillidiidae*). e.g. A. vulgare - The Pill Bug or Woodlouse, 15 mm long; found in dry places under stones, old logs, and in old walls. When it contracts into a ball, the legs and antennae are withdrawn to avoid being seized by predators. Distribution: almost worldwide.

Armeniaca A genus of Trees of the Rose family, *Rosaceae*. e.g. A. vulgaris (= *Prunus* armeniaca) - The Apricot Tree.

Contains tannic and gallic acids; used as an astringent in the form of a decoction (the Bark). From the trunk and branches a dried gummy exudation is prepared (Apricot Gum), which may be used in the preparation of gargles and lotions.

Armeria A genus of plants of the Sea-lavender family, *Plumbaginaceae*. e.g. A. maritima - Sea Pink or Thrift, a perennial plant, 6-20 cm high with fleshy leaves, and pink flowers in solitary terminal heads. Grows on coastal saltmarshes, rocks and cliffs, and also on mountains inland. Distribution: Europe; native in Britain.

Armigeres A genus of Mosquitoes. e.g. A. obturbans - Transmits dengue in Japan.

Armillaria A genus of Mushrooms of the Gill Fungus family, *Agaricaceae*. e.g. A. mellea - The Honey Fungus, an edible species, 5-10 cm high; cap yellowish or pinkish, with dark scales; gills white, later brown; white spores. Grows in clumps on old tree-trunks and also on roots. Found in woods. Distribution: Europe; native in Britain.

Armillifer A former name of a genus of worm-like Arthropods, now known as *Porocephalus*. e.g. A. armillatus (= *Porocephallus* a.) - see **Porocephalus**.

Armoracia A genus of plants of the Cress or Crucifer family, *Cruciferae*, better known by the generic name, *Cochlearia*. e.g. A. rusticana (= A. lapathifolia; *Cochlearia* armoracia) - The Horseradish (see **Cochlearia**).

Arnica A genus of plants of the Daisy family, *Compositae*. e.g. A. montana - Mountain Arnica, Mountain Snuff, or Mountain Tobacco; an aromatic, perennial plant, 30-60 cm high; basal leaves in a rosette; heads large. It has a creeping, underground rhizome. Common on moorland heaths, and rocky slopes in mountain areas. The root (Arnica Rhizome; Arnica Root) has been used as a tincture for local application to bruises and sprains, though of doubtful value.

Place of origin: the mountains of Central Europe.

Arnoglossus A genus of Fishes of the Turbot and allied family, *Bothidae*. e.g. A. laterna - The Scaldfish; these fishes, the left-eye flounders, have a free edge to their operculum, very asymmetrical pelvic fins, and their eyes are on the left side of the body.

Arnoseris A genus of plants of the Daisy family, *Compositae*. e.g. A. minima - Swine's Succory, an annual plant, 5-10 cm high; with hollow scapes or leafless axes, enlarging above, and club-shaped beneath the flower heads. Fringed leaves. Grows by waysides, on sandy banks, waste land, heaths, and in grassy places. Distribution: Europe; native in Britain.

Aromia A genus of Insects of the Longicorn or Long-horned Beetle family, *Cerambycidae*. e.g. A. moschata - The Musk Beetle; colored green to bluish-green, has a strong musky smell, and is found on flowers and felled trees.

Arothron A genus of poisonous Fishes of the Puffer family. e.g. A. tetraodon - The Japanese Puffer Fish or "Deadly Death Puffer", the most poisonous fish in the world. It contains a nerve biotoxin, which is 200,000 times more potent than curare. Known in Japan as the "fugu fish", and considered a great delicacy. It is prepared for eating by removing the poisonous parts without contaminating the rest of the fish (see **Tetraodon**).

Arrhenatherum A genus of plants of the Grass family, *Gramineae*. e.g. A. elatius - Oat-grass, a perennial plant, 50-150 cm high; lower floret usually male, with an awn from the back in the lower third; the upper male floret is usually awnless. Grows commonly in grasslands, sometimes on high ground. Distribution: Europe; native in Britain.

Artamus A genus of Birds of the Wood-swallow family, *Artamidae*. e.g. A. maximus - The Larger Woodswallow, a gregarious tree bird found in the forests of New Guinea; dark slate-grey in color, with a white rump and underparts.

Artemia A genus of Branchiopod Crustaceans of the order *Anostraca*. e.g. A. salina - The Brine Shrimp, 8-11 mm long; with short antennae and stalked eyes. It swims upside-down, and is found in salt-lakes, where the concentration of salt may be higher than in seawater. Once present in the British Isles, but now probably extinct there.

Artemisia A genus of Medicinal and Aromatic plants of the Daisy family, *Compositae*. e.g. A. absinthium - Absinthe or Wormwood, a many-branched, aromatic silvery, perennial plant, 50-100 cm (20-40 in) high, with pendent yellow flowers. Grows in grassy and waste places, on rubbish dumps, in hedgerows and by waysides. The plant contains the lactone, Santonin, a derivative of naphthaline, and was formerly used as an anthelmintic and stomachic. Habitual use or large doses cause absinthism, with restlessness, vomiting, vertigo, tremors and convulsions. The liqueur absinthe is illegal in several countries. Place of origin: Europe; native in Britain.

Arthroderma A genus of dermatophytic Fungi living in the soil, and botanically related to the genus *Trichophyton*.

Arthrographis A genus of Fungi. e.g. A. langeroni - Causes an onchomycosis in man.

Artocarpus A genus of Trees of the Mulberry family, *Moraceae*. The species are the Bread-fruit and the Jack-fruit trees, important economically for their edible fruits.

Artyfechinostomum A genus of Flukes. e.g. A. sufrartyfex - An intestinal fluke from Assam.

Arum A genus of Medicinal and Aromatic plants of the Arum family, *Araceae*. e.g. A. maculatum - The Cuckoo Pint or Lords-and-Ladies, a poisonous plant, 10-50 cm high; with erect,

yellowish-green, cylindrical spadix, having a dull purple upper part. The fruit consists of scarlet berries. Grows in woods and shady hedgebanks. Distribution: Europe; native in Britain.

Aruncus A genus of plants, formerly known by the generic name, *Spiraea*, of the Rose family, *Rosaceae*. e.g. A. dioicus (= A. sylvester; A. sylvestris; *Spiraea* aruncus) - The Goat's Beard or Wood Goatsbeard, a moisture-loving, herbaceous, perennial plant, 90-120 cm high, with dense panicles of creamy-white to yellow flowers. Grows in woods, thickets, and on banks of streams. Place of origin: Northern Hemisphere; introduced into Britain in 1633.

Arundo A genus of plants of the Grass family, *Gramineae*. e.g. A. donax - Spanish Reed, a perennial plant, 2-4 m high; culm up to 2 cm thick, panicle up to 70 cm long; similar to the Common Reed, *Phragmites* communis, but much stouter. Grows at the edges of ponds and in the beds of streams. Distribution: western Europe; Mediterranean region.

Arvicola A genus of Cricetid Mammals of the Vole, Lemming and Muskrat sub-family, *Microtinae*. e.g. A. amphibius - The Water Vole of Western Europe; about 20 cm (8 in) long, with a 10 cm (4 in) tail; coat dark brown. It lives near water, and is a good swimmer and diver, although its feet are not webbed. Over much of the European continent it is replaced by a related species A. terrestris - which is entirely terrestrial and mole-like in its habit of burrowing, but is otherwise similar to A. amphibius.

Asarum A genus of plants of the Birthwort family, *Aristolochiaceae*. e.g. A. europaeum - Asarabacca, Hazelwort, or Wild Nard, a perennial plant, 5-20 cm high; with usually 2 leaves and brownish-red, solitary flowers with short stalks. A poisonous species, it is a powerful emetic and purgative; has been used as an errhine in the treatment of headache, usually as an ingredient of snuffs. Distribution: Europe; native in Britain.

Ascalaphus A genus of Insects of the Snake Fly family, *Ascalaphidae*. e.g. A. libelluloides - 20-25 mm in length, with long, clubbed antennae. Found in warm, temperate climates of southwestern Europe and northwest Africa.

Ascaphus A genus of Amphibians of the primitive Frog family, *Liopelmidae*, found in the cold mountain streams of western North America. e.g. A. truei - This primitive frog has a prolongation of the cloaca, which is used by the male as an intromittent organ, with which it effects fertilization of the female.

Ascaris A genus of Round Worms or Nematodes of the order *Ascaridida*. e.g. A. lumbricoides - The cosmopolitan eel-worm or roundworm, the largest of the common parasitic nematodes of man (up to 35. cm - 13.7 in long and over 5 mm - 0.19 in in diameter). Found in the intestine, especially in children.

Ascidiella A genus of primitive Chordates of the class *Ascidiacea*. e.g. A. scabra - A Sea-squirt, 8 cm long; it has a sack-shaped body with a gelatinous tunic and large siphons. Common on rock and shells on the lower sea-shore. Distribution: the North Atlantic and the Baltic Sea. Native around Britain.

Asclepias A genus of plants of the Milkweed family, *Asclepiadaceae*. e.g. A. tuberosa - Butterfly Weed; Pleurisy Root; formerly used in the treatment of rheumatism and pleurisy. Place of origin: North America; introduced into Europe in 1690.

Ascocotyle A genus of Nematode parasitic worms. e.g. A. pithecophagicola - A species found in the monkey-eating eagle in the Philippines.

Ascophyllum A genus of plants of the class of Brown Algae or Seaweeds, *Phaeophyta*, known as the Wracks. e.g. A. nodosum - The Knotted Wrack, size up to 150 cm; air vesicles large, concep-tacles stalked, round or oval, and dioecious, having male and female organs on separate plants. Found in the seas and along the shores of Europe; native around Britain.

Asellus A genus of Isopods of the Water Louse family, *Asellidae*. e.g. A. aquaticus - The Water Louse, up to 25 mm long; body flat; this crustacean species is found all over the freshwaters of Europe, and also penetrates into brackish water in the Baltic Sea. It is a general scavenger, feeding on organic detritus.

Asilus A genus of Two-winged Robber Flies of the family *Asilidae*. e.g. A. crabroniformis - A large, yellow-and-black, robber fly, found in forests where it hunts various insects, piercing its victims and sucking their juices. Distribution: Europe.

Asimina A genus of North American trees and shrubs, of the family *Caricaceae*. e.g. A. triloba - The Papaw or Pawpaw tree. From the juice of the unripe fruit is expressed a proteolytic enzyme, papain (see **Carica**).

Asio A genus of Birds of the Typical Owl family, *Strigidae*. e.g. A. otus - The Long-eared Owl, 38 cm long; brown, barred plumage; red eyes and large eartufts. A partially migrant bird of prey, feeding on small animals, chiefly voles and mice. It lives in forests, mainly coniferous, and in mixed woods, and has a sensitivity to low light intensities, that is, 50-100 times greater than that of human night vision. Distribution: Europe, Asia, America. Native in Britain.

Asopia A genus of Insects of the family of Pyralid Moths. e.g. A. farinalis - The Meal Moth; acts as the intermediate host of *Hymenolepis* diminuta.

Asparagus A genus of plants of the Lily family, *Liliaceae*. e.g. A. officinalis - The Common Asparagus, a perennial plant, 30-180 cm high; scale-like leaves; needle-like cladodes or modified stems growing in clusters; red berries. The root is used as a mild diuretic. Found in sandy pastures, hillocks and grassy sea-cliffs. Distribution: east Mediterranean region and in central and northwestern Europe. Native in Britain. Some forms are cultivated as vegetables, and naturalized the world over.

Aspergillus A genus of Ascomycetous Fungi of the subclass *Ascomycetes*. e.g. A. flavus - A species yielding the highly toxic and carcinogenous metabolites known as the aflatoxins.

Asperugo A genus of plants of the Borage family, *Boraginaceae*. e.g. A. procumbens - German Madwort, an annual plant, 20-70 cm high; with recurved fruit-stalks. Widespread but rare, growing in waste places and on arable land. Distribution: Europe; introduced and naturalized in Britain.

Asperula A genus of plants of the Bedstraw or Madder family, *Rubiaceae*. e.g. A. cynanchica - Squinancy Wort, a perennial plant, 5-40 cm high; with few-flowered cymes and four-angled, glabrous stems. Grows in calcareous, grassy places and on sand dunes. Distribution: central and southeastern Europe. Native in Britain.

Asphodeline A genus of plants of the Lily family, *Liliaceae*. e.g. A. lutea - Yellow Asphodel or King's Speer, a perennial plant, up to 30 cm high; with yellow flowers in a raceme, and narrow, grass-like leaves. Grows mainly in woods, especially in olive scrubs. Distribution: the Mediterranean region.

Asphodelus A genus of plants of the Lily family, *Liliaceae*. e.g. A. ramosus - Branched Asphodel or King's Rod, a perennial plant, 50-100 cm high; flowers white, with a red midrib, in a large, branched raceme. Grows mainly in woods, especially in olive scrubs. Distribution: the Mediterranean region.

Aspidelaps A genus of Reptiles of the suborder of Snakes, *Ophidia* (= *Serpentes*) order *Squamata*. e.g. A. scu-

tatus - The Shield-nose Snake, a species estimated to have a striking speed of 271 cm/s (106.6 in/s). The human hand by comparison, can record a speed of 328 cm/s (129 in/s) in a quick snatch. Evidently the unexpectedness of a snake's striking movement over a relatively short distance, is the reason why observers usually over-estimate its speed.

Aspidistra A genus of indoor plants of the Lily family, *Liliaceae*. e.g. A. elatior (= A. lurida) - A perennial plant with underground, rhizomatic stems and long large radical leaves with complete, leathery, shiny green lamina. It produces solitary purplish flowers at sea level. Place of origin: Central and East Asia; a native of China.

Aspidogaster A genus of Trematode Worms (the Flukes), of the order *Aspidogastrea*. e.g. A. conchicola - This aspidogastrid trematode is parasitic in freshwater mussels of the genera *Anodonta* and *Unio*, occurring in the pericardial cavity.

Aspidontus A genus of Fishes (subphylum *Vertebrata*). e.g. A. rhinorhynchus - The Sabre-toothed Blenny fish.

Aspidosperma A genus of Trees of the family *Apocynaceae*. e.g. A. quebracho - Quebracho Bark, containing the alkaloids aspidospermine and yohimbine (= quebrachine). Has been used as a bitter, antispasmodic, and febrifuge.

Aspius A genus of Fishes of the Minnow and Carp family, *Cyprinidae*. e.g. A. aspius - The Asp Fish or Rapfen, length 70-80 cm; a favorite game fish with a long body and broad mouth, found in the lower reaches of larger rivers and in backwaters. Distribution: eastern Europe (see **Carassius**).

Asplenium A genus of plants of the Fern or Polypody family, *Aspleniaceae*. e.g. A. ceterach (= *Cetarach* officinarum; *Ceterach* o.) - The Scaly Spleenwort; Rusty-back Fern; a dwarf wall fern, a perennial, evergreen plant, up to 25 cm high, with sage-green fronds covered all over the undersides with silvery scales, which become brown with age. Found on old walls and on rocks. Used for planting in rock gardens. Place of origin: Europe, North Africa, and western Asia.

Aspro A genus of Fishes of the Perch, Walleye, and Darter family, *Percidae*. e.g. A. zingel - The Zingel Fish, length 20-50 cm; a long slender fish found near the bottom of deeper flowing rivers; feeds on invertebrates and small fish. Distribution: The Danube and its tributaries.

Astacilla A genus of higher crustacean animals of the order *Isopoda*. e.g. A. longicornis - An isopod parasite species.

Astacopsis A genus of Crayfishes of the family, *Austroastacidae*. e.g. A. franklini - A crustacean species found in Tasmania, this crayfish reaches a weight of 8 or 9 lb. Related to the northern crayfish of Europe *Astacus* (see **Astacus**).

Astacus A genus of Crustaceans of the Lobster and Crayfish family, *Astacidae*. e.g. A. astacus (= A. fluviatilis; *Potamobius* a.) - The River Crayfish or Common European Crayfish; length 120-160 cm; lives in flowing as well as still waters, burrowing in the banks under roots of trees or hiding under stones. It feeds on small fish and tadpoles, and on carcasses. It has a pointed spike on the forehead and red areas at the leg joints. An edible freshwater species of the River Crayfish family. Distribution: Europe; native in Britain.

Astarte A genus of Molluscs of the Astarte family, *Astartidae*. e.g. A. sulcata - The Furrowed Astarte-shell mollusc; found on the sea-coast where it burrows in mud and gravel down to considerable depths.

Astasia A genus of Phytomast Flagellates of the order *Euglenoidina*. Nutritionally interesting, the cells form paramylum when cultured in media devoid of complex substances, such as sugars and proteins.

Aster A genus of Rock garden plants of the Daisy family, *Compositae* (= *Asteraceae*). e.g. A. alpinus - The Alpine Aster, a low, spreading, hairy, perennial plant, 15 cm (6 in) high, with violet-blue flowers having yellow central disc florets, in large solitary heads, 2.5-5 cm (1-2 in) across. Cultivated forms include white and red flowered varieties. Used in Rock gardens, rocky slopes and banks, etc. Place of origin: mountains of central and eastern Europe.

Asteracantha A genus of plants of the family *Acanthaceae*. e.g. A. longifolia (= *Hygrophila* spinosa) - The Hygrophylla plant; used in India as a decoction diuretic.

Asterias A genus of Starfishes of the family *Asteriidae*. e.g. A. rubens - The Common Starfish; body up to 50 cm in diameter; reddish brown or dark violet, with 5 broad arms (sometimes 4 to 6); it feeds on various molluscs. Distribution: the North Sea and West Baltic down to a depth of 400 m.

Asterina A genus of Starfishes of the class *Asteroidea*. e.g. A. gibbosa - The Starlet or Cushion-star; these echinoderm animals have five arms which are united for almost their entire length. They live on the sides of rocks, etc.

Asterionella A genus of freshwater Diatoms of the class *Bacillariophyta*. e.g. A. formosa - A very common colonial planktonic species, with the extremely narrow cells radiating from the center.

Asterococcus A genus of microorganisms, also known by the generic names, *Astrococcus*, *Bovimyces*, and *Coccobacillus*, and resembling filtrable viruses. e.g. A. mycoides (= *Astrococcus* m.; *Bovimyces* pleuropneumoniae; *Coccobacillus* m.) - These bacteria cause pleuropneumonia in cattle.

Asthenosoma A genus of Sea Urchins of

the class *Echinoidea*. e.g. A. varium - A dangerous species, carrying white bulb-like poisonous glands on its spines. Distribution: along the sandy bottoms of coral reefs; especially abundant in the Red Sea.

Astilbe A genus of plants of the Saxifrage family, *Saxifragaceae*. e.g. A. davidi - A tall species with spikes of small, crowded, crimson flowers up to 60 cm (2 ft) long. A herbaceous perennial plant, with many cultivated hybrids and varieties. Place of origin: Asia and North America.

Astracopsis A genus of freshwater Crustaceans. e.g. A. gouldi - The Crayfish, found in small streams in Tasmania, and said to be the largest freshwater crustacean in the world, reaching a weight of 3.62-4.08 kg (8-9 lb).

Astraea A genus of Archaeogastropod Molluscs of the Topshell family, *Trochaceae*; this species inhabits both American coasts.

Astragalus A genus of plants of the Pea family, *Leguminosae*. e.g. A. diphysus - "Blue Loco" or Blue Loco-weed, poisonous to farm animals, and causing locoism or loco poisoning. These plants contain selenium, and the animals that graze on them act as if crazy (Sp. loco = insane). Distribution: western N. America. Other genera containing selenium include *Hosackia*, *Oxytropis*, and *Sophora*. A. gummifer - The Tragacanth plant; used as a suspending and thickening agent in the manufacture of creams, jellies, and pastes.

Astrangia A genus of Corals of the family *Astrangidae*. e.g. A. danae - A species forming small encrusting colonies along the east coast of the United States, from North Carolina to Massachusetts.

Astrantia A genus of Mediterranean plants of the Carrot family, *Umbelliferae*. e.g. A. major - Great Black or Pink Masterwort, an erect little-branched perennial, 30-90 cm high,

with palmately lobed or cut leaves, and sickly smelling, pinkish or greenish-black colored flowers, surrounded by large pointed bracts or rounded umbels. Grows in scrub and meadows, and used in flowerbeds and rock gardens. Place of origin: Alpine meadows; South and Central Europe.

Astroboa A genus of Echinoderms, the Basketstars, of the class *Ophiuroidea*. e.g. A. nuda - The Basketstar; together with the Brittlestars (genus *Placophiothrix*) and Featherstars (genus *Heterometra*), the Basketstars dominate the underwater scenery of the coral reefs at night, their arms stretched out to strain food from the sea currents; during the daytime they are hidden under ledges and inside crevices on the reef. Distribution: the Coral Reefs of the Red Sea, and other reefs.

Astrococcus A genus of microorganisms resembling filtrable viruses, also known by the generic names, *Asterococcus*, *Bovimyces*, and *Coccobacillus*. e.g. A. mycoides (= *Bovimyces* pleuropneumoniae; etc.) - see **Asterococcus**, **Bovimyces** and **Coccobacillus**.

Astropecten A genus of Starfishes of the family *Asteriidae*. e.g. A. aurantiacus - Body up to 30 cm in diameter; upperside orange, underside yellow; found on sandy sea bottoms at depths of 1-100 m. Distribution: Mediterranean Sea and the Atlantic Ocean, north to Portugal.

Astrophytum A genus of succulent plants or Cacti of the family *Cactaceae*. e.g. A. myriostigma (= *Echinocactus* m.) - Bishop's Cap; a globular, spineless plant with five well-defined ribs like a bishop's cap. Yellow daisy-like flowers. Place of origin: Mexico.

Astroscopus A genus of Fishes (subphylum *Vertebrata*). e.g. A. y-graecum - The Southern Star-gazer fish.

Astrotoma A genus of Basketstar echinoderms of the class *Ophiuroidea*. e.g. A. agassizii - The Antarctic Basketstar, the largest known ophiuroid; has unbranched elongated arms, said to measure as much as 1 m across.

Astylosternus A genus of Amphibians of the Frog family, *Ranidae*. e.g. A. robustus - The Hairy Frog; during the breeding season, the males develop hair-like processes of skin on their flanks.

Ateleopus A genus of deep-sea Fishes of the family *Ateleopidae* (= *Podatelidae*), found off India and Japan. They are characterized by a profound degeneration of their bodies.

Ateles A genus of Mammals of the New World Monkey family, *Cebidae*. e.g. A. ater - The Black-faced Spider Monkey; it has an extremely long prehensile tail which enables it to leap among the trees with remarkable agility. The sensitive tip is also used to grasp food which is beyond reach of the hands. Distribution: Central America, north up to Mexico.

Atelopus A genus of Amphibians of the Tree Frog family, *Atelopidae*. e.g. A. zeteki - Zetek's Frog, a small Central African tree frog found in Panama; very brightly colored, and moving by day under protection of this warning coloration.

Athena See **Athene**.

Athene A genus of Birds, sometimes known as *Athena*, of the Typical Owl family, *Strigidae*. e.g. A. noctua (= A. noctus; *Athena* noctua) - The Little Owl, 20-22.5 cm (8-9 in) long; body small; back spotted; head flat with bright yellow eyes. A resident bird, living near human habitation such as treed avenues, gardens, in buildings and copses. A night bird of prey, feeding on small mammals, birds and insects. Distribution: the Mediterranean region, central and southern Europe, and North Africa; an introduced species in Britain.

Atherestes A genus of Fishes of the Right-eyed Flounder family, *Pleuronectidae*. e.g. A. stomias - The American Arrowtooth Flounder; about 90 cm (3 ft) long; a predator fish, found at times as deep as 600 m (2,000 ft). A

right-eyed flounder, the eye is usually on the right side of the body. A fish of commerce.

Atherina A genus of Silverside Fishes of the family *Atherinidae*. Small edible fishes that often live together in large shoals, often swimming near the surface. They are related to the grey mullets. e.g. A. presbyter - The Sand Smelt, up to 15 cm (6 in) long; translucent greenish color above, and silvery white on the sides. Distribution: the English Channel and the Zuider Zee; the Mediterranean Sea.

Atherinopsis A genus of Silverside Fishes of the family *Atherinidae*. e.g. A. californiensis - The Jacksmelt Silverside fish, a commercially important edible species, harvested off the Pacific coast of North America.

Atherix A genus of Flies. e.g. A. variegata - Found in Europe and North America.

Atherosperma A genus of Trees of the family *Monimiaceae*. e.g. A. moschatum - The Sassafras tree of Australasia, the bark of which is used as a diaphoretic and diuretic.

Atherurus A genus of Old World Porcupines of the family *Hystricidae*, the Brush-tailed Porcupines of southeast Asia and Africa; the tail ends in a tuft of horny hairs simulating spines. The spines or quills on the body are sharply pointed hairs of enormous size; erectile, they serve as organs of defense, controlled by powerful muscles in the skin.

Athous A genus of Click Beetles of the family *Elateridae*. e.g. A. vittatus - A common click beetle found in broad leaved forests in lowlands and foothills. A. haemorrhoidalis - A species of click beetles enduring the highest gravity force encountered in nature (as much as 400+ g), when jack-knifing into the air to escape predators.

Athripsodes A genus of Insects of the Caddis Fly family, *Leptoceridae*. e.g. A. cinereus - Found near slow-flowing waters and ponds; larvae make slender portable cases.

Athyrium A genus of plants of the Fern or Polypody family, *Aspleniaceae*. e.g. A. felix-femina - Lady Fern, a perennial plant, 30-100 cm high; the sori or cluster of fern sporangia are oval or crescent-shaped. Grows in woodlands and hedgebanks. Place of origin: Europe; native in Britain.

Atilax A genus of Mammals of the Mongoose family, *Viverridae*. e.g. A. paludinosus - The Marsh or Water Mongoose; found in Africa south of the Sahara.

Atlanta A genus of Molluscs of the Mesogastropod Limpet superfamily, *Cerithiacea*. These are very active carnivorous animals, which swim upside down. They retain a complete, although very delicate transparent shell.

Atlantisea A genus of Ratite Birds (class *Aves*), of the Rail, Coot, and Moorhen family, *Rallidae*. Also known by the generic name, *Atlantisia*. e.g. A. rogersi - The Flightless Rail, the smallest ratite bird and probably the smallest flightless bird known in existence, living on Inaccessible Island of the Tristan da Cuha group in the South Atlantic. It has degenerate hair-like plumage, and is no bigger than a newly hatched domestic chick. It lives in burrows and is reported to be a very fast runner.

Atlantisia See **Atlantisea**.

Atolla A genus of Jellyfishes of the order *Coronatae*, an inhabitant of deeper waters, with a shallow, dark red bell. It is not usually found above 200 m from the sea surface.

Atractaspis A genus of Snakes of the Viper family, *Viperidae*. The burrowing and oviparous species have an aggressive disposition. In proportion to their length (30 in at most), they have the longest fangs of any snake, and make formidable adversaries. Habitat: Africa.

Atrax A genus of poisonous Funnel-web Spiders of the family *Dipluridae*. e.g. A.

robustus - The Funnel-web spider of New South Wales, Australia; extremely poisonous, and has caused many deaths. Long-lived: as much as 17 years.

Atrichornis A genus of rare Scrub-birds of the family *Atrichornithidae*. e.g. A. clamosus - The Noisy Scrub-bird, believed to have a total world population of less than 100. Length 8.5 in; resembles a wren. Distribution: western Australia.

Atriplex A genus of plants of the Goosefoot family, *Chenopodiaceae*. e.g. A. patula - The Common Orache, an annual plant, 30-90 cm high, with hastate or halberd-shaped fruit-bracheoles. A common and widespread garden weed, growing in waste-land, on rubbish-dumps, by road-sides and paths. Place of origin: Europe; native in Britain.

Atropa A genus of plants of the Nightshade family, *Solanaceae*. e.g. A. belladonna - The Deadly Nightshade, a perennial plant, 60-125 cm high; with entire leaves, bell-shaped, stalked, nodding flowers, and black berries; very poisonous. From the leaf, Belladonna Leaf or Deadly Nightshade Leaf is expressed the alkaloids atropine and hyoscyamine, both parasympatholytic drugs. Distribution: Europe, western Asia, and North Africa. Native in Britain.

Atta A genus of Insects of the Ant order, *Hymenoptera*. These are the Parasolor Harvesting Ants of the Mediterranean region. They live on seeds which they store in underground nests or chambers as a food reserve in times of drought. There they cultivate fungus gardens, the soil of which is made up of chewed leaf fragments.

Attacus A genus of Insects of the Emperor Moth or Giant Silkworm Butterfly family, *Saturniidae*. e.g. A. atlas - The Atlas Moth of southeast Asia, an exceptionally large insect, with a wingspan in one case (a female in the

Dorman Museum) of 264 mm (10.39 in). Among the largest of the world's butterflies.

Attagenus A genus of Insects of the Hide, Carpet, or Museum Beetle family, *Dermestidae*. e.g. A. pellio - The Two-spotted Carpet Beetle, length 4-5.5 mm; bluish-black, with brown legs and antennae. Each elytron has a light spot. The adults and larval forms live on animal hairs. A synanthropic species found in households and warehouses. The larvae are feared pests of furs, fabrics, and zoological collections. Distribution: worldwide; native in Britain.

Attalea A genus of Trees of the Palm family, *Palmae*. e.g. A. funifera - The Pissaba Palm tree of Brazil. From the nut is expressed Babassu Oil, used for making soap, lubricants, diesel oil, and margarine.

Attaphila A genus of Insects of the sub-order of Cockroaches, *Blattodea*. The species are wild, and are associated with ants.

Attelabus A genus of Insects of the Weevil Beetle family, *Curculionidae*. e.g. A. nitens - A Leaf-Roller Beetle, 6 mm long, with red elytra; the female makes rolls with oak leaves, where it deposits its eggs. Distribution: Europe; native in Britain.

Atyaephyra A genus of Crustaceans, found in freshwater. e.g. A. desmaresti - Size up to 20 mm; transparent narrow body; it has pincers on the first and second pairs of walking legs. Place of origin: the Mediterranean region; has been introduced into Holland and the Rhineland.

Aubrieta A genus of Rock-garden plants of the Cress or Crucifer family, *Cruciferae*. Sometimes misspelt *Aubrietia*. e.g. A. deltoide - The Purple Rock Cress, a variable, mat-forming, evergreen perennial, up to 30 cm (1 ft) high; has conspicuous flowers of lilac-mauve deep or red-purple color, growing in broad sheets. Used in rock gardens, cracks in walls or rocks, or in crazy paving. Place of origin: mountains of Sicily, southern Greece, and Asia Minor.

Aubrietia See **Aubrieta**.

Auchmeromyia A genus of Flies. e.g. A. luteola - A fly of the Congo and Nigeria, having a blood-sucking larva, the Congo floor maggot.

Aucuba A genus of indoor plants of the Cornel or Dogwood family, *Cornaceae*. e.g. A. japonica - The Aucuba plant or Spotted Laurel; an evergreen shrub up to 3.6 m (12 ft) high, with 4-petalled flowers in terminal male and female, on separate plants. Clones occur, with variegated foliage and white or yellow berries in clusters. These plants are unisexual, and both male and female plants must be grown in order to obtain the berries. Place of origin: Himalayas; Japan.

Aurelia A genus of True Jellyfish of the family *Ulmariidae*. e.g. A. aurita - The Common Jellyfish, Moon or White Jelly; the umbrella-like body is 400 mm in diameter; violet, reddish or yellowish in color, with numerous short arms; round the mouth are 4 long flat arms. A plentiful species, floating in groups near the shore, and in open waters. Feeds on small animal life such as plankton. Distribution: Mediterranean Sea and Atlantic coast of Europe.

Aurinia A genus of Rock garden plants of the Cress or Crucifer family, *Cruciferae*; formerly included under the genus name, *Alyssum*. e.g. A. saxatilis - Formerly, *Alyssum* saxatile (see **Alyssum**).

Australopithecus A genus of Mammals, the prehistoric Hominids, and an extinct member of the Human family, *Hominidae*, which branched off the main stem of hominid evolution and died about 1 million years ago. e.g. A. africanus - The so-called southern ape of Africa, an early hominid species, the skeletal remains of which have been dug up in the southern Transvaal in South Africa. Thought to have been one of the first Hominid species of African ape-men, preceding the evolution of Man, *Homo* sapiens. These hominids, according to abundant evidence, emerged as an upright-walking, small-brained offshoot of the higher primates, late in the Tertiary Period, which lasted from about 65 million to just under 2 million years ago.

Austrocedrus A genus of Trees of the Cypress family, *Cupressaceae*; also known by the generic name, *Libocedrus*. e.g. A. chilensis (= *Libocedrus* c.) - The Chilean Cedar or Chilean Incense Cedar, an evergreen conifer tree, native to Chile and Argentina, and up to 24 m (80 ft) tall. The foliage is very distinct with flattened pairs of leaves or scales, having bright white bands beneath. Male and female flowers develop into green, ripening to brown cones 8 mm (1/3 in) long, and consisting of 4 scales. Grown in botanical collections.

Austrofundilis A genus of short-lived Fishes (the Killifish), of the suborder *Cyprinodontei*. e.g. A. dolichopterus - The Killifish of South America, found in temporary ponds and drainage ditches. Lives about 8 months in the wildstate.

Austrophlebia A genus of swift Dragon-flies. e.g. A. costalis - A species said to exceed a speed of 57.6 km/h (36 miles/h) in short bursts.

Autographa A genus of Moths of the family *Noctuidae*. e.g. A. gamma - The Silver Y Moth; markings on the forewings resemble the Greek letter gamma (γ). Found in fields, meadows, and gardens. The caterpillar is polyphagous, and when numerous may cause damage to cultivated plants.

Autolytus A genus of Polychaete Segmented Worms of the family *Syllidae*, with species on both sides of the Atlantic. These syllids show a chain of new individuals budded from the hind end of the parent worm.

Automeris A genus of Moths. e.g. A. io - The Io Moth; causes dermatitis by means of irritant hairs on its larvae.

Avahi A genus of Mammals of the Avahi family, *Indriidae*. e.g. A. laniger - The Avahi, a nocturnal, slow-moving, vegetarian mammal with a thick, woolly, grey-brown coat and a 40 cm (16 in) tail, which is 10 cm (4 in) longer than the body. Distribution: the forests of Eastern Madagascar.

Avena A genus of plants of the Grass family, *Gramineae*. e.g. A. fatua - The Common Wild Oat, an annual plant, 70-100 cm high, with 2-3-flowered spikelets, growing on arable and waste land. Distribution: Europe; introduced and naturalized in Britain. Has apparantly been in cultivation since 2,500 B.C. A. sativa - The Common or Cultivated Oat, containing the principle avenine. Not known in the wild state, and probably arose from A. fatua, the Wild Oat.

Avicularia A genus of Arachnid Spiders of the large South American "tarantula" family, *Avicularoideae* (= *Mygalomorphae*). Also known by the generic name, *Mygale*. e.g. A. avicularia (= *Mygale* a.) - One specimen of these Mygale "tarantulas" has been known to measure 203 mm (8 in) across its extended legs.

Axine A genus of Merozoic Cestode Worms of the subclass *Eucestoda* (= *Cestodamerozoa*). e.g. A. belones - The opisthaptor is asymmetrical and bears numerous minute clamps.

Axiolotus A genus of Molluscs of the Freshwater Limpet family. e.g. A. lacustris - The Lake Limpet; 2 mm high; shell is shield-shaped. Distribution: Europe; native in Britain.

Axis A genus of Mammals of the True Deer family, *Cervidae*. e.g. A. axis - The Axis Deer or Chital, with three-tined antlers. Found in India and Sri Lanka, where it associates in large herds. Also known as the Spotted Deer. Its alarm call is a shrill whistle. It is about 1.5 m

(5 ft) long and less than 1 m (3 ft 4 in)
high at the shoulders. Its coat is always
spotted, like that of the fawns of other
species, hence the name.

Aythya A genus of Birds of the Duck,
Goose, and Swan family, *Anatidae*. Also
known as the Aythyini, Diving Duck,
"Bay" Duck, or Pochard Duck. e.g. A.
ferina - The Common Pochard or
Pochard Diving Duck, size 45 cm; male
plumage grey, with brown head and
dark breast; female with blue band on
bill. A freshwater diving duck which
nests in the reedbeds of inland lakes. A
partial migrant bird, foraging mainly
under water. Distribution: Central
Europe; native in Britain.

Azadirachta A genus of plants of the
family *Melicaceae*. Also known by the
generic name, *Melia*. e.g. A. indica (=
Melia azadirachta) - Margosa; Neem;
the dried stembark, root bark, and leaves
are used in India as a bitter and antiperiodic. Neem oil has been used as a hair
tonic and in skin diseases.

Azolla A genus of plants of the Water-
fern family. e.g. A. caroliniana - The
Azolla, a perennial plant, up to 1.2 cm
high; a small floating plant with slender
stems; bluish-green, bilobed, imbricate
leaves borne in two rows; simple roots.
A native of America; naturalized in
Britain, Europe, and other countries.

Azotobacter A genus of microorganisms
of the family *Azotobacteriaceae*. e.g. A.
indicus - Large rods, occurring as free-
living nitrogen-fixing bacteria in the
soil.

Azotomonas A genus of microorganisms
of the family *Pseudomonadaceae*; rod-
shaped tococcoid cells (coccobacilli),
found in the soil, and active in fixation
of atmospheric nitrogen. e.g. A. fluo-
rescens.

B

Babesia A genus of protozoan parasites found in the red blood corpuscles of various domestic animals, and transmitted by ticks. Formerly called by various generic names such as *Piroplasma*, *Pyroplasma* and *Pyrosoma*. Some species are included in the genus *Nuttalia* or *Theileria*. e.g. B. bigemina - the species causing Texas Fever of cattle, and transmitted by the cattle tick, *Boophilus* annulatus (= *Boophilus* bovis; *Margaropus* annulatus; M. bovis) - see **Boophilus**. B. bovis - the species causing Red Water Fever (hemoglobinuria) in cattle, and transmitted by the tick, *Ixodes* ricinus (see **Ixodes**). B. caballi (= B. equi), better known as *Nuttalia* equi - a species causing hemoglobinuric fever of horses in South Africa, and transmitted by the tick, *Rhinencephalus* everti (= *Boophilus* e.) - see **Boophilus**; **Nuttalia**. B. parva (= *Theileria* p.) - a species causing East African Fever in cattle (see **Theileria**).

Bacillus A genus of microorganisms of the family *Bacillaceae*; large gram-positive, spore-forming rods, most spores of which are saprophytes of water and soil, and common laboratory contaminants. e.g. B. anthracis - a species causing anthrax. Large rods with central spore, primarily affecting cattle, sheep and horses. Man is occasionally infected by spores that enter the injured skin or mucous membranes and, less frequently, by inhalation. It produces skin infection (malignant pustule), anthrax pneumonia (wool sorters' disease), septicemia, enteritis and meningitis. B. cereus - a widely distributed aerobic, spore-bearing species producing two types of enterotoxin, a short-acting, heat-stable, emesis-provoking (ST) toxin, and a long-acting, heat-labile (LT) enterotoxin, causing diarrhea and abdominal pain about 12 hours after ingestion of contaminated food, especially cereals. The bacillus may survive in fried rice. B. dysenteriae (= *Shigella* d.) - see **Shigella**.

Bacteriodes See **Bacteroides**.

Bacterionema A genus of microorganisms of the family *Chlamydobacteriaceae*, better known by the generic name, *Leptothrix*. e.g. B. matruchotii (= *Leptothrix* buccalis) - see **Leptothrix**.

Bacterium (a) Any microorganism of the order *Eubacteriales*, not fitting into other formally defined genera. e.g. B. tularense (= *Bact.* t.; *Pasteurella* tularensis) - see **Pasteurella**. B. typhiflavum (= *Chromobacterium* t.) - see **Chromobacterium**. (b) A genus of small anaerobic, gram-negative microorganisms, better known by the generic name, *Dialister*. e.g. B. pneumosintes (= *Dialister* p.) - see **Dialister**.

Bacteroides A genus, also called *Bacteriodes*, of non-sporulating, anaerobic, gram-negative, filamentous or rod-shaped bacteria, classified under the family *Brucellaceae* or *Bacteroidaceae*, and occurring in normal flora in the mouth and the large intestine. Occasionally associated with disease, most often as opportunists. The species are also included under other genera such as *Eikenella*, *Eubacterium*, and *Fusobacterium*. e.g. B. aerofaciens (= *Eubacterium* a.) - see **Eubacterium**. B. corrodens (= *Eikenella* c.) - see **Eikenella**. B. fragilis - a common human pathogen, usually constituting 91-97% of the normal fecal flora, producing β-lactamase, and capable of causing bacteremia, brain abscess and empyema. B. funduliformis (= *Fusobacterium* necrophorus) - see **Fusobacterium**. B. melaninogenicus (= B. nigrescens) - so-named because of its darkly pigmented colonies on blood agar. Part of the normal flora of the mouth, upper respiratory tract, genito-urinary tract, and gastro-intestinal tract. May cause lung abscesses, oropharyngeal infections, and wound infections. B. oralis - part of the normal oral and urethral flora. Not known to be a primary pathogen.

Bactoscilla A genus of saprophytic microorganisms, the Schizomycetes, of the family *Vitreoscillaceae*. The species are similar to those of the genus *Vitreoscilla* (see **Vitreoscilla**).

Balaena A genus of Right Whales of the family *Balaenidae* (order *Cetacea*). e.g. B. mysticetus - The Bowhead Greenland Right Whale, length 15-21 m, occasionally reaching 30.48 m (100 ft). On each side of the upper jaw there are 300-360 whalebones each measuring up to 350 cm in length. The Bowhead is now rare because of overhunting; the blubber is boiled with water to produce high quality whale oil, known as Whale Oil No: 0. Distribution: the Arctic Seas.

Balaeniceps A genus of Birds of the Shoebill family, *Balaenicipitidae*. e.g. B. rex - The Shoebill or Whale-headed Stork, also known as the Whalebird, is a grotesque, stork-like bird living mainly in the marches and swamps along the banks of the Upper White Nile and its tributaries in East Africa. Up to 4 ft tall; the tip of its large broad bill is hooked, and it feeds on lungfishes and other fishes, and even small crocodiles (see **Cochlearis**).

Balaenophilus A genus of Crustaceans of the order *Harpacticoida*. e.g. B. unisetus - This copepod is found in large numbers on the fringed edges of whalebone on the mouths of whales, where it feeds by scraping off algae which grow on the surface of the whalebone.

Balaenoptera A genus of Rorqual or Blue Whales of the family *Balaenopteridae* (order *Cetacea*). e.g. B. physalus - The Common Rorqual Blue Whale, Fin Whale, or Finback Whale, length 18.5-25 m. On each side of the upper jaw there are 320-420 whalebones

which serve as a sieve in foraging for food. Gracefully streamlined and very fast, it has been called the "Ocean Greyhound". Distribution: The Arctic seas. In winter it often swims far south, and is regularly seen in the Mediterranean. Hunted regularly and therefore declining in numbers from year to year.

Balaninus A genus of Weevil insects of the family *Curculionidae*; in the species of this genus the proboscis is greatly attenuated.

Balanocarpus A genus of plants of the family *Dipterocarpaceae*. e.g. B. dammari - The Dammar resin plant; used in the manufacture of varnishes, as a mountant in microscopy, and occasionally as a constituent of plaster-masses. Other genera yielding dammar resins are Hopea and Shorea.

Balanoglossus A genus of Acorn worms (phylum *Hemichordata*), of the family *Ptychoderidae* (class *Enteropneusta*). e.g. B. clavigerus - length 250 mm; elongated worm-like body, colored yellow-brown. The front part has a proboscis, the so-called "acorn". Found on the sea-coast where it forms a U-shaped burrow in the sand; at low tide its presence is revealed by the opening of the tunnel and piles of excrement. Distribution: Atlantic and Mediterranean Sea coast.

Balanophyllia A genus of Solitary Corals of the family *Eupsammidae*. e.g. B. regia - The Scarlet-and-Gold Star Coral, scarlet or orange, with transparent tentacles. Found in California.

Balantidium A genus of Trichostome protozoa (order *Trichostomatida*), of the subclass *Ciliata*, causing balantidiasis (= balantidiosis; balantidosis) in man. e.g. B. coli (= *Blntd.* coli; *Holophyra* c.; *Leucophyra* c.; *Paramecium* c.; *Phagiotoma* c.) - the causative agent of balantidial dysentery.

Balanus A genus of Cirripede Crustaceans of the subclass *Cirripedia*.

e.g. B. improvisus - The Acorn Barnacle; the carapace is highly modified to form a turret in which the trunk can be moved, and from which the cirri can be protruded to catch food.

Balbiana A genus of parasites of the class *Sarcosporidia*. e.g. B. gigantea - A species found in the esophagus of sheep.

Balearica A genus of Birds of the Crane family, *Gruidae*. e.g. B. pavonina - The Crowned Crane of tropical Africa, once known as the Balearic Crane, for it could then be found as far north as the Mediterranean islands of that name. A strikingly beautiful bird with head doubly decorated with a black velvet cap and an occipital tuft of bristle-shaped feathers.

Balistes A genus of Fishes of the family *Balistidae*. e.g. B. capriscus - The Grey Triggerfish, 12 in long; has a laterally compressed body with strong spines on the dorsal, pelvic, and anal fins. Distribution: the Mediterranean, and on the coast of North America.

Balistipus A genus of Fishes (subphylum *Vertebrata*). e.g. B. undulatus - The Undulate Trigger-fish.

Balsamodendron A genus of trees of the family *Palmae*. e.g. B. mukul - A tree of India, producing the fragrant gum-resin bdellium (see **Borassus**).

Baluchitherium A genus of large prehistoric land mammals, a long-necked hornless rhinoceros, which roamed over what is now central and western Asia and East Europe, between 20 and 40 million years ago. e.g. B. grangeri - A species named after its discoverer, Walter Granger; it has an overall length of 10.66-11.27 m (35-37 ft). In life it must have weighed about 16-20 tons.

Bandicota A genus of Old World Rats of the subfamily *Murinae* (family *Muridae*). e.g. B. bengalensis - The Bandicoot Rat, 16 in long (including the tail); very destructive because of its voracious hoarding in deep burrows. Distribution: South Asia.

Banisteria A genus of plants of the family *Malpighiaceae*. e.g. B. caapi - A species yielding an alkaloid banisterine or telepathine, identical with the alkaloids harmine and yageine, which act as psychostimulants and hallucinogens (see **Haemadictyon**; **Peganum**; **Tetrapteris**).

Baptesia A genus of plants of the family *Leguminosae*. e.g. B. tinctoria - A species of herbs.

Barbastella A genus of Bats of the family *Vespertilionidae*. e.g. B. barbastellus - Length of body 42-52 mm; wingspan 240-275 mm. In summer it hides by day in hollow trees or wall crevices, leaving its shelter at dusk. It winters in caves, cellars, and abandoned ruins. Distribution: throughout Europe, Asia Minor, and northwest Africa.

Barbourula A genus of tail-less Amphibians of the Frog or Toad family, *Discoglossidae*. A wholly aquatic and monospecific species of toads found in the Philippines, floating on the surface of streams when undisturbed.

Barbus A genus of Fishes of the Minnow and Carp family, *Cyprinidae*. e.g. B. barbus - The Common Barbel, length 1 m (40 in); lives on the bottom in flowing sections of rivers. Distribution: western and central Europe (see **Carassius**).

Barnea A genus of Bivalve Molluscs of the Piddock family, *Pholadidae*. e.g. B. candida - The White Piddock; length of shell up to 7 cm; valves with radial, spiny ribs, rasp-like especially at front end. Found on the Atlantic and West Baltic coasts. They bore dwelling-holes or tubes in wood, chalk, peat, or firm clay. Native around Britain.

Barosma A genus of plants of the Rue family, *Rutaceae*; also called by the generic name, *Agathosma*. e.g. B. betulina (= *Agathosma* b.) - The Buchu plant (see **Agathosma**).

Bartonella A genus of microorganisms of the family *Bartonellaceae*. e.g. B. bacilliformis - A gram-negative coccobacillus, the causative organism in man, of Oroya Fever (= Carrion's Disease; Infectious Anemia), a severe hemolytic anemia and verruga peruana (a vascular granulomatous skin infection). Occurs in areas of the Andes Mountainsin Peru, Ecuador, and Columbia. Transmitted by the Phlebotomus sandfly.

Bartramia A genus of Birds of the Sandpiper family, *Scolopacidae*. e.g. B. longicauda - The Upland Plover bird of North America.

Basidiobolus A genus of Fungi. e.g. B. haptosporus - A species causing a subcutaneous phycomycosis, a fungus infection in man.

Basilosaurus A genus of extinct whales of great size, also known as *Zeuglodon*, which swam in the seas over what are now the states of Alabama and Arkansas, U.S.A., about 38 to 54 million years ago. One skeleton was found to measure 18.28 m (60 ft) long, had a 1.83 m (6 ft) long skull, with an estimated living weight of 180,000 lb (80.3 tons).

Bassariscus A genus of Mammals of the Raccoon family, *Procyonidae*. e.g. B. astutus - The Cacomistle, Cunning-cat Squirrel, or Ring-tailed Cat, over 1 ft long with a one foot long, beautifully banded black and white tail; the fur is known as the ring-tail cat; a nocturnal and arboreal mammal, somewhat resembling a marten, and living in Central America and the southern United States.

Bassia A genus of trees of the Sapodilla family, *Sapotaceae*. e.g. B. latifolia (= *Madhuca* indica) - the Madhuca or Mowrah tree; medicines are prepared from the bark, leaves, and seeds in India, and are used as astringents, emollients, and expectorants.

Bassogigas A genus of deep-sea Fishes of the family *Brotulidae*, regarded as the deepest-living vertebrates. e.g. B. profundissimus - A specimen of this species was once recovered by the Royal Danish research vessel "Galathea" in the Sunda Trench, south of Java, at a depth of 7,120 m (23,392 ft) in 1951.

Bathycrinus A genus of Sea-lilies of the class *Crinoidea*. e.g. B. gracilis - The smallest extant crinoid or sea-lily; it has a total height of 70-80 mm (2.75-3.14 in).

Bathyergus A genus of African Mole Rats of the family *Bathyergidae*. e.g. B. suillus - The Mole Rat, up to 10 in long; mole-like in shape and habits. It has immense fore claws, and the projecting incisors are powerful and are used in digging. It lives among the coastal sand dunes of South Africa.

Bathynomus A genus of Isopod Crustaceans of the suborder *Flabellifera*. e.g. B. giganteus - the largest isopod of this suborder, 27 cm long, pale lilac in color, and living at considerable depths in the ocean.

Bathypterois A genus of deep-sea Fishes, related to the Lantern-fishes. Some species have been sighted and photographed at a depth of 7,010 m (23,000 ft) (from a bathyscaphe).

Batocera A genus of Longhorned Beetles of the family *Cerambycidae*. e.g. B. wallacei - An extremely long species of long-horned beetles found in New Guinea. It has been measured up to 266 mm (101/2 in) in length (including its 190 mm [71/2 in] long antennae).

Batrachoseps A genus of Worm Salamanders of the family *Plethodontidae*. These amphibians are extremely long and thin, and are found under rotten wood, or burrow deep into the ground. Distribution: United States.

Batrachuperus A genus of Amphibians of the Frog family, *Leptodactylidae*; these frogs contain two aquatic species from high-altitude South American lakes.

Batte Bacillus (= *Mycobacterium* intracellulare) - Nonphotochromogenic acid-fast bacillus that is part of group III "anonymous" or "atypical"

Mycobacteria. Causes pulmonary disease indistinguishable from tuberculosis (see **Mycobacterium**).

Bax A genus of Fishes (subphylum *Vertebrata*). e.g. B. boops - A species of Salt-water fish.

Bdella A genus of Mites. e.g. B. cardinalis - The Snout Mite, parasitic on other insects.

Bdellostoma A genus of Hagfishes of the order *Hyperotreti*. Also called by the generic name, *Heptatretus* or, *Eptatretus*. These species have poorly developed eyes and are parasitic on larger fishes; they have 6-14 gill openings. Distribution: the Pacific Ocean, from California, U.S.A. to Japan (see **Eptatretus**; **Heptatretus**).

Bedsonia A genus of microorganisms of the family *Chlamydiaceae*, better known by the generic name, *Chlamydia*. e.g. B. psittaci (= *Chlamydia* p.) - the causative organism of psittacosis (see **Chlamydia**).

Begonia A genus of flowering plants of the family *Begoniaceae*. e.g. B. semperflorens - an herbaceous perennial species with pink, red or white flowers on axillary cymes. Used in flowerbeds, borders and pots in greenhouses and gardens. Place of origin: Brazil; introduced into Britain in 1829.

Belamcanda A genus of plants of the family *Iridaceae*. e.g. B. chinensis - Blackberry Lily or Leopard Flower plant, a tuberous-rooted herbaceous perennial, growing up to 1 m (3 ft) in height, with sword-shaped basal leaves, and 6-segmented orange colored flowers, spotted with red or brown, in loose clusters. Used in flower-beds, borders, and isolated clumps. Place of origin: China; introduced into Britain in 1823.

Belascaris A genus of ascarid worms. e.g. B. cati - A species found in the intestine of cats.

Bellis A genus of flowering plants of the family *Compositae*. e.g. B. perennis ("Flore Pleno") - The Daisy; English Daisy - a rosette-forming perennial plant. Flowers are 2.5 cm (1 in) across, yellowish disks with white or pinkish ray-florets. Used for edging borders and as potted plants. Place of origin: Europe.

Belone A genus of Garfishes of the family *Belonidae*. e.g. B. belone - The Garfish or Garpike, 80-100 cm long; weight 1 kg. Travels in shoals on the open sea. Its bones are colored green. Distribution: the North Sea; Baltic, Black, and Mediterranean Seas.

Beloperone A genus of plants of the family *Acanthaceae*. e.g. B. guttata - The Shrimp plant; a small twiggy shrub 75 cm - 1 m (2-3 ft) high, with bright green leaves, and creamy flowers with conspicuous pinkish-brown bracts. Place of origin: Mexico; introduced into Europe in 1936.

Benaratherium A genus of large prehistoric land mammals, remains of which have been found in South Georgia, U.S.S.R. They are said to have lived between 20 and 40 million years ago.

Beneckea A genus of microorganisms of the family *Achromobacteraceae*; small to medium-sized rods found in soil and water. e.g. B. indolthetica. B. vulnifica (= *Vibrio* vulnificus) - see **Vibrio**.

Benthamia A genus of shrubs and bushes of the family *Cornaceae*. Also known by the generic name, *Dendrobenthamia*. e.g. B. japonica (= *Dendrobenthamia kousa*) - The Cornus Kousa or Japanese Dogwood plant, a large deciduous shrub or small tree up to 6 m (20 ft) in height with small, inconspicuous, purplish-green flowers like a button surrounded by four showy creamy-white bracts. Strawberry-like scarlet fruits. Place of origin: Japan, Korea, and Central China; introduced into Europe in the mid-ninteeth century.

Bentheuphausia A genus of Euphausiid Crustaceans of the order *Euphausiacea*. The marineshrimp-like species extend down to depths of well over 1,000 m.

Berardius A genus of deep diving

Whales. e.g. B. bairdii - The Berardius Whale of the North Pacific.

Berberis A genus of plants of the family *Berberidaceae*. e.g. B. vulgaris - Barberry Bark; contains the alkaloid berberine (acid salt). It is given orally as a bitter; used in India as injections in the treatment of oriental sore (see **Coptis**).

Bergenia A genus of flowering plants of the family *Saxifragaceae*. Also known by the genus name, *Megasea*. e.g. B. crassifolia (= *Megasea* c.) - The Leather Bergenia plant (= Pig-Squeak; Saxifrage), a coarse perennial with a thick woody rootstock; pinkish-red flowers in panicles on stout stems. Used in rockgardens and flower beds. Place of origin: Siberia; introduced into Europe in 1765.

Beroë A genus of Comb Jellies of the order *Beroidea* (= *Beroida*). e.g. B. cucumis - Length 160 mm. Pink in color; the front end is rounded, with a wide mouth. It floats freely in the sea and feeds on other comb jellies. Distribution: the Mediterranean and Atlantic.

Bertiella A genus of Tapeworms. e.g. B. studeri - A species found occasionally in man and in higher apes in Mauritius, Cuba, India, and the Phillippines.

Beryx A genus of Alfonsino Fishes of the family *Berycidae* (order *Berycomorphi* = *Beryciformes*). e.g. B. alfonsino - A deep-sea species with large eyes; brilliantly colored. It has a single dorsal fin with four to seven spines. Distribution: the warmer parts of the Atlantic.

Bethesda Ballerup group - a genus of microorganisms, now included in the genus *Citrobacter* (see **Citrobacter**).

Betonica A genus of plants of the family *Labiata*. e.g. B. officinalis - Wood Betany; the root is used as an emetic and carthartic.

Betta A genus of Fishes (subphylum *Vertebrata*). e.g. B. splendens - The Siamese Fighting-fish; these fishes choose their spawing sites carefully, and defend them, often until death.

Betula A genus of trees of the family *Betulaceae*. e.g. B. pubescens - The Birch tree, source of birch tar oil.

Bibio A genus of Flies of the family *Bibionidae*. e.g. B. marci - St. Mark's Fly; a short, heavy, slow-flying black species common in Europe in the spring.

Bibos A genus of Wild Cattle of the family *Bovidae*. The wild ox of the genus include the Gayal, Gaur, and Bantang, which are all Asiatic and differ from the European Domestic Ox (*Box* taurus) mainly by the lumps on their backs. e.g. B. gaurus - The Gaur Ox; 6 ft 4 in at the shoulder, with long horns arched over the head. It inhabits the wooded and mountainous regions of India, Burma, and southeast Asia, generally in inaccessible regions. It is not easily domesticated though attempts have been made.

Bielzia A genus of gastropod Molluscs of the family *Limacidae*. e.g. B. coerulans - length 10-12 cm; a well-known mountain slug, colored blue, violet, to almost black; commonly found in forests under the bark of old trees, fallen leaves and stones. Distribution: Europe, in the Carpathian and Sudeten mountain regions.

Bifidobacterium A genus of microorganisms, better classified under other genera, such as *Actinomyces* and *Lactobacillus*. Examples are *Actinomyces* eriksonii, A. parabifidus, and *Lactobacillus* bifidus. These are gram-positive anaerobes; they are intestinal organisms, predominant in breast-fed infants, that disappear on weaning. May also be isolated from the natural orifices of adults. Most strains are non-pathogenic, but on occasion, the organism is associated with abscesses and chronic pulmonary disease (see **Actinomyces**; **Lactobacillus**).

Bigelovia A genus of plants of the family

62

Compositae. e.g. B. veneta - A species affording the tonic damiana.

Bilharzia Former name of the genus *Schistosoma.* e.g. B. hematobium (see **Schistosoma**).

Billbergia A genus of indoor flowering plants of the family *Bromeliaceae.* e.g. B. rhodocyanea - The Aechmea Fasciata plant, an evergreen epiphyte with rosy-pink flowers, the individual blooms being blue at first, then rose, and enclosed by spiny pink bracts. Used as an indoor plant.

Biorhiza A genus of Gall Wasps of the family *Cynipidae.* e.g. B. pallida - The Oak-apple Gall-fly. This gall wasp produces two types of gall: on oak branches and on the roots. Those on the branches, called oak apples are shaped like a potato, up to 40 mm in diameter. Inside this gall the sexual generation develops. After pairing, the female burrows into the ground, laying her eggs which develop into roundish galls, on slender oak roots. Distribution: Europe, Asia Minor, and North Africa.

Birgus A genus of land hermit Crabs of the family *Coenobitidae.* e.g. B. latro - The Coconut or Robber Crab, the heaviest and largest land crustacean known. When full grown it does not carry a shell. It lives on tropical Indo-Pacific islands and atolls, has a leg-span of 914 mm (3 ft), and can reach as much as 4.08 kg (9 lb) in weight.

Bison A genus of now very rare Wisents of the family of Wild Cattle, *Bovidae.* e.g. B. bison - The North American Bison or Buffalo; heavy forequarters and a massive head; large bulls stand 6 ft at the shoulder and may weigh nearly 11/2 tons. This species formerly existed in enormous herds on the prairies of western Canada and the United States; their extermination began towards the end of the 18th century. By 1935 there were only about 20,000 bison in North America, more than half being in Canada. Today the species is protected

and their numbers are large enough to guarantee their survival. B. bonasus - The European Bison, Wisent, or Auroch; length of body 310-350 cm; height at shoulder 85-200 cm. Found in forests with lush undergrowth. Was abundant in Europe in the late Cainozoic era, but the species has now been almost exterminated. Today found only in nature preserves and in zoos.

Bispira A genus of segmented Worms of the family *Sabellariidae.* e.g. B. volutacornis - The Sabellid fan and feather-duster worm; found on exposed rocky coasts, massed on rocks and boulders in honeycomb-like colonies.

Biston A genus of Moths of the family *Geometridae.* e.g. B. betularia - The Peppered Moth; 21-32 mm. Most abundant in its black form, known as "carbonaria", and found in industrial centers. It has another form, known as "ochrearia", with ochre-colored wings. Distribution: Europe and Asia.

Bitis A genus of Reptiles of the Adder or Viper Snake family, *Viperedae.* e.g. B. atropos atropos - The Berg or Mountain Adder of South Africa. Its venom is both cytotoxic causing necrosis, and neurotoxic causing (reversible) ophthalmoplegia, ptosis and fixed widely dilated pupils, not reacting to light.

Bitoma A genus of Insects of the Beetle family, *Colydiidae.* e.g. B. crenata - Length 2.6 - 3.5 mm. Found commonly under the bark of old stumps. Distribution: the Palaearctic region.

Bittium A genus of marine Gastropod Molluscs or Sea-snails of the order *Neogastropoda.* e.g. B. reticuatum - The Needle Shell, length 9-12 mm; the shell has fine lattice markings. A common sea-snail species. Distribution: Europe; native around Britain.

Blaniulus A genus of Diplopod Arthropods of the Millipede (= Millepede) family, *Blaniulidae.* e.g. B. guttulatus - Length 7.5-16 mm. Body composed of several segments, with 103 pairs of legs. Whitish to yellowish-grey with a row of carmine-red spots running along the sides of the body, with one pair to each segment. It is herbivorous, and is found in gardens and greenhouses. Distribution: much of Europe and Canada.

Blarina A genus of Mammals of the Shrew family, *Soricidae*. e.g. B. brevicauda - The Big Short-tailed Shrew, the red-toothed North American variety.

Blastocaulis A genus of microorganisms of the family *Pasteuriaceae*; pear-shaped or globular cells found in freshwater areas. e.g. B. sphaerica.

Blastocerus A genus of True Deer of the family *Cervidae*. e.g. B. benzoarticus - The Pampas Deer of South America. Its antlers tend to be rudimentary, with only three points.

Blastocystis A genus of vegetable "non-pathogenic amebae", related to the yeast. e.g. B. hominis.

Blastomyces A genus of yeastlike fungi, morphologically similar to *Saccharomyces*; pathogenic to man and animals. e.g. B. braziliensis (= *Paracoccidiodes* b.) - The fungus causing South American blastomycosis. B. coccidiodes (= *Coccidiodes* immitis; *Mycoderma* immite). B. dermatitidis (= *Endomyces* epidermidis; *Mycoderma* dermatitidis; *Zymonema* capsulatum; Z. dermatitidis; Z. gilchristi) - A species causing North American blastomycosis. B. farciminosus (= *Zymonema* farciminosum) - A species causing blastomycotic epizootic lymphangitis in horses.

Blastophaga A genus of Insects of the Fig Insect or Bark Beetle family, *Agaontidae*. e.g. B. psenes - The best-known fig insect; introduced into California to make possible there the cultivation of Smyrna figs.

Blatta A genus of Insects, also called *Blatella*, of the Cockroach family, *Blattidae*. e.g. B. orientalis (= *Blatella* o.) - The Common Cockroach, length 18-30 mm; occurs in human habitations;

male has 2 pairs of wings shorter than the abdomen, female has only wing stumps. These roaches soil food and other provisions and may transmit diseases. Adults live up to one year. Distribution: worldwide.

Blatella See **Blatta**.

Blennius A genus of Perciform Bony Fishes of the scaleless Blenny family, *Bleniidae*. e.g. B. pholis - The Shanny fish, length up to 16 cm; occurs near rocky coasts, burying itself in mud or sand. Distribution: the northeast Atlantic.

Blepharidopterus A genus of Heteropterous Insects of the family *Miridae*. These are terrestrial bugs or *Geocorisae*. e.g. B. angulatus - A beneficial orchard insect, delicately built with an elongated oval firm, which feeds on the injurious red-spider mites.

Bletia A genus of plants of the family *Orchidaceae*. e.g. B. hyacinthina - The Bletilla Striata, a terrestrial orchid, with rich purple flowers on erect stems. Place of origin: China; introduced into Europe in 1802.

Blicca A genus of Fishes of the Minnow and Carp family, *Cyprinidae*. e.g. B. bjoerkna - The Silver or White Bream fish, length 35 cm; weight 1 kg. Very deep-bodied and markedly flattened from side to side. Found in the lower reaches of larger rivers, and in pools, inlets, and ponds. An edible fish. Distribution: all over Europe (see **Carassius**).

Blissus A genus of Heteropterous terrestrial bugs or *Geocorisae*, of the family *Lygaeidae*. e.g. B. leucopterous - The Chinch Bug, a plant-feeding insect pest of cereal crops, found in North America.

Boa A genus of Boa snakes of the family *Boidae*. e.g. B. canina - The Emerald Boa of South America; 6 ft long; one of the most magnificent boas; the adult is greenish yellow in color with white rings. Arboreal in habit it destroys many birds, particularly parrots, and monkeys.

Boaedon A genus of Colubrid Snakes of the family *Colubridae*. e.g. B. lineatum - The Domestic Snake of tropical Africa; it catches mice and is known as a good ratter.

Bodo A genus of flagellate microorganisms, also known by the generic name, *Cystomonas*, of the family *Bodonidae*. e.g. B. caudatus (= *Cystomonas* c.) - The common coprozoic flagellate found in human feces.

Boerhaavia A genus of plants of the family *Nyctaginaceae*. e.g. B. diffusa (= B. repens) - The Punarnava or Punarnaba plant; contains an alkaloid punarnavine, and is used in India as a diuretic. The name *Punarnava* is also used for the white variety of *Trianthema* portulacastrium (see **Trianthema**).

Boiga A genus of poisonous Colubrid Snakes of the subfamily *Boiginae*. e.g. B. dendrophila - The Mangrove Snake; a tree-dweller, 5 ft long; lives among the long, outside roots of mangrove trees near coasts and estuaries. Glossy black color with gold bands. It has an aggressive nature and is sometimes called the "Cat Snake". Distribution: Asia.

Bolina See **Bolinopsis**.

Bolinopsis A genus of Comb Jellies of the order *Lobata*; Also called by the generic name, *Bolina*. e.g. B. infundibulum (= *Bolina* i.) - This lobate ctenophore is found in the colder waters of the North Atlantic and North Pacific oceans.

Bolitoglossa A genus of plethodontid Salamanders of the family *Plethodontidae*. They are lungless amphibians and are found from Mexico through Central to South America, where some species may live at heights up to 11,000 ft.

Bombina A genus of tail-less amphibians (Frogs and Toads) of the family *Discoglossidae*. e.g. B. bombina - The Fire-bellied Toad, length 4.5 cm; lives in calm clear waters, and feeds on small aquatic animals. Distribution: central and eastern Europe.

Bombus A genus of Bees of the family *Apidae*. e.g. B. lapidarius - The Red-tailed Bumblebee, length of female 20-24 mm; lives in colonies. Distribution: Central Europe.

Bombycilla A genus of Perching Birds of the Waxwing family, *Bombycillidae*. e.g. B. garrulus - The Bohemian Waxwing, length 18 cm; inhabits forests with dense shrub vegetation, and feeds chiefly on mosquitoes in summer and berries in winter. A migrating bird, living in the north polar region, and travelling southward every winter to Central Europe. The secondary feathers have reddish, waxy-looking tips, giving rise to the name "wax wing".

Bombylius A genus of Bee Flies of the family *Bombyliidae*. e.g. B. major - Length 8-13 mm; resembles a small bumblebee; has a strikingly long proboscis. The larvae are parasites in the nests of certain solitary bees such as *Andrena*, *Colletes*, etc. Distribution: the Palaearctic and Nearctic regions.

Bombyx A genus of Moths of the family *Bombycidae*. e.g. B. mori - The Silkworm Moth, originally from China, creamy-white in color; introduced into many countries and cultivated on a large scale in commercial silk production. Each cocoon consists of a single thread of silk which when unravelled, is about 1,000 yd long.

Bonasa A genus of Birds of the Ptarmigan or Grouse family, *Tetraonidae*. e.g. B. umbellus - The Ruffed Grouse; a resident forest species found in mixed and deciduous woodlands, where it is a popular game bird known to hunters as the partridge. It perches firmly on a log or stump and beats its wings in a vertical position making a "drumming" noise, apart of its courtship-territorial procedure. Distribution: most of Canada and the United States.

Bonellia A genus of Echiuroid Worms of the phylum *Echiuroidea*. e.g. B. viridis - An unusual species; the male 3 mm long, lives as a parasite in the meter-long female. Found in the Mediterranean region.

Boocercus A genus of Antelopes of the subfamily *Antilopinae*. e.g. B. eurycerus - The Bongo, a hollow-horned ruminant, only found in the deep recesses of equatorial forests, from West Africa to the Aberdare Mountains of Kenya.

Boophilus A genus of Cattle Ticks, also variously known as *Margoropus*, *Rhinencephalus*, *Rhinicephalus*, and *Rhinocephalus*, of the Tick and Miteorder, *Acari* (= *Acarina*). e.g. B. annulatus (= B. bovis) - The southern U.S.A. cattle tick transmitting the protozoan parasite, *Babesia* bigemina, the causative agent of Texas fever.

Borassus A genus of Trees of the family *Palmae*. e.g. B. flabelliformis - A palm tree from Africa, producing a gum, bdellium (see **Balsamodendron**).

Bordetella A genus of microorganisms of the family *Brucellaceae*, the species of which were formerly placed in the genera *Haemophilus* or *Hemophilus*, *Alcaligenes* or *Alkaligenes*, *Bacillus*, *Brucella*, etc. Small gram-negative coccobacilli, or rods, which produce a dermonecrotic toxin. e.g. B. bronchiseptica (= *Haemophilus* b.; *Hemophilus* b.; *Alcaligenes* b.; *Alkaligenes* b.; *Bacillus* bronchicanis; *Bacillus* bronchisepticus; *Brucella* bronchiseptica) - May be part of the normal flora in the respiratory tract of man, dogs, rabbits, and other animals. Can cause pneumonitis or a pertussis-like disease. B. pertussis (= *Hemophilus* p.; *Haemophilus* p.; *Bacillus* p.) - The causative agent of pertussis or whooping cough, and an occasionally associated interstitial pneumonia.

Boreogadus A genus of Fishes of the Cod family, *Gadidae* (Cod subfamily, *Gadinae*). e.g. B. saida - The Arctic Cod; a commercially important fish, covered with cycloid scales; found in the cold waters of the North Atlantic, and in the summer, around the edges of the Arctic ice.

Boreus A genus of Insects of the order of Scorpion-flies, *Mecoptera*. The species feed on moss, in contrast to the other genera of this order, which are essentially carnivorous insects.

Borophyrne A genus of Fishes (subphylum *Vertebrata*). e.g. B. apogon - The Deep-sea Angler fish.

Borrelia A genus of microorganisms of the family *Treponemataceae*, some species of which were formerly classified under the generic names, *Entomospira*, *Spironema*, and *Spiroschaudinnia*. These spirochetes exist in many species; some are part of the normal oral or genital flora that, on occasion, cause Vincent's angina, and several are known to cause relapsing fever. e.g. B. burgdorferi - Causes Lyme disease in man. Vector - mites. Animal reservoir of the infected mite: cattle, dogs, deer, etc. B. duttoni (= B. duttonii; B. recurrentis; *Spironema* d.; *Spiroschaudinnia* d.) - A species causing West African Relapsing Fever (= African Tick Fever). Relapsing fever is caused by many other species of spirochetes, including B. aegyptium, B. berbera, B. carteri, B. caucasica, B. hermsii, B. hispanica, B. kochi, B. novyi, B. parkeri, B. persica, B. rossi, B. turicatae, and B. venezuelensis. B. buccalis and B. vincentii - Isolated from normal mouths around the gums and from mucous membranes of the respiratory tract. Associated with Vincent's angina and fusospirochetal disease. B. glossinae (= *Entomospira* g.).

Bos (i) A genus of Horned Ungulates of the family *Bovidae* - (L.) Cow. e.g. B. taurus - The Ankole cattle of Africa. From this animal is produced Beef Suet and Beef Tallow. (ii) An infection: e.g.

B. sinicus - A schistosoma infection in cattle.

Boselaphus A genus of Hollow-horned Ruminants of the Antelope subfamily, *Antilopinae*. e.g. B. tragecamelus - The Nylghaie or Blue Bull, the biggest Asian antelope, with a height of 55 in to the withers, lives on the open plains of India. It has points of resemblance to the ox, deer, goat, and camel. The horns, found in males only, are small and slightly curved forwards.

Boswellia A genus of plants of the family *Burseraceae*. e.g. B. carterii - From this plant is prepared Olibanum (= Encens, Frankincense, Incenso, Thus, or Weihrauch) - used as an ingredient of incense and some fumigating powders.

Botaurus A genus of Birds of the Egret, Bittern, and Heron family, *Ardeidae*. e.g. B. stellaris - The Eurasian Bittern, length 76 cm; a nocturnal bird with three longitudinal black stripes on a chestnut breast; found on large beds of reeds and low, shrubby willows and alders near small bodies of clean shallow water. The booming cry of the male Bittern, audible at a three-mile range, if unforgettable. Distribution: Europe, Asia, and Africa.

Bothops See **Bothrops**.

Bothriocephalus A genus of Tapeworms, also called *Dibothriocephalus*, and today known as *Diphyllobothrium*. e.g. B. latus (= *Dibothriocephalus* latus; *Diphyllobothrium* latum) - The Fish Tapeworm of man (see **Diphyllobothrium**).

Bothrioplana A genus of Tubellarian worms or Flatworms of the suborder *Cyclocoela* (order *Protricladida*). e.g. B. semperi - This proticlad, like the other marine species of this order, has a triclad of pharynx and a single undivided intestine.

Bothrops A genus of South American serpents, also called *Bothops*, of the Pit Viper family, *Crotalidae*. e.g. B. atrox - The Fer-de-lance, a large venomous snake of Central and South America. B. jararacussu (= B. jararaca) - The Tararacucu snake of Brazil. It yields great amounts of venom, more than 200 mg (0.006 oz) dry weight of the peptide, known as teprotide - used in the treatment of hypertension, due to its inhibiting action on the angiotensin converting enzyme (Ace), which is reponsible for the conversion of angiotensin I to angiotensin II. From teprotide, diacid drugs such ascaptopril and enalaprilat have been synthesized and are now used in the lowering of blood pressure.

Botia A genus of Fishes of the Loach family, *Cobitae*. e.g. B. macrocanthus - The Clown Loach, a beautiful, brilliantly colored, small freshwater, Indonesian fish, with an orange-red ground color and three velvet blade bands. The fins are blood red. It lives on the water bottom and feeds on insects and worms.

Botrychium A genus of plants of the Adderstongue or Adder's Tongue family, *Ophioglossaceae*. e.g. B. lunaria - The Common Moonwort, a perennial plant, 8-25 cm high; it has a sterile blade about the middle of the leaf. Grows widespread in meadows, heath and woodland. Place of origin: Europe; native in Britain.

Botryllus A genus of Sea-squirts of the class *Ascidiaceae*. These chordate animals develop colonial forms in which numerous small individuals are invested by a common tunic. e.g. B. schlosseri - The Golden-stars Tunicate or Golden Star Sea-squirt; each animal is up to 2.5 mm long; 6 to 20 individuals are grouped in a star-like colony and grow on boulders exposed to the sea, embedded in a common gelatinous test. Native in Britain (see **Amaroucium**).

Botrytis A genus of fungi causing disease in insects; also affects the skin of grapes and other fruits and vegetables. e.g. B. bassiana - Causing the disease muscaridine in silk-worms. B. cinerea - The grey rot fungus, developing on grapes and other fruits.

Bougainvillea A genus of climbing and spreading plants of the family *Nyctaginaceae*. Named after Louis Antoine de Bougainville who sailed round the world in 1767 to 1769. e.g. B. glabra - The Bougainvillea or Paper Flower plant; a vigorous climber with thick oval leaves and showy clusters of small white flowers surrounded by brilliant cerise bracts. Varieties with orange, lemon and pink bracts exist; also a double and variegated leaved form. Used as a showy and long-flowering conservatory plant. Place of origin: Brazil.

Bourletiella A genus of Springtail Insects (suborder *Symphypleona*), of the family *Sminthuridae*. e.g. B. hortensis - A common symphypleonan insect of Europe and North America, feeding mainly on fungal hyphae or organic debris.

Bouvardia A shrubs and bushes of the family *Rubiaceae*. e.g. B. longiflora - The Bouvardia plant, a low growing, evergreen shrub or herb, much valued in the tropics for its long flowering season. It has terminal corymbs of white, tubular, 4-lobed, fragrant flowers, individually resembling jasmine. Used as a greenhouse pot or garden plant or for cut flowers. Place of origin: Mexico; introduced into Europe in 1827.

Bovimyces A genus of bacteria resembling filtrable viruses. e.g. B. pleuropneumoniae (= *Astrcoccus* mycoides) - A species causing pleuropneumonia in cattle.

Brachinus A genus of Ground Beetles of the family *Carabidae*. These so-called Bombardier beetles defend themselves by audibly emitting puffs of a volatile irritant secretion from glands near the posterior end of the body.

Brachionus A genus of microscopic worms, the Rotifers of the class *Rotifera*. These metazoa belong to the order *Monogononta*, which is characterized by the possession of a single ovary and vitellarium; there is an external cuticle that covers most of the body, but no body wall as such, only an ectoderm and isolated bands of muscle.

Brachiosaurus A genus of prehistoric reptiles or Dinosaurs, known as the Shoulder Lizards. These are the heaviest of all prehistoric animals and the heaviest land vertebrates of all times. They lived in what is now East Africa and Tanzania, and in Colorado and Oklahoma, U.S.A. between 135 and 165 million years ago. Some species weighed an estimated 78.26 tonne (77 tons) when alive, one specimen (the skeleton) measuring 22.7 m (74 ft 6 in) in total length.

Brachycephalus A genus of Amphibians of the family *Atelopidae*. The species are small Treefrogs of Central America.

Brachymystax A genus of Fishes of the family *Salmonidae*, the species of which are the Iconnus, the large fishes in the mountain streams of Siberia.

Brachyplatystoma A genus of large Catfishes (subphylum *Vertebrata*). e.g. B. filamentosum - The Lau-lau Catfish, a very long species found in the rivers of Guayana, as well as the Amazon river in South America, with an average length of 2.43 m (8 ft).

Bradicebus A genus of Mammals of the Loris or Slow-lemur family, *Lorisidae*. e.g. B. tardigradus - A large, slow-moving lemur of India and the East Indies.

Bradypus A genus of slow-moving terrestrial mammals (class *Mammmalia*). e.g. B. tridactylus - The Ai or Three-toed Sloth of central and tropical America, the slowest terrestrial mammal, with a ground speed of 1.83-2.43 m/min (6-8 ft/min). In the trees this speed may be increased to 609 mm/sec (2 ft/sec) or 2.19 km/h or 1.36 miles/h (see **Choleopus**).

Bragada A genus of blood-sucking insects. e.g. B. picta - A species attacking man in India.

Branchinecta A genus of Branchiopod

Crustaceans of the order *Anostraca*. e.g. B. paludosa - A common circum-arctic species of which there are relict populations left behind after the Ice Ages. One such population is found in the Tatra Mountains of Czechoslovakia, and another, on the Medicine Bow Mountains, Wyoming, in the United States.

Branchiostoma A genus of small fish-like chordate animals, the Lancelets, of the subphylum *Cephalochordata*. Formerly known by the generic name, *Amphioxus*. There are some two dozen species; the largest of all, 6 in long, is found in the shallow waters around the coast of Amoy in southern China. They bury themselves in the sand, with only the headend including the mouth, exposed above the surface. They feed on organic particles and small plankton.

Branhamella A genus of microorganisms. e.g. B. catarrhalis.

Branta A genus of True Geese of the Duck family, *Anatidae*. e.g. B. bernicla - The Brant or Brent Goose; length 58 cm; black and brown with a white breast. A migratory bird found on the arctic coasts, mainly near river estuaries. Distribution: northern Palaearctic region.

Brassica A genus of plants of the family *Cruciferae*. Also known by the genus name, *Sinapis*. e.g. B. alba (=, *Sinapis a.*) - The White Mustard plant; it contains the glycoside sinalbin and the enzyme myrosin. Used as a condiment and emetic, and externally as a counter-irritant.

Brayera A genus of plants of the family *Rosaceae*. e.g. B. anthelmintica (= *Hagenia* abyssinica) - Cousso; Flos Koso; Kousso - the pistillate flowers have been used in anthelmintic infusions.

Brephidium A genus of Butterflies. e.g. B. barberae - The Dwarf Blue butterfly of South Africa, the world's smallest known butterfly. It has a wingspan of 14 mm (0.55 in) and weighs less than 10 mg (0.00034 oz).

Breviceps A genus of small tail-less Amphibians of the order of Frogs and Toads, *Anura* or *Salientia*. The species are found in South Africa; they use the special digging shovels in their hind feet to dig into the soft earth and sink into the soil.

Brevoortia A genus of Fishes (subphylum *Vertebrata*). e.g. B. tyrranus - The Atlantic Menhaden fish.

Brodiaea A genus of plants of the Lily family, *Liliaceae*; formerly known by the generic names, *Triteleia* or *Tritelia*. e.g. B. laxa - The Brodiaea plant (= Grassnut; Ithuriel's Spear); produces pale violet-mauve funnel-shaped flowers. Place of origin: western United States.

Brontosaurus A genus of extinct animals, the Dinosaurs. Also known as *Apatosaurus*, a large theropod or Carnosaur ("meat reptile").

Brosme A genus of Fishes of the subfamily *Lotinae*. e.g. B. brosme - The Cusk Fish of the North Atlantic. A deep-water fish with only one dorsal fin.

Brucea A genus of plants. e.g. B. sumatrama - A species growing in southeast Asia. The seeds, known as kosam are used in the treatment of dysentery and uterine hemorrhage. Ya Yan Tzu - A Chinese species of Brucea, the seeds of which are used in the treatment of amebic dysentery.

Brucella A genus of microorganisms of the family *Brucellaceae*; small gram-negative coccobacilli capable of infecting many animal hosts, including man, and producing brucellosis (Undulant fever; Malta fever). e.g. B. melitensis (= Br. melitensis; Br. melitensis; *Bacillus m.*) - A species found in goats; one of the 3 main species (besides Br. abortus and Br. suis) causing Brucellosis in humans. Man is infected through ingestion of infected milk products, droplets onto mucous membranes, or skin contact with infected products or tissues. B. abortus (= Br. abortus) - A species found in cattle (see B. melitensis). B. suis (Br. suis) - A species found in swine (see B. melitensis). B. canis (= Br. canis) - A species found in dogs. Can infect man. B. neatomae (= Br. neatomae) - A species found in the desert rat. Can infect man.

Bruchus A genus of Beetles of the family *Bruchidae*. e.g. B. pisorum - Length 4-5 mm. A common pest of the pea plant, appearing in the fields in spring after hibernating in stored seeds or outdoors. The eggs are laid in young pea pods and covered with a sticky substance by the female to prevent washing off by rain. Distribution: worldwide.

Brugmansia A genus of plants of the family *Solanaceae*, now called by the generic name, *Datura*. e.g. B. sauveolens (= *Datura* s.) - see **Datura**.

Brunfelsia A genus of South American plants. e.g. B. hopeana (= *Franciscea* h.) - A Brazilian plant yielding manaca, used in the treatment of gout and rheumatism.

Bryobia A genus of insects. e.g. B. praetiosa - The red spider or spinning mite, found in clover.

Bryonia A genus of plants of the family *Cucurbitaceae*. e.g. B. dioica (= B. alba) - Root of Bryony or False Mandrake; used as a cathartic.

Bubalus A genus of mammals of the cattle family, *Bovidae*. e.g. B. bubalis - The Indian or Water Buffalo, also known as the Arnaor Carabao, a domestic buffalo found also in Europe (with shorter horns), and other parts of Asia. It has the longest horns grown by any animal; these have been measured at 4.24 m (13 ft 11 in) from tip to tip along the curve across the forehead. There is also a record of a single horn measuring 2 m (6 ft 6½ in) on the outside curve.

Bubo A genus of Typical Owls of the family *Strigidae*. e.g. B. bubo - The Eagle Owl, a night bird of prey; the largest European owl; length 66-71 cm. Found in lowlands, nesting in hollow trees, or in mountain regions building its nest in rock formations. Distribution: most of Europe, Asia, and North Africa.

Buccinum A genus of gastropod molluscs of the family *Buccinidae*. e.g. B. undatum - The Common Whelk; shell up to 110 mm high, 65 mm wide, with 6-8 whorls and a rich surface sculpture. Found on sandy and muddy bottoms on the seashore or at depths of up to 200 m. A scavenger, it feeds on both dead and living animals. A plentiful species, collected for (sea) food. Distribution: the Atlantic and North Sea coasts.

Bucephala A genus of waterfowl of the Duck family, *Anatidae*. e.g. B. clangula - The Goldeneye duck, length 41-45 cm (male 45 cm; female 41 cm). The nest is built in tree hollows, perhaps as far as 2 km away from water. Distribution: northern Europe, Asia, and North America.

Bucephalus A genus of trematode parasitic worms. e.g. B. papillosis - A species found in freshwater fish.

Buceros A genus of Birds of the Hornbill family, *Bucerotidae*. e.g. B. bicornis - The Great Hornbill, 5 ft long, an ungainly-looking bird with a huge bill, nesting in tree hollows. Found in India, Malaya, and Indonesia.

Buddleia See **Buddleja**.

Buddleja A genus of Shrubs of the family *Buddlejaceae*; also known by the generic name, *Buddleia*. e.g. B. davidii (= B. variabilis; *Buddleia* d.; *Buddleia* v.) - The Buddleia or Butterfly Bush; a vigorous deciduous shrub, 4-5 m (12-15 ft) high with mid-green, lanceolate leaves and long slender panicles of closely packed, fragrant, lilac to purple flowers. Used as a summerflowering garden shrub. Place of origin: Central and Western China; introduced into Europe by the French missionary David in 1864.

Budorcas A genus of Goat-antelopes of

70

the subfamily *Caprinae*. One species, the Takin, is remarkable in that it combines features of the ox, goat, andantelope. Its Roman nose, eyes high in the head, and horns placed at the top of the forehead give it an extremely unusual appearance. Distribution: the highlands of Central Asia.

Bufo A genus of Amphibians of the Toad family, *Bufonidae*. e.g. B. bufo - The Common Toad, length 12 cm. Nocturnal, hiding under stones by day. From the skin glands was first isolated a hallucinogenic indole alkaloid, bufotenine, now available from other sources. Distribution: all Europe, Asia, and northwest Africa (see **Piptadenia**). B. viridis - From the skin glands of this toad, is extracted the cardiac poison viridobufagin ($C_{23}H_{34}O_5$).

Bugula A genus of Sea-mats of the family *Bicellariidae* (phylum *Polyzoa*). e.g. B. neritina - A moss animal up to 100 mm high; looks more like a branching tuft of marine grass. Lives in large colonies, each individual forming a shrub-like zoarium (each individual in such a colony is called a zooid). Found chiefly on the hulls of ships and on buoys in shallow and deep water. Distribution: the Atlantic Ocean, North and Mediterranean Seas.

Buistia A genus of viruses. e.g. B. pascheni - Elementary "inclusion bodies" allied to the Vaccinia or Cow-pox virus.

Bulinus A genus of Snails or Molluscs, several species of which are the intermediate hosts of *Schistosoma* hematobium and *Clonorchis*. e.g. B. contortus (see **Schistosoma**).

Bulla A genus of Bubble Shell Sea Slug Molluscs of the order *Cephalaspidae*, also known by the generic name, *Bullomorpha*. A carnivorous animal species inhabiting sandy or muddy sea-bottoms.

Bulweria A genus of Birds of the family *Procellariidae*, the Gadfly Petrels. Little is known about them due to their pelagic haunts and rather secretive breeding habits. A migratory and wholly marine family.

Bungarus A genus of extremely venomous land snakes, the Kraits of Southeast Asia and the Malay Archipelago, of the family *Elapidae*. e.g. B. caeruleus - The Blue Krait or Common Indian Krait, 4 ft long; a particularly venomous species; about 50% of bites being fatal, even with anti-venom treatment. The lethal dose for man is 2-3 mg (0.00007-0.000105 oz) of venom.

Bupalus A genus of Moths of the family *Geometridae*. e.g. B. piniaria - The Bordered White Moth, length 19-22 mm. Found in pine and mixed coniferous forests. Distribution: northern and central Europe.

Buphagus A genus of Oxpecker Birds of the family *Sturnidae*. The species are found in Africa, always in association with large game mammals, especially cattle. They perch on the back of game, feeding on the ectoparasites in the mammal's coat, and even on the larvae embedded in the skin. As soon as any predator approaches, the birds swirl upwards calling, so giving warning of danger to their host.

Buprestis A genus of Metallic Wood-borer Beetles of the family *Buprestidae*. e.g. B. rustica - Length 12-18 mm; found in open coniferous forests. Distribution: most of Europe.

Burhinus A genus of Field or Shorebirds of the Thick-knee family, *Burhinidae*. e.g. B. oedicnemus - The Stone Curlew or Thick-knee; length 41 cm; found in large fields, sandy areas, and steppe with sparse vegetation. A migrating bird. Distribution: temperate regions of Europe, southwest and southern Asia, and North Africa.

Burretis A genus of long-living Beetles. e.g. B. splendens - The Metallic wood-boring beetle, a very long-living species, said to live more than 30 years.

Busycon A genus of Molluscs of the order *Neogastropoda*; these are large scavenging animals, the Whelks; they are all carnivorous, with an eversible proboscis.

Butea A genus of plants of the family *Leguminosae*. e.g. B. frondosa (= B. monosperma) - The seeds yield an anthelmintic oil (moodooga oil).

Buteo A genus of Birds of the Buzzard and Hawk subfamily, *Accipitrinae* (= *Buteoninae*), and of the family *Accipitridae*. e.g. B. buteo - The Common Buzzard, length 51-56 cm, a wingspan of 150 cm; found in woodlands interspersed with fields and meadows. Builds its nest in tall trees. Some birds are resident, while others are migrant or dispersive. Distribution: Europe and Asia.

Buthacus A genus of venomous Scorpions. e.g. B. occitanus - A species found in the Western Desert of North Africa; can be fatal to man.

Buthus A genus of Scorpions (order *Scorpiones*). e.g. B. quinquestriatus - A dangerous species found in Egypt; about 55 mm in length.

Byctiscus A genus of Beetles of the family *Attelabidae*. e.g. B. populi - Length 3.8-6 mm; metallic green, sometimes blue or black. Found on aspen and poplar trees. Distribution: Eurasia.

Byrrhus A genus of Pill Beetles of the family *Byrrhidae*. e.g. B. pilula - Length 7.5-11 mm; when attacked or in danger, it feigns death, changing itself into a motionless ball or pill, by withdrawing its antennae and legs into crevices on the underside of its body. Found in fields and forests. Distribution: the Palaearctic region.

Bythnia A genus of Snails. e.g. B. longicornus (= *Alocimua* longicornis).

Byturus A genus of Beetles of the family *Byturidae*. e.g. B. tomentosus - The Raspberry Beetle; length 3.2-4 mm; yellow in color; its elongated larva is commonly known as the raspberry fruitworm. Found chiefly in the fruit of the raspberry. Distribution: the Palaearctic region.

C

Cabassous A genus of Mammals of the Armadillo family, *Dasypodidae*. e.g. C. unicinctus - The Tatouay Armadillo, found in Central and South America; is distinguished by having a tail covered only by delicate scales, instead of the usual strong armored scales in other genera of the same family.

Cacatua A genus of Birds of the order of Parrots and Parakeets, *Psittaciformes* (family *Psittacidae*). e.g. C. alba - The White-crested Cockatoo, 45 cm (11/2 ft) long, white all over except for the pale yellow underside of the wings and tail. The white crest or cockade is composed of broad feathers arranged in a row one behind the other. Habitat: the jungles of the Molucca Islands in the Malay Archipelago.

Cacajao A genus of Mammals of the New World Monkey family, *Cebidae*. e.g. C. melanocephalus - The Black-headed Uakari, a brightly colored monkey with a short bushy tail. Gregarious in nature, it lives high in the trees of the Amazonian forests, confined to a comparatively small tract in the Amazon basin.

Cactoblastis A genus of Insects of the Moth family, *Pyralididae*. e.g. C. cactorum - The Cactus Moth; this species was deliberately introduced into Australia to destroy the dense growth of prickly pear in New South Wales and Queensland.

Caenorhabditis A genus of Nematodes or Roundworms (class *Nematoda*). e.g. C. elegans - A nematode species with 4 larval stages; 1 mm in length. Development from embryo to adult is complete in about 63 hours.

Caesalpinia A genus of plants of the Pea family, *Leguminosae*. e.g. C. sepiaria - The Mysore-thorn plant, a climbing shrub, up to 4.5 m (15 ft) tall, with bip-innate leaves and erect racemes of canary-yellow, cup-shaped flowers. Place of origin: India; introduced into Europe in 1857.

Caiman A genus of Reptiles of the Crocodile order, *Crocodylia*. e.g. C. crocodilus - The Spectacled Caiman or Cayman, an alligator found in tropical America, from the basin of the Amazon and Orinoco rivers. This species attains a maximum length of 2.4 m (8 ft).

Cairina A genus of Birds of the Duck family, *Anatidae*, belonging to the Cairinini tribe of perching ducks, living in forests and nesting in trees. e.g. C. moschata - The Barbary Duck or Muscovy Duck, originally from Central and South America. Metallic dark green with white spots, and fleshy excrescences at the base of the back. Black glossed with green in the wild state, but domesticated birds have a good deal of white and grey in the plumage.

Cakile A genus of drift-line plants of the Cress or Crucifer family, *Cruciferae*. e.g. C. maritima - The Sea Rocket, a perennial plant, 15-45 cm high; fleshy leaves, lilac to pink flowers, and 2-jointed fruits. Grows on sandy and shingly shores.

Calamagrostis A genus of plants of the Grass family, *Gramineae*. e.g. C. varia - The Variable Small-reed, a perennial grass, from 30-100 cm high, leaves up to 7 mm wide; rough, pale-violet glumes, twice as long as hairs at base of lemma.

Calamintha A genus of plants of the Thyme family, *Labiatae*. e.g. C. ascendens - The Common Calamint, a perennial plant, 25-80 cm high; leaves up to 4 cm long; stalked flowers in whorls of 6 to 10. Found in open deciduous woods, bushy slopes, and dry banks. Distribution: central and southern Europe.

Calamoichthys A genus of river fishes of the order *Cladista* (= *Polypteriformes*). These reed fishes have elongated worm-like forms, and are found in the Niger delta in Africa.

Calamus A genus of Trees of the Palm family, *Palmae*. e.g. C. ciliaris - An Indian palm with pinnate leaves up to 75 cm (30 in) long, each having from 80 to 100 hairy leaflets.

Calandrella A genus of Birds of the Lark family, *Alaudidae*. e.g. C. brachydactyla - The Short-toed Lark, 14 cm long; breastlight-colored without stripes; closely related to the Skylark (*Alauda arvensis*). A migrant bird inhabiting waste ground, lightly tree-clad slopes, and woods.

Calanus A genus of Crustacean animals of the order *Calanoida*. An interesting feature of this order is the possession of a heart, which is absent from other copepods of this subclass (*Copepoda*). e.g. C. finmarchicus - A copepod crustacean living in the sea from the Artic Ocean to New Zealand; especially abundant in the northern seas, it constitutes the main food of the Herring.

Calathea A genus of indoor plants of the family *Marantaceae*. e.g. C. makoyana - The Calathea or Makoy Calathea plant; a herbaceous rhizomic plant, about 50 cm (20 in) high, with oval, rounded leaves on long stems, the lower blades being blotched in red. Place of origin: Tropical America (Brazil and Peru).

Calcarius A genus of Birds of the Bunting family, *Emberizidae*. e.g. C. lapponicus - The Lapland Bunting, length 15 cm (6 in); with conspicuous head and nape markings; found in tundra and on the summits of mountain ranges above the tree line. A migrant bird overwintering in warmer regions. Distribution: the Arctic region and Scandinavia.

Calceolaria A genus of indoor flowering plants of the Snapdragon family, *Scrofulariaceae*. e.g. C. x herbeo-hybrida - A species of half-hardy hybrids, from 15-45 cm (6-18 in) tall, with tremendous variation; large or small,

slipper-like flowers, mostly in shades of yellow, red, or spotted orange. Place of origin: a hybrid from various cross-pollinations, the genus originated from South America, especially from Chile and Peru.

Caldesia A genus of plants of South and Central Europe. e.g. C. parnassifolia - The Caldesia, a perennial plant, 10-40 cm high, with long-stalked, veined, heart-shaped, ovate leaves.

Calendula A genus of medicinal and aromatic plants of the Daisy family, *Compositae*. e.g. C. officinalis - The Pot Marigold plant, an annual plant, 30-50 cm (12-20 in) high, with orange to yellow flowers variable in shade. Marigold flowers contain a yellow substance, calendulin; used in flavoring, coloring, and as a tonic; also used externally in herbal remedies to heal sprains and wounds. Place of origin: southern Europe.

Caleonas A genus of Birds of the Dove and Pigeon family, *Columbidae*. e.g. C. nicobarica - The Nicobar Pigeon, greenish-bronze in color; one of the most beautiful birds of southeast Asian forests. Found in the Nicobar Islands and New Guinea.

Calepina A genus of plants of the Cress or Crucifer family, *Cruciferae*. e.g. C. irregularis - The Calepina, an annual plant, 30-100 cm (12-40 in) high, with small white flowers. Found mainly as weeds in arable and waste land. Distribution: south, central, and western Europe.

Calicotyle A genus of Tapeworms of the subclass *Eucestoda* (= *Cestoda Merozoa*). e.g. C. kroyeri - In this cestode the loculate opisthaptor has one pair of hooks, marginal vitellaria, and central arrangement of follicles.

Calidris A genus of Birds of the Sandpiper or Wader family, *Scolopacidae*. e.g. C. alpina - The Dunlin, length 18 cm; a coastal or beach bird, with a black patch on the lower breast; a partial migrant, seen during winter in fish ponds along the coast and in water reservoirs and marshes. Distribution: northern Europe, Asia, and America.

Caliroa A genus of Insects of the typical Sawfly family, *Tenthredinidae*. e.g. C. cerasi - Length 5 mm; development takes place on cherry trees. The larva is black and smells of ink, and pupates in the ground. Distribution: Eurasia, North and South Africa, and North America.

Calla A genus of plants of the Lily family, *Liliaceae*; better known by the generic name, *Zantedeschia*. e.g. C. aethiopica (= *Zantedeschia* a.) - The Calla Lily, Arum Lily, or Lily-of-the-Nile; a herbaceous perennial, 60-75 cm (2-21/2 ft) high; pure white flowers, 7-25 cm (3-10 in) long, on stout, leafless stems. Place of origin: South Africa; introduced into Europe in 1731 (see **Zantedeschia**).

Callaeas A genus of Birds of the Wattled Crow family, *Callaeidae*. e.g. C. cinerea - The Wattled Crow or Kokako, New Zealand's best song-bird. It lives in forests and feeds on leaves, buds, and berries. It has a sharp-pointed, slightly down-curved beak, long tail and legs, anda velvety-like plumage with brightly colored wattles.

Calliactis A genus of Sea Anemones of the order *Actiniaria*, previously known also by the generic name, *Adamsia*. e.g. C. parasitica (= *Adamsia* p.) - The "Parasitic" Anemone, height 70-100 mm; white, yellow, as well as brown. A commensal animal, living attached to the shells of whelks or marine molluscs (often *Buccinum* or *Murex*), very often these being inhabited by the Hermit Crab, *Eupagurus bernhardus*. If the crab abandons the shell and moves to another, the anemone moves with it, the crab providing food and locomotion, and the anemone protecting the crab with its stinging tentacles. Distribution: the shallow and deep waters of the Mediterranean.

Callianira A genus of Comb Jellies of the order *Cydippida*. Members of this order are the least modified of the comb jellies; they are generally globular, with two long tentacles bearing colloblasts.

Callianthemum A genus of plants of the Buttercup family, *Ranunculaceae*. e.g. C. kerneranum - The Callianthemum, a perennial plant, 5-20 cm high; 2 to 3-pinnate leaves; grows in alpine grasslands up to 2,400 m and more. Distribution: southern Alps.

Callicebus A genus of New World Monkeys of the family *Cebidae*. e.g. C. torquatus - The White-collared Titi, a small monkey no bigger than a large squirrel, with glossy black fur and a white face, collar, and hands. Distribution: South America, from northern Brazil to Ecuador and Peru.

Callichthys A genus of Armoured Catfishes of the family *Callichthyidae*. These fishes have their armour arranged in pairs of plates, covered with denticles meeting at the lateral line, reminiscent of those of sharks.

Callidium A genus of Insects of the Long-horned or Longicorn Beetle family, *Cerambycidae*. e.g. C. violaceum - Length 8-16 mm; the elytra are blue, green, or reddish-violet. Larvae found in house timbers and even in dead wood. Distribution: the Palaearctic region.

Callimorpha A genus of Insects of the Tiger Moth family, *Arctiidae*. e.g. C. dominula - The Scarlet Tiger Moth, size 21-27 mm; found in woodland meadows and near water. Distribution: Europe, the Urals and Caucasus.

Callinectes A genus of Crustaceans of the Crab family, *Portunidae*. e.g. C. sapidus - The Blue Crab, an edible crustacean, found near the sea-shore and estuaries of rivers on the Atlantic coast of North America.

Callionomys A genus of Fishes of the Dragonet family, *Callionymidae* (see **Callionymus**).

Callionymus A genus of Fishes, also known by the generic name,

Callionomys, of the Dragonet family, *Callionymidae*. e.g. C. boekei (= *Callionomys* b.) - The Coral Dragonet Fish; has small terminal proctatile mouth with large lips, large eyes placed near the top of the head, no scales, and a prolonged pre-opercular that carries 2 to 4 spines, the dorsal spine being above the anal.

Calliostoma A genus of Molluscs. e.g. C. conuloide - The Common Top Shell, size up to 3 cm; conical, marbled; has heavy mother-of-pearl layers showing on the surface.

Callipepla A genus of Birds of the Pheasant and Quail family, *Phasianidae*. e.g. C. squamata - The Scaled Quail, a New World quail species of Mexico and the southwestern United States.

Calliphora A genus of scavenger Insects of the Blow-fly family, *Calliphoridae*, which deposit their eggs in wounds or decaying matter. e.g. C. vomitoria - The Common Blowfly or Bluebottle fly; the larvae feed on carrion, and also on living tissues. Under bad hygienic conditions, the larvae can penetrate inside the human body and cause serious damage. Distribution: worldwide.

Calliptamus A genus of Insects of the Locust family, *Catantopidae*. e.g. C. italicus - The Italian locust, length 14-34 mm; tibia red, forewings striped, head roundish; occurs in dry meadows and fallowland, feeding on broadleaved plants. Distribution: Eurasia and North Africa.

Callista A genus of Bivalve Molluscs of the Venus and Carpet Shell family, *Veneridae*. e.g. C. chione - The Smooth Venus, length 80-110 mm, height 65-85 mm, thickness 41-48 mm; surface dirty grey to brown, almost smooth, with darker radiating bands. Lives embedded in clean sand, offshore down to almost 130 m. A tasty, edible species. Distribution: the Mediterranean and Atlantic coasts.

Callistemon A genus of Shrubs of the Myrtle family, *Myrtaceae*. e.g. C. speciosus - The Bottle-brush of Australia, a tall shrub with narrow leaves and cylindrical groups of dark red flowers. The fruit is a capsule with minute seeds.

Callistephus A genus of late blooming plants of the Daisy family, *Compositae* (= *Asteraceae*). e.g. C. chinensis - The China Aster, an annual plant, height 45 cm (18 in), with solitary, dark purple flowers, set off by a ring of green sepals. Flower varieties appear in pink, white, mauve, and violet colors. Place of origin: China, Japan; introduced into Europe in the 18th century.

Callithrix A genus of Mammals of the Marmoset family, *Callithricidae*. e.g. C. jacchus - The Common Marmoset, 9 in long, with a bushy tail over 1 ft long. This primate, more primitive than the Cebidae monkey, inhabits the mouth of the Amazon, where it lives high up among the trees; the offspring are carried hanging to the ventral fur of the males, later on the hip, and finally when older, on the shoulders. The mother merely suckles the young.

Callitriche A genus of plants of the Water-starwort family. e.g. C. stagnalis - The Common Water-starwort, a perennial plant, 25-100 cm high; ovate leaves; 1.5 mm-sized fruit with winged lobes.

Callitris A genus of Trees found in Africa. e.g. C. quadrivalvis - A species yielding a transparent resin, sandarac, used in dentistry as a separating fluid and as a preservative varnish for plastic casts.

Callitroga A genus of Insects of the Fly family, *Calliphoridae*. e.g. C. hominivorax - The Screw-worm Fly of North and South America, a wound-infesting insect injurious to cattle and other livestock.

Callocalia A genus of Birds, also known by the generic name, *Collocalia*, of the Swift family, *Apodidae*. e.g. C. inexpectata (= *Collocalia* i.) - The Edible-nest Swiftlet, a tiny bird with large salivary glands, producing saliva in such quantities that it uses it as the sole material for nest building. The cup-shaped nests, built on the face of tall sea cliffs or in dark caves, are made wholly of saliva. Distribution: Asia and Oceania.

Callophyrs A genus of Butterflies of the Blues, Coppers, and Hairstreaks family, *Lycaenidae*. e.g. C. rubi - The Green Hairstreak, size 15-17 mm, wings colored a striking green on the underside. Distribution: Europe, Asia Minor, North Africa.

Callorhinus A genus of Mammals of the Eared Seal family, *Otariidae*. e.g. C. ursinus (= C. alascanus) - The Alaskan Fur Seal; has external ears, and is not a true seal. The bulls are black, with a grey cape covering the swollen neck; up to 7 ft long and weighing up to 613 lb, while the cows are grey, under 5 ft long and weigh about 138 lb. Distribution: The Pribilof Islands off Alaska, in the Bering Sea, where it visits once a year to breed.

Callorhynchus A genus of Cartilagenous Fishes known as the Chimaeras, of the family *Chimaeridae*. They are variously known as elephant fishes, rabbit fishes, rat fishes, and spook fishes. e.g. C. callorhynchus - The Bottlenose Chimaera from Patagonia, has a peculiar extended snout, part of which is turned back, below and in front of the mouth; this (snout) is a sensory pad well supplied with nerves.

Calluna A genus of Shrubs of the Heath family, *Ericaceae*. e.g. C. vulgaris - The Heather or Ling, an erect, branched, evergreen, perennial shrub with slender, wiry stems up to 90 cm (3 ft) high. Thin petal-like purple flowers. The fruit is a small capsule containing few seeds. Place of origin: Europe, western Asia.

Calocedrus A genus of Trees of the Cypress family, *Cupressaceae*. Sometimes known by the generic name, *Librocedrus*. e.g. C. decurrens (= *Librocedrus* d.) - The Incense Cedar, a native of the western United States from Oregon to California. A columnar evergreen conifer, up to 45 m (150 ft) tall, with flattened branchlets covered with dark-green, tiny, scale-like leaves. The cylindrical pendulous cones are 1.9 cm (3/4 in) long, and have six scales. The bark is red brown, ridged and covered with scales and the durable wood is soft and fragrant.

Calopteryx A genus of Insects of the order of Damselflies, *Odonata*; family *Agriidae*. Also known by the generic name, *Agrion*. e.g. C. virgo (= *Agrion* v.) - The Demoiselle Agrion or Blue Damselfly, size 5-6 cm; the male has dark blue wings, the female dull greyish green. Found by the banks of still and slow-flowing waters. Distribution: Europe, Asia. Native in Britain (see **Agrion**).

Calosoma A genus of Insects of the Ground beetle family, *Carabidae*. e.g. C. sycophanta - Length 24-30 mm; blackish blue, wide body, golden green elytra; one of a handsome species of iridescent, adephaganor gluttonous, ground beetles introduced into the United States from Europe in order to combat the caterpillars of the gipsy/gypsy moth, *Lymantria* dispar (see **Lymantria**).

Calotermes A genus of Insects of the Termite order, *Isoptera*. e.g. C. flavicollis - A termite species found round the Mediterranean, and one of the very few living outside the tropics. They live chiefly inside rotting trees, but will also attack planks and beams in buildings. Termites found in the United States are similar to them in appearance.

Calotes A genus of Reptiles of the Lizard family, *Agamidae*. e.g. C. versicolor - The Harlequin Lizard, also called the Indian Bloodsucker, because its throat changes from pale yellow to scarlet when it is excited. About 40 cm (16 in) long; it lives near human habitations. A gentle and agreable reptile, the male may be seen in the breeding season, dancing before the female, holding himself upright on his hind feet, swaying his head to and fro, opening and shutting his mouth, and changing color rapidly - hence the name versicolor. Distribution: the forests of India, Sri Lanka, the Malay peninsula and Indonesia.

Calothamnus A genus of Shrubs of the Myrtle family, *Myrtaceae*. e.g. C. sanguineus - An evergreen shrub from western Australia. It has narrow leaves and attractive lateral clusters of red flowers.

Calotropis A genus of plants of the Milkweed family, *Asclepiadaceae*. e.g. C. procera - The Mudar Plant; the dried root bark is used in India as an expectorant and emetic. Deaths have been reported from its use as a snuff and as an abortifacient.

Caltha A genus of plants of the Buttercup family, *Ranunculaceae*. e.g. C. palustris - The Marsh Marigold or Kingcup, a smooth perennial plant, 22-37 cm (9-15 in) high, with golden yellow, buttercup-like flowers on branching stems. Common in marshes, damp woods and other wet places. Place of origin: northern temperate and arctic regions.

Calures A genus of Birds of the Trogon family, *Trogonidae*. The species are found in America, and are mostly sexually dimorphic; they are essentially forest-dwelling birds and are known as trogons, belonging to the order *Trogoniformes*.

Calycanthus A genus of Shrubs of the family *Calycanthaceae*, also known by the generic name, *Chimonanthus*. e.g. C. floridus (= *Chimonanthus* f.) - The Caroline Allspice or Common Sweetshrub, an aromatic deciduous shrub, 2-3 m (6-9 ft) tall, with pointed oblong-oval leaves and reddish-purple, aromatic, fragrant flowers. Place of origin: the southern United States; introduced into Europe in 1726. C. praecox (= C. fragrans; *Chimonanthus* fragrans or praecox) - the Winter Sweet, a medium-sized deciduous shrub, 2-3 m (5-8 ft) tall, with ovate-lanceolate leaves, and extremely fragrant, pale yellow flowers stained purplish-red inside. Place of origin: China; introduced into Europe in 1766.

Calymmatobacterium A genus of Microorganisms of the family *Brucellaceae* - Non-motile, gram-negative coccobacilli or rods, with prominent polar granules ("safety pin" appearance). Also known by the generic name, *Donovania*. e.g. C. granulomatis (= *Donovania* g.) - causally related to granuloma inguinale in man, it produces ulcerating, granulomatous lesions in the skin and subcutaneous tissues. May be a part of normal fecal flora.

Calyptomena A genus of primitive perching Birds of the Broadbill family, *Eurylaimidae*. e.g. C. viridis - The green broadbill, a brightly colored frugivorous bird, dark grass-green in color, with black bars on the wings, and a small bill covered by a tuft of feathers at the base. Distribution: the forested areas of Borneo and the other islands of the Malayan archipelago.

Calyptraea A genus of Mesogastropod molluscs of the family *Calyptraeidae*. e.g. C. chinensis - Chinaman's Hat; shell 7 mm high, 29 mm wide, thin-walled, milky white; apex in the center. Found at shallow depths attached to small stones, shells, and other solid objects. Distribution: European coast of the Atlantic, Mediterranean sea.

Calystegia A genus of Shrubs of the Convolvulus family, *Convolvulaceae*. e.g. C. sepium - The Bellbine, Great Bindweed, or Hedge Binewood, a variable perennial climbing plant, 10-30 cm high, widespread in north temperate regions. It has a creeping rhizome and long twining stems, with solitary white or pink flowers possessing a large funnel-shaped corolla.

Calyx A genus of Sponges of the phylum *Parazoa*. e.g. C. nicaeensis - A short, round, thick-walled sponge. these animals often grow their distinctive cups around a piece of projecting debris on the sea bed.

Cambarellus A genus of eastern North American Crustaceans of the Crayfish subfamily, *Cambarinae*. e.g. C. shufeldtii - The diminutive eastern North American Cambarine Crayfish; this crustacean rarely reaches a carapace length of 25 mm (1 in).

Cambarus A genus of River crustaceans, the eastern North American Crayfishes, of the subfamily *Cambarinae*. Also known by the generic name, *Orconectes*. e.g. C. affinis (= C. affinus; *Orconectes limosus*) - Up to 20 cm (8 in) long; with dark red transverse bands on the abdominal segments; a long-tailed decapod species, found east of the Rocky Mountains, and introduced into Europe and Japan.

Camelina A genus of plants of the Cress or Crucifer family, *Cruciferae*. e.g. C. sativa - The Gold of Pleasure, an annual plant, 10-80 cm high; stems and leaves usually glabrous; fruits 7-9 mm, with strongly convex valves. Found as weeds in corn, flax, and lucerne fields. Distribution: eastern Europe.

Camellia A genus of Trees and Shrubs of the Tea family, *Theaceae*. e.g. C. sinensis (= C. thea) - The Tea Plant; the tea leaves and leaf-buds contain caffeine and tannin. It is widely cultivated in moist tropical regions. Place of origin: China.

Camelus A genus of Mammals of the Camel family, *Camelidae*. e.g. C. dromedarius - The Arabian Dromedary or One-humped Camel; measures about 71/2 ft to the top of the hump; fawn or beige, matching the color of its environment. A domestic animal adapted for life in the desert: the broadened feet, eyelashes that protect the eyes from wind-blown sand, nostrils that can be closed, lips thickened to withstand the coarsest plants, the hump with its

reserve of fat, and the ability to survive for long periods without water. it is also known as the Racing Camel, as opposed to the Arabian Baggage Camel which is much heavier in build and can carry a load of 350 to 450 lb forty miles aday. Distribution: desert regions of Asia and Africa. Successfully introduced into the arid regions of Australia.

Campanula A genus of plants of the Bellflower family, *Campanulaceae*. e.g. C. glomerata - The Danesblood or Clustered Bellflower, an erect hairy perennial plant, 10-90 cm (4-36 in) high. Sessile, funnel-shaped, deep blue to purple flowers. Place of origin: Europe, temperate Asia.

Campephaga A genus of Birds of the Cuckoo-shrike and Minivet family, *Campephagidae*. e.g. C. phoenicea - The Red-shouldered Cuckoo-shrike of Africa; about 9 in long, and mainly black with a red shoulder patch.

Campodea A genus of diplurid or japygid Insects of the family *Campodeidae*. e.g. C. fragilis - Length 3.5 mm; a small, blind, fragile, whitish species living in the soil or plant debris. It has two long filamentous, terminal abdominal processes, the cerci or caudal setae. Distribution: cosmopolitan.

Camponotus A genus of hymenopterous Insects of the Ant family, *Formicidae*. e.g. C. ligniperda - Length 7-14 mm; one of the largest European ants; the nest is located in stumps and in the living wood of conifers, chiefly spruce trees. The chambers are excavated along the annual rings. Distribution: central and northern Europe.

Campostoma A genus of Fishes of the Carp, Minnow, Tench, and allied family, *Cyprinidae*. e.g. C. anomalum - The Stoneroller Fish; this North American species feeds on mud and has an exceedingly long intestine arranged in 15 coils around its air bladder. Its fins are black and orange and the breeding tubercles prominent on the head and back (see **Carassius**).

Campsis A genus of climbing plants of the family *Bignoniaceae*. e.g. C. x tagliabuana - The Trumpet Creeper; a collective name given to a group of hybrids between C. grandiflora and C. radicans, of which one called "Madame Galen" is outstanding. Various deciduous climbers reach lengths of 6 m (20 ft) or more; they have orange-scarlet flowers in huge, showy inflorescences.

Camptoptera A genus of Fishes (class *Insecta*). e.g. C. papaveris - The Fairyfly, a very minute insect with remarkable hair-fringed wings; it virtually "swims" through the air when flying, because of its small size and weight. It is probably parasitic on the eggs of insects living in the seed capsules of the poppy plant, *Papaver* rhoeas.

Campylobacter A genus of Microorganisms, also called *Helicobacter*, causing enteric infection (see **Helicobacter**). e.g. C. fetus (subspecies *Jejuni*, called C. jejuni) - A species pathogenic to man, causing severe diarrhea (enterocolitis); previously called *Vibrio* fetus. C. pylori - A species associated with peptic ulcer disease.

Cananga A genus of tropical Trees. e.g. C. odorata - The Ylang-Ylang, a tree of the Malayan islands; its flowers afford a fragrant volatile oil.

Canarium A genus of plants of the family *Burseraceae*. e.g. C. luzonicum (= C. commune) - The Elemi; Manila Elemi - a species yielding resins and oleoresins.

Cancer A genus of short-tailed Crustaceans. e.g. C. pagarus - The Edible Crab; carapace up to 30 cm wide; front edge of shell crenated; pincer tips black. Found mainly on rocky shores under stones and seaweed. Distribution: the North Sea and Atlantic coasts.

Candida A genus of yeast-like Fungi producing mycelia, of the family *Cryptococcaceae*. Other names for this genus are: *Cryptoccocus, Endomyces, Monilia, Myocandida, Mycotoruloides, Oidium, Parasaccharomyces, Saccharomyces, Syringospora, Torula, Torulopsis*, and *Zymonema*. e.g. C. albicans (= ca; Can. albicans; *Endomyces* a.; *Zymonema* a.) - causes candidiasis or thrush in man. C. ashfordi (= *Monilia* psilosis; *Parasaccharomyces* a.). C. utilis (= *Torulopsis* u.) - Torula Yeast.

Canis A genus of Carnivorous Mammals of the Dog, Fox, Jackal and Wolf family, *Canidae*. e.g. C. adustus - The Side-striped Kackal of South and East Africa, the largest of the jackals, measuring 17 in at the shoulder. Jackals hunt mainly at night, feeding on carrion and small animals. C. dingo - the Australian Dingo or Wild Dog, believed to have reverted into this state from domestic dogs brought over from India by natives years ago. C. familiaris - The Domestic Dog; this worldwide species is classified as (dog) breeds according to employment of the dog. They are boxers; bulldogs, dachsunds, gun dogs, hounds, poodles; sheepdogs, spaniels, terriers, toy dogs, watchdogs, etc. C. latrans - The Coyote, Bush Wolf, or Prairie Wolf; this species still haunts the western plains of North America. C. lupus - The European or Common Wolf, length of body 100-165 cm; fur grey; found only in the wilder parts of Scandinavia, western and eastern Europe. A powerful and intelligent animal, living and hunting in packs. It has been greatly feared and ruthlessly slaughtered to protect livestock.

Canella A genus of plants of the family *Canellaceae*. e.g. C. alba (= C. winterana) - White Cinnamon; Wild Cinnamon Bark; has been used as an emmenagogue.

Canna A genus of plants of the family *Cannaceae*. e.g. C. indica - The Canna, Canna Lily, or Indian Shot, a perennial plant with tuberous roots; 11/2 m (5 ft) high; with red pink, orange, or yellow showy flowers. The rhizomes of this plant were once collected and widely used in the Americas as food in pre-columbian times. Place of origin: tropical America and West Indies; introduced into Europe in 1570.

Cannabis A genus of plants of the Hemp and Hop family, *Cannabinaceae.* e.g. C. sativa - The Hemp Plant, an erect annual with deeply palmate-lobed leaves. It originated in central Asia but has been cultivated for centuries in eastern countries for the narcotic resin it contains. The dried flowering tops yield the resin cannabinone, containing the active principles, cannabinol and tetrahydro-cannabinol (THC), psychoactive hallucinating drugs causing euphoria, mental confusion, and motor excitement. The initial phase of inebriation is succeeded by irritability and somnelescence, and after some hours by a comatose sleep. Cannabis is also variously known by some 100 names or more (!), such as: A-Bomb (a cigarette prepared from a mixture of marijuana and heroin); Ace (cigarette - now obsolete); Angel Dust (a mixture of cannabis with phencyclidine); Baby; Bar; Bee; Bhang; Black Russian; Blond Hashish; Bomber (m. cigarette); Boo; Bush; Can; Canadian Black; Canamo Indiano; Cancelled Stick (m. cigarette); Cannabis Indica; Chanore Indien; Charas; CS; Dagga; Duby; Dynamite (concentrated m.); Finger (hashish; marijuana); Gangster (m.); Ganga (m.); Gold (concentrated m.); Grass; Green; Grefa; Greta; Grifa (m. cigarette); Grüner Türke; Guaza; Gunga; Hanfkraut; Haschischol; Hash; Hashish; Hay; Hemp; Herb; Indian Hay; Indian Hemp; J (m. cigarette); Jay (m. cigarette or joint); Joint (m. cigarette); Juanita; Kick Stick (m. cigarette); Kief (cannabis); Kif (cannabis); Killer Weed (potent m.); Kilter; Lebanese (hashish); Loco; Macon (cannabis); Majoon (candy containing hashish or cannabis);

Mariahuana; Mariajuana; Marihuana; Marijuana; Mary Ann; Mary Jane; Mexican Brown; Mexican Green; MJ (Mary Jane/m.); Moon (hashish/peyote); Moota; Mota; Mu; Muggle (m. cigarette); Muta; Number (m. cigarette); Oil (hashish extract high in THC - tetrahydrocannabinol); Panama; Panama Red; Pin (m. cigarette); Pot; Rag (ragweed - dilute m.); Ragweed (dilute m.); Red (Panama Red/m.); Reefer (m. cigarette); Root (m. cigarette); Rope; Roter Libanese; Schwarzer Afghane; Sinsemilla (a seedless marijuana species, produced through intensive cultivation of only the female plant - Spanish American); Snop; Stick (m. twig); Sweet Lucy; Tea; Texas Tea; THC (tetrahydrocannabinol - a major active ingredientof marijuana); Thumb (m. cigarette); TJ (m. purchased in Tijuana, Mexico); Twist; Weed; Wheat; Yerba; Z Purple (m. grown in Zacatecas, Mexico).

Cantharellus A genus of Fungi or Mushroom plants of the family *Agaricaceae.* e.g. C. cibarius - The Chanterelle, found in beech and oak woods during the autumn. It is deep yellow, from 1-4 in a cross, and funnel-shaped with a wavy margin. The thick shallow gills are often branched. An edible species, with an odor of apricots when fresh.

Cantharis A genus of Soldier Beetles of the family *Cantharidae* (= *Meloidae*); also known by the generic name, *Lytta.* e.g. C. vesicatoria (= *Lytta* v.) - The Spanish Fly (= Blister Bug; Blistering Beetle; Cantharides; Cantharis; Insectes coleopteres heteromeres; Lytta; Meloides; Russian Fly), source of cantharidin, the active principle of cantharides. Preparations of cantharides have been employed externally as counter-irritants, rubefacients, and vesicants.

Canthocamptus A genus of Crustaceans of the order *Harpacticoida.* e.g. C.

staphylinus - One of the commonest freshwater species of harpacticoids; abundant in water, but during summer it makes a gelatinous cyst in which it remains inactive when temperatures are high and the oxygen content of the water is low.

Capella A genus of Birds of the Snipe family, *Scolopacidae*. e.g. C. media - The great snipe, 28 cm long; long, straight bill, with a lot of white on the sides of the tail; a migrant bird with zigzag flight, making a peculiar drumming noise with its outer tail feathers. Distribution: a native of Great Britain and much of Europe.

Capillaria A genus of Nematodes parasitic in birds and mammals. e.g. C. hepatica - The Capillary Liver Worm; seen in North America; may infest man.

Capparis A genus of Rock Garden Plants of the Caper family, *Capparidaceae*. e.g. C. spinosa - The Caper-bush, a spiny, deciduous shrub, 1-11/2 m (3-5 ft) high, of spreading habit. Large, white, solitary flowers, tinged reddish or lilac with numerous projecting stamens. Place of origin: southern Europe and the Mediterranean region.

Capra A genus of Mammals of the Goat subfamily, *Caprinae* (family of horned ungulates, bovidae). e.g. C. hircus (= C. hircus aegagrus) - The Wild Goat, Pasang, or Persian Wild Goat; is believed to be the ancestor of the domestic goat (of the same species, C. hircus). About 11/2 m (5 ft) at the shoulders; horns scimitar-shaped; the male has a long beard. It lives in the mountains of Asia Minor up to heights of 13,000 ft, and the islands of the Mediterranean. The Swiss or Alpine goats and the native English goats (renowned for their milk), and the Kashmir and Angora goats (for their wool), are races belonging to the same species (Capra hircus) of domestic goats.

Caprella A genus of Crustaceans of the Shrimp family, *Caprellidae*. e.g. C. aequilibra - It has a long slender body and few limbs; the first free segment carries a well-developed pair of pincers which it uses to catch small animals as they swim by. Like other caprellids, it moves about on hydroids and seaweed in a slow, deliberate manner.

Capreolus A genus of Mammals of the true Deer family, *Cervidae*. e.g. C. capreolus - The Roe Deer; length of body 95-135 cm; height at shoulder 65-75 cm; in summer the coat is reddish-brown, in winter greyish-brown; it has short, branched antlers which grow almost vertically from the forehead. Found in all types of forests from lowlands to mountains. There are three races: 1. C. c. capreolus - The European Roe Deer. 2. C. c. pygargus - The Siberian Roe Deer. 3. C. c. bedfordi - The Manchurian Roe Deer.

Capricornis A genus of Mammals of the Goat-antelope subfamily, *Caprinae*. These are the serows, mountain animals closely related to the gorals, and distributed throughout the Himalayas to China, Taiwan, and Japan, and south to Malaya and Sumatra.

Caprimulgus A genus of Birds of the Nightjar family, *Caprimulgidae*. e.g. C. europaeus - The European Nightjar or Goatsucker, a large bird, 27 cm (11 in) long; found on trees and in bushes, living in woods and forests. It is mottled, with brown, red and black, simulating the coloring of dead leaves. This is an excellent camouflage in its natural habitat, but it can be heard at night making a noise like a distant two-stroke engine. A migrating bird, it winters in southern Asia and Africa. Distribution: Europe and central Asia.

Capromys A genus of very primitive Rodents of the Coypu family, *Echimyidae*. e.g. C. pilorides (= C. pilarides) - The Hutia-Conga or Cuba; this mammalian species is nocturnal and arboreal; 45 cm (18 in) long.

Capsella A genus of plants of the Cress or Crucifer family, *Cruciferae*. e.g. C. bursa-pastoris - Shepherd's Purse; Herba Bursae Pastoris; a cosmopolitan weed of cultivation, 2-80 cm high; the flattened triangular fruit contains over 20 seeds. The dried ariel parts are used in the countries of what was formerly the USSR, as a uterine stimulant.

Capsicum A genus of plants of the Nightshade family, *Solanaceae*. e.g. C. annum (other cultivated varieties are C. frutescens and C. minimum) - African Pepper; Capsici Fructus; Cayenne Pepper; Chilli (es); Paprika; Piment Rouge; Pimentao; Red Pepper; Spanish Pepper - a variable annual plant unknown in the wild state, from 1-6 ft high; with ovate, entire leaves, and white or greenish-white flowers. The fruit is a berry with numerous seeds, first green, and ripening to red. The small forms have the strongest taste and are dried and ground as pepper. They contain capsaicin (capsicin), used as a carminative, and externally as a counter-irritant.

Capulus A genus of Limpet Molluscs of the order *Mesogastropoda*. These meso-gastropod limpets or snails are known as Cap-of-Liberty Shells. They appear in solitary chains, with the males living above and fertilizing the females beneath.

Carabus A genus of Insects of the Ground Beetle family, *Carabidae*. e.g. C. violaceous - The Violet Ground Beetle, length 18-34 mm; has a striking narrow, violet or greenish flattened rim round the edges of the elytra and scutum. Locally common, chiefly in damp forests, fields and gardens. Distribution: Europe.

Caracal A genus of Mammals of the Cat family, *Felidae*. e.g. C. caracal - The Caracal or Desert Lynx. Closely related to the serval (*Leptailurus* serval), it is remarkably swift and a great hunter. Long-legged, its coat is a uniform fawn,

shading to white below and to black at the tips of the ears. It inhabits the deserts and savannas of North Africa and western Asia. Its agility is so great that it was used in pigeon-catching contests by eastern princes.

Caracara A genus of Birds of the Falcon family, *Falconidae* - The Caracaras or Carrion Hawks of the subfamily *Polyborinae*. e.g. C. cheriway - The Carancho or Crested Caracara, a species from Central and South America; primarily a ground bird with strong legs; a scavenger. Has been adopted as the national bird of Mexico.

Caralluma A genus of plants of the Milkweed family, *Asclepiadaceae*. e.g. C. europaea - A succulent, cactus-like, Mediterranean plant with fleshy, four-angled stems. The flowers are pale yellow with purple bands and are clustered at the top of the stem.

Caranx A genus of Fishes of the Horse-mackerel, Jack, and Scad family, *Carangidae*. In this family the scales of the lateral line, especially those near the tail, have comb-like extensions, and there are two spines just in front of the anal fin. e.g. C. bartholomaei - The Yellow Jack Fish.

Caraproctus A genus of Fishes of the Lumpsucker family, *Cyclopteridae*; these have suckers formed from the pelvic fins, and they lay their eggs under crabshells.

Carapus A genus of Fishes of the Pearlfish family, *Carapidae*. e.g. C. bermudensis - The Bermuda Pearlfish; has an elongated body and spineless fins, and is a commensal on molluscs.

Carassius A genus of Fishes of the Carp, Goldfish, and Minnow family, *Cyprinidae*; the principal European genera of this family include the Barbels (*Barbus*), Bitterlings (*Rhodeus*), Bleaks (*Alburnus*), Breams (*Abramis*), Carps (*Cyprinus*), Chubs (*Scardinius*), Dace (*Leuciscus*), Gardons (*Gardonus*), Goldfishes (*Carassius*), Minnows

(*Phoxinus*), Nase (*Chondrostoma*), Orfes (*Idus*), Roaches (*Rutilus*), Silverbreams (*Blicca*), Tenches (*Tinca*), and others. Many species of these genera cross breed with others and the hybrids so formed are fertile. e.g. C. auratus - The Goldfish; of Asiatic origin, from China, Korea, and Japan. By controlled breeding an infinite variety of forms have been produced, some beautiful, others grotesque. The goldfish kept in aquariums are domesticated forms (C. auratus auratus). The original goldfish was first brought to Europe in the 17th century.

Carausius A genus of Insects of the order *Phasmida*. e.g. C. morosus - The Stick Insect or Indian Stick Insect; of oriental origin; has a twig-like appearance and exhibits both parthogenesis and regeneration, and is able to adjust its color fairly rapidly to that of its background. It is a close relative of the American Walkingstick (insect).

Carbenia A genus of plants of the Daisy family, *Compositae*, now called by the generic name, *Cnicus* (see **Cnicus**).

Carcharhinus A genus of Fishes of the Requiem or Requin Shark family, *Carcharhinidae*. e.g. C. leucas - The Ground Shark; a heavy slow-swimming species, almost 12 ft long, with sharp teeth serrated on both edges and very destructive. Rarely seen far from land, and sometimes enters freshwater on occasion.

Carcharias A genus of Fishes of the Sand Shark family, *Carchariidae*. These are primitive sharks, most of them very destructive. The only genus of the family now living, the other genera having become extinct. e.g. C. lamia - The Cub Shark of Florida; a small broad-headed form, very voracious and destructive, living near sandy shores, and sometimes, but not often, attacking man. The tail is unequal and not keeled and the teeth are slender, often with small cusps

at their base. C. littoralis (= *Odontaspis* l.) - a common Atlantic sand shark.

Carcharodon A genus of Fishes of the Shark suborder, *Galeoidea*. e.g. C. carcharias - The White Shark.

Carcinus A genus of Crustaceans of the Crab family, *Portunidae*. e.g. C. maenas - The Common Shore Crab of Northern Europe or the Green Crab of North America; the commonest British crab. Carapace up to 10 cm. It has a very limited ability to swim, and this is reflected in the slight widening of the last leg.

Cardamine A genus of plants of the Cress or Crucifer family, *Cruciferae*. e.g. C. pratensis - The Cuckoo Flower or Lady's Smock; 15-50 cm high; a perennial plant, growing mainly in damp situations; extremely variable, with pink or sometimes white flowers. The long, narrow fruit opens violently, flinging the seeds up to 6 ft away. distribution: Britain; North America.

Cardaminopsis A genus of plants of the Cress or Crucifer family, *Cruciferae*. e.g. C. arenosa - The Sand Rock-cress, an annual to biennial plant, 10-50 cm high; white or reddish flowers 7-11 mm long; leaves with forked hairs. Common on cultivated fields, by waysides, and on wasteland. Distribution: most of Europe.

Cardaria A genus of plants of the Cress or Crucifer family, *Cruciferae*. e.g. C. draba - The Hoary Cress, 30-90 cm high; a widely distributed Mediterranean species, now a pernicious weed in Britain with a long tap-root. The small white flowers are borne in dense clusters, and the short fruit, containing 1 or more seeds, does not open.

Cardium A genus of bivalve Molluscs of the Cockle family, *Cardiidae*. e.g. C. ovale - The Banded Cockle; length of shell up to 11/2 cm, with thin valves and about 25 regular, delicate, radial ribs.

Carduelis A genus of Birds of the Finch (Darwin's finches) family, *Fringillidae*. e.g. C. carduelis - The European Goldfinch, a small bird, 12 cm long, very attractively colored; head black; wingbands in the male, white and red, in the female black and yellow; a stocky body, free toes, and a short, conical bill. Adapted to a seed-eating habit, its chief food being thistle seeds (Carduus). A partially migrant bird. Distribution: Europe, western Asia, and northwest Africa; has been introduced into parts of eastern North America and Bermuda.

Carduus A genus of plants of the Daisy family, *Compositae*. e.g. C. benedictus - The Blessed Thistle Plant (see **Cnicus**). C. nutans - The Musk Thistle, a biennial plant, 30-100 cm high; heads usually solitary, drooping; up to 50 mm-wide bracts, strongly reflexed and spine-tipped.

Caretta A genus of Reptiles of the Marine Turtle family, *Chelonidae*. e.g. C. caretta - The Loggerhead Turtle, up to 1 m (3 ft) long; brown in color; legs and feet like fins; carapace on back with five pairs of main shields. A carnivorous species. One of the four marine turtles common to all tropical and subtropical seas. Its limbs are adapted for swimming, with clawless flippers that contain the digits, and its aquatic habit makes for a diet of fishes, molluscs, and crustaceans.

Carex A genus of plants of the Sedge family, *Cyperaceae*. e.g. C. flava - A true sedge, a perennial plant, seldom more than 2 ft high, with a single male spike, and from 2-4 female spikes. Widespread in North America, but rare in Britain. C. hordeistichos - The Barley Sedge, a perennial plant, 10-20 cm high; stem overtopped by the flat, leaf-like bracts; female spikes 3-4. Distribution: south, southeast, and central Europe.

Cariama A genus of Birds of the Cariama or Seriema family, *Cariamidae*. e.g. C. cristata - The Crested Cariama or Seriema, related to the crane; has a greyish plumage striped with brown with brownish underparts; the long legs are red. Lives on open grasslands and pampas, from central Brazil to Paraguay and northern Argentina.

Carica A genus of Tropical Trees of the family *Caricaceae*. In North America, a close relative is the genus *Asimina*. e.g. C. papaya - The Papaya or Pawpaw Tree, a native of tropical America, now widely cultivated in the tropics for its edible fruit. A rapidly growing, unbranched tree, up to 6 m (20 ft) tall, with large, palmately lobed leaves. The papaya or pawpaw fruit (as in Asimina) contains the proteolytic enzyme, papain, used as a meat tenderizer and in the clarification of beverages; also as a digestive for protein in chronic dyspepsia and gastritis. In recent years, papain has been introduced for intraspinal use in cases of prolapsed intervertebral disc (see also **Asimina**).

Carlina A genus of plants of the Daisy family, *Compositae*. e.g. C. acaulis - Smooth Carline, Stemless Carline Thistle, or Great Carline Thistle, a low perennial plant, 20 cm (8 in) high. Usually solitary flowers, 5-13 cm (2-5 in) across, disk-florets white or reddish-white, surrounded by spreading silvery-white, spiny, lance-shaped bracts. Used as a weather guide, the flowers (bracts) closing under humid conditions. The root was once used to cure plague in mediaeval times. Place of origin: mountains of Europe.

Carludovica A genus of Palm-like plants of the family *Cyclanthaceae*. e.g. C. atrovirens - A species from Columbia, with large, dark green leaves arising from the base of the plant. The small unisexual flowers are borne in close, spirally arranged groups on a cylindrical spike enclosed in bracts.

Carpesium A genus of plants of the Daisy family, *Compositae*. e.g. C. cernuum - The Carpesium, an annual to biennial plant, 20-80 cm high, with lanceolate leaves; found in open woodland. Distribution: South and Central Europe.

Carpinus A genus of Trees of the Hazel

family, *Corylaceae*. e.g. C. betulus - The Common Hornbeam, an attractive tree, up to 30 m tall, with a grey, characteristically grooved trunk. The greenish, ovoid, ripe fruits are attached to large trilobed bracts. Distribution: Europe, western Asia.

Carpiodes A genus of Fishes of the Sucker family, *Catastomidae*. These are known as the Carpsuckers; they are the largest of the sucker fishes and bear an external resemblance to the carp. Their dorsal fins are long, with many rays. Distribution: North America, east of the Rocky Mountains.

Carpocapsa A genus of Insects of the Moth family, *Eucosmidae* (= *Olethreutidae*). e.g. C. pomella - The European Codling Moth; a serious pest of apples and other fruit; widely distributed in North America, South Africa, Australia and New Zealand.

Carpodacus A genus of Birds of the Finch family, *Fringillidae*. e.g. C. erythrinus - The Scarlet Grosbeak, size 14.5 cm; male with red crown, breast and rump, and no white wing-bars; the female brown, streaked. Found nesting in bushes. A migrant bird. Distribution: large areas of Europe.

Carterocephalus A genus of Insects of the Skipper Butterfly family, *Hesperiidae*. e.g. C. palaemon - The Chequered Skipper, size 13-14 mm; found in forest margins, damp meadows, and even on mountains (in the Alps up to 1,500 m). A hibernating species. Distribution: Europe, Asia, and North America.

Carpoglyphus A genus of Mites. e.g. C. passularum - A tyroglyphid mite, infesting dried fruit and causing dermatitis in fruit handlers.

Carum A genus of plants of the Carrot family, *Umbelliferae*. e.g. C. carvi - Caraway; Kummel - a much-branched, erect, biennial plant, up to 60 cm (2 ft) high, with umbels of small white flowers, and strongly scented fruits.

Distribution: Europe; Asia. C. copticum (= *Trachyspermum* ammi) - Ajowan; Ptychotis (see **Trachyspermum**). C. petroselinum (= *Apium* p.; *Petroselinum* crispum; P. sativum) - The Parsley Plant (see **Petroselinum**).

Carya A genus of Trees of the Walnut family, *Juglandaceae*. e.g. C. alba - The Mockernut, an American species; of importance for its wood and fruit.

Caryophyllia A genus of solitary stony Corals of the family *Caryophyllidae*. e.g. C. smithii - The Devonshire Cup Coral, a whitish animal with brownish tentacles when expanded; when contracted the radiating ribs of the theca can be seen. Found in the colder waters of southwest England.

Caryophyllus A genus of plants of the Myrtle family, *Myrtaceae*. e.g. C. aromaticus (= *Eugenia* caryophyllus; E. aromatica; *Jambosa* c.) - Cloves; the dried flower-buds are used as a carminative in flatulence and dyspepsia.

Caryopteris A genus of late blooming plants of the family *Verbenaceae*. e.g. C. incana (= C. mastacanthus) - The Blue Spiraea or Common Bluebeard, a deciduous shrub, 90-150 cm (3-5 ft) tall, delicate-looking, with pale blue flowers in clusters in terminal and axillary cymes. Place of origin: China, Japan; introduced into Europe in 1844.

Carysomyia A genus of Flies. e.g. C. bezziana - maggot-producing fly (Myiasis) of India and Africa.

Casarca A genus of Birds of the Duck family, *Anatidae*. e.g. C. ferruginea - The Ruddy Shelduck, one of the Tadornini, 63 cm in size, the most primitive tribe of the subfamily *Anatinae*. Plumage rust-red, head light-colored, a partially migrant bird. As all the other species of this genus, these are the typical shelducks of the old world.

Casmerodius A genus of Birds of the Bittern, Eegret, and Heron family, *Ardeidae*; better known by the generic name, Egretta. e.g. C. albus egretta -

The White American Egret (see
Egretta).

Caspiomyzon A genus of Fishes of the
Lamprey family, *Petromyzonidae*. e.g.
C. wagneri - Wagner's Lamprey; length
up to 55 cm; lives in the Caspian Sea,
and swims up the Ural, Terek, Volga,
and other rivers to spawn in sandy parts
of these river beds.

Cassia A genus of plants of the Pea family, *Leguminosae*. e.g. C. senna - The
Senna Plant; senna fruit and senna leaf
yield the active principle sennatin, containing the glycosides sennoside A and
B. Used as a cathartic.

Cassida A genus of Insects of the Leaf
Beetle family, *Chrysomelidae*. These are
curiously shaped tortoise beetles, whose
flat, spiny larvae adorn themselves with
their cast skins and excrement. E.g. C.
viridis - Length 7-10 mm; colored bright
green; a herbivorous species.
Distribution: the palaearctic region.

Cassidaria A genus of Molluscs. e.g. C.
echinophora - up to 11 cm high; the
shells have few whorls, and the edge of
the aperture is thickened and polished.

Cassiope A genus of Shrubs of the Heath
family, *Ericaceae*. e.g. C. hypnoides -
Moss Heather, an evergreen shrub, up to
10 cm high, with long-stalked flowers;
found in rocky moorland. Distribution:
central Europe and the arctic tundra.

Cassiopeia A genus of Jellyfishes of the
order *Rhizostomeae*. This animal is a
rhizostome medusa, shaped like a flattened ball, with a marginal ridge giving
the appearance of a saucer. It is sluggish, and lies upside-down in shallow
lagoons with the exumbrella facing
downwards. It has eight branching
mouth arms with suctorial mouths for
drawing in food and oxygen. Commonly
found in Florida and the West Indies.

Cassis A genus of Molluscs or Sea Snails,
of the order *Neogastropoda*. These are
the rounded "cameo-shells", which also
include the great helmet shells. These
animals are all carnivorous scavengers

and move along just below the sand surface. e.g. C. rufa - A species from the
Indian Ocean.

Castanea A genus of Trees of the Beech
family, *Fagaceae*. e.g. C. sativa - The
Sweet or Spanish Chestnut; a tall tree,
up to 20 m (65 ft) high, with a dark
brown, spirally grooved bark. The large
oblong leaves have pointed marginal
teeth. From 1-3 large brown nuts are
enclosed in a green spiny capsule which
opens by several valves. Place of origin:
a native of southern Europe, Asia Minor,
and the Caucasus. Widely planted and
fully naturalized in southeast England.

Castellanella A genus of Sporozoan
Parasites of the family
Trypanosomatidae, now better known
by the generic name, *Trypanosoma*. e.g.
C. gambiense (= C. castellani;
Trypanosoma g.) - causing sleeping
sickness (see **Trypanosoma**).

Castor A genus of Rodents of the Beaver
family, *Castoridae*. e.g. C. canadensis -
The American or Canadian Beaver, a
large rodent, 73 cm (29 in) long; lives in
streams and shows great ingenuity in
building lodges, dams, and canals.
Much hunted for its fur. C. fiber - The
European Beaver, length of body 70-100
cm; tail 30-35 cm; brown coat; broad,
flat, scaly tail; small ears and shortlegs.
Unlike the American Beaver, this old
world European Beaver is chiefly a
bank-dweller, living in underground
holes dug in banks, with entrances
below the water level, or in so-called
ledges. It builds dams to raise the water
level, constructed of mud, stones, and
tree-trunks felled and trimmed by gnaw-
ing with its sharp incisors. It has been
almost exterminated in western Europe
and is now protected by law. From the
dried preputial follicles is prepared castor or castoreum, a darkbrown or grey
mass with an empyreumatic odor, which
is chiefly used as a fixative in perfumery. In the past, tincture of castor

was used in the treatment of hysteria and dysmenorrhea.

Casuarina A genus of Trees of the family *Casuarinaceae*. e.g. C. equisetifolia - An evergreen species found mainly in Australia and Malaya. It has slender green branches and minute whorls of leaves, giving it the appearance of a horsetail. The minute flowers are unisexual and are borne in slender catkins. The hard timber, often known as she-oak, is of commercial value.

Casuarius A genus of Birds of the Cassowary family, *Casuariidae*. e.g. C. casuarius - The Common Cassowary, about 41/2 ft high; the most colorful of the large ratite or flightless birds. The blue, naked skin of the head and neck contrasts with vivid red wattles, and the head is topped with a horny casque. Distribution: northern Australia, New Guinea and parts of Polynesia.

Catabrosa A genus of plants of the Grass family, *Gramineae*. e.g. C. aquatica - The Water Whorl-grass, a perennial plant, 10-60 cm high; culm geniculate at base, panicle loose, and spikelets small; 1-3 flowered. Found on river-banks, in ditches and wet places. Distribution: most of Europe.

Catalpa A genus of Trees of the Bean family, *Bignoniaceae*. e.g. C. bignonioides - The Indian Bean Tree or Catalpa, a deciduous tree up to 50 ft high with a dense spreading crown, and large clusters of trumpet-shaped yellow flowers with yellow and purple spots. The long narrow fruits, about 15 in long, are podlike and contain numerous seeds which have a tuft of white hairs at each end. These seeds have been used in the treatment of bronchial asthma. Distribution: a native of the eastern United States.

Catamblyrhynchus A genus of Birds of the Plush-capped Finch subfamily, *Catamblyrhynchinae*. e.g. C. diadema - Finch-like, 6 in long; the sole member of this subfamily; remarkable for its erect, golden-brown crown feathers which are like a stiff velvet pile. Distribution: tropical South America.

Catasetum A genus of plants of the Orchid family, *Orchidaceae*. e.g. C. maculatum - An orchid species native to Mexico, Guatemala and Venezuela. The greyish flowers are purple spotted and have a large inflated labellum.

Catastomus A genus of Fishes of the Sucker family, *Catastomidae*. e.g. C. commersoni - The Common Sucker; has a round toothless suctorial mouth with fleshy lips, by means of which the fish sucks up its food. Distribution: North America.

Catha A genus of plants of the family *Celastraceae*. e.g. C. edulis - Kat; Kath; khat; Abyssinian, African or Arabian Teaplant - the leaves contain cathine (d-nor-pseudoephedrine), choline, tannins, and inorganic salts. An excitant of the central nervous system; used in northern and eastern Africa as a stimulant. Addiction to catha has been reported, addicts experiencing a dreamy euphoria with loquacity, followed by apathy, depression, and anorexia. Also causes palpitations and sweating.

Catharacta A genus of Birds of the Jaeger and Skua family, *Stercorariidae*. e.g. C. skua - The Great Skua, 2 ft long, found in both polar regions, off the Faroe Islands and Iceland and most abundantly, in the antarctic and subantartic regions. There is no indication that the polar populations meet, although both migrate towards the equator in winter.

Catharanthus A genus of flowering plants of the Periwinkle family, *Apocynaceae*; formerly known by the generic name, *Lochnera*, and recently, the more popular name, *Vinca*. e.g. C. roseus (= *Lochnera* rosea; *Vinca* rosea) - The Madagascar Periwinkle, a widespread plant in the world's tropics with large, flat, rose-pink or white flowers with red centers. From the plant are extracted the alkaloids vinblastine, vincristine, vinleurosine, vinrosidine, etc., used as cytotoxic agents in the treatment of Hodgkin's disease and chorionepithelioma. Place of origin: the tropics; introduced into Europe in 1756.

Catharsius A genus of Insects of the Lamellicorn and Scarab Beetle family, *Scarabaeidae*. The species concerned are the sacred Scareb Beetles, which attained great significance as talismans and as part of the religious mythology in ancient Egypt. They all feed on dung which they roll into balls and bury beneath the ground. In some of these balls the female lays an egg in which the larva completes its development.

Cathartes A genus of Birds of the new world Vulture family, *Cathartidae*. e.g. C. aura - The Turkey Vulture, a new world scavenging bird, the most widespread and common member of the family, found from southern Canada to the Magellan Straits of South America. Also known in the U.S.A. as the Turkey Buzzard, as its almost naked red head and neck resembles a turkey, and its soaring flight is similar to that of the birds known in England as buzzards.

Catla A genus of Fishes of the family *Cyprinidae*. e.g. C. catla - The Catla; an Indian cyprinid species, about 5-6 ft long and weighing over 100 lb (see **Carassius**).

Catocala A genus of Insects of the Butterfly family, *Noctuidae*. e.g. C. sponsa - The Dark Crimson Underwing Moth, size 30-33 mm; found in oak woods, parks, and gardens. The caterpillar feeds on the leaves of the oak, growing to a length of 70 mm. Distribution: Europe.

Catoptrophorus A genus of Birds of the Plover, Sandpiper, and allied family, *Scolopacidae*. e.g. C. semipalmatus - The Willet of the American continent, closely related to the Sandpiper, *Tringa*.

Cattleya A genus of plants of the Orchid family, *Orchidaceae*. e.g. C. labiata - A

hybrid species growing in numerous forms, with deep velvety crimson flowers, 20-25 cm (8-10 in) across, of various shades. Place of origin: South and Central America and the West Indies; introduced into Britain from Brazil in 1818.

Caucalis A genus of plants of the Carrot family, *Umbelliferae*. e.g. C. platycarpos - The Small Bur-parsley, an erect annual plant up to 30 cm (1 ft) high, with dissected leaves and compound umbels of white or pink flowers. The fruit bears rows of hooked spines. Distribution: Europe; western Asia.

Caudamoeba A genus of Protozoans of the order *Amoebina*. e.g. C. sinensis - A species of amoeba found in dysenteric stools in China. The posterior end of the amoeba is drawn out in a tail-like caudostyle.

Caularchus A genus of Fishes of the order of Clingfishes, Cornish Suckers, and Skittle Fishes, Xenopterygii (= *Gobiesociformes*). Small fishes with a small terminal mouth, and eyes on top of the head; found mainly in tidal pools in the warmer seas.

Caulobacter A genus of rod-shaped Bacteria (Schizomycetes) of the family *Caulobacteraceae*, found in fresh or salt water. e.g. C. vibrioides.

Caulophyllum A genus of plants of the family *Berberidaceae*. e.g. C. thalictroides - Blue Cohosh, Papoose, or Squaw Root. The rhizome and roots of this plant have been used as a diuretic and emmenagogue; usually in the form of a liquid extract.

Causus A genus of Reptiles of the Viper family, *Viperidae*. These African snakes are known as the Night Adders; they have very large venom glands extending beyond the head down the front part of the body. e.g. C. rhombeatus - A Night Adder with a chain of lozenge-shaped markings on its body; found in rubbish dumps near human dwellings.

Cavia A genus of Rodents of the Cavy

family, *Caviidae.* e.g. C. cobaya (= C. porcellus) - The Domestic Cavy, Peruvian Cavy, or Laboratory Guinea Pig; from 20-30 cm (8-12 in) long; a small, tailless, short-eared, nocturnal, fast-maturing and prolific rodent, originating in South America, with a stocky rounded body; introduced into Europe in the 16th century. The domesticated form is known as the Guinea Pig, which has a wide range of colors and several different types of hair. These tame rodents, exclusively vegetarian in habit, are kept not only as children's pets, but also as valuable laboratory animals used in medical and biological research.

Cavochalina A genus of horny Sponges of the order *Keratosa.* e.g. C. bilamellata - This animal species is a gelatinous seawater sponge, thin and two-leafed, with a skeleton that consists entirely of horny fibers. These fibers are interlocking, and contain a silk-like material, known as spongin. Distribution: the Antilles (West Indian Archipelago, from Cuba to Trinadad).

Cebuella A genus of Mammals of the Marmoset family, *Callithricidae.* e.g. C. pygmaea (= C. pygmaeus) - The Pigmy Marmoset of Brazil, one of the smallest and most primitive of primates. They are smooth-brained, and their digits have long claws instead of flat nails like other primates. Their thick silky fur ranges in color from brown and yellow to white.

Cebus A genus of Mammals of the new world Monkey family, *Cebidae.* e.g. C. capucinus - The White-throated Capuchin Monkey, white on the head, shoulders and chest, with a dark brown cowl, and a pale, flesh-colored face. While predatory and vicious in the wild state, capuchin monkeys live well in captivity, are cunning, and very rapidly learn to imitate man's actions. Distribution: Central and South America.

Cecropia A genus of Trees growing on the Atlantic coast.

Cecropis A genus of Birds of the Martin and Swallow family, *Hirundinidae.* e.g. C. daurica - The Red-rumped Swallow of southern Europe and Africa; an insectivorous migrating bird, about 4-9 in long, and highly gregarious.

Cecrops A genus of Crustaceans of the order *Caligoida* (subclass *Copepoda*). e.g. C. latreillii - Over 30 mm (1 in) long, and relatively broad; this copepod is parasitic on the ocean sunfish, *Mola mola.*

Cedrus A genus of Trees of the Pine family, *Pinaceae.* These are the north temperate conifers. e.g. C. libani - The Cedar of Lebanon, a graceful tree with horizontal branches, reaching a height of 36 m (120 ft); an evergreen, with needle-like green leaves and erect, brown, woody cones up to 7.5 cm (3 in) long.

Ceiba A genus of Trees of the family *Bombacaceae.* e.g. C. pentandra - The Silk-cotton Tree, a large tree up to 30 m (100 ft) high, with spreading basal flanges. Kapok is obtained from the hairy seeds.

Celerio A genus of Insects of the Hawk Moth family, *Sphingidae.* e.g. C. galii - The Bedstraw Hawk Moth, so-called because the larva feeds on bedstraws; length 7 cm; the fore and hind wings have a wide, yellowish band. Widely distributed in Europe and Asia.

Cellfalcicula A genus of Microorganisms of the family *Spirillaceae.* Short rods or spindle-shaped cells. e.g. C. fusea.

Cellvibrio A genus of Microorganisms of the family *Spirillaceae.* Straight or slightly curved rods. e.g. C. vulgaris.

Celosio A genus of late blooming plants of the family *Amaranthaceae.* e.g. C. argentea var. cristata - Cockscomb or Cock's Comb, an annual plant, a cultivated variety, with showy flower-heads in brilliantly red colored hues; height 45-60 cm (11/2-2 ft). Place of origin:

90

tropical Asia and Africa; introduced into Europe in 1570.

Cenolophium A genus of plants of the Carrot family, *Umbelliferae.* e.g. C. fischeri - The Cenolophium, a perennial plant, up to 150 cm high; 2-5 pinnate leaves, with absent bracts and numerous bracteoles. Growing in still and running water, sometimes also in damp meadows. Distribution: most of Europe.

Centaurea A genus of plants of the Daisy family, *Compositae.* e.g. C. cyanus - Bachelor's Button (= Blue-bottle; Cornflower) - A slender, branching, annual plant, 30-60 cm (1-2 ft) high; flowers in round heads at tops of stems; blue, purple, pink or white. Has been widely distributed as a cornfield weed. Place of origin: Europe, including Britain; naturalized in North America.

Centaurium A genus of plants of the Gentian family, *Gentianaceae*; also known by the generic name, *Erythrace.* e.g. C. erythraea (= C. minus; C. umbellatum; *Erythraea* centaurium) - The Common or Lesser Centaury, an annual plant, 2.5-50 cm (1-20 in) high, with elliptical leaves in a basal rosette, and dense clusters of pink flowers forming a dense corymb-like cyme. Used in medicine as a bitter. Distribution: Britain, Europe, and western Asia. Has become naturalized in North America.

Centella A genus of plants of the Carrot family, *Umbelliferae*; also known by the generic name, *Hydrocotyle.* e.g. C. asiatica (= *Hydrocotyle* a.) - Hydrocotyle; Indian Pennywort; a perennial plant, containing the active principle asiaticoside. The fresh and dried leaves and stems are used in India in the treatment of skin diseases and as a diuretic. It has long been a popular remedy in India for leprosy and syphilis. Large doses are said to have a narcotic action.

Centetes A genus of Mammals of the Tenrec family, *Tenrecidae*, found on the island of Madagascar. It is related both to the hedgehog and the shrew. e.g. C. ecaudatus - The Common Tenrec, the largest of the insectivora, the body being 30-40 cm (12-16 in) long. It is covered with a mixture of hair, bristles, and a few spines; the tail is rudimentary.

Centranthius A genus of Rock Garden plants, also known as *Kentranthus*, of the family *Valerianaceae.* e.g. C. ruber (= *Kentranthus* r.) - Red Valerian, a smooth, upright, perennial plant, 30-100 cm (12-40 in) high, with pointed ovate leaves which emit a cat-like smell when bruised. Pink or red flowers in rich clusters. Place of origin: Europe; naturalized in Britain.

Centriscus A genus of Fishes of the Cornetfish and Snipefish subfamily, *Aulostomidinae.* These fishes are characterized by the tube-like snout of its members. They are mainly marine, tropical and subtropical fishes, and swim in an upright position like seahorses (genus *Hippocampus*) (see also **Fistularia**).

Centrocercus A genus of Birds of the Grouse or Ptarmigan family, *Tetraonidae.* e.g. C. urophasianus - The Sage Grouse; an aggressive species.

Centrocestus A genus of Flukes. e.g. C. cuspidatus - A Fluke occurring in Taiwan.

Centrolenella A genus of Amphibians of the family *Centrolenidae.* These frogs are found in Central and South America, from sea level into the high mountains; the eggs are laid in situations where they are kept moist by the spray from waterfalls and other natural agencies.

Centronotus A genus of Fishes of the Butterfish family. e.g. C. gunellus - The Gunnel Fish, 30.5 cm long; 9-13 eyespots along the dorsal fin; eggs are laid in a clump on the shore and are guarded by the parent fish, even through low tide. Inhabits northern waters.

Centropelma A genus of grebe Birds of the order *Podicipediformes.* e.g. C. micropterum - The Short-winged Grebe of South America, a flightless bird often found on reed-fringed lakes and ponds, nesting in masses of floating vegetation.

Centropristes A genus of Fishes (subphylum *Vertebrata*). e.g. C. striatus - The Black Sea Bass.

Centropus A genus of Birds of the Cuckoo family, *Cuculidae*. e.g. C. phasianus - The Pheasant Coucal, a non-parasitic member of this family. The hen lays three round, white eggs and both parents share the incubation which lasts two weeks.

Centrostephanus A genus of Sea-urchins of the class *Echinoidea*. e.g. C. rodgersi - A Sea-urchin found off the shores of South Australia. These echinodermic animals move about freely with the aid of their spines or cling to rocks, sometimes living in the hollows they have excavated, or burrowing into the sand.

Centruroides A genus of Arachnids of the Scorpion order, *Scorpiones*. e.g. C. suffusus - A poisonous Scorpion found in Mexico and the southwestern United States; the venom, a neurotoxin produced by two glands lying in the base of the sting, causes paralysis of the cardiac and respiratory muscles and can be fatal in children and in the elderly.

Cenurus A generic name formerly given certain tapeworm larvae; also known as Coenurus. e.g. C. cerebralis (= *Coenurus* c.) - The larva of Multiceps multiceps, found in the brain of sheep and goats.

Cepaea A genus of gastropod Molluscs of the garden Snail family, *Helicidae*. e.g. C. nemoralis - The Brown-lipped or Larger Banded Snail; shell 17-18 mm high, 22-23 mm wide; shell coloration very variable, yellow, white pink or brown, with or without up to 5 bands; mouth reddish-brown to black. Found in gardens, parks and open woodlands. Distribution: western and central Europe.

Cephaelis A genus of plants, also known by the generic name, *Uragoga*, of the family *Rubiaceae*. e.g. C. ipecacuanha (= *Uragoga* i.) - The Ipecacuanha Plant. From Ipecacuanha root are extracted the alkaloids emetine and cephaeline. Used in small doses as an expectorant in acute bronchitis; also giving relief in the dry cough of laryngitis and tracheitis. Combined with opium as in Dover's powder (Pulv. Doveri), it is of value as a diaphoretic. Large doses are emetic and are used to induce vomiting.

Cephalanthera A genus of plants of the Orchid family, *Orchidaceae*. e.g. C. longifolia - The Long-leaved Helleborine, a perennial orchid, 20-60 cm tall, with a creeping rhizome; lance-olate, folded leaves, and erect spikes with distant pure white flowers. Found in woods and shady places. Distribution: Europe; Asia.

Cephalodiscus A genus of Marine Chordate Animals of the class *Pterobranchia*; these pterobranchids grow in colonies, new individuals budding and sharing a common chitinous investment.

Cephalopterus A genus of Birds of the Chatterer family, *Cotingidae*. e.g. C. ornatus - The Umbrella-bird or Cotinga, a jet black bird; the size of a crow, with a long feathered lappet and a huge umbrella-like crest on its head that projects beyond the lip of the bill.

Cephaloscyllium A genus of Fishes of the Cat-shark suborder, *Galeoidea*. These swell sharks, which are fairly widespread off the coasts of California, Chili, Australia and Japan, have a habit of filling their stomachs with water and floating belly upwards on the surface.

Cephalosporium A genus of Fungi. e.g. C. granulomatis - A sporotrichum-like fungus causing gumma-like lesions in man. C. salmosynnematum (= *Emericellopsis* s.) - A mold from which is prepared the anti-microbial substance, cephalosporin N (= adicillin; aminocar-boxybutylpenicillin; penicillin N; synnematin B).

Cephalothrix A genus of ribbon Worms

of the order *Palaeonemertini*; the common littoral forms of the most common primitive nemertine or ribbon worms. Their body wall has two or three muscle layers.

Cephenomyia A genus of Insects of the Warble fly family, *Oestridae*. The larvae of these flies live in the tissues and body spaces of mammals. In some species, the larvae perforate through the skin of these animals, forming swellings or "warbles". e.g. C. stimulator - The Deer Bot Fly, a hairy black-and-yellow species, resembling a small bumble-bee. These flies deposit their small larvae inside the nostrils of the roe deer. The larvae then crawl into the animal's throat and bite into its soft palate and root of the tongue. If they do not cause the deer's death, they are sneezed out in summer and pupate in the ground.

Cephus A genus of Monkeys, now classified under the generic name, *Cercopithecus* (see **Cercopithecus**).

Cepphus A genus of Birds of the Auk family, *Alcidae*. e.g. C. grylle - The Black Guillemot, length 35 cm; found near the coast and sometimes on islands with thick vegetation. Distribution: circumpolar in the far north, wintering on the coast of the North and Baltic Seas, sometimes also on the French coast of the Atlantic.

Cerambyx A genus of Insects of the Long-horned or Longicorn Beetle family, *Cerambycidae*. e.g. C. cerdo - The European Longicorn, length 3-5 cm; one of the largest European beetles. Occurs in old, broadleaved, chiefly oak forests, where the larvae burrow in the wood and may be serious pests. Adults fly at dusk and at night, making a considerable noise. Distribution: Europe.

Ceramium A genus of Seaweeds belonging to the Red Algae. e.g. C. rubrum - The Red Seaweed or Common Red Ceramium, up to 15 cm tall; the plants are jointed, with pincer-shaped tips, and act as hosts to the jellyfishes

Craterlophus and *Haliclystus*, which anchor themselves to the branches. Found in the North Sea and the West Baltic.

Cerastes A genus of Reptiles of the Viper family, *Viperidae*. e.g. C. cerastes - The Horned Viper; this true viper lives in the desert regions of Africa, burying itself in the sand; it has a pair of horn-shaped protuberances above the eyes, giving it a ferocious appearance in keeping with the virulence of its poison.

Cerastium A genus of plants of the Pink family, *Caryophyllaceae*. e.g. C. arvense - The Field Mouse-ear Chickweed, a common perennial herb, 10-30 cm tall, widespread in the northern hemisphere, with downy linear leaves, white flowers and upright fruit-stalks. Distribution: Europe.

Cerastoderma A genus of bivalve Molluscs of the Cockle family, *Cardiidae*. e.g. C. edule - The Common Edible Cockle, length 30-50 mm; height and thickness about 30 mm striking shell with broad ribs crossed by darker concentric stripes. Lives shallowly, embedded in sandy bottoms, and capable of jumping with the aid of its foot. In winter it is a popular commodity in fish markets. Distribution: European seas, Canary Islands, North America.

Ceratitis A genus of acalyptrate Insects of the family *Trypetidae*. e.g. C. capitata - The Mediterranean Fruit Fly, a notorious and widespread fruit and cultivated plant pest of the Mediterranean littoral.

Ceratium A genus of Dinoflagellate Protozoans of the order *Dinoflagellata*. The species are mostly marine, though freshwater forms occur. Diameter 100-700 1/2; they possess elaborate armouring, with the epicone or covering on the top half of the body having one process, and the hypocone covering the lower half, three. Food is captured by means of a fine pseudopodial network.

Ceratodus A genus of Lungfishes of the suborder *Ceratodei*. Also known by the generic name, *Neoceratodus*. e.g. C. forsteri (= *Neoceratodus* f.) - The Burnett Salmon, Australian Lungfish, or Monopneume (due to its possessing a single lung), found in the Mary and Burnett rivers of Queensland, Australia; over 4 ft long; weight about 20 lb. They thrive in stagnant waters, where oxygen depletion kills off other fishes; they rise to the surface and breathe in air through the nostrils every 45 minutes. When river conditions return to normal gill respiration is resumed and only rarely is air inhaled.

Ceratonia A genus of Trees of the Pea family, *Leguminosae*. e.g. C. siliqua - The Carob Tree or Locust Bean Tree, up to 6 m (20 ft) tall, with leathery leaves, reddish brown beneath, and long, brownish-violet pods. From the seeds is extracted the mucilage, carob gum or ceratonia gum. Grows wild on mountains, but is widely cultivated. Powdered ceratonia is sometimes known as cheshire gum. Used for the dietetic treatment of diarrhea in infants, children and adults, and also for regurgitation and vomiting in infants. Distribution: the Mediterranean region.

Ceratophrys A genus of Amphibians of the Horned Toad family, *Leptodactylidae*; these are the so-called "horned frogs" of South America. They are very vividly-colored frogs with horn-like protuberances above the eyes, and usually remain hidden in sand or mud, with only the head showing. Their protective coloring makes them almost invisible as they lie in wait for their prey. e.g. C. varia - A horned toad species, up to 25 cm (10 in) in size, and capable of puffing itself up to a great width. It is patterned light green, dark red and dark brown, and can swallow a comparatively large prey.

Ceratophyllum A genus of plants of the Hornwort family. e.g. C. demersum - The Common Hornwort, a perennial plant, up to 2 m tall; rigid, forked, dark green leaves; fruit with 2 spines at base when ripe. Unisexual flowers. A submerged aquatic plant found in still and slow-flowing water.

Ceratophyllus A genus of Insects of the Flea order, *Siphonaptera*. Now known by the generic name, *Nosopsyllus*. e.g. C. fasciatus (= *Nosopsyllus* f.) - The Common Rat Flea, vector of murine typhus (see **Nosopsyllus**). C. gallinae -A species attacking chickens and man. C. silantiewi - A species attacking the tarbagan (a marmot or rodent of the genus arctomys), a natural reservoir, with the plague (see **Arctomys**).

Ceratostigma A genus of late blooming plants, also known by the generic name, *Plumbago*, of the family *Plumbaginaceae*. e.g. C. plumbaginoides (= *Plumbago* larpentae) - Blue Ceratostigma or False Plumbago, a shrubby plant, with rich purplish blue flowers in terminal and axillary clusters. Height 30-45 cm (1-11/2 ft). Place of origin: China; introduced into Europe in 1846.

Ceratotherium A genus of Mammals of the Rhinoceros family, *Rhinocerotidae*. The largest species of this genus is the white rhinoceros of southern Asia, standing over 1.8 m (6 ft) at the shoulder, and weighing up to 4 tons (see **Diceros**).

Ceratozamia A genus of Trees of the family *Cycadales*. e.g. C. mexicana - A species found in southeast Mexico; the plants are unisexual and the ovules, which develop into seeds after fertilization, are borne marginally on the scales of the female cones.

Cerbera A genus of Shrubs of the family *Apocynaceae*. e.g. C. odollam - An evergreen shrub from the tropical regions of Asia. A poisonous plant, with elliptical leaves and white funnel-shaped flowers.

Cerceris A genus of Hymenopterous Insects of the Digger Wasp family, *Sphecidae*. e.g. C. arenaria - The Sand Tailed Digger Wasp, length 9-17 mm;

found in sandy localities; hunts the larvae of various weevils (*Curculionidae*) as food for its offspring. Distribution: much of Europe.

Cercis A genus of Trees of the Pea family, *Leguminosae*; a member of the subfamily *Caesalpiniodeae*, which has also been treated as a separate family *Caesalpinaceae*. e.g. C. siliquastrum - The Judas Tree, up to 8 m (26 ft) tall; a small deciduous tree with alternate, heart-shaped leaves, and clusters of reddish-purple flowers. The fruit is a compressed red pod about 12.5 cm (5 in) long. A woody plant growing wild on mountains in the Mediterranean region and also widely cultivated.

Cercocebus A genus of old world Monkeys of the family *Cercopithecidae*. e.g. C. albigena - The Crested or Grey-cheeked Mangabey Monkey, widespread from Cameroun to Uganda; has a long snout with a brush of long hair on the head, black in the middle, merging into grey.

Cercomela A genus of Birds (class *Aves*). e.g. C. melanura - The Blackstart, found in desert areas.

Cercomonas A genus of Flagellate Protozoa found in feces, also classified under generic names such as, *Chilomastix*, *Giardia*, and *Trichomonas*. e.g. C. hominis (now known as *Trichomonas* h. and *Chilomastix* mesnili) - Found in man (see **Trichomonas**; **Chilomastix**). C. intestinalis (= *Giardia* lamblia) (see **Giardia**).

Cercopis A genus of Homopterous Insects of the Spittle-bug or Frog-hopper family, *Cercopidae*. e.g. C. vulnerata - Length 9.5-11 mm; occurs in abundance in various plants in June and July. The larvae, enclosed in foamy matter, live on plant roots. Distribution: all of Europe.

Cercopithecus A genus of Guenon Old World Monkeys of the family *Cercopithecidae*. Formerly known by the generic name, *Cephus*. e.g. C.

cephus cephus (= *Cephus* cephus) - The Moustached Monkey, found in Africa from eastern Nigeria to the Congo, with an olive-green to blue-grey body, a bright blue face, and a white mark like a moustache across the upper lip.

Cercosphaera A genus of small-spored ringform fungi, also called by the generic names, *Microsporon* or *Microsporum*. e.g. C. addisoni (= *Microsporon* audouini; *Microsporum* audouini) - Causes ringworm of the scalp, especially in children (see **Microsporum**).

Cercosporalla A genus of microscopic fungi. e.g. C. vexans - A fungus causing skin eruptions.

Cercotrichas A genus of Birds of the Thrush family, *Turdidae*. e.g. C. galactotes - The Rufous Warbler or Rufous Bush Robin, length 15 cm; brownish in color; the tip of the tail has a black and white margin; nests in hedges; a migrant bird. Found in South Spain and the Balkans.

Cerdocyon A genus of Mammals of the Wild Dog, Wolf, and Jackal family, *Canidae*. e.g. C. magellanicus - The Cordillera Fox, 90 cm (3 ft) long without the tail; found from as far north as Ecuador to the extreme southern areas of South America.

Cerebratulus A genus of Ribbon Worms (phylum *Nemertina*, or *Rhynchocoela*) of the order *Heteronemertini*. They have three muscle layers, the inner and outer ones being longitudinal, and the middle layer circular. These worms are long, flat creatures in which it is difficult to distinguish the head from the tail. Along the seashore they look like shiny, wet coils of seaweed.

Cereopsis A genus of Birds of the Duck family, *Anatidae*, belonging to the primitive shel geese tribe of Tadornini of the subfamily *Anatinae*. e.g. C. novae hollandiae - The Cape Barren Goose; slighter in build than the Domestic Goose (*Anser* anser domesticus); greyish-brown incolor, with a greenish-yellow bill. It keeps to dry land, and evades water. Habitat: the forests of South Australia, Tasmania and New Zealand.

Cereus (1) A genus of Sea Anemones of the animal order *Actiniaria*. e.g. C. pedunculatus - The Daisy Anemone, height 60-90 mm; yellowish-brown to white, upper third of the body furnished with numerous suction pads. Approximately 700 tentacles arranged in 8 rows, inside tentacles larger than those on the outside. Lives attached to stones, rocks, shells, etc. at shallow depths. A familiar, very abundant animal, viviparous, with a flower-like appearance of its expanded oral disc and margin of tentacles. Found on rocky shores. Distribution: worldwide; especially numerous along the Mediterranean and Atlantic coasts. (2) A genus of Cactus plants of the family *Cactaceae*. e.g. C. peruvianus monstrosus - The Curiosity Plant or Peruvian Torch, an arborescent plant with large white flowers surrounded by red sepals; and with fleshy edible fruits. Place of origin: southeast coast of South America.

Cerianthus A genus of Thorny Corals of the order *Cerianthidea* (= *Ceriantharia*). e.g. C. membranaceus - The Vestlet; length 200 mm; tube-like, elongate, variously colored (from white to brown to violet); body furnished at upper (mouth) end with about 130-long tentacles arranged in 4 rows. Occurs along the sea coast at depths of 1-35 m. Distribution: Mediterranean Sea.

Cerinthe A genus of plants of the Borage family, *Boraginaceae*. e.g. C. glabra - The Alpine Waxyflower, a perennial plant, 30-45 cm high, corolla with 5 recurved teeth; grows in alpine meadows, scrub and open woodlands up to 2,660 m. Found in Central Europe.

Cerithium A genus of Molluscs. e.g. C. vulgatum - Thick-walled, rough, conical shell; up to 7 cm high; edge of aperture dentate. Found in the Mediterranen region.

Ceropegia A genus of plants of the Milkweed family, *Asclepiadaceae*. e.g. C. radicans - A succulent, cactus-like plant from South Africa, with small fleshy, ovate leaves and 5-lobed, erect flowers having long, slender corolla tubes. Insects enter the tube through gaps between the lobes and effect pollination.

Certhia A genus of Birds of the Tree Creeper family, *Certhiidae*. e.g. C. familiaris - The Tree Creeper or Brown Creeper of North America, length 13 cm; a small bird with a long down-curved beak, reddish-brown plumage with black and white streaks on its upper parts and a light grey belly. It has long claws which enable it to grip the bark firmly, while it searches the crevices for insects. Distribution: Europe, Asia, southern parts of North America.

Cerura A genus of Insects of the Prominent Moth family, *Notodontidae*. e.g. C. vinula - The Puss Moth, size 28-36 mm; found along streams, in meadows and in parks. The striking, large caterpillar grows to a length of 7-8 cm. Distribution: the Palaearctic region.

Cervus A genus of Mammals of the True Deer family, *Cervidae*. e.g. C. braziliensis - The Brazilian Deer, whose skin and hair was once used in homeopathic medicine. C. canadensis - The North American Elk or Wapiti, related to the Red Deer of Europe (C. elaphus), but is larger, standing 165 cm (51/2 ft) at the shoulders, with its antlers 150 cm (5 ft) long. Found in Canada and the northern United States.

Ceryle A genus of Birds of the Kingfisher family, *Alcedinidae*, belonging to the fishing subfamily, *Cerylinae*. e.g. C. rudis - The Pied Kingfisher of Africa and southwest Asia; it specializes in taking crabs and crayfishes out of the water, and pounding them to pieces as a preliminary to eating them.

Cestrum A genus of Shrubs and Bushes

96

of the family *Solanaceae*. e.g. C. pur-
pureum (= C. elegans) - The Bastard
Jasmine or Jessamine, a slender evergreen plant, 2.1 m (7 ft) high, with
downy pendulous shoots carrying dense
terminal panicles of 2.5 cm (1 in), funnel-shaped purplish-red flowers with 5-
pointed lobes. Place of origin: Mexico;
introduced into Europe in 1840.

Cestum A genus of Comb Jellies of the
order *Cestida*. e.g. C. veneris - Venus's
Girdle; the species are restricted to tropical and subtropical waters and to the
Mediterranean. Some species may also
be seen around Florida.

Cetarach A genus of plants (see
Ceterach).

Cetengraulis A genus of Fishes (subphylum *Vertebrata*). e.g. C. mysticetus -
The Anchoveta fish.

Ceterach A genus of Rock garden plants,
also known by the generic names,
Asplenium or *Cetarach*, of the Fern or
Polypody family, *Aspleniaceae*. e.g. C.
officinarum (= *Asplenium* ceterach) -
The Scaly Spleenwort or Rusty-back
Fern (see **Asplenium**).

Cetomimus A genus of Fishes of the
Whalefish family, *Cetomimidae*; the
species look like miniature whales, with
enormous heads; the mouth is large with
a big gape, the body is compressed and
scaleless; there is no adipose fin, and the
eyes are rudimentary.

Cetonia A genus of Insects of the Scarab
or Lamellicorn Chafer Beetle family,
Scarabaeidae, the brilliant metallic
green cockchafer scarabaeids. e.g. C.
aurata - The Rose Chafer, length 14-20
mm; metallic or golden-green elytra
with white spots; found in spring on
flowering roses, elder, etc.; larvae generally found in rotting wood of beech and
oak stumps, in garden compost heaps,
and in ant nests. Distribution: Europe,
Asia Minor, Siberia.

Cetorhinus A genus of Fishes of the
Basking Shark family, *Cetorhinidae*.
e.g. C. maximus - The Basking Shark,

almost 12 m (40 ft) long; sluggish in
habit, with broad gill openings which
almost meet under the throat. The teeth
are small and weak and the food is fil-
tered out of the water by gill rakers as it
passes over the gills. Commonly found
in the North Sea.

Cetraria A genus of plants, also known
by the general generic name, *Lichen*, of
the family of Lichens, *Parmeliaceae*.
e.g. C. islandica (= *Lichen* islandicus;
Lobaria islandica) - The Iceland Moss
(= Cetraria; Iceland Lichen; Lichen
d'Islande; Lichen Islandicus); a perennial plant with more or less erect reddish-
brown thallus lobes, about 10 cm (4 in)
high with rows of marginal black spines.
One of the few edible lichens, it is a
small, slow-growing plant, capable of
surviving prolonged desiccation. It
flourishes on the ground in northern
heather moors and open mountain
woods. It has demulcent properties and
is used in the treatment of lung and
bowel disorders. Its bitter taste is due to
Cetraric Acid which can be removed by
maceration in water, weak potassium
carbonate solution, alcohol, etc. Place of
origin: Europe; native in Britain (see
Lobaria).

Cettia A genus of European Birds of the
Old World Warbler family, *Sylviidae*.
e.g. C. cetti - Cetti's Warbler, length 14
cm; upper parts dark reddish brown;
rounded tail; difficult to see in waterside
under growth, but has a characteristic
song. A resident European bird.

Ceuthophilus A genus of Insects of the
Cricket family, *Stenopelmatidae*. These
are the camel crickets of North America.

Chaenomeles A genus of plants, also
known by the generic name, *Cydonia*, of
the Rosefamily, *Rosaceae*. e.g. C. speciosa (= C. lagenaria; *Cydonia* japonica)
- Japanese Quinceor Japonica, a deciduous thorny shrub, 2 m (61/2 ft) high,
with spiny branches and crimson flowers with yellow stamens. Also pink,
white and scarlet cultivars. Yellow or

yellowish-green, rounded or pear-shaped, fragrant fruit, from which is prepared marmalade. Place of origin: probably Japan; introduced into Europe in 1796.

Chaenorhinum A genus of plants of the Figwort family. e.g. C. minus - The Small Toadflax, an annual plant, 5-60 cm high; flowers with a spur about 9 mm long, growing in loose racemes. Found commonly in cultivated fields, by roadsides, and on waste ground.

Chaerophyllum A genus of plants of the Carrot family, *Umbelliferae*. e.g. C. bulbosum - The Tuberous-rooted Chervil, a biennial to perennial plant, 50-200 cm high; root tuberous, stem glabrous or with scattered hairs on leaf-bases, and glabrous bracteoles. Widespread in waste and grassy places, by hedgerows and at edges of woods.

Chaetodipterus A genus of Fishes (sub-phylum *Vertebrata*). e.g. C. faber - The Atlantic Spade-fish.

Chaetodon A genus of Fishes of the Butterflyfish family, *Chaetodontidae*. e.g. C. capistratus - The Foureye Butterflyfish, a small, brightly colored fish of tropical waters with flattened bodies possessing an easily recognizable spot on each side, thin fins partly covered with scales, and small brush-like teeth on the jaws.

Chaetogaster A genus of Oligochaete freshwater segmented Worms of the family *Naididae*; the species are found in the silt at the bottom of ponds, lakes and quiet streams. The Naidids are minute semi-transparent creatures that reproduce vegetatively.

Chaetonotus A genus of minute Worms known as the Gastrotrichs, of the order *Chaetonotoidea*. These are freshwater worms, commonly found at the bottom of ponds, lakes, and even in aquarium tanks or protozoan cultures, where they feed on detritus, bacteria, diatoms or protozoans by means of a long sucking pharynx, and swim around by means of cilia at the front end. Distribution: worldwide.

Chaetophractus A genus of Mammals of the Armadillo family, *Dasypodidae*. e.g. C. villosus - The Hairy Armadillo or Peludo; about 50 cm (1 ft 8 in) long; its carapace is relatively flat, and it cannot curl up like the rest of the species. It digs long tunnels even in hard, stony ground and is omnivorous, eating insects, small vertebrates, and vegetable matter alike. A nocturnal animal. Distribution: the pampas of Argentina, Bolivia and Uruguay.

Chaetopterus A genus of Segmented Worms of the Chaetopterid family, *Chaetopteridae*. e.g. C. variopediatus - An extremely delicate annelid, almost colorless, and luminescing in the dark; remarkable for extreme differentiation of the body segments and by its special feeding mechanism. A marine animal with a worldwide distribution.

Chaetorhynchus A genus of Birds of the Drongo family, *Dicruridae*. e.g. C. papuensis - The Papuan Mountain Drongo, an aggressive bird with twelve tail feathers.

Chaetura A genus of Birds of the Swift family, *Apodidae*. e.g. C. pelagica - The Chimney Swift of North America; its flight is faster than any other bird. These birds sometimes spend a whole night on the wing, and frequently copulate in the air. They winter in South America.

Chalcides A genus of Lizards of the Skink family, *Scincidae*. e.g. C. chalcides - The European Skink, length over 40 cm (16 in); found on grassy hillsides, hiding under stones, in walls, and under leaves. Distribution: northwest Africa and southwestern Europe.

Chalcites A genus of Birds of the Cuckoo family, *Cuculidae*. e.g. C. lucidus - The Bronzed or Shining Cuckoo; breeds on the islands off New Zealand, laying its eggs in the nests of flycatcher birds (subfamily *Muscicapinae*), and after laying, the adults depart. The young, brought up by these foster-parents, follow a month later the same path taken by their parents, making a non-stop flight of some 1,200 miles to Australia, and then northwards another 1,000 miles to the Bismarck and Solomon Islands.

Chalcophora A genus of Insects of the suborder *Polyphaga* (order of Beetles, *Coleoptera*). e.g. C. mariana - This species of Beatles has a beautiful bronze color, but is usually covered with a fine layer of white "sawdust", which renders it practically invisible against the trunk of the pine-tree. The larvae tunnel in the live wood of these trees. Distribution: the United States.

Chalcosoma A genus of Insects of the Dung, Scarab or Lamelliform Chafer Beetle family, *Scarabaeidae* (subfamily, *Dynastinae*). e.g. C. atlas - The Indian or Indonesian Scarab Beetle; black in color with a greenish sheen. The most beautiful specimens live in the Celebes, and the largest in Java and Sumatra.

Chamaea A genus of Birds of the Wrentit subfamily, *Chamaeinae* (family *Muscicapidae*). e.g. C. fasciata - The Wren-tit, the single species of this subfamily; a small brown bird, with a long barred tail, living in low scrub on insects and fruit; it recalls both the wrens and the tit mice. Distribution: western North America.

Chamaebuxus A genus of plants of the Milkwort family, *Polygalaceae*. e.g. C. alpestris - The Boxleaved Milkwort, an evergreen trailing shrub with yellowish flowers, which grows in stony places in the Pyrenees, Alps, Apennines, and the Carpathian Mountains.

Chamaecyparis A genus of Trees of the Pine family, *Pinaceae*. e.g. C. obtusa - The Hinoki Cypress, a native of central and southern Japan, where it is used extensively for forest planning. An evergreen tree up to 36 m (120 ft) tall, with a straight trunk and reddish-brown bark. The dense branches are spreading out and the stalked spherical cones are about 12.5 mm (1/2 in) across.

Chamaedaphne A genus of dwarf shrubs of the Heath family, *Ericaceae*. e.g. C. calyculata - The Chamaedaphne, an evergreen shrub, up to 1 m (39 in) high; leaves with rust-brown glands, whitish green beneath; bell-shaped white flowers. Usually grows on high moorlands. Place of origin: northern Europe.

Chamaedorea A genus of indoor plants, also known by the generic names, *Collinia* or *Neanthe*, of the Palm family, *Palmae*. e.g. C. elegans (= *Collinia* e.; *Neanthe* bella) - Camedorea or Feathers Palm, a small palm, growing to a maximum height of 1-2 m (3-6 ft), with bright green pinnate leaves. Used as an indoor, porch, or greenhouse plant. Dioecious flowers and shining scarlet fruits. These fruits are eaten by Mexicans, as well as the young shoots, like asparagus. Place of origin: Mexico and Central America.

Chamaeleo A genus of Reptiles of the Chameleon family, *Chamaeleonidae* (= *Chameleontidae*). e.g. C. chamaeleon - The Common Chameleon, African Chameleon, or Mediterranean Chameleon, found in Africa, Madagascar and the Mediterranean; length 23-28 cm; body flatenned from side to side; long prehensile tail; fore and hind feet each have two digits opposed to the other three, forming a gripping device and making it an excellent tree-climber. The club-shaped tongue is prehensile and can be extended about the length of the reptile itself. In the chameleons color change reaches its highest development among the vertebrates, the granular skin containing black, yellow, and red color cells together with others that have a whitish pearly appearance. The reptile is noted for its ability to change the color of its skin from almost white to black. The species is oviporous. Distribution: southern

Europe, Asia Minor, the Middle East, North Africa, and Madagascar.

Chamaemelum A genus of plants, known also by the generic name, *Anthemis*, of the Daisy family, *Compositae*. e.g. C. nobile (= *Anthemis* nobilis) - The Common or True Chamomile of Britain, a perennial plant, 10-40 cm high; receptacle-scales oblong and blunt; strongly aromatic; from its capitula is extracted Oil of Chamomile. Place of origin: Britain, central and western Europe (see **Anthemis**).

Chamaenerion A genus of Medicinal and Aromatic plants, better known by the genus name, *Epilobium*, of the family *Oenotheraceae*. e.g. C. augustifolium (= *Epilobium* a.) - The Rosebay Willow-herb or Fireweed plant (see **Epilobium**).

Chamaeorchis A genus of plants of the Orchid family, *Orchidaceae*. e.g. C. alpina - The Alpine False-orchid, a perennial plant, 5-12 cm high; with about 8 basal leaves, and small flowers with ovate to rhombic lips. Grows in mountain meadows and stony pasture-land. Place of origin: South Central and Central Europe; Scandinavia.

Chamaepericlymenum A genus of plants of the Cornel or Dogwood family, *Cornaceae*. e.g. C. suecica - The Dwarf Cornel, a perennial plant, 10-15 cm high; red-brown flowers with 4 white involucral bracts; red berries. A rare member of the Cornel family, found on moors along mountain ranges in Arctic Europe, and extending southwards to north Germany. Native in Britain.

Chamaerops A genus of shrubs and bushes of the Palm family, *Palmaceae*. e.g. C. humilis - The Dwarf Fan Palm, the only native European palm, a woody evergreen perennial plant and a dwarf species, 2-2.6 m (6-8 ft) high; flat, fan-shaped, palmate leaves on long stems, and yellow flowers in dense heads. Place of origin: Mediterranean region; introduced into Britain in 1731.

Chamaesaura A genus of Snake Lizards of the order *Squamata*. These reptiles have spiky scales, and very greatly reduced limbs.

Chamaespartium A genus of Shrubs of the Pea family, *Leguminosae*. e.g. C. sagitalle - The Hare's-foot Greenweed, 12-25 cm high; broadly winged stem; grows in dry meadows, heaths, thickets and open woods. Place of origin: South and Central Europe.

Chaos A genus of Protozoan microorganisms, now included in the genus *Entamoeba*. e.g. C. diffluens (= *Amoeba* proteus; *Entamoeba* p.) (see **Entamoeba**).

Charabdea A genus of Sea Wasp Jellyfishes of the order *Cubomedusae*; these jelly-fishes are generally colorless, with single tentacles at the cuboid bell margin, and are noted for their powerful and often lethal sting, being greatly feared by swimmers off the coastal waters of South Japan, the Philippines and Australia, where they occur.

Charadrius A genus of Birds of the Plover and Lapwing family, *Charadriidae*; also called the Sand larks or Stone-runners, probably because of their preferance for sandy or pebbly beaches. e.g. C. alexandrinus - The Kentish Plover or the Snowy Plover as it is known in America. An active little bird, no bigger than a lark, length 15 cm, with a dark spot on the side of the breast; black legs and bill; a partially migrant bird, nesting along European coasts on sandy beaches, including the Kent coast. Distribution: worldwide.

Chasmorhynchus A genus of Birds of the Cotinga or Chatterer family, *Cotingidae*. e.g. C. niveus - The White Bellbird, an ornamental species from the Guianas; snow-white in color with a black fleshy caruncle nearly 7.6 cm (3 in) long rising from its forehead. When excited, it erects the spike-like caruncle and utters a sound like a bell.

Chauliodus A genus of Bony Fishes of the Viperfish family, *Chauliodontidae*.

e.g. C. sloani - The Viperfish, length 25 cm; a deep-sea fish with sharp, fang-like teeth. Noted for its vertical migration: by night it comes up to the surface, and by day it keeps to depths between 450 and 2,800 m. Distribution: worldwide, on all seas from the equator to the poles.

Chaulmoogra A genus of plants, also known by the generic name, *Gynocardia*, of the family *Bixaceae*. e.g. C. odorata (= *Gynocardia* o.) - A species, at one time believed to produce Chaulmoogra Oil, which was used in the treatment of leprosy, but later (in 1954), was described as being expressed from the seeds of the plant *Hydnocarpus* kurzii (*Taraktogenos* k.) (see **Hydnocarpus**).

Chauna A genus of Birds of the Screamer family, *Anhimidae*. e.g. C. torquata - The Crested Screamer, a hornless species living in the lagoons, swamps, and pampas in Paraguay, southern Brazil, Uruguay, and north eastern Argentina. Its nest, lightly built of rushes, stands in the water.

Cheilanthes A genus of plants of the Fern or Polypody family, *Aspleniaceae*. C. maranthae - A tree fern growing in dry rocky places in southern Europe. It has a creeping rhizome and stalked, pinnate leaves up to 30 cm (1 ft) long, which are covered beneath with pale brown scales.

Cheilomastix A genus of Protozoa, also known by the generic names, *Chilomastix*, *Trichomonas*, etc. e.g. C. mesnili (= *Chilomastix* m.; *Trichomonas* hominis, etc.) - An intestinal flagellate (see **Chilomastix**; **Trichomonas**).

Cheilotrema A genus of Fishes of the Croaker family, *Sciaenidae*. e.g. C. saturnum - The Black Croaker fish of California, a marine fish found in the warmer waters of the Pacific Ocean, widely distributed and commercially important.

Cheiranthus A genus of flowering plants of the Cress or Crucifer family, *Cruciferae*. e.g. C. cheiri - The Wallflower, an erect biennial to perennial, showy plant, 23-60 cm (9 in 2 ft) high, with variously colored, fragrant flowers, according to the strain, from creamy-white, through shades of yellow, brown, red, pink and purple, growing in close spikes. Used as decoration for balconies, terraces, and flower-beds. Place of origin: Europe, especially the eastern Mediterranean; naturalized in Britain.

Cheirogaleus A genus of Mammals of the Lemur family, *Lemuridae*. e.g. C. pusillus - The Mouse Lemur, a rat-sized nocturnal, forest-dwelling primate, living at the tops of the highest trees in a nest like that of a crow, and subsisting on insects and vegetable foods.

Chelidonium A genus of plants of the Poppy family, *Papaveraceae*. e.g. C. majus - The Greater Celandine, a perennial herb, 30-70 cm high, with pinnate leaves and an orange latex which contains poisonous alkaloids. It has bright yellow flowers. The fruit is an elongated capsule containing shiny black seeds. Widespread and common in hedgerows and by walls. Distribution: Britain, Europe, northern Asia, and parts of North America.

Chelifer A genus of Arachnids of the family *Cheliferidae* (the False Scorpion or Pseudoscorpion order, *Pseudoscorpiones*). e.g. C. cancroides - The Common False or Book Scorpion, length 2.6-4.5 mm; a species of pseudoscorpions found in books, human dwellings, and in old warehouses; also encountered sometimes, clinging to the legs of flies, a dispersal activity known as phoresy. They bear a superficial resemblance to scorpions, having a wide flattened body and conspicuously large pincers, but no tail or terminal sting. They feed chiefly on mites. Distribution: worldwide.

Chelodina A genus of Reptiles of the Snake-necked Turtle family, *Chelidae*. e.g. C. longicollis - The Australian

Long-necked or Snake-necked Swamp Tortoise or Turtle, a carnivorous species, preying mainly on fishes, and characterized by its extremely long neck; glossy-brown in color, and found in the swamps and marshlands of Australia.

Chelon A genus of Fishes of the Mullet family, *Mugilidae*. e.g. C. labrosus - The Thick-lipped or Grey Mullet, up to 66 cm long; small mouth; upper lip very thick; two widely separated dorsal fins. Distribution: the Mediterranean, southern North Sea.

Chelonia A genus of Reptiles of the Marine Turtle family, *Chelonidae*. e.g. C. midas - The Green or Edible Turtle, an edible species in danger of extinction, about 1.2 m (4 ft) long, and weighing 300-400 lb. Carapace on back, with four pairs of costal shields. A Sea Turtle, it feeds on algae and other marine plants, and is found in tropical waters and warm oceans.

Chelydra A genus of Reptiles of the Snapping Turtle family, *Chelydridae*. e.g. C. serpentina - The American Common Snapping Turtle; up to 60 cm long; it has a reduced shell, lateral spine and a long tail, is very aggressive and has powerful hooked jaws which can inflict severe injury. Habitat: the rivers and marshes of North America, east of the Rockies, and southwards to northern South America.

Chelys A genus of Reptiles of the Snake-necked Turtle family, *Chelidae*. e.g. C. fimbriata - The Matamata, a side-necked carnivorous tortoise, inhabiting the rivers and swamps of Brazil, Guianas, and Venezuela; its carapace is thrown up into a mass of bosses and bumps. When motion lesson the bottom of a stream it resembles a stone. The head is flat-tenedand triangular, and is extended to form a proboscis.

Chenopodium A genus of plants of the Goosefoot family. e.g. C. urbicum - The Upright Goosefoot, an annual plant, 50-100 cm high; leaves with shallowly toothed margins, and upright axillary and terminal inflorescences. Grows on waste ground, rubbish-dumps, barn-yards, and in cultivated fields. Distribution: most of Europe.

Cheyletiella A genus of Arachnid Worms. e.g. C. parasitovorax - An ascarine living on the cat; may cause adermatosis in man.

Chiasmodon A genus of Fishes (subphylum *Vertebrata*). e.g. C. niger - The Black Swallower fish.

Chiasognathus A genus of Lamellicorn Beetles of the suborder *Polyphaga*. e.g. C. grantii - A tropical Stag Beetle from Chile; a large and striking insect, having a mandible of fearsome dimensions - longer than its body.

Chilocorus A genus of Insects of the Ladybird Beetle family, *Coccinellidae*. e.g. C. renipustulatus - Length 3-41/2 mm; distinguished by the large rounded red spot on each elytron; often found on tree trunks covered with scale insects (family *Diaspididae*) on which it feeds. Distribution: the Palaearctic region.

Chilodon A genus of Ciliate Protozoans of the suborder *Cyrtophorina*. e.g. C. dentatus - A ciliate species sometimes causing dysentery.

Chilodonella A genus of Ciliate Protozoans of the suborder *Cyrtophorina*. The ventral mouth is visible from the dorsal side through the cell, and a line of cilia, the dorsal brush, runs from the edge of the cell to a region over the mouth.

Chilomastix A genus of Protozoans of the class of Flagellates, *Zoomastigophorea*, known by other generic names such as *Cercomonas*, *Cheilomastix*, *Macrostoma*, *Tetramitus*, and *Trichomonas*. e.g. C. mesnili (= *Trichomonas* hominis, etc.) - An intestinal flagellate (see **Trichomonas**).

Chilomonas A genus of Protozoans of the order *Cryptomonadina*; these species of cryptomonads are small cells, 15-40 μ long, usually having two flagella arising in a pit, and one or two yellow or brown chloroplasts. The cells are flattened in section, and grow so readily that they are used frequently as food for particle feeders, such as ciliates.

Chilomycterus A genus of Fishes of the Burrfish family, *Diodontidae*; this is the Common Burrfish or Rabbitfish of the North American Atlantic coast; it does not have a bony case, and the fins have weak short spines.

Chimabacche A genus of Insects of the order of Butterflies and Moths, *Lepidoptera*. e.g. C. fagella - A greyish-brown moth, about 15 mm (3/5 in) long, found on trees and houses near woods. The females have partly stunted wings.

Chimaera A genus of Cartilaginous Fishes of the Chimaera family, *Chimaeridae*. e.g. C. monstrosa - The Rabbit Fish, length about 1 m; weight 21/2 kg. A deep-water species with large eyes, a single erectile appendage on the forehead, and a long, saw-toothed poison spine along its back. It lives on the sea-bottom at depths of 200-500 m. Distribution: along the European and North American Atlantic coast.

Chimaphila A genus of plants of the Wintergreen family. e.g. C. umbellata - The Umbellate Wintergreen, 7-15 cm high; inflorescence almost umbellate; found in dry woods and scrub. Place of origin: north and central Europe.

Chimonanthus A genus of Shrubs, also known by the generic name, *Calycanthus*, of the family *Calycanthaceae*. e.g. C. praecox (= *Calycanthus* p.) - The Winter Sweet plant (see **Calycanthus**).

Chinchilla A genus of Mammals of the Rodent Chinchilla family, *Chinchillidae*. e.g. C. laniger - The Chinchilla, the most valued species of this family, much valued for its fur, which is pearl-grey or silver-grey, and extremely beautiful, having more hairs per square inch than any other animal. It is about 30 cm (1 ft) long - (about the size of a squirrel). It

was much hunted for its valuable fur and almost became extinct, and is now rarely found in the wild state outside Chile, where it lives high up in the Andes Mountains at altitudes of 8,000-12,000 ft. It is now widely bred for its fur on farms in North America and Europe.

Chionactis A genus of Reptiles of the Snake suborder, *Ophidia*. e.g. C. occipitalis annulata - The Shovel-nose Snake, an inhabitant of the Californian desert. It is vividly colored, white, black, and orange, with annular rings on its body; its scales are shiny, and it is only 30 cm (1 ft) long. It lives in the sand, the shape of its head allowing it to "shovel" or burrow through it, hence the name.

Chionaspis A genus of Armored Scale Insects of the family *Diaspididae*. e.g. C. salicis - Length 1.9-2.5 mm; polyphagous and very common on alder, willow, mountain ash, blueberry, and other trees. The red flat females are covered by a white scale. Distribution: the Palaearctic region.

Chionis A genus of Birds of the Sheathbill family, *Chionididae*. e.g. C. alba - A primarily terrestrial bird, with white plumage, small wings, and unwebbed feet. Distribution: Antarctic and subantarctic regions.

Chirocephalus A genus of Crustaceans of the order *Anostraca* (= *Phyllopoda*); the oldest and most primitive of all crustaceans. e.g. C. grubei - Almost 2.5 cm (1 in) long; these anostracans have eleven pairs of foliaceous limbs which form a filtering mechanism for small food particles. The eggs in the brood-pouch have a tough coat, and when liberated are capable of withstanding being dried and frozen. They live in freshwater pools in Central Europe.

Chirolophis A genus of Fishes of the Blenny family, *Bleniidae*. e.g. C. galerita - Yarrell's Blenny; up to 15 cm long; has orbital tentacles on forehead. Inhabits northern waters.

Chironectes A genus of Mammals of the American Opossum family, *Didelphidae.* e.g. C. minimus - The Water Opossum, of Brazil and Guatemala, the only marsupial adapted to an aquatic life. Its hind feet are webbed, and its bare tail is scaly and non-prehensile. It burrows in river banks and feeds entirely on other aquatic animals.

Chironomus A genus of Insects of the Nematoceran family, *Chironomidae.* These are also known as the non-biting midges; the larvae contain hemoglobin which is used for respiratory purposes when the oxygen supply is low; from their red color these larvae are often known as bloodworms.

Chiropotes A genus of Mammals of the New World Monkey family, *Cebidae*; these include the Saki monkeys, small animals no bigger than large squirrels, with long non-prehensile tails, such as the Black Saki and the White-nosed Saki species. These are found in the forests of South America, chiefly in Brazil.

Chiropsalmus A genus of Sea Wasp Jellyfishes of the order *Cubomedusae.* e.g. C. quadrigatus - A species found in northern Australian waters (see **Charabdea**).

Chiropterotriton A genus of Amphibians of the plethodontid Salamander family, *Plethodontidae*; a lungless animal ranging from Mexico through Central to South America, where some species may occur at heights up to 11,000 ft.

Chiroxiphia A genus of Birds of the Manakin family, *Pipridae.* e.g. C. linearis - The Fandango Bird, a small, brightly colored bird of South America. In this Manakin, the central pair of tail feathers are very long, and male courtship of the female involves a remarkable exhibition of acrobatics.

Chiton A genus of Molluscs of the class *Amphineura.* These animals belong to the subclass *Polyplacophora* (= *Loricata*). The species have soft bodies covered by a number of plates, and they are able to coil and uncoil freely. They have neither eyes nor feelers, and live on the ocean bed, feeding on organic matter with a mouth situated on the underside of the body. Chitons are considered to be among the most primitive of Mollusc animals.

Chlamydia A genus of microorganisms of the family *Chlamydiaceae* (order *Rickettsiales*). These are small, intracellular, gram-negative bacteria of the Psittacosis - Lymphogranuloma venereum (LGV) - Trachoma-inclusion conjunctivitis (TRIC) group. Once they were considered viruses because of their small size. They have also been classified under former generic names, such as *Bedsonia* and *Miyagawanella.* e.g. C. psittaci (= *Bedsonia* p.; *Miyagawanella* p.) - Causing psittacosis (ornithosis), a disease of birds which may be transferred to man, resulting in mild respiratory symptoms to severe pneumonia with toxicity and septicemia. The organism can be isolated from the blood or sputum. Many infected persons will have a biologic false-positive (BFP) test for syphilis. C. lymphogranulomatis (= *Bedsonia* l.; *Miyagawanella* l.) or LGV Chlamydia - produces the venereal disease, lymphogranuloma venereum (LGV or lymphopathia venereum). May be isolated from pus, buboes, or biopsy material. C. trachomatis and C. oculogenitalis (TRIC agents), the former causing trachoma and the latter, inclusion conjunctivitis (inclusion blennorrhea), neonatal ophthalmia, non-specific urethritis, pulmonary infections, and lymphogranuloma venereum. These agents have also been isolated from the joints and the urethra of patients with Reiter's disease.

Chlamydomonas A genus of Protozoans of the order *Phytomonadina*, clearly animals nearest to the plant kingdom. The chloroplasts contain a bright green chlorophyll and the cell walls, firm and resistant to distortion, are made of cellulose or a close chemical relative. These are typical phytomonads, about 20 μ in size, with two flagella, a pigment spot or stigma, and are found in freshwater.

Chlamydophrys A genus of Protozoa. e.g. C. stercorea - Found in the feces of man and various animals.

Chlamydosaurus A genus of Reptiles of the Agamid or Lizard family, *Agamidae*. e.g. C. kingi - The Australian Frilled Lizard, 90 cm (3 ft) long, an arboreal and insectivorous reptile; it has a fold of skin or frill that can be spread out like a ruff when the mouth is opened in aggressive display.

Chlamydoselachus A genus of primitive notidanid Sharks of the family *Hexanchidae* (sub-order *Notidanoidea = Hexanchiformes*); sometimes placed in a separate suborder, *Chlamydoselachoidea*. e.g. C. anguineum - The Frill Shark, found off Japanese waters; each gill opening is bordered by a broad frill of skin, and there is only one dorsal fin (see **Hexanchus**).

Chlamydotis A genus of Birds of the Bustard family, *Otididae*. e.g. C. undulata - Macqueen's Bustard; 63.5 cm long; short head-crest and long, black and white feathers on neck; a resident bird, nesting on the steppes and in large fields. A gregarious species, feeding on many plant and animal foods. Habitat: North Africa.

Chlamyphorus A genus of Mammals of the Armadillo family, *Dasypodidae*. e.g. C. truncatus - The Fairy Armadillo or Pichiciego, 12.7 cm (5 in) long excluding the tail; the shell is attached to the body only along the mid-dorsal line, and its fore feet have digging claws. It lives underground and as a burrower has often been compared to a mole. Distribution: the sandy plains of western Argentina.

Chlamys A genus of Molluscs also known by the generic name, *Aequipecten*, of the Scallop family, *Pectinidae*. e.g. C. opercularis (= *Aequipecten* o.) - The Queen Scallop (see **Aequipecten**).

Chlidonias A genus of Birds of the Tern family, *Laridae*. e.g. C. niger - The Black Tern, length 25 cm, the only tern with completely black breeding plumage, except for the white under the forked tail coverts. Builds floating nests on inland lakes. Inhabits in land waters; a migratory bird wintering in tropical Africa. Distribution: Europe, Asia, North America.

Chloëphaga A genus of Birds of the Duck, Goose, and Swan family, *Anatidae*. e.g. C. melanoptera - The Andean Goose, black and white in color; extremely aggressive in the nesting season. Habitat: the mountain ranges of South America.

Chloris A genus of Birds of the Bunting family, *Emberizidae*. e.g. C. chloris - The Greenfinch, length 14.5 cm; green in color except for its wings, which are a mixture of grey, black, and yellow. A partial migrant, it is found in gardens, shrubberies, and farmlands. The male makes a characteristic note like someone whistling with an indrawn breath. Distribution: Europe.

Chlorobacterium A genus of microorganisms of the family *Chlorobacteriaceae* (order *Pseudomonadales*) - Non-motile, rod-shaped cells growing symbiotically on the outside of protozoa. e.g. C. symbioticum.

Chlorobium A genus of microorganisms of the family *Chlorobacteriaceae* (order *Pseudomonadales*) - Spherical to rod-shaped cells.

Chloroceryle A genus of Birds of the Kingfisher family, *Alcedinidae*. Belongs to the subfamily of fishing Kingfishers, *Cerylinae*. e.g. C. amazona - The Crested Amazon Kingfisher, the largest of the New World genera of this family.

Chlorochromatium A genus of microorganisms of the family *Chlorobacteriaceae* (order *Pseudomonadales*) - Ovoid to rod-shaped cells. e.g. C. aggregatum.

Chlorohydra A genus of Hydrozoons of the freshwater Hydra family, *Hydridae*. e.g. C. viridissima - The Green Hydra, the best-known species of this genius, found in Britain and the United States. It is capable of controlling the amount of water entering the body due to the differences in the ionic concentration of cells and freshwater.

Chlorophonia A genus of Birds of the Tanager subfamily, *Thraupinae*. e.g. C. occipitalis - The Blue-crowned Chlorophonia, a vividly colored species, found in the New World.

Chloropsis A genus of Birds of the Leafbird family, *Irenidae*. e.g. C. aurifrons - The Golden-fronted Leafbird, a very colorful bird, green above, greenish yellow below, with a golden-red forehead, blue throat, and black patches on head and breast. A fruit-eater living in the tree-tops in forests from southern Asia to the Philippines.

Chlorostigma A genus of plants. e.g. C. stuckertianum - A plant from the Argentine, used as a galactogogue.

Choanotaenia A genus of Tapeworms. e.g. C. infundibulum - Found as a parasite in chickens and turkeys.

Choeropsis A genus of Mammals of the Hippopotamus family, *Hippopotamidae*. e.g. C. liberiensis - The Pigmy Hippopotamus, the only species in this genus, living largely on dry land, by the rivers of West Africa, in Sierra Leone, Liberia, Guinea and Nigeria, and feeding on shoots and other forest vegetation. Only one tenth the size of the Common Hippopotamus, *Hippopotamus* amphibius, it is up to 1.5 m (5 ft) long, 75 cm (21/2 ft) high, and weighs no more than 1/4 ton (550 lb). It is more terrestrial than the latter, and lives singly or in pairs.

Choeropus A genus of Mammals of the Bandicoot family, *Peramelidae*. These are the pig-footed bandicoots belonging to the pouch-bearing marsupial order, *Marsupialia*. They have a jumping gait which, though less developed, recalls that of a kangaroo (see **Isodon**).

Choiromyces A genus of Mycelia or Fungi of the Truffle family. e.g. C. meandriformis - A rare truffle producing large, yellowish-brown sporophores on the soil surface in woods.

Choisia A genus of Shrubs of the Rue family, *Rutaceae*. e.g. C. ternata - The Mexican Orange Blossom, an evergreen shrub up to 3 m (10 ft) high, with terminal clusters of scented white flowers.

Choleopus A genus of Mammals of the Sloth family, *Bradypodidae*. e.g. C. didactylus - The Two-toed Sloth or Unau of Brazil, a smaller form than the Ai or Three-toed Sloth (see **Bradypus**); it moves less slowly, is more responsive, more sociable and more easily kept incaptivity than the three-toed Ai.

Chologaster A genus of Fishes of the family *Amblyopsoidae*. e.g. C. cornutus - The Swampfish, a small cave-dwelling fish, with its anal opening on the throat, and unlike many members of this family, not blind, but with normal eyes.

Chondrodendron A genus of Shrubs of the family *Menispermaceae*. e.g. C. tomentosum - A species of shrubs producing tubocurarine chloride; used chiefly as an adjuvant to anesthesia to obtain greater muscular relaxation in surgical operations and in orthopedic manipulations.

Chondrilla A genus of plants of the Daisy family, *Compositae*. e.g. C. chondrilloides - Alpine Gum-succory, a perennial plant, 15-30 cm high; rosette-leaves with cartilaginous teeth, glabrous scapes, and heads in an umbellate panicle. It grows on gravel and sand of streams in the Alps, up to 1,500 m, the Apennines, and Corsica.

Chondromyces A genus of Saprophytic

microorganisms of the family of Schizomycetes, *Polyangiaceae*; found in soil and decaying organic matter.

Chondrosteus A genus of prehistoric, now extinct fishes of the Jurassic periods, wherein existed a great reduction of bone in the skeleton, as seen in present-day fossils.

Chondrostoma A genus of freshwater Fishes of the family *Cyprinidae*. e.g. C. nasus - The Nase, length 40 cm; weight 1 kg; has a vertical mouth; the lips are covered with horny skin and have sharp edges. The species is commercially valueless. Distribution: Rivers that flow into the Baltic Sea from the south and into the Black Sea from the north and west (see **Carassius**).

Chondrus A genus of Red Algae, plants of the family *Gigartinaceae*. e.g. C. cripsus - Carrageen, Carragheen, Dried Seaweed, or Irish Moss; fan-shaped, tough, leathery frond-clusters, 15 cm high, and violet-red to dark purple-brown in color. Used as an emulsifying, suspending, and gelling agent; also used in the production of agar for use in constipation. Place of origin: the North Sea, particularly near Heligoland; at times also in the Baltic (see **Gigartina**).

Chonopeltis A genus of Branchiuran Crustaceans of the Fish lice subclass, *Branchiura*. These are parasites attaching themselves to the fish host with the aid of an exceptionally well-developed pair of maxillules with hooked spines.

Chorda A genus of plants of the family of Brown Algae. e.g. C. filum - Sea Lace, a species of brown algae up to 300 cm high, with cord-like fronds up to 5 mm thick.

Chordeiles A genus of Birds of the Nightjar family, *Caprimulgidae* (subfamily of Nighthawks, *Chordeilinae*). e.g. C. minor - The Common Nighthawk, found in North America.

Chordodes A genus of Hairworms of the family *Chordodidae*. The adults are freeliving and the larvae are parasitic to insects. Human infection is accidental.

Chorictis A genus of Birds of the Bustard family, *Otididae*. e.g. C. australis - The Australian Bustard, up to 32 lb in weight, with a wingspan of 7 ft, and probably the heaviest flying bird in existence today. It looks ostrich-like, and though the body is very heavy and it flies clumsily, it is able to run with great speed, hence bustards are also known as running birds.

Chorisochismus A genus of Fishes of the order of Cornish Suckers, Clingfishes, and Skittlefishes, *Xenopterygii* (= *Gobiesociformes*). e.g. C. dentex - The Rock Sucker fish from South Africa; about 30 cm (1 ft) long; a small fish with a terminal mouth and strong teeth. The eyes are on top of the head; the body is scaleless, and there is no air-bladder. The vertical sucker is formed partly by the muscles, with some parts of the internal skeleton.

Choristoneura A genus of Insects of the Bell Moth family, *Tortricidae*. e.g. C. fumiferana - The Spruce Budworm Moth; their caterpillars have caused considerable damage to Canadian conifer forests. The adult moths are known as bell moths from the outline of their folded wings.

Chorthippus A genus of Orthopterous Insects of the Short-horned Grasshopper and Locust family, *Acrididae*. The species are common in Europe.

Chortoicetes A genus of Orthopterous Insects of the Short-horned Grasshopper and Locust family, *Acridadae*. These species include the plague locusts found in the dry interior regions of Australia, where they periodically devastate the crops and natural vegetation.

Chromobacterium A genus of microorganisms of the family *Rhizobiaceae*. These are small, pleomorphic, gram-negative, flagellated rods forming blue, violet, or yellow pigments. They are most often saprophytic and are found in soil and water. Some species are also given the generic name, *Bacterium*. e.g. C. typhiflavum (= *Bacterium* t.) - Forms yellow colonies. Has been recovered, especially from sputum, but is only rarely pathogenic. C. violaceum - Produces violet pigments or colonies; has been associated with septicemia and fatal granulomatosis.

Chromulina A genus of Protozoans of the Chrysomonad order, *Chrysomonadina*; a solitary species, it has two stages, a flagellate stage with one flagellum and one dominant chloroplast, and an ameboid stage in which the flagellum is absent, and locomotion is by means of a single pseudopodium (a blunt-edged lobopodium).

Chrosomus A genus of Fishes of the family *Cyprinidae*. These are colored fishes inhabiting North America (see **Campostoma**; **Carrassius**).

Chrysanthemum A genus of late blooming plants of the Daisy family, *Compositae*; formerly also called by the genus name, *Tanacetum*. e.g. C. x hortorum - The Chrysanthemum, a well-known, woody-stemmed perennial plant, much grown for the cut flower trade and for autumn bedding. Single or double, round, daisy-like flowers, in a wide range of colors. Place of origin: China and Japan. Cultivated varieties reached Europe in 1789, having been grown in China since 500 B.C. C. vulgare (= *Tanacetum* v.) - Buttons; Common Tansy; Tansy; anaromatic, perennial herb, 60-90 cm (2-3 ft) high, with rounded, bright golden yellow flowers growing in compound fat-topped clusters. The leaves and stems are poisonous, but when the flower heads are steeped in liquid, the flavored fluid can be used for cheeses, cakes, and puddings. The flowering heads and dried leaves are used to make infusions and a liquid extract as a home treatment for intestinal worms, and as an insecticide and vermicide for flies and lice. Place of origin: Europe, including Britain. C. cinerariaefolium - The dried flower-heads, known as Pyrethrum Flower (= Pyrethri. Flos.; Insect Flowers; Dalmation Insect Flowers; Chyrantheme Insecticide; Insektenblute; Piretro), contain not less than 1% of pyrethrins and cinerin esters, of which not less than one-half consists of pyrethrin I. Used widely as a domestic and agricultural insecticide.

Chrysaora A genus of True Jellyfishes of the family *Pelagiidae*; also known by the generic name, *Chrysoara*. e.g. C. hyoscella (= *Chrysoara* h.) - The Compass Jellyfish, body about 30 cm (1 ft) in diameter; the margin of the fairly flat, shallowly arched, umbrella-like disc bears 32 lobes and 24 marginal arms, and the sexual organs are white, yellow, or red. The yellowish-white disc is patterned with 16 brownish, forked, radiating arms or stripes. It feeds on fish, other medusae, arrow-worms, etc. Distribution: in open water in the Mediterranean Sea, the Atlantic coast of Europe, and the North Sea.

Chrysemys A genus of Reptiles of the Freshwater Tortoise or Turtle family, *Emydidae*. e.g. C. picta - The Painted Terrapin or Painted Turtle, barely 15 cm (6 in) long; one of the most widespread turtles in the United States; greenish-brown color with prominent yellow bands and plastron. The marginal plates around the edge of the carapace are usually scattered with red.

Chrysis A genus of Insects of the Cuckoo Wasp family, *Chrysididae*. e.g. C. ignita - The Ruby-tailed Wasp, length 7-10 mm, beautiful metallic color; coloration variable, the front part of the body being blue-green to blue, abdomen gold, reddish-gold or purple. Larvae are parasitic on the larvae of various wasps and bees (families *Apidae*, *Eumenidae*, *Sphecidae*, and *Vespidae*). Distribution: much of the Palaearctic region.

Chrysoara A genus of True Jellyfishes of the family *Pelagiidae* (see **Chrysaora**).

Chrysochloris A genus of Mammals of the Golden Mole family, *Chrysochloridae*. e.g. C. asiatica - The Cape Golden Mole of South and Central Africa; a burrowing animal; its fur has a coppery sheen, and one of its nails is greatly developed, making its hand pick-like for digging.

Chrysococcyx A genus of Birds of the Cuckoo family, *Cuculidae*, found in Africa. Like the other cuckoos, the species are arboreal and insectivorous.

Chrysocyon A genus of Mammals of the Wolf family, *Canidae*. e.g. C. brachyurus - The Maned Wolf, the largest of the South American wild dogs; found on the pampas of Argentina, southern Brazil, and Paraguay, and resembling a giant fox. It is 51 in long, with a 16 in tail, and has a mane on the nape of the neck and down the back. Large enough to kill sheep, it hunts them alone, not in packs.

Chrysolophus A genus of Birds of the Cock Pheasant family, *Phasianidae*. e.g. C. amherstiae - Lady Amherst's Pheasant, a very handsome bird, introduced into England by Lady Amherst. The crest of the cock is red and the collaret is silvery. The back is green and the underparts white, while the wings and tail are black and white. Place of origin: Tibet, China, and North Burma.

Chrysomela A genus of Insects of the Leaf Beetle family, *Chrysomelidae*. e.g. C. aenea - Length 6.5-8.5 mm; coloring is variable, usually green, sometimes blue or golden-red. Beetles and larvae found on the leaves of alder. Distribution: Europe, Asia.

Chrysomitra A genus of Hydras, the small medusae released by the Hydra genus, *Velella* of the order *Athecata*. This was thought originally to be a separate genus. The medusae have no mouth and die after releasing gametes. They are now included under the generic name, *Velella* (see **Velella**).

Chrysomyia A genus of Insects of the Fly family, *Calliphoridae*. e.g. C. bezziana - An Asian fly frequently found in wounds of man and animals.

Chrysopa A genus of Insects of the family *Chrysopidae* (order of Lacewings and Ant Lions, *Planipennia*). e.g. C. perla (= C. vulgaris) - The Green Lacewing, length 10 mm; wingspan 25-30 mm; colored blue-green; very common in forests; adults and larvae are predaceous, feeding on aphids. Distribution: Europe.

Chrysopelea A genus of Reptiles of the poisonous Colubrid Snake subfamily, *Boiginae* (family *Colubridae*). e.g. C. ornata - The Indo-Malayan Golden or "Flying" Tree Snake, a tree-dwelling colubrid snake, taking off from the top branches of trees and gliding to a landing at an angle of about 45 degrees. Habitat: the Malayan and Indo-Australian forests.

Chrysophanus A genus of Butterflies and Moths of the Blues family, *Lycaenidae*. e.g. C. hippothoe - Size 3 cm; male reddish-brown; female blackish-brown and reddish-yellow.

Chrysophris A genus of Fishes of the family *Cyprinidae*. e.g. C. aurata - The Gilt-head Bream, length 50-60 cm; yellow spot on the forehead, and a red spot on the gill-covers. A sea fish species native to the Mediterranean. Much prized as a commercial fish (see **Carassius**).

Chrysopogon A genus of plants of the Grass family, *Gramineae*. e.g. C. gryllus - Brush-grass, 50-100 cm high; spikelets with golden red hairs at the base. Place of origin: the Mediterranean region.

Chrysops A genus of Insects of the Deerfly or Horsefly family, *Tabanidae*. e.g. C. caecutiens - The Thunder Fly or Gold-eye, a tabanid fly, size 7.5-11 mm; wing with a dark transverse band; common in summer in damp locations. A blood-sucking species, the female sucking the blood of man, cattle, and horses.

Distribution: Europe, Siberia. Native to Britain. C. silacea - A West African species; spotted or branched wings; an intermediate host, transmitting the parasitic filarial worm, Loa loa to man. The females are blood-suckers of man, cattle and horses.

Chrysospalax A genus of Mammals of the Golden Mole family, *Chrysochloridae*. e.g. C. trevelyani - The Giant Golden Mole of South Africa, a burrowing animal nearly 22.5 cm (9 in) long (see **Chrysochloris**).

Chrysosplenium A genus of plants of the Saxifrage family. e.g. C. oppositifolium - The Opposite-leaved Golden Saxifrage, a perennial plant, 5-20 cm high, with opposite leaves; basal leaves are truncate or wedge-shaped at base. Grows in woods. Place of origin: Western Europe and parts of Central Europe.

Chrysozona A genus of Insects of the Horsefly family, *Tabanidae*, formerly known by the genus name, *Hematopota*. e.g. C. italica (= *Hematopota* i.) - A tabanid species common in Europe.

Chthonerpeton A genus of Caecilian Amphibians of the family *Caeciliidae*. e.g. C. indistinctum - A species said to be capable of using its cloaca as a sucker, to enable the animal to remain in one place against the pull of a strong current of water.

Chthonius A genus of Arachnids of the False Scorpion family, *Chthoniidae*. e.g. C. ischnocheles - Length 1.6-2.5 mm. It has claw-like pedipalps containing the poisonous glands, and large chelicerae which it uses to cut up its food. Unlike the true scorpions there is no "tail" or sting. Found under stones and litter. Distribution: Europe (see **Chelifer**).

Chunga A genus of Birds of the Cariama or Seriema family, *Cariamidae*. e.g. C. burmeisteri - Burmeister's Seriema, an omnivorous bird found in sparse bushy forests, restricted to northern Argentina and parts of Paraguay.

Cicada A genus of Homopterous Insects, related to the plant bugs, of the Cicada family, *Cicadidae*. e.g. C. plebeja - Up to 5 cm (2 in) long; pronotum and elytra have a yellow rear edge. The males produce a chirping noise like grasshoppers by vibrating a drum-like organ called the tymbal. The females, unlike most insects, can hear this. Cicadas live in warm regions.

Cicadella A genus of Homopterous Insects of the Leaf-hopper family, *Cicadellidae* (= *Jassidae*). e.g. C. viridis - Length 5-9 mm; color green; found in damp locations and meadows. Distribution: the Palaearctic and Nearctic regions.

Cicadetta A genus of Homopterous Insects of the Cicada family, *Cicadidae*. e.g. C. tibialis - The Blackhorn Cicada, about 2.5 cm (1 in) long; found on vegetation in sunny places. The most northerly species of this genus, found in Central Europe (see **Cicada**).

Cicendia A genus of plants of the Gentian family, *Gentianaceae*. e.g. C. filiformis - The Slender Cicadia, an annual plant, 1-12 cm high; simple or branched stem; golden yellow, long-stalked, tetramerous flowers; small leaves. Grows on moors. Place of origin: Europe; native in Britain.

Cicerbita A genus of plants of the Daisy family, *Compositae*. e.g. C. alpina - Blue Sow-thistle, a perennial plant, 50-200 cm high; with stiff reddish, glandular hairs; found in damp woodlands and sometimes in moist alpine rocks up to 2,000 m. Place of origin: most of Europe.

Cichlasoma A genus of Freshwater Fishes of the family *Cichlidae*. e.g. C. biocellatum - The Jack Dempsey, a perciform fish, inhabiting sluggish streams and brackish waters from the Amazon River basin and the Rio Negro. It is up to 20 cm (8 in) long, brownish with shiny dots on almost every scale and on parts of the fins. The dorsal fin is bordered with black.

Cichorium A genus of plants of the Daisy family, *Compositae*. e.g. C. intybus - The Chicory plant, a perennial with a long, thick, tap root, Cherry Root, containing the polysaccharide, Alant Starch or Inulin. The erect, grooved stems, up to 1.2 m (4 ft) high, are sometimes hairy and bear numerous bright blue, ray-like, ligulate florets, surrounded by two rows of green bracts. Place of origin: England and Wales, Europe, and western Asia; it has become naturalized in most temperate regions, including the United States (see **Inula**).

Cicindela A genus of Insects of the Tiger Beetle family, *Cicindelidae*. e.g. C. campestris - The Green Tiger Beetle, length 12-15 mm; dull grass-green elytra with usually five light spots; found in forest and sandy localities. A carnivorous species, feeding on other insects and larvae. Distribution: Europe; North Africa.

Ciconia A genus of Birds of the Jabiru and Stork family, *Ciconiidae*. e.g. C. ciconia - The White Stork, length up to 102 cm, and a wingspan of almost 210 cm; a migratory bird, inhabiting open country with bodies of water. It has a strong body, a long, straight, conical bill with cutting edges, white plumage with black wing coverts and flight feathers, and may weigh up to 4 kg or more. Distribution: Eurasia, North Africa.

Cicuta A genus of plants of the Carrot family, *Umbelliferae*. e.g. C. virosa - The Cowbane or Water-hemlock, a very poisonous perennial plant, 1 m (39 in) high, widespread in still water (lakes, ditches and ponds), in canals, fens, marshes, and by brooks and springs. Two to three-pinnate leaves and a tuberous root-stock with transverse cavities. Native in Britain.

Ciliata A genus of Fishes of the Cod family, *Gadidae*. e.g. C. mustela - The Five-bearded Rockling, length 25 cm; one barbell on lower jaw, four on the upper. Inhabits the Atlantic coasts.

Cimbex A genus of Insects of the Sawfly family, *Tenthredinidae*; a leaf-eating relative of bees and wasps. e.g. C. femorata - The Birch Sawfly; size 2.5 cm (1 in); knobbed antennae; head widens behind the eyes. The grubs feed on birch trees.

Cimex A genus of Hemipterous Insects of the Blood-sucking and Bed-bug family, *Cimicidae*. Also variously known by the generic names, *Acanthia*, *Clinocoris*, *Clinophilus*, or *Klinophilus*. e.g. C. lectularius (= *Acanthia* lectularia; *Clinocoris* lectularius, etc.) - The Common Bedbug, length 3.5-8 mm; well-known annoying species that sucks the blood of humans, an itching red spot appearing round the point of puncture. Found primarily in unclean, poorly kept dwellings, hotels, and army barracks. Female successively deposits some 100-200 eggs. Distribution: worldwide.

Cimicifuga A genus of plants of the Buttercup family, *Ranunculaceae*. e.g. C. racemosa - Black Snakeroot, Black Cohosh, or Bugbane, a perennial plant, containing the resin, cimicifugin. Place of origin: East and Central Europe.

Cinchona A genus of Tropical and Subtropical Trees and Shrubs of the Bedstraw or Madder family, *Rubiaceae*. e.g. C. succirubra - The Quinine plant or Cinchona Bark, from which are extracted the cinchona alkaloids, cinchonine, cusconidine, cusconine, quinine, etc.

Cinclus A genus of Birds of the Dipper or Water-ouzel family, *Cinclidae*. e.g. C. cinclus - The European Dipper, length 18 cm; dark brown plumage; throat and breast snow-white; short tail; found along streams and rivers in foothills and mountains. Hunts under water for aquatic insects, crustaceans and molluscs, small fishes, salmon eggs, and other food, running on the stream bed using its wings to help move against the current. A resident and dispersive bird.

Distribution: almost all of Europe, also the Middle East, central Asia and Africa.

Cinnamomum A genus of Trees or Shrubs of the Laurel family, *Lauraceae*. e.g. C. camphora - The Camphor tree or Camphor wood. The bark and foliage yield aromatic oils from which camphor is obtained. Place of origin: southeast Asia. C. zeylanicum - The Cinnamon tree; from the bark is obtained Cinnamon.

Ciona A genus of primitive Chordates of the class *Ascidiacea*. e.g. C. intestinalis - The Sea-squirt or Sea Vase, a sedentary chordate animal, size up to 12 cm; cylindrical, yellowish-white body; found especially on pier-supports and on buoys. It has a test of tunicin which is closely related to cellulose. Distribution: North Atlantic, and Baltic Sea.

Cionus A genus of Insects of the Weevil Beetle family, *Curculionidae*. e.g. C. scrophulariae - Length 4-5 mm; locally abundant on figwort in summer. Distribution: Europe, Asia Minor, the Middle East.

Circaea A genus of plants of the Willow-herb family. e.g. C. lutetiana - The Common Enchanter's Nightshade, a perennial plant, 30-70 cm high; half-heart-shaped leaves; found in deciduous woods. Place of origin: most of Europe; native in Britain.

Circaëtus A genus of Birds of the Eagle subfamily, *Accipitrinae* (family *Accipitridae*), also known as the Serpent Eagles because they hunt and devour snakes and other reptiles. e.g. C. gallicus - The Short-toed Eagle, size 63-65 cm; underwings and breast almost pure white; a migratory bird arriving in France in May and leaving in September. Found in southern Europe, North Africa, and parts of Asia; they favor forest clearings and ponds, and stalk all kinds of rodents, reptiles, amphibians and insects.

Circulifer A genus of Homopterous Insects of the Leaf-hopper family, *Cicadellidae*. e.g. C. tenellus - A leaf-hopper pest of sugar beet in the United States, transmitting virus diseases to the host plants, thus injuring them seriously.

Circus A genus of Birds of the Eagle subfamily, *Accipitrinae* (family *Accipitridae*). e.g. C. cyaneus - The Hen Harrier or Marsh Hawk, length 47 cm; with a conspicuous white spot on the rump; found in swamp and marshland, peatmoors, as well as large damp meadows. A partially migrant bird, nesting on the ground, and feeding mostly on small rodents. Distribution: Europe, North Africa, Asia, and North America.

Cirratulus A genus of Segmented Worms of the family *Cirratulidae*. e.g. C. cirratus - This cirratulid species is found in rather foul mud, full of rotting vegetation or amongst the roots of eelgrass; it has remarkable powers of regeneration, for at times the body breaks up spontaneously into fragments, each becoming a new individual. It is found on both sides of the Atlantic.

Cirsium A genus of flowering plants of the Daisy family, *Compositae*. e.g. C. arvense - The Creeping or Canada Thistle, a perennial plant with long white roots, giving rise to annual erect stems from 30-90 cm (1-3 ft) high. There are rose-purple capitula and the fruits have a radiating tuft of feathery white hairs. It grows by the wayside, on waste land, dry meadows, pastures and in gardens. Distribution: Britain, Europe, and Asia; has also been introduced into North America.

Cistus A genus of Mediterranean plants of the Rockrose family, *Cistaceae*. e.g. C. incanus (= C. villosus) - The Pink Rockrose, a densly-branched evergreen shrub, 30-150 cm (1-5 ft) high, with large, 5 cm (2 in), pink to rosy-purple flowers with numerous bright stamens, growing in prolific terminal clusters. Source of Labdanum, a bitter gum used in perfumery and medicine. Place of origin: Europe; Mediterranean region; introduced into Britain in 1650.

Citellus A genus of Rodents of the Tree Squirrel family, *Sciuridae*. These are the Ground Squirrels. e.g. C. beecheyi - The ground squirrel of California, this species being one of the natural reservoirs of tularemia and plague (see **Lignognathoides**). C. citellus - The Souslik ground squirrel of central Europe, or the European Souslik; length of body 195-240 mm, tail 60-70 mm; the largest species of this genus of ground squirrels. Found in cultivated steppes, meadows and clearings. Lives in colonies, riddling the ground with deep burrows, and hibernating in winter. Distribution: Europe, Asia, and North America.

Citrobacter A genus of microorganisms of the family *Enterobacteriaceae*. Gram-negative rods, including bacteria formerly of the Bethesda-Ballerup group; closely resemble Salmonella and Arizona organisms. e.g. C. freundii - Type species; has been associated with enteritis, septicemia, urinary tract infections, etc.

Citrullus A genus of plants of the Cucumber family, *Cucurbitaceae*. e.g. C. colocynthis - The Colocynth or Bitter Apple plant; from the dried pulp of the fruit, is prepared Colocynth (= Bitter Apple; Colocynth Pulp; Colocynthis; Coloquinte; Coloquintidas; Koloquinthen) - This has a drastic purgative action, and hence is seldom used alone, but is usually combined with hyoscyamus to counteract the griping.

Citrus A genus of Trees and Shrubs of the Rue family, *Rutaceae*. Some species are sometimes classified under other generic names such as, *Aegle* or *Poncirus*. e.g. C. acida - The Lime tree. C. amara - The Seville Orange tree. C. aurantium - The Orange tree. C. bergamia (= C. bergama) - The Bergamot or Bergamotte tree. C. limon

(= C. limonia; C. limonum) - The Lemon tree. C. medica - The Citron tree; considered to be the parent of the Lime and Lemon trees. C. nobilis - The Mandarin Orange tree. C. paradisi - The Grapefruit tree. C. sinensis - The Sweet Orange tree. C. trifoliata (= *Aegle* sepiaria; *Poncirus* t.) - The Hardy Orange or Japanese Bitter Orange, a deciduous shrub with small, round fruits like small greenish yellow oranges. Sometimes used for drinks. Place of origin: North China; introduced into Europe in 1850 (see **Aegle**; **Poncirus**).

Civettictis A genus of Mammals of the Civet, Genet, and Mongoose family, *Viverridae*. e.g. C. civetta - The African Civet; it has a grey coat with black markings, an erectile mane and a short tail; overall length 120 cm (4 ft). Lives in bushy regions. Civets have often been kept in captivity to obtain their scent, a musk-like perfume, which is expressed from the glands about twice a week with a small spatula.

Cladium A genus of plants of the Sedge family, *Cyperaceae*. e.g. C. mariscus - The Prickly Cladium, a stout perennial plant, 0.60-3 m (2-10 ft) high, with long sharp leaves, usually bisexual flowers, and dark brown, ovoid fruit. Widespread, growing on riverbanks and in meadows. Place of origin: the British Isles.

Cladognathus A genus of Lamelliform Beetles of the suborder *Polyphaga*. e.g. C. giraffa - A large, striking, tropical species found in India and Java, having mandibles of fearsome dimensions.

Cladonia A genus of Lichens or Lichen plants. e.g. C. rangiferina - The Reindeer Lichen or Reindeer Moss, a greyish-mauve perennial lichen plant, occurring in the wetter parts of heaths and moors. Formerly used as a stomachic and expectorant. Native in Britain. C. impexa - Also known (as above) by the name of Reindeer Lichen, a plant 10 cm high, with branched bush-shaped podetia in greyish-green tufts. Native in Britain.

Cladorchis A genus of Trematode Worms, also known by the generic names, *Amphistoma* and *Watsonius*. e.g. C. watsoni (= *Amphistoma* w.; *Watsonius* w.) - A pear-shaped parasitic trematode worm causing diarrhea in Africa.

Cladorhynchus A genus of Birds of the Stilt and Avocet family, *Recurvirostridae*. e.g. C. leucocephala - The Banded Stilt, an Australian mono-typic species, especially at home in sea-water or freshwater. In flight it looks like a miniature stork.

Cladosporium A genus of Fungi. e.g. C. herbarum - Causes "black spot" on meat in cold storage.

Clamator A genus of Birds of the Cuckoo family, *Cuculidae*. e.g. C. glan-darius - The Great Spotted Cuckoo, size 40 cm; grey crest and long tail, edged with white; a migrant bird, parasitic on members of the Crow family, especially on Magpies. Habitat: Spain and Portugal.

Clangula A genus of Birds of the Duck family, *Anatidae*. e.g. C. hyemalis - The Long-tailed Sea Duck; size, male 53 cm, female 40 cm; body checkered, with pointed tail in the male, and face white in the female. A migrant bird, nesting in the tundra, by lakes or on moors in the north.

Clatochloris A genus of microorganisms of the family *Chlorobacteriaceae* - spherical cells in chains and trellis-like aggregates. e.g. C. sulphurica.

Clathrina A genus of Ascon Sponges of the suborder *Homocoela*. These are cal-careous sponges where the whole of the cloacal cavity is lined with collared cells (see also **Leucosolenia**).

Clathrocystis A genus of microorgan-isms, now known by the generic name, *Lamprocystis*, of the family *Thiorhodaceae* (order *Pseudomonadales*) - Spherical toovoid

cells in a common capsule. e.g. C. roseopersicina (= *Lamprocystis* r.).

Clathrulina A genus of Protozoans of the Helizoan order, *Helizoa*. These are freshwater organisms with silaceous skeletons, found in peat and other acid pools. Reproduction is by repeated bina-ry fission.

Clathrus A genus of Gastropod Molluscs, also known by the generic names, *Epitonium* or *Scalaria*, of the family *Epitoniidae*. e.g. C. clathrus (= *Epitonium* clathrum; *Scalaria* clathra) - The Common Wentletrap; brownish-yel-low shell about 30 mm high, 12 mm wide, composed of 12-15 whorls, and decorated with very tall, striking ridges crossing each whorl at right angles. The oval mouth is closed by an operculum. Found on muddy and sandy sea bottoms at depths of as much as 100 m. Distribution: coasts of the Atlantic Ocean, Mediterranean, North and Baltic Seas.

Clausilia A genus of Gastropod Molluscs of the Doorshell family, *Clausiliidae*. e.g. C. dubia - The Door-shell, 12-13 mm high, 2.8-3 mm wide; elongate, spindle-shaped, sinistral (left-spiralling) shell, found in moist locations, ancient ruins, old tree stumps, hedgerows and meadows. Distribution: western Europe.

Clava A genus of Hydroids or Sea firs of the family *Clavidae*. e.g. C. squamata - This animal produces pink colonies on seaweeds, particularly the wracks, such as the knotted wrack, *Ascophyllum* nodosum.

Clavaria A genus of Fungi, belonging to the group *Basidiomycetes*, of the Fairy Club Fungi family. e.g. C. flava - The Crusted Fairy Club; fruit-body 10-12 cm long; branches firm, yellow, later yel-lowish-brown at the tips; palatable when young.

Clavelina A genus of Sea-squirts of the class *Ascediacea*; this chordate develops a tadpole-like larva with a well-devel-oped notochord in the muscular tail, which is lost at metamorphosis. The adults develop a common stolon or root stalk upon which new individuals are budded. e.g. C. lepadiformis - Up to 3 cm; usually orange in color. Distribution: Baltic and Atlantic coasts.

Claviceps A genus of parasitic Fungi of the family *Hypocreaceae*, which infest the seeds of various plants. e.g. C. purpurea - The ascospores of the common ergot, infecting the flowers of grasses, particularly the rye; the fruit is later replaced by a hardened mass of mycelium, the sclerotium. This is the source of the common ergot, the fungus developing in the ovary of the rye, *Secalecereale (Gramineae)*. From the sclerotium are extracted the poisonous ergotoxine alkaloids, which can cause serious illness if they get into the flour. Used in the treatment of post-partum hemorrhage.

Clelia A genus of Colubrid Snakes (family *Colubridae*), of the snake-eating subfamily, *Boiginae*. e.g. C. clelia - This species lives in Mexico, Central America and northern South America, and appears to be completely immune to the venom of the snakes that form its food.

Clematis A genus of climbing plants of the Buttercup family, *Ranunculaceae*. e.g. C. flammula - Virgin's Bower, a strong deciduous climbing plant, 3-6 m (10-20 ft) long, forming a dense tangle of cinnamon colored stems clothed in green bipinnate leaves, with white, 4-sepalled, small almond scented flowers, in loose panicles all over the plant. Place of origin: southern Europe; cultivated in England since the late 16th century.

Clemmys A genus of Reptiles of the Freshwater Tortoise or Turtle family, *Emydidae*. e.g. C. leprosa - The Spanish Terrapin, very common in ponds and rivers in southwest Europe, the Iberian peninsula, and northwest Africa. This turtle frequently suffers from a skin dis-ease inside the carapace caused by freshwater algae.

Cleome A genus of flowering plants of the family *Capparidaceae*. e.g. C. spinosa - The Spider Flower, a strongly scented annual plant, sticky to the touch, 0.60-1.2 m (2-4 ft) high, with white, flesh, rose-pink to purple flowers in large and showy racemes. These have long, 5-7.5 cm (2-3 in) extending stamens. Place of origin: Tropical America; introduced into Britain in 1817.

Clerodendrum A genus of Shrubs of the Verbena family, *Verbenaceae*. e.g. C. bungei (= C. foetidum) - Clerodendrum, a medium sized, deciduous shrub, 2-3 m (6-8 ft) high, with fragrant, rose-pink, star-shaped flowers in flat terminal heads, 10-13 cm across. Place of origin: China; introduced into Europe in 1844.

Clethrionomys A genus of Rodents of the Vole subfamily, *Microtinae* (Mouse family, *Cricetidae*). e.g. C. glareolus - The Bank Vole, length of body 85-110 mm; tail 35-55 mm; upper side of body red-brown; inhabitant of woodlands and dense vegetation. Makes two types of holes, one serving as a hiding place, the other for breeding. Differs from true voles in having rooted molars. A very common rodent in Europe.

Clevelandella A genus of Spirotrich Protozoans of the order of Heterotrichs, *Heterotrichidae*; the species are invertebrate gut symbionts, with the somatic cilia more or less well represented and of uniform size. The oral cilia are several times larger than the somatic cilia and may be fused into cirri.

Climacia A genus of Endopterygote Insects of the family *Sisyridae*; small, brownish insects whose larvae feed on freshwater sponges.

Climacteris A genus of Birds of the Australian Tree Creeper family, *Climacteridae*. e.g. C. erythrops - The Red-browed Tree Creeper; it has long legs and toes, with strong claws; the bill is long and decurved; there is a noticeable sexual dimorphism. Habitat: eastern Australia.

Clinocoris A genus of Insects of the family *Cimicidae*, better known by the generic name, *Cimex*. e.g. C. lectularius (= *Cimex* l.) - The Common Bedbug (see **Cimex**).

Clinophilus A genus of Insects of the family *Cimicidae*, better known by the generic name, *Cimex*. e.g. C. lectularius (= *Cimex* l.) - The Common Bedbug (see **Cimex**).

Clinopodium A genus of plants of the Thyme family, *Labiatae*. e.g. C. vulgare - The Wild Basil, a perennial plant, 25-80 cm high; leaves, stem and calyx woolly; flowers red or white, in whorls of 10-20; found in woods and hedges, sometimes also in grassland. Place of origin: Europe; native in Britain.

Cliona A genus of Calcareous Sponges of the suborder *Astrosclerophora*. In these animals the spicules are microscleres based on rays that emanate from a common center, and are known as euasters. e.g. C. celata - The Sulfur Sponge; round bore-holes 2-3 mm in diameter; a Boring Sponge which will bore into limestones or shells, especially oysters, which it often reduces to a crumbly texture.

Clivia A genus of plants of the family *Amaryllidaceae*. e.g. C. miniata - The Kaffir Lily or Natal Lily, a fine bulbous plant with evergreen strap-shaped foliage, and funnel-shaped orange-red flowers with yellow throats. Place of origin: South Africa.

Cloeon A genus of Exopterygote Insects of the order of Mayflies, *Ephemeroptera*, of the family *Baetidae*. The species swim actively in mid-winter with their hair-fringed cerci.

Clonorchis A genus of Digenetic Flukes or Trematode Worms of the order *Digenea*, formerly known under the generic names, *Distoma* or *Opisthorcis*. e.g. C. sinensis (= C. endemicus; Clon. sinensis; Clon. endimicus; *Distoma* s.;

D. japonicum; *Opisthorchis* s.) - The Human Liver Fluke, a species of Asian flukes, endemic to man in China and Japan. It encysts in many species of freshwater fishes, some of which are ingested and serve as vectors of parasitic disease.

Clonothrix A genus of microorganisms of the family *Crenotrichaceae*; cylindrical cells showing false branching. e.g. C. putealis.

Clossiana A genus of Insects of the Butterfly or Moth family, *Nymphalidae*. e.g. C. euphrosyne - The Pearl-bordered Fritillary Butterfly, size 21-25 mm; found on flowers in forest clearings, glades, and margins. Distribution: Europe, and temperate parts of Asia.

Clostridium A genus of microorganisms of the family *Bacillaceae* (= *Schizomycetaceae*); these are anaerobic, large gram-positive rods with spores; most are soil saprophytes, while some are a part of the normal intestinal flora. Many species decompose protein and/or form toxins. Symbol: C; Cl. More than 150 species exist:
(a) Species causing Gas Gangrene (in order of decreasing frequency): Cl. perfringens (= Cl. welchii) - Has 5 toxin types: A, B, C, D, & E. Type A is associated with Gas gangrene. Some strains produce a powerful enterotoxin that causes "food poisoning" and severe diarrhea. Presenting the genital tract of 5% of normal females; also part of the normal fecal flora; has been isolated from milk and soil. Cl. novyi (Cl. oedematiens). Cl. bifermentans (Cl. sordellii) - Cl. sordellii is the name given to more violent strains of Cl. bifermentans. Commonly isolated from normal feces and soil. Cl. tertium. Cl. tetani - see (c): species causing Tetanus. Cl. septicum (= *Vibrion* septique) - see **Vibrion**. Cl. putrificum; Cl. butyricum; Cl. cochlearium; Cl. fallax; Cl. capitovale; Cl. histolyticum; Cl. multifermentans; Cl.

sphenoides; Cl. hastiforme; Cl. regulare; Cl. tetanomorphum; Cl. paraputrificum.
(b) Species causing Botulism: Cl. botulinum (= C. botulinum) - includes types A through F, based on immunologic specificity of toxin. Types A, B, and E are most frequently associated with human disease and produce one of the most highly toxic substances known. Botulism occurs when toxin in foods is ingested; the toxin is heat labile. Worldwide distribution in soil and occasionally, in animal feces.
(c) Species causing Tetanus: Cl. tetani (C. tetani; Cl. tetani; *Bacillus* tetanus) - the cause of tetanus in man and domestic animals. The bacillus has a large terminal spore that produces a drumstick appearance, and several types exist. Worldwide distribution in soil and animal feces. Tetanus results from atoxemia, with toxin acting on nerve tissue, the spinal cord, and peripheral nerves, with resulting lockjaw, risus sardonicus (sardonic smile), etc. This species may also be one of the causes of Gas gangrene.

Clupea A genus of Fishes of the Herring family, *Clupeidae*. e.g. C. harengus harengus - The Atlantic Herring, length 25-40 cm; a slender pelagic fish, with pelvic fins behind leading edge of dorsal fins; conjugating in huge shoals that stay in deep water by day and ascend to the surface by night. One of the most important sea fishes of commerce. Distribution: The northern Atlantic, where the warm southern ocean currents mingle with the cold northern currents.

Clymenella A genus of Segmented Worms of the Bamboo Worm family, *Maldanidae*. e.g. C. torquata - Like the other malanids, these worms have long and narrow segments, rather like the joints of a bamboo cane. They lie buried vertically in the mud or sand within a thin gritty tube, and with the head downwards. Found commonly near Woods Hole, Massachusetts.

Clytia A genus of Thecate Hydroids of the family *Campanulariidae*; these species produce delicate branching colonies on red weeds, releasing free medusae.

Clytocybe A genus of Fungi or Molds. e.g. C. gigantea - A mold from which is extracted clytocybine, which has a bacteriolytic effect on *Mycobacterium* tuberculosis.

Clytus A genus of Insects of the Longhorned or Longicorn Beetle family, *Cerambycidae*. e.g. C. arietis - The Wasp Beetle; black with yellow bands; mimies wasps of the genus *Vespula*, resembling them not only in color but also in its jerky movements and quivering antennae. Distribution: common in Britain.

Cnemidocoptes A genus of Sarcoptid Mites of the order of Mites and Ticks, *Acarina*. e.g. C. gallinae - This insect species causes the depluming of fowls.

Cnicus A genus of plants of the Daisy family, *Compositae*; formerly known by the generic names, *Carbenia* or *Carduus*. e.g. C. benedictus (= *Carbenia* benedicta; *Carduus* benedictus) - The Blessed Thistle or Holy Thistle plant; it contains the volatile oil, cnicin, used as a diaphoretic and emmenagogue.

Cnidium A genus of plants of the Carrot family, *Umbelliferae*. e.g. C. dubium - The Cnidium, a biennial plant, 30-60 cm high; leaves 2-pinnate, with linear segments and numerous subulate bracteoles. Grows mainly in fens and damp meadows. Place of origin: Europe.

Cobaea A genus of late blooming plants of Phlox family, *Polemoniaceae*. e.g. C. scandens - The Cup and Saucer plant, a vigorous climbing perennial, about 6 in (20 ft) tall; with large, 7 cm (3 in) across, deep purple, cup-shaped flowers and green, prominent, saucer-like calyces. Place of origin: Mexico; introduced into Europe in 1787.

Cobitis A genus of Fishes of the Loach family, *Cobitidae*. e.g. C. taenia - The

Spotted Weatherfish or Spined Loach; length 6-10 cm; a slender fish with head and body laterally compressed. Has 6 bar-bells round the lower jaw and a movable hinged spine beneath each eye. Found in calm and slow-flowing fresh waters. Indigenous to European rivers. Also found in Asia and North Africa.

Coccidioides A genus of pathogenic yeast-like Fungi of the family *Endomycetales*, also known by other generic names, such as *Blastomyces* and *Mycoderma*. e.g. C. immitis (= *Blastomyces* coccidioides; *Mycoderma* immite) -causes Coccidioidomycosis (see **Blastomyces; Mycoderma**).

Coccidium A generic name formerly given to a group of microorganisms (order *Sporozoa*), resembling the malarial parasites both in schizogony as well as sporogony. e.g. C. hominis (= *Isospora* h.); causes a severe diarrhea (Coccidiosis) in man (see **Isospora**).

Coccinella A genus of Beetles of the Ladybird family, *Coccinellidae*. e.g. C. septempunctata - The Seven-spotted Ladybird, length 5.5-8 mm; elytra red, with a total of 7 black spots; one of the commonest and most beautiful of beetles; they feed on aphids or green flies and other plant lice, and are injurious agricultural and horticultural pests of north temperate countries. Distribution: Europe, Asia, North Africa. C. bipunctata (= *Adalia* b.) - The Two-spot Ladybird (see **Adalia**).

Coccobacillus A genus of microorganisms resembling filtrable viruses; better known by the generic name, *Asterococcus* (see **Asterococcus**).

Coccothraustes A genus of Birds of the Finch family, *Fringillidae*. e.g. C. coccothraustes - The Hawfinch, a relatively large bird, 17 cm long, with a large cone-shaped bill occupying most of its face. It has a rich brown color with metallic tints, and bold white shoulder patches. Fond of cracking open cherrystones with its powerful bill, and eating the kernels. Mostly a resident bird, nesting high up in trees. Distribution: Europe, Asia, northwest Africa.

Coccus A genus of Hemipterous Insects of the family *Coccidae*, also known by the generic name, *Dactylopius*; source of cochineal, kermes and lac. e.g. C. cacti (*Dactylopius* coccus) - From the dried female insect is extracted Cochineal (Coccinilla), used as a red coloring agent.

Coccytius A genus of Insects of the Hawk Moth family, *Sphingidae*, species of which are found in Central America; up to 25 cm (10 in) long; large and striking insects, capable of strong flight around disc, hovering over the flowers from which they feed with their long probosces.

Coccyzus A genus of Birds of the Cuckoo family, *Cuculidae*. e.g. C. erythrophthalmus - The Yellow-billed Cuckoo of North America; non-parasitic, building its own nest of twigs, and unlike the Common Cuckoo (*Cuculus*), not encroaching upon the nests of other birds.

Cochlearia A genus of plants of the Cress or Crucifer family, *Cruciferae*. e.g. C. armoracia (= *Armoracia* lapathifolia; A. rusticana) - The Horseradish Plant or Root contains the glyceride sinigrin, used as a condiment and stomachic (see **Armoracia**). C. officinalis - Common Scurvy-grass, a biennial to perennial plant, 10-25 cm high; basal leaves orbicular to kidney or heart-shaped at base; stem leaves coarsely toothed, clasping the stem. Found on sea-coasts. Native in Britain.

Cochlearius A genus of Birds of the Boatbilled Heron or Boatbill family, *Cochleariidae*, considered to be an aberrant member of the Heron family (*Ardeidae*). e.g. C. cochlearius - The Boatbill, the only species of this family, whose bill resembles that of the shoebill with its peculiar large broad "boat" or "shoe"-shape, except that the upper mandible is not hooked. It inhabits the swamps of Guyana and Brazil (see **Balaeniceps**).

Cochliomyia A genus of Flies of the family *Calliphoridae*; also known by the generic name, *Compsomyia*. e.g. C. hominivorax (= *Compsomyia* macellaria) - The American screw-worm fly.

Cochlodina A genus of Gastropod Molluscs of the family *Clausiliidae*. e.g. C. orthostoma - Shell 12-13 mm high, 3 mm wide, and marked with fine ribs. A common species found on moss-covered rocks and on the trunks of sycamore trees and beeches in broadleaved forests. Distribution: Europe.

Cocos A genus of Trees of the Palm family, *Palmae* (= *Arecaceae*). One species is now more correctly called by the generic name, *Microcoelum*, namely M. weddeliana (= *Cocos* w.). e.g. C. nucifera - The Coconut or Coconut Palm, a native of the Cocosand Keeling Islands, now widespread in the tropics. It is a slender palm up to 30 m (100 ft) high, with pinnate leaves 1.8-6 m (6-20 ft) long. The copious sap is used fresh or fermented to produce palmwine or toddy, the fruit yields Coconut Oil, and the dried white flesh of the seed is used to make copra, while the fibrous outer layer of the fruit is coir. C. weddeliana (*Microcoelum* w.; *Sygarus* w.) - The Weddel Palm (see **Microcoelum**).

Codiaeum A genus of indoor plants of the family *Euphorbiaceae*. e.g. C. variegatum - The Croton plant, an evergreen shrub, varying greatly in leaf-shape and coloring in red, green and yellow, orange ornear black colors. Height 2-3 m (6-10 ft); grows in tropical gardens; also a shorter variety used as a pot plant, 30-45 cm (12-18 in) high. Place of origin: Malayan archipelago and Pacific islands; introduced into Europe in the late 19th century.

Codium A genus of Green Algae, the species of which act as host plants to the common European Sea Slug, *Elysia* viridis of the same color (see **Elysia**).

Coelacanthus A genus of Deep-bodied Fishes of the order *Actinistia* (= *Coelacanthiformes*), known to have lived about 300 million years ago. In the Carboniferous period, this genius lived in freshwater. The species of this order had a three-lobed diphycercal tail and is now believed to be extinct.

Coelagyne A genus of plants of the Orchid family, *Orchidaceae*. e.g. C. cristata - A small epiphytic orchid species from the Himalayas, about 45 cm (18 in) high. It has smooth ovoid pseudo-bulbs and long leathery, bright green leaves. The beautiful, white fragrant flowers are borne in drooping spikes.

Coeloglossum A genus of plants of the Orchid family, *Orchidaceae*. e.g. C. viride - The Frog Orchid, a perennial plant, 10-30 cm high; 2-3 ovate basal leaves; lip 3.5-6 mm long, parallel-sided, hanging vertically, and 3-lobed near tip. Grows especially on calcareous soil, also in meadows and woods. Native in Britain.

Coeloplana A genus of Tentaculate Comb Jellies of the order *Platyctenea*, found originally in the Red Sea and later off Japan; a flattened oval animal, which although possessing tentacles, lacks comb plates and is ectocommensal on species of soft coral (order *Alcyonacea*).

Coelosomides A genus of Trichostome Protozoans of the order *Trichostomatida*; a free-living ciliated organism, this animal feeds upon algae and bacteria.

Coenagrion A genus of Insects of the order of Damsel-flies, *Odonata*; belonging to the family *Coenagriidae*. e.g. C. puella - The Common Coenagrion, size 3.5 cm; a freshwater fly; in the male the body is sky-blue with black joints, and has a U-shaped mark on the second segments; in the female the back is black, and the sides apple-green. Capable of folding the wings over the back when at rest. Found by the banks of still and slow-flowing water.

Coendou A genus of Mammals of the New World Porcupine family, *Erethizontidae*. e.g. C. prehensilis - The Prehensile-tailed Porcupine of South America, an arboreal species having a long prehensile tail with stiff bristles at its root, that assist by gripping the tree trunk when the animal is climbing.

Coenogonimus A former name, now *Heterophyes*, given to a genus of minute Trematode Worms. e.g. C. heterophyes (= *Heterophyes* h.) - see **Heterophyes**.

Coenonympha A genus of Butterflies of the Browns family, *Satyridae*. e.g. C. pamphilius - Small Heath, size 5 cm (2 in); wings orche with a dark margin; caterpillars on soft grasses; hibernates in the caterpillar stage.

Coenurus A genus name, also *Cenurus*, formerly given certain Tapeworm Larvae. e.g. C. cerebralis (= *Cenurus* c.) - The larva of *Multiceps* multiceps (see **Cenurus**).

Coffea A genus of plants of the Bedstraw or Madder family, *Rubiaceae*. e.g. C. arabica - The Coffee plant; the kernel of the dried ripe seed, Coffee Seed, contains about 1-2% of caffeine. It is employed in the form of an infusion or decoction as a stimulant and very powerful beverage. A decoction is used as a flavoring agent in some pharmaceutical preparations e.g. Caffeine Iodide Elixir, BPC.

Cola A genus of plants of the family *Sterculiaceae*. e.g. C. nitida - The Kola or Cola plant; the dried cotyledons, Cola Seeds (= Kola Nuts; Embryo Colae) contain about 1.5-2.5% of caffeine and traces of theobromine. Used as a beverage drink, a stimulant, and a flavoring agent. Has been used with phenazone in the treatment of migraine.

Colacium A genus of Euglenoid Protozoans of the order *Euglenoidina*. A common freshwater animal, each individual cell forming stalks of mucilage. While in the pseudo-colonial phase there is no large flagellum, but when the cells separate they can grow a single flagellum and can swim around like other flagellates.

Colaptes A genus of Birds of the True Woodpecker subfamily, *Picinae* (family *Picidae*). e.g. C. auratus - The Yellow-shafted Flicker, widespread in the eastern United States, where it is also known as the Yellow hammer or Golden wing. Mainly a terrestrial bird, it feeds almost exclusively on ants, using its strong bill and sticky tongue to extract them from their nests.

Colchicum A genus of late blooming plants of the Lily family, *Liliaceae*. e.g. C. autumnale - The Meadow Saffron, a crocus-like, medicinal, perennial plant, named after the land of Colchis at the eastern tip of the Black Sea; 5-30 cm high, with solitary, pale purple flowers, occurring in deep meadows and woods. It yields the alkaloid, colchicine, used in the treatment of gout. Distribution: England and Wales, central and southern Europe. Often cultivated in the US.

Coleonyx A genus of Lizards of the Gecko family, *Gekkonidae*. e.g. C. variegatus - The Banded Gecko, a species living in the deserts of California and southern Texas; it has, unlike most other species, free eyelids, and does not possess adhesive foot pads.

Colesiota A genus of microorganisms of the family *Chlamydiaceae* (order *Rickettsiales*). e.g. C. conjunctivae - Causing infectious ophthalmia in sheep.

Colettsia A genus of microorganisms of the family *Chlamydiaceae* (order *Rickettsiales*). e.g. C. pecoris - A parasite found in the conjunctiva of domestic animals.

Coleus A genus of indoor plants of the Thyme family, *Labiatae*. e.g. C. blumei - The Coleus, a hybrid, evergreen, subshrubby, perennial plant with ornamental foliage of various colors and patterns

- green, yellow, crimson, mauve and pink. Leaves are nettle-shaped and velvety to the touch, with dentate margins. Place of origin: India, Java, tropical Asia.

Colias A genus of Insects of the White and Yellow Butterfly family, *Pieridae*. e.g. C. croceus - The Clouded Yellow Butterfly, size 2.5-5 cm; wings yellowish-red with marginal band; a migrant species found in woods and meadows. Distribution: most of Europe; also the Middle East and North Africa. Native in Britain.

Colinus A genus of Birds of the Pheasant family, *Phasianidae*. e.g. C. virginianus - The Bobwhite, a New World Quail, also called a partridge in the southern United States. A common game bird.

Colletes A genus of Insects of the Solitary Bee family, *Apidae*. e.g. C. bombylius.

Collinia A genus of plants, better known by the generic name, *Chamaedorea*, of the Palm family, *Palmae*. e.g. C. elegans (= *Chamaedorea* e.) - The Feathers Palm (see **Chamaedorea**).

Collocalia A genus of Birds of the Swift family, *Apodidae*; also called *Callocalia* (see **Callocalia**).

Collotheca A genus of Rotifer Metazoa of the suborder *Callothecacea*; in these carnivorous animals the anterior end of the sessile forms is expanded into a funnel, which is used for the capture and ingestion of large prey.

Colobocentrotus A genus of Echinoderm animals of the class of Sea-urchins, *Echinoidea*. e.g. C. atratus - A species of Sea-urchin from Peru; purple in color; the tip is quite smooth, the central mouth has five strong, chisel-shaped teeth, which extend into the musculature as calcareous rods, into a special formation known as "Aristotle's lantern".

Colobus A genus of Mammals of the Old World Monkey family, *Cercopithecidae*. e.g. C. abyssinicus - The Guereza, a black monkey, with white round the face, and long white hairs on flank and tail. The young are white all over. Found in Central Africa.

Colocasia A genus of plants of the Arum family, *Araceae*. e.g. C. esculenta var. antiquorum (= C. antiquorum) - Elephant's Ear or Taro, an ornamental aroid plant, used for indoor pools or as a potplant close to water, with large edible tubers and huge, 60 cm (2 ft), ovate, cordate leaves shaped like the ears of an elephant, and on short stems. Pale yellow flowers. Colocasias have edible rhizomes when, after boiling, they lose their poisonous properties. A staple article of diet throughout the Pacific, where they are widely cultivated. Place of origin: Tropical Asia.

Colpoda A genus of Protozoans of the order *Trichostomatida*. These trichostomes are freshwater animals, kidney-shaped and about 75 μ long. The mouth is on the indented side, and the vestibular cilia are a little shorter than the somatic cilia.

Coluber A genus of Reptiles of the Colubrid or Grass Snake family, *Colubridae*. e.g. C. jugularis - The Large Whip Snake, 2-3 m long; color light brown; the largest European snake, found in steppe and scrub country; feeds on various small mammals, birds, lizards, and other snakes. Distribution: southeastern Europe and western Asia.

Columba A genus of Birds of the Dove and Pigeon family, *Columbidae*. e.g. C. livia - The Rock Dove, length 33 cm; has two black wingbands; all breeds of the Domestic Pigeon originated from this dove. Found in rocky, sparsely vegetated country from the coast to inland areas. The Domestic pigeons live in areas of human habitation. The Roman pigeons are one of the domestic races closest to the ancestral stock. Distribution: southern Europe, North Africa, the Middle East, southern Asia, and Japan.

Columbigallina A genus of Birds of the

Dove and Pigeon family, *Columbidae*. e.g. C. passerina - The Ground Dove, a small, sparrow-like bird, spending most of its time on the ground searching for seed, and sometimes even nesting there. Distribution: the Americas.

Columnea A genus of indoor flowering plants of the family *Gesneriaceae*. e.g. C. gloriosa - Columnea or Flying Gold Fish plant, an evergreen epiphyte with slender, pendulous branches and bright scarlet, tubular flowers with yellow throats. An ideal hanging basket plant. Place of origin: Central America; introduced into Europe in 1915.

Colutea A genus of Shrubs of the Pea family, *Leguminosae*. e.g. C. arborescens - The Bladder Senna, a deciduous shrub from the Mediterranean region. About 3.6 m (12 ft) high, with small racemes of yellowish flowers. Its leaves are used to adulterate senna which is obtained from *Cassia* sp.

Colymbetes A genus of Insects of the carnivorous Water Beetle family, *Dytiscidae*. e.g. C. fuscus - Size 15 cm; upper side brown; edges rust-red; long, flattened, long-haired hind legs. Similar to the Great Water Beetle, *Dytiscus* marginalis, but smaller. Native in Britain.

Colymbus A genus of Birds, better known by the generic name, *Podiceps*, of the Grebe family, *Podicepedidae* (= *Colymbidae*). A widely distributed family of freshwater diving birds which visit the sea when migrating during winter. They have lobed toes and are excellent swimmers and divers, but are ungainly on land, walking upright like penguins. e.g. C. cristatus (= *Podiceps* c.) - The Great Crested Grebe (see **Podiceps**).

Combretum A genus of plants of the family *Combretaceae*. e.g. C. micranthum (= C. altum; C. raimbaultii) - The Kinkeliba plant (= Combrete Folium), reputed to be of value in the treatment of blackwater and other fevers.

Commiphora A genus of plants of the family *Burseraceae*. e.g. C. molmol -

This plant yields an oleo-gum-resin, Myrrh; it is carminative and, during excretion, mildly expectorant, diuretic, and diaphoretic. It is astringent to mucous membranes and is occasionally used as a tincture in mouthwashes and gargles for ulcers in the mouth and pharynx.

Compsomyia A name formerly given a genus of flies, now better known as *Cochliomyia*. e.g. C. macellaria (= *Cochliomyia* hominovorax) - The American Screw-worm fly (see **Cochliomyia**).

Condylura A genus of Mammals of the Mole family, *Talpidae*. e.g. C. cristata - The Star-nosed Mole, with an extraordinary fleshy, star-shaped, sensitive outgrowth of 22 feelers encircling the snout, forming a highly developed sensory organ. It lives in colonies, especially in marshy areas, and it spends much time in winter, being a good swimmer.

Conger A genus of Fishes of the Conger Eel family, *Congridae*. e.g. C. conger - The Conger Eel, a marine fish, up to 3 m (10 ft) long, and weighing almost 45.36 kg (100 lb). It has no scales, and its nostrils are placed laterally, the caudal fin being continuous with the dorsal and anal fins. It feeds on many other fishes and on the octopus, being one of the largest fishes caught off the sea shore. Distribution: northern Mediterranean, the Baltic, and widely in the oceans around North America.

Conioselinum A genus of plants of the Carrot family, *Umbelliferae*. e.g. C. vaginatum - Sheathed Hemlock-parsley; a perennial plant, up to 1.5 m high; 2 to 3-pinnate leaves, 8-winged fruit. Grows in woods and on bushy slopes. Place of origin: Europe.

Conium A genus of plants of the Carrot family, *Umbelliferae*. e.g. C. maculatum - The Conium or Hemlock tree, an annual to biennial plant, 50-200 cm high; furrowed, purple-spotted stem, with smell ofmice. Conium Fruit or

Hemlock Fruit contains the alkaloid coniine which causes coniine poisoning or coniism. Distribution: widespread; native in Britain.

Conochilus A genus of Rotifer Metazoa of the suborder *Flosculariacea*. e.g. C. hippocrepis - A species found in Britain, encased in gelatinous sheaths, and sometimes fused to form a colony.

Conocybe A genus of Mushrooms, the species of which yield the Mexican Indian drug, Teonanacatyl (see **Psilocybe**; **Stropharia**).

Conolophus A genus of Reptiles of the Iguana family, *Iguanidae*. e.g. C. subcristatus - The Galapagos Land Iguana, up to 120 cm (4 ft) long, living in sandy burrows, and feeding on plants and grasshoppers. Found only on the Galapagos Islands, and in danger of extinction because of domestic dogs and cats released on these islands.

Conophytum A genus of ornamental plants of the family *Aizoaceae*, sometimes known as the *Mesembryanthemaceae*. e.g. C. meyeri - A succulent species with showy white flowers, growing in arid stony areas. Place of origin: South Africa.

Conorhinus A former generic name applied to Insects of the family *Reduviidae*, now known as *Panstrongylus* and *Triatoma*. e.g. C. megistus (= *Panstrongylus* m.; *Triatoma* m.) - The Brazil Bug, transmitting trypanosomiasis (Chagas' disease) - see **Panstrongylus**; **Triatoma**.

Conringia A genus of plants of the Cress or Crucifer family, *Cruciferae*. e.g. C. orientalis - The Hare's-ear Cabbage, an annual plant, 15-80 cm high; petals 10-13 mm long; yellowish-white, 4-angled fruit; found mainly as weeds in arable and waste land; may occur as casuals or be widely naturalized. Place of origin: Mediterranean region.

Constrictor A genus of Reptiles of the Boa Snake family, *Boidae*. e.g. C. constrictor - The Boa Constrictor, one of the best-known species of this genus of snakes, found in Central and South America and the West Indies; almost 3.6 m (12 ft) long. A series of tight coils thrown round the prey kills by constriction. It gives birth to live young. Not dangerous to man.

Contopus A genus of Birds of the Tyrant Flycatcher or Kingbird family, *Tyrannidae*. e.g. C. virens - The Eastern Wood Pewee, a species found in Canada and the U.S.

Conuropsis A genus of Birds of the Parrot and Parakeet family, *Psittacidae*. e.g. C. carolinensis - The Caroline Parakeet, a small parrot with a long tail, once found living in the northern U.S., but is now extinct.

Conus A genus of Neogastropod Molluscs of the family *Conidae*. e.g. C. ventricosus (= C. mediterraneus) - The Mediterranean Cone Shell, a poisonous toxiglossid predacious carnivore, living in a shell up to 57 mm high, of variable color (brownish-yellow to olive-brown), common on rocky coasts amidst vegetation, and pouncing on its prey which it poisons and then swallows whole.

Convallaria A genus of plants of the Lily family, *Liliaceae*. e.g. C. majalis - Lily of the Valley, a charming perennial, 15-25 cm high, with slender creeping rhizomes, 2 radical oval-lanceolate leaves, and drooping bell-shaped, 6-lobed, very fragrant white flowers. The berries are red and poisonous. Place of origin: Europe, western Asia.

Convoluta A genus of Turbellerians or Flatworms of the order *Archoophora*. These forms are without an intestine.

Convolvulus A genus of plants of the Bindweed family, *Convolvulaceae*; also known by the genus name, *Ipomoea*. e.g. C. scammonia (= *Ipomoea* orizabensis) - The Ipomoea or Scammony plant; from the root, Ipomoea Root or Scammony Root, is extracted Scammony, a resinous gum which is anthelmintic and cathartic. Place of origin: Europe.

Conyza A genus of plants of the Daisy family, *Compositae*. e.g. C. canadensis - Canadian Fleabane, an annual to biennial plant, 10-100 cm high; heads 3-5 mm wide, white or reddish. Grows on dry grassland, rocky slopes, and in arable and waste land. Place of origin: a native of North America; well-established throughout Europe.

Cooperia A genus of parasitic worms, the cooperids. e.g. C. oncophora - Sometimes found in cattle.

Copaifera A genus of plants of the Pea family, *Leguminosae*. e.g. C. langsdorffii (= C. lansdorfii) - Copaiba Balsam, formerly used in the treatment of gonorrhea.

Copeina A genus of Fishes of the subphylum *Vertebrata*. e.g. C. arnoldi - The Characin fish.

Copernicia A genus of Trees of the Palm family, *Palmae* (= *Arecaceae*). e.g. C. cerifera - The Caranda Tree (= Carnauba Tree; Brazilian Wax Palm) - From this species is extracted Caranda Wax (= Carnauba Wax; Brazilian Wax), containing myricyl cerotate; used in the manufacture of polishes.

Coprinus A genus of Fungi of the family *Agaricaceae*. e.g. C. comatus - The Shaggy Cap or Lawyer's Wig, a common roadside fungus, with a white, hollow stipe up to 25 cm (10 in) long, and a cap which, cylindrical at first, becomes bell-shaped.

Copris A genus of Insects of the Scarab Beetle family, *Scarabaeidae*, belonging to the group of sacred scarab beetles of ancient Egypt. e.g. C. hispanus - The Spanish Copris, up to 20 mm long; body bluish-black with a horn; makes pear-shaped breeding pellets. Distribution: Mediterranean region (see **Catharsius**).

Copromastix A genus of flagellate microorganisms. e.g. C. prowazeki - A Brazilian species found in the feces.

Copromonas A genus of flagellate

microorganisms. e.g. C. subtilis - Found in the species of frogs, and sometimes in man.

Coptis A genus of plants of the Buttercup family, *Ranunculaceae*. e.g. C. teeta - An Asiatic plant species containing the alkaloid berberine. Used in India orally as a bitter, and in the form of subcutaneous injections (of berberine sulfate) in the treatment of oriental sore (see **Berberis**).

Coracia A genus of Birds of the Crow family, *Corvidae*. e.g. C. graculus - The Yellow-billed or Alpine Chough, size 38 cm; yellow bill; plumage bluish-black; a resident bird living in mountainous forests, rock-crevices, and ruins. Distribution: Europe.

Coracias A genus of Birds of the Roller family, *Coraciidae*. e.g. C. garrulus - The Eurasian Common Roller, a jay-like bird, 30 cm long, with a greenish-blue or aquamarine head and front, yellow back, purplish-blue wing tips, and greyish-brown tail. Nests in tree holes and cavities in walls. A migratory bird, wintering in East Africa. Distribution: most of Europe, southwest Asia, northwest Africa.

Coracina A genus of Birds of the Cuckoo-shrike and Minivet family, *Campephagidae*. e.g. C. lineata - The Barred Cuckoo-shrike, 24 cm (91/2 in) long, mainly grey with underparts barred in black and white. Distribution: Australia to the Solomon Islands.

Coragyps A genus of Birds of the New World Vulture family, *Cathartidae*. e.g. C. atratus - The Black Vulture, also known as the Urubu or Gallinazo, smaller than the 180 cm (6 ft) wingspan, King Vulture, and entirely black. A scavenger bird, having become quite domesticated in certain areas of the central United States and South America.

Corallina A genus of Red Algae plants of the family *Gigartinaceae*. e.g. C. officinalis - The Corallina, up to 10 cm high; a calcareous seaweed, its cell walls storing calcium carbonate, and with opposite club-shaped branches, 1-2 mm thick. Found in the North Sea, chiefly on rocky shallows, and in the West Baltic.

Corallium A genus of Horny Corals of the order *Gorgonacea* (= *Metazoa*), of the subclass *Octocorallia*. e.g. C. rubrum - The Red Coral, a colony-forming precious coral found off the coast of the Mediterranean Sea and Japan, whose red, solid axis of fused spicules (red calcareous skeleton) is used to make coral jewellery. It has become very rare almost in all regions because of over-collection.

Corallorhiza A genus of Saprophyte plants of the Orchid family, *Orchidaceae*. e.g. C. trifida - The Coral-root, a perennial plant, 7-30 cm high; the rhizome is fleshy, much-branched and coral-like; the lip is short and white, with red dots. Native in Britain.

Corallus A genus of Reptiles of the Boa Snake family, *Boidae*. e.g. C. caninus - The Emerald Tree Boa, a beautiful species, this snake has white markings on its green body, which is strongly compressed laterally. About 1.8 m (6 ft) long, it lives among the branches of trees in the forests of tropical South America. It feeds on birds.

Corchorus A genus of plants of the Jute or Lime family, *Tiliaceae*. e.g. C. olitorius - The Jute plant, native to tropical Asia and cultivated especially in East Bengal and Bangla Desh. Widely used for cotton baling and sacking, cordage and paper. Formerly used in the preparation of surgical dressings.

Corcorax A genus of Birds of the Magpie-lark family, *Grallinidae*. e.g. C. melanorhamphus - The White-winged Chough; several birds combine to build a bowl-shaped nest out of mud, in which the females lay their eggs when the mud-bowl has been completed.

Cordulegaster A genus of Insects of the Hawker Damselfly or Dragonfly family (order *Odonata*). e.g. C. bidentatus - Size 7-8.5 cm; eyes touch each other; a swift-flying insect hunter found by brooks and rivers, preying on large insects such as butterflies. Dragonflies cannot fold their wings; at rest they are held at right angles to the body; in flight the pairs of wings are used independently, and make a rustling sound.

Cordulia A genus of Insects of the family of Hawker Dragonflies (order *Odonata*). e.g. C. aenea - The Downy Emerald Dragonfly, size 5.5 cm; body golden green, without any yellow spot; rather short abdomen. A swift flier, with a well-defined beat over the pools and ponds where it hunts insects, such as butterflies. It spreads its wings out horizontally when at rest. These insects may sometimes be seen well away from water, occasionally in large numbers on migration.

Cordyline A genus of Shrubs of the family *Agavaceae*. e.g. C. indivisa - The Cabbage Palm or Dracena (U.S.); this plant has a stout upright trunk, 3-8 m (10-25 ft) tall, topped by long, dark green, sword-shaped leaves. Hay-like, scented, off-white flowers are followed by round purple berries. Place of origin: New Zealand.

Cordylobia A genus of Flies. e.g. C. anthropophaga - A species of African flies, the larvae of which cause myiasis in man and animals.

Cordylophora A genus of Sponges (phylum *Porifera*). e.g. C. caspia - Size 5 cm; colony with many branching tubes. These coelenterates feed on smaller animals which they subdue with poisonous stings. Native in Britain.

Cordylus A genus of Reptiles of the Girdle-tailed Lizard family, *Cordylidae*. e.g. C. giganteus - The Giant Girdle-tail or Zonure Lizard, nearly 60 cm (2 ft) long, feeding mainly on grasshoppers, and living in rocky districts in South Africa. The scales are ossified, those of the back and tail being developed into spines.

Coregonus A genus of Fishes of the Whitefish family, *Coregonidae*; identified with the Salmon. e.g. C. lavaretus - The Freshwater Houting or Powan fish, length 50-70 cm; weight 3-4 kg; a deepwater fish with small eyes, inferior mouth, and 20-34 gill-rakers, that enters shallow water during the spawning season in November. Originally native to Lake Miedwie in Poland, it was introduced to many European ponds in the late 19th century. Native in Britain. C. artedii - The Cisco or Lake Herring.

Coreopsis A genus of plants of the Daisy family, *Compositae*. e.g. C. grandiflora - Coreopsis or Tickseed, a smooth, leafy, perennial plant, 30-60 cm (1-2 ft) high, with bright yellow, daisy-type flowers. Place of origin: southern United States; introduced into Europe in 1826.

Corethra A genus of Mosquitoes. e.g. C. plumicornis - The Phantom, length 6 mm; hairy, unspotted wings; proboscis not adapted for piercing as in other mosquitoes. Found in still or slow-flowing freshwater. Native in Britain.

Coriandrum A genus of plants of the Carrot family, *Umbelliferae*. e.g. C. sativum - The Coriander plant; the dried ripe fruits of Coriander (= Coentro; Coriand.; Coriander Fruit; Coriander Seed; Fruto de cilantro), are aromatic and carminative, and Powdered Coriander, containing not less than 0.2% v/w of volatile oil, is added to purgative medicines to prevent griping. Coriander Oil (= Oleum Coriandri; Ol. Coriand.), a volatile oil obtained by distillation from coriander, contains about 65-80% of (+)-linalol (coriandrol); it is aromatic, stimulant, and carminative.

Coris A genus of Fishes of the Wrasse family, *Labridae*. e.g. C. julis - The Rainbow Wrasse, 25 cm long; occurs near coastal rock formations down to depths of about 120 m. The young differ in color from the adults. Distribution: northwest Atlantic, Mediterranean Sea.

Corixa A genus of Insects of the Corixid or Water Boatman family, *Corixidae*. e.g. C. punctata - The Lesser Water Boatman, a bug-like insect up to 14 mm long; front of back with 15-16 transverse lines; middle legs the longest; a freshwater dweller, swimming by rowing movements of the legs, rising to the surface to breathe. Native in Britain.

Cornufer A genus of Amphibians of the Frog family, *Ranidae*, found from New Guinea to the Philippines. The various species live wholly in forest plants, where the eggs are deposited, or on the forest floor where the eggs are laid.

Cornus A genus of Shrubs of the Cornel or Dogwood family, *Cornaceae*; also known by the generic name, *Cynoxylon*. e.g. C. florida "Rubra" (= *Cynoxylon floridum*) - Pink Flowering Dogwood, the common dogwood of North America, a deciduous shrub or small tree, 3-5 m (10-15 ft) tall, insignificant greenish flowers surrounded by 4 large, red or pink bracts. Used as an astringent, antiperiodic, and tonic. Place of origin: Eastern United States; introduced into Europe c. 1730.

Coronella A genus of Reptiles of the Colubrid Snake family, *Colubridae*. e.g. C. austriaca - The European Smooth Snake, length up to 75 cm (30 in); its scales are not keeled but smooth and reddish in color; it has dark lengthwise stripes on the sides of the head and neck. Bites when in danger. Found on dry, sunny, stony hillsides covered with vegetation. Ovoviviparous method of reproduction. Frequently mistaken for the venomous viper. Distribution: almost all of Europe and in southern England.

Coronilla A genus of Herbs of the Pea family, *Leguminosae*. e.g. C. varia - The Common Crown-vetch, a poisonous perennial herb, 30-120 cm high, with heads of purple, pink, or white flowers.

Grows in meadows and bushy places. Place of origin: central and southern Europe.

Coronopus A genus of plants of the Cress or Crucifer family, *Cruciferae*. e.g. C. squammatus - The Common Wart-cress, an annual plant, 2-30 cm high; flowers white, and in short racemes; fruits apiculate, ridged or warty, and kidney-shaped. Grows in dry pastures, on walls, banks, roadsides, and in cultivated fields and waste places. Place of origin: Europe; a widespread species, native in Britain.

Coronula A genus of Crustaceans of the order *Thoracica*; a commercial species, attaching themselves to the surface of whales, although they are not true parasites.

Corophium A genus of Crustaceans, widespread in European coastal waters. e.g. C. volutator - The Burrowing Shrimp, size 6 mm; eyes dark red; male has very long second antennae. Digs holes for itself in muddy shallows. Native in Britain.

Corrigiola A genus of plants of the Pink family, *Caryophyllaceae*. e.g. C. littoralis - The Strapwort, an annual to biennial plant, 3-30 cm high; small flowers, in crowded terminals and axillary inflorescences; sepals green or red in center, with broad white margins. Grows in damp places. Place of origin: from southwest Europe to Denmark; native in Britain.

Cortaderia A genus of plants of the Grass family, *Gramineae*; also known by the generic name, *Gynerium*. e.g. C. selloana (= C. argentea; *Gynerium argenteum*) - Pampas Grass, a grass species widely grown on account of its stately habit. Forms huge mounds of tough, saw-edged, grassy leaves, 1-3 m (3-9 ft) long and huge silky panicles of silvery-white or pink flowers at the top. Place of origin: South America; introduced into Europe in 1848.

Cortusa A genus of plants of the Primrose family, *Primulaceae*. e.g. C. matthioli - The Alpine-sanicle, a perennial plant, 15-40 cm high; widespread, growing in meadows and woods, but rare on the mountains of Europe (in the Alps up to 1,900 m) and Asia. The terminal clusters of dull red or purple flowers are borne on slender stems, 3-12 in each umbel.

Corvus A genus of Birds of the Crow, Magpie, and Jay family, *Corvidae*. e.g. C. splendens - The House Crow, a perching bird, 30 cm long; ows its scientific name to the metallic tints of its plummage. Very common in built-up areas in towns and villages. Distribution: India and Sri Lanka.

Coryaria A genus of plants.

Corycaeus A genus of Crustaceans of the Copepod subclass, *Copepoda*; in these species two of the ocelli develop large lenses so that the animal appears to have two enormous eyes.

Corydalis A genus of plants of the Fumitory family, *Fumariaceae*. Closely related to the Poppy family, *Papaveraceae*. Some species also known by the generic names, *Dicentra* or *Dielytra*, are allied to *Corydalis*. e.g. C. canadensis (= *Dicentra* c.; *Dielytra* c.) - Squirrel or TurkeyCorn; the tubers of this plant contain a number of alkaloids, some of which are physiologically active, and one of them, Bulbocapnine, has been used as a sedative in postencephalitic conditions, Meniere's syndrome, and in tremor of various origins (see **Dicentra**; **Dielytra**). C. lutea - Yellow Corydalis or Yellow Fumewort, a branched, spreading, hairless, perennial plant, 20-30 cm (8-12 in) high; with soft 2-pinnate leaves and tubular, spurred, yellow flowers; growing in rock gardens and other cultivated areas. Place of origin: Europe, including Britain.

Corydalus A genus of Insects of the Dobson-fly family, *Corydalidae*. e.g. C. cornutus - A species from North

America, reaching a wingspan of over 10 cm (4 in); commonly found on vegetation near water, the larvae being aquatic with long, segmented gills arranged along each side of the abdomen. The male possesses very powerful mandibles.

Corylus A genus of plants of the Hazel family, *Corylaceae*. e.g. C. avellana - The Hazel Nut or Cob Nut, a shrub 1-6 m tall, producing long male catkins and bud-like groups of female flowers very early in the year. The fruit is a woody nut surrounded by a thin-toothed bract. Grows in open woodlands and scrubs. Distribution: Europe, Asia Minor. Native in Britain.

Corymbites A genus of Insects of the Click Beetle family, *Elateridae*. e.g. C. purpureus - Length 8-14 mm; has characteristic raised, longitudinal ribs on the reddish-purple elytra, and cannot be mistaken for any similar species. Found on various deciduous shrubs in foothills. Distribution: central and southern Europe, Asia.

Corymorpha A genus of True Jellyfishes of the order *Scyphozoa*. e.g. C. nutans - Size 5-6 cm; bell-pointed, with one tentacle. Found in the Atlantic, North Sea, and the West Baltic.

Corynactis A genus of Sea Anemones of the order *Corallimorpharia*; a small solitary anemone, more closely related to a coral than an anemone, growing in groups. It is a delicate green animal, about 0.5-1.5 cm high, but may be orange, grey or scarlet, with numerous knobbed tentacles.

Corynanthe A genus of plants of the Bedstraw or Madder family, *Rubiaceae*; also known by the generic name, *Pausinystalia*. e.g. C. yohimbi (= *Pausinystalia* yohimbe) - The Yohimbe tree; it yields the alkaloid yohimbine. Has been used for its alleged aphrodisiac properties but without convincing results.

Coryne A genus of Hydrozoans of the family *Corynidae*; this animal has rose-colored polyps bearing knobbed tentacles, and is found on larger seaweeds or rocks.

Corynebacterium A genus of microorganisms of the family *Corynebacteriaceae*; gram-positive, straight to slightly curved, club-shaped rods; generally aerobic, but may be microaerophilic or even anaerobic; often appearing in palisade arrangements (Chinese letters; picket fence). Widely distributed in nature in soil, plants, and animals. Many species are part of the normal flora of the skin, respiratory tract, and mucous membranes, but several species are pathogenic for humans, animals and plants. Some species also have other generic headings, such as *Haemophilus*, *Nocardia* and *Propionibacterium*. e.g. C. acnes (= *Propionibacterium* a.) (see **Propionibacterium**). C. aquaticum - Has been isolated from blood, cerebrospinal fluid, sputum and water sources. Its appearance has been confused with that of *Listeria* monocytogenes (see **Listeria**). C. diphtheriae (= *Bacillus* d.) - The Diphtheria bacillus or Klebs-Loeffler bacillus, the causative agent of diphtheria; its potent exotoxin and pseudomembrane produce many effects. There are 3 types: mitis, intermedius, and gravis. C. diphtheriae gravis tends to give rise to the most severe infections. May exist in healthy carriers in the respiratory tract, conjunctiva, and the skin or in wounds. C. equi - Usually a contaminant; has been isolated from blood, urine, cerebrospinal fluid, lung, female genital tract, wounds, and abscesses. C. hemolyticum - Has been associated with throat, blood, and skin affections. C. minutissimum (= *Nocardia* minutissima) - The causative agent of erythrasma (see **Nocardia**). C. pseudodiphtheriticum (= C. hofmanii; diphtheroids) - May be apart of the normal throat flora. C. ulcerans - Has been isolated from cases of tonsillitis, but rarely in the United States. C. vaginale (= *Haemophilus* vaginalis) - Has been isolated fromthe female urogenital tract and is associated with venereally transmitted vaginitis, urethritis, puerperal fever, and septicemia (see **Haemophilus**). C. xerosis - Has been isolated from both normal and infected conjunctivae. Probably nonpathogenic.

Corynephorus A genus of plants of the Grass family, *Gramineae*. e.g. C. canescens - Grey Hair-grass, a perennial plant, 15-50 cm high; the species is greyish green, with silvery grey, tufted panicles; awn of lemma club-shaped at tip. Grows on waste and arable land. Native in Britain.

Coryphaena A genus of Fishes of the Dolphin family, not to be confused with the True Dolphins, which are mammals. e.g. C. hippurus - The Dolphin or Dorado, length up to 1 m; body color very variable; long dorsal fin extending from top of head. A fast-swimming predaceous species found mainly in the warmer waters of the world, living at a depth of 200 m.

Coryphella A genus of Sea Slugs of the subclass *Opisthobranchia*. In these marine gastropods the brightly colored elongated body carries a series of projecting "cerata", in which are stored sting cells (nematocytes) obtained from the coelenterates on which it feeds. In the absence of a shell these sting cells furnish protection.

Coscinasterias A genus of Starfishes of the class *Asteroidea*. e.g. C. tenuispina - A species with as much as 8 spiny arms, feeding on lampshells, opening them in the same way as other starfishes open bivalve molluscs.

Coscinia A genus of Insects of the Tiger Moth family, *Arctiidae*. e.g. C. cribraria - The Speckled Footman Moth, size 15-21 mm; found over heaths and in pinewood clearings. Distribution: Europe; temperate parts of Asia.

Coscinium A genus of plants of the family *Menispermaceae*. e.g. C. fenestratum - Ceylon Calumba, False Calumba, or Tree Tumeric plant; the dried stem is used extensively in India and the Far East as a bitter and as a substitute for calumba in atonic dyspepsia (see **Jateorhiza**).

Coscinodiscus A genus of Diatoms of the class *Bacillariophyta*. e.g. C. radiatus - A diatom species with a shape like a pill-box.

Coscinoscera A genus of Insects of the Silkworm Moth family, *Saturniidae*. e.g. C. hercules - A species found in Australia, and one of the largest moths in the world.

Coscoroba A genus of Birds of the Duck family, *Anatidae*. e.g. C. coscoroba - The Coscoroba Swan, a small species found in the waters of South America, with a long flexible neck, able to swim and paddle, but cannot dive like other swans. It is white, with black wing-tips and slightly smaller than the Swans.

Cosmarium A genus of Green Algae of the class *Chlorophyta*. These species are the Desmids, beautiful unicellular green algae, particularly common in the water of bogs and tarns.

Cosmos A genus of flowering plants of the Daisy family, *Compositae*. e.g. C. bipinnatus - Cosmos, a half-hardy annual plant, up to 1 m (3ft) high, with large, rose, purple, or white flowers with yellow discs. Place of origin: Mexico; introduced into Europe in 1799.

Cosmotriche A genus of Insects of the order of Butterflies and Moths, *Lepidoptera*. e.g. C. potatoria - The female moth is larger, light yellow in color and the smaller male is brownish-red. Found in damp meadows. Place of origin: Central Europe.

Cossus A genus of Insects of the Goat Moth family, *Cossidae*. e.g. C. cossus - The European Goat Moth, size up to 9.5 cm; ash-grey wings; a big insect whose very large, long-lived, pinkish-brown larvae burrow in and feed on the wood of the oak, elm, ash or willow. It has a strong goat-like smell. Native in Britain.

Cothurnia A genus of Protozoan animals of the class *Ciliata*. These ciliate species, size about 0.0075 mm (3/10,000 in), live attached to various objects in the water, forming a protective cup on a stalk.

Cotoneaster A genus of plants of the Rose family, *Rosaceae*. e.g. C. pyracantha (= *Crataegus* p.; *Pyracantha* coccinea) - The Firethorn plant (see **Crataegus**). C. salicifolia - Willowleaf Cotoneaster, a graceful evergreen shrub or small tree, up to 4.5 m (15 ft) tall, with spreading branches, turning reddish with age. Small, pinkish flowers in compact woolly corymbs, succeeded by attractive bright red berries in late autumn. Place of origin: China; introduced into Europe in 1908.

Cottus A genus of Fishes of the Bullhead or Sculpin family, *Cottidae*. e.g. C. bubalis - The Long-spined Sea Scorpion fish, length 18 cm; during spawing time the male has a cherry-red belly covered with brownish spots. Distribution: European seas, common especially in the north and west. C. gobio - The Miller's Thumb or Bullhead, a not very active, freshwater fish, length 15 cm; fins large and flexible; scales along lateral line only, midway down the body. Common in clear mountain streams and rivers. Distribution: Europe, except the south.

Cotula A genus of plants. e.g. C. coronopifolia - The Buttonweed, an annual to perennial plant, 8-50 cm high; strongly aromatic; leaves deeply toothed to pinnatifid; heads with tubular florets. Place of origin: a native of South Africa; now widely distributed in Europe.

Coturnix A genus of Birds of the Partridge, Pheasant, and Quail family, *Phasianidae*. e.g. C. coturnix - The Common Quail, a game-bird, 17 cm long; the smallest fowl-like bird inhabiting Europe. It is something like a miniature partridge, with a striped head, and a greyish brown body variegated with black, white, and yellow spots. The nest is located on the ground. The only migratory member of this family in Europe. It winters in Africa. Distribution: temperate and warm regions of Europe; Asia, east to Japan.

Cotylaspis A genus of Flukes of the family *Aspidogastridae*. These trematode worms are mainly endoparasites of molluscs and cold-blooded vertebrates.

Cotyledon A genus of Shrubs of the Crassula family, *Crassulaceae*. e.g. C. undulata - A shrubby species from South Africa; it has beautiful leaves with wavy edges, covered with a white mealy deposit. The flowers are cream-colored, with a red stripe.

Cotylogaster A genus of Flukes of the family *Aspidogastridae*. Similar to the genus *Cotylaspis* (see **Cotylaspis**).

Cotylogasteroides A genus of Flukes of the family, *Aspidogastridae*. As in the genus *Cotylaspis* (see **Cotylaspis**).

Coua A genus of Birds of the Cuckoo family, *Cuculidae*. e.g. C. cristata - The Crested Coua, a large ground-dwelling, noisy cuckoo found only in Madagascar. These birds build their own nests and seek safety by running instead of flying.

Councilmania A genus of Amebae. e.g. C. dissimilis - A species resembling *Entameba* histolytica.

Coura A genus of Birds of the Dove and Pigeon family, *Columbidae*; these are species of three-crowned pigeons, up to 82.5 cm (33 in) long, found in New Guinea. They possess a fan-like crest of feathers on the head, which can be erected or lowered at will. These birds wander in forests in small groups, gathering up seeds and fruits fallen from the trees.

Cowdria A genus of microorganisms of the family *Rickettsiaceae*. e.g. C. ruminantium - A species causing Heartwater disease of sheep, goats, and cattle.

Coxiella A genus of microorganisms, also known by the generic name, *Rickettsia*, of the family *Rickettsiaceae*. e.g. C. burnetii (= C. burnettii; *Rickettsia* diaporica) - A species transmitted to animals by ticks, causing a fever. In man it causes Q fever, which is transmitted by inhalation of aerosols containing the microorganisms, by ingestion of raw milk, or rarely, by tick bites. Now widespread in the United States in cattle. May be isolated from blood and urine. Clinical findings include bronchopneumonia and occasionally, liver granulomas or endocarditis (see **Rickettsia**).

Crabro A genus of Insects of the Digger Wasp family, *Sphecidae*. e.g. C. cribrarius - The Slender-bodied Digger Wasp, length 11-17 mm; found in summer on flowering umbelliferous plants. Distribution: the Palaearctic region.

Cracticus A genus of Birds of the Bellmagpie, Australian Butcherbird, and Piping Crow family, *Cracticidae*. e.g. C. torquatus - The Grey Butcherbird, a species of bell-magpies found in Australia and Tasmania. This non-migratory bird is chiefly arboreal, building its nest on trees, and using its strong, hooked bill to kill large insects and lizards as well as small birds and mammals.

Craigia A genus of flagellate Protozoans, formerly known as Parameba or Paramoeba. e.g. C. hominis (= *Parameba* h.; *Paramoeba* h.) - A species causing dysentery.

Crambe A genus of plants of the Cress or Crucifer family, *Cruciferae*. e.g. C. maritima - The Sea-kale, a perennial plant, 40-60 cm high; very large, long-stalked, glaucous, cabbage-like leaves; white flowers; 2-jointed fruits. Grows on shores and on cliffs along the Atlantic and Baltic coasts and around the Black Sea. Native in Britain.

Crangon A genus of Crustaceans of the Shrimp family, *Crangonidae*. e.g. C. crangon - The Common Shrimp, 40-50 mm long; greyish-yellow, yellow or greenish, sometimes dark brown on the upper side; no rostrum, but sharp point between the eyes. The shells are inflexible and shed periodically with growth. A nocturnal creature, found along the seacoast up to a depth of 10-50 m. Consumed as food, used as bait, and also for making fishmeal. Distribution: Mediterreanean, coast of western Europe, eastern coast of North America.

Craspedacusta A genus of Hydrozoans of the order *Limnomedusae*. e.g. C. sowerbii (= C. sowerbyi) - A large hydrozoan medusa, up to 20 mm in diameter, an uncommon freshwater species; resembling a small jellyfish, and found in certain ponds and lakes in America. It has a minute polyp, formerly called *Microhydra*, before its relation to the medusa was discovered. These Coelenterates feed on smaller animals which they subdue with poisonous stings. Also native in Britain (see **Microhydra**).

Craspedella A genus of Turbellarian Flatworms of the order *Temnocephala*; these are leach-like forms with anterior tentacles.

Crassostrea A genus of Bivalve Molluscs of the edible Oyster family, *Ostreidae*. e.g. C. gigas - The Japanese Oyster, a very numerous and important species of commercial bivalves, with an elongated cupped shell.

Crassula A genus of succulent plants of the Crassula family, *Crassulaceae*. e.g. C. lycopodioides - Club Moss Crassula, an unusual succulent plant, resembling a club moss, with up to 30 cm (12 in), stringy, branching stems sheathed by 4 rows of tiny, scale-like, oval, pointed, closely set leaves. Minute, whitish, sessile flowers are situated in the upper leaf axis. Place of origin: southwest Africa.

Crataegus A genus of Hawthorn plants of the Rose family, *Rosaceae*; also known by the generic names, *Cotoneaster* or *Pyracantha*. e.g. C.

pyracantha (= *Cotoneaster* p.;
Pyracantha coccinea) - The Firethorn, a
large evergreen shrub or small tree, with
thorny branches and white flowers, followed by dense bunches of rich red
fruits all along the branches. Place of
origin: southern Europe and Asia Minor;
introduced into Britain in 1629.

Craterellus A genus of Fungi or fungal
plants of the Cantharell or
Chanterellefamily. e.g. C. cornucopiodes
- Horn of Plenty; cap up to 12 cm; funnel-shaped.

Craterolophus A genus of Stalked
Jellyfishes of the order *Stauromedusae*.
e.g. C. convolvulus - The Stalked
Jellyfish, a sessile animal living fastened
to seaweeds or stone by the exumbrella
surface, and feeding on small crustaceans. The bell is trumpet-shaped with
a flattened subumbrella surface bearing
tentacles, and with a central mouth. The
short stalk-like body is anchored by an
adhesive disc (see **Ceramium**).

Crax A genus of Birds of the Curassow
family, *Cracidae*. e.g. C. rubra - The
Great Curassow, 90 cm (3 ft) long; the
male is black with a yellow knob on top
of the beak, the female is reddish-
brown, with no beak knob. An arboreal
rooster, but a ground-feeder, found in
the Americas between Texas in the north
and the Argentine in the south.

Crenilabrus A genus of Fishes of the
Wrasse family, *Labridae*. e.g. C. melops
- The Gold Sinny or Corkwing Wrasse,
length 15-25 cm; variably colored, but
all have a black spot at the base of the
caudal fin. Common along the shores of
Europe as far north as the coast of
Norway; also in the Mediterranean.

Crenothrix A genus of microorganisms
of the family *Crenotrichaceae*; disc
shaped to cylindrical cells occurring in
attached trichomes. e.g. C. polyspora.

Creophilus A genus of Insects of the
Rose Beetle family, *Staphylinidae*. e.g.
C. maxillosus - Length 15-25 mm;
predaceous like many other rove bee-

tles. Found in dung, piles of rubbish and
on carrion, where it hunts insect larvae.
Distribution: the Palaearctic, Nearctic,
and Oriental regions.

Crepidula A genus of Mesogastropod
Molluscs of the family *Calyptraeidae*.
e.g. C. fornicata - The Slipper Limpet;
shell 20 mm long; up to 47 mm wide,
and dish-shaped with a plate inside.
These animals live in piles or chains, the
lowest individuals being females, and
the highest (and thus youngest) males,
which fertilize the females below. Those
in the middle are hermaphrodites. Found
at shallow depths, they are often a pest
of oyster beds, smothering the oysters
and competing with them for food.
Distribution: along the coast of Great
Britain, and the Atlantic coast of
Europe. Introduced from America.

Crepis A genus of plants of the Daisy
family, *Compositae*. e.g. C. aurea -
Golden Hawksbeard, a perennial plant,
5-30 cm high; solitary head; the fruits
have 20 ribs. Grows in alpine meadows
and turfs, and on scree slopes; also in
tall herb meadows from 1,200 to 2,700
m. Place of origin: South and
Southcentral Europe.

Crex A genus of Birds of the Coot,
Gallinule, Moorhen, and Rail family,
Rallidae. e.g. C. crex - The Corncrake,
also known as the Land Rail, a migratory land bird, 27 cm long; upper parts
rusty brown, below yellow-brown; bill
and feet yellow; living on dry land in
damp meadows from lowland to hilly
country, sometimes even in mountains.
Its cry, like the sound of a stick being
drawn across a comb, is heard chiefly in
the evening. It winters around the
Mediterranean Sea and in tropical and
southern Africa. Distribution: Europe
(except the north, Italy, and the Iberian
Peninsula); Asia (as far as Lake Baikal).
Native in Britain.

Cricetomys A genus of Mammals of the
Old World Rat and Mouse subfamily,
Murinae. e.g. C. gambianus - The

Gambian Pouched Rat or Giant Bamboo-rat, almost 90 cm (3 ft) long (of which half is the tail), Africa's largest rat. An inoffensive creature, but very destructive because of its voracious hoarding in deep burrows.

Cricetulus A genus of Mammals of the Mouse family, *Cricetidae*. e.g. C. migratorius - The Migratory or Grey Hamster, 11-13 cm long; upper side grey; coat not very bright; ears large; tail short. Builds its nest in the ground. Distribution: the Balkans and South Russia.

Cricetus A genus of Mammals of the Hamster and New World Rat, Mouse, and Vole Cricetid subfamily, *Cricetinae*. e.g. C. cricetus - The Common Hamster, length of body 210-340 mm, tail 25-65 mm; coat brightly colored, underside black; a solitary, nocturnal, rat-like rodent that lives in underground burrows, expanding into a nesting chamber and a food chamber. In winter it hibernates at a depth of up to 2 m ($6^1/2$ ft). A species used extensively as a laboratory experimental animal. Distribution: Europe, from France southward; Asia as far as the Yenisei River.

Crinia A genus of Amphibians of the family *Leptodactylidae*. These species of frogs are found in the more arid parts of Australia, where their eggs may be laid on land or in the water.

Crinitaria A genus of plants of the Daisy family, *Compositae*. e.g. C. linosyris - Goldilocks, a perennial plant, 30-60 cm high; leaves 2 mm wide; uncommon or rare; grows on dry grassland, bushy places, and limestone cliffs. Place of origin: South and Central Europe.

Crinum A genus of plants of the Lily family, *Amaryllidaceae*. e.g. C. bulbispermum (= C. longifolium) - The Cape Lily or Jamaica Crinum; an evergreen plant with long-necked, ovoid bulbs; grey-green fleshy, floppy leaves, and 30 cm (1 ft) fleshy stalks, each carrying 6-12 fragrant white flowers which are pink-flushed on the outside. Place of

origin: Natal, Transvaal; introduced into Britain in 1750.

Crisia A genus of Moss Animals or Polyzoa, also known as Bryozoa, Ectoprocta, or Sea Mats, of the order *Cyclostomata*. These are small, inconspicuous, highly complex, hermaphrodite organisms, where both sexual and asexual reproduction occurs. The species reproduce by embryonic fission, the fertilized egg developing into an embryo in a modified zooecium, known as an ovicell, giving rise to numerous secondary embryos, each of which becomes a free-swimming larva. They form numerous, large, sessile colonies. e.g. C. eburnea - Size up to 2.5 cm (1 in); cells cylindrical, mostly with a long spine; colonies branch out like a tree. Distribution: worldwide.

Cristatella A genus of Moss Animals or Polyzoa, also known as *Bryozoa* or *Ectoprocta* of the class *Phylactolaemata*. These are partially sedentary freshwater animals that can creep slowly over the surfaces of water plants, aided by a gelatinous secretion which serves the same purpose as the slimetrack of slugs. Colonies may divide into two by a constriction in the middle, the two parts then continuing to grow to full size.

Cristispira A genus of microorganisms of the family *Spirochaetaceae*, found in the intestinal tracts of molluscs. e.g. C. anodontae.

Cristivomer A genus of Fishes of the Salmon and Char family, *Salmonidae*, now known by the generic name, *Salvelinus*. e.g. C. namaycush (= *Salvelinus* n.) - The Great Lake Char or "Trout" of North America (see **Salvelinus**).

Crithidia 1. A genus of Zoomast Protozoans of the order *Protomonadina*. These protomonads resemble the trypanosoma, and have only 1 or 2 flagella, but the species are otherwise probably unrelated. e.g. C. gerridis - A protozoan parasite of the water bug, *Gerris*, found in crithidial, leptomonad, and leishmanial stages. 2. A developmental form of the trypanosome in its insect host. e.g. C. cunninghami (= *Leishmania* tropica).

Crocethia A genus of Birds of the Sandpiper or Wader family, *Scolopacidae*. e.g. C. alba - The Sanderling, length 20.5 cm; winter plumage very light-colored; shoulders black. A migrant bird, nesting in the tundra. Winter visitors on the west and south coasts of Europe.

Crocidura A genus of Mammals of the Shrew family, *Soricidae*. e.g. C. russula (= C. suaveolens) - The Greater or Common White-toothed Shrew; length of body 64-95 mm; tail 33-46 mm; upper side brownish-grey, underside yellowish-grey, not sharply contrasting; found chiefly in human habitations, fields, gardens, and cemetries. Also in mountains up to elevations of 1,600 m. Distribution: central Europe and the Mediterranean region; also central Asia and northeastern Siberia.

Crocodylus A genus of Reptiles of the Crocodile family, *Crocodylidae*. e.g. C. niloticus - The Nile Crocodile, up to 4.5 m (15 ft) long; once very common in Africa but is becoming rare because of unrestricted hunting. Still found in South and Central Africa and Madagascar, but has disappeared from many of its former habitats, particularly in North Africa.

Crocosmia A genus of garden plants of the Iris family, *Iridaceae*; also known by the generic names, *Monbretia*, *Montbretia*, or *Tritonia*. e.g. C. x crocosmiiflora (= *Monbretia* c.; *Montbretia* c.; *Tritonia* c.) - The Garden Montbretiia, a hybrid herbaceous plant derived by crossing *Crocosmia* durea with *Crocosmia* pottsii; it has sword-shaped leaves and many orange-scarlet tubular flowers with flaring petals. This hybrid has become naturalized in parts

of England, making wide drifts in open woodland.

Crocus A genus of flowering plants of the Iris family, *Iridaceae*. e.g. C. sativus - The Saffron, Safran, or Azafran, a perennial plant, 8-30 cm high, with flared to funnel-shaped solitary flowers of brilliant red, pink, and violet colors; the throat is bearded at base. From the dried stigmas and tops is extracted Saffron (Acafrao; Azafran; CI Natural Yellow 6; Colour Index No: 75, 100; Crocus; Estigmas deazafran; Safran; Stigma Croci). Used as a food dye and flavoring agent. Was widely used as a glycerine, syrup and tincture for coloring medicines, but it has been largely replaced for this purpose by solutions of orange G and tartrazine. The Saffron plant is a native of the Mediterranean region; naturalized in Central and West Europe; introduced and naturalized in Britain.

Crocuta A genus of Mammals of the Hyena family, *Hyaenidae*. e.g. C. crocuta - The Spotted Hyena or Spotted Hyaena, a direct descendant of the Cave Hyaena (*Hyaena* spelaea), whose fossil bones and dung are found in Paleolithic beds. The coat is reddish grey with brown spots. A scavenging carnivore, adapted for carrion-feeding, it travels in packs to attack sheep and other domestic animals when food is scarce. It has a wide vocal range. Habitat: East and South Africa.

Crossandra A genus of indoor flowering plants of the family *Acanthaceae*. e.g. C. infundibuliformis - The Crossandra, a small evergreen shrub with showy salmon-pink flowers in 4-cornered spikes. Place of origin: India.

Crossomys A genus of Mammals of the Australasian Water Rat subfamily, *Hydromyinae* (Rat family, *Muridae*). e.g. C. moncktoni - The New Guinea Water Rat, with large webbed hind feet, small fore feet, vestigial ears, and a tail fringed with hairs for use in swimming.

Distribution: from the Philippines to New Guinea and Australia.

Crossoptilon A genus of Birds of the Pheasant family, *Phasianidae*; also known by the generic name, *Crossoptilum* (see **Crossoptilum**).

Crossoptilum A genus of Birds, also called *Crossoptilon*, of the Pheasant family, *Phasianidae*. e.g. C. mantchuricum (= *Crossoptilon* m.) - The Brown-eared Pheasant; in this species the sexes are similar, and in both, the cheek feathers project like ears. Habitat: Manchuria.

Crotalus A genus of Reptiles of the Pit Viper family, *Crotalidae*. e.g. C. horridus - The Timber Rattlesnake, the common species of the eastern United States. It varies in length from 90-180 cm (3-6 ft), and its color pattern is highly variable, consisting of transverse brown bands on a yellowish background. The rattle at the end of the tail is formed by a series of loosely connected horny segments. When the snake moves, these segments make a noise like that of a hand-rattle.

Crotaphytus A genus of Reptiles of the Iguanid Lizard family, *Iguanidae* (order *Squamata*). e.g. C. collaris - The Collared Lizard, 30 cm (1 ft) long; with black markings on the neck and shoulder; can run bipedally to escape enemies, a feature unusual among iguanid lizards.

Croton A genus of plants of the family *Euphorbiaceae*. e.g. C. tiglium - The Croton plant; from the seed is extracted Croton Oil (Oleum Crotonis; Olium Tiglii), containing an exotoxin or phytotoxin, crotin. Croton Oil is so violently purgative that it is now rarely employed, though it has been used in very severe constipation. Externally, it is a powerful counter-irritant and vesicant.

Crotophaga A genus of the Cuckoo family, *Cuculidae*; known as the Anis. The species range from Mexico to Argentina; with glossy black plumage, long tail and laterally compressed bill, an antithesis of what is expected of members of the cuckoo family.

Cruciata A genus of plants of the Bedstraw or Madder family, *Rubiaceae*. e.g. C. chersonensis - The Crosswort, a perennial plant, 15-70 cm high; leaves hairy on the veins. Grows in open woodlands, hedges, pastures and by roadsides. Native in Britain.

Crynchia A genus of Molluscs of the class *Cephalopoda*. e.g. C. scabra - A cephalopod with a high degree of complexity in its structure. This squid is exclusively a marine animal.

Cryphalus A genus of Insects of the Bark Beetle family, *Agaontidae* (also known as the Fig Insect family). e.g. C. piceae - Size 1.5-2 mm; lives in firs; main borehole small and room-like; eating pattern star-shaped; lays its eggs under the bark inside niches in a main borehole. The larvae do not bore deeply into the wood, but make a series of tunnels which may be seen when the bark is stripped off a tree or log (see **Blastophaga**).

Cryptanthus A genus of Trees of the Pineapple family, *Bromeliaceae*. e.g. C. zonatus - A species from Brazil; it has a rozette of beautifully banded leaves and a small central cluster of white flowers.

Cryptobranchus A genus of Amphibians of the family *Cryptobranchidae*. The Hellbinder, a heavily built newt or salamander, up to 45 cm (11/2 ft) long, with no external gills and living in mountain streams. Habitat: the eastern U.S.

Cryptocellus A genus of Arachnid Insects of the order *Ricinulei*. e.g. C. albosquamatus - A recently discovered Ricinuleid species, less than 1 cm long, living in the rotting humus and moist litter on the floor of dense rain forests or in caves. Like other representatives of this rare order, little is known about its biology. Habitat: several parts of tropical South America.

Cryptocephalus A genus of Insects of the Leaf Beetle family, *Chrysomelidae*.

e.g. C. moraei - Length 3-5 mm; glossy black with large yellow patches across the elytra. Found on blossoms in sunny areas. Distribution: Europe.

Cryptocercus A genus of Insects of the Cockroach suborder, *Blattodea*. e.g. C. punctulatus - A North American species of wild cockroaches, which feeds on dead wood, digesting it with the aid of symbiotic protozoans inhabiting its gut.

Cryptochiton A genus of Molluscs of the Chiton subclass, *Polyplacophora* (= *Loricata*). e.g. C. stellari - The largest of all the chitons; a species found in the waters of the northern Pacific (see **Chiton**).

Cryptococcus 1. A genus of Molds or Yeast-like Fungi of the family *Cryotococcaceae*; formerly classified under the generic names, *Torula* and *Torulopsis*. e.g. C. hominis (= C. histolytica; C. meningitidis; C. neoformans; *Torula* histolytica, et al.; *Torulopsis* h., et al.) - It causes the infection, cryptococcosis or torulosis in humans. C. epidermis (= *Saccharomyces* epidermica) (see **Saccharomyces**; also **Candida**). 2. A genus of Homopterous Insects of the family *Eriococcidae*. e.g. fagi - Length 0.8-1 mm; forms large colonies on the trunks of beech trees. Locally very abundant. Distribution: Europe, Asia Minor, North America.

Cryptogamma A genus of plants of the Fern or Polypody family, *Aspleniaceae*. e.g. C. crispa - The Parsley-fern, a perennial plant, 15-30 cm high; yellowish-green sterile frond; fertile frond with cylindrical inrolled terminal segments. Grows on acid-rock scree, mainly in mountainous areas. Native in Britain.

Cryptolucilia A genus of Flies. e.g. C. caesarion - A bright green fly breeding in cow-manure.

Cryptoprocta A genus of Mammals of the Civet, Ginet, and Mongoose family, *Viverridae*. e.g. C. ferox - The Fossa, the largest carnivore in Madagascar, 90 cm (3 ft) long with a 60 cm (2 ft) tail, and 150 cm (5 ft) in overall length. It has short legs, as in all the Viverridae. Though largely a terrestrial animal, it climbs trees to hunt lemurs. Its coat is a rich warm brown.

Cryptotis A genus of Mammals of the Shrew family, *Soricidae*. e.g. C. parva - The Short-tailed Shrew, a North American species; these are the red-toothed shrews, though the color is really more of a chestnut brown.

Crypturellus A genus of Birds of the Tinamou order, *Tinamiformes*. e.g. C. variegatus - The Variegated Tinamou, a species inhabiting the dense rain forests of South America; its color is cryptic to afford protection. The eggs are laid on the ground, their shiny surfaces blending into the wet, glistening moss and leaves of the forest floor, giving the appearance of glazed porcelain in pale pastel grey, lilac and primrose, or olive and dark reds, purples and even blacks. These (eggs) are incubated by the male.

Crystalogobius A genus of Fishes of the Goby family, *Gobiidae*. e.g. C. linearis - Size 48 cm; a completely transparent fish, with 3-5 small spots along the side. An active sea fish living in tidal pools. Native in Britain.

Ctenocephalides A genus of Insects of the Flea family, *Pulicidae*; also known by the generic names, *Ctenocephalus*, *Ctinocephalus*, or *Pulex*. e.g. C. canis (= Ct. canis; *Ctenocephalus* c.; *Ctinocephalus* c.; *Pulex serraticeps*) - the dog flea, a wingless insect, externally parasitic on dogs, and capable of transmitting the dog tapeworm (*Dipylidium caninum*) and other animal tapeworms to man (see **Pulex**).

Ctenocephalus A genus of Insects of the Flea family, *Pulicidae*, better known by the generic name, *Ctenocephalides* (see **Ctenocephalides**; **Pulex**).

Ctenodactylus A genus of Rodents (class *Mammalia*). e.g. C. gondi - A North American rodent, being a common reservoir of the protozoan parasite,

Toxoplasma gondi, the cause of Toxoplasmosis in man (see **Toxoplasma**).

Ctenolabrus A genus of Fishes of the Wrasse family, *Labridae*. e.g. C. rupestris - The Blunt-snouted Gold Sinny, length 18 cm; blackspot on tip of tail; partly spiny, long dorsal fin; thick lips which fold inside the mouth. Lives mostly amidst marine rock formations, feeding on bottom-dwelling fauna. Distribution: the northeast Atlantic, the Mediterranean, and the Black Sea.

Ctenolepisma A genus of Insects of the Silverfish family, *Lepismatidae*. e.g. C. longicaudata - The Australian Silverfish, a domestic species; occurs as a minor insect pest in warm, dry places such as restaurants or centrally heated premises.

Ctenomys A genus of very primitive Rodents of the New World, of the Coypu family, *Echimyidae*. e.g. C. mendocinus - The Tucotuco, a large rodent, about 22.5 cm (9 in) long; adapted to a burrowing habit; it has a cylindrical body, with short limbs bearing large digging claws, and small eyes and ears. Distribution: common in Chile, Argentina, and some parts of Brazil, where it makes extensive burrows in open country.

Ctenophthalmus A genus of Fleas. e.g. C. agrytes - The European mouse flea.

Ctenophyryngodon A genus of Fishes of the Carp and Minnow family, *Cyprinidae*. e.g. C. idella - The Grass Carp, length 1 m (39 in); weight 32 kg (701/2 lb); native to the Amur River in northern China, from where it was introduced into European ponds before several decades, where it became established. Feeds chiefly on aquatic vegetation (see **Carassius**).

Ctenopsylla See **Ctenopsyllus**.

Ctenopsyllus A genus of Fleas found on rats and mice, also known by the generic names, *Ctenopsylla* and *Leptopsylla*. e.g. C. segnis (= *Ctenopsylla* s.;

Leptopsylla musculi) - The common flea of the house rat.

Ctenosaura A genus of Reptiles of the Iguana family, *Iguanidae*. The species are the black iguanas of Central America. Only the tail has rows of spines which can wound potential aggressors. These reptiles can run in a bipedal fashion. They feed on mice, birds, and plants, and may reach a length of over 60 cm (2 ft).

Ctinocephalus (= *Ctenocephalus*) (see **Ctenocephalides**; **Pulex**).

Cucubalus A genus of plants of the Pink family, *Caryophyllaceae*. e.g. C. baccifer - The Berry Catchfly, a perennial plant, up to 3 m tall; flowers usually solitary; petals with coronal scales; the fruit is a black berry. Grows in scattered localities in woods. Place of origin: Central and South Europe.

Cuculia A genus of Insects of the order of Butterflies and Moths, *Lepidoptera*. e.g. C. umbratica - The Shark Moth; its grey wings resemble decaying wood.

Cuculus A genus of Birds of the Cuckoo family, *Cuculidae*. e.g. C. canorus - The Common Cuckoo, length 33 cm; a lively, agile species; plumage chiefly grey. The characteristic "cuckoo" call is made only by the male; the female has a bubbling call. Found in open, mixed groves with dense undergrowth of thickets. A migrant bird, it winters in tropical and South Africa. Distribution; all over Europe, Asia, most of Africa.

Cucumaria A genus of animals of the Sea Cucumber class, *Holothuroidea*. e.g. C. planci - Size 15 cm; feathery tentacles; related to the star-fish, and lives on muddy sea-beds. Distribution: Mediterranean; also more northerly waters.

Cucumis A genus of plants of the Cucumber or Gourd family, *Cucurbitaceae*. e.g. C. sativus (= C. sativa) - The Cucumber plant, a native of southern Asia, but cultivated varieties are worldwide. The seeds are diuretic.

Cucurbita A genus of plants or vines of the Cucumber or Gourd family, *Cucurbitaceae*. e.g. C. citrillus - The Water-melon, containing the sterol cucurbitol, and the extract cucurbocitrin. Has been used as a tenicide. C. lagenaria - The poisonous tula de mate. C. maxima - The fresh seeds have been used as a tenicide. C. pepo - The Pumpkin or North American Squash. The fresh seeds known as Cucurbita (= Abobora; Kurbissame; Melon Pumpkin Seeds; Pepo; Semena de Courge), have been used as a tenicide.

Culex A genus of Insects of the Mosquito family, *Culicidae*. e.g. C. fatigans - The culicine mosquito, transmitting elephantiasis (filariasis) to man.

Culicoides A genus of Biting Fleas of the family *Heleidae*. e.g. C. austeni - Intermediate host of *Acanthocheilonema perstans*.

Cultellus A genus of Molluscs of the Razor Shell family, *Solenidae*. e.g. C. pellucidus - The Transparent Razor Shell, 3.9 cm long; 1 cm wide; thin greenish-white shell. Lives in vertical holes on sandy shores. Although dead shells may often be found, it is rare to catch a live Razor Shell, as these animals can dig faster into the sand than most people.

Cuminum A genus of plants of the Carrot family, *Umbelliferae*. e.g. C. cyminum - The Cummin plant. Cummin fruit yields the volatile Cummin Oil (Cumin oil). Used as a carminative.

Cuniculus A genus of Mammals of the Paca family, *Cuniculidae*. e.g. C. paca - The Sooty Paca or Spotted Cavy of Mexico and Brazil, a large tailless, heavy-bodied rodent, about 70 cm (28 in) long. It has a brown coat, ornamented on each side by 3-5 lines of pale spots.

Cunila A genus of plants of the Thyme family, *Labiatae*. e.g. C. mariana - A North American species. It is diuretic and diaphoretic.

Cunina A genus of Hydrozoans of the order *Narcomedusae*; in these species the planula becomes parasitic, attaching itself to the trachymedusa, *Geryonia*, and developing as an elongated stolon which then buds off medusae.

Cunninghamia A genus of Trees of the Pine family, *Pinaceae*. e.g. C. lanceolata - The Chinese Fir or China-Fir Tree, an evergreen tree up to 45 m (150 ft) tall, from Central and South China. Each side of the ovoid, mature cone bears 3 flattened seeds.

Cuon A genus of Mammals of the Wild Dog, Wolf, and Jackal family, *Canidae*. e.g. C. alpinus - The Siberian Wild Dog, 90 cm (3 ft) long (without the tail); reddish above and whitish below; looks like a shaggy Alsation but differs from domestic dogs in having only 40 teeth (instead of 42 or 44), and more numerous teats. These creatures hunt in small packs.

Cupressus A genus of Trees of the Pine family, *Pinaceae*. e.g. C. sempervirens - The Mediterranean Cypress, the cypress tree of classical literature; grows wild on the mountains of the eastern Mediterranean region, and is widely planted in southern Europe, particularly Italy. It reaches 45 m (150 ft) in height. The shiny brown or grey globose cone is about 2.5 cm (1 in) across.

Curculio A genus of Beetles of the Weevil family, *Curculionidae*. e.g. C. venosus - Length 5-7 1/2 mm; these insects are found on oaks, their larvae developing in acorns which, as a result, fall prematurely. They pupate on the ground. Distribution: Europe, Asia Minor.

Curcuma A genus of plants of the Ginger family, *Zingiberaceae*. e.g. C. longa - The Turmeric plant; the dried rhizome, Turmeric (= Curcuma; Indian Saffron; CI Natural Yellow 3; Color Index No: 75300) contains a yellow pigment, curcumin. Turmeric is used principally as a constituent of curry powders and other condiments. It has been employed in the treatment of chronic cholecystitis. The tincture is used for thepreparation of turmeric paper as a test for boric acid and borates; it has also been used as a yellow coloring agent, but the color is fugitive in solution. C. zeodaria - The Zedoary turmeric plant of East India; from the rhizome, which resembles ginger, is obtained an aromatic substance, used in perfumes and medicines.

Cursorius A genus of Birds of the Pratnicole and Courser family, *Glareolidae*. e.g. C. cursor - The Cream-colored Cursor or Courser of Africa, also known as the Desert Runner; length 23 cm; with black and white eye-stripe. Its cream color is a perfect camouflage in the sandy desert and it can run with incredible speed; it also flies well from district to district. Distribution: the deserts of North Africa; also on the sand dunes of southern France and other Mediterranean coasts of Europe; southeast Asia.

Cuscuta A genus of plants of the Bindweed or Convolvulus family, *Convolvulaceae*. e.g. C. europaea - The Large or Greater Dodder, a non-green, annual, parasitic, twining plant found particularly on the Stinging Nettle, *Urtica* dioica and the Hop, *Humulus* lupulus. It has slender reddish stems which are attached to the host by suckers or haustoria. The small pinkish-white flowers are in clusters, and the capsule contains 4 seeds. Place of origin: Europe; Asia; has been introduced into North America. Native in Britain.

Cuspidaria A genus of Molluscs of the subclass *Septibranchia*; these are small carnivores confined to the sea, with a characteristic siphonal shell extension occurring in moderate depths on muddy gravel.

Cyanea A genus of True Jellyfishes of the order *Semaeostomeae*; they have a diameter of 30-200 cm, and are often colored in patches and streaks; these animals can be identified by the lobed bell margin with 8 bundles of tentacles arranged around it. They are found in Polar regions, and off the coasts of Britain and America. e.g. C. capillata - The Red or Arctic Jellyfish; disc diameter up to 1 m (39 in); umbrella flat, with enteron sexual organs, and gastric pouches yellowish brown to reddish yellow. The tentacles are up to 2 m (61/2 ft) long. Habitat: open water in the North Sea and the Atlantic Ocean.

Cyanerpes A genus of Birds of the Hawaiian Honeycreeper family, *Drepanididae*. e.g. C. cyaneus - The Yellow-winged Blue Honeycreeper; the males are brilliantly colored with beautiful tones of blue, blue-green and torquoise blue. The females are more sober with tones of green. The males moult into a female-like plumage during the non-breeding season.

Cyaniris A genus of Insects of the Blue, Copper, and Hairstreak Butterfly family, *Lycaenidae*. e.g. C. semiargus - The Mazarine Blue Butterfly, length 16-18 mm; the male wings are blue-violet on the upper side, the female wings brown. Occurs also in mountains (in the Alps at about 2,500 m). Distribution: Europe, Asia.

Cyanocitta A genus of Birds of the Crow, Magpie, and Jay family, *Corvidae*. e.g. C. cristata - The Blue Jay; a noisy blue-plumed bird living in North America, and commonly found east of the Rocky Mountains, from southern Canada to the Gulf Coast.

Cyanocorax A genus of Birds of the Crow, Magpie, and Jay family, *Corvidae*. e.g. C. yncas - The Crestless Green Jay; it has a blue head, black cheeks and bib, and yellow outer tail feathers composing its otherwise green plumage.

Cyanopica A genus of Birds of the Crow, Magpie, and Jay family, *Corvidae*. e.g. C. cyanus - The Azure-winged Magpie, length 34 cm; black cap; blue wings and tail; a resident bird, building its nest in trees. Distribution: Spain and Portugal.

Cyanopsis A genus of plants of the Daisy family, *Compositae*. e.g. C. tetragonaloba - The Guar or Jaguar plant. From the ground endosperms of the seeds, is prepared Guar Gum (= Guar Flour; Jaguar Gum), used as a thickening and suspending agent and emulsion stabilizer in the manufacture of various foods. A 1-1½% mucilage is used as a pill excipient and as a binding and disintegrating agent in tablets.

Cyanosylvia A genus of Birds of the Thrush, Robin, Nightingale and Old World Blackbird subfamily, *Turdinae* (family *Muscicapidae*). e.g. C. svecica - The Bluethroat; this bird has a lovely azure blue throat and breast, with a central white patch and a tri-colored collar. Distribution: Europe.

Cyathea A genus of Trees of the Fern or Polypody family, *Aspleniaceae*, which grow in tropical and subtropical regions. The species have an erect, straight trunk covered with roots and leaf bases, and some reach a height of 18-21 m (60-70 ft).

Cyathura A genus of Crustaceans of the family *Anthuridae*. e.g. C. polita - Like other crustaceans in this family, the body is elongated, and there is a pair of statocysts in the telson. The species live in brackish and even freshwater, and have been found in densities up to 4,000 per m² in the Pocasset River, Massachusetts.

Cybister A genus of Insects of the family of the True or Carnivorous Water Beetles, *Dytiscidae*. e.g. C. laterali marginalis - Size 30-35 mm; hind feet with one claw; tibia of hind leg very short. Feeds on many types of prey. Found in pools and ponds, and may also cause considerable damage in fish hatcheries. Distribution: Europe.

Cycas A genus of indoor plants of the family *Cycadaceae*. e.g. C. revoluta - The Cycas plant, sometimes known as the Sago Palm; it has a cylindrical trunk growing to 2.1 m (7 ft), covered by the remains of leaf bases. Leaves consist of flat, slightly sickle-shaped pinnules, rolled at the edges, with sharp tips. Place of origin: China and Japan.

Cychrus A genus of Insects of the Ground Beetle family, *Carabidae*. e.g. C. caraboides - Length 15-23 mm; found in damp forests from lowlands to mountainous elevations. A predaceous species, hiding under logs, stones, and beneath the bark of tree stumps. It does not chew or swallow its prey, but decomposes it first by regurgitating a small amount of digestive juice after which it can suck up its meal in liquid form. Distribution: much of Europe.

Cyclamen A genus of Mediterranean plants of the Primrose family, *Primulaceae*. e.g. C. europaeum (= C. purpurascens) - The European Cyclamen or Purple Sowbread, a perennial plant, 6-15 cm high; with a large globular or flattened corm. Long-stalked, heart- or kidney-shaped leaves, silvery-zoned above and often purplish below. Sweetly scented, rich carmine flowers, occasionally white or pink, with reflexed petals. It has an acrid cathartic root, from which is extracted the glycoside cyclamin. Used in domestic medicine as a purgative and painkiller. The corms are regarded as a favorite food for swine in parts of southern Europe, hence the name Sowbread. The plant grows in deep mountain forests. Place of origin: Central and Southern Europe; introduced into Britain in 1613.

Cycleptus A genus of Fishes of the Sucker family, *Catastomidae*; these are the Carpsuckers, also known as the Missouri Sucker or Blackhorse; they are jet black, large fishes, bearing an external resemblance to the carp, and move up the smaller tributaries of the Mississippi and Ohio in spring.

Cyclestheria A genus of Crustaceans of the order *Conchostraca*. e.g. C. hislopi -

This conchostracan species is circum-tropical, and is exceptional in producing a second type of egg, which will develop quickly in the brood-pouch of the mother, and give rise to a miniature of the adult without first having to pass through a nauplius stage.

Cyclopes A genus of Mammals of the Anteater family, *Myrmecophagidae*. e.g. C. didactylus - The Two-toed or Silky Anteater, a species about the size of a squirrel, about 30 cm (12 in) long, an arboreal edentate, with long fur and an extremely elongated snout. Habitat: the hottest parts of Brazil, Guiana, and Venezuela.

Cyclops A genus of minute Crustaceans of the family *Cyclopidae* (subclass *Copepoda*). e.g. C. strenuus - This species of copepods acts as host to the Fish Tapeworm, *Diphyllobothrium latum*.

Cycloptera A genus of Insects of the orthopteran order of Grasshoppers, Crickets, and their allies, *Orthoptera*; in these species the color, shape, and pattern closely imitate the surrounding leaves, producing a perfect camouflage to the environment.

Cyclopterus A genus of Bony Fishes of the Lumpsucker family, *Cyclopteridae*. e.g. C. lumpus - The Lumpfish, Sea Hen, or Lumpsucker Fish; length 60 cm; male mainly red, female mainly blue; ventral fins modified to forma sucking disc. Lives near the sea bottom from shallows to depths of about 300 m. Distribution: both sides of the North Atlantic; native in Britain.

Cyclospora A genus of pathogenic proto-zoa, first identified in 1977. e.g. C. cayetanensis - A species of intestinal protozoan parasites, causing waterborne outbreaks of gastroenteritis and entero-colitis, with vomiting, severe intestinal cramps, and diarrhea. Can be treated with antibiotics.

Cyclothone A genus of Bony Fishes of the Bristlemouth family, *Gonostomatidae*. e.g. C. signata - Length 5 cm (2 in); member of a family that is the most numerous of deepsea fishes. Eyes silver, and light organs arranged in rows along the sides of the body. Distribution: the depths of all seas and oceans.

Cydia A genus of Insects of the Moth family, *Tortricidae*. e.g. C. pomonella - The Codling Moth; size 7-9 mm; the pink caterpillars are pests, living in apples and pears, and growing to a length of about 20 mm. Distribution: cosmopolitan.

Cydonia A genus of plants, better known by the generic name, *Chaenomeles*, of the Rose family, *Rosaceae*. e.g. C. oblonga (= C. vulgaris) - The Cydonia (= Cognassier; Cydoniae Semen; Quince Seed; Quittenkern), contain about 20% of mucilage (cydonine) and is used as a demulcent and suspending agent. C. japonica (= *Chaenomeles* speciosa) - Japanese Quince (see **Chaenomeles**).

Cygnopsis A genus of Birds of the Duck, Goose, and Swan family, *Anatidae*. e.g. C. cygnoides - The Swan Goose; inhabits parts of Siberia, the Altai and Mongolia, where it is found in its original wild form. It has been domesticated by the Chinese, and this variety is known as the Chinese Goose. The bird roosts in trees and is extremely shy.

Cygnus A genus of Birds of the Duck, Goose, and Swan family, *Anatidae*. e.g. C. cygnus - The Whooper or Wild Swan, an all-white European swan; length 1.5 m (5 ft); the basal half of its black bill is lemon-yellow and the neck is held straight; it is found on open expanses of water and in swamps in the tundra, in river estuaries, and on mountain lakes. A migratory bird and a winter visitor from the tundra, it travels across Central Europe between early October and April. Distribution: Iceland, northern Scandinavia, and northern Asia. Native in Britain.

Cylindrogloea A genus of microorganisms of the family *Chlorobacteriaceae* (order *Pseudomonadales*) - Ovoid to rod-shaped cells. e.g. C. bacterifera.

Cymbalaria A genus of plants of the Figwort family. e.g. C. muralis - The Ivy-leaved Toadflax, a perennial plant, 10-80 cm high; blue flowers with 2 yellow spots on palate. Commonly grows on walls. Place of origin: South Europe; naturalized in Central and North Europe, including Great Britain.

Cymbidium A genus of plants of the Orchid family, *Orchidaceae*. There are numerous species or cultivars, with 15-25 blooms per stem on a well-grown plant, and up to 200 sprays. Place of origin: Asia, Australia; in Europe since c. 1838. e.g. C. insigne - A beautiful, variable species from Assam. The flowers are pinkish-red with darker spots.

Cymbopogon A genus of plants of the Grass family, *Gramineae*. e.g. C. nardus - The Citronella plant; distillation of the plant produces a pale to deep yellow oil with a pleasant characteristic odor, Citronella Oil (= Oleum Citronellae; Ol. Citronell.), containing geraniol and citronellal. There are two main types of Citronella Oil, differing in odor and composition and known as Ceylon Oil and Java Oil. Chiefly used as a perfume, particularly for soaps and brilliantines; also as a constituent of insect repellents.

Cynara A genus of plants of the Daisy family, *Compositae*. e.g. C. scolymus - The True or Globe Artichoke, a native of southern Europe; has been widely cultivated for the sake of its large globular capitula. The numerous fleshy involucral bracts or leaves, Artichoke Leaf (= Alcachofra; Artichaut; Cyanara) contain cynarin, polyphenolic acidic substances, and flavonoids, and are said to have diuretic and choleretic properties. The leaves are eaten with the tip of the stem when young, before the florets are exposed.

Cynips A genus of Insects of the Gall Wasp family, *Cynipidae*. Also known by the generic name, *Adleria*. e.g. C. gallae tinctoriae (= *Adleria* g.t.) - The Gall Wasp (see **Adleria**).

Cynocephalus A genus of Mammals of the order of Flying Lemurs, *Dermoptera*. e.g. C. variegatus - The Coluga, Cobego, or Flying Lemur; once classified as an insectivore, but now placed in a separate order. Its gliding membranes show affinities with lemurs and bats, and it has been classified as a carnivore on account of its dentition. It resembles bats also in having only one young, which clings to its mother's breast or belly. There is only a single species, ranging over Malaya, South China, Indonesia, and the Philippines.

Cynodon A genus of plants of the Grass family, *Gramineae*. e.g. C. dactylon - Bermuda-grass, a perennial plant, 20-40 cm high; spikes in clusters of 3-6 at tip of culm; growth widespread, especially in warmer or dry areas. Introduced and naturalized in Britain.

Cynogale A genus of Mammals of the Civet, Genet, and Mongoose family, *Viverridae*. e.g. c. bennetti - The Otter Civet, 75 cm (21/2 ft) long; an aquatic animal, although still partly arboreal; found in Malaya, Sumatra, and Borneo. It has soft thick fur and a long moustache. The feet are webbed and it feeds largely on fishes and frogs.

Cynomys A genus of Mammals of the Squirrel, Prairie dog, and Marmot family, *Sciuridae*. e.g. C. ludovicianus - The Prairie Dog, one of the North American ground squirrels; 35 cm (2 in) long, with a tawny brown coat and a dirty white underside, living in huge colonial burrows called "dog-towns", with each member having its own den. It was given its common name by the early Canadian and American trappers because of its yelping and barking.

Cynoglossum A genus of plants of the Borage family, *Boraginaceae*. e.g. C. officinale - Hound's Tongue, an erect, bristly, biennial plant, from 30-90 cm

142

(1-3 ft) high; the leaves are covered on both sides with silky hairs, there are small, dull red flowers, and the flattened nutlets are covered with short spines. It smells of mice. The dried root, Cynoglossum Root (= Hound's-tongue Root) has been used as a demulcent and sedative in coughs and diarrheas. It grows in dry, arable and waste places, in short grasslands and by roadsides. Distribution: Europe, Asia; native in Britain.

Cynopithecus A genus of Mammals of the Old World Monkey family, *Cercopithecidae*. e.g. C. niger - The Black Ape of the Celebes; it has a crest of stiff hairs on the top of its head which can be erected at will. Although known as an ape, this animal is in fact, a monkey. It is a shy, gentle creature, and lives in coastal mangrove forests, feeding upon fruit, leaves, and sea foods.

Cynops A genus of Amphibians of the Salamander family, *Salamandridae*. These newts range from China to the United States and Europe.

Cynoscion A genus of Fishes (subphylum *Vertebrata*). e.g. C. regalis - The Weakfish.

Cynosurus A genus of plants of the Grass family, *Gramineae*. e.g. C. cristatus - The Crested Dog's-tail, a perennial plant, 25-60 cm high; panicles one-sided; dense spikelets in clusters of two kinds, fertile and sterile. Common in meadows, pastures, and sandy places. Place of origin: Europe; native in Britain.

Cynoxylon A genus of Shrubs of the Cornel or Dogwood family, *Cornaceae*; better known by the generic name, *Cornus*. e.g. C. floridum (= *Cornus florida* "Rubra") - The Pink Flowering Dogwood (see **Cornus**).

Cynthia 1. A genus of Insects of the Tortoiseshell Butterfly family, *Nymphalidae*; often known by the generic name, *Vanessa*. e.g. C. cardui (= *Vanessa* c.) - Painted Lady, size 27-55

mm; wings yellowish-red, with black and white spots; flies in dry meadows, steppes, and fields, and avoids forests. A migrant that arrives in central Europe from the south in early April. Distribution: almost worldwide; absent only in South America. 2. A genus of Sea-squirts of the class *Ascidiacea*. e.g. C. papillosa - Size 12 cm; this animal has a beautiful red, tuberous tunic. Distribution: the Mediterranean coast (see **Clavelina**).

Cyperus A genus of grasslike reed, rush, and sedge plants of the Sedge family, *Cyperaceae*. e.g. C. articulatus - a grasslike species growing in the West Indies, with a tonic, antiemetic, and anthelmintic root (adrue). C. papyrus - The Egyptian Paper-plant, Paper Reed, or Papyrus, a noble, perennial plant with large, round "mop-heads" of greenish-brown inflorescences on short, triangular, bare stems, up to 4-6 m (13-19 1/2 ft) tall in its native habitat. Grows in marshy and damp places beside ponds and in ditches. The pith of this plant supplied the paper or papyrus of the ancient Egyptians, and these were the "bull rushes" that sheltered the infant Moses. Place of origin: Egypt, tropical Africa, Syria; introduced into Europe in 1803.

Cypraea A genus of Molluscs of the class of Gastropod animals, *Gastropoda*. e.g. C. pyrum - Size 4 cm; shell brownish golden. Habitat: the Mediterranean coast.

Cyprinodon A genus of Fishes of the suborder *Cyprinodontei*. e.g. C. mascularius - The Desert Pupfish; this species can live in hot springs where the water temperature reaches 50° C (122° F).

Cyprinus A genus of Bony Fishes of the Carp and Minnow family, *Cyprinidae*. e.g. C. carpio - The Common Carp, length 120 cm; weight more than 30 kg; mouth terminal, upper lip with 4 barbels, and scales large and regularly arranged; most popular fish of European pond culture. The most common variety bred in ponds is the Mirror Carp, whose body is irregularly covered with scales of different sizes. The species originated in central Asia, China, and Japan, and was transported by man, becoming worldwide and an important food of commerce (see **Carassius**).

Cypripedium A genus of plants of the Orchid family, *Orchidaceae*. e.g. C. calceolus - Lady's Slipper, Mocassinflower, or Common Slipper Orchid, a perennial plant, 15-45 cm (6-18 in) high; lip large, inflated, slipper-shaped. A terrestrial orchid with a thick, short-jointed, creeping rhizome carrying numerous roots. Solitary flowers with bright yellow pouches and brownish-purple sepals and petals about 8 cm (3.2 in) across, with a fruity smell. Place of origin: Europe, North Asia, and North America.

Cypris A genus of Crustaceans of the subclass of Ostracods or Seed Shrimps, *Ostracoda*. As with other ostracods, the species have a bivalved carapace into which they can withdraw all their limbs. They are about the size of a cherry, and the antennae are protruded between the two shell valves, being used for swimming.

Cypselurus A genus of Fishes of the order of Skippers and Flyingfishes, *Synentognathi* (= *Beloniformes*), belonging to the family *Exocoetidae*. Also known by the generic name, *Cypsilurus*. e.g. C. californicus (= *Cypsilurus* c.) - The California Flying-fish. C. cyanopterus (= *Cypsilurus* c.) - This species has an unusual barbel development, which appears when the fish is about 19 mm (0.75 in) long. When it is 5 cm (2 in) long, these barbels are longer than the body; however when the fish is fully grown (about 25 cm or 10 in long), the barbels have completely disappeared.

Cypsilurus See **Cypselurus**.

Cyrtodactylus A genus of Reptiles of the Gecko family, *Gekkonidae*. e.g. C. kotschyi - Kotschyi's Gecko, a lizard-like reptile, 10 cm long; lives amidst large piles of stones, on rocks and in ruins. Also found on the walls of buildings near lamps or lanterns, where it captures insects attracted to the light. Distribution: southern Europe and western Asia.

Cyrtonyx A genus of Birds of the Pheasant family, *Phasianidae*. e.g. C. montezumae - The Harlequin Quail, a species of New World quails, having picturesque plumes arising from the crown of their heads.

Cysticercus A genus name once given to larval forms of the Tapeworm, but now denoting a common juvenile form usually known as a Bladder worm because the inverted developing scolex lies within a vesicle filled with liquid. Some forms of cysticerci were discovered and named before the corresponding adults were known, as in C. pisiformis. e.g. C. bovis - The bladder worm corresponding to *Taenia* saginata or the adult Beef Tapeworm. C. pisiformis - Corresponding to the adult *Taenia* pisiformis.

Cystomonas A genus of flagellate microorganisms of the family *Bodonidae*; better known by the generic name, *Bodo* (see **Bodo**).

Cystophora A genus of Mammals of the Early Seal family, *Phocidae* (subfamily *Cystophorinae*). e.g. C. cristata - The Hooded Seal or Bladdernose; the males are over 240 cm (8 ft) long, and the females about 180 cm (6 ft). The males can inflate their nostrils to form a hood or crest. Body light grey; head and flippers black. Habitat: the Arctic waters and drift ice from Spitsbergen to Canada.

Cystopteris A genus of plants of the Fern or Polypody family, *Aspleniaceae*. e.g. C. fragilis - The Brittle Bladder Fern, a perennial plant, 10-40 cm high; tufted, 2-3-pinnate leaves; spores prickly.

Grows on old walls and rocks, and also found in mountainous regions. Place of origin: Southern Europe; native in Britain.

Cytherella A genus of Ostracod Crustaceans or Seed Shrimps of the order *Platycopa*. The species lives in deep water and has a heavy shell which probably leaves it unable to swim; it has branched antennae and lacks a heart and eyes.

Cytisus A genus of Shrubs and Rock garden plants of the Pea family, *Leguminosae*, the various species of which are classified under different generic names, such as *Genista*, *Laburnum*, and *Sarothamnus*. e.g. C. laburnum (= *Laburnum* anagyroides; L. vulgare) - Golden Rain, one of the most poisonous trees growing in Great Britain; all parts are toxic. The toxic principle is Cytisine (= Baptitoxine; Laburnine; Sophorine; Ulexine) - A highly toxic alkaloid causing poisoning (cytisism), with nausea, vomiting, dizziness, mental confusion, muscular incoordination and weakness, convulsions, respiratory paralysis, and ultimately death by asphyxia (it resembles nicotine in its actions) (see **Laburnum**). C. radiatus (= *Genista* radiata) - The Royal Broom, a densely branched shrub up to 45 cm (11/2 ft) high; yellow flowers, silky-haired outside, in dense terminal clusters. Place of origin: southeastern France, central and eastern Europe; introduced into Britain in 1750 (see **Genista**). C. scoparius (= *Sarothamnus* s.) - The Scorparium plant, from which is obtained the dibasic alkaloid, sparteine (see **Sarothamnus**).

Cytopspermium A former name given to a genus of microorganisms, the *Coccidia*, now called by the generic name, *Isospora*. Other genus names used are *Coccidium*, *Eimeria*, etc. e.g. C. hominis (*Isospora* h.; *Coccidium* bigeminum; *Eimeria* stiedae; etc.) (see **Isospora**).

D

Daboia A genus of venemous Snakes. e.g. D. elegans - The Cobra-monil or Ticplonga, an extremely venemous serpent of Sri Lanka and India.

Dacelo A genus of Birds of the Kingfisher family, *Alcedinidae* (subfamily *Daceloninae* - the tree, forest, or wood kingfishers). e.g. D. novaeguineae - The Laughing Jackass or Kookaburra, the largest representative of the family, a powerful bird with a large head and a heavy bill, attaining the size of a Raven, almost 50 cm (1 ft 8 in) long. It ows its popular name to its cry, which sounds like human laughter. Often seen in towns, where it scavenges and eats snails and slugs. It makes its burrow in termite nests. Distribution: Australia, Tasmania, and New Guinea.

Dactylis A genus of plants of the Grass family, *Gramineae*. e.g. D. glomerata - Cock's-foot or Orchard Grass, a coarse, tufted, perennial plant, 30-125 cm high, with keeled leaves, long pointed ligules, and panicles with short-stalked compressed spikelets crowded at the ends of the branches. Grows in meadows, roadsides, and on waste ground. Distribution: widespread in Europe, Asia, North America, and North Africa. Native in Britain.

Dactylochalina A genus of Sponges (phylum *Porifera*) of the class *Silicea* - the horny or siliceous sponges. e.g. D. cylindrica - A South African species of keratosa or horny sponges, with a widely branched body.

Dactylogyrus A genus of Flukes of the order of monogenetic trematodes *Monogenea*. e.g. D. vastator - A gill parasite, infecting the carp fish less than one month old. The monogenetic lifecycle is so-called because it occurs within a single host.

Dactylopius A genus of Insects of the family *Coccidae*, better known by the generic name, *Coccus*. e.g. D. coccus (= *Coccus* cacti) (see **Coccus**).

Dactylopterus A genus of Fishes (subphylum *Vertebrata*). e.g. D. volitans - The Flying Gurnard fish.

Dactylorchis A genus of plants of the Orchid family, *Orchidaceae*. e.g. D. fuchsii - The Common Spotted Orchid, a species with spotted leaves and white to reddish-purple flowers growing on erect spikes. Occurs widely in wet non-acid soils. Distribution: Europe.

Dactylorhiza A genus of plants of the Orchid family, *Orchidaceae*. e.g. D. sambucina - Elder-scented Orchid, a perennial plant, 10-40 cm high, with purple or yellow flowers smelling of elders. Grows in meadows and on grassy hills. Distribution: South and Central Europe.

Dacus A genus of Acalyptrate Insects of the Fly family, *Trypetidae*. e.g. D. oleae - The Olive Fly, a pest of the Olive fruit, found in the Mediterranean countries and South Africa.

Daedalea A genus of Bracket Fungi belonging to the group of *Basidiomycetes*. They have a separate mycelium, and reproduction involves the formation of a basilium which produces usually four spores in small projections. e.g. D. confragosa - A species with elongated pores, growing on the fallen branches of deciduous trees.

Daedolacanthus A genus of Shrubs of the large tropical Acanthus family, *Acanthaceae*. e.g. D. nervosus - A prickly shrub of the East Indies, from 60-180 cm (2-6 ft) high, with ovate leaves and terminal spikes of blue, and two-lipped flowers about 2.5 cm (1 in) across.

Dahlia A genus of plants of the Daisy family, *Compositae*. e.g. D. variabilis - One of the 20 species of Garden Dahlias, a tuberous-rooted perennial plant with large, showy flowers. From the bulb is obtained the polysaccharide Inulin or Alant starch. Place of origin:

Mexico; introduced into Europe in 1789 (see **Inula**).

Dalechampia A genus of plants of the Spurge family, *Euphorbiaceae*. e.g. D. roezliana - An erect shrub, up to 1.2 m (4 ft) high, with large, coarsely toothed, lanceolate leaves and small dense groups of yellow unisexual flowers, surrounded by two large, pale red bracts. A native of Mexico.

Dallia A genus of Fishes of the order of Pikes and Mudminnows, *Haplomi*. These are the blackfishes found in bogs and swamps in Alaska, living in banks of sphagnum moss. They can be frozen and still survive.

Dama A genus of Mammals of the True Deer family, *Cervidae*. e.g. D. dama - The Fallow Deer, up to 1.5 m (5 ft) long, height at shoulder up to 90 cm (3 ft), and weighing about 68 kg (150 lb). It has broad, flattened, beautifully palmated antlers; its summer coat is reddish-brown with whitish spots, changing in winter to a dark greyish-brown. Found in broadleaved woodlands and meadows at lower elevations. It is native to the Mediterranean region and Asia Minor, but even there it is rare in a truly wild state. In Britain, western and central Europe, the Fallow Deer exists as a park animal since medieval times.

Dammara A genus of Trees of the family, *Dipterocarpaceae*. e.g. orientalis - This species produces dammar, a transparent resin used as a mountant in microscopy and in the manufacture of varnishes and plasters.

Danaus A genus of Insects of the Butterfly family. e.g. D. plexippus - The Monarch Butterfly, a famous insect migrant. In summer it occurs over the whole of the U.S. and southern Canada, migrating southwards in September in great swarms and overwintering in semi-hibernation on selected "butterfly trees" in Florida and California. Some 200 specimens of this butterfly have been reported from Britain; some may have been blown across the Atlantic on strong westerly winds, but most of them probably went by ship.

Daphne A genus of plants of the Daphne family, *Thymelaeaceae*. e.g. D. mezereum - The February Daphne or Mezereon plant, an erect, deciduous shrub, 1.5 m (5 ft) high, with wedge-shaped leaves, and deep pink to purple, fragrant flowers in threes, on each branch. It has round, scarlet, fleshy berries or fruits, and grows in wooded areas, mainly in mountain forests. The dried bark, Mezereon Bark (= Bois Gentil; Seidelbastrinde) is used as an external stimulant and vesicant. Distribution: widespread in Europe and western Asia; introduced into Britain in 1561.

Daphnia A genus of freshwater Crustaceans (order of Water-fleas, *Cladocera*), of the family *Daphnidae* or *Daphniidae*. e.g. D. magna - A large species in which reproduction is prolific, females producing broods of up to 100 eggs at intervals of 3 days. These eggs do not need to be fertilized, and each egg can give rise to a female which becomes mature in 8 days. In unfavorable conditions such as low temperatures or shortage of food, some of the eggs develop into males, and some females then produce another type of egg which needs fertilization.

Daphnis A genus of Insects of the Hawk-moth family, *Sphingidae*. e.g. D. nerii - The Orchid Hawk Moth; size up to 8.5 cm; broad body with beautiful green, whitish rose, and violet-colored wings. The caterpillars feed on oleander. Found in central and western Europe.

Dascyllus A genus of Fishes of the Damsel Fish family. e.g. D. marginatus - The Marginate Damsel Fish, a small, aggressive, territorial species, living among corals in groups, the size of which depends on the size of the coral. Habitat: the Coral Reefs of the Red Sea region.

Dasogale A genus of Mammals of the group of Tenrec Insectivores. e.g. D. fontoynonti - The rarest insectivore, a tenrec living in eastern Madagascar.

Dasyatis A genus of Fishes of the order of Rays and Skates, *Rajiformes*. e.g. D. pastinaca - The Common Sting Ray, up to 2.5 m long; the tail has a dorsal spine. Lives on the seabed, feeding mainly on molluscs. Found in the Mediterranean Sea.

Dasychira A genus of Insects of the Tussock Moth family, *Lymantriidae*. e.g. D. pudibunda - The Pale Tussock Moth, size 20-29 mm; forewings greyish with darker, variable pattern. Found on the edges of broadleaved forests, in parks and gardens. Distribution: western and central Europe, Siberia, and as far east as Japan.

Dasypeltis A genus of Reptiles of the Colubrid Snake family, *Colubridae*. e.g. D. scabra - The African Egg-eating Snake; although no thicker than a man's finger, it can swallow an egg three or four times the size of its own head.

Dasyphyllia A genus of Sea Anemones or Actiniaria (large marine polyps) of the subclass *Hexacorallia*. e.g. D. echinulata - A coral species found in the Indian Ocean.

Dasypoda A genus of Insects of the Bee family, *Apidae*. e.g. D. altercator (= D. hirtipes) - The Hairy-legged Mining Bee, length 13-15 mm; found on flowering hawkweed, field scabious, etc. The nest consists of a number of underground cells. Distribution: the Palaearctic region.

Dasyprocta A genus of Mammals of the Agouti Rodent family, *Dasyproctidae*. e.g. D. aguti - The Golden Agouti, a South American rodent resembling the hare in size and shape; brown in color with a tawny or yellow rump. Its fur is valued locally, in Brazil and Guiana. It inhabits open and virgin forest from Mexico to the Amazon valley.

Dasypus A genus of Mammals of the Armadillo family, *Dasypodidae*. e.g. D. novemcinctus - The Nine-banded Armadillo, about 75 cm (21/2 ft) long, and weighing up to 6 3/4 kg (15 lb). It has nine mobile bands between the scapular and pelvic shields - strong, bony plates covered with horn. The top of the head and the exposed parts of the limbs are also covered with scales, and the tail is also armored. Habitat: Central and South America, and the southern United States.

Dasyurops A genus of Mammals of the Dasyurus Cat family, *Dasyuridae*. e.g. D. maculatus - The Tiger Cat or Large Spotted-tailed Native Cat, nearly 1.2 m (4 ft) long, with a tail almost as long as the head and body. It is a handsome, arboreal marsupial or pouch-bearing mammal, with a reputation for savagery and destructiveness, spending the daytime in trees and bringing forth its young there. The Tiger Cat inhabits eastern and southeastern Australia.

Dasyurus A genus of Mammals of the Dasyurus Cat family, *Dasyuridae*. e.g. D. quoll - The Australian Eastern Native Cat, as big as a Domestic Cat; a terrestrial, carnivorous marsupial, savage and destructive on the farm, and holding on the Australian continent the same ecological position as weasels, wild cats, wolves, and other placental animals do elsewhere. It has a very short gestation period, as little as eight days sometimes.

Datura A genus of Shrubs and Bushes of the Nightshade family, *Solanaceae*, formerly known by the generic name, *Brugmansia*. e.g. D. metel - The seeds of this species have been erroneously named "Ololiuqui" (see **Rivea**). D. stramonium - The Stramonium plant (= Datura Officinal; Estramonio; Jamestown Weed; Jimson Weed; Stechapfelblätter; Stramoine; Stramon; Stramonii Folia; Stramonii Herba; Stramonium Leaves; Thornapple Leaves; Thornapple Plant) - An annual plant, 30-150 cm high, with sinuate leaves and solitary white flowers up to 7 mm long. The fruit is an ovoid capsule about 5 cm (2 in) long, usually covered with long sharp spines, and the plant, which is very poisonous, contains the alkaloids atropine, hyoscyamine, hyoscine, and scopolamine. These alkaloids are extracted and used to control bronchospasm in asthma, intestinal colic, salivation, and muscular spasms in the treatment of paralysis agitans and postencephalitic parkinsonism. D. suaveolens (= *Brugmansia* s.) - The Angel's Trumpet, a tree or large shrub, 3-4.5 m (10-15 ft) high, with huge, pendulous, musks cented, trumpet-shaped flowers up to 25 cm (10 in) in length. These are pure white at first but mellow to cream-yellow with age. Place of origin: Mexico.

Daubentonia A genus of Mammals of the Aye-aye family, *Daubentoniidae*. e.g. D. madagascariensis - The Aye-aye, the only member of this family. About the size of a cat, this primate which outwardly resembles a lemur, has a thick silvery fur and a long bushy tail. Its arms are much shorter than its legs, and a striking feature is the slenderness of its bony fingers. It is a nocturnal tree-climber, and not much is known of its habits.

Daucus A genus of plants of the Carrot family, *Umbelliferae*. e.g. D. carota - Wild Carrot or Queen Anne's Lace, an erect, biennial plant, up to 90 cm (3 ft) high; with ternate or pinnatifid bracts, flat or convex umbels, and white flowers: the central flower is often red. The ovoid fruits are covered with rows of hooked bristles. Grows widespread in meadows and pastureland. Distribution: Europe, Asia. Native in Britain; naturalized in America.

Davainea A genus of Cyclophyllidean Tapeworms of the family *Davainiidae*. A well-known species of merozoic cestode tapeworms found in poultry.

Davidia A genus of Trees of the family *Nyssaceae*. e.g. D. involucrata - The Chinese Dove Tree, a deciduous tree, up to 15 m (50 ft) tall, with alternate, ovate, toothed leaves and small flowers growing in dense heads and surrounded by two large bracts round each head. Place of origin: China.

Decticus A genus of Insects of the Bush Cricket family, *Tettigoniidae*. e.g. D. verrucivorus - The Wart-biter; length 24-44 mm; green or brown body with "horse" head; found in meadows, fields, heaths, etc. Distribution: Eurasia; native in Britain.

Deilephila A genus of Insects of the Hawk Moth family, *Sphingidae*. e.g. D. elpenor - The Elephant Hawk Moth, size 25-32 mm; an accomplished insect flier, with rose-colored wings, capable of travelling long distances at relatively high speeds, and of hovering above the flowers from which it feeds. The full-grown larva measures 7-8 cm, and changes into a yellow-brown, dark-spotted pupa. Distribution: the Palaearctic region.

Deinocheirus A genus of prehistoric mammals, the Carnosaurs. This extinct animal is said to have been truly colossal; a pair of forelimbs discovered in the south Gobi desert by a Polish-Mongolian expedition in the summers of 1963-5, were found to measure 2.59 m (8 ft 6 in) in length.

Deinosuchus A genus of now extinct Crocodiles. e.g. D. hatcheri (= D. riograndensis; *Phobosuchus* h.) - The largest known crocodile, now extinct, lived in the lakes and swamps of what are now the states of Montana and Texas, U.S.A. about 750 million years ago. From the remains of this gigantic saurian reptile discovered in southern Texas in 1940, the total length of the crocodile was calculated as measuring 15.24 m (50 ft).

Delesseria A genus of plants of the Red Algae family, *Gigartinaceae*. e.g. D. sanguinea - Delesseria, up to 15 cm high; a delicate, leaf-like, crimson-red plant with a thick midrib and lateral veins. Found in the West Baltic and North Seas.

Delia A genus of Two-winged flies (order *Diptera*) of the family *Anthomyidae*. e.g. D. brassicae - The Cabbage-root Maggot Fly, length 5.5-7.5 mm. Larvae live in the ground, feed on the roots of wild and cultivated plants, and finally tunnel into the host plant, where they pupate. Destructive to crops. Distribution: Europe, North America.

Delichon A genus of Birds of the Martin and Swallow family, *Hirundinidae*. e.g. D. urbica - The House Martin, length 12.5 cm (5 in); found in human habitations on roofs, and telephone wires. It builds a mud nest on houses, usually against a vertical wall. It has long pointed wings, darkish plumage, a forked tail, and a pure white rump. A palaearctic migrant, it winters in Africa and Asia. Distribution: northwest Africa, Europe, and Asia.

Delospermum A genus of plants of the family *Aizoaceae*, sometimes also known as the *Mesembryanthemaceae*. e.g. D. echinatum - A herb found in South Africa, with prickly, elliptical, fleshy leaves and yellow flowers. The leaves contain water storage tissue, and the species often grow in arid stony areas. Cultivated as an ornamental plant.

Delphinapterus A genus of Mammals of the Beaked or Toothed Whale family, *Ziphiidae*. e.g. D. leucas - The White Whale or Beluga. Although belonging to this family, it has no beak, and looks more than a porpoise than a dolphin. It lives in Arctic waters. In the adult state it is completely white, and measures up to 5.1 m (17 ft) long. The White Whale is hunted in the northern coasts of Europe, Asia and America. Its flesh, fat, and leather (sold as "porpoise hide") are of commercial value.

Delphinium A genus of plants of the Buttercup family, *Ranunculaceae*. e.g.

D. staphisagria - The Stavesacre or Paparraz plants; the seeds, Stavesacre Seeds or Paparraz Seeds contain the alkaloids delphinine, delphinoidine, and delphisine. Used as a pediculicide and externally to relieve pain in neuralgia and rheumatism. Distribution: worldwide.

Delphinus A genus of Mammals of the Dolphin family, *Delphinidae*. e.g. D. delphis - The Common Dolphin, up to 2.4 m (8 ft) long; has a very slender body and pointed beak; one of the swiftest cetacean mammals, pursuing shoals of herring and sardines, and devouring them in vast quantities. A highly intelligent creature and a very popular performer in marine shows. Dolphins are found in seas throughout the world, being particularly abundant in the North Sea and the Mediterranean, at times in schools numbering as much as ten thousand animals. Does not occur in the Arctic seas.

Demodex A genus of Arachnids of the order of Mites, *Acari*. e.g. D. folliculorum - A species found in hair follicles and in sebaceous secretions, especially of the face and nose. Infestation is known as demodicidosis or follicular mange.

Demoixys A genus of Crustaceans of the Cyclopoid family, *Notodelphyidae* (subclass *Copepoda*). These copepods are parasites of the sea-squirts (class *Ascidiacea*), and the females have an internal brood-pouch which can be distended into a globular shape.

Dendroaspis A genus of Reptiles of the Cobra, Mamba, and Krait Snake family, *Elapidae*. e.g. D. polylepis - The Black Mamba, a tree-dwelling snake, probably the fastest-moving land snake in the world, confined to equatorial and southern Africa. It has a slender body, and its olive green or grey color gives effective camouflage in its normal habitat. It is extremely poisonous and its venom can kill a man within ten minutes.

Dendroaster A genus of Sea-urchins of the class *Echinoidea*. e.g. D. excentricus - The Sand Dollar, a sea-urchin living off the Pacific coast of North America. It has extremely small spines and its test or shell is flattened, and may even be flexible. When exposed at low tide it drops flat and quickly covers itself with sand, a means of escape from the starfish that feeds on it.

Dendrobates A genus of Amphibians of the Frog family, *Ranidae*. These are the South American arrow-poison frogs, brightly patterned with warning coloration, living both on the forest floor and in the trees, moving actively and fearlessly by day. The male of the various species carries the tadpoles on his back.

Dendrobium A genus of plants of the Orchid family, *Orchidaceae*. There are several hybrid evergreen and deciduous species. These orchids are handsome and epiphytic, possessing leathery leaves and pseudobulbs. The flowers are variable in size, color, and markings. e.g. D. nobile - One of the best-known species in cultivation, with stout fleshy pseudobulbs, 60-90 cm (2-3 ft) high, and 6-10 bright green leaves. The fragrant flowers are reddish-pink in color. Place of origin: North India, the Far East, Australia, and the islands of the South Pacific. Introduced into Europe in 1836.

Dendrocalamus A genus of plants of the Grass family, *Gramineae*. e.g. D. giganteus - The Giant Bamboo, one of the largest and highest species of grasses, reaching 30 m (100 ft) in height. Place of origin: Java, southeast Asia.

Dendrochilum A genus of plants of the Orchid family, *Orchidaceae*. e.g. D. glumaceum - A species of orchids growing in the Philippines, with broad leaves and a pointed spike of white or creamy-white flowers, borne on a slender, curved stalk.

Dendrocoelum A genus of Turbellarian Flatworms of the Planarian or Triclad order, *Tricladida*, in which the intestine has three main branches, one directed forward and the two others backward. e.g. D. lacteum - Length 26 mm; milky white in color; the head has short movable tentacles and two black eyes. A freshwater species found under stones in clean, slow-flowing or still waters; also in brackish waters. Distribution: Europe, as far north as Scandinavia.

Dendrocopos A genus of Birds of the Woodpecker family, *Picidae*. e.g. D. major - The Great Spotted Woodpecker, sometimes called the Pied Woodpecker, length 23 cm (9 in), a typical species of the true woodpeckers (subfamily *Picinae*). It has a white patch on each side below the black nape-band; the male has a red spot on the nape. Using its large robust head, the woodpecker hammers away at the bark of the tree trunk with its strong wedge-shaped bill to extract insect larvae. Found in coniferous forests, treed-avenues, parks and large gardens. Distribution: Europe, Asia, northwest Africa.

Dendrocygna A genus of Birds of the Duck family, *Anatidae* (subfamily of Whistling Ducks or Tree Ducks, the, *Dendrocygnini*). e.g. D. bicolor - The Fulvous Tree Duck, found mostly in the tropics of the New World and Africa. Unlike the typical duck it shows no sexual dimorphism.

Dendrodoa A genus of Sea-squirts of the class *Ascidiaceae*. e.g. D. grossularia - The Gooseberry Sea-squirt; size up to 15 mm; red bodies with two prominent siphons. Dense masses of these animals may become fused to common bases. Found on the lower seashore. Habitat: on the Baltic and some Atlantic coasts; native in Britain.

Dendrogyra A genus of Sea Anemones (order *Actiniaria*) of the large subclass of marine polyps, *Hexacorallia*. e.g. D. cylindrus - A coral species from the Bahamas.

Dendrohyrax A genus of Mammals of the Hyrax order, *Hyracoidea*. These are the true hyraxes, about the size of marmots. They belong to only one family, *Procaviidae*, and are found in Africa, Arabia and Syria, living in large colonies, much like rabbits. They have incisor teeth like rodents and live on grasses, buds, seeds and fruit.

Dendroica A genus of Birds of the Wood-warbler family, *Parulidae*. e.g. D. dominica - The Yellow-throated Warbler, a bright-colored bird of the Americas, breeding in the more temperate regions and migrating to the tropical areas for winter. In spring, migration in and to the United States is often spectacular, with myriads of these birds passing through in a relatively short period of time.

Dendrolagus A genus of Mammals of the Kangaroo family, *Macropodidae*. e.g. D. lumholtzi - The True Kangaroo, a marsupial found in northern Queensland and New Guinea. Closely related to the other kangaroos, it has become adapted to arboreal life, with roughened feet pads for climbing and a thick brush at the end of the tail to serve as a balancer.

Dendrolimus A genus of Insects of the Moth family, *Lasiocampidae*. e.g. D. pini - The Pine Lappet Moth, size 25-36 m; coloring ranges from light to dark brown; there is a snow-white moon-shaped spot on the forewings. Found in pine forests where it causes considerable damage. Distribution: much of Europe; China; Japan.

Dendromus A genus of Mammals of the common Rat and Mouse family, *Muridae*, and the Banana Mouse subfamily, *Dendromyinae*. These are the Banana or Tree Mice, delicately built with a long prehensile tail. They are light fawn-buff in color, with a dark stripe down the back, and live in Africa, where they hide in the folds of banana leaves or among the fruit.

Dendronephyta A genus of Octocorals living in the Red Sea region, similar to the genus *Akabaria* (see **Akabaria**).

Dendronessa A genus of Birds of the Duck family, *Anatidae*, belonging to the tribe of perching ducks, the Cairinini. e.g. D. galericulata - The Mandarin Duck of China, a popular ornamental bird. It is vividly colored with reds, greens, blues, yellows and browns, contrasting with the pure white of its front. Males and females show great mutual affection, which has made the species a symbol of conjugal fidelity in China.

Dendrostoma A genus of Sipunculoid Animals of the phylum *Sipunculoidea*. Worm-like marine animals with a long, cylindrical, unsegmented body. Sedentary in habit, it lives in permanent or semi-permanent burrows in mud or sand, in rocky fissures, under stones, or amongst algal holdfasts (see **Golfingia**; **Phascolosoma**).

Dentalium A genus of Gastropod Molluscs found on the seashore and seabeds. e.g. D. cutalis - Tusk Shell, length 5 cm; a burrowing animal with a shell like an elephant's tusk. Native in Britain.

Dentex A genus of Bony Fishes of the Porgy, Sea Bream, and Snapper family, *Sparidae*. e.g. D. dentex - The Dentex, length more than 1 m; with a large, sickle-shaped caudal fin. A predaceous marine fish feeding chiefly on other fish; occurs over rocky bottoms at depths of 10-200 m. A popular food fish. Distribution: the Mediterranean; eastern parts of the Atlantic, as far north as the Bay of Biscay.

Deporaus A genus of Insects of the Weevil Beetle family, *Attelabidae*. e.g. D. betulae - The Birch Leaf Roller Weevil, a leaf-rolling beetle, 2.5-4 mm long, with a shining black body. Common on birch. The female makes striking leaf cases, cutting through the apical part of the leaf up to the midrib and then rolling the two flaps into a loose cigar-like structure in which eggs are laid and the larvae develop. Distribution: much of Europe, Siberia, Mongolia, and North Africa.

Deraeocoris A genus of Insects of the Capsid True Bug family, *Miridae*. e.g. D. ruber - Length 6.5-7.5 mm; variable in color; found on deciduous trees and herbaceous plants. Distribution: Europe, North Africa, North America.

Dermacentor A genus of Ticks, also known by the generic name, *Dermatocentor*, transmitting disease. e.g. D. andersoni - A species transmitting *Rickettsia* rickettsii, the cause of Rocky Mountain Spotted Fever. D. nitens - Now known as *Otocentor* n. (see **Otocentor**).

Dermacentroxenus A genus name formerly given to the microorganisms parasitic in ticks, now called *Rickettsia*. e.g. D. rickettsi (*Rickettsia* rickettsii) (see **Rickettsia**).

Dermanyssus A genus of Mites. e.g. D. avium et gallinae - The Bird Mite, Poultry Mite, or Chicken Louse, sometimes infesting man.

Dermatobia A genus of Insects of the Warble Fly family, *Oestridae*. e.g. D. hominis - The Human Botfly of South America, whose larvae are parasitic on the skin of man, mammals, and birds.

Dermatocentor See **Dermacentor**.

Dermatophagoides A genus of Arachnids of the order of Mites, *Acari*. e.g. D. pteronyssinus - The House-mite, found in house dust, being one of the causes of allergic (bronchial) asthma. A vaccine from these mites has been prepared for use in desensitization in patients found allergic to this pest.

Dermatophilus A genus of Arachnids of the order of Mites, *Acari*. e.g. D. penetrans.

Dermestes A genus of Insects of the Carpet, Hide, and Museum Beetle family, *Dermestidae*. e.g. D. lardarius - The Bacon, Larder, or Skin Beetle, length 7-9.5 mm; has a grey band with dark spots on the elytra. Originally it lived in the

wild, and even today may be found occasionally in the nests of certain birds. In time however, it gradually became a domestic pest of man, and is now found in his vicinity - in dwellings, warehouses, museums, dove-cotes, beehives, etc. The beetle itself is harmless but the larvae feed on fabrics, leather, badly prepared skins, zoological and entomological collections in museums, etc. Distribution: worldwide.

Dermochelys A genus of Reptiles of the Leathery or Leather-backed Turtle family, *Dermochelidae* (= *Dermochelyidae*). e.g. D. coriacea - The Leathery Turtle, Leatherback, or Luth, length 2.0 m; weight 600 kg, the typical marine turtle of the open sea and the world's largest and heaviest chelonian. It looks heart-shaped from above and is formed simply of dermal ossicles which are not fused to the vertebrae and ribs. The skin has the appearance of brown leather and bears seven longitudinal ridges. It always comes ashore to breed. Distribution: Atlantic, Pacific, Indian Oceans, and the Mediterranean Sea.

Deroceras A genus of Gastropod Molluscs of the family *Limacidae*. e.g. D. reticulatum - The Field Slug, length 50-60 mm. Ground color greyish, yellowish-white, or brownish-red, and marked with a dark network pattern. Found in forests, meadows, fields and gardens, and other moist locations. An omnivore, it causes much damage to gardens and greenhouses. Distribution: worldwide.

Derocheilocaris A genus of Crustaceans of the subclass *Mystacocarida*. A blind animal, with a worm-like body, feeding by filtering small particles from the water between sand grains.

Deschampsia A genus of plants of the Grass family, *Gramineae*. e.g. D. cespitosa - Tufted Hair-grass, a perennial plant, 50-150 cm high; leaves flat, with translucent veins; lemma with short awn from near base. Widespread and common in moorland, meadows, and heaths. Native in Britain.

Descurainia A genus of plants of the Cress or Crucifer family, *Cruciferae*. e.g. D. sophia - Flixweed, an annual to perennial plant, 10-90 cm high, with pale yellow flowers and fruits curved upwards. Widespread, growing as weeds in arable and waste land, and on roadsides. Place of origin: Europe; introduced and naturalized in Britain.

Desmacidon A genus of Sponges (phylum *Parazoa*). e.g. D. fruticosus - A siliceous sponge, with its main skeleton a network of needle-shaped spicules (oxea), living in the shallow seas of the eastern North Atlantic. A prawn lives in its cloacal cavities and feeds on its tissues.

Desmana A genus of Mammals of the Desman or Mole family, *Talpidae* (subfamily *Desmaninae*). e.g. D. moschata - The Russian Desman; length of body 215 mm, tail 170 mm. It has a long snout and no external ears. Found near large rivers, chiefly along backwaters with thick vegetation. An aquatic fur animal, with a long and extremely mobile snout, smelling strongly of musk. Lives in burrows with an underwater entrance. Distribution: native to Russia, along the lower reaches of the Volga, Don, and Ural Rivers.

Desmodus A genus of Mammals of the Blood-sucking Bat family, *Desmodontidae*. e.g. D. rotundus - The Great Blood-sucking Common Vampire, or Great Vampire Bat, a small mammal, the head and body being under 7.5 cm (3 in) in length. The teeth are almost reduced to a pair of razor-sharp upper canines and incisors for puncturing the skin of its prey before sucking the blood. The most dangerous bat of tropical and subtropical America, it not only drinks the blood of its victims but also transmits disease, including the paralytic rabies virus (hydrophobia), which is almost 100% fatal to livestock and man.

Distribution: Central and South America, from Mexico to Paraguay.

Desmognathus A genus of Amphibians of the Plethodontid Salamander family, *Plethodontidae*. These are lungless species found in North America; some are terrestrial, others are aquatic animals.

Desulfovibrio A genus of microorganisms of the family *Spirillaceae*. Actively mobile, slightly curved rods. e.g. D. desulfuricans.

Desvoidea A genus of Mosquitoes. e.g. D. obturbans - Transmits dengue.

Deutzia A genus of plants of the family *Philadelphaceae*, sometimes also known by the name *Hydrangeaceae*. e.g. D. scabra - Deutzia, a deciduous flowering shrub, a bushy species reaching a height of 2-3 m (6-10 ft), with opposite, ovate leaves and large paniculate clusters of white or pinkish, 5-petalled flowers. Place of origin: Japan and China; introduced into Europe in 1822.

Diadema A genus of Sea-urchins of the class *Echinoidea*. e.g. D. setosum - The Hat-pin Sea-urchin, an echinoderm with long, slender, numerous spines which move in relation to changes in the intensity of light. Found on barrier reefs, these sea-urchins live in company with corals; their spines are poisonous and are used as a means of defence.

Diadumene A genus of Anthozoan Coelenterates of the order of Sea Anemones, *Actiniaria*. e.g. D. luciae - Size 1.4 cm; thickness 5 mm; with 8-20 orange, vertical stripes; tentacles up to 2.5 cm long and non-contractile. Found on firm seabeds on rocks, breakwaters, mussels, etc. in the North Sea.

Diaea A genus of Arachnids of the Crab Spider family, *Thomisidae*. e.g. D. dorsata - Length 5-7 mm; colored green, only the top of the abdomen is brown. Found in shrubby oak stands and coniferous forests. Distribution: Europe, the Caucasus, Turkestan.

Dialister A genus of minute, anaerobic,

gram-negative cocco-bacilli found in the respiratory tract. e.g. D. pneumosintes (=*Bacterium* p.) - Found in the nasopharynx in epidemic influenza.

Diamanus A genus of Insects of the order of Fleas *Siphonaptera*. e.g. D. montanus - A species found on rodents in the United States; has been implicated in the transmission of sylvatic plague.

Dianthus A genus of plants of the Pink family, *Caryophyllaceae*. e.g. D. caryophyllus - Carnation, a plant with a long history of cultivation, with a woody base, evergreen grey-green leaves and large, double, fragrant flowers. Color of these flowers vary from white and yellow, through shades of pink and red. Carnation flowers are the source of an essential oil used for high grade perfumery. Place of origin: southern Europe.

Diapensia A genus of plants of the Diapensia family. e.g. D. lapponica - Diapensia, a perennial plant, 3-5 cm high; the matted stems form a "cushion"; it has solitary, yellowish-white flowers and grows in Arctic Europe and the hill-tops of Scotland. Native in Britain.

Diapheromera A genus of Insects of the order of Plasmid Stick Insects, *Phasmida*. e.g. D. femorata - The North American Walking Stick, an elongate, usually wingless, twig-like species.

Diaphus A genus of Fishes of the Lanternfish family, *Myctophidae*. These species live in deep water and come to the surface at night; they have large eyes and at the edge of the snout, a large luminous organ rather like a headlight. Their bodies are covered by small, luminous spots.

Diaptomus A genus of Copepod Crustaceans. e.g. D. vulgaris - A copepod host to the hexacanth larvae of *Diphyllobothrium* latum.

Dibothriocephalus - Former generic name for *Diphyllobothrium* (see **Diphyllobothrium**).

Dibranchus A genus of Fishes of the suborder *Antennarioidei*. e.g. D. atlanticus - A species related to the Batfishes of the same suborder; it has only two pairs of gills.

Dicamptodon A genus of Amphibians, also known as *Ensatus*, of the Salamander family, *Ambystomatidae*. There is only one species, known as Dicamptodon or Ensatus, the Pacific Giant Salamander, up to 30 cm (1 ft) long, found from British Columbia to California. On land it occurs under logs in moist places, and probably lays its eggs in similar situations and near water.

Dicentra A genus of plants, also known by the generic name, *Dielytra*, of the Fumitory family, *Fumariaceae*. e.g. D. canadiensis (= *Dielytra* c.) - Squirrel Corn or Turkey Corn; an elegant, slender-stemmed plant containing the alkaloid bulbocapnine. Used as a sedative in post-encephalitic parkinsonism and in Meniere's syndrome (see **Corydalis**).

Diceromonas - A former generic name for *Giardia*. e.g. D. muris (= *Giardia* lamblia) (see **Giardia**).

Dicerorhinus A genus of Mammals of the Rhinoceros family, *Rhinocerotidae*. e.g. D. sumatrensis - The Sumatran Rhinoceros, a two-horned mammal found in the forests of East Asia - in Sumatra, Thailand, Borneo, and the Malay peninsula. Smaller than other species of rhinoceroses, it is only about 135 cm (41/2 ft) at the shoulder.

Diceros A genus of Mammals of the Rhinoceros family, *Rhinocerotidae*. e.g. D. simus - The African White or Square-lipped Rhinoceros, slate-grey in color, with a square upper lip and flattened nose, the name "white" being a misnomer, being derived from the Africaans, meaning "wide". It is the largest species, being 4.8 m (16 ft) long, 195 cm (61/2 ft) tall, and weighing from 3-4 tons, with formidable horns. The anterior horn is the longest, measuring up to 11/2 m (5 ft). It is an inoffensive animal, unless provoked. Formerly common in South Africa, it has been hunted so widely that it is now confined to reserves in Zululand and other protected territories in central and southern Africa.

Dichonemertes A genus of Enoplous Ribbon Worms (phylum *Nemertina* or *Rhynchocoela*), of the order *Hoplonemertini*. In these hermaphrodite species of nemertines or ribbon worms, the proboscis is armed with one or more stylets.

Diclodophora A genus of Tapeworms of the subclass of Merozoic Cestodes, *Eucestoda* or *Cestoda Merozoa*. e.g. D. merlangi.

Dicranocephalus A genus of Insects of the Cockchafer or Scarab Beetle family, *Scarabaeidae*. e.g. D. dabryi - A Chinese beetle of the large subfamily of Rose Chafers, *Cetonidae*, about 3 cm (1.2 in) long. Distribution: Sechuan province of China.

Dicranum A genus of Moss plants of the group *Bryophyta*. These are small plants without roots, and with woody, water-conducting strands. e.g. D. polysetum - The Woodland Moss; the crowded, pointed leaves develop in a spiral arrangement.

Dicranura A genus of Insects of the Prominent or Puss Moth family, *Notodontidae*. e.g. D. vinula - The European Puss Moth, a large greyish-colored insect. The caterpillar is found on willows, poplars and aspens; it is green and purple on top; when in danger it projects two red, motile appendages in the rear part of its body which it raises aloft, frightening off a possible attacker.

Dicrocoelium A genus of Flukes or Flatworms of the order of Digenetic Trematodes, *Digenea*. e.g. D. dendriticum - The Lancet Fluke, a lancet-shaped flatworm, the larvae of which pass through land snails before achieving adulthood in their mammalian host livers; in sheep, cattle, deer, etc. Also found in humans.

Dicrurus A genus of Birds of the Drongo family, *Dicruridae*. e.g. D. paradiseus - The Greater Racket-tailed Drongo of India and Malaya, an arboreal bird and a good flyer, black in color, with a shaggy coat and a forked tail, with "rackets" at its tip as the common name indicates. Drongos are fearless in attacking hawks and crows, and smaller birds often nest in the same tree for protection.

Dictamnus A genus of plants of the Rue family, *Rutaceae*. e.g. D. albus - The Burning Bush, a perennial plant, 40-100 cm high; lemon-scented, white, pink or purplish flowers with dark veins. It secretes so much volatile oil that on a warm, windless day it can burst into flame at the touch of a match, hence its name. Distribution: south, central, and eastern Europe; Asia.

Dictyoptera A genus of Insects of the Beetle family, *Lycidae*. e.g. D. accrora - Length 7-13 mm; orange-red carapace; found in forest undergrowth and on stumps. Distribution: Europe, across Siberia to Korea and Japan.

Didelphis A genus of Mammals of the American Opossum family, *Didelphidae*. e.g. D. virginiana - The Common or Virginian Opossum, one of the commonest marsupials in the U.S., ranging from the south and east United States to the Argentine. It lives chiefly in trees and is active at night, frequenting fields, orchards and farmyards. It grows bigger than a cat, up to 50 cm (1 ft 8 in) long without the tail, and is rat-like with a pointed snout. At birth the young are only 1.25 cm (1/2 in) long, and are kept in the mother's marsupium or ventral pouch, which contains the teats of the milk glands. When older, they are carried on the mother's back.

Didinium A genus of Protozoans of the Gymnostome suborder, *Rhabdophorina*. A carnivorous microscopic organism found on the surface of freshwater

156

where it feeds on dinoflagellates. It can ingest a whole Paramecium, although the prey is several times larger than itself.

Didunculus A genus of Birds of the Dove and Pigeon family, *Columbidae*. e.g. D. strigirostris - The Manumea or Tooth-billed Pigeon, an almost extinct native of the islands of Samoa. A terrestrial bird, about the same size as a Domestic Pigeon (*Columba* livia). The plumage is a dark, shining green (see **Columba**).

Dieffenbachia A genus of plants of the Arum family, *Araceae*. e.g. D. amoena - Diffenbachia; Dumb Cane; Tuftroot; a polymorphous species with dark green, lanceolate leaves having irregular cream marks between the veins. Used as an indoor plant. Place of origin: Tropical America; introduced into Europe in 1880.

Dielytra A genus of plants, better known by the generic name, *Dicentra*, of the Fumitory family, *Fumariaceae*. e.g. D. canadensis (= *Dicentra* c.) (see **Dicentra**; **Corydalis**).

Diemictylus A genus of Amphibians of the Newt or Salamander family, *Salamandridae*. e.g. D. viridescens - The American Newt; the land (early adult) form is known as the Red Eft, spending the first 2 to 3 years of its life on land. In later adult life it turns into the red-spotted green water phase before entering the water to breed, the full adult becoming the completely aquatic newt. The land and water phases of *Diemictylus* are so dissimilar that at one time they were considered different creatures. Habitat: the United States.

Diendamoeba The former name for the genus *Dientamoeba* (see **Dientamoeba**).

Dientamoeba A genus of Protozoans, for-merly called *Diendamoeba*, of the order *Rhizomastigina*. These are small (5-18 μ) limax-type amoebae with two nuclei, sometimes found in the human intestine. e.g. D. fragilis - A pathogenic species, sometimes causing acute dysentery.

Diervilla A genus of plants, better known by the generic name *Weigela*, of the Honeysuckle family, *Caprifoliaceae*. e.g. D. florida (= *Weigela* f.) - The Weigela plant (see **Weigela**).

Difflugia A genus of Protozoans of the order of Testacean Amoebas, *Testacea*. The species have a single symmetrical round test or shell, cemented around for added protection by grains of sand, with a pointed apex and a rim around the opening. The pseudopodia are blunt.

Digaster A genus of Segmented Worms (phylum *Annelida*) of the order of Earthworms, *Oligochaeta*. e.g. D. longi-mani - A giant species of Oligochaete worms found in Queensland and New South Wales. The largest specimen has been measured at 1,651 mm (5 ft 3 in) in length and over 25 mm (1 in) in thickness when suspended alive by the tail.

Digenea A genus of plants of the family of Red Algae, *Gigartinaceae*. e.g. D. simplex - A species of Red Algae, from which is obtained the ascaride Digenic acid (= Digenin; Helminal; Kainic acid). Used in conjunction with Santonin in the treatment of *Ascaris* lumbricoides in children.

Digitalis A genus of plants of the Figwort or Snapdragon family, *Scrophulariaceae*. e.g. D. purpurea - The Foxglove, a biennial plant, 60-150 cm (2-5 ft) high; large, bell-shaped, pink-colored or white flowers with dark spots. The leaf, Foxglove Leaf, contains a number of glycosides, including digi-talin, digitoxin, gitaloxin and gitoxin. Digitalis increases the force of myocar-dial contraction, slows the heart rate by depressing conduction in the atrioven-tricular bundle, increases the cardiac output, decreases cardiac enlargement, reduces venous pressure and improves renal function. It is especially used in the treatment of congestive cardiac fail-ure. Also, in North Wales a dye from the leaves is used to color stonework. Place of origin: Western Europe, including Britain. Extensively naturalized in northwestern North America.

Digitaria A genus of plants of the Grass family, *Gramineae*. e.g. D. sanguinalis - Hairy Finger-grass, an annual plant, an uncommon species, 8-50 cm high; the leaf-sheaths are usually hairy, with five racemes. Grows as weeds on arable and waste land. Distribution: Europe; introduced and naturalized in Britain.

Digramma A genus of Cestodes or Tapeworms. e.g. D. brauni - An avian species of tapeworms.

Dileptus A genus of Protozoans of the Gynostome suborder, *Rhabdophorina*. These are carnivorous animals, remarkably voracious feeders, with extended trichocyst and cilia-bearing anterior processes, giving a comb-like appearance to the region in front of the mouth.

Diloba A genus of Insects of the Owlet-moth or Noctuid family, *Noctuidae*. e.g. D. caeruleocephala - One of over 10,000 species of this family, found the world over. The caterpillar can cause serious damage to crops.

Dimastigamoeba A genus of Coprozoic Amoebae which has both an amoeboid and a flagellate stage in its life history. e.g. D. gruberi (= *Amoeba* g.; A. tachypodia; *Naegleria* g.; N. punctata; *Vahlkamfia* punctata;, *Wasielewskia* g.).

Dimorphotheca A genus of plants of the Daisy family, *Compositae*. e.g. D. sinuata (= D. durantiaca) - African Daisy; Cape Marigold; Star of the Veldt; a half-hardy perennial plant, with thick radical leaves, entire, alternate and rough to the touch with cut edges; large, showy, deep orange flowers with darker centers. Produces two different kinds of seed, hence the generic name. Place of origin: South Africa; introduced into Europe in 1774.

Dinomys A genus of Mammals of the Pacarana Rodent family, *Dinomyidae*. e.g. D. branickii - Sole species of this family, the False Paca or Pacarana, one of the largest rodents, found in the mountain forests of the eastern Andes. It measures 90 cm (3 ft) including the tail. Like the Paca (see **Cuniculus**) it has a large head, is more heavily built, is black in color, with whitish stripes instead of spots on its back, and lives in burrows and among rocks.

Dinoponera A genus of Insects of the order, *Hymenoptera*. e.g. D. gigantea - The Ponerine Ant of Brazil, the world's largest ant; worker ants of this species have been measured up to 33 mm (1.31 in) in length.

Dinornis A genus of Extinct Birds. e.g. D. giganteus - The Flightless Moa of North Island, New Zealand, which was exterminated by the Maoris quite recently in human history. It attained a maximum height of 3.96 m (13 ft) and had an estimated weight of 245 kg (520 lb).

Dioctophyma A genus of Parasitic Roundworms, also called by the generic name, *Dioctophyme* (class *Nematoda*), of the order *Enoplida* (= *Enoploidea*). Formerly the genus was also called *Eustrongylus* or *Strongylus*. e.g. D. renale (= *Dioctophyme* r.; *Eustrongylus* r.; *Strongylus* r.) - The Giant Kidney Worm, the largest-known nematode, 30-90 cm (1-3 ft) long and 1 cm (0.4 in) in diameter; commonly found in the kidneys of dogs and other animals. There are two intermediate hosts - fish and a fish parasite, the branchiobdellid annelid.

Dioctophyme See **Dioctophyma**.

Diodon A genus of Fishes of the Burrfish and Porcupinefish family, *Diodontidae*. e.g. D. hystrix - The Common Porcupinefish or Porcupine Globefish, widely distributed in the warmer waters of all seas. This fish does not have a bony case, and the body is covered with spines. When in danger it inflates itself, the spines become erect, and though not poisonous can inflict severe pain on the intruder.

Diodora A genus of primitive gastropod

Molluscs of the order of Limpets, *Archaeogastropoda*. e.g. D. aspera - The Keyhole Limpet, with a solitary apical opening, found off California on the American West Coast.

Diogenes A genus of Crustaceans of the order *Decapoda*. e.g. D. pugilator - A long-tailed Crustacean Decapod, up to 3 cm long with a small body, short, feather-like, second antennae, a very large left pincer, and varied coloring. Lives on muddy bottoms and flat beaches. Habitat: the Mediterranean.

Diomedea A genus of Birds of the Albatross family, *Diomedeidae*. e.g. D. exulans - The Wandering Albatross of the southern oceans, alarge, long-winged bird, the largest of this family, with a wingspan of 3 m (10 ft) or more and a weight of 9 kg (20 lb). The flight of the Albatross is aerodynamically similar to that of a seaplane, neck retracted, so that the head lies along the body axis, with wings spread out and the undercarriage drawn up.

Dioon A genus of Trees of the family *Cycadales*. This family represents the remnants of a large group of plants that flourished in the Mesozoic era about 200 million years ago. e.g. D. edule - A species from Mexico, producing abundant pollen in the male cones, which are usually large and woody. The plant is unisexual and the ovules, which develop into seeds after fertilization, are borne marginally in the scales of the female cones.

Dioscorea A genus of plants, the Yams. e.g. D. hirsuta - The source of dioscorine. D. mexicana - The source of botogenin.

Dioscoreophyllum A genus of plants. e.g. D. cumminsii - The Serendipity plant, a West African species; its red berries contain monellin, a natural sweetener, about 2,500 times as sweet as sucrose.

Dipetalonema A genus of filarial nematode worms, better known by the generic name, *Acanthocheilonema*. e.g. D. perstans (= *Acanthocheilonema* p.) (see **Acanthocheilonema**).

Diphylla A genus of Mammals of the Blood-sucking Bat family, *Desmodontidae*. e.g. D. ecaudata - The Lesser Blood-sucking or Lesser Vampire Bat. Like the Great Vampire Bat (see **Desmodus**), this species is found from Mexico to Paraguay, in Central and South America. These bats can puncture the skin of an animal unobserved, and it has been suggested that some anesthetic process is involved.

Diphyllobothrium A genus of Merozoic cestodes or tapeworms (subclass *Cestoda Merozoa* or *Eucestoda*) of the order *Pseudophyllidea*, formerly called by the generic names, *Bothriocephalus* or *Dibothriocephalus*. e.g. D. latum (= *Bothriocephalus* latus; *Dibothriocephalus* latus) - The Broad Tapeworm or Fish Tapeworm of man; as much as 18 m (60 ft) long, and comprising 3,000 to 4,000 proglottides. The respective hosts in its life-cycle are first, a copepod (*Cyclops*), a second intermediate host, the fish and finally, man or some other fish-eating vertebrate, in whom maturity occurs.

Diplacanthus A genus of Cestode Worms, also known as *Hymenolepis*. e.g. D. nanus (= *Hymenolepis* nana), the Dwarf Tapeworm found in the human intestine (see **Hymenolepis**).

Diplobacillus A genus of microorganisms, also known by the generic name, *Diplobacterium*. Short rod-shaped organisms occurring in pairs. e.g. D. liquefaciens (= *Diplobacterium* l.) - A species associated with conjunctivitis.

Diplobacterium See **Diplobacillus**.

Diplococcus A genus of microorganisms of the family *Lactobacillaceae*. e.g. D. pneumoniae (= *Pneumococcus*; *Streptococcus* p.) - The common cause of lobar pneumonia.

Diplodocus A genus of extinct prehistoric animals, the longest of the Dinosaurs, which ranged over what is now western

North America about 150 million years ago. e.g. D. carnegii - A species of sauropod or "double-beam" dinosaurs, skeletons of which have been discovered in the Utah mountains, U.S.A. They have measured some 26.67 m (87 1/2 ft) in total length, nearly the length of three London double-decker buses, and weighed an estimated 10.56 tonne/tons when alive.

Diplodus A genus of Bony Fishes of the Porgy, Sea Bream, and Snapper family, *Sparidae*. e.g. D. annularis - The Annular Porgy, length 12-20 cm; one of the commonest fishes off rocky coasts and a popular commercial fish; often found also in brackish water. Distribution: the Mediterranean, eastern Atlantic and the Black Sea.

Diplogonoporus A genus of Tapeworms. e.g. D. grandis - A common parasite in whales. Has been found in man in Japan.

Diplolepis A genus of Hymenopterous Insects of the Gall Wasp family, *Cynipidae*. e.g. D. rosae - Robin's Pincushion Gall-fly or the Rose Gall-wasp; length 3.7-4.3 mm; produces a single type of gall, up to 5 cm (2 in) in diameter, called robin's pin cushion or bedeguar gall, on roses. It varies in size, and is green at first, turning to yellow and red later. The gall is composed of several chambers, and is also inhabited by large numbers of other hymenopterous insects that have varying relationships to the rightful inhabitants. Distribution: Europe; native in Britain.

Diploneis A genus of plants of the class of Diatoms, *Bacillariophyta*. e.g. D. crabro - A species of Diatoms with a narrow groove or raphe, lying along the middle of each valve, and filled with cytoplasm.

Diplopylidium A genus of Tapeworms better known by the generic name, *Dipylidium*. e.g. D. caninum (= *Dipylidium* c.) - The Double-pored Dog Tapeworm (see **Dipylidium**).

Diploria A genus of Sea Anemones (*Actiniaria*) of the large subclass of Marine polyps, *Hexacorallia*. e.g. D. cerebriformis - Neptune's Brain, a coral species from the Bahamas, resembling the human brain in both size and shape.

Diplotaxis A genus of plants of the Cress or Crucifer family, *Cruciferae*. e.g. D. muralis - The Common Wall-rocket or Stinkweed, an annual to perennial plant, 12-60 cm high; leaves confined to a basal rosette; yellow flowers. Grows on waste land and on walls. Distribution: South and Central Europe. Introduced and naturalized in Britain.

Diplozoon A genus of Flukes of the order of monogenetic Trematodes, *Monogenea*. The monogenetic life-cycle is so called because it occurs within a single host. e.g. D. paradoxum - This species inhabits the gills of minnows and other freshwater fishes. The fertilized egg, after hatching as a larva takes a peculiar form (diporpa) which mates with another individual, remaining united in an X-shape as a permanent copula; each individual has four pairs of sessile clamps. It is parasitic on the gills of Cyprinoid (carplike) fishes.

Dipodomys A genus of Mammals of the Kangaroo Rat family, *Heteromyidae*. e.g. D. deserti - The Desert Kangaroo Rat, of the southwestern deserts and semi-deserts of the United States. It has elongated limbs and proceeds by a series of long leaps, using its tail as a balancer. Like the gophers (genus *Thomomys*), it has capacious cheek-pouches which it fills with seeds, enabling it to carry large quantities to its burrow.

Diprion A genus of Hymenopterous Insects of the family *Diprionidae*. e.g. D. pini - The Pine Sawfly, length 7-10 mm; found in pine forests where there are weak and diseased trees. Distribution: central and northern Europe, Spain, North Africa.

Dipsacus A genus of plants of the

Scabious family, *Dipsacaceae*. e.g. D. pilosus - The Shepherd's Rod, a biennial plant, 50-200 cm high; stem with spines on the angles; flower-heads about 2 cm wide; showy pale-yellow flowers borne at the end of tall stalks. Grows in woodlands, hedge-banks, roadsides and pastures. Was once thought effective as a remedy for scabies. Distribution: Europe; native in Britain.

Dipteryx A genus of Trees. e.g. D. odorata - The Tonka Bean tree, a species found in North America. The seed contains coumarin, and is used as a flavoring agent and to disguise odors.

Dipus A genus of Mammals of the Jerboa Rodent family, *Dipodidae*. e.g. D. sagitta nogai - The Northern Three-toed Jerboa, 17 cm long with a 21 cm tail; a rodent with small ears and very short forelegs. It lives in holes dug in the ground, with oval entrances to these burrows. Distribution: the steppes of southwest Russia.

Dipylidium A genus of Merozoic Cestodes or Tapeworms (subclass *Eucestoda*), also known by the generic name, *Diplopylidium*, of the order *Cyclophyllidea*, belonging to the family *Dilepididae*. e.g. D. caninum (= *Diplopylidium* c.) - The Double-pored Dog Tapeworm; it has a two-host cycle involving one intermediate host, in fleas (*Ctenocephalus* canis) and lice (*Trichodectes* canis) - the larval stage, and the final host in cats and dogs. This post-larval form may then be transferred to man (see **Ctenocephalides**).

Diretmus A genus of Fishes of the Diretmid family, *Diretmidae*. e.g. D. argentens - The only species of this family; it has cycloid aswell as ctenoid scales, and teeth only on the jaws.

Dirofilaria A genus of small nematodes or filarian worms. e.g. D. conjunctivae - A species found in the conjunctivae. D. inmitis - The Heartworm; a species found in animals.

Discoglossus A genus of Amphibians of the Frog family, *Discoglossidae*. e.g. D. pictus - The Painted Frog, a Mediterranean species found in two distinct phases, occurring near water, in which the females may spawn several times during the year.

Discoma A genus of Anthozoan Coelenterates of the order of Sea Anemones, *Actiniaria*. These species are the largest known sea-anemones, found on the Great Barrier Reef off Queensland, Australia. Its expanded oral disc has been found to measure up to 609 mm (2 ft) in diameter.

Disculiceps A genus of Merozoic Cestodes or Tapeworms (subclass *Eucestoda*) of the order *Cyclophyllidea*. In these species the entire scolex is able to penetrate into the host's mucosa, there to fuse with the tissues in more or less permanent attachment.

Discus A genus of Gastropod Molluscs of the family *Endodontidae*. e.g. D. rotundatus - The Rounded Snail; shell 2.4-2.8 mm high, 5.8-7 mm wide; yellow-brown, marked with reddish-brown stripes. A common species found in forests amidst fallen leaves, under old wood, stones, etc. Distribution: Europe, North Africa.

Dispholidus A genus of Reptiles of the Colubrid Snake family, *Colubridae* (subfamily *Boiginae*). e.g. D. typus - The Boomslang, an arboreal species, one of the few colubrid snakes potentially dangerous to man. It lives in South Africa, particularly in the Cape area. Although its venom is a potential danger, it rarely bites humans, because its fangs are far in the rear of its mouth.

Distichopora A genus of Coelenterates of the class of Polyps, *Hydrozoa*. These are coral-like living growths which have acquired a calcareous skeleton, and are found among the colonies on coral reefs. e.g. D. violacea - A species found in the Pacific, colored a beautiful violet.

Distoma A former name of a genus of trematode worms, also called *Distomum*,

now named under different genera. e.g.
D. haematobium (= D. hematobium)
(see **Schistosoma**). D. hepaticum (=
Fasciola hepatica) - The sheep liver
fluke (see **Fasciola**). D. ringeri (=
Paragonimus westermani) - The lung
fluke (see **Paragonimus**). D. sinensis (=
Distomum japonicum; *Clonorchis* s.) -
The human liver fluke (see **Clonorchis**).
Distomum See **Distoma**.
Ditylenchus A genus of Roundworms
(class *Nematoda*) of the order
Tylenchida. These are the Stem and
Bulb Eelworms, nematode parasites of
plants. The stylet is used for sucking the
cell contents of the living plant.
Dizygotheca A genus of plants, also
known by the generic name, *Aralia*, of
the family *Araliaceae*. e.g. D. elegantis-
sima (=*Aralia* e.) - False Aralia, a slender evergreen shrub with elegant digital-
ly composed leaves, composed of 7-11
narrow, toothed, bronze-green leaflets.
Place of origin: New Caledonia; intro-
duced into Europe in 1873 (see **Aralia**).
Dochmius A genus of nematode worms,
better known by the generic name,
Ancylostoma. e.g. D. duodenalis (=
Ancylostoma duodenale) (see
Ancylostoma).
Dodecatheon A genus of plants of the
Primrose family, *Primulaceae*. e.g. D.
meadia - A handsome species, which
has been in cultivation for many years.
The flowers have reflexed petals.
Dolichonyx A genus of Birds of the
American Blackbird and Oriole family,
Icteridae. e.g. D. oryzivorus - The
Bobolink, the oriole found over most of
North America. The male is black
below, with white patches on the wings
and back, and has a yellowish nape. A
sociable bird, with several pairs nesting
in the same area, and singing continual-
ly.
Dolichotis A genus of Mammals of the
Cavy family, *Caviidae*. e.g. D. patagon-
um (= D. patagonica) - The Patagonian
Cavy or Mara, a rodent about the size of

a hare, about 50 cm (1 ft 8 in) long,
sometimes known as the Patagonian
Hare, inhabiting the pampas of
Patagonia and the Argentine. A good
runner, it can outdistance the dogs used
by the natives to hunt it.
Dolichovespula A genus of
Hymenopterous Insects. e.g. D. saxonia
- The Saxon Wasp; a true wasp belonging to the group of Aculeata.
Doliolum A genus of Chordate Animals
of the order *Doliolida*. These belong to
the class *Thaliaceae*, and have a complex method of asexual reproduction.
The species are the planktonic salps,
which have a transparent body with
opaque muscle bands like barrel-hoops.
The mouth and the atrial opening are at
opposite ends of the body, and the animals swim and feed by means of these
muscle bands which pump water
through the digestive canal. By the contraction and relaxation of these bands
the barrel-like body pulsates as water is
drawn in at the front and forcibly ejected at the rear. Most of the species are
transparent or translucent, some are
luminescent, and since they tend to congregate in large shoals they are one of
the commoner causes of phosphorescence on the surface of the sea at
night.
Dolomedes A genus of Arachnids of the
Swamp Spider family. e.g. D. fimbriatus
- It has a brown body with yellowish-
white edges; size of male 12 mm, of
female 20 mm. Lives by freshwater
regions and swamps. Distribution:
Europe; native in Britain.
Dolomys A genus of Mammals of the
Cricetid Mouse, Rat, and Vole family,
Cricetidae. e.g. D. bogdanovi -
Martino's Snow Vole, a rodent 14 cm
long, with a 12 cm-long tail; coat light
grey with very dense, silky hairs;
whitish underside. Lives in the dwarf-
tree zone of the mountains of
Yugoslavia.
Dolycoris A genus of Insects of the fami-

ly of Plant Bugs (order of True Bugs, *Heteroptera*). e.g. D. baccarum - The Sloe Bug; size up to 12 mm; brown body with horny elytra from base to center, the remaining half being membranous. A common Plant Bug which lives by sucking plant juices. Distribution: Europe; native in Britain.

Dombeya A genus of Tropical Trees of the family *Sterculiaceae*. e.g. D. wallichii - A tree from Madagascar, up to 9 m (30 ft) tall, with velvety, heart-shaped leaves and dense, pendulous clusters of scarlet flowers.

Donax A genus of Bivalve Molluscs of the Wedge Shell family, *Donacidae*. e.g. D. vittatus - The Banded Wedge Shell; length 35 mm, height 14-18 mm, thickness 10-12 mm. Highly polished shell, white to yellowish-brown; surface marked with concentric lines and radiating ribs. Lives in sand, from low-water mark down to depths of up to 20 m. Distribution: the Mediterranean Sea; European coast of the Atlantic Ocean; off the Canary Islands. Native in Britain.

Donovania A genus of microorganisms, the Schizomycetes of the family *Parvobacteriaceae* or *Brucellaceae*, better known by the generic name, *Calymmatobacterium*. e.g. D. granulomatis (= *Calymmatobacterium* g.) - Causative agent of granuloma inguinale in man; also called the Donovan body or organism (see **Calymmatobacterium**).

Dorcadion A genus of Insects of the Longicorn or Long-horned Beetle family, *Cerambycidae*. e.g. D. pedestre - Length 11-17 mm; color black; differs from most cerambycids by having a robust body and short antennae as well as different habits, for the larva does not live in wood but feeds on the roots of various grasses. Distribution: eastern part of Central Europe and the Balkans.

Dorcus A genus of Insects of the order of Beetles, *Coleoptera*. e.g. D. parallelopipedus - The Dorcus Beetle, length 1.2-

2.2 cm, body dull black with very wide thorax; male has elongated mandibles. Adults feed on exuding sap of trees. Distribution: Europe; native in Britain.

Dorema A genus of plants of the Carrot family, *Umbelliferae*. e.g. D. ammoniacum - From the stem is produced a gum-resin Ammoniacum (ammonia solution); used as an expectorant.

Dorippe A genus of Crustaceans of the order *Decapoda*. e.g. D. lanata - A short-tailed Crab, 2-2.5 cm long; with a rounded reddish-yellow body and small pincers. The first and second pairs ofwalking legs are very long. Lives among sea-grass, on rocky bottoms, or in shingle. Distribution: the Mediterranean.

Dorocidaris A genus of Sea Urchins of the class *Echinoidea*. e.g. D. papillata - A Pencil Urchin found in the Mediterranean; size 4-4.5 cm, with thick spines up to 9 cm long.

Doronicum A genus of plants of the Daisy family, *Compositae*. e.g. D. pardalianches - The Common Leopardsbane, a perennial plant, 30-90 cm high; with deeply heart-shaped basal leaves present at flowering time. Grows mainly in mountain forests. Distribution: southwest and south central Europe.

Dorotheanthus A genus of plants, formerly known by the generic name, *Mesembryanthemum*, of the family *Aizoaceae*. e.g. D. bellidiflorus (= D. bellidiformis; *Mesembryanthemum criniflorum*) - The Livingstone Daisy, an annual plant forming a low mat, 2.5 cm (1 in) high, with thick, almost cylindrical, succulent leaves, and bright, daisy-like, crimson to yellow, to pale mauve flowers. Place of origin: South Africa; introduced from the Cape Province into Britain circa 1880.

Dorstenia A genus of Trees or Shrubs of the Mulberry family, *Moraceae*. e.g. D. radiata - A herbaceous species, growing in tropical areas of Africa and America, with minute unisexual flowers borne in separate pits on the upper surface of a thick circular receptacle, growing on a long stalk, so that the whole complex inflorescence resembles a single flower. Cultivated as an ornamental plant.

Dorycnium A genus of plants of the Pea family, *Leguminosae*. e.g. D. hirsutum - The Hairy Dorycnium, a perennial plant, 20-60 cm high, with greyish-white hairs, reddish-white flowers, and a white or pink keel. Grows in stony pastureland and grassy places. Native of the Mediterranean region.

Doryichthys A genus of Fishes of the order of Snipefishes and Seahorses, *Solenichthyes* (suborder *Syngnathoidei*). The bodies are covered in bony rings and the brood-pouches of all male members of the species are on the belly.

Dorylus A genus of Insects of the Driver Ant family, *Dorylinae*. e.g. D. helvolus - The Driver Ant of South Africa. The huge wingless queen of the species is sometimes nearly 51 mm (2 in) in length. When searching for food these ants move off in highly organized columns numbering anything up to 150,000 individuals, and will devour any animal that is too slow to get out of the way.

Dosidicus A genus of Cephalopod Molluscs of the order *Decabrachia*; also known by the generic name, *Ommastrephes*. e.g. D. gigas (= *Ommastrephes* g.) - The most dangerous Squid of the Humboldt Current off Peru, which may reach a total length of 3.65 m (12 ft) and a weight of 158.7 kg (350 lb). This extremely aggressive animal is greatly feared by native fishermen.

Dosinia A genus of Bivalve Molluscs of the Venus and Carpet Shell family, *Veneridae*. e.g. D. lupinus - Smooth Artemis; length 25-30 mm, height 25-30 mm, thickness 12-14 mm. Shell whitish to yellowish, smooth with only the growth lines discernible. It burrows deeply in sand, mud or shell-gravel

down to 125 m. Distribution: Canary Islands, Mediterranean Sea, Atlantic Coast of Europe.

Draba A genus of plants of the Cress or Crucifer family, *Cruciferae*. e.g. D. tomentosa - Downy Whitlow-grass, a perennial plant, 3-12 cm high; stem, leaves and fruit downy. The species form "cushions", growing on scree and in rock-crevices on mountains of South and Central Europe; also in the Balkans.

Dracaena A genus of plants of the Lily family, *Liliaceae*. e.g. D. draco - The Dragon Tree, a native of the Canary Islands. It is densely branched and woody, from 9-18 m (30-60 ft) tall, and has crowded, sword-shaped leaves up to 60 cm (2 ft) long. D. fragrans - Fragrant Dracaena, a species with large, green, glossy leaves up to 90 cm (3 ft) long, very fragrant, yellowish flowers growing in clusters, and orange-red berries. Clones also occur, with gold leaf-edges and central bands. Used as an indoor ornamental plant. Place of origin: Tropical Africa; Guinea. Introduced into Europe in 1768.

Dracocephalum A genus of plants, also known by the generic name, *Physostegia*, of the Thyme family, *Labiatae*. e.g. D. virginianum (= *Physostegia* virginiana) - The False Dragon-head or Obedient Plant, an erect perennial, herbaceous plant with lanceolate sharply-toothed leaves, and thick spikes of flesh-pink to purple, sessile, tubular flowers. These can be moved at will, and will stay as placed, hence the name "obedient". Place of origin: North America; introduced into Europe in 1683.

Dracontium A genus of Tapeworms, also known by the generic name, *Symplocarpus*. e.g. D. foetidum (= *Symplocarpus* foetidus).

Dracunculus 1. Animal Kingdom. A genus of Parasitic Nematodes or Roundworms of the order *Dracunculoidea*. e.g. D. medinensis -

The Guinea-worm, Medina worm, Medusa worm, Dragonworm; a filarial nematode, parasitic to man and animals, 1 m long, the fiery serpent of the Old Testament, which is removed from the human host by winding it round a stick. The larvae hatch within the female worm, which makes its way to the subcutaneous tissue of the human host and forms a small ulcer to which the nematode applies her genital opening. When the ulcer comes in contact with water large numbers of juvenile larvae are set free and swim around, until they are eaten by copepod crustaceans of the genus *Cyclops*. Within the crustacean the parasites develop to their infective stage and when humans accidentally swallow water containing the crustaceans, they become infected. 2. Vegetable Kingdom. A genus of plants of the Arum family, *Araceae*. e.g. D. vulgaris - The Common Dragon-plant, a perennial plant, up to 100 cm high; the spathe is purplish red, 25-40 cm long. It grows mainly in cultivated fields and grassy places. Distribution: the Mediterranean region.

Drascius A genus of Fishes of the suborder of Scorpionfishes, *Scorpaenoidea*. These belong to the family *Agonidae*, in which the heads are completely encased in a body shield of many plates. e.g. D. sachi - The Japanese Pogge or Japanese form of the Armed Bull-head fish. It has greatly enlarged dorsal and anal fins.

Drawida A genus of Segmented Worms of the family *Moniligastridae*. The species are the so-called Giant Earthworms. The family belongs to the class of earthworms or oligochaetes (class *Oligochaeta*). e.g. D. grandis - The Giant Earthworm of India, which reaches a length of 108 cm (3 ft 61/2 in).

Dreissena A genus of Bivalve Molluscs of the family *Dreissenidae*. e.g. D. polymorpha - The Zebra Mussel; length of shell 20-30 mm, height 13-15 mm, thickness 17-20 mm; pointed, with a keel extending from the front to the hind edge. It is attached by the aid of byssal threads to objects such as stones, wooden posts, etc. in rivers, lakes and canals. One of the few freshwater molluscs to produce planktonic larvae. Distribution: the rivers of Europe, of the Black and Caspian Sea regions.

Drepana A genus of Insects of the Butterfly family, *Drepanidae*. e.g. D. falcataria - The Pebble Hooktip Butterfly, size 16-18 mm; found in deciduous groves, clearings, heaths and alongside streams. Distribution: Europe.

Drepanidum A genus name for the larval stage of certain protozoa. e.g. D. ranorum - A cytozoon of frog's blood.

Dromaius A genus of Birds, also known by the generic name *Dromiceius*, of the Emu family, *Dromaiidae*. These belong to the superorder of Walking Birds, *Paleognathae*. e.g. D. novaehollandiae (= *Dromiceius* n.) - The Common Emu, the second largest living bird after the Ostrich (*Struthio* camelus), the cock often standing over 1.8 m (6 ft) high. Its neck is partly feathered, the plumage is deep brown, and it has three toes. It lays 15 or more dark green eggs, each nearly 15 cm (6 in) long, which the male alone incubates for 60 days. A sturdy walking bird, it is found on the grassy planes, dry open forests, and deserts of Australia. It feeds on low-growing plants and small animals.

Dromas A genus of Birds of the Crab-plover family, *Dromadidae*. e.g. D. ardeola - The Crab-plover, a large noisy bird, with mainly white plumage, long greenish-blue legs and a bill like that of a Tern. It lives on the coast of East Africa and the northern and western shores of the Indian Ocean. The sole member of this family.

Dromia A genus of Crustaceans of the primitive Crab family, *Dromiidae*. e.g. D. vulgaris - The Sponge Crab, a marine dromiid crab; size up to 15 cm; body strongly arched and covered with short, dense bristles. Its pincers are very powerful and the last legs are modified so that they bend upwards, and can be used to hold mollusc shells or pieces of sponge on its back, letting them grow there, as protection from predators such as the Octopus. Mainly a Mediterranean species.

Dromiceius A genus of Birds of the Emu family, *Dromaiidae*; better known by the generic name, *Dromaius*. e.g. D. novaehollandiae (= *Dromaius* n.) - The Common Emu (see **Dromaius**).

Drosera A genus of plants of the Sundew family, *Droseraceae*. e.g. D. rotundifolia - The Common or Round-leaved Sundew, or Rorela, a perennial plant, 5-25 cm high; circular, long-stalked leaves, densely glandular and fringed with long glandular hairs, forming a rosette round the stalks. The species is insectivorus. Insects are attracted by the leaves and get caught on the sticky hairs. The struggling insect stimulates the hairs to bend over, and when it dies they secrete a digestive juice which breaks down the soft parts of the body. The soluble products thus formed are absorbed and utilized by the plant. The small white flowers grow on one side of a long thin stalk which uncurls from the center of the rosette. The plant grows on wet heaths and moors. It is used in the treatment of chronic bronchitis and asthma. Distribution: Europe; native in Britain.

Drosophila A genus of Insects of the Fruit-fly family, *Drosophilidae*. e.g. D. melanogaster - The Vinegar-fly, length 2 mm; found in fields, gardens, and households. The species is much used in genetic experiments. Distribution: worldwide.

Dryas A genus of plants of the Rose family, *Rosaceae*. e.g. D. octopetala - Mountain Avens, a circumpolar, arctic-alpine, perennial evergreen, creeping under-scrub, 2-15 cm high; with dark-green, oblong leaves, downy beneath; white flowers, solitary, axillary, long-stalked, with 8 petals. The fruit is a dry head of achenes, with long persistent, feathery styles. Grows on mountains and the tundra. Distribution: North and Central Europe; native in Britain.

Dryocopus A genus of Birds of the Woodpecker and Wryneck family, *Picidae*. e.g. D. martius - The Black Woodpecker, length 45 cm; black plumage; the male has red crown markings. The largest European woodpecker, found chiefly in vast coniferous forests at both lowland and mountain elevations. For nesting it seeks out old trees in which it excavates a deep hole with its beak. A resident bird. Distribution: Europe, Asia.

Dryomys A genus of Mammals of the Dormouse Rodent family, *Gliridae*. e.g. D. nitedula - The Forest or Asiatic Dormouse; length of body 75-100 mm (3-4 in); bushy tail, 70-95 mm (2.8-3.8 in) long. It has a typical black eye-stripe running across the face, extending from the nose to the ear. Found commonly in broadleaved, mountain forests, clearings and gardens. A nocturnal animal, feeding on nuts, fruit, and invertebrates. Distribution: central and eastern Europe; southwest and central Asia.

Dryophis A genus of Reptiles of the Colubrid Snake family, *Colubridae*. e.g. D. prasinus - The Whip Snake, an excessively long and slender snake found in Malaya, and an expert tree-climber. Even the pupil ofthe eye is elongated in the horizontal, in conformity with the long head.

Dryopithecus A genus of prehistoric Mammals of the Anthropoid Ape family, *Pongidae*. The species lived in the Cainozoic era.

Dryopteris A genus of plants of the Fern or Polypody family, *Aspleniaceae* (= *Polypodiaceae*). e.g. D. filix-mas - The Male Fern or Aspidium, a perennial plant, up to 1 m high, with densely scaly stalks. Grows in woodlands and also in hedgebanks. Contains the glycosides, filicin, and is used as an anthelmintic for the expulsion of tapeworms and flukes. Distribution: most of Europe and temperate Asia. Native in Britain.

Duboisia A genus of plants of the Nightshade family, *Solanaceae*. e.g. D. myoporoides - From the species, is extracted a mixture of alkaloidal sulfates (duboisinine), chiefly hyoscyamine and hyoscine. Used as a mydriatic.

Dugesia A genus of Turbellarian Flatworms, formerly called by the generic name, *Planaria*, of the Planarian order, *Tricladida*. e.g. D. gonocephala (= *Planaria* g.) - Length up to 25 mm. Body elongate, brownish, blackish-brown or greyish-brown, with ear-like growths at the front end. A freshwater species of flatworm found in clean, flowing water. Because the food canal does not have an anal opening, the undigested food remnants are expelled through the mouth. Distribution: the Palaearctic region.

Dugesiella A genus of Arachnids of the order of True Spiders, *Araneae*. e.g. D. crinita - An enormous "tarantula", living in Mexico, with a body length of 85 mm (3.32 in) and weighing 54.7 g (1.97 oz).

Dugong A genus of Mammals of the Dugong Sea-cow family, *Dugongidae*. e.g. D. dugon - The Dugong, one of the four species of sea-cows. It frequents the shores of the Indian Ocean; about 2.4 m (8 ft) long, it has a bulky, whale-like body which ends in a horizontal tail-fin. There is no trace of hind limbs and the fore limbs are flattened, paddle-shaped and lack external digits. A slow-moving and almost defenceless creature, it subsists chiefly on underwater vegetation.

Dulus A genus of Birds of the Palm-chat family, *Dulidae*. e.g. D. dominicus - The Palm-chat, the single species in this family of starling-sized birds, is localized on the West Indian islands of Hispaniola and La Conave. These birds live in flocks in trees and build communal nests for about four pairs. They feed on flowers and berries.

Dumetella A genus of Birds of the Catbird, Mockingbird, and Thrasher family, *Mimidae*. e.g. D. carolinensis - The Catbird. Its chief call is a distinctive, cat-like mewing note, usually heard issuing from the thickets. It does not repeat its song, as in the Thrasher (genus *Toxostoma*). Habitat: ranges from southern Canada southwards to all but the extreme south of South America.

Dunckerocampus See **Dunkerocampus**.

Dunkerocampus A genus of Fishes, also known by the generic name, *Dunckerocampus*, of the Pipe Fish and Sea Horse family, *Syngnathidae*. e.g. D. caullervi chapmani (= *Dunckerocampus* c. c.) - The Banded Pipe-fish, a species characterized by a small mouth at the end of a tube-like snout. It lives on the sea-bottom among weeds and algae on coral reefs.

Duttonella A genus of Trepanosomes or Sporozoan parasites, now called *Trypanosoma* (see **Trypanosoma**).

Dyamena A genus of Thecate Hydroids or Sea-firs, also known by the generic name, *Dynamena*, and more commonly as *Sertularia*, of the family *Sertulariidae*. e.g. D. pumila (= *Dynamena* p.; *Sertularia* p.) (see **Sertularia**).

Dynamena See **Dyamena**.

Dynastes A genus of Insects of the Scarabaeid or Lamellicorn Beetle family, *Scarabaeidae* (subfamily *Dynastinae*). e.g. D. hercules (= D. herculeus) - The Hercules Beetle, over 12.5 cm (5 in) long, a ground insect living in timber, and found in Central and South America. The wing-sheaths are a light greenish-blue, irregularly spotted with black. It is considered to be the world's largest beetle.

Dytiscus A genus of Insects of the True Water Beetle family, *Dytiscidae*. e.g. D.

marginalis - The Great Diving Beetle or Great Water-beetle, 35 mm long; with yellow-edged pronotum and elytra, yellow underside, and hind legs fringed with long hairs for swimming, with two claws. A carnivorous water beetle, found mostly in freshwater ponds and also in garden pools. It hunts small animals and even feeds on carrion. When full-grown it climbs out of the water, flies from one pond to another, and later burrows in the soil of the shore, where it pupates in a cocoon. Adults live to an age of about 18 months. Distribution: Europe, Siberia, Japan, and North America. Native in Britain.

E

Eberthella A genus of microorganisms now known by the generic name, *Salmonella*. e.g. E. typhosa (= *Bacillus* typhosus; E.typhi; *Salmonella* t.) - the causative organism of typhoid fever (see **Salmonella**).

Ecballium A genus of plants of the Cucumber, Gourd, or Melon family, *Cucurbitaceae*. e.g. E. elaterium - The Squirting Cucumber, a rough, annual plant, 50-100 cm high, bluish-green in color, with a stiff-hairy surface and ripe oblong fruit, squirting its seeds when touched, to a distance of some 3-7.5 m (10-25 ft). It contains the active principle elaterin (momordicin), a powerful hydragogue purgative. Place of origin: the central and eastern Mediterranean region.

Echeneis A genus of carnivorous Fishes (subphylum *Vertebrata*). e.g. E. naucrates - The Sucker Remora or Sucker Shark, a carnivorous fish inhabiting warm regions such as the Red Sea and the Gulf of Eilat (Akaba), which attaches itself by a sucking disk or sucker, situated above the mouth, with a vacuum-type mechanism, to sharks and other large fishes, to reptiles such as turtles, and to ships. It feeds on the "leftovers" of the host fish.

Echeveria A genus of succulent plants of the Crassula family, *Crassulaceae*. e.g. E. setosa - A species found growing on rockeries. It has rosette leaves covered with long white hairs, and orange flowers. Place of origin: Mexico.

Echidna A genus of Fishes of the Moray Eel family, *Muraenidae*. e.g. E. nebulosa - A species found in the Red Sea, attaining a length of over 2 m (61/2 ft) and a diameter of 15 cm (6 in). Their teeth are very strong and sharp, and their vice-like grip is feared by divers. Yellowish-white in color with large brown spots, they live in crevices in the coral reefs of the Gulf of Eilat (Akaba), but are sometimes seen on the seabed.

Echidnophaga A genus of Insects of the Flea order, *Siphonaptera*. e.g. E. gallinacea - The Sticktight Flea.

Echinacea A genus of plants of the Daisy family, *Compositae*. Sometimes known by the generic name, *Rudbeckia*. e.g. E. augustifolia - A species used in medicine as a tonic. E. purpurea (= *Rudbeckia* p.) - The Purple Cone-flower or Rudbeckia, a strong growing perennial with oval-lanceolate leaves and purple-crimson flowers on 1-1.2 m (3-4 ft) stems. The cone-shaped, prominent centers are deep mahogany red. The black roots are edible.

Echinaster A genus of Echinoderm animals of the class of Starfishes, *Asteroidea*. e.g. E. sepositus - Size up to 25 cm; vermilion-colored with five thick arms, and skin without spines. Lives on muddy seabeds. Distribution: the Mediterranean.

Echinocactus A genus of plants of the Cactus family, *Cactaceae*. e.g. E. grusonii - The Gold Ball Cactus or Golden Barrel Cactus, a succulent plant grown for its attractive golden spines which cover the whole rounded body, attaining a diameter of up to 1 m (3 ft). There are yellow flowers at the top of the plant, but are seldom produced in cultivation. The fruits are thin-skinned, egg-shaped and woolly. Place of origin: Mexico.

Echinocardium A genus of Irregular Sea-urchins of the class *Echinoidea*. e.g. E. cordatum - The Heart-urchin or Sea Potato, 4 cm long and 3 cm wide, an Echinoderm species found in shallow waters off the coasts of Britain, and although the dead and empty heart-shaped shells are a common feature, the live animal is seen only by digging it out of the sand. It lies buried up to 20 cm (8 in) deep, with a vertical shaft to the water above for breathing purposes. The species are numerous and short, looking almost like a coat of short fur. Distribution: the rocky shores of Europe. Native in Britain.

Echinochloa A genus of plants of the Grass family, *Gramineae*. e.g. E. crusgalli - Cockspur-grass, an annual plant, 30-100 cm high, with panicles of few to many clustered racemes. Grows as a weed on arable and waste land. Introduced and naturalized in Britain.

Echinococcus A genus of Tapeworms of the class *Cestoda* (subclass of Merozoic Cestodes *Eucestoda* or *Cestoda* merozoa), belonging to the family *Taeniidae*. e.g. E. granulosus - The Liver Tapeworm, a most dangerous species of tapeworm, 8 mm (0.31 in) long, the larvae of which cause Hydatid disease or Echinococcosis (helminthic hydatidosis), a malady of global importance in carnivores or omnivorous mammals, including man, which are infected by ingesting "hydatid sand", the grains of which are microscopic eggs or minute juveniles of this parasite, which adhere to the hair of dogs and make their way into human tissue via the mouths of dog owners and others who pat these canines.

Echinocyamus A genus of Irregular Seaurchins (phylum *Echinodermata*) of the class *Echinoidea*. e.g. E. pusillus - The Green Sea Urchin, size up to 1 cm; with an ovate flattened body, living at depths of 10-50 m in burrows on rocky shores. Distribution: the shores of Europe from Scandinavia southwards.

Echinometra A genus of Echinoderms of the class of Sea-urchins, *Echinoidea*. e.g. E. mathaei - A red-colored seaurchin found in the Red Sea, inhabiting the sandy bottoms and coral reefs of the Gulf of Eilat (Akaba). It possesses a calcareous box-like exoskeleton with movable spines used for walking.

Echinomyia A genus of Insects of the order of Two-winged Flies, *Diptera*. e.g. E. grossa - The Black Spiny Fly, size 25 mm; black body with thick spiky bris-

tles; head and wing-base reddish yellow. The larvae are parasitic in domestic animals.

Echinoprocta A genus of Mammals of the New World Porcupine family, *Erethizontidae*. e.g. E. rufescens - The porcupine of Columbia, an arboreal mammal with a prehensile tail. It sleeps in the trees during the day and feeds at night on leaves, buds and bark.

Echinops A genus of plants of the Daisy family, *Compositae*. e.g. E. sphaerocephalus - The Great Globe-thistle, a stout perennial plant, 50-200 cm high; pinnated leaves with spiny teeth, white-downy beneath. The florets are bluish-grey, and the fruit has a pappus of short scales. Grows in dry bushy places, open woods, and on waste land. Distribution: southeast Europe and southwest Asia. Introduced and naturalized in Britain.

Echinorhinus A genus of Fishes of the suborder of Angel Sharks and Dogfish Sharks, *Squaloidea*, of the family *Squalidae* or *Scymnorhinidae*. These are the Bramble Sharks, large fishes with no anal fin; the dorsal fins are very small and without spines. The teeth in each jaw have several cusps, but they are so oblique that they form a continuous cutting edge. Fairly widespread in southern oceans.

Echinorhynchus A genus of Thorny-headed Worms of the phylum *Acanthocephala*. e.g. E. hominis - A parasitic acanthocephalan species occasionally found in man.

Echinosorex A genus of Mammals of the Hedgehog family, *Erinaceidae*. e.g. E. gymnura - The Moon rat or Raffle's Gymnura, a long-tailed, somewhat rat-like animal, lacking spines, the largest known insectivore, over 30 cm (1 ft) long without the tail, which is 20-221/2 cm (8-9 in). Found in India and southeast Asia.

Echinostoma A genus of parasitic Flukes of the family *Echinostomatoides*. These are intestinal parasites, formerly known as *Fascioletta*. e.g. E. ilocanum - Garrison's Fluke, found in man in Java and the Philippines.

Echinostrephus A genus of Sea-urchins of the class *Echinoidea*. e.g. E. molaris - A species found in the shallow waters of the Indo-Pacific Oceans, burrowing several inches into rock. It has needle-sharp spines with a so-called poison bag at the tip. Penetration of the skin by these spines is a very painful experience.

Echinus A genus of Regular Sea-urchins of the family *Echinidae*. e.g. E. esculentus - The Common or Edible Mediterranean Sea-urchin, a bulky spherical-shaped animal, 10-12.5 cm (4-5 in) in diameter, somewhat globular, red, and studded with a great number of spines. It can climb readily up the glass wall of the aquarium using only its tube-feet. Locally very abundant, it is found along the coast at depths of up to 50 m, on rocks, and amidst seaweed. Distribution: Mediterranean Sea and the Atlantic (from Portugal to the North Sea).

Echis A genus of Reptiles of the Viper Snake family, *Viperidae*. e.g. E. carinatus - The Carpet or Saw-scaled Viper, the most dangerous snake in the world, which ranges across Africa north of the Equator, through the Middle East to India. It is an extremely prolific species and its venom is unusually toxic to man. The lethal dose has been estimated as small as 3-5 mg (0.000105-0.000175 oz).

Echium A genus of plants of the Borage family, *Boraginaceae*. e.g. E. vulgare - Viper's Bugloss, a hairy, biennial plant, 25-120 cm high, with long leaves and narrow blue, pink or white corolla. Widespread and locally common in arable and waste places, short grasslands, and on roadsides. Distribution: Europe; common in Britain.

Echiurus A genus of Echiuroid Worms of the phylum *Echiuroidea*. e.g. E. echiurus - Length up to 12 cm; it has a limb-less body. A browsing worm, it feeds by means of a spathulate, extensile proboscis extending from its head. Distribution: the seashores of Europe; native in Britain.

Eciton A genus of Insects of the Ant family (order *Hymenoptera*). These are the Army Ants of South America. When searching for food they march in highly organized columns numbering up to 150,000 individuals and devouring any animal that is too slow to get out of the way.

Ectobius A genus of Insects of the Cockroach family, *Blattidae*. e.g. E. lapponicus - The Dusky Cockroach, length 7-10 mm, common in woodlands. Distribution: Europe; western Siberia.

Ectopistes A genus of Birds of the Pigeon family, *Columbidae*. e.g. E. migratorius - The Passenger Pigeon, a carrier pigeon once very numerous, has become totally extinct since 1914 as a result of massive slaughter and epidemics. It has been estimated that there were between 5 and 9 billion of these birds before 1840, forming 20-40% of the total bird population of the United States.

Edwardsiella A genus of microorganisms of the family *Enterobacteriaceae*; gram-negative rods, once classified under what was originally called, the "Bartholomew group". e.g. E. tarda - A species isolated from normal stools and from cases of diarrhea, acute gastroenteritis, abscesses, septicemia, and meningitis.

Egeria A genus of plants of the Frogbit or Waterweed family, *Hydrocharitaceae*. e.g. E. densa - The Dense-leaved Pondweed, a perennial plant, up to 3 m tall, with mostly 4 leaves in a whorl. Place of origin: South America; naturalized in Europe.

Egretta A genus of Birds formerly known by the generic name, *Casmerodius*, of the Egret, Heron, and Bittern family *Ardeidae*. e.g. E. alba (= *Casmerodius* albus; C. albus egretta) - The Common, Great, or White American Egret, or Great White Heron, a tall, slender, elegant bird, size 89 cm, with brilliant white plumage, more distinguished-looking than the heron, and found nesting in trees or among reeds in heronries, and almost always near water. A partial migrant, rarely found in Europe except near the Danube and in southern Russia. Other subspecies are cosmopolitan, and are found all over the world.

Ehrlichia A genus of microorganisms of the family *Rickettsiaceae*, non-pathogenic to man. e.g. E. canis - A species causing disease in dogs.

Eichhornia See **Eichornia**.

Eichornia A genus of plants, also known as *Eichhornia*, of the Water Hyacinth family, *Pontederiaceae*. e.g. E. crassipes (= E. speciosa; *Eichhornia* c.) - The Water Hyacinth, a showy aquatic plant, floating freely aided by large air sacs formed at the base of the leaf stalks. Owing to its rapid vegetative reproduction, it becomes a serious pest in clogging waterways. It has pale violet-blue flowers with conspicuous blue and gold peacock markings on the upper petals. Place of origin: South America; introduced into Europe in 1879.

Eidolon A genus of Mammals of the order of Bats, *Chiroptera*. e.g. E. helvium - The Straw-colored Fruit Bat, a long living species, surpassing the age of 21 years.

Eikenella A genus of microorganisms of the family *Brucellaceae*, once included in the genus *Bacteroides*. Gram-negative rods, facultatively anaerobic, and considered part of the normal mouth flora. e.g. E. corrodens (= *Bacteroides* c.) - A species isolated from infected tonsils, bronchial secretions, pleural fluid, and dental abscesses; from the blood immediately after dental extractions, from wounds inflicted by human bites, and in cases of pharyngitis, pneumonia, empyema, otitis media, sinusitis, liver and neck abscesses, osteomyelitis, urinary tract infections, gastroenteritis, meningitis, and bacterial endocarditis.

Eimeria A genus of Protozoans of the suborder *Eimeriidea* (class *Sporozoa*), sometimes found in the liver of man. e.g. E. avium - A species found in birds. E. stiedae - A species found in man (see **Isospora**). E. tenella (= E. necatrix) - A species causing fatal "coccidiosis" in chickens.

Eisenia A genus of Chaelopod segmented worms or Annelids of the Earth-worm family, *Lumbricidae*. e.g. E. foetida - A species found in urine. E. lucens (= E. submontana) - Length 90-135 mm; width 5-6.5 mm; segments marked with brown stripes. Found under the bark of old broadleaved trees and tree stumps (chiefly beech), in rotting wood, and in moss in submontane and mountainous regions. Capable of luminescence. A plentiful species. Distribution: Central Europe; Russia.

Eiseniella A genus of Segmented Worms or Annelids of the Earthworm family, *Lumbricidae*. e.g. E. tetraëdra - The Square-tailed Worm, 3-5 cm long; the middle and rear parts of the body are four-angled, hence the name. The clitellum or centrally situated belt-like segment, by which the worm unites during mating, is long. Widely found on muddy beds. Common in Europe; native in Britain.

Elaeagnus A genus of plants of the Oleaster family, *Elaeagnaceae*. e.g. E. augustifolia - The Narrow-leaved Oleaster, a large, thorny, deciduous shrub or small tree, up to 8 m (26 ft) tall, with leaves silvery-white beneath, and berry-like, reddish-yellow, oval fruit covered with silvery scales. Place of origin: Europe and Asia; especially the Mediterranean region.

Elaeis A genus of Trees of the Palm family, *Palmae*. e.g. E. guineensis - The Oil Palm, a palm tree species with a thick trunk, up to 9 m (30 ft) tall, and supported at the base by a mass of fibrous roots. It has dark green, pinnate leaves 3-4.5 m (10-15 ft) long. Widely cultivated for the oil which is extracted from the fruits, it yields Palm-kernel Oil or Palm-nut Oil, used in the manufacture of soaps and margarine. Place of origin: tropical West Africa.

Elanoides A genus of Birds of the Old World Vulture family, *Accipitridae* (fork-tailed kite subfamily, *Perninae*). e.g. E. forficatus - The Swallow-tailed Kite, a North American species 60 cm (2 ft) long, with two deeply sunk eyes, a hooked beak and claws shaped into talons. The plumage on the head, neck, breast and underparts is snow white, while the rest of the body is black, with a metallic sheen. The wings cross over each other when at rest, and it has a forked tail. A great destroyer of insects.

Elanus A genus of Birds of the Old World Vulture family, *Accipitridae* (white-tailed kite subfamily, *Elaninae*). e.g. E. leucurus - The American White-tailed Kite, a white and grey resident bird flying over marshes, river valleys, and well-watered foothills where it preys on rodents, lizards, and large insects. Distribution: southern Europe, central Asia, southwestern United States, southern South America, and Australia.

Elaphe A genus of Reptiles of the Colubrid Snake family, *Colubridae*. e.g. E. scalaris - The Ladder Snake, a colubrine species with a prominent, pointed snout, and a pattern of black spots along its back resembling a ladder. It is 1.2 m (4 ft) long, ill-tempered and vicious. Found in vineyards, hedges, and on heaps of stones; it hunts rodents, birds and lizards. Distribution: France.

Elaphurus A genus of Mammals of the True Deer family, *Cervidae*. e.g. E. davidianus - The Milu or Père David's Deer, named after the French missionary who observed them in the gardens of the Emperor's palace near Peking. Up to 130 cm (4 ft 4 in) high at the shoulders and 220 cm (7 ft 4 in) long, it differs from all the Deer by having a long tail, and antlers forming long forks with no brow tine. They were destroyed during the Boxer rebellion, and the species now exists only in zoos, especially in Woburn Park in England.

Elaps A genus of Reptiles of the venomous Snake family, *Elapidae*. e.g. E. dorsalis - The Striped Dwarf Garter Snake, the shortest venomenous snake in the world, averaging 152 mm (6 in) in length. Found in South Africa. E. fulvius (= *Micrurus* f.) - The Harlequin Snake (see **Micrurus**).

Elasmosaurus A genus of Extinct Reptiles (class *Reptilia*), the long-necked Plesiosaur which swam in the seas over what is now the state of Kansas, U.S.A., about 130 million years ago. Estimated as measuring up to 14.32 m (47 ft) in total length, of which the flexible neck accounted for 7.62 m (25 ft).

Elater A genus of Insects of the Click Beetle family, *Elateridae*. e.g. E. sanguineus - Size 11-13 mm; blood-red elytra. Found in fields, gardens and parks. Native in Britain.

Elatine A genus of plants of the Waterwort family. e.g. E. alsinastrum - The Whorled Waterwort, a perennial plant, 3-50 cm high, with 8-16 submerged leaves and 3 aerial leaves in each whorl, and greenish flowers. Grows in ponds and small lakes, also on wet mud. Distribution: most of Europe.

Electrophorus A genus of Fishes of the Electric Eel family, *Gymnotidae*. e.g. E. electricus - The Electric Eel, up to 2.4 m (8 ft) long, and weighing 27 kg (60 lb). It can produce a powerful electric shock of about half an ampere at 200-300 volts, capable of killing a human being, from its electric organs which extend about four-fifths of the sides of the body. It is an elongated cylindrical fish found in the rivers of northeast South America, and the Amazon basin.

Eledone A genus of Marine Cephalopod Molluscs, also known by the generic name, *Ozaena*, of the Octopus family, *Octopodidae*. e.g. E. moschata (= *Ozaena* m.) - The Musk Octopus, size up to 40 cm; it has eight arms or tentacles, each with a single row of suckers and joined by a web of thin skin which enables it to grasp its victim more firmly. Distribution: a northern species found in the Mediterranean and other European waters; native around the coast of Britain.

Eleginus A genus of Fishes of the Cod, Haddock, Hake and Whiting family, *Gadidae*. These species of Cod Fishes are the so-called Navagas, common in North European waters and in the northern waters of the Far East.

Eleocharis A genus of plants of the Sedge family, *Cyperaceae*. e.g. E. palustris - The Common Spike-rush, a perennial plant, 8-60 cm high, stem terete or cylindrical, spikelet up to 20 mm; 2 stigmas. Grows on wet, sandy and muddy places at margins of lakes and ponds, damp peaty places on moors and fens, and marshes and ditches. Distribution: most of Europe; native in Britain.

Elephantulus A genus of Mammals of the family of Jumping or Elephant Shrews, *Macroscelididae* (see **Macroscelides**).

Elephas A genus of Mammals of the Elephant family, *Elephantidae* (order *Proboscidea*). e.g. E. maximus (= E. maximus indicus) - The Asiatic or Indian Elephant, height under 3 m (10 ft) and weighing 4.5 tonne/ton; markedly smaller than its African cousin. It has a concave forehead, knobby head, small ears, and a single appendix or "lip" at the end of the trunk. It is the longest-lived land mammal after man; some have been said to live more than 80-100 years. It has the longest of all mammalian gestation periods, with an average of 606 days or just over 20 months. There are four recognized races of Asiatic elephant. They are: E. m. indicus - above; E. m. maximus - of Sri Lanka; E. m. ceylanicus- of Sri Lanka; and E. m. sumatricus - of Sumatra. The species is found in the wild state in India, Sri Lanka, Assam, Burma, Thailand, Campuchia, Laos, Vietnam, Malaysia, Sumatra and Borneo.

Elephe A genus of Reptiles of the Colubrid Snake family, *Colubridae*. e.g. E. guttata - The Corn Snake; average length 1 m (39 in).

Elettaria A genus of plants of the Ginger family, *Zingiberaceae*. e.g. E. cardamomum var. miniscula - The fruit of this species, Cardamom Fruit is carminative and purgative.

Eleutherodactylus A genus of Amphibians of the Frog family, *Leptodactylidae*. e.g. E. ricordi planirostris - The Greenhouse Frog, a diminutive anuran with a snout-vent length of only 31 mm (11/4 in). Found in the United States.

Eliomys A genus of Mammals of the Dormouse family, *Muscardinidae* (= *Gliridae*). e.g. E. quercinus - The Garden Dormouse, a small animal; length of body 105-147 mm, tail 80-135 mm; with tawny grey fur above and whitish grey below, and a black patch on each side of the head. The end of the tail carries a black and white tuft of hair. Found in gardens, parks, and broadleaved woodlands. Distribution: central and southern Europe, North Africa.

Elodea A genus of plants of the Frogbit or Waterweed family, *Hydrocharitaceae*. e.g. E. canadensis - Canadian Waterweed or Pondweed, a submerged, branched, aquatic, perennial plant, up to 3 m high, with whorls of darkgreen, simple leaves, mostly 3 in a whorl; long-stalked flowers. Place of origin: North Africa; naturalized in Europe, including Britain.

174

Elphidium A genus of Protozoans of the order *Foraminifera*. The species are spirally arranged on seaweeds in littoral areas. A slowly creeping animal, it can climb the glass sides of a tank.

Elymus A genus of plants of the Grass family, *Gramineae*, also known by the generic name, *Leymus*. e.g. E. arenarius (= *Leymus* a.) - Leyme-grass, a perennial plant, 100-150 cm high, rhizomatous with bluish grey leaves, 3-6-flowered spikelets usually growing in pairs, and densely hairy lemmas (the lower two scales enclosing the flower). Found along the seacoast of north and western Europe. Native in Britain.

Elysia A genus of Sea Slugs of the order *Sacoglossa*. e.g. E. viridis - The Common European Sea-slug, named after its green color, and often found feeding on the green algae, its host plant, of the genus *Codium* (see **Codium**).

Emarginula A genus of Archeogastropod Molluscs of the family *Fissurellidae*. e.g. E. huzardi - The Slit-limpet; shell 13-20 mm long; whitish and shaped like a pointed cap, with a network pattern of radiating ribs crossed by finer concentric grooves. Found on the hard sea bottom at shallow depths, generally under stones. Feeds on algae. Distribution: the Mediterranean Sea.

Embadomonas A genus of intestinal flagellates, now known by the generic name, *Retortamonas*. e.g. E. intestinalis (= *Retortamonas* i.) (see **Retortamonas**).

Embelia A genus of East Indian climbing plants of the Myrtle family, *Myrtaceae*. e.g. E. ribes - In this species, the dried fruit contains embelin, used in India and in the Far East as a tenicide.

Emberiza A genus of Birds of the Bunting family, *Emberizidae*; formerly placed in the Finch family, *Fringillidae*. e.g. E. citrinella - The Yellowhammer or Yellow Bunting, a beautiful sparrow-like bird, length 17 cm, the commonest member of this family, with its head and neck a fine lemon yellow, and reddish-brown hump. A partial migrant, it lives on farmlands, roadsides, and in open country. Distribution: Europe. In America the Woodpecker, known as the Yellow-shafted Flicker is often called the Yellowhammer.

Embiotica A genus of Fishes of the Perch family, *Percidae*. e.g. E. jacksoni - The Black Sea-perch fish.

Emericellopsis A genus of Molds or Fungi, better known by the generic name, *Cephalosporium*. e.g. E. salmosynnematum (= *Cephalosporium* s.) - The mold species, from which is prepared the antibiotic, Adicillin (see **Cephalosporium**).

Empetrum A genus of plants of the Crowberry family, *Empetraceae*. e.g. E. nigrum - The Crowberry, a small, creeping, evergreen shrub, 15-60 cm high, with needle-shaped leaves and a white keel (sharp ridge resembling the keel of a boat). Its fruit is a small black berry. Widespread on heaths. Native in Britain.

Empicoris A genus of small predatory Insects (class *Insecta*). e.g. E. vagabunda - A species living in human dwellings, but is useful because it attacks and kills flies and other harmful insects.

Empusa A genus of parasitic fungi. e.g. E. muscae - A species developing in the bodies of flies.

Emys A genus of Reptiles of the Freshwater Terrapin, Tortoise or Turtle family, *Emydidae*. e.g. E. orbicularis - The European Pond Tortoise or Terrapin, a carnivorous animal about 30 cm (1 ft) long. It has a dark brown carapace marked with yellow lines, and a flattened shell and limbs which though bearing claws, are web-footed and serve as swimming organs. It has a long life-span, some individuals reaching an age of more than a hundred years. Found in muddy ponds and ditches. Distribution:

southern Europe, North Africa and southwest Asia.

Enantiothamnus A genus of fungi. e.g. E. braulti - Causes development of pustular nodules.

Encephalitozoon A genus of microsporidian parasites, also known by the genus name, *Encephalocytozoon*. e.g. E. hominis (= *Encephalocytozoon* h.) - A species isolated from granulomatous encephalomyelitis.

Encephalocytozoon See **Encephalitozoon**.

Enchelyopus A genus of Fishes of the Cod family, *Gadidae* (Ling subfamily, *Lotinae*). These marine species, the Rocklings, have a small first dorsal fin which is often merely a single ray followed by a small fringe-like band.

Enchytraeus A genus of Segmented Worms of the family *Enchytraeidae*. e.g. E. albidus - Length 20-40 mm; yellowish-white; lives in compost, by the banks of ponds and brooks, amidst vegetation, under stones, and on the seacoast. Feeds on the roots of houseplants in flower-pots. Cultivated as food for aquarium fish and cage birds. Distribution: worldwide.

Encope A genus of Sea-urchins of the class *Echinoidea*. e.g. E. grandis - A species living in the ocean near the Antilles; it has a shield-shaped skeleton, and its body is covered with spines.

Endamoeba A genus of amoebic parasites in the intestines of invertebrates. e.g. E. blattae - A species found in the cockroach.

Endodermophyton A genus of Fungi, now named *Trichophyton*. e.g. E. castellanii (= *Trichophyton* c.) - Causes tinea intersecta (see **Trichophyton**).

Endolimax A genus of non-pathogenic intestinal amebae, also called by the generic names, *Amoeba* and *Entamoeba*, of the class *Rhizopoda*. e.g. E. nana (= *Amoeba* limax; *Entamoeba* n.) - A harmless commensal found in man (see **Entamoeba**).

Endomyces A name formerly given a genus of fungi, now known as *Candida*, *Blastomyces*, or *Zymonema*. e.g. E. albicans (= *Candida* a.; *Zymonema* a.) (see **Candida**). E. epidermidis (= *Blastomyces* dermatitidis; *Zymonema* capsulatum) (see **Blastomyces**).

Endomychus A genus of Insects of the Beetle family, *Endomychidae*. e.g. E. coccineus - Length 4-6 mm; red-colored carapace with four large, round, black spots. Found in beech woods under the moldy bark of felled trees or in tree fungi. Distribution: Europe.

Endromis A genus of Insects of the Emperor Moth or Butterfly family, *Saturniidae*.g. e. versicolor - The Kentish Glory Moth, a quickly growing species; in temperate regions it changes into pupae either the same year or after hibernation.

Endrosa A genus of Insects of the order of Butterflies and Moths, *Lepidoptera*. e.g. E. aurita - Size 3-3.4 cm; wide veins; the eggs are laid on rock-lichens. Distribution: Europe; found mainly in the Alps.

Endymion A genus of plants of the Lily family, *Liliaceae*. e.g. E. hispanicus (formerly called *Scilla* campanulatus; S. hispanica; *Urginea* maritima; U. scilla) - The Spanish Bluebell or Spanish Squill (= Bulbo de escila; Cila; Meerzwiebel; Scilla; Scillae Bulbus; Scille; Squill; White Squill) - A stout bulbous plant, 45 cm (18 in) high, with smooth strap-shaped flowers varying in color from pale china-blue to deep blue or indigo. Grown in woodlands and shrubberies. It contains a number of glycosides, especially scillarin A and scillarin B, which have a digitalis-like action on the heart. Squill has been used in the treatment of cardiac edema and as an expectorant in chronic bronchitis. Place of origin: the Mediterranean region, especially Spain and Portugal.

Engraulicypris A genus of African Cyprinid Fishes of the family

Cpyrinidae. These include barbels, carps, minnows, etc.

Engraulis A genus of Fishes of the Anchovy family, *Engraulidae.* e.g. E. encrasicholus - The European Anchovy Fish, a small herring-like fish, rarely more than 16 cm long, with a snout projecting in front of the mouth. Pelagic, they travel in large shoals and feed on marine zooplankton. They have a wide distribution and are suitable for canning in oil. Distribution: along the European Atlantic coast, in the Mediterranean, and in the Black Sea.

Enhydra A genus of Mammals of the Badger, Otter, Skunk and Weasel family, *Mustelidae.* e.g. E. lutris - The Sea Otter, the smallest totally marine animal, 1.2 m (4 ft) long with a 30 cm (1 ft) tail; weighing up to 20 kg (90 lb). Both morphologically and in its habits it forms a transition between the land carnivores and the seals. The skin is loose and its fur is greatly valued. The Sea Otter has been extensively hunted almost to extermination, but protective measures have led to its numbers increasing, especially along the Californian coast, and the Aleutian Islands in the extreme North Pacific.

Enicurus A genus of Birds of the family *Muscicapidae* (Thrush, Nightingale, Robin subfamily, *Turdinae*). e.g. E. leschenaulti - The White-crowned Forktail, a black bird with a white crown, rump and abdomen, and a white bar on the wings and white tips to the long, black tail feathers. Distribution: from the Himalayas to the China Sea.

Enkianthus A genus of Shrubs of the Heath family, *Ericaceae.* e.g. E. campanulatus - The Redvein Enkianthus, an erect shrub up to 6 m (20 ft) tall, with whorls of ovate, finely toothed leaves. The small bell-shaped, creamy-white flowers appear in pendulous clusters during May. The fruit is a capsule. Place of origin: Japan; introduced into Europe in 1880.

Ensatus A genus of Amphibians, also known by the generic name, *Dicamptodon*, of the Salamander family, *Ambystomatidae* (see **Dicamptodon**).

Ensis A genus of Bivalve Molluscs of the Razor Shell family, *Solenidae.* e.g. E. ensis - The Sword Razor; length 115-160 mm, height 18-21 mm, thickness 12-16 mm; shell strikingly long and narrow, slightly curved with blunt ends, glossy yellowish-white or greenish-brown, with reddish-brown bands marking the growth lines. Lives in vertical holes in fine or silty sand bottoms. Edible, often harvested. Distribution: from the Mediterranean to the North and Baltic Seas.

Entameba or **Entamoeba** A genus of parasitic amebae in the intestines of vertebrates, of the order *Amoebina*. Formerly called by various generic names such as: *Amoeba*; *Amoebus*; *Chaos*; *Endolimax*; *Iodamoeba*; and *Loeschia* (in 1912). e.g. E. buetschlii (= *Iodamoeba* b.) (see **Iodamoeba**). E. coli - Found in the large intestine; non-pathogenic. E. histolytica (= *Amoeba* dysenteriae; E. tropicalis) - The causal organism of amoebiasis. E. invadens - A species causing great damage in snakes and other reptiles; does not infect man. E. nana (= *Endolimax* n.) (see **Endolimax**). E. proteus (= *Amoeba* p.; *Chaos* diffluens) - The most commonly studied amebic organism in schools. E. tropicalis (= E. histolytica) - above.

Entelurus A genus of Fishes of the Pipefish and Seahorse family, *Syngnathidae.* e.g. E. aequoreus - The Snake Pipefish, up to 60 cm long; it has a very small caudal fin, and is a feeble swimmer, seeking the protection of seaweed in the lower tidal and offshore zone. Habitat: the Atlantic coast and the North Sea.

Enterobacter A genus of microorganisms of the family *Enterobacteriaceae*. These are gram-negative rods, belonging to the tribe *Klebsielleae*, and tending to be less invasive than the closely related Klebsiella group of organisms. The species are also classified under other generic headings, such as *Aerobacter, Serratia*, etc. These organisms are present in soil, water, dairy products, and the human intestinal tract. Most infections are hospital-acquired and associated with altered host resistance. e.g. E. aerogenes (= *Aerobacter* a.) - A species causing urinary tract infections (usually in debilitated individuals or after genito-urinary tract instrumentation), endocarditis, pneumonia, and bacteremia (see **Aerobacter**). E. agglomerans (Herbicola-Lathyri Group) - Gram-negative bacilliforming part of the *Erwineae* tribe, but now a proposed species of the genus *Enterobacter*. Part of the normal skin. An opportunist organism implicated in wound infections, urinary tract infections, bacteremia, meningitis, brain abscess, and septicemia from contaminated intravenous fluids. E. liquefaciens (= *Serratia* l.) (see **Serratia**).

Enterobius A genus of Nematode Intestinal Worms of the family *Oxyuridae*, formerly called by the generic name, *Oxyuris*. e.g. E. vermicularis (= *Oxyuris* v.) - The Pinworm, Seatworm or Thread-worm, a parasite in the large intestine of man; length of male 0.5-1cm, of female 1.5-2 cm. At night it emerges, remaining around the anus, where the female lays as many as 12,000 microscopic eggs. Inadequate personal hygiene results in continued reinfection. The disease caused by pinworms is called Oxyuriasis, which is widespread especially in countries with poor hygiene. Distribution: worldwide.

Enterococcus A genus of microorganisms of the Group D Streptococci normally residing in the human intestine. The Enterococci are part of the normal flora of the intestines, but may cause disease when removed from this habitat into the blood stream (septicemia and subacute bacterial endocarditis), urinary tract, or

meninges. They are occasionally associated with "food poisoning". e.g. E. faecalis (= *Streptococcus* f.) (see **Streptococcus**).

Enteromonas A genus of Flagellate Protozoa, formerly also called *Tricercomonas*. e.g. E. hominis (= *Tricercomonas* h.) - A rare parasite in the human intestine.

Enteromorpha A genus of plants of the family of Green Algae. e.g. E. compressa - Sea Grass, up to 30 cm high and 1 cm wide; compressed and tubular; narrowed at base. Distribution: the North Sea; native around Britain.

Entomobrya A genus of Apterygote or Primitive wingless Insects of the suborder *Arthropleona*. Among members of this suborder are the *Entomobryoidea*, from which the genus *Entomobrya* is derived. The species are brownish, white or mottled in color, without wings, and with a smooth, sometimes scaly integument.

Entomospira A genus of Spirochaetes, now included in the genus *Borrelia*. e.g. E. glossinae (= *Borrelia* g.) (see **Borrelia**).

Entriatoma A genus of Insects of the order of True Bugs, *Heteroptera*, better known by the generic name, *Triatoma*. e.g. E. sordida (= *Triatoma* s.) (see **Triatoma**).

Eozapus A genus of Mammals of the Jumping Mouse family, *Zapodidae*. These species are the Szechwan jumping mice of China.

Ephedra A genus of low, evergreen Shrubs of the family *Gnetaceae*. e.g. E. equisetina - A species containing the sympathomimetic alkaloid ephedrine. E. major - A species about 1.8 m (6 ft) high, with dense, dark-green branches bearing scale leaves. Distribution: from the Mediterranean region to northern India. E. vulgaris - A species found in China, from which is extracted the Chinese drug, ma huang.

Ephemera A genus of Exopterygote

Insects of the order of Mayflies, *Ephemeroptera* (family *Ephemeridae*). e.g. E. danica - The Mayfly, length of body 15-24 mm; a fragile insect in which the large forewings have a complete set of veins, and the wings have a slightly pleated appearance resembling the folds of a fan. The development of the Mayfly is associated with an early phase in insect evolution. The larvae frequent flowing water in hilly country. Distribution: most of Europe.

Ephestia A genus of Insects of the Moth family, *Pyralidae*. e.g. E. kuehniella (= E. kuhniella) - The Mediterranean Flour Moth, 10-12 mm long, found chiefly in mills, storehouses and homes. The larvae generally live in flour. A severe pest, the cocoons soil the flour and the surroundings, sticking the flour on which the insect feeds, into small lumps. Distribution: worldwide; introduced into Europe in the 19th century.

Ephippiorhynchus A genus of Birds of the Stork family, *Ciconiidae*. e.g. E. senegalensis - The Senegal Jabiru or Saddle-billed Stork, a giant, colorful bird found in Africa, with a beak "covered with a horse's saddle", hence the name. It has a brilliant black head, neck, wings and tail, while the rest of the plumage is pure white. The iris and cere are golden yellow, the bill red with a black median band, and the feet greyish brown marked with red at the root of the toes. It nests in trees, living on the banks of large rivers and feeding on fish, amphibians and insects. Distribution: Africa.

Ephoron A genus of Insects of the order of Mayflies, *Ephemeroptera*. e.g. E. virgo - Size 1.6 cm; whitish wings with three tail bristles. Spends most of its life in the larval stage in freshwater; after emergence the adults mate and die within a few hours.

Ephydatia A genus of Sponges of the phylum *Porifera*. e.g. E. fluviatilis - The River Sponge, a parazoan animal, with a body often pale in color when not exposed to light. Forms encrusting growths in plants, wood or stones in the water. A freshwater species. Native in Britain.

Epicrates A genus of Reptiles of the Boa family, *Boidae*. e.g. E. cenchris (= E. cenchria) - The Rainbow Boa, a species of Boa snakes found in Central and South America, up to 90 cm (3 ft) long, similar to the constrictor but with a smaller head and narrower body. Its golden brown to red-brown skin is strikingly iridescent, with all the hues of the rainbow, particularly after sloughing, hence the name (see **Constrictor**).

Epidendrum A genus of plants of the Orchid family, *Orchidaceae*. e.g. E. falcatum - A species derived from Mexico, with a stout rhizome, thick pseudobulbs and fleshy, tapering leaves from 15-30 cm (6-12 in) long. The flowers are creamy-white, each possessing five narrow, spreading perianth segments.

Epidermophyton A genus of dermatophytic Fungi, keratinophilic molds causing ringworm (tinea or dermatophytosis). e.g. E. cruris (= E. floccosum) - The causative organism of dermatomycosis - tinea cruris, tinea pedis, and tinea ungium.

Epidinium A genus of Protozoans of the order of Entodiniomorphs or Spirotrichs, *Entodiniomorphida*. An extremely complex ciliate species, with a macronucleus, a micronucleus, a permanent cytoproct, and two contractile vacuoles. The AZM (adoral zone of membranellae) is the only feeding ciliature.

Epilobium A genus of Medicinal and Aromatic plants, also known by the generic name, *Chamaenerion*, of the Willow-herb family, *Oenotheraceae*. e.g. E. augustifolium (= *Chamaenerion* a.) - The Fireweed; Rosebay Willow-Herb; an erect, little-branched perennial, 50-150 cm high, with long, narrow, lance-shaped leaves and rose-purple, 4-

petalled flowers in long, leafless, terminal spikes. Common in clearings at altitudes of up to 2,400 m. The leaves are used for making tea in parts of Russia. Place of origin: Europe; native in Britain.

Epimedium A genus of plants of the Barberry family, *Berberidaceae*. e.g. E. alpinum - The Barrenwort, a perennial plant, 20-30 cm high, with numerous stems each carrying one 2-3-ternate leaf. Distribution: Europe, especially northeast Italy, Austria and the Balkans. Introduced and naturalized in Britain.

Epimys A genus of Mammals, better known by the generic name, *Rattus*, of the Old World Rat and Mouse family, *Muridae*. e.g. E. norvegicus (= *Rattus* n.) - The Brown Rat or Ship Rat. E. rattus (= *Rattus* r.) - The Black Rat or Indian Plague Rat (see **Rattus**).

Epinephelus A genus of Fishes of the Grouper and Sea bass family, *Serranidae*. e.g. E. fasciatus - The Banded Grouper, a species of night fish found in the Red Sea. These are solitary predators preying on live fish and dwelling at the entrance of small caves in the coral reefs off the Gulf of Eilat (Akaba), and particularly active at night. They have large eyes, characteristic of all night fish.

Epipactis A genus of plants of the Orchid family, *Orchidaceae*. e.g. E. purpurata - The Violet Helleborine, a perennial plant, 15-50 cm high; leaves as long as the internodes, often violet beneath; greenish flowers. Grows scattered in shady woods. Distribution: Europe; native in Britain.

Epiphyllum A genus of plants of the Cactus family, *Cactaceae*. Also known by the generic names, *Rhipsalidopsis* and *Schlumbergera*. e.g. E. gaertneri (= *Rhipsalidopsis* g.; *Schlumbergera* g.) - The Easter Cactus plant, an epiphytic species with spreading, pendant, joined, flattened stems and starry scarlet flowers at the tips of the branches. There are many cultivated forms with beautiful flowers. Place of origin: Tropical America; Brazil (see **Rhipsalidopsis**; **Schlumbergera**).

Epipogium A genus of plants of the Orchid family, *Orchidaceae*. e.g. E. aphyllum - The Ghost Orchid, a rhizomatous, non-green, saprophytic, perennial plant living on decaying organic matter, 5-20 cm high, with coral-like, fleshy rhizomes and yellowish or reddish, drooping, spurred flowers. A very rare species. Place of origin: from northern Europe to the Himalayas. Native in Britain.

Episyrphus A genus of Insects of the Hover Fly family, *Syrphidae*. e.g. E. balteatus - The Hover Fly, length 11-12 mm; very abundant in gardens, parks, and forests. Settles on flowers and pollinates them. Distribution: the Palaearctic and Austro-oriental regions.

Epitonium A genus of Gastropod Molluscs, better known by the generic name, *Clathrus*. e.g. E. clathrum (= *Clathrus* clathrus) - The Common Wentletrap (see **Clathrus**).

Epizoanthus A genus of Sea Anemones of the order *Zoanthiniaria*. These animals resemble small anemones, but with a different arrangement of septa. They are generally found in warm, shallow waters, growing as encrusting organisms on sponges, corals and shells. Distribution: Off southwest Britain and Bermuda.

Epomorphorus A genus of Mammals of the suborder of Fruit-sucking Bats, *Megachiroptera*. These fruit bats are the epauletted bats, so called from the tufts of white hair on the shoulders of the males. Distribution: Central and South Africa.

Eptatretus A genus of Fishes of the order of Hagfishes, *Hyperotreti*; also called by the generic names, *Bdellostoma* or *Heptatretus*. These are marine species with 6-14 gill pouches or openings, and are found only in the Pacific in

California, Chile, Patagonia, South Africa, and Japan (see **Bdellostoma**; **Heptatretus**).

Eptesicus A genus of Mammals of the Long-eared Bat family, *Vespertilionidae*. e.g. E. fuscus - The Big Brown Bat, a fast-flying, mouse-eared species, hibernating in caves, and in summer is found in the attics of old houses and farm buildings. Distribution: North America.

Equisetum A genus of plants of the Horsetail family, *Equisetaceae*. e.g. E. arvense - The Common or Field Horsetail, or Herba Equiseti, a perennial plant, up to 50 cm high, with distant sheaths of fertile stems bearing 6-16 teeth. A common weed, it causes poisoning in horses that eat it with hay. The stem contains the alkaloids nicotine and palustrine. It has weak diuretic properties. Grows in damp places, on fields and in meadows. Distribution: Europe; native in Britain.

Equus A genus of Mammals of the Ass, Horse, and Zebra family, *Equidae*. e.g. E. caballus - The Wild Horse, now represented by only one race, the Mongolian Wild Horse, Przewalski's Horse, or the Przewalsky Horse (E. c. przewalskii), a herbivorous animal with short ears, round undivided hooves, and tail hairs growing the whole of its length. Very short, it is only 1.35 m (41/2 ft) at the withers. The species was very common in Europe in prehistoric times, finally dying out in South Russia in 1851, but back-breeding from primitive types of wild horses has produced a type much like the wild ancestor. It has been domesticated since prehistoric times as a beast of burden, a draught animal, and for riding and racing. The domestic breed today has a flowing mane and a tail of coarse hair, and is much taller and larger than its wild predecessor.

Eragrostis A genus of plants of the Grass family, *Gramineae*. e.g. E. minor - Small Eragrostis, an annual plant, 15-50 cm high; ring of hairs at junction of leaf-sheath and blade; spikelets often blackish-violet. Distribution: Mediterranean region; introduced and naturalized in Britain.

Eranis A genus of Butterflies and Moths of the Geometer Moth or Looper family, *Geometridae*. e.g. E. defoliaria - The Mottled Umber Moth, length 22-26 mm; it has several color forms, and lives in gardens and forests. The caterpillars live on forest trees and are known as loopers or measuring worms, as they move by humping their body into a loop, then throwing themselves forward. Distribution: Europe; native in Britain.

Eranthis A genus of plants of the Buttercup family, *Ranunculaceae*. e.g. E. hyemalis (= E. hiemalis) - The Winter Aconite, a perennial herb with a tuberous rhizome, 4-15 cm high; it has three stem leaves arranged in a whorl just beneath the solitary flowers. Distribution: South Europe; introduced and naturalized in Britain.

Erebia A genus of Butterflies and Moths of the Brown family, *Satyridae*. e.g. E. medusa - The Woodland Ringlet Moth, size 23-24 mm; found in glades and clearings up to elevations of 2,500 m. Distribution: Europe, Asia Minor.

Erebus A genus of Insects of the Noctuid or Owlet-moth family, *Noctuidae*. e.g. E. aggrippina - The Agrippina Moth from South America, a huge insect with a wingspan of up to 25 cm (10 in). E. odora - The Black Witch Moth, a heavily built, dull-colored nocturnal species, with a wingspan of almost 12.5 cm (5 in); commonly attracted as the other noctuid moths to lights. It migrates regularly northwards from tropical America.

Eremias A genus of Reptiles of the Lizard family, *Lacertidae*. e.g. E. arguta deserti - The Steppe Racer, length 20 cm; noted for its ability to bury itself in the sand with lightning rapidity when danger threatens. Distribution: sandy regions of eastern Rumania as far as southern Russia.

Eremicaster A genus of Echinoderms of the class of Starfishes, *Asteroidea*. e.g. E. tenebrarius - A deep-sea starfish living in the great depths of the Central Pacific.

Eremophila A genus of Birds of the Lark family, *Alaudidae*. e.g. E. alpestris - The Shore Lark, size 16.5 cm; a migrant bird possessing striking head markings with "horns". Distribution: Europe.

Eremurus A genus of plants of the Lily family, *Liliaceae*. e.g. E. robustus - A stout perennial plant from Turkestan, with bright green leaves from 60-150 cm (2-5 ft) long and up to 10 cm (4 in) wide, and handsome spikes of many peach-colored flowers, reaching heights of 270 cm (9 ft) or more.

Eresus A genus of True Spiders (order *Araneae*), of the family *Eresidae*. e.g. E. niger - The Black Spider; length of male 8-11 mm, of the female 9-16 mm; the sexes differ in coloration: the male is black with the upper surface of the abdomen red with four large black spots, and the female is blackish-brown. Distribution: Europe, the Caucasus and Turkestan.

Erethrizon A genus of Mammals of the New World Porcupine family, *Erethizontidae*. e.g. E. dorsatum - The Canadian Porcupine, 1.05 m (31/2 ft) long including 15 cm (6 in) of tail, with long brownish-black fur sprinkled with long white hairs concealing short barbed spines. The hind foot has a well-developed big toe. Distribution: From Alaska across Canada and, in the United States, to central California, New Mexico and eastwards to Virginia.

Eretmochelys A genus of Reptiles of the Marine Turtle family, *Chelonidae*. e.g. E. imbricata - The Hawksbill Turtle, the smallest of the species; its carapace is about 90 cm (3 ft) long, it has a hooked beak, and it lays its eggs by day or by night on beaches along sheltered bays.

Distribution: In warm waters, but may be carried by currents into the Mediterranean and to the European and American shores of the North Atlantic.

Erica A genus of plants of the Heath family, *Ericaceae*. e.g. E. tetralix - The Bog Heather, Cross-leaved Heather or Heath, an evergreen shrub, often cultivated, 10-60 cm high; flowers in umbel-like clusters. Grows on the bogs and moors of West and Central Europe, northwards to South Scandinavia. Native in Britain.

Ericulus A genus of Mammals of the Tenrec family, *Tenrecidae*, related both to the hedgehogs and the shrews. e.g. E. setosus - The Hedgehog Tenrec, an insectivore in which the whole of the back and the short tail are covered with spines. It lives as much as 10-11 years. Distribution: Madagascar, where the animals are hunted for their flesh.

Erigeron A genus of plants of the Daisy family, *Compositae*. e.g. E. alpinus - The Alpine Fleabane, a perennial plant, 2-20 cm high; outer disc-florets slender. Grows in stony alpine pastures, on scree and rocks. Distribution: southwest and southeast Europe.

Erignathus A genus of Mammals of the True or Earless Seal family, *Phocidae*. e.g. E. barbatus - The Bearded Seal, size up to 3.1 m; brown skin; upper lip has very long light-colored vibrissae or whiskers, hence the name. The species lives among the drift ice of the Arctic Ocean.

Erinaceus A genus of Mammals of the Hedgehog family, *Erinaceidae*. e.g. E. europaeus - The European Hedgehog or Urchin, a short-legged animal, 30 cm (1 ft) long, covered by an armature of spines all over the back and sides, which can be erected by powerful skin muscles. The spiny coat is dark brown and grey, and the ears are small. The creature can roll up into a ball, thus protecting the belly, head and legs. A feature of the species is its relative immunity against snake bites. A dose of poison that could be fatal to another animal of its size, leaves it completely unaffected and enables it to kill its adversary with impunity. It is found in woods among undergrowth, on shrub-covered hillsides, in gardens and parks in cities and villages. Distribution: western Europe, in Bohemia and the Alps. Native in Britain.

Erinus A genus of plants of the Figwort or Snapdragon family, *Scrophulariaceae*. e.g. E. alpinus - The Fairy Foxglove, a perennial plant, 10-20 cm high; basal leaves in a rosette; violet flowers in corymb-like racemes. Grows in stony mountain pastures, on gravelly slopes, on scree and in crevices. Distribution: South, West and Central Europe, at 1,500 to 2,350 m. Introduced and naturalized in Britain.

Eriobotrya A genus of plants of the Rose family, *Rosaceae*, also known by the generic name, *Mespilus*. e.g. E. japonica (= *Mespilus* j.) - The Japanese Plum Tree or Loquat Tree, an evergreen with handsome, broad, dark, glossy green foliage; under-surface brown and deeply ribbed. Yellowish, fragrant flowers growing in woolly clusters and producing edible, pear-shaped fruits. Place of origin: China, Japan; introduced into Europe in 1787.

Eriocheir See **Eriocheiris**.

Eriocheiris A genus of Decapod Crustaceans of the family, *Grapsidae*, also known by the generic name, *Eriocheir*. e.g. E. sinensis (= *Eriocheir* s.) - The Chinese Mitten Crab; length 7cm, width 9 cm. Has large pincers, those of males thickly covered with felty hairs. A marine crab, it travels against the current up rivers, but reproduces only in salt water. Distribution: The Yellow Sea. At the beginning of the century it was introduced from China into Europe, where it spread up the Elbe, Ems and Weser rivers, and has now reached the lower regions of the

Vltava River. Also found in the Atlantic Ocean north from the English Channel, and in the North and Baltic Seas.

Eriodictyon A genus of plants of the family *Hydrophyllaceae*. e.g. E. californicum (= E. glutinosum) - An hydrophyllaceous species, from which is prepared yerba santa or "mountain balm", once used for the treatment of bronchitis. The dried leaf yields a fluid extract used as a vehicle for dispensing drugs.

Eriophorum A genus of plants of the Sedge family, *Cyperaceae*. e.g. E. angustifolium - The Common Cottongrass, a rhizomatous perennial plant, 20-60 cm high; 3-7 heads with smooth stalks. The fruits bear conspicuous white cottony hairs. Grows in wet bogs and damp peaty places, in arctic and north temperate regions. Native in Britain.

Eristalis A genus of Insects of the Hoverfly family, *Syrphidae*. e.g. E. tenax - The Drone Fly or Dronefly, a syrphid species of flies, covered with hair, with a robust body, and dark coloring resembling a bee, for which it is often mistaken. A medium-sized, flower-hunting insect, 15-19 mm long, the larvae are aquatic scavengers found in heavily polluted water, and are provided with long respiratory siphons which extend to the surface from deep below, serving as telescopic breathing tubes. Distribution: worldwide; native in Britain.

Erithacus A genus of Birds of the Babbler, Thrush and Warbler family, *Muscicapidae* or *Turdidae* (Nightingale, Robin, and Thrush subfamily, *Turdinae*). e.g. E. rubecula - The European Robin; length 14 cm; it has a bright-red throat and breast. A solitary, partially migrant bird, it lives in woodlands, parks and heaths, and builds a rather crude cup-shaped nest on or near the ground. The Robin is the national bird of Great Britain. A short-lived species, its average life-expectancy is 13 months in the wild state. Distribution: Europe, the Middle East, Asia Minor, northwest Africa.

Eritrichium A genus of plants of the Borage family, *Boraginaceae*. e.g. E. nanum - The Dwarf Fairy-borage, a perennial plant, 2-5 cm high, cushion-like with glossy silky hairs. Grows in rock-crevices, on scree and alpine grass-land, at 2,500 to 3,390 m. Habitat: Alps and Carpathian mountains.

Erodium A genus of plants of the Cranesbill or Geranium family, *Geraniaceae*. e.g. E. cicutarium - The Common Storksbill or Alfileria, an annual plant, up to 60 cm high; with bright, rosy purple petals, often with blackish spot at base; umbel-like inflo-rescences. Widespread and common in arable and waste land, and by roadsides. Distribution: most of Europe; Asia, and North Africa. Native in Britain.

Erolia A genus of Birds of the Sandpiper family, *Scolopacidae*. Closely related to the Dunlin species of the genus *Calidris* (see **Calidris**).

Erophila A genus of plants of the Cress or Crucifer family, *Cruciferae*. e.g. E. verna - Common Whitlow-grass, an annual plant, 2-15 cm high; fruits are oblanceolate or elliptical. Widespread and local on rocks, walls, and in dry places. Distribution: most of Europe; native in Britain.

Erpobdella A genus of Segmented Worms, the Annelids, of the class of Leeches, *Hirudinea*. The genus belongs to the family *Pharynchobdellae* (= *Erpobdellidae*), in which the species have entirely lost the power of piercing the tissues of other animals and of suck-ing their blood. They feed instead on whole insect larvae, small earthworms, etc. e.g. E. punctata - One of the com-monest and most widespread species in North America. They have 3 or 4 pairs of eyes arranged in 2 transverse rows.

Erucastrum A genus of plants of the Cress or Crucifer family, *Cruciferae*.

e.g. E. gallicum - The Hairy Mustard, a biennial plant, 10-60 cm high; erect sepals, pale or golden-yellow petals, fruits curving upwards. Grows on arable and waste land. A native of central and southwestern Europe; also found in the Alps. Introduced and naturalized in Britain.

Erwinia A genus of microorganisms resembling members of the family *Enterobacteriaceae*; these are gram-neg-ative rods. Primarily a plant pathogen, but has been isolated from wounds, uri-nary tract infections, skin infections, etc., and has been responsible for sever-al cases of bacteremia traced to intra-venous infusion apparatus.

Eryngium A genus of plants of the Carrot family, *Umbelliferae*. e.g. E. campestre - The Field Eryngo, a peren-nial plant, 10-90 cm high; yellowish, almost spherical heads with spiny leaves, bracts and bracteoles. Grows on arable and waste land, in dry grassy places, and by waysides. Distribution: South, West and Central Europe. Native in Britain. Also found in Iran and North Africa.

Erynnis A genus of Insects of the Skipper Moth family, *Hesperiidae*. e.g. E. tages - The Dinger Skipper Moth, length 13-14 mm; flies over sunny paths in glades, and sometimes also in gar-dens. Distribution: southern and central Europe, northern Asia.

Erysimum A genus of plants of the Cress or Crucifer family, *Cruciferae*. e.g. E. canescens - A Treacle-mustard plant, containing the glycoside Erysimin (Helveticoside), whose action resembles that of Strophanthin-K, and is used in the treatment of circulatory insufficien-cy.

Erysipelothrix A genus of microorgan-isms of the family *Corynebacteriaceae*; gram-positive rods present worldwide and widely distributed in nature in water and on fish surfaces. e.g. E. insidiosa (= E. rhusiopathiae) - A species causing erysipeloid (erysipelas) in animals and man - a skin infection, particularly of the hands, with lymphadenopathy and occasionally, arthritis. Has also caused rapidly fatal septicemia and endocarditis. Infection occurs from contact with infected materials such as meat, fish, poultry, manure, hides, and bones. E. muriseptica (= Ery. muriseptica; *Bacillus* murisepticus) - A species causing an epizootic septicemia in mice.

Erythraea A genus of red-flowered plants of the Gentian family, *Gentianaceae*, better known by the generic name, *Centaurium*. e.g. E. centaurium (= *Centaurium* minus) - The Common or Lesser Centaury plant (see **Centaurium**).

Erythrina A genus of plants of the Pea family, *Leguminosae*. e.g. E. crista-galli - The Cockscomb, Cockspur, Coral Bean or Coral Tree, up to 1.5 m (5 ft) tall, having trifoliate, prickly leaves with ovate leaflets. The red flowers have large standard petals and grow indense terminal racemes, looking like waxen sweet peas. Place of origin: Brazil; introduced into Europe in 1771.

Erythrinus A genus of Fishes of the Characin family, *Characinidae*. These are carnivorous fishes with a scaly body, normal pharyngeal bones and an adipose fin. They are interesting in that they are capable of breathing air.

Erythrobacillus A genus of small aerobic non-pathogenic bacteria, producing red or pink pigments. Now known by the generic name, *Serratia*. e.g. E. indica (= *Serratia* i.) (see **Serratia**).

Erythrocebus A genus of Mammals of the Old World Monkey family, *Cercopithecidae*. e.g. E. patas - The Patas or Hussar Monkey, a non-arboreal animal related to the guenon monkeys. It is orange above, white below, and has a black face. Preferring the ground, it seeks refuge in trees only when alarmed. Its many races are distributed from Senegal to Ethiopia and East Africa, in dry savannahs where trees are within reach for emergencies.

Erythronium A genus of plants of the Lily family, *Liliaceae*. e.g. E. dens. canis - The Dog's-tooth Violet, European Fawn-lily, or Trout-lily, an erect, bulbous, perennial plant, with oval pointed leaves and solitary, drooping, pinkish-purple flowers borne on short, erect stems. Place of origin: Europe, Asia; introduced into Britain in 1596.

Erythrophloeum A genus of Trees of the Pea family, *Leguminosae*. e.g. E. guineense - A species yielding casca, mancona, or sassybark, an African poison.

Erythroxylon See **Erythroxylum**.

Erythroxylum A genus of South American plants, also cultivated in Indonesia, of the family *Erythroxylaceae*. Sometimes called by the generic name, *Erythroxylon*. e.g. E. coca (= *Erythroxylon* c.) - A shrub 3-4.5 m (10-15 ft) high; a native of the Andes, of Peru and Bolivia. The leaf is known as the Bolivian or Huanuco Leaf. E. truxillense - Peruvian or Truxillo Leaf, also known in Indonesia as the Java Leaf. These shrubs or trees have simple opposite leaves and small bisexual flowers. There are usually 10 stamens more or less fused basally. The fruit is usually succulent. The dried leaves of these species are known as Coca, which contains the alkaloid ecgonine, from which is prepared synthetically the crystalline alkaloid, methyl benzoyl ecgonine or Cocaine - the oldest local anesthetic known to man, causing systemic toxic effects (cocainism) and addiction. Its medical use today is almost entirely restricted to local application (local anesthesia) in ophthalmic surgery and surgery of the ear, nose and throat. Cocaine is also known by other names, abbreviations, and popular or slang expressions, such as: Bernie; Bolivian marching powder; Bouncing powder;

Burese; C.; C&H - cocaine and heroin; Charles; C&M - cocaine and morphine mixture; Co.; Coke; Crack; Dynamite; Flake; Frisco speedball - mixture of heroin, cocaine and LSD-25; FS (= frisco speedball); gold dust; heaven dust; her; HMC - heroin, morphine and cocaine; incentive; jam; lady snow; M&C- morphine and cocaine (drug addiction); mojo - cocaine, morphine or heroin; rock, sniff; snort; speedball (mixture of cocaine and heroin); star dust; white lady (name also used for heroin).

Eryx A genus of Reptiles of the Boa Snake family, *Boidae*. e.g. E. jaculus - The Sand Boa, length 80 cm; found in warm regions in stony places, usually under stones or in holes in the ground. It kills its victims, such as lizards and mice, by coiling itself around their bodies and suffocating them. Distribution: southeastern Europe, western Asia, and North Africa.

Escherichia A genus of microorganisms of the family *Enterobacteriaceae*; gram-negative short rods, widely distributed in nature and the human intestinal tract. Has been used for research into fundamental genetical problems in recent years. Some species are also known by other generic names, such as *Citrobacter*. e.g. E. aurescens - A species characterized by yellow-orange pigment production. Has been isolated from feces, eye infections, and contaminated water supplies. E. coli (= EC; E. coli communis; *Bacillus* c.) - A part of the normal intestinal flora, and frequently used as an indicator of fecal contamination in water. Also appears as an enteropathogenic form (= EEC or Enteropathogenic *Escherichia* coli), and the most frequent cause of urinary tract infections, such as cystitis, pyelitis, and pyelonephritis. May also cause puerperal sepsis, cholecystits, appendicitis, summer diarrhea, diarrhea of travellers (especially to Asia, Central and South America, etc.), and epidemic diarrhea of the newborn. There are approximately 150 "O" antigenic groups, eleven of which contain EEG (enteropathogenic) types that produce infantile diarrhea, i.e. O111; O55; O26; O86; O112; O119; O124; O125; O126; O127; and O128. E. freundii (*Citrobacter* f.) - see **Citrobacter**.

Eschichtius A genus of Mammals of the Whale order *Cetacea*. e.g. E. gibbosus - The Pacific Gray Whale, the largest marine animal ever held in captivitiy. A female specimen, when returned to the sea after being captured, measured 8.22 m (27 ft) long, and weighed an estimated 6,350 kg (14,000 lb) in 1972. It was at that time one year old, and was rapidly outgrowing her surroundings in San Diego, California.

Eschscholtzia A genus of plants, also called *Eschscholzia*, of the Poppy family, *Papaveraceae*. e.g. E. californica (= *Eschscholzia* c.) - The California Poppy, a brilliant annual plant with divided leaves and solitary dark-red to cream-colored flowers. A favorite garden species. Also a hypnotic and anodyne used in medicine. Place of origin: California; introduced into Europe in 1790.

Eschscholzia See **Eschscholtzia**.

Esox A genus of Fishes, also called *Essox*, of the Pike, Pickerel, and Muskellunge family, *Esocidae*. e.g. E. lucius (= *Essox* l.) - The Northern Pike, length 150 cm (5 ft), weight 35 kg (77 lb); a swift, powerful, predatory fish with an elongated body and a snout protruding like a duck's bill; mouth slit to below the eyes, numerous pointed teeth, and a dorsal fin far back on the body. Popular with anglers and much esteemed as food in many parts of Europe. Found in the lower reaches of rivers, in pools and backwaters overgrown with vegetation. Planted as a supplementary fish in carpponds.

Distribution: throughout Europe, most of Asia, and North America.

Essox See **Esox.**

Estrilda A genus of Birds of the Grassfinch, Mannikin, and Waxbill family, *Estrildidae*. These species of Old World birds are the waxbills, best known in northern countries as attractive cage-birds. The males are predominantly a buffish grey with white spots and often have scarlet backs, heads, or bills.

Etheostoma A genus of Fishes of the Perch family, *Percidae*. These are the darter fishes, small and many of them brilliantly colored. e.g. E. microperca - The Least Darter, a very tiny fish.

Ethusina A genus of Crustaceans of the order *Decapoda*. e.g. E. abyssicola - A deep-living marine crab, some specimens of which have been recovered from a depth of about 4,267 m (14,000 ft) in the Pacific Ocean.

Euastacus A genus of Crustaceans of the Lobster family, *Astacidae*. e.g. E. armatus - The Murray River Lobster, a large freshwater crustacean, found in South Australia, of impressive size, and credited with a maximum weight of 2.72 kg (6 lb).

Eubacterium A genus of microorganisms, also known by the genus name, *Bacteroides*, the species of which are anaerobic, gram-negative rods normally found in soil, feces and mucous membranes of the upper respiratory, gastrointestinal, and genito-urinary tracts. e.g. E. aerofaciens (= *Bacteroides* a.) - A species capable of causing empyema, peritonitis, postoperative wound infections, furuncles, etc.

Eubalaena A genus of Mammals of the Right Whale family, *Balaenidae*. e.g. E. glacialis - The Biscayan or Atlantic Right Whale, a rare species with a total length of 15-18 m (50-60 ft). The color is black and the jaws and lips are curiously built, the two halves of the upper jaw forming a "V", and at the same time being sharply arched. The lower jaw is also V-shaped, but more or less horizontal.

Eucalia A genus of Fishes of the Stickleback family, *Gasterosteidae*. e.g. E. inconstans - The North American Brook Stickleback, a freshwater species living in clean streams from New York to Indiana and Minnesota.

Eucalyptus A genus of Trees of the Myrtle family, *Myrtaceae*. e.g. E. globulus - The Blue Gum, a tree attaining a height of 60 m (200 ft). From the leaves is distilled Eucalyptus Oil, used as an antiseptic, deodorant and expectorant. Place of origin: tropical America and Australia. Cultivated in many countries all over the world.

Eucarya A genus of Trees of the Sandalwood family, *Santalaceae*. Also known by the generic name, *Santalum*. e.g. E. spicata (= *Santalum* spicatum) - Australian Sandalwood; formerly used as a urinary antiseptic (see **Santalum**).

Eucharis A genus of plants of the Daffodil family, *Amaryllidaceae*. e.g. E. grandiflora - The Amazon Lily, a daffodil species with a spherical bulb, and leaves about 60 cm (2 ft) long. The erect stem bears from 3-6 star-shaped flowers. Place of origin: Columbia.

Euchlanis A genus of minute Metazoans (phylum *Aschelminthes*), of the class of Rotifers, *Rotifera*. The species are minute, unsegmented, freshwater animals, less than 0.5 mm long. The anterior end is flattened into what is known as a trochal disc, which is ciliated for locomotion and food collection. The posterior end terminates in an appendage that is pincer-like, and the rotifer can anchor itself with this.

Euchlora 1. A genus of Comb Jellies or Sea Walnuts (= Ctenophores) of the phylum *Ctenophora*. e.g. E. rubra - A species of small, marine, carnivorous animals with a cosmopolitan distribution, generally found in the plankton. The gelatinous body is composed of two cell layers, ectoderm and endoderm, separated by the non-cellular mesogloea. They swim by means of transverse rows of fused cilia and are bioluminescent: after being placed in the dark for some time, light can be seen in the region of the comb plates, originating from the canal system lying below them. 2. A genus of Insects of the Chafer and Dung Beetle family, *Scarabaeidae*. e.g. E. dubia - Length 12-15 mm; a chafer beetle, green in color with a brown carapace, found in sandy and sandy-loamy soil. Distribution: Europe, from southern Italy to southern Sweden and Finland.

Eucinostomus A genus of Fishes (sub-phylum *Vertebrata*). e.g. E. gula - The Silver Jenny Fish.

Eucomis A genus of plants of the Lily family, *Liliaceae*. e.g. E. regia - A bulbous plant with a globose bulb, and 6-8 leaves up to 45 cm (18 in) wide and 10 cm (4 in) across. The stout central stem bears a dense mass of green flowers and above this, a terminal crown of 12-20 leaves. Place of origin: tropical and South Africa.

Euctimsna A genus of Arachnids of the True Spider order, *Araneae*. e.g. E. tibialis - An aggressive species, the most dangerous spider found in Australia; can be fatal to man.

Eudendrium A genus of Thecate Hydroids or Sea-fir animals, colonial relatives of the Sea Anemones. e.g. E. ramosum - Size 5-10 cm; the polyp colony is axial and the polyps do not possess cup-shaped hydrotheca. Found on seaweed and rocks. Distribution: the Mediterranean Sea.

Eudia A genus of Insects of the Emperor Moth family, *Saturniidae*. e.g. E. pavonia - The Lesser Emperor Moth, widely distributed throughout Europe. The male has a hyperacute sense of smell exhibited in nature, and can pick up inconceivably minute traces of the scent of its female counterpart at distances of even

3.2 km (2 miles) in the 40,000 sensory nerve cells of its feather-like antennae.

Eudontomyzon A genus of Chordate animals of the family of Lampreys, *Petromyzonidae*. e.g. E. danfordi - The Danube Lamprey, length 20-30 cm; a freshwater species parasitic on fish, attaching itself to their bodies with its sucktorial mouth and feeding on their blood and muscles. Breeds in the shallow waters of rivers and streams. Distribution: the tributaries of the Danube River and in rivers flowing south of the Danube into the Black Sea.

Eudorina A genus of plants of the class of Algae, *Flagellata*. e.g. E. elegans - Single celled or colonial microorganisms, having one flagellum and forming green, chlorophyll-containing colonies.

Eudromia A genus of Birds of the order of Tinamous, *Tinamiformes*. e.g. E. elegans - The Crested Tinamou, a species found on the open pampas of South America. Its coloring is cryptic to afford protection. The eggs, incubated by the male, are laid on the ground, their shiny surfaces blending into the wet, glistening moss and leaves of the pampas floor.

Eudromias A genus of Birds of the order of Auks, Gulls and Shorebirds, *Charadriiformes*, and belonging to the Plover family and its allies, *Charadriidae*. e.g. E. morinellus - The Dotterel, length 22 cm; it nests in stony and swampy northern tundras and in high mountains. A migratory bird, it winters in North Africa and the Middle East. Distribution: is discontinous, in Europe, in the mountains of Scotland, Scandinavia, the southern Alps and the Rumanian Carpathians; in northern Siberia, and in the high mountains of Central Asia.

Eudyptes A genus of Birds of the order of Penguins, *Sphenisciformes*. e.g. E. crestatus - The Rockhopper Penguin, found in New Zealand and neighboring islands. A marine, flightless bird, its webbed toes and wings are reduced to form strong flippers that cannot be folded, and it is capable of swimming underwater as fast as a seal. The penguin order is gregarious and monogamous, and the individuals nest in large colonies, the male taking an active role in rearing the young.

Eudyptula A genus of Birds of the Penguin order, *Sphemsciformes*. e.g. E. minor - The Fairy Penguin, a small species, also confined to the southern hemisphere (see **Eudyptes**).

Eugenia A genus of Trees of the Myrtle family, *Myrtaceae*. The species have other generic names such as *Caryophyllus*, *Pimenta*, etc. e.g. E. caryophyllus (= *Caryophyllus* aromaticus) - The Clove tree or shrub (see **Caryophyllus**). E. chequen - The Cheken plant. E. myrtifolia - The Australian Brush Cherry, a tree up to 24 m (80 ft) tall, with lanceolate leaves and white flowers. The red, ovoid fruits have a sharp flavor and are used in making jellies. Often used for clipped hedges in California. E. pimenta (= *Pimenta* officinalis) - The Pimenta tree (see **Pimenta**).

Euglena A genus of Protozoans of the order of infusorian animals, the Euglenoids (order *Euglenoidina*). The species range from 30 μ to 400 μ in length. e.g. E. halophila - A marine species, most tolerant of very high concentrations of salt. E. rubra - A reddish species, commonly occurring in late summer as the scum on stagnant waters rich in organic matter, such as farmyard ponds. E. viridis - A green-colored species, found in stagnant pools.

Eulalia A genus of Segmented Worms, the Polychaetes (class *Polychaeta*), of the Phyllodocid family, *Phyllodocidae*. e.g. E. viridis - A common European species living in crevices on rocky shores; about 5-7.5 cm (2-3 in) long and brilliant green in color, with a slender body composed of a large number of segments. The family is distinguished by the dorsal leaf-like extensions of the parapodia.

Eulota A genus of Gastropod Molluscs of the Snail family, *Helicidae*. e.g. E. fruticum - The Glass Shell Snail, width 22 mm; height 17 mm; the shell has a large deep navel, is uniformly grey or reddish, and the inside of the lip is whitish. Widespread in hedgerows, on walls, stones and in meadows. During rain or at nightfall, this damp-loving animal leaves its hiding-place to feed on leaves and soft fruit. Distribution: Europe.

Eumeces A genus of Reptiles of the Skink family, *Scincidae*. e.g. E. fasciatus - The Five-lined Skink, a lizard species common in the United States, in which the young have distinct light stripes down the back and a bright blue tail. In the adult the stripes are very indistinct and the tail becomes brownish like the rest of the body.

Eumenes A genus of Insects of the Wasp family, *Eumenidae*. e.g. E. pomiformis - The Potter Wasp, length 10-16 mm; has a variable yellow pattern on the dark body. The female is an excellent builder, making a flask-like nest with a narrow neck, attaching it to a plant, to a twig, or under a bark, and stocking it with several small caterpillars which later serve as food for the larvae. Distribution: the Palaearctic and Neoarctic regions.

Eumenis A genus of Insects of the Browns Butterfly family, *Satyridae*. e.g. E. semele - The Grayling Moth or Butterfly, size 5.5 cm; upperside dark brown with yellowish bands; the forewings have two eye-spots, the hindwings one. Eggs are laid on grasses. Found in open country and deciduous forests. Distribution: Europe.

Eumetopias A genus of Mammals of the Eared Seal family, *Otariidae*. e.g. E. jubatus - Steller's Sea-lion, a large animal, the male being up to 4 m (13 ft) long and weighing up to 585 kg (1,300 lb). Due to its great weight, this creature moves on land with difficulty. The species was discovered in the 18th century on the Pribilof Islands by Bering, and was first described by Steller, hence its name. Found all over the Pacific Coast from San Francisco to Kamchatka, but it has been greatly reduced in numbers by excessive hunting, especially in the 19th century.

Eumops A genus of Mammals of the Mastiff Bat family, *Molossidae*. e.g. E. californicus - The Californian Mastiff Bat, the largest bat in the United States. A thick-set, snub-nosed animal in which, when the wings are folded the forearm can be used for walking.

Eunectes A genus of Reptiles of the True Boa family, *Boidae*. e.g. E. murinus - The Giant Anaconda, a largely aquatic species, usually living in swamps and along the banks of pools and slow-moving rivers, preying on mammals and birds that come there to drink. An aquatic arboreal member of the Boa family, the Giant Anaconda may reach a length of up to 9 m (30 ft). Its color pattern consists of black spots and rings against an olive-green background. The age of such large snakes is estimated at 50 years. It is the world's largest snake, and is found in the northern parts of South America and Trinidad.

Eunice A genus of Segmented Worms or Annelids of the Ragworm family, *Nereidae*. e.g. E. harassi - A Ragworm species, 20-25 cm long; numerous body segments; parapodia with red gill tufts. A predaceous worm, feeding on many kinds of small animals including the tentacles of tubeworms. Found in the North Sea and the Baltic.

Eunicea A genus of Octocoral Metazoa of the order of Horny Corals, *Gorgonacea*. e.g. E. anceps - An animal species forming bushes whose branches appear to be made of cork. Found in the warm waters of the West Indies.

Eunicella A genus of Horny Corals of the order *Gorgonacea*. These animals are found in cold water areas of the sea; they are whitish in color and are often dredged around the southwest coast of Britain. e.g. E. verrucosa - The Common Sea Fan, a species found in European waters, growing like a plant with a short main trunk fastened to the ground, and lateral branching stems with numerous polyps arranged around an internal axis, giving the animal a feathery appearance.

Euonymus A genus of Trees or small Shrubs of the Spindle-tree family, *Celastraceae*. e.g. E. europaeus - The European Spindle-tree, a small deciduous tree or shrub, up to 6 m (20 ft) tall, whose hard wood is used to make skewers and spindles. The branches are green and 4-angled. The inconspicuous green flowers produce deep pink, or red, 4-lobed, or 4-valved capsules (the fruit), enclosing seeds with orange arils (fleshy envelopes). It usually grows on alkaline soils. Distribution: widespread in Europe and western Asia. Native in Britain.

Eupagurus A genus of Decapod Crustaceans (order *Decapoda*) of the Hermit Crab family, *Paguridae*. e.g. E. bernhardus - The Common Hermit Crab; size up to 10 cm (4 in); it has unequal pincers, the larger protecting the shell aperture; the hind part of the body is soft-skinned. It is also known for its habit of keeping its large soft abdomen within the shells of gastropod molluscs such as the colonial hydroid, *Hydractinia*. Sea-anemones (genus *Calliactis*) are often found growing on the shells of these crustacean crabs, living in symbiosis with them. Distribution: generally in littoral areas; along the seacoast of Europe; native along British shores and the deeper waters of the North Sea and the Atlantic Ocean (see **Calliactis**; **Hydractinia**).

Eupatorium A genus of plants of the Daisy family, *Compositae*, also known by the generic name, *Hebeclinium*. e.g. E. ianthinum (= *Hebeclinium* i.) - The

Eupatorium Sordidum, an attractive plant with woody stems, 1 m (3 ft) high; with oval-lanceolate leaves and large, shiny, terminal corymbs of mauve-purple fragrant flowers. Place of origin: Mexico; introduced into Europe in 1849. E. triplinerve (= E. ayapana) - A species containing ayapanin and ayapin; said to have diaphoretic, hemostatic, and stimulant properties. E. urticaefolium - The White Snakeroot, a species causing a disease known as "milk sickness" in persons taking milk or milk products from cattle or sheep, made ill by eating this plant. The animal becomes weak and may stumble and fall; the disease is known as "trembles" or "slow". In humans, milk sickness can be fatal. Another genus of plants, *Aploppus* e.g. A. heterophyllus and A. fructicosus, can also cause this condition in cattle and humans (see **Aploppus**).

Eupetes A genus of Birds of the Babbler family, *Muscicapidae* (subfamily *Timaliinae*). e.g. E. macrocerus - The Malay Rail-babbler, a ground-babbler living in the dense Malayan jungle and moving fast on the ground, difficult to spot. These jungle babblers are dull-covered birds, looking very much like warblers; they are also known as babbling thrushes.

Euphagus A genus of Birds of the American Blackbird and Oriole family, *Icteridae*.g. *e.* cyanocephalus - Brewer's Blackbird, essentially a bird of the prairies and meadows, particularly of the western United States. It nests either in bushes or the lower branches of trees, and even on the ground. The male is black with a purple sheen on the head, and greenish over the rest of the body; the eyes are white.

Eupharynx A genus of Fishes of the order of Gulper Eels, *Lyomeri* (= *Saccopharyngiformes*), belonging to the family *Eupharyngidae*. The species have extraordinary eel-like forms, living in deep water at depths of 1,800 to 2,700

m (6,000 to 9,000 ft), and eating invertebrates.

Euphorbia A genus of plants of the Spurge family, *Euphorbiaceae*. e.g. E. exigua - The Dwarf Spurge, an annual plant, 3-30 cm high; linear, stalkless, glaucous leaves up to 4 mm wide with pointed tips. Grows as weeds in gardens and cultivated fields and on waste ground. Distribution: widespread and common throughout Europe; native in Britain.

Euphrasia A genus of plants of the Figwort or Snapdragon family, *Scrophulariaceae*. e.g. E. rostkoviana - Meadow Eyebright, an annual plant, 4-50 cm high; calyx with glandular hairs; white flowers with violet upper lip. Widespread in meadows, on bushy slopes, and in open woodlands. Distribution: most of Europe; native in Britain.

Euphydryas A genus of Insects of the Tortoiseshell Butterfly family, *Nyphalidae*. e.g. E. maturna - A Tortoiseshell Butterfly species, size 4.5 cm; it has dark brown wings with a row of light yellow spots along the outer edge. Common in large areas of Europe.

Euplagia A genus of Insects of the Tiger Moth family, *Arctiidae*. e.g. E. quadripunctaria - The Jersey Tiger Butterfly, size 26-30 mm; found on rocky and stony hillsides and valleys (chiefly with limestone substrate). Distribution: Europe, Asia.

Euplectella A genus of Marine Sponges (phylum *Parazoa*) of the suborder *Hexasterophora*. e.g. E. aspergillum - Venus's Flower Basket; a hexactinellid (six-rayed), deep-sea Sponge, living in the ocean depths near the Philippines and the coasts of equitorial Africa. The naked siliceous skeletons are extremely beautiful and ornamental (see **Aphrocallistes**).

Euplotes A genus of Protozoans of the order of Hypotrichs, *Hypotrichida*. These are active ciliate organisms of medium size, 70-150 μ in length, occurring in mud and rotting vegetation. The macronucleus is horse shaped, and adjacent to the center, and on its outer surface is a single micronucleus.

Eupomotes A genus of Freshwater Fishes of the Perch family, *Percidae*. e.g. E. gibbosus - The Common Sunfish, length 10-15 cm; very high body, laterally compressed; it has a small mouth edged with small hooked teeth. A tasty edible species, introduced into Europe from America, and now well established in ponds and lakes.

Euproctis A genus of Insects of the Tussock Moth or Browntail family, *Lymantriidae*. e.g. E. chrysorrhoea (= E. phaeorrhoea) - The Brown-tail Moth or Butterfly, size 17-22 mm; white-winged with forewings marked sometimes with small black dots. Brown-colored posterior end of abdomen, hence the name. In winter the early larval stage is spent in a communal silk cocoon, where the larval hairs can cause skin rashes (dermatitis) in humans. Found in gardens and parks. Distribution: Europe, Asia Minor, North Africa.

Euproctus A genus of Amphibians of the family *Salamandridae*. In these species, the males clasp the females during mating, using their tails as nooses to hold their mates, so that their cloacae are very close together.

Euptasia A genus of Coelenterate Animals of the order of Sea Anemones, *Actiniaria*. Living in large clusters, the species of this genus are the Hexacorallia or Hexacorals, closely related to the stony corals, and have soft, hollow bodies, with a mouth surrounded by tentacles containing thousands of stinging cells that erupt when touched, paralyzing the prey (usually fish and crustaceans), which is then brought into the mouth to be digested. Habitat: the Red Sea.

Eurotium A genus of Fungi or Molds. e.g. E. repens - A species growing on bread and preserved fruits. E. malignum - A species occasionally found in the human ear.

Eurycea A genus of Amphibians of the Plethodontid Salamander family, *Plethodontidae*. They are lungless species from the United States, and are entirely neotenic (with a tendency to remain in a larval state, although gaining sexual maturity).

Eurydema A genus of Insects of the True Bug family, *Pentatomidae*. e.g. E. oleracea - The Brassica Bug, length 5-7 mm; ground color metallic green or blue with red, yellow, white or orange spots. Found mostly in coniferous forests. Distribution: Europe, Asia.

Eurylaimus A genus of Birds of the Broadbill family, *Eurylaimidae*. e.g. E. ochromalus - The Black-and-yellow Broadbill, found in the islands of the Malayan archipelago. A white-colored bird, with black wings with yellowish-white barring, and rather vinous-colored underparts. The males perch for long periods singing a loud melodious song.

Eurypelma A genus of Arachnids of the order of True Spiders, *Araneae*. e.g. E. hentzii - The American Tarantula, an extremely venomous spider and a long-living species, living as much as 20 years.

Eurypyga A genus of Birds of the Sunbittern family, *Eurypygidae*. e.g. E. helias - The Sunbittern of Central and South America, 45-52.5 cm (18-21 in) long; it flies very little but walks slowly and deliberately on its orange-colored, heron-like legs, its long, snake-like neck held parallel to the ground. It lives in dense tropical forests and swamps, usually near water.

Eurystomus A genus of Birds of the Roller family, *Coraciidae*. e.g. E. orientalis pacificus - The Broad-mouthed Dollarbird, breeding in northern and eastern Australia, nesting in tree-holes and sometimes using the old nests of other species. Noted for performing aerial acrobatics, rolling, zig-zagging, rocketing up with closed wings and diving, and making long swoops during display flights.

Euschemon A genus of Insects of the Skipper Butterfly family, *Hesperiidae*. e.g. E. rafflesia - The Australian Regent Skipper, a black and yellow butterfly, probably the most archaic butterfly existing today. A medium-sized species with a characteristic darting, erratic flight.

Euscorpius A genus of Arachnids of the True Scorpion family, *Chactidae*. e.g. E. flavicaudus - Size up to 4 cm; body blackish brown; abdomen with poisonous sting which reaches and kills its prey when the abdomen is twisted upwards and forwards. Found in the Mediterranean region. E. italicus - The Black Scorpion of Europe and North Africa.

Eusimulium A genus of Insects of the Fly family, *Simulidae*, found in Mexico and Central America. Various species of these flies are common hosts of the microfilarial worm, *Onchocerca* volvulus, causing what is locally known as coast erysipelas (see **Onchocerca**).

Euspongia A genus of Calcareous Sponges, also known by the generic name, *Spongia*, of the family *Spongiidae*. e.g. E. officinalis (= *Spongia* o.) - The Bath, Toilet or Turkey Sponge; diameter 15-20 cm or more; soft, one of the best-known forms, it has a horny skeleton, being composed of spongin fibers colored red, brown, green, greenish-violet to black. Found near the coast at shallow depths of 10-50 m, firmly anchored to one spot, and forming encrusting growths on plants, wood, or stones in the water. It is collected, dried, and sold commercially on the market. Distribution: the Mediterranean Sea (see **Spongia**).

Euspongilla A genus of Calcareous Sponges of the Freshwater Sponge family, *Spongillidae*; better known by the generic name, *Spongilla*. e.g. E. lacustris (= *Spongilla* l.) - The Pond Sponge. The body has finger-like projections and is colored yellowish grey or brown. The size of the sponge depends on the size of the object on which it lives. Clumps of this species smell of iodine or mud. Found in freshwater lakes and ponds (see **Spongilla**).

Eustrongylus Former name of a genus of nematodes, now called *Dioctophyma*. Also formerly classified under the genus *Strongylus*. e.g. E. gigas (= *Dioctophyma* renale) (see **Dioctophyma**).

Eutamias A genus of Mammals of the Squirrel family, *Sciuridae*. e.g. E. sibiricus - The Siberian Chipmunk or Borunduki. This rodent species is a North Asiatic ground squirrel. It lives on the ground and in subterranean dens in not too dense coniferous forests, but it is also a proficient climber. Together with the tail it measures 25 cm (10 in). Its fur is a tawny yellow with alternating black and light yellow stripes on the back, and the underside of the body is white. Distribution: northern Asia, United States and Canada.

Eutrombicula A genus of Arachnids of the order of Mites, *Acari*; also known by the generic name, *Trombicula*. e.g. E. alfreddugesi (= *Trombicula* irritans) - The Common Chigger Mite of the United States; also known as the Harvest Mite or Red Bug (see **Trombicula**).

Euxenura A genus of Birds of the Stork and Jabiru family, *Ciconiidae*. e.g. E. galeata - The Maguari Stork of South America, 1 m (40 in) long; white with black on the wings and upper tail coverts; red feet. The tail is slightly forked.

Evasterias A genus of Echinoderms of the Starfish family, *Asteriidae*. e.g. E. echinosoma - A large marine species found in the North Pacific which measures as much as 96 cm (37.79 in) from arm tip to arm tip, and weighs up to 5 kg (11 lb).

Evernia A genus of Lichens, small, undifferentiated, slow-moving, branching plants, capable of surviving prolonged desiccation. e.g. E. prunastri - A species of Lichens with a forked green thallus, growing commonly on the twigs of bushes and trees.

Eviota A genus of Fishes of the Goby family, *Gobiidae*. e.g. E. zonura - The Marshall Islands Goby, the shortest recorded marine fish, measuring 12-16 mm (0.47-0.62 in). Mature specimens weigh only 2 mg (0.0007 oz).

Excalfactoria A genus of Birds of the Partridge, Pheasant, and Quail family, *Phasianidae*. e.g. E. chinensis - The Chinese Painted Quail, one of the smallest Galliformes or game birds, living in China, India, Africa, Indonesia, New Guinea and Australia.

Exochorda A genus of Shrubs and Trees of the Rose family, *Rosaceae*. e.g. E. korolkowii - A deciduous shrub from Turkestan. It reaches 4.5 m (15 ft) in height and has simple obovate leaves and white flowers, more than 2.5 cm (1 in) across.

Exocoetus A genus of Bony Fishes of the Flying Fish family, *Exocoetidae*. e.g. E. volitans - The Flying Fish, length 18 cm; distinguished by the large and long, wing-like pectoral fins and conspicuously elongated lower lobe of the tail fin. It usually swims in shoals just below the surface. When in danger it flies with outspread fins above the water, taking advantage of the air currents, as long a distance as 200 m. Found in all tropical waters; also in the western parts of the Mediterranean Sea.

Exogone A genus of Segmented Worms of the Syllid family, *Syllidae*. The species have quite short and numerous cirri or appendages on the head and along the sides of the body. Most of these worms are very small and unlikely to be noticed by the casual observer.

Exogonium A genus of plants, better known by the generic name, *Ipomoea*, of the Bindweed or Convolvulus family, *Convolvulaceae*. e.g. E. purga (= *Ipomoea* p.) - Jalap; Jalapa; Jalapa Root - contains Jalap resin, a powerful purgative (see **Ipomoea**).

Extastosoma A genus of Insects of the Stick Insect family, *Phasmatidae*. e.g. E. tirartum - The Spiny Stick Insect of Australia, a giant plasmid beetle, with a long, thin, stick-like body and reaching a bodyweight of 30 g or more after heavy summer feeding.

F

Fabriciana A genus of Insects of the Butterfly and Moth family, *Nymphalidae*. e.g. F. adippe - The High Brown Fritillary Moth, size 29-34 mm; found on flowers in glades, clearings and forest rides. Also found in mountains. Distribution: Europe; temperate parts of Asia; Japan; North Africa.

Fagopyrum A genus of plants of the Dock or Knotgrass family, *Polygonaceae*. e.g. F. esculentum - The Buckwheat, an erect, annual plant with loose clusters of flowers, from which is obtained Rutin or Rutoside, used in the treatment of capillary bleeding. Can cause poisoning (fagopyrism). Place of origin: Central Asia; cultivated in many countries for green fodder and as a grain crop.

Fagus A genus of plants of the Beech family, *Fagaceae*. e.g. F. sylvatica - The Beech Tree, up to 45 m (150 ft) tall; leaves fringed when young; the fruit is a 3-sided nut enclosed in a 4-valved scaly capsule. From the wood is obtained Beech Tar (= Pix Fagi), used as an antipruritic in chronic skin disease such as eczema and psoriasis. Place of origin: West and Central Europe; widespread in Asia; native in Britain.

Falcaria A genus of plants of the Carrot family, *Umbelliferae*. e.g. F. vulgaris - The Longleaf, a perennial plant, 25-80 cm high; 1-2-ternate leaves with narrow, sickle-shaped segments. Grows on arable and waste land, dry grassy places and waysides. Distribution: central and southern Europe.

Falco A genus of Birds of the Falcon family, *Falconidae*. e.g. F. peregrinus - The Peregrine Falcon, length 47.5 cm (1 ft 7in); a bird of prey commonly found all over Europe, Asia, and the Americas, where it is known as the Duck-hawk. The adult falcon has dark "moustaches" on white cheeks; its beak is short, powerful, and curved down to the base making it a very powerful and merciless hunter of animal pests and smaller birds, such as pigeons. It nests at the edge of forests and on cliffs. The falcons living in Europe migrate to equatorial Africa every winter, returning home the next summer. Native in Britain.

Fannia A genus of Insects of the House-fly family, *Muscidae*. e.g. F. canicularis - The Lesser House Fly; small, greyish; size up to 6 mm; the male thorax has three dark lengthwise stripes. It flies characteristically around hanging lamps inside houses. The spiny maggots or larvae, and the eggs infest man (causing myiasis). Distribution: worldwide.

Fasciola A genus of large Flukes or Digenetic Trematodes (order *Digenea*), of the phylum of Flatworms, *Platyhelminthes*. Also called by the generic name, *Distoma*. e.g. F. hepatica (= *Distoma* hepaticum; F. humana; F. venarum) - The Common Liver Fluke, found in sheep and other herbivorous animals; occasionally also affecting humans. The adult is about 1.25 cm (1/2 in) long and 1.25 cm (1/2 in) wide, flat, oval and leaf-like, and at least one stage of its life-history is spent in an invertebrate host or hosts, generally a mollusc.

Fasciolaria A genus of Marine Molluscs of the class *Gastropoda*. e.g. F. gigantea - The Horse Conch, a marine slug or snail, found off the Florida Keys and the West Indies; a very large species weighing up to 2.27 kg (5 lb) and having a shell measuring 30-60 cm (1-2 ft) in length.

Fascioletta A former name given to a genus of parasitic flukes. e.g. F. iliocana (= *Echinostoma* ilocanum) (see **Echinostoma**).

Fascioloides A genus of Flukes or Trematode worms. e.g. F. magna - The large American liver fluke found in herbivorous animals.

Fasciolopsis A genus of large Flukes or

Digenetic Trematodes of the order *Digenea*. e.g. F. buski - The Human Intestinal Fluke; also infests pigs. Acquired by eating the tubers of plants on which snails feed and cercariae encyst.

Fatsia A genus of plants of the Ivy family, *Araliaceae*. e.g. F. japonia - An evergreen shrub, 4.5 m (15 ft) tall, with thick smooth stems and large alternate leaves with broad, palmately-lobed blades and umbels of white flowers. Place of origin: Japan.

Faucaria A genus of plants of the family *Aizoaceae* (= *Mesembryanthemaceae*). e.g. F. lupina - A succulent, perennial herb with fleshy leaves, growing in opposite pairs and with bristly margins; containing water storage tissue and growing in arid stony areas. Large attractive flowers. Cultivated as an ornamental plant. Place of origin: South Africa.

Favia A genus of True or Stony Corals of the order *Scleractinia*. These animals are the astraeid corals, which are reef-building corals developing into massive forms. Found on the coral reefs of the Red Sea coasts.

Feijoa A genus of plants of the Myrtle family, *Myrtaceae*. e.g. F. sellowiana - The Feijoa or Pineapple Guava, a large evergreen shrub or small tree of bushy habit, with opposite, dark green, ovate leaves. The flowers have white petals and red stamens. The egg-shaped fruits or berries are edible, up to 10 cm (2 in) across, and with a strong aromatic flavor. Place of origin: Brazil, Uruguay; introduced into Europe in 1898.

Felis A genus of Mammals of the Cat family, *Felidae*. e.g. F. domestica (= F. catus) - The Domestic Cat; little is known for certain about the origin of domestic cats but they interbreed with other species. A well-known breed is the Siamese Cat, with light-colored, close fur, a black-brown face, and blue eyes. F. leo (= *Leo* l.) - The Lion, one of the largest members of the cat family. F. onca (= *Jaguarius* o.; *Panthera* o.) - The Jaguar (see **Panthera**). F. tigris - The Tiger, the largest feline, its size varying in the eleven or so existing subspecies. e.g. F. tigris longipilis - The Siberian Tiger, with a length of 2.7 m (9 ft), a tail up to 1.1 m (3 ft 8 in), and weighing as much as 292 kg (650 lb).

Fennecus A genus of Mammals of the Dog family, *Canidae*. e.g. F. zerda - The Fennec, a small desert fox, the smallest known, about 40 cm (16 in) long with a tail of about 20 cm (8 in). It is sandy in color and has enormous round eyes and large sensitive ears, 10 cm (4 in) long, so typical of nocturnal animals. It feeds on birds, jerboas, lizards, locusts, rodents and fruit. Fennecs are easily domesticated and become affectionate pets. Natural habitat: North Africa and the Sahara, the deserts of Arabia.

Ferocactus A genus of plants of the Cactus family, *Cactaceae*. e.g. F. horridus - The Fish Hook Cactus, a large and formidable rounded cactus plant with 12 ribs bearing groups of 8-12 white, slender, spreading outer spines. One in each group is flat and hooked and may be up to 15 cm (6 in) long. Place of origin: Mexico.

Ferula A genus of plants of the Cactus family, *Umbelliferae*. e.g. F. assafoetida (= F. asafoetida) - The Asafetida plant; from the rhizome and root is obtained an oleo-gum-resin, asafetida (= Asant or Devil's Dung); it has a carminative and expectorant action and has also been used in nervous disorders due to the psychological response to its objectionable odor and taste.

Festuca A genus of plants of the Grass family, *Gramineae*. e.g. F. pratensis - The Meadow Fescue or Meadow-grass, a perennial plant, 40-100 cm high, growing in meadows, pastures and by roadsides. It has auricles of hairless leaves, compact panicles, and awnless lemmas. Common throughout Europe; native in Britain.

Ficedula A genus of Birds of the Old World Flycatcher family, *Muscicapidae.* e.g. F. hypoleuca - The Pied Flycatcher, length 13 cm; a migrant bird wintering in North and Central Africa. Found in open, broad-leaved woodlands, parks, and gardens. Nests in tree-holes. Distribution: most of Europe, western Siberia, northeast Africa.

Ficus A genus of trees or shrubs of the Mulberry family, *Moraceae.* e.g. F. carica - The Common Fig Tree, a small tree or shrub, 2-8 m high, with 3-5-lobed palmate leaves, soft-hairy beneath. Conical edible fruit. Grows wild in mountains in the Mediterranean region and also widely cultivated. Place of origin: probably in western Asia.

Filago A genus of plants of the Daisy family, *Compositae.* e.g. F. germanica - The Common Cudweed, an annual plant, 7-40 cm high, with 10-30 heads in each cluster and straight-pointed bracts. Grows in cultivated fields and grassy and sandy places, dry pastures and by waysides. Distribution: South and Central Europe; native in Britain.

Filaria A former generally applied generic name for members of the Nematode or Roundworm superfamily, *Filaroidea.* e.g. F. bancrofti (= *Wuchereria* b.). F. diurna (the larval form of *Loa* loa). F. immitis (= *Dirofilaria* i.). F. loa (= *Loa* loa). F. medinensis (= *Dracunculus* m.). F. ozzardi (= *Mansonella* o.). F. perstans (= *Acanthocheilonema* p.). F. volvulus (= *Onchocerca* v.).

Filipendula A genus of plants of the Rose family, *Rosaceae.* e.g. F. vulgaris - The Dropwort, a perennial plant, 15-80 cm high; roots with ovoid tubers; pinnate leaves. Widespread in Europe; native in Britain.

Fimbriaria A genus of Tapeworms. e.g. F. fasciolaris - A species infecting wild and domestic fowl.

Fissurella A genus of Mesogastropod Molluscs of the order *Anaspidea.* e.g. F. costata - The Keyhole Limpet, size up to 5 cm (2 in); it has a cup-shaped shell with a hole for a breathing tube. A sea animal. Distribution: Europe; native around British waters.

Fistularia A genus of Fishes of the Snipefish and Cornetfish subfamily, *Aulostomidinae.* These are the Cornetfishes, marine fishes found in tropical waters, usually in shallow water. Like the Snipefishes (genus *Centriscus*), they swim in an upright position. e.g. F. petimba - The Flute Fish, a species found in the Gulf of Eilat (Akaba) in the Red Sea.

Fittonia A genus of prickly plants of the family *Acanthaceae.* e.g. F. arguroneura (= F. verschaffeltii "Argyroneura") - A small, herbaceous, perennial plant, up to 15 cm (6 in) high, with erect spikes of insignificant flowers subtended by overlapping ovate bracts. Often grown in hot-houses for its beautiful heart-shaped, velvety-green leaves with conspicuous silvery-white veins, giving them a nettled appearance. Place of origin: Peru; introduced into Europe in 1867.

Flavobacterium A genus of nonfermentative gram-negative, pigmented bacilli widely distributed in nature and most often recovered from sputum of debilitated patients. Composed of three main groups: 1. F. meningosepticum or *Flavobacterium* - group I - Species of gram-negative rods that produce pronounced, yellow-orange pigments in culture. Natural habitat is soil and water. Highly virulent for newborn and premature babies, and capable of causing outbreaks of neonatal meningitis and septicemia in hospital nurseries. Has been found in the water supply in drinking fountains, in hospital equipment exposed to air and containing water, on nipples of nursing bottles, and on delivery-room faucets. Rarely isolated from adults and then usually asymptomatic. 2.

Flavobacterium - group II - Saccharolytic coccobacilli, producing yellow or orange-yellow colonies. Frequently isolated as opportunistic pathogens. 3. *Flavobacterium* - group III - Nonsaccharolytic species that produces grey-green or yellow-green colonies. Not known to cause disease.

Flustra A genus of Moss Animals or Bryozoans of the Sea-mat family, *Bicellariidae*. These are small, inconspicuous, but highly complex animals, which form numerous large, sessile colonies. e.g. F. foliacea - The Hornwrack, size up to 20 cm; seaweed-like colonies in leaf-shaped lobes with rounded corners. Distribution: the seas of Europe.

Fockea A genus of plants of the Milkweed family, *Asclepiadaceae.* e.g. F. capensis - A plant with a thick root and thin untidy branches with small leaves, and grey-green flowers. Place of origin: the desert regions of South Africa.

Foeniculum A genus of plants of the Carrot family, *Umbelliferae*. e.g. F. vulgare - The Fennel, a smooth, glaucous perennial plant, up to 1.5 m (5 ft) high, with linear-segmented, many-divided leaves, umbels of yellow flowers, and ovoid fruits. It has a characteristic strong smell, and grows on waste places and sea cliffs. Place of origin: the Mediterranean region; widely naturalized in temperate countries, including Britain.

Fomes A genus of plants, the Telephora of the Bracket Fungus family, *Polyporaceae*. Also called by the generic name, *Polyporus*. e.g. F. officinalis (= *Polyporus* o.) - A species found growing on the trunks of birch trees. From the dried stroma of this fungus is prepared Agaric acid (= Agaricin; Agaricinic acid), used as an astringent in hyperhydrosis, and in large doses as a purgative.

Fontinalis A genus of plants of the True Moss family. e.g. F. antipyretica - The Willow Moss, a perennial plant, 40 cm long; densely leafy, growing entirely submerged and floating in freshwater. Distribution: Europe; native in Britain.

Forcipomyia A genus of Insects (class *Insecta*); these are midget species with a rapid wingbeat. The highest wingbeat frequency so far recorded for any insect under natural conditions, was on one specimen, by Dr. Olavi Sotavalta, the Finnish entomologist. He obtained a speed of 1,046 cycles/sec (i.e. 62,760 cycles/min).

Forficula A genus of Insects of the Earwig family, *Forficulidae* (order *Dermaptera*). e.g. F. auricularia - The Common or European Earwig, length up to 2 cm; brown-colored body; a common, omnivorous, nocturnal insect which has been introduced into the Americas, South Africa and Australasia. It has strong, curved, forceps-like jointed processes or cerci at the end of the abdomen, and is generally considered a pest in parts of the United States. Distribution: worldwide.

Formica A genus of Insects of the Ant family, *Formicidae* (order of Ants, Bees, and Wasps, *Hymenoptera*). e.g. F. rufa - The European Ant, Wood Ant, or Red Ant, 5-11 mm long; an active predator, destroying large numbers of various harmful larvae, and important in maintaining the balance of nature. The thorax and scale are red; the abdomen is black. It builds large ant-hills, especially in coniferous woods and spruce forests. It feeds on the honeydew of aphids (*Lachnus* roboris), a sugary liquid exuded from the anus of this insect, and lives in close association with it. There are two castes in the species, the sexual forms (male and female), and the worker ants. The workers are always wingless, sterile females, and are often very numerous in large colonies. Distribution: Europe; (native in Britain); the Caucasus, Siberia, North America (see **Lachnus**).

Formicaleon A genus of Insects of the order of Alder-flies, Ant-lion flies, Lacewings and allies, *Neuroptera*. e.g. F. tetragrammicus - The Ant-lion fly, a predacious species, with biting mouthparts and membranous wings. The larvae bear powerful jaws and make snares for trapping their prey.

Forsythia A genus of Shrubs of the Olive family, *Oleaceae*. e.g. F. suspensa - A deciduous shrub, up to 3 m (10 ft) tall, with arching branches and simple opposite, toothed leaves about 10 cm (4 in) long. There are shortly stalked, yellow flowers and the fruit is a capsule. Frequently cultivated. Place of origin: China.

Fossa A genus of Mammals of the Civet, Genet and Mongoose family, *Viverridae*. This is the Madagascan Civet, not to be confused with the ferocious carnivore of Madagascar, the Fossa (*Cryptoprocta ferox*) (see **Cryptoprocta**).

Fragaria A genus of plants of the Rose family, *Rosaceae*. e.g. F. vesca - The Wild or Woodland Strawberry, a perennial plant, 4 to 25 cm high; lateral leaflets, stalkless or nearly so; white petals and fleshy fruits. Common in woods, scrubs, thickets and grassland. The common Cultivated Strawberry is a hybrid between F. virginiana from North America and F. chiloensis from Chile. The Wild Strawberry grows in Europe, western Asia and eastern North America. It is native in Britain.

Franciscea A genus of South American plants, better known by the generic name, *Brunfelsia*. e.g. F. hopeana (= *Brunfelsia* h.) (see **Brunfelsia**).

Francisella A genus of microorganisms, also known by the generic name, *Pasteurella*, of the family *Brucellaceae*. e.g. F. tularensis (= *Pasteurella* t.; *Bacterium* tularense) - A species of small gram-negative coccobacilli that often show bipolar staining and cause the disease tularemia. Has been isolated from many sources including wild animals, foods, water and insects, and from clinical specimens in local lesions, blood, sputum, gastric aspirates, conjunctival swabs and pleural fluid. There are different categories of tularemia depending on the site of infection - ulceroglandular, with a local skin lesion and regional lymphadenitis; oculoglandular, with conjunctivitis or conjunctival ulcer and regional lymphadenitis; pulmonary, with pneumonitis. Bacteremia, toxemia, pharyngitis, and miliary necrosis in almost all organs may occur. Transmitted to man by handling of infected animals, such as rodents; by bloodsucking insects; via the respiratory or gastrointestinal tract; or occasionally by contact with contaminated water (see **Pasteurella**).

Francolinus A genus of Birds of the Pheasant family, *Phasianidae*. e.g. F. leucoscepus - The Yellow-necked Francolin, a partridge-like bird found in Kenya; the male has spurs and sometimes the female too. It inhabits bushy grasslands and is closely related to the Partridge (see **Perdix**).

Frangula A genus of plants of the Buckthorn family, *Rhamnaceae*. e.g. F. alnus - The Alder Buckthorn, a shrub or small tree up to 7 m (23 ft) tall; it has pentamerous or 5-parted flowers, and black, unpalatable berries. Distribution: Europe; native in Britain.

Fratercula A genus of Birds of the Auk, Guillemot and Puffin family, *Alcidae*. e.g. F. arctica - The Common or Atlantic Puffin, or so-called Sea-parrot, a migrant bird about 30 cm (1 ft) long, with a laterally compressed parrot-like bill, slate-grey at the base and ornamental during the breeding season with vivid red and white; the hood is black. Puffins breed in colonies, the female laying a single speckled white egg. Distribution: northern Europe, northern parts of North America, Greenland.

Fraxinus A genus of Trees of the Olive family, *Oleaceae*. e.g. F. ornus - The

European Flowering Ash or Manna Ash, a tree up to 8 m (26 ft) tall; it has pinnate leaves mostly with 7 leaflets and narrow, showy panicles of white, petalled flowers. The blooms are heavily scented. From this species is extracted Manna or Manna sugar, the dried saccharine juice containing 40 to 60% mannitol, employed as a laxative for infants and children. Place of origin: a native of South Europe and Asia Minor; introduced and naturalized in many parts of the world, including Britain before 1700.

Freesia A genus of plants of the Iris family, *Iridaceae*. e.g. F. refracta - The Freesia, a slender plant about 45 cm (18 in) high, with bright-green, linear leaves arising from small corms and one-sided spikes of richly scented, greenish-yellow, tubular flowers with six divisions. These are basically yellow but forms exist having white, mauve, purple, pink, red and violet flowers. All are extremely fragrant. Place of origin: southern Africa; introduced into Europe since 1875.

Fregata A genus of Birds of the Frigatebird family, *Fregatidae*. e.g. F. aquila (= F. magnificens) - The Ascension or Magnificient Frigate-bird, found only on Ascencion Island, and believed to be one of the swiftest of large birds. Its body is black with a metallic sheen. The male has a bright red pouch of naked skin under the throat, used for inflation during courtship and before mating. The bird has a 2.1 m (7 ft) wingspan and weighs up to 11/2 kg (31/2 lb); it is very aggressive, and powerful in flight, being the fastest-flying seabird. It is also called at times, the Man-o'-war Bird. Frigate-birds are indigenous to the tropical waters of the south Atlantic, the Pacific and the Indian oceans.

Fringilla A genus of Birds of the Finch and Darwin's Finches family, *Fringillidae*. e.g. F. coelebs - The Chaffinch, 15 cm (6 in) long; a small,

stockybird, frequently seen in woods, orchards and farmlands. It has a pinkish-brown front, blue-grey hood, brown wings with double white bars, brown tail, and greenish rump. A really handsome and lively bird, it has a fine song. A partial migrant living in Europe, western Asia, and northwest Africa. Native in Britain.

Fritillaria A genus of temperate plants of the northern hemisphere, of the Lily family, *Liliaceae*. e.g. F. verticillata - The Fritillary plant, with erect buds and fruit, and pendulous open flowers. It contains a crystalline alkaloid, verticine. Grows in damp meadows and pastures. Cultivated for its ornamental flowers, like the rest of the genus. Distribution: Europe; native in Britain.

Fromia A genus of Starfishes of the class *Asteroidea*. e.g. F. ghardaquana - The Starfish, a red-colored echinoderm with five arms, and a carnivore that feeds on molluscs, aided by suction of its tiny tube-feet, with which it engulfs its prey by everting its stomach. Found on the coral reefs of the Red Sea.

Frontonia A genus of Protozoan animals (phylum *Protozoa*). e.g. F. leucas - A type species.

Fuchsia A genus of plants of the family *Onagraceae*. e.g. F. triphylla var. "Mantilla" - The Fuchsia, a garden hybrid from F. triphylla, much-esteemed for hanging basket work, 30-60 cm (1-2 ft) high, with few-branched stems; bronzed, stemmed, opposite or verticillate, entire, oval lanceolate leaves and deep carmine flowers, borne in pendulous terminal racemes. Place of origin: species from Haiti and Santo Domingo; introduced into Britain in 1872.

Fucus A genus of plants of the Brown Algae family, *Fucaceae*. e.g. F. vesiculosus - The Bladder Wrack (= Bladderwrack; Kelpware; Seawrack), a Brown Algae plant, up to 1 m (40 in) long, with pairs of air vesicles or conceptacles (hollow cases covering the reproductive organs) - male or female. These are dioecious, with male and female organs on separate plants. From the dried plant is extracted the gelatinous substance, Algin, used in the treatment of obesity. It contains the carotinoid Zeaxanthin ($C_{40}H_{56}O_2$). The species grows on rocks, stones and wood along the seacoast. Place of origin: Europe; native in Britain.

Fulica A genus of Birds of the Coot, Gallinula, Moorhen and Rail family, *Rallidae*. e.g. F. atra - The Common Coot of Europe, a partly resident, partly migratory bird, about 37.5 cm (1 ft 3 in) long; its dark plumage is broken by its white bill and frontal shield, and its toes are edged with scalloped membranes or lobed-webbing, very much like a web. It has a white forehead which gives it a bald appearance from a distance. It is resident in the British Isles, wherever there is a sizeable expanse of freshwater; also found all across Europe, except the far north, central and southern Asia, northwestern Africa, and Australia.

Fulmarus A genus of Birds of the Tubenose family, *Procellariidae* (including the Albatrosses, Fulmars, Petrels and Shearwater birds). e.g. F. glacialis - The Northern Fulmar, Fulmar Petrel, or Foul-gull; 47 cm long; an exclusively marine bird with a thick nape, heavy bill, prominent nasal tubes (hence the family name, Tubenose), and a gull-like appearance. An ill-smelling oil that is spat at intruders at the nest, has earned the Fulmar its common name, Foul-gull. A partial migrant, it is found in the Arctic region and the northern hemisphere. Native in Britain.

Fumana A genus of plants of the Rockrose family, *Cistaceae*. e.g. F. procumbens - The Heath-rose, a perennial plant, 10-20 cm high; it has needle-shaped leaves, and grows on sunny hillocks. Distribution: the Mediterranean region and Central Europe.

Fumaria A genus of plants of the Fumitory family, *Fumariaceae*. e.g. F. officinalis - The Common Fumitory, an annual plant, 7-50 cm high; the bracts are half as long as the fruit-stalks, the corolla 7-9 mm wide, and the fruit is broad and wrinkled. Grows in cultivated fields and waste places. Distribution: Europe; native in Britain.

Fundulus A genus of Fishes of the Guppy, Killifish, Swordtail and Foureyed Fish order, *Cyprinodontidae* (= *Cyprinodontiformes*; *Microcyprini*). e.g. F. heteroclitus - The Mummichog or Killifish, the common or green killifish, a hardy species used in biological research.

Fungia A genus of True or Stony Corals of the order *Scleractinia*. These are the Solitary or Mushroom corals. When the young planula (aquatic invertebrate larva) settles down and becomes attached, the oval end expands into a disc and lays down a calcareous skeleton. This disc breaks off, setting free the adult, which remains unattached. Distribution: the coral reefs of the Red Sea coast.

Funkia A genus of plants, better known by the generic name, *Hosta*, of the Lily family, *Liliaceae*. e.g. F. ovata (= *Hosta ventricosa*) (see **Hosta**).

Furcellaria A genus of plants of the Red Algae family, *Gigartinaceae*. e.g. F. fastigata - Furcellaria, a Red Algae species; size up to 25 cm; blackish, cartilaginous, forked fronds with yellow tips in the male, and club-shaped, black tips in the female. Grows in depth off the seashore. Distribution: Europe; native around Britain.

Furnarius A genus of Birds of the Ovenbird family, *Furnariidae*. e.g. F. rufus - The Red Ovenbird or Baker, a stout little bird with a slender, slightly curved beak almost 2.5 cm (1 in) long, and strong legs suited to its terrestrial habits. Rufous brown in color. It builds its nest of mud strengthened with root-fibers, which when completed has the

shape of a kiln or baker's oven (hence
the name), but with a deeper and narrower entrance. Distribution: southern
Mexico, Central and South America.

Fusarium A genus of Molds or Fungi of
the class *Ascomycetes*. e.g. F. lateritium
- From this mold is extracted the antimicrobial substance Fusafungine, active
against gram-positive and gram-negative
microorganisms and also against
Candida albicans.

Fusidium A genus of Molds or Fungi.
e.g. F. coccineum - From this mold is
extracted the antimicrobial substance
Fusidic Acid, active against gram-positive bacteria and gram-negative cocci.

Fusiformis A genus of microorganisms,
now known by the generic name
Fusobacterium (see **Fusobacterium**).

Fusobacterium A genus of anaerobic,
spindle-shaped or filamentous, gram-
negative microorganisms, also known as
the Fusiform Bacilli, of the family
Bacteroidaceae. Also known by other
generic names, such as *Bacteroides*,
Fusiformis, *Bacillus* and
Sphaerophorus. They occur as normal
flora in the mouth and large intestine.
e.g. F. plauti-vincenti (= F. fusiforme;
Bacillus fusiformis; *Fusiformis* p-v.; The
Plaut-Vincent Bacillus) - Has been isolated from the mucosa in ulcerative gin-
givitis and stomatitis (Trench mouth or
Vincent's angina), from the mucous
membranes of the intestine or respiratory tract in ulcerative or otherwise infected lesions (as in lung abscesses), and
from infected human bite wounds,
rarely progressing to septicemia. Often
co-exists with spirochetes (*Borrelia* sp.)
to form the fusospirochetal complex (as
in Vincent's angina) (see **Borrelia**). F.
necrophorus (= *Bacteroides* funduliformis; *Sphaerophorus* n.) - May be a
part of the normal flora of the mouth,
intestinal and genital tracts. Considered
to be a secondary invader that causes
mouth ulcerations, surgical wound
abscesses, etc.

G

Gadus A genus of Fishes of the Cod, Hake, Haddock and Whiting family, *Gadidae* (Cod subfamily, *Gadinae*). e.g. G. morrhua (= G. callarius; G. morhua) - The Atlantic Cod, a commercial fish up to 1.5 m (5 ft) long and weighing up to 40 kg (90 lb); light-colored body covered with cycloid scales, and a very high fecundity, having more than six million eggs in its ovaries. From its liver is prepared Codliver Oil (= Oleum Morrhuae; Oleum Jecoris Aselli). Due to its high Vitamin D content the oil is a valuable supplement to the diet of infants for the prevention of rickets. Distribution: common along the European coast of the Atlantic as well as far out at sea. Native around the seacoasts of Britain.

Gaffkya A genus of microorganisms of the family *Micrococcaceae*. These are gram-positive cocci occurring in tetrads, with a large encompassing capsule. Also known by the generic name, *Micrococcus*. e.g. G. tetragena (= *Micrococcus* tetragenus) - a part of the normal flora of the upper respiratory tract; pathogenic to mice. Regarded as an opportunist in man, and has been responsible for abscesses, septic arthritis, meningitis, pneumonia, bacterial endocarditis and septicemia.

Gagea A genus of plants of the Lily family, *Liliaceae*. e.g. G. pratensis - The Meadow Yellow-star-of-Bethlehem, a perennial plant, 4-30 cm high, with radical linear leaves, fringed stem flowers and glabrous flower-stalks. Grows mostly in grasslands and woods. Place of origin: widespread all over Europe.

Gaidropsaurus A genus of Fishes of the Cod, Haddock, Hake and Whiting family, *Gadidae*. The species belong to the rockling subfamily, *Lotinae*. These are small marine fishes with two dorsal fins, the first dorsal fin being often merely a single ray followed by a small fringe-like band. e.g. G. vulgaris - The Three-bearded Rockling, 60 cm long; one barbel on the lower jaw and two on the upper. A sea fish inhabiting the Atlantic coasts and the Mediterranean.

Gaillardia A genus of plants of the Daisy family, *Compositae*. e.g. G. aristata - The Blanket Flower or Gaillardia, a herbaceous, perennial plant with erect, 60 cm (2 ft) stems, lancelate or oblong, toothed leaves and yellow and red, large, 7-10 cm (3-4 in), daisy-like flowers. Produces many-colored hybrids in deep red, crimson and cream, tangerine and other shades. Place of origin: western North America; introduced into Europe in 1812.

Galaeorhinus A genus of Fishes of the Requiem Shark family, *Carcharhinidae*; also known by the generic name, *Galeorhinus*. e.g. G. zyopterus (= *Galeorhinus* z.) - The Soupfin Shark, a requiem shark species found off the California coast, so-called because its fins, forming a highly flavored gelatine, are valued for making soup. It is also in much commercial demand because of the high Vitamin A content of its liver.

Galago A genus of Mammals of the Loris family, *Lorisidae*. e.g. G. crassicaudatus - The Bush-tailed or Great Galago of East Africa, over 30 cm (1 ft) long with a tail slightly longer. G. senegalensis - The Senegal Galago, Night-ape, or Bush Baby, body colored a pale greyish-brown and having the size of a small squirrel. The tail is longer than the body and is generally bushy. The ears are bare and membranous, similar to those of the Bats, which can be folded during its daytime sleep. Distribution: the grass woodlands in most of Africa south of the Sahara.

Galanthus A genus of plants of the Daffodil family, *Amaryllidaceae*. e.g. G. woronowii - The Caucasian Snowdrop or Veronov's Snowdrop, a perennial plant, growing in damp meadows and woods of the Caucasus. It has an underground bulb which yields the alkaloid galanthamine hydrobromide, an anticholinesterase used in Russia for the treatment of myasthenia gravis and cerebral palsy.

Galathea A genus of Crustaceans of the Squat Lobster family, *Galatheidae*. The species resemble lobsters, but carry the abdomen flexed forewards under the thorax, and have a small last leg which can be tucked into the branchial chamber and used for cleaning purposes. A marine animal, it is found on the surface of sand or mud beds. e.g. G. squamifera - The Green Squat Lobster; it has a dark brownish, green body up to 8 cm long, and is found under stones on the lower shore of the North Sea and Atlantic coasts.

Galaxea A genus of Celenterate Animals of the subclass of Sea Anemones and Stony Corals, *Hexacorallia* (= *Zoantharia*). e.g. G. lamarci - A stony coral species found in the Red Sea.

Galega A genus of plants of the Pea family, *Leguminosae*. e.g. G. officinalis - Goat's Rue or French Lilac, an herbaceous perennial plant, 90-120 cm (3-4 ft) tall; pinnate leaves with 11-17 finely-pointed leaflets. Bluish-white, pea-shaped flowers growing in dense axillary clusters. Place of origin: southern Europe, Asia Minor. Introduced into Britain in 1568.

Galemys A genus of Mammals of the Mole or Desman family, *Talpidae* (subfamily *Desmaninae*). e.g. G. pyrenaica (= G. pyrenaicus) - The Pyrenean Desman, length of body 110-135 mm; tail 130 mm; an elongated, long-tailed mole with along and extremely mobile snout. An aquatic animal with webbed feet and smelling strongly of musk. It burrows into the banks of watercourses, and is found chiefly in places with lush shoreline vegetation; while the entrance to the burrow is under water, the nesting chamber is always above water level in a dry section. Distribution: along rivers and streams in the Pyrenees and other places in northern Spain and Portugal.

Galeocerdo A genus of Fishes of the order of Sharks, *Selachii*. e.g. G. cuvieri - The Tiger Shark, a carnivorous fish measuring over 6 m (20 ft) and found in tropical and subtropical seas.

Galeodes A genus of Arachnids of the order of "Sun Spiders", *Solpugida*. They are also known as "Wind Scorpions". e.g. G. arabs - A long-legged species, with a legspan of up to 12.5 cm (5 in), one of the fastest-moving solpugids. Found in the warmer, drier parts of the world, from Mexico to India and Turkestan.

Galeopsis A genus of plants of the Mint or Thyme family, *Labiatae*. e.g. G. speciosa - The Large Hemp-nettle, a stout annual plant, up to 1 m high, with hairy, erect stems and ovate, toothed leaves. The pale yellow corolla has a compressed, hooded upper lip and a three-lobed, purple-colored lower lip. Distribution: Europe, Siberia. Found in cultivated land throughout Britain.

Galeorhinus See **Galaeorhinus**.

Galerida A genus of Birds of the Lark family, *Alaudidae*. e.g. G. cristata - The Crested Lark, length 17 cm; it has a prominent crest on the head, and a short tail with brown outer edges. Originally a bird of the steppes, it occurs now in cultivated steppes and fields. A resident bird, it overwinters in the neighborhood of its nesting ground. Distribution: Europe, Asia, North Africa. Absent in Britain.

Galeruca A genus of Insects of the Leaf Beetle family, *Chrysomelidae*. e.g. G. tanaceti - Length 6.5-11 mm; colored black; very abundant from spring to autumn. Found on forest and field paths. Distribution: western parts of the Palaearctic region.

Galeus A genus of Fishes of the class of Cartilaginous Sharks, *Chondrichthyes*. e.g. G. canis - The Tope, a cartilaginous fish, measuring about 2 m (6 1/2 ft). This shark lives in almost every sea, feeding on the smaller marine creatures. It is viviparous, having as many as thirty young at a time.

Galinsoga A genus of plants of the Daisy family, *Compositae*. e.g. G. ciliata - The Hairy Galinsoga or Shaggy Soldier, an annual plant, 6-80 cm high; it has a hairy stem with receptacle-scales finely toothed above. Grows on waste and arable land, and in gardens. Place of origin: Europe; introduced and naturalized in Britain. Also now well established in the New World.

Galipea A genus of plants of the Rue family, *Rutaceae*. e.g. G. officinalis - The Caromy or Cusparia plant. The bark, Cusparia Bark (= Angostura Bark; Caromy Bark), yields an aromatic bitter. Used in South America and the West Indies in the treatment of diarrhea and dysentery.

Galium A genus of plants of the Bedstraw or Madder family, *Rubiaceae*. e.g. G. mollugo - The Hedge Bedstraw, a perennial plant, 20-60 cm high; 4-angled stem and one-veined leaves, 6-8 in a whorl; the corolla-lobes are pointed. Grows commonly in grassland. Place of origin: Europe; native in Britain.

Gallinago A genus of Birds of the Sandpiper and Snipe family, *Scolopacidae*. e.g. G. gallinago - The Common Snipe, length 27 cm (10 2/3 in); common in marshland, around ponds and in damp meadows. Its habits are similar to those of the Woodcock (see **Scolopax**). It migrates to Africa for the winter. Distribution: Europe; Asia; North America, where it was formerly called Wilson's Snipe.

Gallinula A genus of Waterbirds of the Coot, Rail, and Moorhen (Gallinule) family, *Rallidae*. e.g. G. chloropus - The Moorhen, known in America as the Common Gallinule; length 33 cm; it has a conical bill, red with a yellow tip and a red frontal shield, a part of the upper

bill flaring out on the forehead. The plumage is olive green with the rest of the body slate-grey; the feet are green. Found in freshwater regions near pools, ponds and marches, and any stretch of river where weeds and rushes abound. It is either resident (in more southerly parts) or migratory. Distribution: worldwide (except Australia and northern Europe).

Gallionella A genus of microorganisms of the family *Caulobacteraceae*, growing only in iron-containing fresh or salt water. e.g. G. ferruginea.

Gallirallus A genus of Birds of the Coot, Moorhen, and Rail family, *Rallidae*. e.g. G. australis - The Flightless Wood-rail or Weka of New Zealand; as large as a domestic hen. A nocturnal bird found near human habitations. It will eat almost anything, including rats and mice, as well as birds the size of a duck.

Gallus A genus of Birds of the Partridge, Pheasant and Quail family, *Phasianidae*. e.g. G. gallus bankiva - The Red Jungle Fowl, considered to be the ancestor of all domestic breeds (see G. gallus domestica). With its tail feathers the cock measures about 70 cm (2 ft 4 in). The feathers of the back are purple-brown, bright red in the middle and brownish-yellow round the edges. The head and neck are a brilliant yellow. The eye is orange-red, the comb red, the bill brownish, and the feet are a dark slate color. The hen is smaller in size and duller in color, and the comb and wattles little developed. This species lives in the under-growth of jungles from India to Indonesia. G. gallus domestica (= G. bankiva var domesticus) - The Domestic Fowl, one of the most important of all domestic animals; it has been bred by man for more than 4,000 years. Under favorable conditions a single hen may lay as many as 300 eggs a year. Distribution: There are believed to be about 3,500 million (3 1/2 billion) in the world, or nearly one chicken for every member of the human race.

Galtonia A genus of plants, also known by the generic name, *Hyacinthus*, of the Lily family, *Liliaceae*. e.g. G. candicans (= *Hyacinthus* c.) - The Giant Summer Hyacinth or Spire Lily, a summer-blooming herbaceous plant with a large, round, tunicated bulb and long, strap-shaped, basal leaves up to 75 cm (2 1/2 ft) long, and long, slender stems of up to 1.2 m (4 ft) or more, carrying loose racemes of white, scented, pendent, fun-nel-shaped flowers. Place of origin: Natal, South Africa. Introduced into Europe in 1870.

Gambusia A genus of Fishes of the sub-order *Cyprinodontei*; these are the Mosquito-fishes capable of destroying and eating mosquito larvae, thus con-trolling malaria. e.g. G. affinis - A min-now, feeding upon the larvae of the Anopheles mosquito, which has been used in the eradication of malaria in the major affected regions of the world.

Gammarus A genus of Crustaceans, now often called by the generic name, *Rivulogammarus*, of the Freshwater Shrimp family, *Gammaridae*. e.g. G. pulex (= G. pulex fossarum; *Rivulogammarus* p. fossarum) - The Common Freshwater Shrimp (see **Rivulogammarus**).

Garcinia A genus of plants of the family *Guttiferae*. e.g. G. hanburyi - A species from which is extracted a yellow gum-resin, Cambogia or Gamboge, a drastic hydragogue cathartic. G. indica - From the seeds of this species is expressed the solid fat, Kokum Butter (= Oleum Garciniae; Goa Butter; Mangosteen Oil), used as a basis in the preparation of suppositories and as an astringent.

Gardenia A genus of plants of the Bedstraw or Madder family, *Rubiaceae*. e.g. G. jasminoides - The Cape Jasmine or Gardenia, an evergreen shrub, 0.3-2 m (1-6 ft) high; with oval or lanceolate, opposite, darkgreen and shiny leaves

and white, scented flowers with spirally arranged petals. An essential oil used in perfumery is obtained from the flowers, which are also used for scenting tea. The fruit yields a yellow dye sold in some parts of tropical Africa. Place of origin: China; introduced into Europe in 1763.

Gardonus A genus of Fishes (one of the principal European genera) of the Carp, Minnow, Tench and allied family, *Cyprinidae*. These are the Gardons or Gardon Fishes (see **Carassius**).

Gari A genus of Bivalve Molluscs of the order *Veneroida*. e.g. G. fervensis - The Faroe Sunset Shell, size 5 cm (2 in); lat-tice sculpturing in the posterior part of the shell which is highly polished inside. Found in the seas of northern Europe; native around Britain.

Garrulus A genus of Birds of the Crow, Jay and Magpie family, *Corvidae*. e.g. G. glandarius - The European Jay, length 34 cm (1 ft 11/2 in); has a bill of medium length, and gay plumage. The wing coverts are barred with light blue, dark blue and black, and the body is pinkish brown. It lives in woods from lowlands to mountains, feeding on acorns, beechnuts, berries, birds' eggs, worms and small vertebrates. Like oth-ers of its family, it is a great robber of nests. A resident or dispersive bird. Distribution: Europe, Asia, northwest Africa.

Garypus A genus of Arthropods of the order of True Spiders, *Araneae*. e.g. G. beauvoisi - Up to 6 cm long; this spider has four pairs of legs and one pair of pincers; the skin looks finely grained. Found under piles of seaweed thrown up along the seashore. Habitat: the Mediterranean.

Gasteracanthus A genus of Spiders (class *Arachnida*); the species are highly colored creatures from East Africa.

Gasteria A genus of plants, also known by the generic name, *Aloë*, of the Lily family, *Liliaceae*. e.g. G. verrucosa (=

Aloë v.) - The Ox-tongue, an attractive, stemless, succulent plant, whose much thickened, succulent leaves are arranged in two ranks. They are 10-15 cm (4-6 in) long, and are rough to the touch due to raised greyish-white spots on the surface. The swollen red flowers are small (2.5 cm/1 in) and drooping. Place of origin: South Africa (see **Aloë**).

Gasterophilus A genus of Dipterous Insects of the Bot-fly family, *Gasterophilidae*. Also known by the generic name, *Gastrophilus*. e.g. G. intestinalis (= G. equi; *Gastrophilus* e.; *Gastrophilus* i.) - The Common Horse Bot-fly, length 12-15 mm; a parasite of horses and donkeys. The wings are glass-clear with grey transverse bands. The female lays several hundred eggs on the host's body, usually on the legs. When the animal licks its skin the eggs are transferred to the tongue, and from there the larvae infest the stomach and intestines, passing out of the body with the feces. The larva pupates either in the excrement or on the ground. Distribution: worldwide.

Gasteroplecus A genus of Fishes of the Characin family, *Characinidae*. The species are carnivorous, freshwater fishes from America. They have a disc-like arrangement on their thorax and abdomen, which has a bony edge; to this are attached the pectoral muscles which enable the fishes to flap their fins as if they were wings and thus fly.

Gasterosteus A genus of Bony Fishes of the Indostomid, Stickleback, and Trumpetfish family, *Gasterosteidae*. e.g. G. aculeatus - The Three-spine Stickleback, length 5-10 cm (2-4 in); a small fish with three free spines in front of the dorsal fin and bony plates along the sides. A circumpolar species of the cold and temperate zone of the northern hemisphere, equally at home in fresh or salt water (an euryhaline type), living in small stagnant or brackish waters near the estuaries of large rivers flowing into the sea. In the spring mating season the male acquires beautiful green and red colors and builds a fine nest out of aquatic plants, in which it fertilizes the eggs of the female. Distribution: in Europe, from the Black Sea, southern Italy and the Iberian Peninsula to the northern coast of Norway.

Gastrochaena A genus of Bivalve Molluscs of the family *Gastrochaenidae*. e.g. G. dubia - The Flask Shell, length 20 mm, height 10 mm, thickness 8 mm. Shell white to brown, with short anterior end and enlarged, elliptical hind end. It bores holes into various kinds of hard matter, including rocks and dead shells of other molluscs, such as oysters. Distribution: European Atlantic coast, Black Sea, Mediterranean Sea.

Gastrodiscoides A genus of Intestinal Trematodes or Flatworms, also known by the generic name, *Gastrodiscus*. e.g. G. hominis (= *Gastrodiscus* h.) - A fluke found in East Asia, infesting the cecum and large intestine of pigs and occasionally of man.

Gastrodiscus See **Gastrodiscoides**.

Gastropacha A genus of Insects of the Butterfly and Moth order, *Lepidoptera*. e.g. G. quercifolia - The Lappet-Moth, an insect with protective coloring, looking like a dry leaf. Its hairy grey caterpillar is approximately 10 cm (4 in) long and feeds on blackthorn bushes, deciduous bushes, deciduous trees, and in particular, fruit trees.

Gastrophilus See **Gasterophilus**.

Gastrophyrne A genus of Amphibians of the Frog family, *Microhylidae*. A narrow-mouth, toad species in which during amplexus, the male and the female adhere to one another by means of their sticky skin secretions.

Gastrotheca A genus of Amphibians of the Typical Treefrog family, *Hylidae*. The species are specially adapted to arboreal life. The female has an enclosed pouch on its back into which the male pushes the fertilized eggs as soon as they are laid, singly or in pairs.

Gattyana A genus of Segmented Worms or Annelids of the order *Errantia*. e.g. G. cirrosa - The Scale Worm, length 21/2-5 cm; brown body with 15 pairs of transparent scales on the back. A scavenger species found on the seacoasts of Europe. Native around Britain.

Gaultheria A genus of plants of the Heath family, *Ericaceae*. e.g. G. fragantissima - From the leaves is extracted Gaultheria Oil (= Oleum Betulae; Sweet Birch Oil; Wintergreen Oil), containing Methyl Salicylate, and is used externally as a counter-irritant.

Gavia A genus of Birds of the Diver or Loon family, *Gaviidae*. e.g. G. arctica - The Black-throated Diver or Loon, length 70 cm; colored crown-grey with its back checkered with white; found in the region of large northern lakes in the tundra and forest-tundra where it nests on the shore close to the water, often on small islands. A migrating bird, over-wintering on large rivers and lakes as well as the shores of the North, Mediterranean, and Black Seas. Distribution: northern Europe, Asia, North America. Native in Britain.

Gavialis A genus of Reptiles of the Indian Crocodile or Gavial family, *Gavialidae*. e.g. G. gangeticus - The Indian Gavial or Gharial, or Ganges Gavial, the Indian Crocodile, the only genus of this family, living in the waters of the Brahmaputra, Ganges, Indus and Mahanadi river basins; up to 6 m (20 ft) long, with a long, rod-like snout that widens at the nostrils. Not dangerous to man, feeding almost entirely on fishes. The female lays almost 40 eggs in a nest on the river bank. Was once regarded by the Hindus as sacred.

Gazania A genus of plants of the Daisy family, *Compositae*. e.g. G. nivea - The Gazania or Treasure Flower, a perennial plant with rough, orange yellow capitula on peduncles. The central disc of the capitulum is surrounded by a brown ring, which marks the base from which the petals sprout. Place of origin: South Africa; found in Europe since 1892.

Gazella A genus of Mammals of the Horned Ungulate family, *Bovidae* (Antelope subfamily, *Antilopinae*). e.g. G. dorcas - The Dorcas Gazelle, one of the smallest gazelles, barely 60 cm (2 ft) high at the shoulders, found all over North Africa. A beautiful sandy-colored animal, with white rump patches, and V-shaped or lyrate horns with the points generally turned foreward. It has a remarkable speed and is capable of travelling over 40 miles per hour for 15 minutes or more.

Geaster A genus of plants of the class of Fungi or Molds, *Gasteromycetes*. e.g. G. pouzari - A species of Earth-stars, generally occurring on the ground in woods. At first the fungus is attached but a thick outer layer splits from the top downwards into a number of segments which curve back to form a stand, with the spherical spore-containing structure in the center.

Gekko A genus of Reptiles of the Gecko Lizard family, *Gekkonidae*. e.g. G. gecko - The Tokay, Tokay Gecko or Lizard, up to 35 cm (14 in) long; light greyish-purple in color with orange or reddish spots, andbroad, strong toe-pads, enabling it to climb walls and the smoothest surfaces in pursuit of insects and mice. It lives among the trees in the jungle, as well as in houses. Its common name is based on its call, "to-kay, to-kay", repeated several times, and heard over a distance of over 90 m (100 yards). Distribution: one of the commonest species in Bengal (India), southern China, Indonesia and the whole of southeast Asia.

Gelochelidon A genus of Birds of the Tern family, *Laridae*. e.g. G. nilotica - The Gull-billed Tern, size 38 cm; with a short, powerful, black bill and black legs. A migrant bird. It lays its beautifully camouflaged eggs on the bare sand or shingle, and does not build any nest. Habitat: the Mediterranean region, breeding on Danish coasts.

Gempylus A genus of Fishes of the Snake Mackerel family, *Gempylidae*. Often regarded as degenerate mackerels, the dorsal fin is in three parts, the last being a series of branched rays with no membranes in between.

Genetta A genus of Mammals of the Civet, Genet and Mongoose family, *Viverridae*. e.g. G. genetta - The Feline or Small Spotted European Genet, or Weasel Cat; length of body 50 cm (20 in), tail 45 cm (18 in). It is yellowish-grey with black markings and is found on rocky slopes covered with dense thickets. Active mostly at night; by day it sleeps in hollow trees or rock crevices. A good climber. Distribution: throughout Africa; in Europe it inhabits the Iberian Peninsula and southern France. The Viverridae, including the Genet or Weasel Cat is a survivor of a primitive carnivore family which used to live all over the world except Australia and the American continent.

Genista A genus of plants, also known by the generic name, *Cytisus*, of the Peafamily, *Leguminosae*. e.g. G. radiata (= *Cytisus* radiatus) - The Royal Broom (see **Cytisus**).

Gentiana A genus of plants of the Gentian family, *Gentianaceae*. e.g. G. lutea - Yellow Gentian or Great Yellow Gentian, an upright, glabrous, perennial plant, from 1.2-1.8 m (4-6 ft) high, with ovate leaves and dense whorls of golden-yellow, tubular flowers. The root, Gentian Root, contains gentianine, a mixture of gentianic acid and gentiopicrin; it is an aperient, digestive, febrifuge and gastric, and is used in medicine as a bitter tonic, appetizer and antiseptic. Also used to make a liqueur. Place of origin: wet pastures in the mountains of South and Central Europe and Asia Minor.

Gentianella A genus of plants of the Gentian family, *Gentianaceae*. e.g. G. aspera - Rough Gentian, a biennial plant, 4-20 cm high, with short, hairy leaf-margins; violet, lilac or whitish corolla; and arough, hairy calyx. Grows in poor alpine pastures, rock-crevices and high ledges. Place of origin: South and Central Europe.

Geocapromys A genus of Mammals of the Coypu Rodent family, *Echimyidae*. These are the Hutias, very primitive rodents of the New World, found in the West Indies, with one species in Jamaica, one in the Bahamas and a third on Little Swan Island.

Geocarcoidea A genus of Decapod Crustaceans of the order of True Crabs, *Brachyura*. e.g. G. humei - The Land Crab, found in the Malacca Strait. These crabs can wander a long way from the sea, and scavenge on the floor of forests. The gill chambers are modified with spongy walls so that they can breathe in air.

Geochelone A genus of Reptiles of the Tortoise suborder, *Cryptodira* (order *Chelonia*). e.g. G. gigantea - The largest living tortoise of the Indian Ocean Islands of Aldabra, Mauritius and the Seychelles, weighing up to 181 kg (400 lb), and with a carapace over 1 m (34 in) long.

Geococcyx A genus of Birds of the Cuckoo and Road-runner family, *Cuculidae*. e.g. G. californianus - The American Road-runner or Chaparral Cock; the fastest-running carinate or flying bird; chiefly a ground bird, as its name implies, with long powerful legs and a long tail, running fast with its short wings outstretched. It has been clocked at 42 km/h (26 miles/h). A solitary species, it seldom flies, and then poorly, and does not migrate. It lives among desert scrub, where it feeds on lizards and small snakes. Distribution: the plains and deserts of the southwestern United States.

Geodia A genus of Sponges (phylum *Parazoa*) of the four-rayed order, *Tetraxonida*. The species have long four-rayed siliceous spicules (little spikes) as well as needle-shaped spicules (megascleres), and two small sperical microscleres.

Geometra A genus of Insects of the Looper or Geometer Moth family, *Geometridae*. e.g. G. papilionaria - The Large Emerald Moth, size 21-29 mm; found in broadleaved and mixed woodlands. Distribution: Central and North Europe, Asia Minor, Siberia, Japan (see **Alsophila**).

Geomys A genus of Mammals of the Pocket Gopher family, *Geomyidae* (= *Geomydidae*). These species are North American rodents found in the eastern United States. The expression "pocket" is the cheek pouch in which they carry their food for storing underground.

Geonemertes A genus of Nemertine or Ribbon Worms of the class *Enopla*, belonging to the order *Hoplonemertini*, in which the proboscis is armed. The species is found in tropical regions in the southern hemisphere.

Geophilus A genus of Centipede Arthropods (class *Chilopoda*) of the family *Geophilidae*. e.g. G. longicornis - The Necrophloephagus, a common species of centipedes, 20-40 mm long with up to 57 pairs of legs. Yellowish in color; matures after two years. Distribution: Europe, North Africa, Siberia; introduced into North America.

Geotrupes A genus of Insects of the family of Chafer and Dung Beetles, *Scarabaeidae*. e.g. G. silvaticus - The European Dor Beetle, a well-known species found in the forests of continental Europe, with a shiny, blackish-blue body. The female lays her eggs in a ball of dung which she then buries in the ground. Closely related species with similar habits are found in North America.

Geranium A genus of plants of the Cranesbill or Geranium family, *Geraniaceae*. e.g. G. pratense - The Meadow Cranesbill, a perennial herb; it has palmate 5-7-lobed leaves, with a thick rhizome and erect stems up to 90 cm (3 ft) high. The large, violet-blue, cup-shaped flowers grow in axillary pairs. The fruit has five lobes each containing a single seed. Widespread in Britain, Europe, North and Central Asia. Has become naturalized in North America.

Gerardinus A genus of Freshwater Fishes of the Carp and Minnow family, *Cyprinidae*. e.g. G. poeciloides - A species of Minnows used in Central America toe at the larvae of the Anopheles mosquito (see **Carassius**).

Gerbera A genus of plants of the Daisy family, *Compositae*. e.g. G. jamesonii - The Barberton Daisy or Transvaal Daisy, a hairy perennial plant with a woody base and numerous lanceolate, pinnatifed leaves and large, orange, capitulate flowers borne on the ends of long, erect, unbranched stems. Beaked fruits with a pappus of rough hairs. Place of origin: Natal, Transvaal; introduced into Europe in 1887.

Gerbillus A genus of Mammals of the Rat family, *Muridae* (Sand rat subfamily, *Gerbillinae*). These are the Gerbils or Sand rats, nocturnal creatures inhabiting the drier parts of Africa and Asia. Sandy or pale buff above, they have white underparts and feet, and leap on their hind legs like kangaroos. They all hoard their food in their underground burrows. e.g. G. iateronia - The Gerbil or Sand Rat, a small burrowing rodent, native of the South African veldt, and one of the chief agents, transmitting plague to humans.

Gerris A genus of Insects of the family of Pond-skaters, Water-bugs, or Water-striders, *Gerridae*. e.g. G. gibbifer - Length 10-13 mm; a water-bug found commonly on the surface of pools and puddles. The adults hibernate.

210

Distribution: Europe, Middle East, North Africa.

Geryonia A genus of Celenterate animals of the Medusa order, *Narcomedusae* (= *Trachymedusae*). These hydrozoans are a small group with no polyp phase. They have broad flat bells and lack a manubrium; the mouth opens directly into the stomach region. Their life-history is considered by some to be ancestral. They live in the warmer waters of the tropics (see **Liriope**).

Geum A genus of plants of the Rose family, *Rosaceae*. e.g. G. chiloense (= G. coccineum) - The Geum or Scarlet Avens, a herbaceous, perennial plant, 40-60 cm (1 ft 4 in - 2 ft) high, with scarlet, chalice-shaped flowers. Place of origin: Chile; introduced into Europe in 1826.

Giardia A genus of Flagellate Protozoans of the Metamonad order, *Metamonadina* (= *Distomatina*). The species are found in the intestinal tract of man and animals. Also known by other generic names. e.g. G. lamblia (= *Cercomonas* intestinalis; G. intestinalis; *Dicercomonas* muris; *Lamblia* i.; *Megastoma* entericum) - The body has a characteristic kite shape, and is bilaterally symmetrical, with two nuclei. There are six flagella. A parasite found in man, sometimes causing severe diarrhea and intestinal colic, especially in children (giardiasis or lambliasis).

Gibberella A genus of Molds or Fungi. e.g. G. fujikuroi - A species from the cultural filtrates of which, are obtained the gibberellins or plant growth stimulators. Gibberellic acid is the most widely used of the gibberellins.

Gibbula A genus of Gastropod Molluscs of the family *Trochidae*. e.g. G. divaricata - The Variegated Topshell, a greenish-yellow species with lines of carmine red spots; shell 23 mm high, 19 mm across and with 6 bulging whorls. The top whorls are smooth, while the bottom ones have spiral ridges. Found in the sea at shallow depths under stones or amongst seaweed on which it feeds. Distribution: the Mediterranean Sea and the Atlantic coasts, north to the English Channel.

Gigantocypris A genus of Ostracod Crustaceans of the order *Myodocopa*. These are ostracods possessing a heart, this genus being the largest of these animals. The species are marine and planktonic; the branched antennae can be protruded through notches in the carapace and used in locomotion while the shell is lightly closed.

Gigantophis A genus of Extinct Reptiles of the Snake suborder, *Ophidia*. e.g. G. garstini - A python-like reptile, the longest prehistoric snake known, which inhabited what is now Egypt about 50 million years ago. Parts of a spinal column and a small piece of jaw, discovered at El Faiyum, indicate a length of about 11.28 m (37 ft).

Gigantornis A genus of Extinct Prehistoric Birds (class *Aves*). e.g. G. eaglesomei - Probably the largest prehistoric bird to actually fly (in terms of wingspan), which soared over what is now Nigeria about 45 million years ago. It is only known from a breastbone, but the enormous size of this fossil, and its close similarity to the breastbone of the albatross, suggest that the bird had long, narrow wings spanning as much as 6 m (20 ft).

Gigartina A genus of plants of the family of Red Algae, *Gigartinaceae*. e.g. G. stellata - A seaweed species from which is obtained Agar (= Agar-agar; Gelosa; Japanese Isinglass), used in the treatment of constipation (see **Crispus**).

Ginglymostoma A genus of Fishes, also known by the generic name, *Ginglyostoma*, of the Nurse Shark or Carpet Shark family, *Orectolobidae*. e.g. G. cirratum (= *Ginglyostoma* c.) - The Common Nurse Shark; for the most part, a scavenger fish that does not attack man.

Ginglyostoma See **Ginglymostoma**.

Ginkgo A genus of Trees of the family *Ginkgogaceae*. e.g. G. biloba - The Ginkgo, Gingko or Maidenhair Tree, a slender, gymnospermous, deciduous tree, up to 30 m (100 ft) tall, with dioecious flowers (male and female flowers on separate plants), native to China and Japan, the only living representative of the genus. The evolutionary line of the Gingko is a long one, stretching back at least 200 million years, and fossil leaves now found in rocks from the Jurassic period (about 180 million years ago) are indistinguishable from those of the present species, hence the Gingko has been called a living fossil. Gingkos (= Gingkoes, Ginkgos, or Ginkgoes) have been preserved sacred in Japanese temple gardens, and are rarely found in the wild.

Giraffa A genus of Mammals of the Giraffe and Okapi family, *Giraffidae*. e.g. G. camelopardalis - The Giraffe, a species now inclusive of all types of giraffes. It is the tallest of all living animals, reaching a height of up to 6 m (20 ft) at the shoulder. The very long neck results from an increase in the length of the vertebrae and the legs are also disproportionately long in relation to the body. The tail is over 1 m (3 ft) long; the hooves are 15 cm (6 in) high in males and 10 cm (4 in) in females. Weights vary from a half to one ton. Giraffes are found in their natural habitat in Africa on savannas, where they feed on acacias and climbing leguminous plants, from Lake Chad to eastern Africa.

Gladiolus A genus of plants of the Iris family, *Iridaceae*. e.g. G. imbricatus - The Meadow Gladiolus, a perennial plant, 30-60 cm (1-2 ft) high. There are 4 to 10 contiguous, crimson-colored flowers, borne on erect one-sided spikes. Two or three leaves are produced by each corm. Place of origin: eastern Europe and southern Russia.

Glandiceps A genus of Chordate animals of the Acorn worm class, *Enteropneusta* (see **Harrimania**).

Glareola A genus of Birds of the Courser and Pratincole family, *Glareolidae*. e.g. G. pratincola - The Common Pratincole, known in South Africa as the Locust Bird; length 23 cm; it has a light throat-spot with rust-red underwing linings shown in flight, hunting on the wing, and skimming across the grass to catch dragonflies and grasshoppers. Distribution: southern Europe, Africa, and southern Asia.

Glaucidium A genus of Birds of the Typical Owl family, *Strigidae*. e.g. G. passerinum - The Sparrow Owl or Pigmy Owl, the smallest European owl, 16.5 cm (6 1/2 in) long, with yellow-colored eyes, and found in coniferous woods at both lowland and mountain elevations. Feeds on small mammals and perching birds. A resident bird, nesting in treeholes, often in holes made by woodpeckers. Distribution: northern Europe and Asia as far as the Amur region.

Glaucium A genus of plants of the Poppy family, *Papaveraceae*. e.g. G. flavium - The Yellow Horned Poppy or Sea Poppy, an erect biennial or short-lived perennial plant, 30-60 cm (1-2 ft) high, with glaucous, rough, deeply lobed and toothed, fleshy leaves and 4-petalled bright yellow flowers. The sap is yellow. The fruit is a narrow curved capsule up to 30 cm (1 ft) long, splitting into 2 valves. Place of origin: coast-line of the Mediterranean and Western Europe (including Britain), usually on shingle; introduced to and naturalized in North America.

Glaucomys A genus of Mammals of the Squirrel, Marmot and Prarie Dog family, *Sciuridae*. e.g. G. volans - The Flying Squirrel of North America, a nocturnal animal living in pine forests and active throughout the winter. About 12.5 cm (5 in) in body length; it has furred membranes between front and hind legs on each side, which it expands into a kind of parachute as it leaps through the air from one tree to another. Its 10 cm (4 in) bushy tail is used as a rudder. The species has spread from North America south to Guatemala in Central America.

Glaux A genus of plants of the Primrose family, *Primulaceae*. e.g. G. maritima - The Sea Milkwort, a perennial plant, 10-30 cm high, with small fleshy leaves, usually growing on the seashore. Place of origin: European seacoasts; native around Britain.

Glechoma A genus of plants of the Mint or Thyme family, *Labiatae*. e.g. G. hederacea - Ground Ivy, a perennial plant, 2-15 cm high; with crenate, kidney-shaped leaves and bluish violet flowers growing in whorls. Found on the edges of cultivated fields, on village greens, by waysides and in hedgerows. Widespread and common in Europe; native in Britain.

Glenospora A genus of Molds or Fungi. e.g. G. graphii - A species causing otomycosis.

Gliricola A genus of Insects of the order of Lice, *Anoplura* (= *Mallophaga*). e.g. G. porcelli - A biting louse found on guinea pigs.

Glis A genus of Mammals of the Doormouse family, *Gliridae* (= *Muscardinidae*). e.g. G. glis - The Fat or Edible Doormouse, the largest of the Doormice, a rodent intermediate between a squirrel and a rat, length of body 15 cm (6 in), tail 12.5 cm (5 in); bright silvery-grey colored back and yellowish-white underneath. It has a long bushy tail and resembles a small squirrel. Found in broadleaved woodlands, chiefly beechwoods, in parks, gardens and cemeteries. Chiefly nocturnal. It hibernates in winter in a hole lined with moss and dry grasses. Feeds on seeds, nuts and young tree shoots, and is destructive to orchards. Distribution: Europe, Asia.

Glischrochilus A genus of Insects of the Flower Beetle family, *Nitidulidae*. e.g. G. quadripunctatus - Length 3-6.5 mm; found under bark and on injured tree trunks oozing sweet sap. Feeds chiefly on bark beetles. Distribution: Europe, Siberia.

Globicephala A genus of Mammals of the Grampus Whale and Dolphin family, *Delphinidae*. e.g. G. melaena - The Northern Pilot Whale, also called the Northern Blackfish, Caa'ing Whale or Grindhval; length 6-7 m (20-23 ft). It has a globular head; beneath the protuberant forehead the muzzle forms a very small beak. An inoffensive predator, it feeds mostly on cuttle-fish. It lives in schools of several hundred, which appear to obey a leader. Habitat: the Arctic Ocean and northern parts of the Pacific and Atlantic.

Globigerina A genus of Protozoan marine animals of the order *Foraminifera*. The foraminifers taken collectively are important because they concentrate calcium or silica with which they form their tests or shells in which they live. The species form globigerina ooze on the floor of deep oceans in warm regions; they have spherical locula arranged in spirals.

Globularia A genus of plants of the Globe-daisy family, *Globulariaceae*. e.g. G. cordifolia - The Heart-leaved Globe-daisy or Heartleaf Globularia, a mat-forming, dwarf, perennial plant, 3-10 cm high, with flowering, woody stems. The leaves are shiny, dark-green, somewhat fleshy, and rounded-ovate, growing in rosettes. The flowers are in rounded heads, terminal, blue, and occasionally white or rose. Grows on rocks, in stony places and alpine grassland. Place of origin: South and Central Europe; West Asia. Introduced into Britain in 1633.

Glomeris A genus of Arthropods of the class of Millipedes, *Diplopoda*. e.g. G. marginata - The Pill Millipede or Pill

Bug, a diplopod myriapod or millipede, a vegetarian species found in Britain, 7-20 mm long. It has a shining black body, 17 to 19 pairs of legs, according to sex and it rolls itself into a ball the size of a pea or pill. It has two rows of stink glands along the sides of the body which secrete a noxious, colored substance containing hydrocyanic acid, capable of killing or repelling insects. Widespread in damp places in rotten wood, under stones or by paths. Distribution: Europe; native in Britain.

Gloriosa A genus of plants of the Lily plants, *Liliaceae*. e.g. G. rothschildiana - The Gloriosa Lily or Glory Lily, a climbing plant, 1.8 m (6 ft) high, with tuberous roots, ascending by means of "finger tip" tendrils at the end of the smooth, green, lanceolate leaves. The flowers are very showy, with six waxy-edged, crimson petaloid segments, which bend backwards to reveal their yellow bases and long extruding stamens. Place of origin: Tropical Africa.

Glossina A genus of Insects of the Biting or Tsetse Fly family, *Stomoxyidae*. e.g. G. palpalis - The Tsetse Fly of Central Africa; living in the damp forests of this part of the continent, it transmits Trypanosoma gambiense, the causative organism of the dreaded African sleeping sickness in human beings.

Glossiphonia A genus of Leeches (class *Hirudinea*) of the Sucker Leech family, *Glossiphoniidae*. e.g. G. complanata - A small, freshwater leech, up to 3 cm long, with a dorso-ventrally flattened body and two dark, lengthwise stripes along the top, inhabiting European ponds. The anterior sucker is no wider than the head, and it can elongate the front part of its body. Distribution: Europe; native in Britain.

Glossobalanus A genus of Chordate animals of the Acorn worm class, *Enteropneusta* (see **Harrimania**).

Glossodoris A genus of Molluscs (phylum *Mollusca*). e.g. G. quadricolor - The Nudibranch, a colorful, unsegmented animal secreting no shell. It crawls about on a muscular "foot", digging away at sponges and coral. Habitat: the Red Sea.

Glossoscolex A genus of Segmented Worms of the class *Oligochaeta*. e.g. G. giganteus - The Great Earthworm of Brazil, measuring up to 1.26 m (4 ft 2 in) when naturally extended.

Glossus A genus of Bivalve Molluscs of the order *Veneroida*. e.g. G. humanus - The Heart Cockle, a seawater cockle; size of shell up to 12 cm; valves are concentrically striped. Found at depths greater than 4 fathoms (7.2 m = 24 ft). Distribution: the seas of Europe; native around Britain.

Gloxinia A genus of plants, now believed to really belong to the genus *Sinningia* of the family *Gesneriaceae*. About 150 years ago, a plant was introduced into Britain from Brazil under the name *Gloxinia* speciosa. It caused much interest and from it, all the present Gloxinias are derived. The correct name of course is *Sinningia* speciosa. It differs from the true *Gloxinia* spp. in having a tuberous rhizome and 5 separate glands instead of a ring at the base of the ovary (as in the gloxinias) (see **Sinningia**).

Glyceria A genus plants of the Grass family, *Gramineae*. e.g. G. fluitans - Floating Sweet-grass or Manna-grass, a perennial plant, 40-120 cm high; with contracting panicles, shortly-projecting, 2-toothed paleas or upper bracts, and violet anthers (the stamen portions containing the pollen). It grows on muddy or grassy banks, by ponds, freshwater marshes, lakes and rivers. Distribution: Europe; native in Britain.

Glycine A genus of plants of the Pea family, *Leguminosae*; also known by the generic name, *Soja*. e.g. G. soja (= *Soja* hispida) - The Soya Bean, Chinese Bean or Japanese Bean, a plant widely cultivated in eastern Asia for its valuable edible seeds. Also used as a green fodder. From the seed is expressed Soya Oil (= Oleum Sojae; Soya Bean Oil; Soybean Oil), containing the albuminoid globulin, Glycinin. It is used as an edible oil, and has also been tried in the treatment of atherosclerosis.

Glyciphagus A genus of Mites related to the genus *Tyroglyphus*. e.g. G. domesticus (= G. prunorum) - A species infesting sugar and causing an itch (Grocer's itch) (see **Tyroglyphus**).

Glycymeris A genus of Bivalve Molluscs of the order *Veneroida*. e.g. G. glycymeris - The Dog Cockle; length of shell 6-10 cm, thick, almost circular, with numerous teeth. Lives on hard seabeds in the English Channel and along the Atlantic coasts, commonly in large colonies.

Glycyrrhiza A genus of plants of the Pea family, *Leguminosae*. e.g. G. glabra var. glandulifera - The Liquorice plant. Liquorice-root contains the sweet substance, glycyrrhizin, used as a demulcent and flavoring agent.

Glyphesis A genus of Arachnids of the order of True Spiders, *Araneae*. e.g. G. cottonae - A species of diminutive spiders which spin some of the world's smallest webs, each measuring as little as 19-20 mm (0.75-0.78 in) in diameter.

Glyptocephalus A genus of Marine Fishes of the Flatfish family (order *Heterosomata* or *Pleuronectiformes*). e.g. G. cynoglossus - The Witch, up to 50 cm (20 in) long; the right pectoral fin is blackish. The fish lies on the left side, the right eye being uppermost. Habitat: the North Sea. Native around Britain.

Glyptocranium A genus of Arachnids of the order of True Spiders, *Araneae*. e.g. G. gasteracanthoides - The Podadora or Bola-spider, a highly venomous spider of Argentina and Peru, which has been credited with many fatalities.

Gnaphalium A genus of plants, also known by the generic name, *Antennaria*, of the Daisy family, *Compositae*. e.g. G. dioicum (= *Antennaria* dioica) (see **Antennaria**).

Gnathoremus A genus of Fishes of the Mormyrid suborder, *Mormyroidea*. e.g. G. petersi - The Ubangi Mormyrid fish from Africa, a freshwater fish with a very long snout, living in muddy, slow-flowing waters.

Gnathostoma A genus of Nematode Worms, also called *Gnathostomum*, found chiefly in the Far East i.e. Thailand and China. e.g. G. spinigerum (= G. siamense; *Gnathostomum* s.) - A species found in animals and occasionally in man, causing the disease known as Gnathostomiasis. The development host is *Cyclops*.

Gnathostomum See **Gnathostoma**.

Gobiesox A genus of Fishes of the order of Cornish Suckers, Clingfishes and Skittlefishes, *Xenopterygii* (= *Gobiesociformes*). e.g. G. maeandricus - The Northern Clingfish of California, reaching a length of 15 cm (6 in); found mainly in tidal pools in warmer seas. A small fish with its eyes on top of the head. The body is scaleless, and there is no air bladder.

Gobio A genus of Fishes of the Carp and Minnow family, *Cyprinidae*. e.g. G. gobio - The Gudgeon, a carp-like fish, 10-20 cm (4-8 in) long; the mouth has two short barbels, the body is elongated and spindle-shaped, with large scales, and it has short dorsal and anal fins. Found on the bottoms of all types of fresh waters. It is much favored as food in France. Popular as bait for predaceous fish. Distribution: Europe; native around Britain (see **Carassius**).

Gobionellus A genus of Fishes of the Goby family, *Gobiidae*. e.g. G. oceanicus - The High-fin Goby fish (see **Gobius**).

Gobius A genus of Fishes of the Goby family, *Gobiidae*. e.g. G. niger - The Black Goby, length 15 cm (6 in); back brownish, belly yellow-white; a marine fish found over sandy or muddy bottoms close to the seashore and also in tidal pools in river estuaries. If left stranded it is capable of jumping across the sand to find water. Distribution: along the coast of the eastern Atlantic, Baltic, Mediterranean and Black Seas. In the Mediterranean and the Black Sea it is caught for food. Native around Britain.

Godetia A genus of plants of the family *Onagraceae*. e.g. G. grandiflora (= G. amoena whitneyi) - The Godetia or Satin Flower, a compact, bushy, annual plant, 30-37 cm (12-15 in) high, with oblong pointed leaves and satin-textured, funnel-shaped flowers, growing in clusters, white ranging to pink, red and carmine. Place of origin: California; introduced into Britain in 1867.

Golfingia A genus of Sipunculoid Animals of the phylum *Sipunculoidea*. e.g. G. elongata - A worm-like marine animal, resembling an annelid, with a long cylindrical, unsegmented body and an uninterrupted coelomic cavity. The forepart of the body can be retracted into the rear part. Sedentary in habit, it lives in permanent or semipermanent burrows in mud or sand, in rocky fissures, under stones or amongst algal colonies (see **Dendrostoma**; **Phasocolosoma**).

Goliathus A genus of Insects of the Dung, Scarab or Scarabaeid Beetle family, *Scarabaeidae*. e.g. G. caccicus - A species from West Africa, one of the world's largest beetles; about 7.5 cm (3 in) long. It flies among the jungle treetops, feeding on the petals and stamens of palm and other flowers. The larvae live in rotting wood. G. goliatus - The Goliath Beetle of equatorial Africa, the world's heaviest insect. Adult males weigh 70-100 g (2.47-3.53 oz) and measure up to 12 cm (4.72 in) from the tip of the frontal horn to the end of the abdomen. Females are smaller. These beetles have a massive build-up of heavy chitin forming their thorax and anterior sternum, which adds considerably to their weight.

Golofa A genus of Insects of the Dung or Scarab Beetle family, *Scarabaeidae* (subfamily *Dynastinae*). e.g. G. pizzaro - A species from Mexico. It has a rectangular plate, hairy below, at the end of its horn, and a curiously-shaped structure on the top of its head and prothorax.

Gomophia A genus of Echinoderm Animals of the class of Star-fishes, *Asteroidea*. These are carnivorous, marine organisms, feeding on molluscs, possessing five radially symmetrical arms, and a mouth on the dorsal surface of a central disc. The dorsal surface of the arms bears rows of tube-feet. Aided by the suction power of these tiny tube-feet, they can open any shell, engulfing their prey by everting their stomach. Thus their digestive process begins outside the body. Habitat: the coral reefs of the Red Sea and the Gulf of Eilat (Akaba).

Gomphidius A genus of Molds or Fungi of the class *Basidiomycetes*. e.g. G. glutinosus - A species growing in coniferous woods during the autumn. It has a stipe (short stem) from 5-15 cm (2-6 in) long and a convex greyish-brown cap from 5-10 cm (2-4 in) across. A membrane between the edge of the cap and the stem protects the young radiating gills on which the basidia (spore-bearing conidophores) are borne.

Gomphosus A genus of Fishes (subphylum *Vertebrata*). e.g. G. caeruleus - The Bird-Fish, a species of fishes with an elongated snout, somewhat resembling the beaks of certain birds. Also the swimming motion of this fish reminds one of a bird. The male has a dark blue-green color and the female is brown. Found in the Red Sea.

Gomphus A genus of Insects of the Damselfly and Dragonfly family, *Gomphidae*. e.g. G. vulgatissimus - The Club-tail Dragonfly, length 40-55 mm, wingspan 60-70 mm; found near water in clearings, forest-rides and meadows, by brooks and rivers. Distribution: Europe, Asia Minor. Native in Britain.

Gonepteryx A genus of Insects of the White and Yellow family, *Pieridae*. e.g. G. rhamni - The Brimstone, size 3-5 cm, a bright yellow butterfly (male), the female being greenish-white. All four wings have a small orange spot. The caterpillar feeds on Buckthorn (family *Rhamnaceae*). Distribution: Europe, Asia Minor, the Middle East, North Africa (see **Rhamnus**).

Gongylonema A genus of Filarial Nematode Worms. e.g. G. pulchrum - A scutate threadworm commonly found in the esophageal mucosa of domestic animals. Also found on the lips and mouth in humans. It develops in cockroaches and dung beetles.

Goniodes A genus of Insects of the order of Biting Lice, *Mallophaga*. e.g. G. colchicus - A bird-parasite species, infesting pheasants. The female measures only 2.5 mm (1/10 in). This louse lives among the feathers of its host, feeding on horny particles of the skin and on the feathers themselves.

Gonionemus A genus of Sea-firs or Hydras (class *Hydrozoa*) of the order of Medusae, *Linnomedusae*. The species are found in Britain and America. These animals are saucer-shaped with 4 to 6 radial canals and numerous tentacles in sets around the bell margin, together with suckers.

Goniopora A genus of True or Stony Corals of the order *Scleractinia*. Habitat: the Red Sea coast (see **Favia**).

Gonium A genus of Protozoans of the order of Chrysomonads, *Chrysomonadina*. These are yellowish-brown, very small flagellates ubiquitous in fresh, brackish and seawater. In size they seldom exceed 20 μ in length, have one or two flagella and form silica-containing cysts. The species exist in colonies.

Gonococcus A genus of microorganisms classified under the genus *Neisseria*. e.g. G. neisseriae (= *Neisseria* gonorrhoeae) (see **Neisseria**).

Gonodontis A genus of Insects of of the Looper or Geometer Moth family, *Geometridae*. e.g. G. bidentata - The Scalloped Hazel Moth, a species widely distributed in northern Europe and Asia. It lays its eggs, which are bluish in color, later turning reddish-brown, on a variety of larval food-plants such as oak, birch, hawthorn, sallow, etc.

Gonolobus A genus of plants of the Milkweed family, *Asclepiadaceace*; also known by the generic name, *Marsdenia*. e.g. G. condurango (= *Marsdenia* c.) - The Condurango or Eagle-vine; the bark contains a mixture of glycosides known as condurangin. Used as an aromatic bitter and gastric sedative.

Gonorhynchus A genus of Fishes of the Sandfish and Beaked Salmon suborder, *Gonorhynchoidea*. This is an elongated fish with a long pointed snout and a rostral tentacle in front of the lips. It is found in shallow water around the coast of Japan, Australia, New Zealand and South Africa.

Gonyaulaux See **Gonyaulax**.

Gonyaulax A genus of Protozoa of the order of Dinoflagellates, *Dinoflagellata*; also known by the generic name, *Gonyaulaux*. e.g. G. catanella (= *Gonyaulaux* c.) - A red plankton dinoflagellate with armored plates surrounding the body. These plates have a characteristic shape and regularly pitted surfaces. When these animals increase in number they give the sea a reddish tinge and kill off fishes and crustaceans, due to the toxic alkaloid they contain. This substance is known as saxitoxin, and it causes Mussel Poisoning (see **Mytilus**).

Goodyera A genus of plants of the Orchid family, *Orchidaceae*. e.g. G. repens - Creeping Lady's Tresses, a perennial plant, 10-25 cm high; the small, scented flowers have lips like pouches with a projecting tongue.

Grows in pine forests, rarely in deciduous woods or on fixed sand-dunes. Distribution: Europe; native in Britain.

Gopherus A genus of Reptiles of the Land Tortoise family, *Testudinidae*. e.g. G. polyphemus - The Gopher Tortoise of the southeastern United States. It uses its hard plate extensively for burrowing, for it spends much of its life in burrows.

Gordius A genus of Worms of the family *Gordiaceae*, known as the horse snakes, hairworms or horsehair worms. e.g. G. sensu stricto - A species passed per urethra and causing dysuria.

Gorgon A genus of Mammals of the Ungulate or Cattle family, *Bovidae* (Antelope subfamily, *Antilopinae*). e.g. G. taurinus - The Brindled Gnu or Blue Wildebeest, an ungainly antelope with buffalo-like head and horns, dark grey in color with black stripes on the neck and a black tail; its body is over 2.4 m (8 ft) long, and it stands some 1.3 m (4 ft 4 in) at the shoulders, with a 60 cm (2 ft) tail. It has a very keen sense of smell and remarkable sight, enabling it to avoid lions, its greatest enemy. It lives in relatively large numbers in open woodlands, from Kenya to South Africa.

Gorgonaria A genus of Horny Octocorals of the family *Gorgoniidae* (order *Gorgonaceae*). These are the Sea Fans, creatures with fan-like branches and found only in deep water. They have a delicate architecture. Habitat: the coral reefs of the Red Sea and the Gulf of Eilat (Akaba).

Gorgonia A genus of Horny Octocorals (order *Gorgonacea*) of the family *Gorgoniidae*). e.g. G. flabellum - Venus's Fan, a horny coral animal, forming large fan-shaped colonies with a horny skeleton, covered on both sides with a soft layer of retractable polyps. It grows on the ocean bed, especially in the West Indies.

Gorgonocephalus A genus of Echinoderm animals of the class of Basketstars, *Ophiuroidea*. These ophi-

uroids are large Basketstars, which take 20 to 30 years to reach maximum size. They usually live in plankton-rich waters and have a more rapid growth than other (smaller) ophiuroids.

Gorilla A genus of Mammals of the Anthropoid Ape family, *Pongidae*. These are the largest living primates. e.g. G. gorilla - The Gorilla, the only species of this genus, with a number of races (e.g. G. gorilla beringei - the Mountain Gorilla), and the biggest and strongest of all the primates. The males grow to well over 1.8 m (6 ft), and weigh as much as 227 kg (500 lb). Gorillas live in groups and feed on juicy shoots, fruit and other vegetable matter. Habitat: the forests of equatorial Africa.

Gorsachius A genus of Birds of the Bittern, Egret and Heron family, *Ardeidae*. e.g. G. goisagi - The Japanese Night Heron, a timid forest bird, rather like a Bittern in appearance (see **Botaurus**).

Gortyna A genus of Insects of the order of Butterflies and Moths, *Lepidoptera*. e.g. G. micacea - The Rosy Rustic Moth; the larvae feed on a very wide range of hosts, including wild plants like docks, and cultivated ones such as potatoes, sugarbeet, marigolds and cereals.

Gossypium A genus of plants of the Mallow family, *Malvaceae*. e.g. G. herbaceum - The Levant Cotton plant, a stout herb or shrub with alternate, lobed leaves. The flower, which may be purple, yellow or white, has five large bracts surrounding the entire calyx and a funnel-shaped corolla. The fruit is a capsule which on drying, opens to expose the long hairs of the numerous small seeds. From the seed is extracted Cottonseed Oil (= Oleum Gossypii seminis; Ol Gossyp Sem; Cotton Oil). Emulsions of cottonseed oil are given intravenously in the treatment of severe nutritional deficiency. The Cotton plant, of course has its worldwide place in commerce, industry and medicine.

Goura A genus of Birds of the Dove and Pigeon family, *Columbidae*. e.g. G. coronata - The Crowned Pigeon, a large and very attractive bird, 75 cm (2 1/2 ft) long, bluish-grey in color, and with a tall head-crest of feathers. A resident bird; during the day it keeps to the ground and only perches in trees for the night. Distribution: northwestern New Guinea and some of its neighboring islands.

Gracilaria A genus of plants, the Lichens. e.g. G. lichenoides - Ceylon Moss, a seaweed, one of the sources of agar.

Gracula A genus of Birds of the Oxpeeker and Starling family, *Sturnidae*. e.g. G. religiosa - The Talking Mynah, Hill Mynah or Indian Grackle, a widely distributed oriental species, the size of a Jackdaw, frequently imported into western countries as a cagebird. It is black with a white patch on the wings, and a thick orange bill; the yellow legs are short. It is very attractive and has a long, melodious warble. Some mynahs may be taught to talk, hence the name. The Hill Mynah lives in flocks and breeds in cavities in trees.

Graellsia A genus of Insects of the Emperor or Giant Silkworm Moth family, *Saturniidae*. e.g. G. isabellae - A beautiful Saturnid butterfly, pale greenish-blue in color with reddish-brown nervures in the wings. The hind wings are protracted into "tails". The caterpillars dwell on pine trees in the mountain districts of Spain and in the Near East.

Grahamella A genus of microorganisms of the order *Rickettsiales*. e.g. G. talpae - A species infecting moles.

Grallina A genus of Birds of the Magpie-Lark family, *Grallinidae*. e.g. G. cyanoleuca - The Mudlark of Australia; black and white in color. They live along the edges of muddy lakes. Their nest is a mudbowl, strengthened with hair and fur, and lined with feathers and grass.

Grammistes A genus of Fishes of the Grouper and Sea bass family, *Serranidae*, a family of tropical fish common in warm seas, such as the Caribbean and Red Seas. e.g. G. sexlineatus - The Sixline Grouper, a solitary grouper fish with large eyes, particularly active at night, and dwelling at the entrance to small caves in coral reefs. A predator preying only on live fish. Distribution: the coral reefs of Eilat (Akaba) in the Red Sea.

Grammostola A genus of Arachnids of the order of True Spiders, *Araneae*. e.g. G. mollicoma - A giant spider from Brazil, with a legspread measuring as much as 25 cm (10 in).

Grampus A genus of Mammals of the Grampus Whale and Dolphin family, *Delphinidae*. e.g. G. griseus - Risso's Dolphin, not to be confused with the Grampusor Killer Whale, *Orcinus* orca (of the same family), because of its scientific name. It too has a rudimentary beak under a rounded forehead, and long pointed flippers. It is up to 4 m (13 ft) long and lives in large schools, sometimes coming ashore and doing much damage to fishing nets. Habitat: off the coasts of Europe, in the Gulf Stream; from colder regions some migrate to African waters during winter (see **Orcinus**).

Grantia A genus of Sponges of the phylum *Parazoa*. e.g. G. compressa - The Purse Sponge, about 2.5-5 cm (1-2 in) high; grows between tidemarks; a marine animal with non-sexual reproduction, grown for commercial use by artificial propagation of "seed sponges" (mature sponges cut into a number of small ones and returned to the sea).

Graphidostreptus A genus of Arthropods of the class of Millipedes, *Diplopoda*. e.g. G. gigas - The largest known species of millipedes of Africa, measuring up to 28 cm (11 in) long and 2 cm (0.78 in) broad.

Graphiurus A genus of Mammals of the Dormouse family, *Muscardinidae*. e.g. G. ocularis - The Cape Dormouse; it has a bold pattern of black and white on the face. Habitat: Africa, from the Cape to Somaliland and Senegal.

Graphoderus A genus of Insects of the family of True or Carnivorous Water Beetles, *Dytiscidae*. e.g. G. cinereus - A species of Water Beetles found in Britain, occurring only in East Anglia; 15 mm long; colored black; found in pools and freshwater ponds, causing considerable damage in fish hatcheries, and feeding on many types of prey. The beetles trap air for breathing under their wing-cases, coming to the surface tail-first to do so.

Grapholitha A genus of Insects of the Moth family, *Eucosmidae* or *Olethreutidae* (order *Lepidoptera*). e.g. G. molesta - The Oriental Fruit Moth, a serious pest, damaging peaches in the United States.

Graphosoma A genus of Insects of the family of Bugs, *Pentatomidae* (order of True Bugs, *Heteroptera*). e.g. G. lineatum - The Linear Striped Bug, length 9-11 mm; with characteristic longitudinal black stripes on a red ground color. Warmth loving, found on flowering Umbelliferae (Carrot family) and other plants. Distribution: southern and central Europe, Asia Minor, Middle East.

Graptemys A genus of Reptiles of the Tortoise and Turtle order, *Chelonia* (= *Testudines*). e.g. G. pseudogeographica - The False Map Turtle, an aquatic form from the United States.

Graptodytes A genus of Insects of the Carnivorous Water or True Water Beetle family, *Dytiscidae*. e.g. G. pictus - Length 2.3 mm; found in pools with rich vegetation. Distribution: much of Europe.

Gratiola A genus of plants of the Figwort or Snapdragon family, *Scrophulariaceae*. e.g. G. officinalis - The Hedge-hyssop, a perennial plant, 14-40 cm high, with 5-petalled light-pink flowers; used as an emetic, diuretic and purgative. Place of origin: northeast and central Europe.

Gregarina A genus of Sporozoans or Spore-forming animals of the Cephaline suborder, *Cephalina*. The body of the trophozoite (cephalont), which may be up to 400 µ long, has a marked knob (epimerite) on the anterior tip and the remaining part is two-jointed. The species are common parasites in the digestive tract and body cavities of insects. They have been found in some cancers in humans (see **Monocystis**).

Grevillea A genus of plants of the family *Proteaceae*. e.g. G. sulphurea - An Australian shrub, up to 1.8 m (6 ft) tall, with narrow leaves and terminal clusters of pale yellow flowers.

Greyia A genus of plants of the family *Melianthaceae*. e.g. G. sutherlandii - An attractive South African species, with toothed leaves and dense pyramidal spikes of flowers.

Grifolia A genus of Molds or Fungi of the class *Basidiomycetes*. e.g. G. sulphurea - This species of fungi attacks the wood of deciduous trees. The sporophores are produced in groups on the tree trunks. Each is thin and fleshy, with a wavy margin and basidia line the shallow pores on the under surface.

Grindelia A genus of plants of the Daisy family, *Compositae*. e.g. G. camporum - The Gumweed (= Gum Plant; Tar weed). It contains the extract grindelia. Used as an antispasmodic in the treatment of asthma and bronchitis.

Grison A genus of Mammals of the Badger, Otter, Skunk and Weasel family, *Mustelidae*. e.g. G. vittata - The Grison, just over 40 cm (16 in) long, with a 20 cm (8 in) tail. It is grey above and dark brown underneath. Unlike most of the Mustelidae, it is not a climber but keeps to the ground.

Groenlandia A genus of plants of the Pondweed family, *Potamogetonaceae*. e.g. G. densa - Opposite-leaved Pondweed, a perennial plant, up to 1 m (40 in) high; it has opposite leaves and the spike has usually only 4 flowers. It grows in freshwater ponds and in still and slow-flowing water. Place of origin: widespread in Europe; native in Britain.

Grubyella A genus of Molds or Fungi, also known by the generic name, *Microsporon*. e.g. G. ferruginea (= *Microsporon* ferrugineum) (see **Microsporon**).

Grus A genus of Birds of the Crane family, *Gruidae*. e.g. G. grus - The Common Crane, 1.2 m (4 ft) high, with a wingspan of over 21/2 m (8 ft), and weighing over 41/2 kg (10 lb). Its plumage is ashen grey, except for the forehead, nape, throat, remiges and rectrices, which are black. On top of the head is a bald red patch. A migrating bird, as winter approaches, it can be seen flying with the flockin V-formation, heading for Europe (its natural habitat), for the Mediterranean and northeast Africa. These cranes are found in lowland bogs and barren swamps in woods where they nest. Distribution: Europe; Asia as far as eastern Siberia.

Grylloblatta A genus of Insects of the order *Grylloblattodea*. e.g. G. campodeiformis - A species discovered as recently as 1914 in the Canadian Rockies; a pale yellowish, wingless insect, 15-30 mm long, found under stones in high mountainous regions where the temperature is often near freezing point. It lays eggs in the soil or among moss, and takes about 5 years to develop. These creatures are known as "living fossils", relics of an early stage in Orthopteroid evolution (from the Grasshopper and Locust order, *Orthoptera*).

Gryllotalpa A genus of Insects of the Mole Cricket family, *Gryllotalpidae* (order of Crickets and Grasshoppers, *Orthoptera*). e.g. G. gryllotalpa - The Mole-cricket, a well-known European garden and field pest; length 35-50 mm; brownish in color; adapted to life underground (front pair of legs are modified as burrowing limbs or digging shovels), but surfaces during the mating period. Distribution: Europe, western Asia, North Africa, North America (introduced). Rare in Britain.

Gryllus A genus of Insects of the True Cricket family, *Gryllidae*. e.g. G. campestris - The Field Cricket, length 20-26 mm; head black, wings yellow at base; the adult lives in underground chambers which it excavates in warm places, such as meadows and hedgerows. The males give "concerts" in the fields and dry hillsides, stridulating most intensely during the mating period. Distribution: central and southern Europe, southern Great Britain, western Asia, North Africa.

Grystes A genus of Freshwater Fishes of the Perch family, *Percidae*. e.g. G. nigricans - The Small-mouthed Black Bass, length 45 cm; it has small scales and a small mouth, extending to under the eyes. The pelvic fins are set far forward. A commercially valuable, freshwater fish, introduced into Europe from America and now well established in ponds and lakes.

Guaiacum A genus of Trees of the family *Zygophyllaceae*. e.g. G. officinale - The Guaiacum Tree. Guaiacum Wood or Heartwood contains guaiacum resin, and has been used in the treatment of asthma and rheumatism.

Guaree A genus of Shrubs of the family *Meliaceae*. e.g. G. rusbyi - The Coccillana (= Grape Bark; Guapi Bark; Huapi Bark) - used as an expectorant.

Gulo A genus of Mammals of the Badger, Otter, Skunk and Weasel family, *Mustelidae*. e.g. G. gulo (= G. luscus) - The Wolverine or Glutton; length of body over 90 cm (3 ft), tail 18 cm (7 in). It has a short-legged, thick-set body, dark brown fur but paler at the flanks, whitish frontlet and yellowish back markings. The largest and fiercest member of the weasel family, it weighs up to 13.5 kg (30 lb). It feeds chiefly on rodents such as lemmings, but it also attacks hares, foxes, martens, various birds and fishes. It has a voracious appetite, also eating plant food, including various fruits and well deserves the name glutton. Distribution: the northern regions of Europe, Asia, North America.

Guzmania A genus of plants of the Pineapple family, *Bromeliaceae*. e.g. G. cardinalis (probably a natural hybrid of G. sanguinea) - The Guzmania plant, a handsome epiphyte, growing on another plant but physiologically independent, and producing rosettes of evergreen leaves arranged to form a vase-like interior. There are flowering stems with clusters of golden florets at their tops, surrounded by conspicuous scarlet bracts. Place of origin: Colombia.

Gymnadenia A genus of plants of the Orchid family, *Orchidaceae*. e.g. G. conopsea - The Fragrant Orchid, a perennial plant, 10-60 cm high; lip bluntly 3-angled; spur about twice as long as the ovary; scented flowers. Grows commonly in meadows and on grassy hills. Place of origin: widespread in Europe; native in Britain.

Gymnema A genus of Trees. e.g. G. sylvestre - A species found in Africa. The leaves are used to disguise the taste of unpleasant medicines.

Gymnocalycium A genus of plants of the Cactus family, *Cactaceae*. e.g. G. saglionis - The Gymnocalycium, a large, round cactus plant, up to 30 cm (12 in) across, with numerous domed ridges with woolly centers, each having 7 to 12 curved browny-black spines which turn white with age. Pinkish-white flowers with red stamens. Place of origin: Argentine, Bolivia.

Gymnocarpium A genus of plants of the Polypody or Fern family, *Polypodiaceae*. e.g. G. dryopteris - The Oak Fern, a perennial plant, 5-40 cm high; 2-pinnate fronds with lowest pair of pinnae 3-angled. Grows mainly in damp woodlands. Place of origin: Europe; native in Britain.

Gymnocephalus A genus of Fishes of the Darter, Perch and Walleye family, *Percidae*. e.g. G. cernua - The Ruffe of Pope, length 10-15 cm; common in the lower reaches of rivers, in dam reservoirs and in ponds. Feeds on smallfish and invertebrates. Distribution: all of Europe.

Gymnodactylus A genus of Reptiles of the Gecko family, *Gekkonidae*. e.g. G. miluisi - The Naked-toed Gecko, a lizard found in the deserts of Australia, with slender, pointed digits. Its body and head are very dark, with a sprinkling of light-colored spots. It runs well up on its feet in a cat-like fashion.

Gymnodinium A genus of Protozoans of the order of Dinoflagellates, *Dinoflagellata*. Known as a "naked" dinoflagellate animal, it is probably the simplest of this order, without sculptured plates. The species are found in lakes, freshwater ponds and also in the sea. Some are green and holophytic, that is, they make complex organic substances by photosynthesis. Some species have been found off the Florida coast in such large numbers that the fish eating them, have been poisoned. This species is reddish and the plague is known as the "red tide".

Gymnogyps A genus of Birds of the New World Vulture family, *Cathartidae*. e.g. G. californianus - The California Condor, the largest American bird, averaging 2.74 m (9 ft) in wing expanse and weighing up to 9 kg (20 lb); now a rare species, numbering today at an estimated 60 individuals. It incubates a single egg.

Gymnothorax A genus of Teleostean or True Bony Fishes of the family of Moray Eels, *Muraenidae*. e.g. G. grisea - A Moray Eel species, attaining a length of over 2 m (61/2 ft), and although resembling a snake it is a fish. The teeth are very strong and sharp and the mouth has a vice-like grip. It lives in crevices on coral reefs, but sometimes also on the seabed. Habitat: the coral reefs of the Gulf of Eilat (Akaba) and the Red Sea.

Gynerium A genus of plants, better known by the generic name, *Cortaderia*, of the Grass family, *Gramineae*. e.g. G. argenteum (= *Cortaderia* selloana; C. argentea) - Pampas Grass (see **Cortaderia**).

Gynocardia A genus of plants, better known by the generic name, *Chaulmoogra*, of the family *Bixaceae*. e.g. G. odorata (= *Chaulmoogra* o.) (see **Chaulmoogra**).

Gypaëtus A genus of Birds of the Old World Vulture, Hawk and Harrier family, *Accipitridae* (Vulture subfamily, *Gypaëtinae*). e.g. G. barbatus - The Lammergeier or Bearded Vulture, the single species of the subfamily, a resident bird, and unlike other vultures, it has feathers on the head and neck, and an eagle-like bearing. It measures up to almost 2.7 m (9 ft) with expanded wings. Its name Gyps (vulture) and aetos (eagle) emphasises that it is an eagle that has taken to eating carrion. It lives alone or in pairs, and nests in remote, rocky regions in high mountains. It is gradually disappearing from Europe through persecution by man, but still inhabits Africa and Asia as far as western China.

Gyposophila See **Gypsophila**.

Gyps A genus of Birds of the Old World Eagle, Harrier, Hawk and Vulture family *Accipitridae* (Vulture subfamily, *Aegypiinae*). e.g. G. fulvus - The Griffon Vulture, length 1 m (3 1/4 ft), the second largest vulture in Europe (the largest is the Black or Hooded Vulture, *Aegypius* monachus); it has a wingspan of about 2.4 m (8 ft). The plumage is a lightish sandy-brown. A carrion-eater, it nests in holes and on rocky ledges. A resident bird; though it prefers rocky areas, it is also found in the plains. Habitat: southern Europe, North Africa and western Asia.

Gypsophila A genus of plants, also known by the generic name, *Gyposophila*, of the Pink family, *Caryophyllaceae*. e.g. G. muralis (= *Gyposophila* m.) - The Wall Gypsophila or Chalkwort, an annual plant, 4-18 cm high; pink flowers with dark veins. Place of origin: Europe; introduced and naturalized in Britain. G. paniculata (= *Gyposophila* p.) - Maiden's Breath, an annual plant bearing tiny white flowers on thread-like stalks, and containing saponaria or saponin, known as Saponariae Allbae Radix (Hungarian Pharm) - used in medicine as an expectorant and diuretic.

Gyrinocheilus A genus of Fishes of the Gyrinocheilid family, *Gyrinocheilidae*; an unusal family characterized by having the upper part of the gill slit separated from the lower half by a partition of skin, water reaching the gills through the upper opening and passing out by the lower. These aperatures are so small that the respiratory movements are very rapid (as many as 250 per minute), thus ensuring an adequate supply of oxygen. The species may reach a length of 45 cm (11/2 ft), and are found in the torrential streams of southeast Asia and Borneo.

Gyrinophilus A genus of Amphibians of the Salamander family, *Plethodontidae*. These species of newts are found in the United States, and are entirely neotenic (retaining their larval characteristics).

Gyrinus A genus of Insects of the family of Whirligig Scavenging Beetles, *Gyrinidae*. e.g. G. natator - The Whirligig Beetle, length 5-7 mm; the eye is divided by a partition into 2 parts, an upper part for seeing over the surface and lower for seeing down into the water. It has two long prehensile legs and four short "swimming" legs. Occurs abundantly on the surface of ponds. Distribution: Europe, Siberia, Mongolia, North Africa. Native in Britain.

Gyrocotyle A genus of Monozoic Cestode worms of the order *Gyrocotylidea*. These are small animals having a deep, funnel-like organ with a frilled margin at one end of the body and a sucker at the other. The larva or lycophore has 5 pairs of hooks, distinguishing them from the tapeworm larvae which have 3 pairs.

Gyrodactylus A genus of Flukes or Flatworms of the class *Trematoda*. The species belong to the monogenetic order of trematodes, *Monogenea*, in which the fertilized egg develops into an embryo within the body of the parent worm.

Gyromitra A genus of Molds or Fungi of the Helvel family. e.g. G. esculenta - The Spring Helvel; the cap is up to 12 cm (5 in) wide, tawny thin coffee-brown, and irregularly undulated. Grows in coniferous woods. Place of origin: Europe; native in Britain.

Gyropus A genus of Lice. e.g. G. ovalis - A biting louse found on guinea-pigs.

Gyrostoma A genus of Anthozoan Celenterates of the order of Sea Anemones, *Actiniaria*. The species are stinging animals with beautiful glowing pink colors. Most fish avoid any contact with them except for the tiny clownfish (genus *Amphiprion*) that lives in close symbiotic relationship, completely immune to their paralysing sting. Habitat: the Red Sea.

H

Haberlea A genus of plants of the family *Gesneriaceae*. e.g. H. rhodopensis - A small, softly hairy perennial plant with thick leathery, ovate leaves. The calyx is 5-toothed, and the pale mauve corolla has 5 unequal lobes. The 4 stamens are included within the corolla tube and bear from 2 to 5 flowers. A white-flowered form, var. virginalis, is also known. Place of origin and distribution: a small rocky area of Thrace in the southeastern Balkan peninsula (shared by Greece, Turkey and Bulgaria since 1878).

Habronema A genus of Nematode Worms parasitic to horses. e.g. H. megastoma - A species causing the disease habronemiasis in horses.

Hacquetia A genus of plants of the Carrot family, *Umbelliferae*. e.g. H. epipactis - The Hacquetia plant, a small perennial shrub, 10-25 cm high, with yellow flowers growing in a single head-like umbel, surrounded by 5 to 8 serrate bracts. Found in the deciduous woods of Europe, in the eastern Alps and West Carpathians.

Hadogenes A genus of Arachnids of the order of Scorpions, *Scorpionida*. e.g. H. troglodytes - A very large species, the males measuring up to 17.5 cm (7 in), found in South Africa.

Haemadictyon A genus of plants of the Periwinkle family, *Apocynaceae*. e.g. H. amazonicum - This species yields the alkaloid yageine, identical with the alkaloids banisterine, harmaline, harmine and telepathine, which act as psychostimulants and hallucinogens (see **Banisteria**; **Peganum**; **Tetrapteris**).

Haemadipsa A genus of Segmented Worms or Annelids of the class of Leeches, *Hirudinea*. e.g. H. ceylonica - A species of terrestrial or land leeches, common in Sri Lanka (Ceylon), annoy-ing to man and animals because of its painful bite.

Haemagogus A genus of Mosquitoes transmitting yellow fever in the jungles of South America.

Haemanthus A genus of plants of the Daffodil family, *Amaryllidaceae*. e.g. H. puniceus - The Blood Flower or Blood Lily, a handsome plant with round, thick bulbs about 5 cm (2 in) across by 5-10 cm (2-4 in) long; wavy-edged and nerved, brilliant green leaves and dense round heads of bright scarlet flowers with long anthers on 30 cm (1 ft) stems. Place of origin: South Africa.

Haemaphysalis A genus of Arachnids of the Tick subfamily, *Ixodoidae*. e.g. H. humerosa - The Bandicoot tick, one of the vectors of *Coxiellaburnetii*. H. leachi - The common Dog tick of South Africa, transmitting the protozoan parasite of tick-bite fever and malignant jaundice.

Haemapium A genus of plasmocytes. e.g. H. riedyi - A species found in the erythrocytes of salamanders.

Haematobium A genus of Insects of the order of Two-winged Flies, *Diptera*. e.g. H. irritans - The Horn Fly, a species troublesome to cattle.

Haematomyzus A genus of Insects of the order of Biting Lice, *Mallophaga* (suborder *Rhynchophthirina*). The species are wingless, rather flat insects with bristly bodies and without compound eyes. They are ectoparasites found on elephants and warthogs.

Haematopinus A genus of Insects of the order of Sucking Lice, *Anoplura*. The species infest cattle, horses and swine. e.g. H. suis - The Swine Louse, a large, wingless, parasitic insect, about 0.5 cm (1/5 in) long, with rudimentary eyes, and mouthparts adapted for sucking or biting. It infests the skin of the domestic pig.

Haematopota A genus of Insects of the family of Horse Flies, *Tabanidae*. Also known by the generic names,

Hematopota and *Chrysozona*. e.g. H. pluvialis - The Clegg, Horse-fly or Rain fly, length 8-12 mm; the wings have white spots. Found on pathways and in forests; the female attacks warm-blooded animals such as horses (also man), and sucks their blood. Distribution: the Palaearctic region (see **Chrysozona**).

Haematopus A genus of Birds of the Oyster catcher family, *Haematopodidae*. e.g. H. ostralegus - The European Oyster catcher or Sea-pie, 50 cm (20 in) long, black and white, with a long orange-red beak and pink legs and feet. It has a loud, shrill cry. A coastal bird, partly migrant, it feeds on shellfish such as mussels, limpets, and more rarely oysters. A long-living creature, it has been known to reach an authenticated age of over 35 years. Its name comes from its habit of decorating its nest, a little hollow on the shore, with fragments of seashells or oysters. In some parts of England it is known as the Musselpecker. Distribution: the coasts of all continents, including Europe.

Haematoxylon A genus of plants of the Pea family, *Leguminosae*. e.g. H. camphechianum - The Heartwood or Logwood plant. It contains haematoxylin and tannin, and has been used as an astringent in diarrhea. It is now used as a red dye in the textile, fur, and leather industries.

Haementaria A genus of Segmented Worms or Annelids of the class of Leeches, *Hirudinea*. Also known by the generic name, *Hementaria*. e.g. H. officinalis (= *Hementaria* o.) - A species used commonly for medicinal purposes in Central and South America.

Haemobartonella A genus of microorganisms of the order *Rickettsiales*; these are parasitic organisms affecting lower animals. e.g. H. canis - A species infecting dogs.

Haemodipsus A genus of Insects of the order of Sucking Lice, *Anoplura*. e.g. H.

ventricosus - The sucking louse transmitting tularemia in rabbits.

Haemogregarina A genus of Sporozoan parasites (order *Eucoccidia*, suborder *Adeleidea*), found in the blood cells of amphibians, reptiles (such as turtles), and segmented worms (such as leeches).

Haemonchus A genus of Nematode worms. e.g. H. contortus - The Sheep Wire Worm; it may also infect humans.

Haemophilus A genus of microorganisms of the family *Brucellaceae*. Also known as *Hemophilus*; many species are classified under other generic names such as *Bordetella*, *Corynebacterium*, *Moraxella*, etc. These are minute gram-negative rods, most of whom are part of the normal mucosal flora of the upper respiratory tract. In children, they commonly cause sinusitis, nasopharyngitis, conjunctivitis, acute epiglossitis (croup), obstructive laryngitis (stridor), pneumonia, meningitis, etc. They grow best in the presence of hemoglobin. e.g. H. aegyptius (= H. conjunctivitidis; Koch-Weeks bacillus) - A species causing a purulent and highly contagious conjunctivitis (pinkeye). H. bovis (= *Moraxella* b.) - see **Moraxella**. H. ducreyi (= Ducrey's bacillus) - A species producing lesions of chancroid (soft chancre). Also isolated from sputum, cerebrospinal fluid and blood. Usually transmitted by direct contact or by surgical instruments or dressings. H. duplex (= *Moraxella* lacunata; *Haemophilus* of Morax-Axenfeld) -A species causing blepharo-conjunctivitis (see **Moraxella**). H. haemoglobinophilus (= H. canis) - A species found in dogs; can cause human endocarditis. H. influenzae - May be part of the normal respiratory tract; isolated from sputum, urine, cerebrospinal fluid, blood and pus. A species once thought to be the cause of epidemic influenza. The most common cause of bacterial meningitis in infants and young children. Also causes obstructive laryngitis, sinusitis, otitis, epiglossitis, conjunctivitis, etc. H. pertussis (= *Bordetella* p.) - A species causing whooping-cough (see **Bordetella**). H. suis and a virus - Both necessary agents of swine influenza. No known human infections. H. vaginalis (= *Corynebacterium* vaginale) - see **Corynebacterium**.

Haemophoructus A genus of Insects of the Blood-sucking Fly family, *Heleidae*.

Haemopis A genus of Segmented Worms of the class of Leeches, *Hirudinea* (group of jawed leeches, *Gnathobdellae*). The species belong to the Jawed Leech family of the so-called medicinal leeches, *Hirudinidae*. e.g. H. sanguisuga - The Horse Leech, 6-10 cm long; greenish-brown in color. It feeds on the blood of various soft-bodied vertebrates, and is found infesting the nasal passages of horses. Was once used for bloodletting in humans. Found in ditches, ponds and lakes throughout Europe and North America (see **Hirudo**).

Haemoproteus A genus of Sporozoan parasites of the suborder *Haemosporidia*, found in birds and reptiles. e.g. H. columbae - An amoeboid sporozoan parasite, found in the red blood cells of pigeons.

Haemulon A genus of Fishes (subphylum *Vertebrata*). e.g. H. flavolineatum - The French Grunt Fish.

Hagenia A genus of plants of the Rose family, *Rosaceae*, better known by the generic name, *Brayera*. e.g. H. abyssinica (= *Brayera* anthelmintica) - The Cousso plant (see **Brayera**).

Haideotriton A genus of Amphibians of the Salamander family, *Plethodontidae*. These are blind newts found in deep wells and caves in Georgia, U.S.A.

Halarachnion A genus of plants of the Red Algae family, *Gigartinaceae*. e.g. H. ligulatum - Size 15 cm; a strap-shaped plant, ligulate-fringed at the margins. Distribution: the North Sea.

Halcyon A genus of Birds of the Kingfisher family, *Alcedinidae*. e.g. H. smyrensis - The Smyrna Kingfisher, a river and lake bird, with a large, crested head, a dagger-shaped bill and brilliant plumage. Seen near water reservoirs, plantations and humid places. Distribution: Asia Minor.

Halecium A genus of Hydroids or Sea-firs (class *Hydrozoa*), of the order *Siphonophora*. e.g. H. halecinum - The Herring-bone Polyp; size up to 10 cm (4 in); these animals form polypoid, fan-shaped colonies with a sturdy central stem. These Sea Firs are found on the Atlantic Coast, in the North Sea, and more rarely in the West Baltic; occasionally also in the Mediterranean.

Halesia A genus of plants of the Storax family, *Styracaceae*. e.g. H. carolina (= H. tetraptera) - The Carolina Silverbell, Snowdrop Tree, or Silver Bell Tree; a large deciduous shrub or tree up to 9 m (30 ft) high, with ovate-lanceolate, downy leaves and pure white, pendulous, bell-shaped flowers growing in clusters on slender stalks. Place of origin: southeastern United States; introduced into Europe in 1756.

Haliaeetus See **Haliaëtus**.

Haliaëtus A genus of Birds of the Old World Harrier, Hawk, and Vulture family, *Accipitridae* (Eagle and Sea-eagle subfamily, *Accipitrinae*). Also spelled *Haliaeetus*. e.g. H. leucocephalus (= *Haliaeetus* l.) - The Bald Eagle or American Eagle, so-called from its white head and neck, a species resident in America, the national bird or emblem of the United States. It has a large, strong, sharply hooked beak which it uses to kill and dismember its prey. Found now in numerous numbers only in Alaska and Florida. Even in Florida its numbers have markedly dropped in recent years due to indiscriminate hunting and persecution.

Haliastur A genus of Birds of the Old World Harrier, Hawk, and Vulture family, *Accipitridae* (Old World Vulture subfamily *Aegypiinae*). e.g. H. indus - The

Brahminy Kite, a common species found in Australasia. It is a slim bird with long narrow wings, and an elongated tail. By nature it is a scavenger, and although fast in flight, it is not swift enough to catch other birds on the wing.

Halichoeres A genus of Fishes of the Wrasse family, *Labridae*. e.g. H. centriquadrus - The Fourspot Wrasse, a brilliantly, light-colored marine fish, with spiny fins and prominent thick lips, found in abundance in warm coastal waters. The juveniles are often completely different from the adults, both in color and form. Especially frequent in the Red Sea.

Halichoerus A genus of Mammals of the Earless Seal family, *Phocidae*. e.g. H. grypus - The Grey or Atlantic Seal, up to 2.7 m (9 ft) long and weighing about 227 kg (500 lb); a grey and spotted pinniped, the underside being paler. It has flippers modified for swimming, and no external ears. Distribution: the northern coast of Great Britain, off the coast of Scotland to the Orkneys and the Shetlands, the Scilly Isles, and the Faroe Islands.

Halichondria A genus of Gelatinose Sponges of the suborder *Sigmatosclerophora*. e.g. H. panicea - The Siliceous, Bread-crumb or Crumbo'-bread Sponge; size up to 10 cm (4 in). The microscleres or oscula (large pores through which water is pumped out during feeding), are based on C- or S-shapes (tower-shaped). The sponge forms encrusting masses on rock, the bases of seaweed fronds, etc. These animals are common around the coasts of Europe, Australia and the Pacific coast of Canada.

Haliclona A genus of Gelatinose Sponges (phylum *Parazoa*) of the suborder *Sigmatosclerophora*. These animals develop as in the genus *Halichondria*. e.g. H. oculata - Size up to 30 cm (12 in); the stems grow like antlers; the species is found commonly around the

coasts of Europe, and elsewhere as far away as Australia and the Pacific coast of Canada (see **Halichondria**).

Haliclystus A genus of Jellyfishes (class *Scyphozoa*) of the Stalked Jellyfish order, *Stauromedusae*. e.g. H. auricula - A stalked jellyfish with 8 dark patches; generally greenish or rose-pink in color, measuring several centimeters in diameter.

Halicobacter A genus of microorganisms, recently discovered as pathogenic to man. e.g. H. pylori - A species now regarded by some, as being the bacterial cause of duodenal ulcer.

Halictus A genus of Insects of the Bee family, *Apidae*. e.g. H. quadricinctus - Length 15 cm (6 in); a long, narrow, dark body with clearly marked abdominal segments. After mating the females of the species lead a solitary life and build simple or branched nests in the ground or in soil banks. Distribution: Europe.

Halidrys A genus of plants of the family of Brown Algae, *Fucaceae*. e.g. H. siliquosa - The Sea Oak, size up to 3 m (10 ft); with pod-shaped, mucronate air vesicles. Found in the seas of Europe and around Britain.

Halieutaea A genus of Fishes (subphylum *Vertebrata*). e.g. H. retifera - The Torpedo Bat-fish.

Halimione A genus of plants of the Goosefoot cm high; with stalkless fruit and usually 3-lobed bracteoles. Grows in salt-marshes and wet meadows by the North Sea and the Baltic. Native around Britain.

Haliotis A genus of Molluscs of the family of Ear Shells, *Haliotidae*. e.g. H. tuberculata - The Green Ormer or Sea Ear, a marine gastropod; the shell is 80-90 mm long, 60 mm high, flat, and marked with radial and concentric grooves. At the outer edge is a row of about 5-7 holes through which the animal extends tactile filaments. The inside of the shell is lined with a thick layer of mother-of-pearl. This snail has a broad foot, which is furnished with a large number of lateral filamentous outgrowths. Found in shallow water where it adheres firmly to the underside of stones. An edible species, gathered mainly for the valuable mother-of-pearl. Distribution: the Atlantic coast, Mediterranean Sea, Canary Islands and Azores. Does not occur in the North Sea.

Halistemma A genus of Hydras, Hydroids or Sea Firs, and Siphonophores (class *Hydrozoa*), of the order *Siphonophora*. These siphonophores form swimming or floating colonies composed of both polypoid and medusoid forms. These animals are found in all seas, but prefer warmer waters. Relatively small and transparent, they are often unnoticed.

Halobacterium A genus of Halophilic (requiring salt for optimal growth) microorganisms of the order *Pseudomonadales*. They occur as rod-shaped cells. e.g. H. salinarium. H. halobium - A species producing bacteriorhodopsimarine gerrid found far from land on the larger oceans.

Haloporphyrus A genus of deep-sea Fishes; the species have been photographed at depths of 7,010 m (23,000 ft).

Hamamelis A genus of plants of the Witch-hazel family, *Hamamelidaceae*. e.g. H. virginiana - The North American Witch-hazel, a large to 4.5 m (15 ft) high, bearing yellow flowers with narrow, crumpled petals. The dried leaves and bark yield an extract containing tannin and gallic acid. The bark contains a natural sugar hamamelose. Used as an astringent and hemostatic, mainly in the treatment of hemorrhoids. Place of origin: eastern North America.

Hammarbya A genus of plants of the Orchid family, *Orchidaceae*. e.g. H. paludosa - The Bog Orchid, a perennial plant, 5-15 cm high; the flowers grow in many-flowered racemes, with acute, erect lips with the bases clasping the column. Found on moorlands and in peat bogs, mostly in North and Central Europe. Native in Britain.

Hapalochlaena A genus of Marine Cephalopod Molluscs of the Octopus family, *Octopodidae*. Better known by the generic name, *Octopus*. e.g. H. lanulata (= *Octopus* rugosus) - The Blue-ringed Octopus (see **Octopus**).

Hapalodermes A genus of Birds of the Trogon family, *Trogonidae*. These trogons are essentially forest-dwelling birds, characterized by sexual dimorphism. They lead solitary lives, making little noise and concealing themselves in the densest and darkest places. Distribution: the forests of Africa.

Haplobothrium A genus of Merozoic Cestode tapeworms (subclass *Eucestoda* or *Cestodamerozoa*), of the order *Haplobothrioidea*. e.g. H. globuliforme - A strobilate cestode tapeworm, a parasite of the Bowfin fish, *Amia* calva (see **Amia**).

Haplochilus A genus of Fishes (subphylum *Vertebrata*). e.g. H. panchax - A small fish used in fishponds to eat the larvae of *Anopheles* mosquitoes, in Southeast Asia.

Haplopterus A genus of Birds of the Plover family, *Charadriidae*. e.g. H. spinosus - The Spur-winged Plover, length 26.5 cm; plumage black and white, with spurs on the wings. A migrant bird from North Africa and Asia; nests from time to time in Greece.

Haplosporangium A genus of Fungi or Molds. e.g. H. parvum - A fungus causing granulomatous lesions in the lungs of rats.

Harpactes A genus of Birds of the Trogon family, *Trogonidae*. A sexually dimorphic species, the male and female are distinctly and beautifully colored. Distribution: the forests of Malaysia (see **Hapalodermes**).

Harpalus A genus of Insects of the

Ground Beetle family, *Carabidae*. e.g. H. rufipes (= H. pubescens) - A black-colored species, 14-16 mm long; the elytra are thickly covered with yellow-ish hairs. Common in fields, on field paths and under stones. Distribution: the Palaearctic region.

Harpia A genus of Birds of the Old World Harrier, Hawk, and Vulture fami-ly, *Accipitridae* (Eagle and Sea-eagle subfamily *Accipitrinae*). e.g. H.harpyja - The Harpy Eagle of South America, the largest, heaviest, and most powerful eagle in the world, and an aggressive predator; up to 9 kg (20 lb) in weight. The wings are short but very broad, with a span of more than 1.83 m (6 ft). Distribution: South America from Mexico to eastern Bolivia, southern Brazil and northern Argentina.

Harpium A genus of Insects of the Long-horned or Longicorn Beetle family, *Cerambycidae*. e.g. H. inquisitor - A long, brownish-grey colored species, 2 cm (4/5 in) long, with two dark, oblique stripes on its wings. Found most com-monly in coniferous forests of Central Europe and North America. The larvae develop under the bark of dead conifer-ous trees. Before turning into pupae, they build a kind of cradle out of dry pine needles in which the pupa lives until the following spring.

Harpodon A genus of Fishes of the Lanternfish family, *Myctophidae*. The species of this genus are the Bunmallow fishes, found in the Indian Ocean; they are brilliantly phosphorescent all over, having specific light-producing organs. They have large mouths and arrow-shaped teeth. When dried these fishes are called by the well-known name, "Bombay ducks".

Harpyrynchus A genus of Mites para-sitic on birds.

Harrimania A genus of Chordate ani-mals of the class of Acorn Worms, *Enteropneusta*. They have a worm-shaped body and an acorn-shaped pro-boscis. They occur in all seas and live in sand or mud, ranging from 5-50 cm (2-20 in) in length (see **Balanoglossus**).

Hartmanella A genus of free-living Protozoans, the Amoebae, of the class *Rhizopoda*. e.g. H. hyalina - A coprozoic amoeba found in human feces. May cause meningoencephalitis.

Hartmannia A genus of plants of the family *Onagraceae*, better known by the generic name, *Oenothera*. e.g. H. spe-ciosa (= *Oenothera* s.) - The Showy Evening Primrose or White Sundrop plant (see **Oenothera**).

Haverhillia A genus of microorganisms of the family of gram-negative strepto-bacilli, *Bacteroidaceae*. Better known today by the generic name, *Streptobacillus*. e.g. H. multiformis (= *Streptobacillus* moniliformis), now called S. murisratti, found in cases of Haverhill fever, with bacteremia, abscesses, arthritis and endocarditis, and what is also known as Erythema arthriticum epidemicum (see **Streptobacillus**).

Hebeclinium A genus of plants of the Daisy family, *Compositae*, better known by the generic name, *Eupatorium*. e.g. H. ianthinum (= *Eupatorium* i.) - The Eupatorium Sordidum plant (see **Eupatorium**).

Hedera A genus of plants of the Ivy fam-ily, *Araliaceae*. e.g. H. helix - The Ivy plant or "Gold Heart", a woody ever-green climber with adventitious roots, flowers in umbels and blue-black berries. Characterized by a conspicuous dash of golden-yellow at the center of each lobed, heart-shaped leaf, hence the name. Place of origin: Europe, Asia Minor; native in Britain. Widely planted in North America.

Hedychium A genus of plants of the Ginger family, *Zingiberaceae*. e.g. H. gardnerianum - The Ginger Lily, an erect, rhizomatous perennial plant, up to 1.8 m (6 ft) high, with large, stiff stems and large, sessile or stalked, lanceolate leaves up to 37 cm (15 in) long. Pale yellow flowers in bold spikes, with red filaments to the stamens. Place of origin: Northern India; introduced into Europe in 1819.

Hedysarum A genus of plants of the Pea family, *Leguminosae*. e.g. H. hedysaroides - The Alpine Hedysarum, a perennial plant, 1-25 cm high; the leaves have 5-9 pairs of leaflets; there are dense racemes with jointed pods. Grows in mountainous regions. Place of origin: South and Central Europe.

Helarctos A genus of Mammals of the Bear family, *Ursidae*. e.g. H. malayanus - The Malayan Bear or Sun Bear; it has a black coat with a yellowish chest patch. An extremely active mammal, it is very agile in climbing trees. It is just over 1.2 m (4 ft) tall, and not dangerous to man. Distribution: Southeast Asia, including Burma, Malaya, Thailand, Indo-China and parts of Indonesia.

Helcosoma A genus of flagellate protozoan parasites, now known by the generic name, *Leishmania*. e.g. H. tropicum (*Leishmania* tropica) - A species causing cutaneous leishmaniasis (see **Leishmania**).

Helenium A genus of plants of the Daisy family, *Compositae*. e.g. H. autumnale - The Sneezeweed or Sneeze Wort, a coarse, herbaceous, perennial plant, 1-1.8 m (4-6 ft) high; with smooth, stout, leafy and winged stems, branched near the tops. The leaves are alternate, entire and lanceolate, and the flowers plentiful, rayed, yellow, with darker disk-florets. There are many hybrids with copper-orange, clear yellow and red-crimson flowers. Place of origin: North America; introduced into Europe in 1729.

Heliactin A genus of Birds of the Hummingbird family, *Trochilidae*. e.g. H. cornuta - A hummingbird species of tropical South America, a minute and beautiful creature. The rapid vibration of its wings, the fastest recorded wingbeat of any bird, with a frequency of 90 beats a second, produces a humming sound and allows it to hover, and feed on insects and nectar, while on the wing.

Helianthemum A genus of plants of the Rock-rose family, *Cistaceae*. e.g. H. chamaecistus (= H. nummularium) - The Common Rockrose or Sun-rose, a spreading, semi-shrubby, evergreen perennial plant, 30-60 cm (1-2 ft) high; it has a woody stem, opposite leaves with lanceolate stipules and yellow to rose, coppery red and crimson flowers. Widespread on basic grassland and in scrub. Place of origin: Europe, including Britain.

Helianthus A genus of plants of the Daisy family, *Compositae*. e.g. H. annuus - The Common Sunflower, a stout, hairy, annual plant, from 0.9-3.6 m (3-12 ft) high, with broadly ovate leaves up to 30 cm (1 ft) long. The terminal capitulum, produced between July and September, is up to 30 cm (1 ft) across, and even broader in cultivation. It has several rows of involucral bracts and a large number of yellow ray florets. From the fruit is expressed Sunflower Oil (= Oleum Helianthe; Sunflower-seed Oil), used in food and pharmaceutical preparations. Place of origin: a native of the western United States; introduced into Europe in 1596. H. tuberosus - The Jerusalem Artichoke, a herbaceous plant, 2-3 m (7-10 ft) high; with well-stemmed, almost all opposite, oval leaves with large teeth; capitula with yellow peripheral ligules in a row, and a similarly yellow disc. From the root is prepared Inulin or Alant Starch. Place of origin: North America; introduced into Europe in 1617 (see **Inula**).

Heliaster A genus of Echinoderm animals of the class of Starfishes, *Asteroidea*. e.g. H. helianthus - A species found on the coast of Chile. These echinoderms have separate sexes and reproduce by means of eggs which are fertilized in the water. They are predators and cause considerable damage in oysterbeds.

Helicella A genus of Gastropod Molluscs of the Snail family, *Helicidae*. e.g. H. itala (= H. ericetorum) - The Heath Snail; shell 7-8 mm wide; whitish-yellow, usually marked with pale brown, translucent bands. Found on dry hillsides, chiefly with a calcareous substrate. Distribution: western and central Europe.

Helichrysum A genus of plants of the Daisy family, *Compositae*. e.g. H. sanguineum - An "Everlasting", perennial plant, known in Israel as the "Blood of the Maccabeans"; about 30 cm high; a protected wild species, with everlasting red flowers blooming in the hills in late spring.

Helicigona A genus of Gastropod Molluscs of the Snail family, *Helicidae*. e.g. H. lapicida - The Lapidary Snail; shell 7.5-8.5 mm high, 16-17 mm wide, lentil-shaped, greyish-brown, usually covered with reddish-brown spots. Found chiefly in foothills on beech trunks, palings, old walls and rocks. Distribution: Europe, including Britain.

Helicobacter A genus of microorganisms, better known as *Campylobacter*, one of whose species has been said to be the cause of peptic ulcer disease. e.g. H. pylori (= *Campylobacter* p.) - The newly-discovered source of hypergastrinemia and duodenal ulcer (see **Campylobacter**).

Helicodonta A genus of Gastropod Molluscs of the Snail family, *Helicidae*. e.g. H. obvoluta - The Cheese Snail; shell 5 mm high, 11 mm wide, flat, covered with hairs; the mouth is 3-lobed. A warmth-loving species found on calcareous substrates, among fallen leaves and under stones and old wood. Distribution: western and central Europe, including Britain.

Helicotrichon A genus of plants of the Grass family, *Gramineae*. e.g. H. pratense - Meadow Oat-grass, a perennial plant, 30-100 cm high, with glabrous leaf-sheaths and stiff, glaucous leaves. Widespread and common in meadows. Distribution: Europe; native in Britain.

Heliometra A genus of Echinoderm animals of the class of Feather-stars, *Crinoidea*. e.g. H. glacialis (= H. glacialis maxima) - The Unstalked Feather-star, the largest of the crinoids; almost 90 cm (3 ft) across, found from the Okhotsk Sea (an inlet of the northwestern Pacific on the coast of Khabarovsk Territory, Russia, southwards to the Korean Straits, at all depths down to 1,800 m (6,000 ft).

Heliopais A genus of Birds of the Finfoot or Sun-grebe family, *Heliornithidae*. e.g. H. personata - The Asian Sun-grebe, 50 cm (20 in) long; brown above, with black head and throat; yellow bill and green legs, and a white stripe behind the eye.

Heliornis A genus of Birds of the Finfoot or Sun-grebe family, *Heliornithidae*. e.g. H. fulica - The American Sun-grebe, the smallest bird of the Sun-grebe family, about 30 cm (1 ft) long, olive-brown above and whitish below, with a white stripe behind the eye. The bill is scarlet and the feet are banded with yellow and black.

Heliosperma A genus of plants of the Pink family, *Caryophyllaceae*. e.g. H. alpestris - The Alpine Rayseed, a perennial plant, 5-25 cm high; with linear-lanceolate leaves up to 9 mm wide and 4-6-toothed petals. Grows in damp stony places, on scree and on gravel by streambeds. Distribution: Europe, in the eastern Alps and northern Balkans.

Heliothrips A genus of Insects of the Thrips order, *Thysanoptera*. e.g. H. haemorrhoidalis - A widespread greenhouse pest, feeding on plant juices with its piercing mouthparts.

Heliotropium A genus of plants of the Borage family, *Boraginaceae*. e.g. H. europaeum - The Common Heliotrope, an annual plant, 20-30 cm high; with soft-hairy leaves and whitish flowers. Distribution: South Europe; naturalized or casual elsewhere.

Helipora A genus of Octocorals of the order of Soft Corals, *Alcyonacea*. This is the Blue Coral; it forms a massive non-spicular skeleton of an amorphous type. The polyps and living tissue overlie the lobed skeletal mass and are brownish in color. The skeleton itself is blue, due to the incorporation of iron salts.

Helix A genus of Gastropod Molluscs of the family of Land Snails, *Helicinidae* (order *Stylommatophora*). e.g. H. pomatia - The Roman or Edible Snail, a hermaphrodite, terrestrial animal, one of the most common of the gastropods, having two pairs of tentacles with eyes on the tips of the hind pair. The shell is up to 4 cm wide and 4 cm high, a dirty straw color with a thin-lipped aperture, cross-grooved whorls and 5 dark stripes. It is a herbivorous creature, highly specialized and capable of breathing by a kind of lung. It has an average lifespan of two to three years, living on chalky soil and attaching itself to the barks of trees or burrowing into the ground to lay its eggs. Very common in gardens where it is a pest of vegetables. Distribution: Western Europe and the Mediterranean region; introduced and naturalized in Britain.

Helkesimastix A genus of coprozoic flagellate organisms. e.g. H. fecicola - A species grown from human feces.

Helleborus A genus of plants of the Buttercup family, *Ranunculaceae*. e.g. H. niger - The Black Hellebore, Christmas Rose, or Melampodium, a perennial plant, 10-30 cm high; with basal, pedate, dark green, toothed leaves having 7-9 perianth-segments and beautiful flowers, pure white or tinged with pink. A poisonous species. From the dried rhizome and root are expressed three glycosides, helleborein, helleborin and hellebrin - used as a powerful hydr-

agogue purgative and emmenagogue. It grows mainly on calcareous soil in mountain forests and alpine grasslands, up to 2,150 m. Distribution: central and southern Europe.

Helobdella A genus of Segmented Annelids of the class of Leeches, *Hirudinea*. e.g. H. algira - A species of aquatic, suctorial leeches, which transmit *Trypanosoma* inopinatum in the frog.

Heloderma A genus of Reptiles of the Lizard family, *Helodermatidae*. These heloderms are venomous lizards inhabiting the deserts of Arizona, Nevada, Utah and Mexico. e.g. H. suspectum - The Gila Monster, a poisonous lizard, up to 50 cm (20 in) long; the dull-black body has a rounded appearance, with variously-shaped reddish-orange markings, and is entirely covered with coarse, bead-like tubercles. These tubercles are many-colored and form designs of orange-red or salmon-colored spots on a dark brown background. There are about ten long pointed and grooved teeth in each jaw, carrying an extremely painful, neurotic venom, though the bite is not usually lethal to man. Distribution: the southwestern United States, including Arizona and New Mexico; also Mexico.

Helophilus A genus of Insects of the Hover Fly family, *Syrphidae*. The larvae of the species may cause nasal and intestinal myiasis (infection with flies).

Helvella A genus of plants of the family of Cup Fungi. e.g. H. crispa - The Common Helvel, a cup fungus with a cap 2-5 cm wide. The cap has 3 undulate lobes; the stem is deeply grooved and pitted, with longitudinal ribs. An edible species, growing amongst grass, usually in woods. Distribution: Europe; native in Britain.

Helwingia A genus of plants of the Cornel or Dogwood family, *Cornaceae*. e.g. H. japonica - A small deciduous shrub, up to 1.2 m (4 ft) high; it has alternate, oval leaves with finely toothed margins. The minute male and female flowers are produced on different plants during May. The rounded fruit is about 6 mm (1/4 in) across. Place of origin: Japan.

Hemaris A genus of Insects of the Hawk Moth family, *Sphingidae*. e.g. H. tityus - The Narrow-bordered Bee Hawk Moth, size 4 cm, having large transparent wings with brown edges. Found in meadows and on hill-sides where there are plenty of flowers. Distribution: Europe; native in Britain.

Hematopota A genus of Insects, now classified under the generic name, *Chrysozona* of the Tabanid or Horse-fly family, *Tabanidae*. e.g. H. italica (= *Chrysozona* i.) (see **Chrysozona**).

Hementaria A genus of Annelids, also known by the generic name, *Haementaria*, of the class of Leeches, *Hirudinea*. e.g. H. officinalis (= *Haementaria* o.) (see **Haementaria**).

Hemerocallis A genus of plants of the Lily family, *Liliaceae*. e.g. H. lilio-asphodelus (= H. flava) - The Lemon Day Lily, a widely distributed perennial plant, 60-120 cm high; with linear, radical leaves in 2 ranks, ascending, spreading and recurving, and numerous large, funnel-shaped, yellow to reddish-orange, pleasantly scented flowers. Fibrous fleshy roots and short underground rhizomes. Place of origin: temperate zones of East Asia, China and Japan. Also now naturalized in S., S.E. and C. Europe. Introduced into Britain in 1570.

Hemerocampa A genus of Insects of the Tussock Moth family, *Lymantriidae*. e.g. H. leucostigma (= H. leukostigma) - The White-marked Tussock Moth, common in North America. Its larvae can cause severe urticaria.

Hemicentetes A genus of Mammals of the Tenrec family, *Tenrecidae*. e.g. H. semispinosus - The Streaked Tenrec, a small, insectivorous, tailless species, commonly found in Malagasy (formerly Madagascar). The fastest-developing mammal in the world; it is weaned after only 5 days, and females are capable of breeding 3-4 weeks after birth.

Hemiclepsis A genus of Segmented Annelids of the class of Leeches, *Hirudinea*. e.g. H. marginata - This species of leeches acts as an intermediate host of *Trypanosoma* granulosum, a sporozoan parasite in the soil.

Hemidactylis See **Hemidactylus**.

Hemidactylus A genus of Reptiles, also known by the generic name, *Hemidactylis*, of the Gecko Lizard family, *Gekkonidae*. e.g. H. turcicus (= *Hemidactylis* t.) - The Disc-fingered or Turkish Gecko, 8-10 cm long, with wide toes having adhesive discs and more than half the length of the body. The eyes are large. The lizard is found in rock crevices, on walls and under stones, and is partial to sunny regions. Distribution: the Mediterranean area and North Africa, India and Pakistan; introduced by man to the American continent.

Hemidesmus A genus of plants of the Milkweed family, *Asclepiadaceae*. e.g. H.indicus - The Anantamul or Indian Sarsaparilla; the dried root of this species is used in India in place of sarsaparilla as a vehicle and flavoring agent for medicaments.

Hemigymnus A genus of Fishes of the Wrasse family, *Labridae*. e.g. H. fasciatus - The Barred Thicklip, a solitary, brilliantly colored, marine fish with spiny fins and prominent, thick lips. Found in warm seas such as the Red Sea region.

Hemimerus A genus of Insects of the order of Earwigs, *Dermaptera* (suborder *Hemimerina*). A blind ectoparasite, it is found on bamboo-rats (genus *Cricetomys*) in Africa. A wingless, viviparous, rather bristly insect whose cerci are soft, flexible structures not modified into forceps (see **Cricetomys**).

Hemiprocne A genus of Birds of the

Swift family, *Apodidae*. e.g. H. mystacea - The Crested Tree Swift, a very small species of birds, which builds minute nests (just bigger than the size of a thimble). Distribution: Malaysia and Australia.

Hemispora A genus of Fungi or Molds. e.g. H. stellata - A fungus found in mycosis, resembling sporotrichosis (in man).

Hemitragus A genus of Mammals of the Cattle family, *Bovidae* (Goat-antelope subfamily, *Caprinae*). These are the Tahrs, goat-antelopes with very short horns, one species found in the Himalayas and two others, in southern India and southeast Arabia.

Hemophilus A genus, also called *Haemophilus*, a name formerly given to a group of microorganisms growing best in the presence of hemoglobin (see **Haemophilus**).

Hemoproteus A genus of Sporozoan parasites, also known as *Haemoproteus*, of the suborder *Haemosporidia*. e.g. H. columbae (= *Haemoproteus* c.) - A species parasitic in pigeons (see **Haemoproteus**).

Hendersonula A genus of non-dermatophytic Molds, causing both nail infections and skin changes resembling tinea.

Heniochus A genus of Fishes of the Butterfly Fish family, *Chaetodontidae*. e.g. H. intermedius - The Antenna Butterfly Fish, a vividly colored, spiny-finned, marine, tropical fish, a beautiful member of the family, found throughout the world, but particularly common in the Red Sea.

Heodes A genus of Insects of the Blue, Copper and Hairstreak Butterfly family, *Lycaenidae*. e.g. H. tityrus - The Sooty Copper moth, size 15-17 mm; the male and female differ in color. Common in fields and meadows, also in the Alps up to elevations of 2,000 m. Distribution: Europe, Asia Minor. Absent in Britain.

Hepatica A genus of plants of the Buttercup family, *Ranunculaceae*. e.g.

H. nobilis - The Noble Liverleaf, a small perennial plant, 4-15 cm high; it has 3-lobed, evergreen leaves heart-shaped at the base, bluish-purple flowers and hairy flower- and leaf-stalks. Distribution: widespread in Europe, except in the north. Introduced and naturalized in Britain.

Hepaticola A genus of Nematode Worms found in the liver of rats. e.g. H. hepatica - A species of nematode parasites found in the liver of man in India, causing the disease hepaticoliasis.

Hepatozoon A genus of parasites found in blood cells. e.g. H. muris - A species found heavy creature, the female of the species reaching 50-60 g in weight after heavy summer feeding.

Hepialus A genus of Insects of the Swift Moth or Butterfly family, *Hepialidae*. e.g. H. humuli - The Ghost Swift moth; size 20-35 mm; male differ in size and coloring of the wings. The female deposits its eggs in flight. Distribution: much of the Palaearctic region (north of the Arctic Circle), the Middle East.

Heptatretus A genus of Fishes of the order of Hagfishes, *Hyperotreti*. These are species of fishes with 6-14 gill openings, also known by the generic names, *Bdellostoma* or *Eptatretus*. Distribution: the Pacific Ocean, from California to Japan (see **Bdellostoma**; **Eptatretus**).

Heptranchias A genus of Fishes of the order of Cow Sharks and their allies, *Squaliformes* (suborder *Notidanoidea* or *Hexanchiformes*). e.g. H. maculatus - A seven gill shark found in the Pacific Ocean; a notidanid species.

Heracleum A genus of plants of the Carrot family, *Umbelliferae*. e.g. H. sphondylium - The Hogwood, Cow Parsnip or Keck, a perennial plant, 60-120 cm high; the stem is angled, furrowed, rough and with short hairs; the flowers are white, the outer ones being extremely unequal, and the fruits are flattened. Grows widespread in meadows and pastureland; also by hedges, roadsides and in woods. Distribution: Europe, Asia; native in Britain. Has been introduced into North America.

Herellea A genus of microorganisms now known by the generic name, *Acinetobacter*, of the family *Brucellaceae*. e.g. H. vaginicola (= *Acinetobacter* anitratus) (see **Acinetobacter**).

Hermetia A genus of Insects of the Soldier Fly family, *Stratiomyidae*. e.g. H. illucens - The Soldier Fly, the larvae of which cause intestinal myiasis in man.

Herminium A genus of plants of the Orchid family, *Orchidaceae*. e.g. H. monorchis - The Musk Orchid, a perennial plant, 8-30 cm high; 2 to 3 oblong or lanceolate leaves with very small flowers. Grows in meadows, grassy slopes, hillsides, and mountainous areas up to 1,800 m. Distribution: Europe; native in Britain.

Hermodactylus A genus of plants of the Iris family, *Iridaceae*; also known by the generic name, *Iris*. e.g. H. tuberosus (= *Iris* tuberosa) - The Snake's Head Iris, an iris-like perennial plant, with a slender, hollow stem, 20-40 cm (8-16 in) high and swollen tuberous roots. The leaves are rush-like, lanceolate and erect, and the solitary flowers striking, greenish-yellow and with blackish or brownish-purple reflexed segments, overarched by a long green spathe. Grows in rocky and stony places. Place of origin: central and eastern Europe; introduced into Britain in medieval times.

Hernanda A genus of Trees. e.g. H. sonora - From the fruit (Netherlands East Indies) is expressed an oil used in lamps. Intoxication caused by the use of this oil in food, is known in Malaya by the term "Moentjang tina".

Herniaria A genus of plants of the Pink family, *Caryophyllaceae*. e.g. H. glabra - The Smooth Herniary or Smooth Rupture-wort, a perennial plant, 5-30 cm high; a glabrous species with numerous prostrate shoots. An extract of the leaves and flowering tops is used as an astringent and diuretic. Distribution: Europe; native in Britain.

Herpestes A genus of Mammals of the Civet, Genet and Mongoose family, *Viverridae*. e.g. H. edwardsi (= H. edwardii) - The Indian Grey Mongoose or Common Indian Mongoose, iron grey in color, under 50 cm (20 in) long and fairly widely distributed in southern Asia. It is adept at killing snakes because of its agility in dodging the fangs. After a kill it will eat the whole snake, including the poison glands. Its eyes have long, oval-shaped irises. Habitat: India, Iran, Sri Lanka and Arabia.

Herpeton A genus of Reptiles of the Colubrid Snake family, *Colubridae* (subfamily *Boiginae*). e.g. H. tentaculatum - The Tentacled Water Snake, a poisonous, aquatic colubrid of Indo-China and Thailand. The head of this small snake has two small appendages which possibly attract the snake's prey and earn it its common name.

Herpobdella A genus of Segmented Worms, the Leeches, of the family *Herpobdellidae*. e.g. H. octoculata - Length 60 mm; width 8 mm; variable brown to dirty green. The body segments are marked with yellow or yellow-white spots. Found in calm and slow-flowing waters, feeding on various aquatic animals. Distribution: Europe.

Herse A genus of Insects of the Hawkmoth family, *Sphingidae*. e.g. H. convolvuli - The Convolvulus Hawk Moth, size 11.5 cm; the hind wings have black, transverse stripes and the rose-colored abdomen has grey or black stripes. This insect blends excellently with the background on which it rests - the trunks of trees, wooden poles and pillars. It migrates from the south and occasionally its caterpillars and chrysalids are found in western and central Europe. Native in Britain.

Hesperis A genus of plants of the Cress or Crucifer family, *Cruciferae*. e.g. H. tristis - The Night-scented Rocket, a biennial plant, 30-60 cm high, with purple-red or white flowers and curved fruits. Grows in waste land, hedgerows and grassy places. Place of origin: East and Central Europe.

Hesperoloxodon A genus of Prehistoric Mammals of the Elephant family, *Elephantidae*. e.g. H. antiquus germanicus - The Straight-tusked Elephant, a prehistoric mammal, which lived in what is now northern Germany about 2 million years ago. Its average length (in adult bulls), has been estimated at 5 m (161/2 ft), and it possessed some of the largest tusks of any prehistoric animal.

Heterakis A genus of Nematode parasites. e.g. H. gallinae - A species found in wild and domestic fowl.

Heteralocha A genus of Birds of the *Aeidae*. e.g. H. acutirostris - The Huia. The sexes show a marked difference in the beak; the male has a straight, sharp beak and the female, a long, down-curved bill.

Heterocentrotus A genus of Echinoderms of the class of Sea Urchins, *Echinoidae*. These animals possess a calcareous boxlike exoskeleton with movable spines used for walking. e.g. H. mammillatus - The Slate-Pencil Urchin, a relatively harmless species, with numerous blunt, reddish-brown, pencil-like arms. Found in tropical waters on coral reefs, such as the Red Sea coast.

Heterocephalus A genus of Mammals of the African Mole Rat family, *Bathyergidae*. e.g. H. glaber - The Naked Mole Rat of northern Kenya and Somaliland. It is 10 cm (4 in) long, with a tail 2.5 cm (1 in) long, and is entirely naked except for a few scattered hairs. Blind and without external ears, it shuns the light and burrows in the sandy soil,

where its presence can usually be detected by the numerous small craters it makes in the ground.

Heterodera A genus of Nematodes or Roundworms of the order *Tylenchida*. These are the Cyst Eelworms, important nematode parasites of plants. e.g. H. radicicola - A nematode parasite found on common root vegetables. It has a stylet which it uses for sucking the cell contents of the living plant. Non-pathogenic to man.

Heterodon A genus of Reptiles of the Colubrid Snake family, *Colubridae*. e.g. H. platyrhinos - The Eastern Hog-nose Snake, a brightly-colored, large, non-venomous, colubrine snake from Florida and to the west as far as Texas, with a fearsome appearance and a hard, trowel-shaped projection which enables it to burrow in loose soil. Its bite can be very painful but it is not poisonous. Distribution: the southeastern United States.

Heterodontus A genus of Fishes of the Port Jackson or Horn Shark family, *Heterodontidae*. This archaic form was the dominant shark from the Triassic or earliest period of the Mesozoic era (about 200 million years ago), marked by the dominance of reptiles and the appearance of gymnosperm plants in geological time, and it is suggested that all modern sharks arose from this form. e.g. H. philippi - The Port Jackson shark or Horn shark. These sharks have a high head with projecting eyebrows, the lateral teeth are pad-like, ridged or rounded in many rows and quite different from the front teeth, which are pointed. They are found in the tropical and warm temperate seas of the world except the Atlantic and Mediterranean.

Heterodoxus A genus of Segmented Worms or Annelids of the class of Leeches, *Hirudinea*. e.g. H. longitarsus - A kangaroo-leech found sometimes on dogs.

Heterometra A genus of Echinoderm

animals of the class of Sea-lilies or Feather-stars, *Crinoidea*. The species are Feather-stars, free-swimming crinoids found in deep marine waters. They have 10 radial arms which are stretched out at night, to strain food from the sea currents. During the daytime they lie hidden under ledges and inside crevices on the coral reef, which is their normal habitat. Distribution: warm tropical regions, such as the Red Sea and the Gulf of Eilat (Akaba).

Heterometrus A genus of Arachnids of the order of Scorpions, *Scorpionida*. These are huge scorpions with large stings, capable of injecting a massive dose of venom. The potency of the poison is low and human fatalities are extremely rare. Distribution: Africa, India. e.g. H. swammerdami - A species found in India. The males average 18 cm (7 in), from the tips of their pincers to the end of the sting.

Heterophyes A genus of minute Trematode Worms, formerly called by the generic names, *Coenogonimus* and *Mesogonimus*. e.g. H. heterophyes (= *Coenogonimus* h.; *Mesogonimus* h.) - A fluke worm found in the small intestine in man, and causing abdominal pain with mucous diarrhea (heterophyiasis).

Heteropoda A genus of Arachnids of the order of Spiders, *Araneida*. These are large spiders, sometimes confused with tarantulas. e.g. H. venatoria - A species of spiders associated with shipments of tropical fruit, especially bananas. Its bite is painful but not lethal.

Heteroscelus A genus of Birds of the Curlew, Sandpiper and Woodcock family, *Scolopacidae*. These are the Tattlers, closely related to the Sandpiper birds (see **Tringa**).

Heteroteuthis A genus of Molluscs of the class *Cephalopoda*. e.g. H. dispar - A marine animal having bilateral symmetry. These cephalopods are predacious carnivores with a highly concentrated central nervous system, within a protec-

tive cartilaginous cranium and with sense organs, notably image-forming eyes, equivalent to those of vertebrates.

Heterotis A genus of Fishes of the Arapaima and Bonytongue suborder, *Osteoglossoidea*. e.g. H. niloticus - A large pike-like fish of tropical rivers, armed with hard bony scales; this ancient fish is found in the Nile, and is of considerable economic importance. It is mainly a grubber with a weak mouth and barbels, and an elongated snout.

Heurnia A genus of plants of the Milkweed family, *Asclepiadaceae*. e.g. H. schneideriana - The Heurnia plant, a succulent, cactus species with light green stems up to 20 cm (8 in) long and 5-7-angled. The bell-shaped flowers are pale brownish outside and velvety black within. Place of origin: Nyassaland (now Malawi), Mozambique.

Hevea A genus of plants of the Spurge family, *Euphorbiaceae*. e.g. H. brasiliensis - The Rubber plant (= Caoutchouc, Gum elastic, or India-rubber plant) - Used as a constituent of the bases of self-adhesive plasters.

Hexabranchia A genus of Gastropod Molluscs of the marine Naked Snail suborder, *Nudibranchia*. This animal, the nudibranch, has no shell and no true gills, but often possesses external gills on the back. e.g. H. imperialis - The Spanish Dancer, a marine snail unbelievably red in color, one of the largest (40 cm or 16 in), of the family of Naked Snails. Found below 9 m (30 ft) in regions where coral reefs abound, such as the Red Sea coast and the Gulf of Eilat (Akaba).

Hexactinella A genus of Marine Sponges of the phylum *Porifera* (= *Parazoa*). e.g. H. ventilabrum - a glass sponge species found in the seas around Japan.

Hexagenia A genus of Insects of the order of Flies, *Plectoptera*. These are the Mayflies, lacy-winged, aquatic imago insects with a short adult life (1-2 days) and a lengthy nymph stage (up to 2 years). e.g. H. bilineata - A mayfly species causing bronchial asthma, and found on the shores of Lake Erie (U.S. and Canada).

Hexamita A genus of flagellate parasites found in various animals, causing hexamitiasis. e.g. H. meleagridis - A species found in turkeys.

Hexanchus A genus of Fishes of the order of Sharks, *Squaliformes* (suborder of Notidanid Sharks, *Notidanoidea* or *Hexanchiformes*); family *Hexanchidae*. e.g. H. griseum (= H. griseus) - The Cow Shark, Six-gilled or Brown Shark, the European species of notidanid sharks; length up to 5 m (16 ft); the most primitive form of existing sharks. It has 6 gill clefts or openings on each side. Found in the seas around Europe; also around the West Indies and off the southeastern United States, where it is called the Sixgill Shark.

Hibiscus A genus of plants, also known by the generic name, *Althea*, of the Mallow family, *Malvaceae*. e.g. H. syriacus (= *Althea* frutex) - The Althea Shrub or the Rose of Sharon (see **Althea**).

Hieracium A genus of plants of the Daisy family, *Compositae*. e.g. H. sylvaticum - The Wood Hawkweed, a perennial plant, 25-50 cm high; the stem usually has one leaf and the basal leaves are stalked. Grows commonly on rocks, walls and woods. Distribution: Europe.

Hieraëtus A genus of Birds of the Old World Harrier, Hawk and Vulture family, *Accipitridae* (Eagle subfamily, *Accipitrinae*). e.g. H. pennatus - The Booted Eagle; size 43-45 cm; pure white on the front of the body and dark brown on the back and ear regions; the legs are feathered and the tail has an almost uniform color. It nests in trees. It was once a summer migrant common in the wooded areas of western Europe, but today is found mainly in Central Europe, Asia Minor and Africa.

Hierochloë A genus of plants of the

Grass family, *Gramineae*. e.g. H. odorata - Holy-grass, a perennial plant, 20-50 cm high; it has 3-flowered spikelets with two lower florets in the male, the uppermost florets being bisexual. Found locally, growing on riverbanks, ditches and wet (freshwater) places. Distribution: Europe; native in Britain.

Hildenbrandia A genus of plants of the Red Algae family, *Gigartinaceae*. e.g. H. prototypus - The Hildenbrandia, a skin-like plant up to 15 cm high, blood-colored to brownish-red, forming smooth and firm crusts on stones. Found on flat beaches covering stones, along the North Sea and West Baltic coasts of Europe.

Himantarum A genus of Arthropods of the class of Centipedes, *Chilopoda*. e.g. H. gabrielis - A species found in southern Europe, possessing 171 to 177 pairs of legs when adult, the greatest number of legs in any known centipede. Half of the known species of centipedes only have 15 pairs of legs when adult, but they all start life with about 6 pairs, including H. gabrielis.

Himanthalia A genus of plants of the Brown Algae family, *Fucaceae*. e.g. H. elongata - The Sea Thong, a species of brown algal plants, up to 3 m (10 ft) high when reproductive; vegetatively a button-like structure 4 cm (1 3/5 in) across, with male or female conceptacles, and belt-shaped, branched reproductive parts. Distribution: Europe; native in Britain.

Himantoglossum A genus of plants of the Orchid family, *Orchidaceae*. e.g. H. hircinum - The Lizard Orchid, a perennial plant, 30-80 cm high; with an angled stem, elliptical leaves and greenish-purple flowers, smelling strongly of goats. It has ovoid root tubers and grows on calcareous soil. Place of origin: Central Europe and the Mediterranean region; native in Britain.

Himantopus A genus of Birds of the Avocet and Stilt family,

Recurvirostridae. e.g. H. himantopus -
The Common, Black-necked, or Black-
winged 38 cm long; a monotypic
species recognized by its black and
white coloring, long pink legs, very long
pointed bill and pointed wings, longer
than the body. Equally at home in fresh
or seawater, and in swamps. A partially
migrant bird. In flight it looks like a
miniature stork. Distribution: a resident
bird in southern Europe, Africa and
Asia. Occasionally found in Britain.

Hippeastrum A genus of plants of the
Daffodil family, *Amaryllidaceae*. e.g. H.
equestre - The Amaryllis, Barbados Lily
or Equestrian Star-flower, a handsome
plant with globose and stoloniferous
bulbs; radical, strap-shaped, glaucous
green leaves and funnel-shaped, vividly
scarlet flowers, green at the base. Place
of origin: West Indies; introduced into
Europe in 1725.

Hippelates A genus of Insects of the
order of Two-winged Flies, *Diptera*. e.g.
H. pusio - The "Eye-gnat" of Coachella
Valley in California; transmits a severe
epidemic conjunctivitis.

Hippobosca A genus of Insects of the
order of Two-winged Flies, *Diptera*.
These are species of pupiparous, dipterous, parasitic insects called Winged Tick
Flies and are found on horses and cattle.
e.g. H. rufipes - A tick fly said to transmit the disease galziekte, the South
African name for gall sickness, a disease
of cattle marked by fever, anemia and
jaundice, and caused by *Anaplasma*
marginale (see **Anaplasma**).

Hippocampus A genus of Bony Fishes of
the Pipefish and Seahorse family,
Syngnathidae. e.g. H. hippocampus -
The Common Seahorse, length 10-15
cm; a well-known marine fish found in
places thick with seaweed, both in the
sea and in brackish waters. The body is
shaped like the head and arched neck of
a horse, and is covered in bony rings.
These species are the only fishes possessing a prehensile tail with which they
can hold onto seaweed. Fertilized eggs
are deposited by the female in the brood
pouch of the male, and the fully developed young leave 4-5 weeks later.
Distribution: the Mediterranean Sea.
Related species are found in the Black
Sea, and northeastern Atlantic (e.g. H.
hudsonius or the Atlantic Salmon).

Hippocrepis A genus of plants of the Pea
family, *Leguminosae*. e.g. H. comosa -
The Horse-shoe Vetch, a perennial plant,
10-40 cm high; 5-12 flowers in umbel-
like inflorescences; covered stalk with
horseshoe-shaped pod segments (hence
the name). Grows on dry pastures and
cliffs. Distribution: southern and western Europe; native in Britain.

Hippoglossoides A genus of Fishes of the
American Plaice, Flatfish, or Flounder
family, *Hippoglossidae*. e.g. H. plates-
soides - The Long Rough Dab or
American Plaice, up to 30 cm long; it
has a large mouth, a straight lateral line
and very rough scales. It lies on the left
side and has the right eye uppermost.
Distribution: the North Sea and the West
Baltic.

Hippoglossus A genus of Fishes of the
Halibut or Right-eyed Flounder family,
Pleuronectidae. e.g. H. hippoglossus -
The Atlantic Halibut, a right-eyed flounder, up to 3 m long, the largest member
of the family, this fish has a grey-brown
body, a free edge to the operculum and
the caudal fins are long, concave and
symmetrical. The lateral line is strongly
curved. Of much economic importance;
from the liver is extracted Halibut-liver
Oil (= Oleum Hypoglossi), used as a
means of administering vitamins A and
D. Found over sandy or rocky bottoms
at depths of 50-2,000 m. Distribution:
the northeast Atlantic from the White
Sea to the Bay of Biscay. Also in the
North Sea and occasionally in the
Baltic.

Hippolais A genus of Birds of the
Babbler, Thrush and Warbler family,
Muscicapidae (Old World or True

Warbler subfamily, *Sylviinae*). e.g. H.
icterina - The Icterine Warbler, length
13 cm; an olive-colored migrant bird,
inhabiting parks, gardens and broadleaved woods, and wintering in tropical
and South Africa. These are the Bush-
warblers, experts at imitating the songs
of other birds, and sometimes producing
a regular orchestral mix-up.
Distribution: Europe, western Siberia.

Hippolyte A genus of Decapod
Crustaceans of the order *Decapoda*. e.g.
H. varians - The Chameleon Prawn, up
to 1.5 cm long; the back of the body is
humped; it gets its name from its ability
to change color, and is often missed
because it matches its background in the
seashore perfectly. Distribution: the
western Baltic from the south North
Sea.

Hippomane A genus of Trees. e.g. H.
mancinella - A tree of tropical Africa,
yielding manchineel, a caustic poisonous sap or juice.

Hippophaë A genus of plants of the
Oleaster family, *Elaeagnaceae*. e.g. H.
rhamnoides - The Sea Buckthorn, a
deciduous shrub, 1-5 m tall; willow-like
and thorny, with silvery lanceolate
leaves, minute greenish, dioecious,
inconspicuous flowers (i.e. male and
female on separate plants), and orange-
red, berry-like fruit. Distribution: widespread in Europe and Asia; native in
Britain.

Hippopotamus A genus of Mammals of
the Hippopotamus family,
Hippopotamidae. e.g. H. amphibius -
The Great African Hippopotamus,
almost 4 ft) long and nearly 1.5 m (5 ft)
at the shoulder, but the legs are less than
60 cm (2 ft) long. They are more at
home in water than on land, where they
are almost overpowered by their own
weight, which can amount to some 4
tons. The color of the thick, bare skin
varies from light reddish-pink to brown
and greyish-blue. Distribution:
Hippopotamuses live in herds in the

swamps, rivers and lakes of Central
Africa, chiefly in the inaccessible interi-
or, and in reserves.

Hippospongia A genus of Calcareous
Sponges (class *Calcarea* or
Calcispongiae), of the family
Spongiidae. e.g. H. communis - The
Honeycomb Sponge, related to the Bath
Sponge, is also collected for commerce
but its coarser skeleton makes it better
suited for cleaning school blackboards,
etc. Distribution: this animal species is
found in the Mediterranean Sea (see
Spongia).

Hippotigris A former generic name for
Zebras or Roman horse tigers (so-called
because of their stripes), and which
were used by the Romans in circuses.
Today, Zebras are included in the generic name, *Equus*. e.g. H. burchelli (=
Equus b.) - The Common Zebra or
Burchell's Zebra; the pattern of stripes
varies considerably among individuals
of the same herd, and there are numerous sub species. The Zebra is closer to
an ass than a horse; it has long ears, a
short stiff mane and a tufted tail. Native
to Africa, being distributed between
Ethiopia and the Orange River.

Hippotragus A genus of Mammals of the
Cattle family, *Bovidae* (Antelope subfamily, *Antilopinae*). e.g. H. equinus -
The Roan Antelope, a grey-colored,
horse-like, even-toed ungulate, standing
about 1.5 m (5 ft) at the shoulder and
with long horns curving backwards.
Distribution: chiefly in west and southeastern Africa.

Hippuris A genus of freshwater plants of
the Marestail or Mare's tail family,
Haloragaceae. e.g. H. vulgaris - The
Marestail or Mare's tail plant (pl. mare's
tails; mares' tails), an aquatic perennial,
15-75 cm high, with long stalks covered
with fine leaves, 6-12 in a whorl. Grows
in still or slow-flowing freshwater.
Distribution: Europe; native in Britain.

Hirudo A genus of Segmented Worms or
Annelids (class *Hirudinea*) of the family

of Jawed Leeches, *Hirudidae*. e.g. H.
medicinalis - The Medicinal Leech, 3-5
cm long; the well-known olive-grey
leech, with six reddish longitudinal
stripes, a blood-sucker on mammals,
and formerly used extensively for therapeutic purposes. Widespread in ponds,
lakes and ditches. Distribution: Europe;
native in Britain but now quite rare in
this country.

Hirundo A genus of Birds of the Martin
and Swallow family, *Hirundinidae*. e.g.
H. rustica - The European Swallow or
the American Barn Swallow. A small
insectivorous bird, about 10-20 cm (4-8
in) long, with long pointed wings, darkish plumage and forked tail. A short-
lived species with an average life-
expectancy of 13 months in the wild
state. A migratory bird, breeding in
Europe, Asia and North America and
wintering in Africa, southern Asia and
South America.

Hister A genus of Insects of the Histerid
Beetle family, *Histeridae*. e.g. H.
impressus (= H. cadaverinus) - Length
4-7 mm; one of the commonest members of the histerid beetle family. Feeds
on carrion, but also attracted to decaying
plant remnants, such as old mushrooms,
etc., and fermenting sap that oozes from
fresh tree stumps, or injured and felled
trees. Distribution: the Palaearctic
region.

Histiobranchus A genus of Fishes of the
order of Eels, *Apodes* (=
Anguilliformes). These are marine
species of eels, living in the northern
Pacific in extremely deep water. They
are predatory creatures, eating small
fishes and crustaceans; they have poison
fangs at the bottom of which are poison
glands. In several species of this genus
the venom may be fatal to man.

Histiogaster A genus of Arachnids of the
order of Mites, *Acarina*. e.g. H. ento-
mophagus - An acarid mite, possibly
causing cutaneous vanillism, with der-

matitis and general malaise in raw vanilla handlers.

Histomonas A genus of Protozoans, also
known by the generic name, *Amoe*,
order *Rhizomastigina*. e.g. H. meleagridis (= *Amoeba* m.) - An amoeba with
1 to 4 flagella. It causes hepatitis in
turkeys; the comb and wattles of poultry
infected by this parasite turn black, and
the disease is known as "blackhead".
The parasite is flagellated in the lumen
of the intestine but becomes amoeboid
upon invading the liver (see **Amoeba**).

Histoplasma A genus of Molds or Fungi
Imperfecti. e.g. H. capsulatum - A
species causing splenomegaly and anemia in humans (histoplasmosis).

Histrio A genus of Fishes of the suborder
of Frogfishes or Sargassum-fishes,
Antennarioidei. e.g. H. histrio - The
Sargassum or Frogfish; this is a tropical
seafish generally found near sandbanks
or coral reefs, especially on the rocks
around the West Indies and Indonesia.
About 20-30 cm (8-12 in) long, they
drift about in masses of seaweed, often
indistinguishable from the weed itself.

Histrionicus A genus of Birds of the
family of Ducks, Geese and Swans,
Anatidae. e.g. H. histrionicus - The
Harlequin Duck, a diving sea dlong;
with harlequin or "clownish" markings;
a resident bird nesting gregariously near
rapids. Distribution: Iceland.

Histrix See **Hystrix**.

Hogna A genus of Arachnid arthropods,
also known by the generic name,
Trochosa of the order of true spiders,
Araneae. e.g. H. singoriensis (=
Trochosa s.) - The largest spider species
in the eastern part of Central Europe and
the neighboring warm Mediterranean
region. It inhabits dry hillsides and open
scrub forests.

Holacanthella A genus of Apterygote
Insects of the Springtail order,
Collembola. These are the poduroid
insects (suborder *Poduroidea*), dark-colored with a rough warty integument and

short appendages. e.g. H. spinosa - A springtail species from New Zealand, up to 1 cm long, and a giant of this order.

Holacanthus A genus of Fishes (subphylum *Vertebrata*). e.g. H. ciliaris - The Queen Angle-fish.

Holarrhena A genus of plants of the Periwinkle family, *Apocyanacae*. e.g. H. antidysenterica. The Kurchi plant (= Conessi Bark; Tellicherry Bark); it contains the principal alkaloid, conessine, used in the treatment of amebic dysentery.

Holcus A genus of plants of the Grass family, *Gramineae*. e.g. H. lanatus - The Yorkshire Fog, a perennial plant high; leaves and leafsheaths are densely hairy, awn scarcely projecting, and finally hooked. Widespread and common in grasslands. Distribution: Europe; native in Britain.

Holocentrus A genus of Fishes of the Soldierfish and Squirrelfish family, *Holocentridae*. e.g. H. ascensionis - The Soldierfish or Squirrelfish; it has thoracic pelvic fins, 4 spines on the anal fin and 10-12 spines on the front dorsal fin.

Hololepta A genus of Insects of the Polyphagan Beetle family, *Histeridae*. These are very flat species, up to 1.8 cm long, the largest representative of the family. They are compact, shining, predacious creatures, frequenting dung, carrion, and ant nests or living beneath bark, and in tunnels of wood-boring insects.

Holophrya A genus of Holotrich Protozoan animals of the order of Gymnostomes, *Gymnostomatida* (suborder *Rhabdophorina*). These are simple organisms, oval and about 160 μ in length, found on the surface of freshwater where they feed on dinoflagellates.

Holorusia A genus of Insects of the Crane-fly family, *Tipulidae*. e.g. H. brobdignagius - A daddy-long-leg species of crane-flies, said to measure as much as 22.5 cm (9 in) from the tips of the front legs to the tips of the hind legs,

and with a wingspan of almost 10 cm (4 in). Distribution: tropical South America.

Holospora A genus of certain microorganisms, parasitic on protozoa.

Holosteum A genus of plants of the Pink family, *Caryophyllaceae*. e.g. high; it has white or pale pink flowers, 3-15 in each umbel-like cyme; the fruit-stalks are deflexed. Widespread in Europe.

Holothuria A genus of Echinoderm animals of the family of Sea-cucumbers, *Holothuriidae*. e.g. H. forskali - A species of sea-cucumbers known as the Cotton-spinner. When molested, the holothurian bends the hind part of the body round in the direction of the throat and ejects white threads of the respiratory system, known as the Cuvierian tubules, through the anus. In the water the threads swell and form a sticky mass, engulfing the attacker. These tubules are later generated.

Holothyrus A genus of Arachnids of the superfamily of Ticks, *Ixodoidae*. e.g. H. coccinella - A tick species from Mauritius. It infects gand is poisonous to humans.

Homalomyia A genus of Insects of the order of Two-winged Flies, *Diptera*. e.g. H. canicularis - The Small House-fly, a species which in summer flies tirelessly round the lights in houses. The maggots or larvae sometimes infest the human intestine.

Homarus A genus of Decapod Crustaceans of the Lobster family, *Homaridae* (= *Astacidae*). e.g. H. gammarus (= H. vulgaris) - The Common Lobster, 30-50 cm long; a marine species inhabiting rocky and stony seacoasts to depths of about 40 m. Has large, unequal pincers, a large pincer for crushing and a smaller one for seizing prey. It has a natural lifespan of up to 30 years. Widely captured as a great delicacy. Distribution: Mediterranean Sea, Atlantic coast of Europe, and the North and Baltic Seas.

Homo A genus of Mammals of the Human family, *Hominidae*, and closely related to, and branching off the Anthropoid Ape family, *Pongidae*, probably sometime in the Miocene era (about 30 million years ago). e.g. H. sapiens - Modern man.

Homogyne A genus of plants of the Daisy family, *Compositae*. e.g. H. alpina - The Purpole Coltsfoot, a perennial plant, 10-40 cm high; it has stolons or creeping stems with green leaves beneath, and carries purplish-red tubular flowers above. Grows in damp alpine meadows, scrubland and open woodlands. Place of origin: south and south-central Europe, up to 2,300 m, and the North Balkans.

Honkenya A genus of plants of the Pink family, *Caryophyllaceae*. e.g. H. peploides - The Sea Sandwort, a perennial plant, 5-25 cm high; with acute, ovate, fleshy, yellowish-green leaves and greenish-white, clustered flowers. Grows widely on mobile sand and sandy shingle. Distribution: Europe; native in Britain.

Hopea A genus of resinous plants of the family, *Dipterocarpaceae*, used in the manufacture of varnishes and as a mountant in microscopy. An important constituent of East Indian or Singapore Dammar, a resin compound (see **Balanocarpus**).

Hoplophorus A genus of Crustaceans of the order of Decapods, *Decapoda*.

Hoplopsyllus A genus of wingless Insects (order *Siphonaptera*), of the Flea family, *Pulicidae*. e.g. H. anomalus - A species of fleas found in ground squirrels, transmitting plague.

Hordelymus A genus of plants of the Grass family, *Gramineae*. e.g. H. europaeus - The Wood Barley, a perennial plant, 40-120 cm high; the spikelets are usually in threes at each node of the axis. Grows fairly widespread in some places, but generally uncommon. Distribution: Europe; native in Britain.

Hordeum A genus of plants of the Grass family, *Gramineae*. e.g. H. distichon - The Two-row Barley; the lateral spikelets are more or less aborted so that the mature ear bears only the vertical rows of grains. H. vulgare - The Six-row Barley; all the spikelets are fertile, with the result that there are six vertical rows of grains in the mature ear. These two species form the Common Barley, a cereal grass plant; the grains are partially germinated artificially and then dried. Used for animal food, infant foods, Malt syrup, Malt extract or Byne, brewing, and bread making. One of the oldest of cultivated cereals. There is clear evidence that it was grown and eaten in Egypt and China over 4,500 years ago.

Horminium A genus of plants of the Mint or Thyme family, *Labiatae*. e.g. H. pyrenaicum - The Pyrenean Dead-nettle, a perennial plant, 10-25 cm high, with large crenate, basal leaves. Grows on dry alpine grassland, between 300-2,450 m. Distribution: Europe, in the Pyrenees and Alps.

Hormodendrum A genus of Fungi or Molds. e.g. H. pedrosoi - A species of fungi, causing the disease known as chromoblastomycosis.

Hornungia A genus of plants of the Cress or Crucifer family, *Cruciferae*. e.g. H. petraea - The Rock Hutchinsia, an annual plant, 3-10 cm high with pinnate leaves and white flowers. Usually grows on limestone rocks. Place of origin: South, West and Central Europe; native in Britain.

Hosackia A genus of plants of the Pea family, *Leguminosae*, many of whose species are poisonous to farm animals, and are known as loco (Spanish for insane), because of the drug, selenium they contain (see **Astragalus**).

Hosta A genus of plants, also known by the generic name, *Funkia*, of the Lily family, *Liliaceae*. e.g. H. ventricosa (= *Funkia* ovata) - A species from eastern Asia, with heart-shaped leaves up to

22.5 cm (9 in) long, and lavender-blue flowers.

Hottonia A genus of plants of the Primrose family, *Primulaceae*. e.g. H. palustris - The Water-violet, a perennial plant, up to 60 cm high; comb-like, 1-2-pinnate leaves, and lilac flowers with yellow throats, 3-8 in a whorl. Widespread in still and slow-flowing waters, by ponds and lakes, and in marshes. Distribution: Europe; native in Britain.

Houttuynia A genus of plants. e.g. H. californica - A species known as Mansa; the root or rhizome of this plant is used as a tonic in malaria and dysentery.

Hoya A genus of plants of the Milkweed family, *Asclepiadaceae*. e.g. H. bella - A sft) high, with ovate, dark-green leaves and pure white flowers with a deep red or purple center, borne in short-stalked pendulous clusters. Sometimes grown in greenhouses. Place of origin: India.

Hucho A genus of Fishes of the Char, Salmon and Trout family, *Salmonidae*. e.g. H. hucho - The Danubian Salmon; length up to 1.2 m; weigh kg. These are the Huchen fishes; they have a naked head, complete opercula, forked caudal fin, lateral line and a dorsal fin of moderate length. The Huchen of the Danube is long and slender, and has a flattish snout. It is a very good indicator of water pollution, though its numbers have been rapidly decreasing in many rivers in recent years. Distribution: Europe, in the mountain and foothill streams that feed the Danube and its tributaries.

Humulus A genus of plants of the Hemp or Hop family, *Cannabinaceae*. e.g. H. lupulus - The Hop plant (= Hops; Lupulus), a climbing perennial plant up to 7 m (23 ft) tall; with 3-5-lobed leaves, cone-like infructescences and dioecious flowers (male and female flowers on separate plants). From the plant is prepared an aromatic bitter, used for improving the appetite and diges-tion. Grows commonly on the banks of rivers, in damp thickets and in meadow forests. Distribution: Europe, western Asia. Native in Britain.

Hupertzia A genus of plants of the sub-phylum of Club Mosses, *Lycopsida*. e.g. H. selago - The Fir Clubmoss, a low spreading, vascular, evergreen plant with erect, branched stems from 5-30 cm (2-12 in) high, densely clothed with small lanceolate, dark green leaves. Small, flattened leafy bulbils are produced and these serve as organs of vegetative reproduction. Grows on heaths, moors and mountains. Distribution: north temperate zones.

Huso A genus of Fishes of the Sturgeon family, *Acipenseridae*. Also known by the generic name, *Acipenser*. e.g. H. huso (= *Acipenser* h.) - The Giant Beluga or Russian sturgeon, length 5-6 m; weight more than 1 ton; the largest of the sturgeon fishes. A migratory species, spending most of its life in the sea and spawning in large rivers. Prized not only for its tasty meat but also for its large number of eggs (roe), from which black caviar is processed. Distribution: common in rivers flowing into the Black Sea, Caspian Sea and the Sea of Azov (see **Acipenser**).

Hutchinsia A genus of plants of the Cress or Crucifer family, *Cruciferae*. e.g. H. alpina - The Alpine Hutchinsia or Alpencress, a tufted, hceous, perennial plant, 5-10 cm (2-4 in) high; with stalked, pinna teleaves and hairy fruit-stalks. Very small white flowers in short-stemmed clusters. Grows on basic limestone rocks and screes, and prefers mountain areas up to 3,400 m. Place of origin: South and Central Europe; also the Pyrenees. Introduced into Britain in 1775.

Hutchinsoniella A genus of Crustaceans of the family *Hutchinsoniellidae*. These are very small creatures, only 2-3 mm long, and do not have any trace of eyes. The head bears a short horseshoe-shaped carapace; the body has 20 segments and 8 well-developed limbs. They live in soft marine segments where they feed on organic debris. Distribution: Japan, North America and the Caribbean.

Hyacinthus A genus of plants, better known by the generic name, *Galtonia*, of the Lily family, *Liliaceae*. e.g. H. candicans (= *Galtonia* c.) - The Giant Summer Hyacinth or Spire Lily (see **Galtonia**).

Hyaena A genus of Mammals of the Hyaena or Hyena family, *Hyaenidae* (subfamily of True Hyenas, *Hyaeninae*). e.g. H. hyaena - The Striped Hyaena (Hyena), about 1 m (40 in) long, with a tail some 40 cm (16 in) in length; a carnivorous animal adapted for carrion-feeding. It has dark stripes on a dull grey or yellowish-grey background, and ranges from Central Africa to India.

Hyaenictis A genus of Mammals of the Hyaena family, *Hyaenidae*, one of the now-extinct ancestors of the present-day hyaenas, which roamed the earth in the Miocene epoch in time about 30 million years ago (the later Tertiary period in Geological time).

Hyalinoecia A genus of Ragworms of the class *Annelida*. e.g. H. tubicola - A species with horny tubes, 5-7 cm long and 2-4 mm thick. The animal wanders around within its tube. Distribution: Europe, in the Mediterranean region. Native in Britain.

Hyalomma A genus of Arachnids of the superfamily of Ticks, *Ixodoidae*. e.g. H. aegyptium - A cattle tick of Africa, India and southern Europe. H. mauritanicum - A cattle tick of North Africa (see **Theileria**).

Hyalonema A genus of Sponges of the Glass Sponge order, *Hexactinella* belong to the suborder, *Amphidiscophora*, in which the typical spicule of the glass sponge is an amphidisc. e.g. H. sieboldi - A species of glassy sponges. These animals are found mostly in the seas around Japan, living in the mud at great depths.

Hyalophora A genus of Insects of the order of Butterflies and Moths, belonging to the Emperor Moth family, *Saturniidae*. e.g. H. cecropia - The Cecropia Moth, the largest Emperor Moth in North America.

Hyas A genus of Decapod Crustaceans of the Crab suborder, *Brachyura*. e.g. H. araneus - The Spider Crab, a short-tailed crustacean; the size of the ovate carapace is about 10.5 cm; the first pair of legs have pincers. It camouflages itself with seaweed along the seashore. Distribution: the North Sea, Atlantic coasts and the West Baltic. Native around Britain.

Hydatigera A genus of Tapeworms of the subclass *Cestoda*. These are long, ribbon-shaped parasitic worms, resembling Taenia, found mostly in carnivorous animals. e.g. H. infantis - A species which has been found in humans in Argentina.

Hydnocarpus A genus of plants of the family *Flacourtiaceae*. Also known by the generic name, *Taraktogenos*. e.g. H. kurzii (= *Taraktogenos* k.) - From the seed is expressed Chaulmoogra Oil (= Gynocardia Oil; Hydnocarpus Oil), used in the treatment of leprosy (see **Chaulmoogra**; **Taraktogenos**).

Hydnum A genus of Fungi or Molds of the Hydnum family. e.g. H. imbricatum - The Imbricated Hydnum; the cap is 5-20 cm wide, greyish-brown with greyish-white spines. An edible species. Distribution: Europe; native in Britain.

Hydra A genus of Hydrozoan animals of the order *Athecata* (family *Hydridae*). These are small freshwater, hydrozoan polyps or hydras, having a basal disc attaching it to a stone or plant. Its slender cylindrical body has thread-like, stinging tentacles around the mouth as an aid to catching prey. It is nonmetagenetic, and reproduction is typically by asexual budding, though sexual reproduction does occur. e.g. H. vulgaris -

The Common Hydra, length up to 1 cm; greyish body with 6 retractable tentacles. A freshwater species. Distribution: Europe; native in Britain.

Hydractinia A genus of Sea Firs or Colonial Hydroid animals (class *Hydrozoa*), of the family *Bougainvilliidae*. e.g. H. echinata - A colonial hydroid species, 6 mm high; it forms a thick mat of stolons on shells occupied by hermit crabs (genus *Eupagurus*). It is polymorphic, the main forms being feeding polyps with proctective spines between them, reproductive polyps, and defensive polyps with tentacles only. Distribution: the Atlantic coast of Europe, in the North Sea, and more rarely in the West Baltic. Native in Britain.

Hydraena A genus of Insects of the Beetle family, *Hydraenidae*. e.g. H. riparia - Length 2.2-2.4 mm; a small, elongate water beetle with greatly developed maxillary polypi, which are larger than the antennae located behind them. Distribution: from central Italy to central and northern Europe.

Hydrallmania A genus of Sea Firs or Colonial Hydroid animals of the class *Hydrozoa*. e.g. H. falcata - A species up to 45 cm in length; spirally twisted and the cups or polyps are in rows along the branches. It grows preferably on oyster-beds. Distribution: the North Sea; native around Britain.

Hydrangea A genus of plants of the Saxifrage family, *Saxifragaceae*. e.g. H. arborescens - A Hydrangea species, a deciduous shrub. The tree and roots are used as a diuretic.

Hydrilla A genus of plants of the Frogbit or Waterweed family, *Hydrocharitaceae*. e.g. H. lithuanica - The leaves mostly 5 in a whorl, male and female flowers with long stalks. Place of origin: Central Europe; native in Britain.

Hydrobates A genus of Birds of the Tubenose family, *Procellariidae*. e.g. H. pelagicus - The Storm Petrel, size 15 cm; white rump, blackfeet and square-ended tail. A partial migrant and exclusively marine bird. Breeds along the coasts of Britain and Iceland; also on the islands of the Mediterranean (see **Fulmarus**).

Hydrobia A genus of marine Gastropod Mulluscs or Seasnails of the order *Neogastropoda*. e.g. H. ulvae - Shell 0.5 cm high, with 8 whorls or spirals, the body one as large as the other seven. Very numerous in muddy areas, sometimes so much so, that the whole surface of the mud looks granular. Distribution: Europe; native in Britain.

Hydrochara A genus of Insects of the Water-scavenger Beetle family, *Hydrophilidae*. e.g. H. caraboides - A scavenging beetle, 14-18 mm long lump, slow, pitch-black, freshwater insect with short, club-shaped antennae. It puts its head above water to trap air for breathing under the wingcases. Distribution: Europe; native in Britain.

Hydrocharis A genus of plants of the Frogbit or Waterweed family, *Hydrocharitaceae*. e.g. H. morsus-ranae - The Frogbit, a pehigh. A floating herb with groups of long-stalked leaves, the male with stalks 1-6 cm long, the female smaller and stalkless. The flowers are unisexual. Found in ponds and ditches. Distribution: widespread in Europe and Asia; native in Britain.

Hydrochoerus A genus of Mammals of the Cavy and Guinea-pig family, *Caviidae*. e.g. H. hydrochaeris (= H. capybara) - The Capybara, Carp Water Hog, the world's largest rodent, growing to the size of a small pig, 1.2 m (4 ft) long and 54 kg (120 lb) in weight. Its fur is greyish-brown, thin and coarse. A sort of gigantic guinea-pig, it is essentially aquatic in habit and is easily domesticated. Distribution: by rivers and lakes in tropical South America, from Panama to Brazil.

Hydrocleis A genus of plants of the Flowering Rush family, *Butomaceae*.

e.g. H. nymphaeoides - An aquatic plant
species from tropical South America. It
is a floating herb, having leaves with
long stalks and ovate blades. The flowers have numerous stamens but the outer
ones are sterile.

Hydrocotyle A genus of plants of the
Carrot family, *Umbelliferae*, better
known by the generic name, *Centella*.
e.g. H. asiatica (= *Centella* a.) - The
Indian Pennywort (see **Centella**).

Hydrocyon A genus of Fishes of the
Characin family, *Characinidae*. This
species looks rather like a salmon with
an adipose fin, but it has ferocious-look-
ing teeth, and is called the Tigerfish. It
may grow to a length of 1.5 m (5 ft) and
weigh 30 kg (66 lb). Distribution: in the
large rivers of Africa.

Hydrodamalis A genus of Mammals of
the Dugong Sea-cow family,
Dugongidae (order *Sirenia*); the species
are most closely related to the Elephant.
e.g. H. stelleri - Steller's Sea-cow, a
dugong siren or sirenian, once abundant
in the Bering Sea and off Kamchatka
and the Aleutian Islands when first discovered in 1741. It is now probably
extinct, having been steadily hunted to
extermination by the end of the 18th
century. What we know of it today is
from Steller's original description and
from various museum skeletons. It was
like a dugong, with a rough and crinkled
skin, but had a horny plating over the
gums and palate instead of teeth, and
was considerably larger, reaching a
length of 9 m (30 ft) and weighing up to
4 tons (see **Dugong**).

Hydrogenomonas A genus of microorganisms of the order *Pseudomonadales*;
short rod-shaped cells obtaining energy
from the oxidation of hydrogen. e.g. H.
facilis.

Hydrolagus A genus of Fishes (subphylum *Vertebrata*). e.g. H. several species
of which are found in mountainous
areas of Europe and the High Sierras of
California.

Hydrometra A genus of Insects of the
order of True Bugs, *Heteroptera*. These
are the Water-striders or Pond-skaters,
freshwater bugs which run only on the
surface of the water. e.g. H. stagnorum -
The Water Measurer, a freshwater bug,
up to 12 mm long; it crawls slowly over
the water on six legs. The part of the
head in front of the eyes is twice as long
as that behind. Distribution: Europe;
native in Britain.

Hydrophidus A genus of Reptiles of the
Sea-snake family, *Hydrophiidae*. These
are front-fanged marine species, whose
natural habitat is the tropical parts of the
Indian and Pacific oceans, the open seas,
coastal waters and even estuaries.
Owing to their lung structure and tight
sealing of the nostrils, they can stay
under water for about eight hours.

Hydrophilus A genus of Insects of the
Polyphagan Beetle family,
Hydrophilidae. e.g. H. peceus - One of
the largest beetles found in Britain are
laid in a remarkable cocoon-like structure which may be attached to floating
grass and other objects. They live in
damp and marshy areas.

Hydrophis A genus of Reptiles of the
Sea-snake family, *Hydrophiidae* (=
Hydrophidae). e.g. H. jerdoni - The
Kerril, a front-fanged, venomous sea-
snake found in the Indian Ocean.

Hydropogne A genus of Birds of the
Tern family, *Laridae*. e.g. H. caspia -
The Caspian Tern, 53.5 cm long, the
largest of the terns; the crown and sides
of the head are black in color, and the
species is distinguished by a heavy bill
or beak. A migrant bird, almost gull-like
in flight. Breeds along the sandy coasts
of the Baltic.

Hydroporus A genus of Insects of the
True Water Carnivorous Beetle family,
Dytiscidae. e.g. H. palustris - Size 3.5-4
mm; the upper side is black with yellow
patches. It attacks a wide range of freshwater animals. Distribution: Europe;
native in Britain.

Hydropotes A genus of Mammals of the Deer family, *Cervidae*. e.g. H. inermis - The Chinese Water Deer; length 90 cm (3 ft); the male has no antlers, but its upper canine teeth project downwards. Distribution: Europe; introduced and naturalized in Britain.

Hydrothaea A genus of Insects (Flies) of the class *Insecta*. e.g. H. meteorica - A fly species, which attacks the eyes and nostrils of man and animals.

Hydrous A genus of Insects of the Water-scavenger Beetle family, *Hydrophilidae*. e.g. H. piceus - The Great Silver Water Beetle, length 34-47 mm; large, pitch-black with short, club-shaped antennae. Found chiefly in quiet waters, in ponds, river backwaters, forest pools, etc., with abundant aquatic vegetation, the only food on which the adult beetle feeds. Distribution: the Palaearctic region, including Europe. Native in Britain.

Hydrurga A genus of Mammals of the Pinniped or Earless Seal family, *Phocidae*. e.g. H. leptonyx - The Sea Leopard or Leopard Seal, the large tic seal except for the Sea Elephant. The female is longer than the male and may reach 3.6 m (12 ft). It is very aggressive, hence the name, and feeds mainly on fishes and cuttlefishes, but also on penguins and seabirds, disgorging the feathers when the flesh has been dissolved. A very fast animal in water, it has been known to swim while hunting, at a maximum speed of c. 37 km/h (23 miles/h) (see **Mirounga**).

Hygrobia A genus of Insects of the Scavenging Beetle family (order *Coleoptera*). e.g. H. hermanni - Size 12 mm; brownish-red elytra with black length-wise stripes. Distribution: Europe; native in Britain.

Hygrophila A genus of plants of the Acanthus family, *Acanthaceae*; better known by the generic name, *Asteracantha*. e.g. H. spinosa (= *Asteracantha* longifolia) - The

Hygrophylla plant (see **Acanthus**; **Asteracantha**).

Hyla A genus of Amphibians of the Typical Treefrog family, *Hylidae*. e.g. H. crucifer (= H. crucifer crucifer) - The Spring Peeper, a treefrog abundant in the eastern United States and southeastern Canada. Found in ponds and marshes where it spawns in water. It is a strong jumper, capable of launching itself at a sedentary insect accurately enough to engulf it as it lands. Specially adapted to arboreal life, its digits have expanded discs at the tips, which assist gripping so effectively that the treefrog can cling to a vertical sheet of glass.

Hylaeosaurus A genus of Prehistoric Reptiles, the Dinosaurs (class *Reptilia*). In 1832 Mantell found fragmentary remains of a 9.14 m (30 ft) long armored dinosaur in the Tilside Forest, Sussex, England, and called it Hylaeosaurus or the "Toad Lizard". It was not until 1842 however, that the name Dinosauria or "Fearfully Great Lizard" was given to these reptiles by Professor (later Sir) Richard Owen, the great English anatomist and vertebrate palaeontologist.

Hylarnus A genus of Mammals of the Cattle family, *Bovidae* (Antelope sub-family, *Antilopinae*). These are the Dwarf Antelopes from West Africa. They are related to the Royal Antelope (see **Neotragus**), but are larger than it.

Hylecoetus A genus of Insects of the Beetle family, *Lymexylidae*. e.g. H. dermestoides - Length 6-18 mm. Adult beetles live only 2-4 days. The larvae develop in the wood of oak and beech, chiefly in stumps, where they bore long, narrow tunnels. Distribution: central and northern Europe, Siberia.

Hylemyia A genus of Insects, the flies, the larvae of which infest vegetables. e.g. H. brassicae - The cabbage root maggot.

Hyles A genus of Insects of the Hawk Moth family, *Sphingidae*. e.g. H. gallii -

The Bedstraw Moth, 32-35 mm long; a multicolored insect. The caterpillars are polyphagous but are partial to bedstraw (Gallium). Distribution: temperate regions of Europe and Asia, and the Nearctic region.

Hylobates A genus of Mammals of the Anthropoid Ape family, *Pongidae*. Anthropoid apes are the primates closest to man, anatomically, physiologically and psychologically. e.g. H. lar - The Lar Gibbon, about 90 cm (3 ft) high and weighing up to 6.8 kg (15 lb). It has a dark face fringed with white; the body color is dark grey or brownish-yellow. It is an arboreal creature. The Gibbon is the least anthropoid of the Pongidae. Distribution: the mountainous forests of southeast Asia i.e. Burma, Thailand and the Sunda Islands.

Hylobius A genus of Insects of the Weevil family, *Curculionidae*. e.g. H. abietis - The Pine Weevil, length 8-13 mm; chestnut brown or black elytra with rust-yellow cross spots. Common and abundant in pine forests. Hibernates under felled logs, amidst fallen leaves, in stumps, etc. Its larvae bore tunnels inside the roots of pine and spruce trees and pupate there. The adult insects grow and severely damage the growing shoots of these young trees. Distribution: Europe, Siberia, Japan. Native in Britain.

Hylochoerus A genus of Mammals of the family of Wild Pigs, *Suidae*. e.g. H. meinertzhageni - The Forest Hog, a species of African pigs, first discovered in the forests of Kenya. Now known to range across equatorial Africa to Liberia. An ugly animal with a grey naked skin, misshapen head and a body covered with grotesque warts.

Hylocichla A genus of Birds of the Babbler and Thrush family, *Muscicapidae* (Thrush subfamily, *Turdinae*). e.g. H. mustelina - The Wood Thrush; it has a conspicuous spotting on the breast and sides, and a reddish head.

Found in deciduous woodlands in North America. It usually winters in places like Florida.

Hyloconium A genus of plants of the phylum *Bryophyta*. These are primitive landplants, the liverworts and mosses, characterized by a rhizoid root anda non-vascular moisture-transport system. e.g. H. splendens - A unisexual species with somewhat flattened shoot systems, red stems and pointed, concave leaves. Grows commonly on heaths and moors.

Hyloicus A genus of Insects of the Hawk Moth family, *Sphingidae*. e.g. H. pinastri - The Pine Hawk Moth, size 4-8 cm; forewings greyish-brown with brownish patches and 3 dark streaks. Found in pine forests, where the caterpillars feed on pine needles. Distribution: temperate parts of Europe and Asia; rare in Britain.

Hylomys A genus of Mammals of the Hedgehog family, *Erinaceidae*. e.g. H. suillus - The Lesser or Short-tailed Gymnura, under 15 cm (6 in) long without the tail, which is almost as long. A rat-like species found in India and southeast Asia.

Hylonomus A genus of Prehistoric Extinct Reptiles (class *Reptilia*), the species of which lived about 290 million years ago (the Carboniferous period in Geological time), and whose remains have been discovered in Nova Scotia (Canada).

Hylotrupes A genus of Insects of the Long-horned or Longicorn Beetle family, *Cerambycidae*. e.g. H. bajulus - The House Longicorn, 9-20 mm long; brown or black in color; the wings have 2 whitish haired bands. Found in coniferous forests and in attics, households and cut timber. The larvae live in wood for several years, making oval holes, often in house timbers. Distribution: worldwide; native in Britain.

Hyloxalus A genus of Amphibians of the Frog family, *Pelobatidae*. The species carry their tadpoles about on their backs.

Hymeniacidon A genus of Calcareous

Sponges of the order *Calcarea*. e.g. H. perlevis - A littoral sponge of the gelatinose type with calcite spicules, growing half-buried in the sand. In the littoral it forms low cushions, tan or bright red, but offshore, it tends to assume a larger size with finger-shaped surface processes. These animals are common around the coasts of Europe, South Africa, Australia and the Pacific coast of Canada.

Hymenobolus A genus of plants of the Cress or Crucifer family, *Cruciferae*. e.g. H. procumbens - The Salt-cress, an annual plant, 2-15 cm white flowers with elliptical to obovate fruit. Grows frequently near the sea, mainly in South Europe.

Hymenocallis A genus of plants of the Daffodil family, *Amaryllidaceae*. e.g. H. narcissiflora (= H. calathina) - The Chalice-crowned Sea Daffodil, Peruvian Daffodil or Spider Lily, a globose bulb with an elongated, cylindrical neck; strap-shaped leaves in double rows, and white fragrant flowers with a wide corona, and carried in umbels of 2 to 8. The tube at the back of the flowers is greenish. Place of origin: The Peruvian Andes; found in Europe since 1794.

Hymenoea A genus of plants of the Pea family, *Leguminosae*. The species yield what is known as the Brazilian Copal (see **Agathis**).

Hymenogorgia A genus of Horny Corals of the order *Gorgonacea*. e.g. H. quercifolia - An Octocoral species living in the warm waters of the West Indies. These animals grow in colonies resembling leaves.

Hymenolepis A genus of Merozoic Cestodes or Tapeworms (subclass *Eucestoda*), also known by the generic name, *Diplacanthus*, of the family *Hymenolepididae*. e.g. H. nana (= *Diplacanthus* n.) - The Dwarf Tapeworm, a species about 2.5 cm (1 in) long, found in the adult form in the intestine of man; it can cause colic and diarrhea.

Hymenophyllum A genus of plants of the Filmy-fern family. e.g. H. tunbrigense - The Tunbridge Filmy-fern, a perennial plant, 2-8 cm high; it has moss-like fronds with toothed indusium valves. Grows locally on rocks and felled tree-trunks. Place of origin: Europe; native in Britain.

Hymenopus A genus of Insects of the voracious Praying Mantid suborder, *Mantodea* (order *Dictyoptera*). e.g. H. coronatus - The Flower Mantis from Java, an Indo-Malayan species of praying mantids, brightly colored and simulating the shades and markings of flowers which it frequents (see **Tenodera**).

Hynobius A genus of Amphibians of the primitive Salamander family, *Hynobiidae*. The female enters the water where she lays her egg sacs, while the male spends the winter there as he guards them. The species are found in Asia.

Hyocoris A genus of Insects of the Swimming Bug family. e.g. H. cimicoides - The Saucer Bug, 35 mm long; the body is compressed and the hind legs possess swimming hairs. Found in freshwater. Distribution: Europe.

Hyoscyamus A genus of plants of the Nightshade family, *Solanaceae*. e.g. H. niger - The Henbane, an annual to biennial high; it has sinuate, coarsely toothed leaves and pale yellow flowers with violet veins. A poisonous plant, growing in waste places, on rubbish dumps, by roadsides and in sandy situations. From the dried Henbane Leaves (= Hyoscy. Leaves) and flowering tops are extracted the alkaloids hyoscyamine and scopolamine, anticholinergic drugs with actions similar to those of atropine. Distribution: Europe (native in Britain), western Asia, North Africa and North America. Also abbreviated: Hyo - as in Hyo niger.

Hyostrongylus A genus of Worms. e.g.

H. rubidus - A small red worm found in the pig stomach.

Hyperchiria A genus of *Isaturniidae*. e.g. H. io - The North American Emperor Butterfly, an exceedingly beautiful insect, with vivid blue eyes against a yellow background on its hind legs. This striking effect intimidates would-be attackers.

Hyperia A genus of Crustaceans of the order *Amphipoda* (suborder *Hyperiidea*). e.g. H. galba - The Sandhopper, an active marine plankton commonly found in the subgenital pits of jellyfishes (class *Scyphozoa*) such as *Rhizostoma* octopus. This amphipod has no carapace and possesses enormous eyes and small coxal plates (see **Rhizostoma**).

Hypericum A genus of plants of St. John's Wort family, *Hypericaceae*. e.g. H. perforatum - The Common St. John's Wort or Millepertius, a perennial plant, 20-60 cm high; the stem has two raised lines, the leaves have translucent glandular dots and the sepals are entire. Grows by waysides, at the edges of cultivated fields, in grassland and open woods. Used as an astringent and diuretic. The herb contains a red pigment, hypericin which causes photosensitization. Common throughout Europe; native in Britain.

Hyperoödon A genus of Mammals of the Beaked and Bottle-nosed Whale family, *Ziphiidae*. e.g. H. rostratus - The Bottle-nosed Whale, a Toothed Whale species, 6-9 m (20-30 ft) long, with a conspicuously protuberant forehead; sometimes found stranded on the coasts of western Europe and the north Atlantic. From the blubber is extracted Spermaceti (= Cetaceum; Walrat) - a solid wax and a common ingredient of cold creams (see **Physeter**).

Hypholomoa A genus of plants of the Gill Fungus family, *Agaricaceae*. e.g. H. fasciculare - The Sulphur Tuft, a gill fungus with a cap 3-5 cm wide, yellow, often darker on the top. The gills are yellowish-green. Grows on treestumps. A poisonous species. Distribution: Europe; native in Britain.

Hyphomicrobium A genus of microorganisms of the order *Hyphomicrobiales*; ovoid cells with filaments radiating outwards. e.g. H. vulgare.

Hyphydrus A genus of Insects of the Carnivorous Water Beetle family, *Dytiscidae*. e.g. H. ovatus - Length 4.5-5 mm; the body is very convex above and below. It attacks a wide range of freshwater animals. Distribution: Europe; native in Britain.

Hypobosca A genus of Insects, the Flies. e.g. H. rufipes - A South African fly transmitting *Trypanosomatheileri*.

Hypochoeris A genus of plants of the Daisy family, *Compositae*. e.g. H. radicata - The Common Cat's-ear, a perennial plant, 25-80 cm high; it has sinuate-toothed leaves, the scape usually possesses a solitary head and the outer ligules are deeply toothed. Widespread in meadows, heaths and open woodlands. Distribution: Europe; native in Britain.

Hypocolius A genus of Birds of the Waxwing family, *Bombycillidae*. e.g. H. ampelinus - The Hypocolius; distributed around the northern end of the Persian Gulf, feeding largely on figs, mulberries and dates. Has been considered the sole member of a separate family, *Hypocoliidae*.

Hypoctonus A genus of Arachnids of the order of Whip Scorpions, *Uropygi*. The prosoma is covered by a single large carapace, and the abdomen terminates in a long flagellum or tail, hence the common name of the order (see **Mastigoproctus**).

Hypoderma A genus of Insects of the Warble Fly family, *Hypodermatidae*. e.g. H. bovis - The Ox or Deer Warble Fly, length 13-15 mm; the body is covered with dense black and yellow hairs. The eggs are spindle-shaped, and are laid on the hind quarters and legs of the host's body. Parasitic on cattle, the larvae bore inside and live under the skin. They cause sometimes a creeping eruption of the skin (myiasis) in man. Distribution: the Palaearctic and Nearctic regions. Native in Britain.

Hypomesus A genus of Fishes (subphylum *Vertebrata*). e.g. H. pretiosus - The Surf Smelt fish.

Hyponomeuta A genus of Insects of the order are white with black dots; the caterpillars live communally in webs, feeding on the spindle-tree. Found in meadows and on pastureland. Distribution: Europe; native in Britain.

Hypopachus A genus of Amphibians of the Toad family, *Microhylidae*. A narrow mouthtoad, showing adhesive properties during amplexus, the male and female during this time adhering to one another by means of their sticky skin secretions (see **Gastrophryne; Kaloula**).

Hypophthalmichthys A genus of Fishes of the Carp, Minnow, Tench and allied family, *Cyprinidae*. These are the Silver Carps, Asiatic cyprinid fishes, in which the eyes are low down, and the helical organ is in the branchial cavity. e.g. H. molitrix - The Silver Carp or Tolstol, length up to 1 m; weight 10 kg. Feeds exclusively on plant plankton. Has a very rapid rate of growth. Place of origin: East Asia; has been introduced to and is now well established in Europe (see **Carassius**).

Hypositta A genus of Birds of the Nuthatch family, *Sittidae*. e.g. H. corallirostris - The Coral-billed Nuthatch of Madagascar, the only species of the subfamily, *Hypositinae*. A tree-climbing bird, descending head first; sometimes called the "upside-down" bird.

Hypselosaurus A genus of prehistoric extinct Reptiles, the Dinosaurs (class *Reptilia*). e.g. H. priscus - A sauropod species of reptiles, 9.14 m (30 ft) long, which lived about 80 million years ago (Early Paleocene epoch of the Tertiary Period in geological time). Some specimens of their eggs, the largest known dinosaur eggs, have been found in the Valley of the Durance, near Aix-en-Provence, southern France, in October 1961, with a (long axis) length of 30 cm (12 in) and a (shorter axis) breadth of 25 cm (10 in), giving them a capacity of 3.3 l. (5.77 pints) each!

Hypsignathus A genus of Mammals of the order of Bats, *Chiroptera* (fruit-sucking bat, suborder *Megachiroptera*). e.g. H. monstrosus - The Hammerhead Bat from Central and West Africa, a fruit-sucking bat with a greatly enlarged snout.

Hypsopsetta A genus of Fishes (subphylum *Vertebrata*). e.g. H. guttulata - The Diamond Tcommercial value as food.

Hyssopus A genus of plants of the Mint or Thyme family, *Labiatae*. e.g. H. officinalis - The Hyssop, a perennial plant, 50-150 cm high; a woody, aromatic species with linear-lanceolate leaves. From the leaves and flowering tops is prepared a carminative and expectorant. A native of South Europe, it is cultivated and has been naturalized in other parts, including Britain.

Hysterocrates A genus of Arachnids of the Wolf Spider family, *Pisauridae* (order *Araneae*). e.g. H. hercules - One of the largest African spiders, with a bodylength of 7.5 cm (3 in) and a legspread of 20 cm (8 in).

Hystricopsylla A genus of Insects of the order of Fleas, *Siphonaptera*. e.g. Hcies being found in the United States. The male measures as much as 6 mm (0.23 in) and the female 8 mm (0.31 in) long.

Hystrix A genus of Mammals, also called *Histrix*, of the Old World Porcupine family, *Hystricidae*. e.g. H. cristata (= *Histrix* c.) - The Common or Crested Porcupine, a nocturnal rodent up to 70 cm (28 in) long and weighing 27 kg (60 lb), with a crest of white-tipped grey quills, a similar covering of quills which are hollow and open at the end, and a

short tail. Found in dry flatlands and at
the feet of mountains in dry thickets.
Distribution: southern Europe, Asia
Minor, North Africa.

254

I

Iberis A genus of plants of the Cress or Crucifer family, *Cruciferae*. e.g. I. sempervirens - The Perennial Candytuft; a spreading, evergreen (except in areas of deep forest), perennial plant, 22 cm (9 in) high. The dark green leaves are narrow, flat and blunt. The flowers are pure white, and grow in dense, flat-topped heads, 4-5 cm (11/2-2 in) across. Variegated and double varieties are known. The short, flatenned fruit contains two winged seeds. Place of origin: the mountains of southern Europe; introduced into Britain in 1820.

Ibis A genus of Birds of the Jabiru and Stork family, *Ciconiidae*. e.g. I. ibis - The Wood Ibis or Yellowbill Stork. It has pink-colored plumage, a yellow beak and purplish-red bare spots on the head, long slender legs and short toes, almost totally unwebbed. It nests in colonies in the trees, and is found near water, feeding on small animals. Distribution: Central Africa.

Ichneumia A genus of Mammals of the Civet, Genet and Mongoose family, *Viverridae*. e.g. I. albicauda - The White-tailed Mongoose of southern Arabia and Africa, found south of the Sahara.

Ichthyophthirius A genus of protozoan animals (phylum *Protozoa*). e.g. I. multifiliis - A ciliate protoz the skin of freshwater fish.

Icichthys A genus of Fishes of the order of Ragfishes, *Icosteiformes* or *Malacichthyes*. These are degenerate, deep-sea forms with limp bodies, rather like wet rags, hence their common name. Their skeletons are very poorly developed, and are mainly cartilaginous.

Icosteus A genus of Fishes of the family of Ragfishes, *Icostidae* (see **Icichthys**).

Ictalurus A genus of Bony Fishes, better known by the generic name, *Ameiurus*, of the Catfish family, *Ameiuridae* or *Ictaluridae*. e.g. I. nebulosus (= *Ameiurus* n.) - The Brown Bullhead or Horned Pout, a well-known freshwater catfish of North America (see **Ameiurus**).

Icterus A genus of Birds of the American Blackbird and Oriole family, *Ecteridae* e.g. I. icterus - The Common Hangnest, one of the troupials or American orioles. It builds a woven, pendulous nest suspended under a broadleaf in the groves of trees. It has a striking black and yellow plumage and is a good singer. Distribution: Brazil.

Ictinia A genus of Birds of the Old World Harrier, Hawk, and Vulture family, *Accipitridae* (true kite subfamily, *Milvinae*). e.g. I. mississippiensis - The Mississippi Kite; a great destroyer of insects. It has a slim form, long narrow wings and an elongated tail. Distribution: North America.

Ictiobus A genus of Fishes of the Sucker family, *Catastomidae*. These are the Buffalo fishes, large fishes resembling carps, possessing long fins with many rays. Distribution: the freshwaters of North America and eastern Asia.

Ictonyx A genus of Mammals of the Badger, Otter, Skunk, and Weasel family, *Mustelidae*. e.g. I. striatus - The Zorille or Cape Polecat, a mustelid living in rocky country throughout Africa. It has an obnoxious smell, in contrast to its beautiful coat of glossy black fur, with a white stripe on each side running into the tail.

Idaea A genus of Insects of the Butterfly and Moth family, *Geometridae*. e.g. I. aversata - The ical form has a broad dark band on the wings; yellowish-brown in color. Found in open woodlands and clearings. Distribution: Europe, Transcaucasus, Asia Minor and the Middle East.

Idiacanthus A genus of Bony Fishes of the Sawtailfish family, *Idiacanthidae*. e.g. I. fasciola - The female up to 30 cm; marked sexual dimorphism. The male has atrophied digestive organs, teeth, and light organs which are reduced to a single photophore above the eyes; very short-lived, dying soon after spawning. The female is long-lived. The larvae show no resemblance to the adult fish, with the eyes set entirely off the head on long stalks. Distribution: the depths of all seas and oceans.

Idotea A genus of Crustaceans of the Shrimp suborder, *Natantia*. e.g. I. baltica - A small, edible, marine shrimp, with a flattened body and seven pairs of legs; the tail plate has three spines. A vegetarian species, it subsides on aquatic plants. Widespread in European coastal waters, especially in the North Sea and the Baltic.

Idus A genus of Fishes of the Carp, Minnow, Tench and allied family, *Cyprinidae*. e.g. I. idus - The Ide, 30-75 cm long, with a terminal mouth and scales on the lateral line. A freshwater fish found almost throughout Europe (see **Carassius**). I. melanotus - A species of Orfe fishes transmitting the Siberian liver fluke, *Opisthorcis* felineus to cats, dogs, pigs, and man (see **Carassius**; **Opisthorcis**).

Iguana A genus of Reptiles of the Iguana Lizard family, *Iguanidae*. e.g. I. iguana - The Common or Green Iguana, a beautiful, green or black-colored arboreal lizard, found in the tropical forests of South America. It measures up to 1.8 m (6 ft) in length; a vegetarian, it feeds on leaves, shoots and fruits of plants. It is fond of water and is an accomplished swimmer. It lays its eggs hidden in the sand on the banks of rivers. The natives hunt the iguana for its tasty flesh.

Iguanodon A genus of prehistoric animals, the now extinct bipedal dinosaurs, the fossilized teeth of which were first described in 1825 by this name (meaning "iguana-tooth"), by Dr. Gideon Mantell and his wife Mary, who discovered them in the district of Cuckfield in 1822. This dinosaur was estimated to be 9.14 m (30 ft) long, and was herbivorous; it was supposed to have stalked across what is now southern England about 130 million years ago (during the Lower Cretaceous period of the Mesozoicera).

Ijimaia A genus of Fishes of the deep-sea fish family, *Podatelidae* (order *Chondrobrachii*). These are deep-sea fishes living in Japanese waters, and are characterized by a profound degeneration of the body. The skeleton is partly cartilaginous, some of the cranial bones are missing, and the supra-temporal bone is attached to the skull only by ligaments.

Ilex A genus of plants of the Holly family, *Aquifoliaceae*. e.g. I. paraguariensis - The Mate or Paraguay Tbush or small tree, with alternate, very leafy, simple leaves and flowers arranged in small, axillary clusters. The fruit is fleshy and contains several seeds. From the plant is extracted caffeine (0.2-2%), as well as the alkaloid yerbine (resembling caffeine). Used as a beverage in South America; it is a diuretic and diaphoretic.

Ilia A genus of Crustaceans of the short-tailed, ten-legged Crab order, *Decapoda*. e.g. I. nucleus - A species of Crabs, up to 3 cm long, with an almost spherical body, and eyes and antennae situated well forward. A marine creature, it lives in the Mediterranean among seagrass, on rocky bottoms or in shingle.

Illecebrum A genus of plants of the Pink family, *Caryophyllaceae*. e.g. I. verticillatum - The Coral Necklace, an annual plant, 2-30 cm high; the white flowers grow in clusters with scarious, silvery bracteoles and thread-like petals. Found in damp places. Habitat: West and Central Europe; native in Britain.

Illicium A genus of plants of the Magnolia family, *Magnoliaceae*. e.g. I. verum - A magnolia species, native to Asia and North America; a tree with large alternate leaves and large terminal white, pink or yellowish waxy flowers. From this species is obtained the Star Anise Fruit, source of Anise Oil or Aniseed Oil, used as a carminative, expectorant, and a flavoring agent.

Ilybius A genus of Insects of the True Water Beetle family, *Dytiscidae*. e.g. I. fenestratus - Length 11-12 mm; common in river back-waterpools, and small forest springs. Distribution: northern part of southern Europe, central and northern Europe as far as Lapland, Siberia, North America.

Ilyocoris A genus of Insects of the Creeping Water-bug family, *Naucoridae*. e.g. I. cimicoides - The Saucer Bug, length 15 mm; inhabits calm and slow-flowing waters. The eggs are deposited in the tissues of aquatic plants. Distribution: Europe, the Caucasus.

Impatiens A genus of plants of the Balsam family, *Balsaminaceae*. e.g. I. balsamina - The Balsam plant, a herbaceous annual, up to 50 cm (20 in) high, with a brittle, succulent stem and toothed leaves. The flowers are irregular, double or single, red, pink or white. Numerous cultivars vary in form of leaf, size, and color of flower. The genus receives its name from the Latin for impatient, the seeds being ejected at the slightest touch. Place of origin: India, Malaya, China; introduced into Europe in 1596. The flowers are used in parts of Asia instead of henna for dyeing fingernails.

Inachis A genus of Insects of the Butterfly and Moth family, *Nymphalidae*. e.g. I. io - The Peacock Butterfly, size 27-35 mm; it emerges from winter shelter in spring and basks in the sun. The caterpillar feeds on nettles. Distribution: Europe (absent in northern Scandinavia), temperate parts of Asia, to Japan.

Inachus A genus of Crustaceans of the order *Decapoda*. e.g. I. scorpio - The Spider-crab, a small ten-legged decapod,

found in the Mediterranean. An omnivorous species.

Incarvillea A genus of plants of the Bean family, *Bignoniaceae*. e.g. I. grandiflora - The Trumpet Flower plant, a herbaceous, dwarf perennial with radical, pinnate leaves and stout stems, up to 45 cm (18 in) high, carrying one or two large (10 cm = 4 in), funnel-shaped, 2-lipped flowers of rich rosy-red color, with throat markings and yellow tubes. Place of origin: W. China; introduced into Britain in 1898.

Indicator A genus of Birds of the Honey-guide family, *Indicatoridae*. e.g. I. indicator - The Greater Honey-guide, a small bird found in the tropical forests of Asia and Africa. Particularly well known for its remarkable skill in finding the nests of wild bees, hence its common name. It feeds on the comb, and unique secretions in the stomach enable it to digest the wax.

Indiella A genus of parasitic Fungi or Molds. e.g. I. mansoni - A species of fungus causing white mycetoma.

Indigofera A genus of plants of the Pea family, *Leguminosae*. e.g. I. gerardiana (= I. dosua) - The Himalayan Indigo plant or Indigo Shrub, a deciduous species 1.5-2 m (5-6 ft) high, with very elegant pinnate foliage, having opposite leaflets and racemes of bright purplish-rose, pea-shaped flowers. Place of origin: Northwest Himalayas; introduced into Europe about 1840.

Indri A genus of Mammals of the Sifaka, Indris, and Avahi family, *Indriidae*. e.g. I. brevicaudata - The Short-tailed Indris, the largest le existence, 90 cm (3 ft) long; it has an occasional bipedal gait. The animal has a thick, silky fur, black back, fawn hind-quarters and a black stripe down the legs. Its tail is a mere 2.5 cm (1 in) stump. A vegetarian, it lives in the forests of eastern Madagascar.

Indricotherium A genus of one of the largest prehistoric mammals, and one of the largest mammals ever recorded, and like the genus *Pristinotherium*, the fossil remains were found in Kazakhstan in Central Asia. Closely related to the genus *Baluchitherium*, the largest land mammal on record, it was a long-necked hornless rhinoceros roaming the prehistoric terrain between 20 and 40 million years ago (see **Baluchitherium**).

Ingolfiella A genus of Crustaceans of the order of Amphipods, Scuds and Sideswimmers, *Amphipoda*. These are blind creatures of the suborder *Ingolfiellida*, ranging in length from 1-15 mm throughout the seven known species of the genus. The body is narrow and elongated, and the first 2 legs have a characteristic two-segmented claw which bends back against a larger swollen segment. Separate species have been found in the depths of the ocean, in cave pools in the Congo, in coral sands in the Gulf of Siam, in ground waters of Yugoslavia, and in the gravels off the Eddystone lighthouse in the English Channel.

Inia A genus of Mammals of the River Dolphin family, *Platanistidae* (subfamily *Iniidae*). e.g. I. geoffrensis - The Amazonian Dolphin or Boutu, found in the upper reaches of the Amazon River. This is a very old form of cetacean which has adapted to a freshwater existence, having lost all contact with the sea. It has a long, almost bird-like beak, but is furnished with teeth, and is about 1.8-3 m (6-10 ft) long. It feeds on the ferocious river fish, the Piranha with impunity (see **Serrasalmus**).

Inimicus A genus of Fishes of the Scorpion Fish family, *Scorpaenidae*. e.g. I. filamentosus - The Lump fish, a dangerous member of the scorpionfish family; a marine species with sharp spines containing venom along the back of the body, which is covered with scales; there are robust spiny rays also in the fins. These spines are capable of causing excruciating pain and even death to man, and also of course, to other fishes. The Lump Fish lies in wait completely motionless for many hours on the sea bed in fairly shallow water, for unsuspecting fish to approach; its camouflage is probably the most efficient in the animal kingdom. A poor swimmer, it prefers to crawl. Habitat: the coral reefs of the Gulf of Eilat (Akaba) and the Red Sea. Also widely spread in other ocean waters.

Inocybe A genus of plants of the phylum of Gill Fungi, *Eumycophyta*. e.g. I. patouillardii - The Red-staining Inocybe, a gill-fungus plant; width of the cup is 4-8 cm; the cap is conical or bell-shaped, and split at the margin. A very poisonous species. Found rarely in coniferous woods. Place of origin: Europe; native in Britain.

Inula A genus of plants of the Daisy family, *Compositae*. e.g. I. helenium - The Elecampane plant. Elecampane root is one source of the vegetable starch, Alant Starch, a polymerized form of fructofuranose (a polysaccharide), yielding levulose on hydrolysis, and is used in the preparation of special breads for diabetics; also as a diagnostic agent in the kidney function test, ICT (Inulin clearance test). Other sources of Alant Starch are the bulbs of *Dahlia* variabilis, the root of the Chicory plant, *Cichorium* intybus and the tubers of the Jerusalem artichoke, *Helianthus* tuberosus, all members of the daisy family, *Compositae*.

Iodamoeba A genus of True Amoebas of the order *Amoebina*. e.g. I. butschlii (= I. buetschlii) - A species of nonpathogenic amoebas found in the human intestine. Also called *Endolimax* williamsi, *Entamoeba* butschlii and I. williamsi (see **Endolimax; Entamoeba**).

Iphiclides A genus of Insects of the Swallow, *Papilionidae*. e.g. I. podalirius - The Scarce Swallow-tail Butterfly, size 4-8 cm; it has a black margin on the outer edge and blue moon-shaped spots

on the hind wings. Found in forest margins, fields and gardens, most often on
warm, sunny, limestone hillsides.
Caterpillars feed on blackthorn,
hawthorn and fruit trees. It is protected
by law in many countries of Central
Europe, but its numbers continue to
decline with the eradication of its most
favored habitat (hedgerows with blackthorn and hawthorn). Distribution: much
of Europe (absent in the north and in
Britain), Transcaucasia, Asia Minor, Iran
and western China.

Ipomoea A genus of herbs and shrubs,
comprising some 300 species, many of
whom have other generic names, such
as, *Convolvulus, Exogonium, Pharbitis*,
etc., of the Bindweed family,
Convolvulaceae. e.g. I. batatas - The
Sweet Potato, a native plant of Central
America, and now widely cultivated in
the tropics for its underground edible
root. I. hederacea (= *Pharbitis* nil) - The
Ivy Leaf Morning Glory plant (see
Pharbitis). I. orizabensis (=
Convolvulus scammonia) - The
Scammony plant; from the root is
extracted Ipomoea Resin (= Mexican
Scammony Resin; Scammony Resin);
used as a drastic purgative (see
Convolvulus). I. purga (= *Exogonium*
p.) - The Jalap plant. Jalap Root contains not less than 10% Jalap Resin and
is used as a powerful purgative (see
Exogonium). I. tricolor - From the
black seeds (the "badoh negro" of
Mexico), is extracted the old Aztec hallucinogenic drug known as Ololiuqui
(see **Rivea**). I. violacea - The seeds of
this plant are known as "morning
glory"seeds, and contain lysergic acid
derivatives, with hallucinogenic properties. MGS or morning glory seeds are
purchased and used as hallucinogens,
and many species of *Ipomoea* are sold
by seedsmen under the name "morning
glory". I. violacea by far, surpasses
other species of this genus, in its lysergic acid derivative content.

Ips A genus of Insects of the Bark Beetle
family, *Scolytidae*. e.g. I. typographus -
The Spruce Bark Beetle, length 4.2-5.5
mm; a serious insect pest attacking
weak or injured spruce and other conifer
trees in the forests of central and northern Europe and North Africa. At swarm-
ing time the females bore into the bark,
burrowing short vertical corridors, while
the males help to excavate the wood
dust. Blind white larvae are hatched in
these corridors out of the 20 odd eggs
laid by each female. The larvae then
make their own horizontal burrows as
they eat their way between the wood
and the bark. This gives rise to the typical appearance of these engraved tree
trunks in the affected forest.
Distribution: Europe, Asia Minor,
Siberia, Korea, northern China. Native
in Britain.

Irena A genus of Birds of the Fairy
Bluebird, Leafbird, and Iora family,
Irenidae. e.g. I. puella - The Blue-
backed Fairy Bluebird, a very colorful
bird, iridescent black and ultramarine in
color, a fruit-eater living in the tree-tops
in forests from southern Asia to the
Philippines.

Iridoprocne A genus of Birds of the
Martin and Swallow family,
Hirundinidae. e.g. I. bicolor - The Tree
Swallow, a holenester bird, nestholes
and bird boxes, and very common
through most of North America. During
winter it subsists on the waxy fruit of
the bayberry. It is about 10-25 cm (4-10
in) long, and highly gregarious.

Iris A genus of plants of the Iris family,
Iridaceae. e.g. I. florentina - The Orris
Root or Iridis Rhizoma plant; it contains
a volatile substance known as Concrete
Oil of Orris or Butter of Orris, and is
used as an ingredient in toilet and tooth
powders. I. tuberosa (= *Hermodactylus*
tuberosus) - The Snake's Head Iris plant
(see **Hermodactylus**).

Isatis A genus of plants of the Cress or
Crucifer family, *Cruciferae*. e.g. I. tinc-

toria - The Woad plant, a bhigh; with numerous yellow flowers and oblong, flattened, 1-seeded, drooping fruits. Distribution: most of Europe; introduced and naturalized in Britain.

Ishthyostega A genus of prehistoric extinct amphibians (class *Amphibia*), one of the earliest known, and the first quadruped, which lived about 350 million years ago. Its remains have been discovered in Greenland.

Isidora A genus of Molluscs (class *Amphineura*) of the family of Snails, the species of which are intermediate hosts of *Schistosoma*.

Isis A genus of Coelenterate animals of the order of Horny Corals, *Gorgonacea*. e.g. I. hippuris - A species developing striking colonies of branches with alternating horny and calcareous sections.

Isodon A genus of Mammals of the Bandicoot family, *Peramelidae*. These are the short-nosed bandicoots, omnivorous animals, most of their food being insects and worms; also, roots and tubers. They have short snouts and numerous incisor teeth as well as canines and sharp molars. They sleep underground all day, emerging from their burrows at night (see **Choeropus**).

Isoetes A genus of plants of the Quillwort family. e.g. I. lacustris - The Common Quillwort, a perennial plant, up to 15 cm high; with stiff leaves, and sporangia embedded in the leaf-base below the ligule. A freshwater species found in lake beds in most of Europe. Native in Britain.

Isognomostoma A genus of Gastropod Molluscs of the Snail family, *Helicidae*. e.g. I. isognomostoma - A species of ly flat shells, 6 mm high and 10.5 mm wide, covered with comparatively long hairs. The mouth is three-lobed, a shape caused by the teeth and lamellae. Found in mountains and submontane forests in stony locations. Plentiful also amongst fallen leaves and under logs.

Distribution: the mountains of Central Europe and the Pyrenees.

Isopyrum A genus of plants of the Buttercup family, *Ranunculaceae*. e.g. I. thalictroides - The Isopyrum plant, a perennial species, 10-30 cm high; with 2-ternate, bluish green leaves and solitary, pentamerous (having parts in fives) flowers. Place of origin: Central Europe.

Isospora A genus of Protozoan animals, the *Coccidia*, causing intestinal infection with diarrhea (coccidiasis). Known also by other generic names, such as *Coccidium, Cytospermium, Eimeria*, etc. e.g. I. hominis (= *Coccidium* bigeminum; C. hominis; C. perforans; *Cytospermium* hominis; *Eimeria* stiedae; I. bigemina) - A non-pathogenic coccidian species, sometimes causing diarrhea in man.

Isotoma A genus of Wingless Insects of the primitive Springtail order, *Collembola*. e.g. I. saltans - A Springtail species, size 2.5 mm; the body is covered with dense black hairs. These insects can leap or spring by suddenly straightening the penultimate segment of their bodies. They are primitive creatures, found in Europe up to a height of almost 4,000 m.

Ispidina A genus of Birds of the Kingfisher family, *Alcedinidae* (subfamily *Alcedininae*). e.g. I. picta - The Pygmy Kingfisher, an insectivorous species found in Africa. Its nest, as in the other species of the family, is a burrow in the river bank, and consists of a long entrance passage terminating in a roomy chamber, usually strewn with fish-bones (see **Alcedo**).

Istiophorus A genus of Fishes of the Sailfish family, *Istiophoridae*. e.g. I. platypterus - The highly streamlined Sailfish, a seafish found in all tropical waters, and said to be the fastest fish in the worldover a short distance. The species has been clocked by ichthyologists in a series of speed trials, at a burst of speed equivalent to 109 km/h (68.18

miles/h), as compared to 96 km/h (60 miles/h) for a cheetah!

Isurus A genus of Fishes of the Mackerel Shark family, *Lamnidae*; sometimes known as *Isuridae*. e.g. I. oxyrinchus - The Sharp-nose Mackerel Shark or Mako Shark, one of the fastest-swimming species; from 1.5-3.9 m (5-13 ft) long. A ferocious man eater, it has no cusps on its teeth, and the dorsal fin is set rather back. Found in the seas of the Atlantic Ocean.

Iulus A genus of Millipedes (class *Diplopoda*), of the family *Iulidae*. e.g. I. terrestris - Length 17-23 mm; width 1.5-2.1 mm; has up to 89 pairs of legs. Black or black-brown, elongate body, lighter on the sides, with yellowish-white legs. The head bears 42 ocelli. Common in damp places, amidst fallen leaves in woodlands and gardens, under stones, etc. Distribution: Europe.

Ixia A genus of plants of the Iris family *Iridaceae*. These are South African plants with small corms producing spikes of dainty star-like flowers on thin wiry stems. Height about 30 cm (1 ft). Variously colored blooms with 6 perianth segments. The leaves are linear and somewhat swordshaped. Place of origin: South Africa; introduced into Europe in 1792.

Ixobrychus A genus of Birds of the Bittern, Egret, and Heron family, *Ardeidae*. e.g. I. minutus - The Little Bittern, the smallest European mthis family, a water bird living among reed-beds of small ponds, and widespread throughout the Old World. Up to 38 cm (15 in) long, it has brown, black, yellow and white plumage, giving it perfect camouflage against a background of dry reeds and other thick vegetation. It makes its nest in the reeds above the surface of the water, and is active by night. Migratory, it winters in North Africa. Distribution: Europe (except Britain), central and southern Asia, as far north as Siberia.

Ixodes 1. A genus of Arachnids of the Tick family, *Ixodidae*. e.g. I. ricinus - The Sheep or Castor-bean Tick, length 4 mm; parasitic to man and beast. The female sucks blood before laying its eggs, using its long, barbed, piercing proboscis, and can reach a length of 10 mm after its meal. The male does not suck blood. These creatures live in deciduous woods with dense undergrowth, the young larvae climbing up on shrubs, whence they drop on the first host, generally a bird nesting on the ground or a lizard, sucking its blood for 3-5 days. The larva next changes into a nymph (the next stage), on the ground. The nymph likewise sucks blood, but from a mammal such as the sheep, leaving it after a certain time to change into the adult form or imago. The castor-bean tick transmits a virus encephalomyelitis in cattle, sheep, pigs and mice. It transmits the protozoan parasite, *Babesia* bovis and the virus of louping illness or "tickbite fever". Distribution: worldwide (see **Babesia**). 2. A genus of Spirochaetes. e.g. I. dammini - A spirochaete species causing Lyme arthritis.

Ixodiphagus A genus of Flies (order *Diptera*). e.g. I. caucurtei - A species parasitic on ticks, of the family *Ixodidae* or Hard Ticks.

J

Jabiru A genus of Birds of the Jabiru and Stork family, *Ciconiidae*. e.g. J. mycteria - The Jabiru, 1.4 m (55 in) long, one of the largest flying birds, living in the jungle swamps of Central and South America, ranging from Mexico to Argentina. It is white with a bare black head and upper neck; the lower neck is orange and scarlet. It also makes its nest high up in the trees.

Jacana A genus of Birds of the Jacana or Lily-trotter family, *Jacanidae*. e.g. J. spinosa - The American Jacana, ranging from Texas southwards to Argentina. Its general color is maroon and black; while in courtship it displays its wings to reveal bright yellow patches of feathers.

Jacobinia A genus of plants of the Acanthus family, *Acanthaceae*. e.g. J. carnea - The Jacobinia Carnea plant, an upright shrubby species, with opposite, downy, lanceolate leaves up to 30 cm (1 ft) long. The purplish flowers develop into handsome erect inflorescences. Each flower has a hooded upper lip, two fertile stamens, and a downward-curving lower lip. Place of origin: Brazil.

Jacquinia A genus of plants of the Joewood family, *Theophrastaceae*. e.g. J. smaragdina - A woody greenish flowers. Place of origin: Tropical America.

Jaculus A genus of Mammals of the Jerboa family, *Dipodidae*. e.g. J. jaculus - The Egypboas, found in the dry regions of North Africa, Egypt, Arabia, and Syria. A most gentle and charming rodent, it is adapted for jumping, its hind legs being unusually long. It has a long tail which helps to support it when squatting. Its coat is fawn-colored, harmonizing with the terrain.

Jaguarius A genus of Mammals, better known by the generic name, *Panthera*, of the Cat family, *Felidae*. e.g. J. onca (= *Felis* o.; *Panthera* o.) - The Jaguar, or Leopard of the New World (see **Panthera**).

Jambosa A genus of plants of the Myrtle family, *Myrtaceae*. Better known by the generic name, *Caryophyllus*. e.g. J. caryophyllus - The Clove or Caryophyllum plant (= *Caryophyllus aromaticus*) (see **Caryophyllus**).

Janthella A genus of Sponges of the order of Horny Sponges, *Keratosa*. These are primitive cellular animals, the gelatinous sponges, with a horny or siliceous skeleton that consists entirely of interlocking horny fibers. e.g. J. basta - A beautiful species found on the coral reefs of Australia. This horny sponge has no mineral skeleton. The fibers are interlocking and consist of a silk-like protein material known as spongia.

Janthinosoma A genus of two-winged Insects of the the Mosquito family, e.g. J. lutzi - A species of Mosquitoes, transmitting the eggs of the Bot fly (*Dermatobia*), glued to its abdomen.

Jasione A genus of plants of the Bellflower or Campanula family, *Campanulaceae*. e.g. J. montana - The Common Sheepsbit, a biennial plant, 10-15 cm high; without stolones; the leaves have undulate margins. Widespread in grassland, on heaths, cliffs and shingle. Distribution: Europe; native in Britain.

Jasminium A genus of plants, also known by the generic name, *Jasminum*, of the Olive family, *Oleaceae*. e.g. J. nudiflorum (= *Jasminum* n.) - The Winter Jasmine, a deciduous rambling plant with slender, angled stems, shortly stalked, trifoliate leaves and solitary, axillary flowers, each having a tubular corolla with six spreading lobes. The fruit is a black berry. Place of origin: China and Japan.

Jasminum See **Jasminium**.

Jateorhiza A genus of plants of the family *Menispermaceae*. e.g. J. palmata - The Calumba or Colombo plant; from the root is prepared a bitter, used in the treatment of atonic dyspepsia.

Jatropha A genus of plants of the Spurge family, *Euphorbiaceae*. e.g. J. urens - The Ortiga plant; it bears white hairs which may cause itching.

Johanssonia A genus of Insects of the order of Butterflies and Moths, *Lepidoptera*. e.g. J. acetosae - One of the smallest species of butterflies in existence, with an average wingspan of less than 3-4 mm (0.11-0.15 in) and a body length of 2 mm (0.078 in). Distribution: Britain (see **Nepticula**; **Stigmella**).

Jordanella A genus of Fishes of the Carp family, *Cyprinidae*. e.g. J. floridae - The Tooth Carp of Florida, U.S.A., the bony fish producing the least number of eggs. Mature females deposit only 200 eggs over a period of several days.

Jovibarba A genus of plants of the Crassula family, *Crassulaceae*. e.g. J. hirta - A perennial species with green lanceolate leaves. The erect stem bears a terminal cluster of bell-shaped flowers with erect pale yellow petals. Place of origin: the mountains of Central Europe and the northern part of the Balkans.

Juglans A genus of plants of the Walnut family, *Juglandaceae*. e.g. J. cinerea - The Butternut or Walnut tree. From the root bark is extracted an aperient or mild laxative. From the black walnut tree is derived an antibiotic juglone, active against certain fungi. J. regia - The Common Walnut tree; up to 30 m (100 ft) tall; with a pale grey bark. The pinnate leaves have 7 to 9 hairy leaflets, and the male flowers are borne in dense pendulous catkins. The greenfruit has an outer leathery layer. Has been planted for this edible fruit, the familiar "walnut", and its valuable timber. Place of origin: the Balkans; has become naturalized in many countries, including Britain.

Julus A genus of Diplopods or Millipedes of the order *Julida*. These creatures have up to 70 segments and can be rolled up into a ball. The first leg is modified for copulation, and in the seventh segment there is an extra copulatory apparatus for transferring sperm into the female vagina. One species found in Ambon, an island in the Central Moluccas, is reported to have an exceptionally venomous secretion. e.g. J. terrestris - A burrowing millipede. The female constructs a dome-shaped nest of earth mixed with saliva and lays her eggs through a hole in the top, sealing the nest with the same material. This species is often mistaken for a Wireworm (the larval form of the Click-beetle or Elaterid family, *Elateridae*).

Juncus A genus of plants of the Rush family, *Juncaceae*. e.g. J. subuliflorus - The Common Rush, a perennial plant, 25-100 cm high; the stem has approximately 40 strong ridges, the path is continu-ous, and the inflorescence is unusually dense and head-like. Grows on mountains up to 1,800 m, on banks, in ditches, damp meadows and wet places. Distribution: Widespread throughout Europe; native in Britain.

Juniperus A genus of evergreen shrubs or small trees, widely distributed over the northern hemisphere, of the Cypress family, *Cupressaceae*. e.g. J. communis - The Juniper tree, up to 9 m (30 ft) tall, with short linear leaves in whorls of three. A unisexual plant. The cone is a globose, blue-black, berry-like fruit containing from 1 to 6 seeds. From this berry or fruit is extracted a volatile oil, Juniper Oil, used as a carminative and diuretic; also to flavor gin and some liquers; the wood is used in cabinet-making and in the manufacture of pencils. Widespread in arctic and north temperate regions.

Jurinea A genus of plants of the Daisy family, *Compositae*. e.g. J. cyanoides - The Sand Jurinea, a perennial plant, 25-75 cm high; white-downy stem and pinnatisect leaves. Grows in sandy places, rich in chalk, in Central Europe and in the steppe, on "black earth" in south-eastern Europe.

Jynx A genus of Birds of the Woodpecker family, *Picidae* (Wryneck subfamily, *Jynginae*). e.g. J. torquilla - The Eurasian Wryneck, about 16.25 cm (61/2 in) long. A perching bird, making its nest in natural tree holes and feeding chiefly on ants caught with its long, extensile tongue. The plumage is a cryptic mottle of greys and browns. The wry neck is a primitive form of the woodpecker, and takes its common name from its habit of twisting its neck, rolling its eyes, and writhing from side to side when disturbed. A migrant bird, it inhabits most of Europe, central and northern Asia, and northwest Africa, wintering in the south.

K

Kakatoe A genus of Birds of the Parrot, Cockatoo, Macaw and Lory family, *Psittacidae*. e.g. K. galerita - The Sulfur-crested Cockatoo, a completely white bird, except for its distinctive erectile crest of long yellow feathers. Easily domesticated and popular as pets. Found in Australia, Tasmania, New Guinea, the Philippines and neighboring islands.

Kalanchoe A genus of plants of the Crassula family, *Crassulaceae*. e.g. K. blossfeldiana - The Kalanchoe, a flowering plant, about 30 cm (12 in) high, with light green, fleshy leaves edged in red, broadly ovate with scalloped margins. The flowers are borne in dense clusters from the upper leaf axils, they are richly scarlet in color, and individually small and tubular. Place of origin: Tropical Africa, southern regions of America and Africa.

Kallima A genus of Insects of the Tortoise shell Butterfly or Moth family, *Nymphalidae*. e.g. K. inachus var. formosana - The Dead-leaf Butterfly from Formosa (now Taiwan), an exotic species with perfect protective coloring and shape. The underside is inconspicuously greyish-, reddish-, or greenish-brown, shaded and stippled, and veined in exact imitation of a dead or dry leaf. The upper side is a shiny blue with large orange-yellow patches.

Kalmia A genus of plants of the Heath family, *Ericaceae*. e.g. K. latiflora - The Calico Bush, Sheep Laurel or Swamp Laurel, an evergreen shrub up to 2.7-3.6 m (9-12 ft) tall, with alternate or sometimes verticillate leaves grouped in threes at the tips of the branches, oblong-lanceolate, stemmed, and a beautiful bright green in color; the pinkish-white, large parasol-shaped flowers are borne on long peduncles in terminal corymbs. The leaves are used in the treatment of diarrhea and other inflammatory diseases. Place of origin: eastern North America; introduced into Europe in 1734.

Kalopanax A genus of Trees of the Ivy family, *Araliaceae*. e.g. K. pictus - The Prickly Castor-oil Tree, height 25-28 m (80-90 ft) in the wild, usually much smaller in cultivation; resembles a maple, but has alternate leaves and stout yellow prickles on its branches and suckers. The bark is dark grey and ridged, often knobbly with spines. Place of origin: Japan, East Russia, Korea and China.

Karagassiema A genus of prehistoric Arthropods of the class *Crustacea*. These are the earliest known crustaceans, 12-legged creatures which lived about 650 million years ago. Their remains have been found in the Sayan Mountains in the (former) U.S.S.R.

Karyamoebina A genus of amoebas, in which the peripheral chromatin is clumped into masses. e.g. K. falcata - A species found in the human intestine in California.

Katayama A genus of Gastropod Molluscs of the Snail family, *Helicidae*. e.g. K. nosopohra - An intermediate host of Schistosoma japonicum (Katayama disease or Schistosomiasis japonica) (see **Schistosoma**).

Katsuwonus A genus of Fishes of the family *Scombridae*. e.g. K. pelamis. Head to tail, with a silver underside. Habitat: the Atlantic and Pacific oceans.

Kentranthus A genus of plants of the Valerian family, *Valerianaceae*, better known by the generic name, *Centranthius* (see **Centranthius**).

Keratella A genus of microorganisms, freshwater worms, the Rotifers or Wheel Animalcules. These creatures are about 1/25 in long. They belong to the class *Rotifera*. These are metazoa; the whirling movement of two rings of cilia at the foot of the body propel them through the water, and are also used to convey food to the mouth. e.g. K. quadrata - An ornamental species.

Kernera A genus of plants of the Cress or Crucifer family; the basal leaves grow in a hairy rosette. Found on rocks, stony slopes and gravel. Place of origin: South and Central Europe.

Kerria A genus of plants of the Rose family, *Rosaceae*. e.g. K. japonica - The Japanese Kerria, an erect, slender-branched, deciduous shrub, 1.2-1.8 m (4-6 ft) high, with alternate leaves and solitary yellow, 5-petalled flowers. A double form, var. plenifera ("Plena"), is frequently seen in gardens, and has been known in Europe since 1700. Place of origin: China and Japan; introduced in its single form, by William Kerr, into Europe in 1804.

Kickxia A genus of plants of the Figwort or Snapdragon family, *Scrophulariaceae*. e.g. K. spuria - The Round-leaved Fluellen, an annual plant, 3-50 cm high, with decumbent, woolly flower-stalks. Grows in cultivated fields on light soils. Place of origin: widespread in Europe; native in Britain.

Kinosternon A genus of Reptiles of the Mud and Musk Turtle family, *Kinosternidae*. e.g. K. flavescens - The Yellow Mud Turtle or Terrapin, a freshwater species, about 12.5 cm (5 in) long, found in ponds and marshes of the southern United States and Mexico. It has distinctive yellow markings on the chin and throat, and can enclose itself completely within its shell by pushing the ends of the plastron (the skeletal structure protecting its ventral surface), up against the flattened carapace.

Kinyxis A genus of Reptiles of the family of Land Tortoises, *Testudinidae*. e.g. K. erosa - The Forest Hinged Tortoise, a strange reptile, wshell more than 30 cm (1 ft) long, and a native of tropical West Africa. Its carapace is divided, being bell-shaped in front.

Kitaibelia A genus of plants of the

Mallow family, *Malvaceae*. e.g. K. vitifolia - Leaves and attractive flowers. Place of origin: Bosnia and Serbia.

Kittacincla A genus of Birds of the Thrush, Warbler and allied family, *Mus* (Thrush, Nightingale, Robin, etc. subfamily, *Purdinae*). e.g. K. malabarica - The white-rumped Shama, a black-bird with a chestnut abdomen and a white rump; the outermost feathers are also white. It lives in thickets in the forests, and is regarded as one of the best songbirds in the world. Habitat: the islands of the Indo-Australasian archipelago.

Klebsiella A genus of microorganisms of the family *Enterobacteriaceae*. These consist of short gram-negative rods. They have been isolated from several animals and from inanimate objects. May be part of the normal flora orpathogenic (= Kb; Klebs). e.g. K. pneumoniae (= Friedländer's bacillus; K. friedlandedlanderi; Klebs f.; Kb pneumoniae; Klebs p.; Pneumobacillus Friedländer) - Part of the normal flora in the nose, mouth and intestines. Tends to be more invasive than the closely related (genus) *Enterobacter* organisms and may cause lesions in almost every part of the body - pneumonia, chronic lung abscess, upper respiratory-tract infections, sinusitis, endocarditis, septicemia, meningitis, gastro-enteritis, wound infections, uterine and vaginal infections, and skin and urinary tract infections. In children, may cause severe enteritis. Debilitated persons and cirrhotics are more susceptible to respiratory infections by this organism.

Kleinia A genus of plants, also known by the generic name, *Senecio*, of the Daisyfamily, *Compositae*. e.g. K. cuneifolia (= *Senecio* cuneatus) - A species native to South Africa; the stems are jointed and the leaves are thick and fleshy, with small terminal clusters of cylindrical, rayless capitula (see **Senecio**).

Klinophilus A genus of Insects also

known by the generic name, *Cimex*. e.g.
K. lectularius (= *Cimex* l.) - The
Common Bedbug (see **Cimex**).

Knautia A genus of plants of the
Scabious or Teasel family, *Dipsacaceae*.
With mostly entire, serrate leaves, and
calyx with 8 bristles. Found in woods in
hilly districts. Place of origin: Central
Europe.

Kniphofia A genus of plants, also known
by the generic name, *Tritoma*, of the
Lily family, *Liliaceae*. e.g. K. uvaria (=
Tritoma u.) - The Red Hot Poker or
Torch Lily, a grass-leaved plant with
conspicuous spikes of brilliant coral-red,
drooping, tubular flowers with long stamens. Widely strap-shaped, long, arching leaves. There are many cultivars,
variously colored in reds, oranges,
creams and scarlet-reds. Place of origin:
South Africa; introduced into Europe
since 1707.

Kobresia A genus of plants of the Sedge
family, *Cyperaceae*. e.g. K. simplicius-
cula - The Common False-sedge, a
perennial plant, 5-30 cm high; with
grooved leaf blades. Grows on damp
calcareous ground at heights of 1600-
2800 m. Place of origin: the Pyrenees;
south, central and northern Europe; the
Arctic region. Native in Britain.

Kochia A genus of plants of the
Goosefoot family, *Chenopodiaceae*. e.g.
K. laniflora - The Hairy Summer
Cypress, an annual plant, 14-50 cm
high; the perianth or outer part of the
flower has 5 appendages. Grows on
sandy fields and hillocks, and in pastures. Place of origin: South and Central
Europe.

Koeleria A genus of plants of the Grass
family, *Gramineae*. e.g. K. glauca -
Glaucous or Sand Hair-grass, a perennial plant, up to 45 cm (18 in) high. The
grass stem or culm is swollen at the
base, and the leaves are small and greyish-green. The narrow, shiny inflorescences have compressed spikelets, each
containing 2 to 3 florets. Grows wide-

spread in non-alkaline, usually sandy
soils, and especially dunes. Place of origin: Central Europe; western Asia.

Koelreuteria A genus of Trees of the
Soapberry family, *Sapindaceae*. e.g. K.
paniculata - The Goldenrain Tree or
Pride of India, a deciduous tree from 9-
18 m (30-60 ft) tall; yellow flowers
growing in clusters. The fruit is a red
pod, about 3.7-5 cm (11/2-2 in) long and
contains 3 black seeds. The green leaves
have 9-15 leaflets, and turn yellow in
the autumn. Cultivated for ornamental
usage in large gardens and collections.
Place of origin: China, Korea and Japan.

Koenenia A genus of Arachnids of the
order of Micro-whip Scorpions,
Palpigradi. These palpigrades are
obscure creatures, less than 2 mm long,
about which very little is known. The
body consists of a prosoma covered by a
large anterior carapace. The species are
worldwide.

Krameria A genus of polygalaceous
shrubs of the family *Krameriaceae*. e.g.
K. argentea - The Brazilian Rhatany.
The root is known as Pará rhatany, and
is used as an astringent; also as a sup-
pository for bleeding or prolapsed hemorrhoids.

Krohnius A genus of Fishes of the Rat-
tail family, *Moridae*. These are oceanic
deepsea forms, with two forward projections from the air bladder; the larvae
have extremely long pelvic rays, hence
the common name. They are found
especially in the North Atlantic and
Pacific Oceans, and are related to the
Grenadier fishes (family *Macrouridae*).

Kronosaurus A genus of extinct prehistoric animals of the class of Reptiles,
Reptilia. e.g. K. queenslandicus - A
short-necked pliosaur, the largest marine
reptile ever recorded, which swam in the
seas around what is now Australia about
100 million years ago. It had a 3.04 m
(10 ft) long skull containing 80 spiked
teeth and measured up to 15.25 m (50
ft) in overall length.

Kryptopterus A genus of Fishes of the
suborder of Catfishes, *Siluroidea*. e.g. K.
bicirrhis - The Glass Catfish, a voracious eater with barb and a scale-less
bony-plated body, with a fleshy fin near
the tail. A freshwater edible species.

Kurthia A genus of microorganisms, for-
merly known by the generic name,
Zopfius, of the order *Eubacteriales*.
Long rods with rounded ends found in
decomposing matter. e.g. K. zenkeri (=
Zopfius z.).

Kurtus A genus of Fishes of the
Forehead Brooder or Kurtus family,
Kurtidae. These are flattened, deep-bodied fishes with a large, almost vertical
mouth, with 10 pairs of much enlarged
ribs encapsulating the air bladder. The
common name is due to the eggs being
retained and "brooded" in a special cavity on the forehead until hatched.

Kyphosus A genus of Fishes of the Carp
family, *Cyprinidae*. e.g. K. sectatrix -
The Bermuda Chub fish. It is frequently
found in deep holes in rivers, shaded by
trees; rarely weighs over 5 lbs, and
makes poor food.

L

Labea A genus of Afro-asian Cyprinid Fishes of the family *Cyprinidae*. These include the barbels, carps, minnows, etc. (see **Carassius**).

Labeo A genus of Freshwater Fishes of the Carp, Minnow, Tench and allied faily, *Cyprinidae*. e.g. L. rohita - The Rohu, a large cyprinid fish, up to 1.8 m (6 ft) long and weighing over 45 kg (100 lb), found in rivers and streams in India. Other Indian species are the Mahseer (*Barbus* tor) and the Catla (*Catla* catla). In Asia alone there are nearly 150 genera in India and China, and nearly 60 genera in Siberia, of the family *Cyprinidae* (see **Carassius**).

Labia A genus of Insects of the Earwig family, *Labiidae*. e.g. L. minor - The Small Earwig, leng insect, common in fields, meadows, forest margins and greenhouses. Distribution: almost worldwide.

Labroides A genus of Marine Fishes of the Wrasse family, *Labridae*. e.g. L. dimidiatus - The Cleaner Wrasse, and white fish with spiny fins and prominent thick lips. This species lives in symbiosis with its neighbors, cleaning the parasites off many other fish, and thereby obtaining food while providing a much needed service. It is amazing how the Cleaner Wrasse enters the mouth and cleans the teeth of the most voracious predators. Distribution: warmer seas, such as the Red Sea and the Gulf of Eilat (Akaba) (see **Labrus**).

Labrus A genus of Marine Fishes of the Wrasse family, *Labridae*. These are short fishes in which the spiny part of the dorsal fin is longer than the soft-rayed part. Many of these are amongst the most brilliantly colored fishes anywhere in the world. e.g. L. viridis - The Green Wrasse, 40 cm (16 in) long; the back is greenish with blue dots. A marine species widespread along the coast of Europe from the Mediterranean to the North Sea.

Laburnocytisus A genus of Trees of the Pea family, *Leguminosae*. e.g. L. adamii - Adam's Laburnum, a fascinating tree, originally produced by grafting Dwarf Purple Broom (*Cytisus* purpureus) on to the Common Laburnum (*Laburnum* anagyroides), forming a hybrid or chimaera at the graft junction. Height about 7.5 m (25 ft). The yellow-purple brown flowers occur on sprouts of broom-type foliage and are produced along with flowers of both parent trees. Grown for its curiosity value in Britain, Europe and North America (see **Cytisus**; **Laburnum**).

Laburnum A genus of Trees of the Pea family, *Leguminosae*. e.g. L. anagyroides (= *Cytisus* laburnum; L. vulgare) - The Common Laburnum, Golden Rain, or the Golden Chain Tree, a deciduous tree and one of the most poisonous species in Britain; up to 7 m (23 ft) tall. It has a smooth bark, alternate trifoliate leaves and golden-yellow, papilionate or butterfly-shaped flowers growing in racemes and with a bean-like scent. The flattened pods each contain several poisonous seeds; all parts are toxic, causing Cytisism. The toxic alkaloid is Cystisine (= Babitoxine; Laburnine; Sophorine; Ulexine). Used as a respiratory stimulant. The wood can be used as a substitute for ebony in furniture making. A native of central and southern Europe but is widely cultivated; introduced and naturalized in Britain (see **Cytisus**).

Laccifer A genus of Homopterous Insects of the family of Scale Insects, *Coccidae* e.g. L. lacca - A scale insect species, living on and suck of the stems of various plants, and producing a resinous substance known as Shellac (= Lacca; Lacca in Tabulis). It has the appearance of small scales on the plant's surface, hence the name. Shellac is used as a lacquer in hair sprays, and may cause lung damage (interstitial pulmonary fibrosis or thesaurosis).

Lacerta A genus of Reptiles of the Typical Lizard family, *Lacertidae*. e.g. L. agilis - The European Sand Lizard, up to 20 cm (8 in) long; the female is light-brown or greyish above with 3 longitudinal series of irregularly shaped dark brown or black spots, each with a central white spot. The flanks and underparts of the mature male are bright emerald green. Found mainly on warm hillsides, field boundaries and the fringes of forests, and in gardens, parks and clearings. Distribution: the most common lizard of Central Europe, its range extending eastward as far as the temperate regions of western Asia.

Lachesis A genus of Reptiles of the Pit Viper snake family, *Crotalidae*. e.g. L. muta - The Bushmaster, a large, venomous snake, about 3 m (10 ft) long, living in the scrub and forest regions of tropical and Central America, where it is camouflaged into near invisibility by its dead-leaf color and spotted markings. An oviparous species, the female protects her eggs by coiling round them until they hatch. e.g. L. lancelatus - The Fer-de-lance, a venomous species.

Lachnus A genus of Homopterous Insects of the Fly superfamily, *Aphidoidea*. e.g. L. roboris - The Aphid or Greenfly, one of the most injurious agricultural and horticultural pests of north temperate countries, and serving as vector of several plant viruses. It secretes a sugary liquidor honeydew which is attractive to ants, such as *Formica* rufa (see **Formica**), and evolves a beneficial association with the species.

Lactarius A genus of plants of the family of Gill Fungi, *Agaricaceae*. e.g. L. deliciosus - The Saffron Milk Cap, an edible species 4-10 cm high; the brick-colored cap has darker concentric markings and the milk or sap is a bright saffron-red. Found growing in coniferous

forests and woods. Place of origin: Europe; native in Britain.

Lactobacillus A genus of microorganisms of the family *Lactobacillaceae*. Large gram-positive microaerophilic or anaerobic rods that may be part of the normal flora of the intestinal tract, vagina, or mouth. Mostly regarded as non-pathogenic but some species are considered etiologically related to dental caries. e.g. L. acidophilus (= *Bacillus* a.; Boas-Oppler bacillus; Doderlein's b.) - A lactic acid producing organism found in milk and buttermilk, and normally present in the human intestine. Also part of the normal oral flora. It causes acid fermentation of glucose and lactose, and is commonly used as living cultures in broth, whey, and whole milk. Has been isolated from carious tooth dentin, and from stools of bottle-fed infants or adults on diets high in milk content. L. bifidus (see **Bifidobacterium**). L. bulgaricus (produces the LBF or LB Factor - pantetheine).

Lactophrys A genus of Fishes of the suborder of Trunkfishes, *Ostracoidei*. e.g. L. trigonus - The Common Trunkfish, sometimes called a Cowfish because the dermal prolongations from the head resemble cows' horns. It is up to 30 cm (1 ft) long and has a bony case around the whole body, caused by the fusing together of its scales.

Lactrodectus A genus of Arachnids of the order of True Spiders, *Araneae*. e.g. L. mactans mactans - A poisonous species of spiders with a powerful neurotoxic venom. Only 0.11 mg (0.0000038 oz) of the poison is needed to kill a 20 g (0.706 oz) mouse when injected intravenously, as discovered under laboratory conditions. The lethal dose for man is unknown.

Lactuca A genus of plants of the Daisy family, *Compositae*. e.g. L. virosa - The Greater Prickly Lettuce or Wild Lettuce, an annual to biennial plant, 50-100 cm high, with the stem leaves usually horizontal and spiny-fringed at the margins. Grows mainly in wasteland, by roads and sometimes on walls. An extract is used in the treatment of cough. The dried juice is known as Lactucarium or Lettuce Opium and is used likewise as a sedative in cough. Place of origin: South and Central Europe; native in Britain.

Laelaps A genus of Arachnids of the Tick superfamily, *Ixodoidae*. These are related to mites but are larger. The adult tick has an oval unsegmented body with a movable head, through which it draws blood from man and other animals, after burrowing under the skin. The species of this genus are found on rats and in stables, and their bite causes intense itching.

Laevicardium A genus of Bivalve Molluscs of the Cockle family, *Cardiida*. Their valves are thick and there are about 40 ill-defined ribs. Distribution: Europe; native around Britain.

Lafoea A genus of Hybrids or Sea-firs of the family *Lafoeidae*. These thecate hydroid animals form colonies of parallel stolons giving mechanical support, and polyps growing at right angles to these stolons in elongated thecae, without lids. They are found on algae and stones near the seacoast or between tidemarks.

Lagendelphis A genus of Mammals of the Dolphin family, *Delphinidae*. e.g. L. hosei - Hose's Sarawak Dolphin or Fraser's Dolphin, a species inhabiting the tropical waters of the Pacific and Indian Oceans. A rare marine animal.

Lagenorhynchus A genus of Mammals of the Dolphin and Grampus family, *Delphinidae*. e.g. albirostris - The White-beaked Dolphimal, found in all seas.

Lagerstroemia A genus of plants of the Loosestrife family, *Lythraceae*. e.g. L. indica - The Crape Myrtle, a large deciduous shrub or small tree, with an attractively mottled stem in grey, pink and cinnamon. The flowers grow in 6-petalled, showy, terminal patches, pink to deep-red with crinkled petals and numerous stamens. Place of origin: China, Korea. Introduced into Europe in 1754.

Lagidium A genus of Mammals of the Chinchilla and Viscacha family, *Chinchillidae* e.g. L. viscaccia - The Mountain Chinchilla or Mountain Viscacha, a gregarious mammal living much like a rabbit, in ramifying burrows near arable land. These burrows undermine the ground. As big as a hare, it has a fur of poor quality, unlike the species of the genus *Chinchilla*. Distribution: the foothills of the Andes in South America (see **Chinchilla**).

Lagochilascaris A genus of unsegmented worms of the phylum *Nematoda*. e.g. L. minor - A nematode worm found in the human intestine in Trinidad.

Lagopus A genus of Birds of the Grouse or Ptarmigan family, *Tetraonidae*. e.g. L. mutus - The Rock Ptarmigan, length 37 cm (15 in), a grouse-like resident bird of circumpolar distribution, the northernly part of their range covering Scotland, Sweden, Norway, Lapland, Siberia, and Canada. In summer, distribution is just below the snowline, nestling close to fern-covered and glacial regions. During the winter they may descend to the valleys, where they are confined to rocky areas close to forest. In winter the plumage is white, so are the feathered legs and toes; only the beak and tail are black. The bird is thus well camouflaged against the snow. In summer, red and black-barred feathers of the breeding plumage appear on the wings and the upper parts of the body.

Lagorchestes A genus of Mammals of the Kangaroo family, *Macropodidae*. These are the Hare-wallabies, deriving their name from their resemblance to the hare. Distribution: Australia.

Lagostomus A genus of Mammals of the Chinchilla and Viscacha family,

Chinchillidae e.g. L. maximus - The Plains Viscacha or Vizcacha, about 50 cm (20 in) long; it has a coarse, brownish-grey coat with characteristic paler bands on its large ugly head. These animals build underground burrows, where they breed in large numbers. Distribution: the pampas of Argentina.

Lagothrix A genus of Mammals of the New World Monkey family, *Cebidae*. These are the Woolly monkeys, and are fairly widespread in the northern half of South America. e.g. L. lagothrica - Humboldt's Woolly Monkey, up to 67.5 cm (2 ft 3 in) long, with a tail of the same length; one of the largest of the New World monkeys. A slow-moving animal, it lives in small groups and is called by the natives of South America, where it lives, the Capparo or Barrigudo. It has a grey woolly coat covering the whole body except the palms of the hands, soles of the feet, and the ears. Distribution: the forests of South America, especially in the region of the Orinoco and Amazon rivers.

Lagria A group of Insects of the Coleopterous Beetle family, *Lagriidae*. e.g. L. hirta - Length 7-10 mm; found in damp vegetation in meadow and alongside paths and streams. Distribution: Europe, Siberia.

Lagurus A genus of plants of the Grass family, *Gramineae*. e.g. L. ovatus. 10-30 cm (4-12 in) high, with greyish-green leaves and soft, woolly, ovoid inflorescences. The compressed spikelet contains a single floret. Found in dry areas near the sea. This beautiful grass is often grown for the attractive flowering heads. Place of origin: the Mediterranean region; has become widely naturalized in southern England, Australia, South Africa and South America.

Lama A genus of Mammals, also known by the generic name, *Llama*, of the Camel family, *Camelidae*. e.g. L. guanacos (= L. huanacos; *Llama* g.; *Llama* h.)

- The Guanaco or Huanaco, a wild ruminant mammal, the Camel of the New World, and closely allied to the camel, but smaller (about 1.2 m = 4 ft high), humpless and woolly-coated, and generally colored a red-brown. Distribution: the Cordillera mountain range of the Andes in southern Peru and Chile, extending southwards to Patagonia and Tierra del Fuego. An extremely shy animal which has also been domesticated. L. guanicoe glama (= L. glama) - The Domesticated South American Guanoco, a sturdier species than its wild cousin, standing as much as 1.25 m (4 ft 2 in) at the shoulder. Its fur is white, black, reddish-brown, and various combinations of these. Used as an indispensable beast of burden. Cloth is made from the wool of these animals. It inhabits with man the mountain plateaux of Bolivia, Peru and other South American countries.

Lamblia A genus of Flagellate Protozoans, better known by the generic name, *Giardia*. e.g. L. intestinalis (= *Giardia* lamblia) (see **Giardia**).

Lamellaria A genus of Molluscs of the Periwinkle and Slipper Limpet order, *Mesogastropoda*. These are the slug-like mesogastropods, and unlike the "true" sea slugs, they have only one pair of tentacles.

Lamia A genus of Insects of the Longhorned or Longicorn Beetle family, *Cerambycidae*. e.g. L. textor - Size 2.6-3.2 cm; broad body with very thick first segment of the antennae. The larvae burrow in wood and may be serious pests of forest trees.

Laminaria A genus of plants of the family of Brown Algae, *Fucaceae*, known as the Kelps or Oarweeds (subfamily, *Phaeophyceae*). e.g. L. digitata - The Tangle Seaweed or Oarweed, one of the Kelp Algae, growing in large quantities on rocky coasts, at or slightly below tide mark off the west coast of Scotland and Ireland. Up to 3 m (10 ft) high, the laminae are split into linear, strap-like, leathery segments. From this species is extracted Sodium Alginate (= Algin; Sodium Polymannuronate). Used as a suspending and thickening agent in the preparation of pastes and creams. Laminaria stalks are used surgically to dilate cavities in the form of solid or hollow cylinders known as "Laminaria Tents". Place of origin: Europe; native in Britain.

Lamium A genus of plants of the Thyme family, *Labiatae*. e.g. L. album - The White Dead-nettle, a hairy perennial plant, 15-80 cm high, with white flowers borne in dense axillary whorls; the coralla-tube is slightly curved and has an oblique ring of hairs. Grows in hedgebanks, waste places and by road-sides. Place of origin: widespread from western Europe to the Himalayas and Japan. Native in Britain.

Lamna A genus of Fishes of the Mackerel Shark family, *Lamnidae* (sometimes known as *Isuridae*). e.g. L. nasus - The Porbeagle or Common Atlantic Mackerel Shark, 1.5-3.5 m long, and weighing 100-150 kg. The teeth are slender, narrow and straight with a small cusp on either side; the dorsal fin is far forward. A viviparous shark that hunts chiefly herring and mackerel but also smaller sharks, halibut, cod and octopus near the water surface. Distribution: from the seas off Japan to European waters. Also the western Atlantic (see **Isurus**).

Lampetra A genus of Fishes of the order of Lampreys, *Hyperoartia* or *Hyperortii* (= *Petrozntiformes*). An eel-like fish with tooth-studded mouth, used as a sucker when it preys upon other fish. It is the most primitive extant vertebrate (i.e. still in existence up to the present day). e.g. L. fluviatilis - The Lampern or River Lamprey fish, a freshwater species which leaves the sea and swims far up rivers to their upper reaches. 13-50 cm long; the body is eel-shaped with 7 gill openings on either side, and has

no pelvic or pectoral fins. Distribution: Europe; native in Britain.

Lampranthus A genus of plants of the Mesembryanthemum family, *Aizoaceae*. e.g. L. coccineus - The Ice Plant, a beautiful perennial, succul species, 30-45 cm (1-11/2 ft) high, with masses of bright, red daisy flowers in spring. The leaves are cylindrical, fleshy and grey-green in color. Cultivated as an orna-mental plant. Place of origin: South Africa.

Lampris A genus of Fishes of the Moonfish family, *Lamprididae*. e.g. L. regius - The Moonfish, also called the Opah or Jerusalem haddock, the only species of the genus; 1.8 m (6 ft) long and weighing 270 kg (600 lb); the body is almost as deep as its length, and it is plump and smooth, without scales or bony plates. A rare fish, occasionally seen floating on the surface of the water. Distribution: cosmopolitan.

Lamprocystis A genus of microorgan-isms, formerly known by the generic name, *Clathrocystis*, of the family *Thiorhodaceae*. Spherical to ovoid cells embedded in a common gelatinous cap-sule. e.g. L. roseopersicina (= *Clathrocystis* r.).

Lampropeltis A genus of Reptiles of the Colubrid Snake family, *Colubridae*. e.g. L. zonata - The from the western United States. A beautifully vivid, pastel-col-ored species up to 1.5 m (5 ft) long. It attacks its prey in the same way as the Boa and Python, coiling itself round the victim, and only when the latter is weakened by strangulation does it start to swallow it.

Lampyris A genus of Insects of the Glow-worm Beetle family, *Lampyridae*. e.g. L. noctiluca - The Common Glow-worm, a beetle species widely inhabit-ing many regions, including North America. Length 11-18 mm; the males are winged, the females wingless. Found on the fringes of woods and in damp forest clearings. They are luminescent, grub-like creatures with a complex light organ on the underside of the body or abdomen, which emits a shining bright, green light in the dark in wavelengths of 518-656 mm. Light is produced by a complex chemical process in which oxygen, conveyed by the tracheae, plays an important role. The adult glow-worms are called Fireflies. The larvae, which also emit light, grow to a length of 23 mm. They live in grass and are carnivorous on snails and slugs. Distribution: Europe, from the Mediterranean to Central Scandinavia, the Caucasus, Siberia, China, North America. Native in Britain.

Lamus A genus of predatory Insects of the Cone-nosed Bug family, *Reduviidae*, the species of which are now classified under the generic names, *Panstrongylus* and *Triatoma*. e.g. L. magistus (= *Panstrongylus* m.; *Triatoma* megista) (see **Panstrongylus**; **Triatoma**).

Lanice A genus of Segmented Worms of the Terebellid family, *Terebellidae*. e.g. L. conchilega - The "Sand Mason" Terebellid worm, up to 30 ft long; familiar in Europe. It constructs shelly tubes with frilly tops projecting about 5 cm (2 in) out of the sand. Distribution: North European coasts; native in Britain.

Lanius A genus of Birds of the Shrike or Butcherbird, and allied family, *Laniidae*. e.g. L. excubitor - The Great Grey Shrike, length 25 cm (10 in), a species found in Europe and America. In the New World it is known as the Northern Shrike. It is the largest European shrike, inhabiting olive groves, woods, hedges, and isolated trees. It has an ash-grey-hood, pinkish-white belly and black wing feathers edged with white. An aggressive and carnivorous bird, its bill is strong, conical and hooked, and it has strong, sharp-pointed claws. The shrikes or butcher-birds are thus well equipped for hunting, but unlike other birds of prey they carry their victim in their beaks, not in their claws. A partial

migrant bird. Distribution: Europe (not found in the north or Italy and the Balkans); Asia; North Africa; North America.

Lankesterella A genus of parasitic protozoans of the class *Sporozoa*. e.g. L. ranarum - A sporozoan parasite of the red blood cells of the frog.

Lankesteria A genus of parasitic protozoans of the class *Sporozoa*. e.g. L. culicis - A gregarine sporozoan parasite found in the intestine of the mosquito, *Aedes* aegypti.

Lantana A genus of plants of the Verbena or Vervain family, *Verbenaceae*. e.g. L. camara - The Lantana, 1.8-3 m (6-10 ft) high, with heads of pink to orange flowers and round black fruits. The infusion from the leaves is used in South America as a tonic and stimulant. Place of origin: tropical South America and Jamaica; introduced into Britain in 1692.

Laothoe A genus of Insects of the Hawk Moth or Hawk Butterfly family, *Sphingidae*. e.g. L. populi - The Poplar Hawk Moth, 8.5 cm long; the wings are ash-grey, the margins indented, and the forewings have a dark band. Distribution: widespread in temperate Europe; native in Britain.

Laphria A genus of Insects of the dipterous family of Robber Flies, *Asilidae*. e.g. L. flava - Length 16-25 mm; it waits for the arrival of its prey on logs of plants, where it hunts various insects, piercing its victims and sucking their juices. Distribution: Europe.

Laportea A genus of plants of the Nettle family, *Urticaceae*. e.g. L. gigas - A tree with poof fruits resembling raspberries. Place of origin: Queensland, Australia.

Lappula A genus of plants of the Borage family, *Boraginaceae*. e.g. L. myosotis - The Common Stitchseed, an annual plant, 10-80 cm high; the stem is weak and the nutlets have hooked spines on the winged edges. Found commonly in dry waste and arable land. Place of origin: Europe; introduced and naturalized in Britain.

Lapsana A genus of plants of the Daisy family, *Compositae*. e.g. L. communis - The Nipplewort, an annual plant, 20-90 cm high; the basal leaves have large terminal lobes; there are numerous heads with a few ray-florets. Grows by waysides, on sandy banks, waste land, heaths and in grassy places. Place of origin: Europe; native in Britain.

Laptonychotes A genus of Mammals of the Earless Seal family, *Phocidae*. e.g. L. weddelli - The Weddell Seal, the deepest-diving pinniped and the world's most southerly mammal. Found along the Antarctic mainland and neighboring islands. Adult bulls regularly descend to 274-305 m (900-1,000 ft) in search of food like fish, squid and crustaceans, staying there for 20 minutes to an hour or more.

Laria A genus of Insects of the order of Beetles, *Coleoptera*. e.g. L. lentis - A pest of North America, this beetle attacks lentils. Its larvae eat their way inside a lentil and there pupate. The adult insect then eats its way out again.

Larix A genus of Trees of the Pine family, *Pinaceae*. e.g. L. decidua (= L. europaea) - The Common or European Larch, a deciduous tree up to 45 m (150 ft) tall, with clusters of narrow leaves or needles borne on lateral short shoots. The soft pinkish cones are pinkish at the time of pollination, but when mature they are brown and woody. From the tree is extracted the oleoresin, Venice turpentine (= Larch turpentine; Terebenthina Laricina), used as an ingredient of ointments and plasters. Place of origin: this conifer grows wild in forests on central and south European mountains east of Siberia, but is extensively planted all over for its valuable timber.

Larosterna A genus of Birds of the Gull and Tern family, *Laridae*. e.g. L. inca - The Inca Tern, a monotypic bird. The bill is pointed and it has very long and narrow pointed wings. Sometimes called the sea-swallow because of its forked tail. Distribution: South America.

Larus A genus of Birds of the Gull and Tern family, *Laridae*. e.g. L. marinus - The Great Black-backed Gull, about 75 cm (21/2 ft) long, with a wingspan of 1.5 m (5 ft). The largest of the gull family and a partial migrant. It is white, with black on top of the wings and body. Habitat: around the seacoast and on the estuaries of large rivers in northern Europe, Asia and North America, nesting on the cliffs or on the ground. Native in Britain.

Laser A genus of plants of the Carrot family, *Umbelliferae*. e.g. L. trilobum - The Common Laser, a perennial plant, 60-120 cm (2-4 ft) high; the leaflets are roundish and smell of caraway. Place of origin: Central Europe; introduced and naturalized in Britain.

Laserpitium A genus of plants of the Carrot family, *Umbelliferae*. e.g. L. latifolium - The Herb-frankincense or Broad-leaved Laserwort, a perennial plant, 60-125 cm high; pinnate leaves with ovate segments, which are heart-shaped at the base. The fruit is 8-winged. Grows in hilly districts. Place of origin: Europe.

Lasiocampa A genus of Insects of the Lepidopterous Butterfly and Moth family, *Lasiocampidae*. e.g. L. quercus - The Oak Eggar Butterfly, 26-37 mm long; a species found in oak and mixed woodlands or over heaths and moors. There are several color forms. Distribution: southern and central Europe, Asia Minor, Transcaucasia, Siberia. Native in Britain.

Lasiodora A genus of Arachnids of the order of True Spiders, *Araneae*. e.g. L. klugi - A long-haired species, the heaviest spider ever recorded. It measures up to 24 cm across the legs and weighs almost 85 g. It is most formidable looking, but is deadly only to small animals such as mice. Found in Brazil and other countries in South America.

Lasiohelia A genus of Insects of the blood-sucking Fly family, *Heleidae.*

Lasiorhinus A genus of Mammals of the Wombat family, *Vombatidae.* Hairy-nosed wombats; they are somewhat badger-like, but their dentition is exactly found in rodents. They are omnivorous animals, though vegetable foods predominate. They are like rodents in that they live socially and riddle the earth with their burrows.

Lasius A genus of Insects of the Ant family, *Formicidae.* e.g. L. niger - The Black Ant, length 3-5 mm; blackish-brown in color; it feeds on seeds, small animals and honeydew milked from plantlice. Distribution: Europe; native in Britain.

Laspeyresia A genus of Insects of the Bell Moth, Coddling Moth or Tortoise family, *Tortricidae.* e.g. L. pomonella - The Coddling Moth, length 2.8 cm; the forewings have rust-red transverse lines. It is an orchard pest, the caterpillars living in apples. Distribution: Europe; native in Britain.

Laterallus A genus of Birds of the Coot, Moorhen, and Rail family, *Rallidae.* e.g. and the smallest rail; black with a reddish nape and white bars on the back. Distribution: temperate North America.

Laternaria A genus of Insects of the Lantern Fly family, *Fulgoridae.* e.g. L. phosphorea - The Central American Lantern Fly, a tropical insect with a hollow anterior prolongation of the head.

Lates A genus of Fishes of the Perch family, *Percidae.* e.g. L. niloticus - The Nile Perch, Africa's only giant freshwater fish, measuring up to 1.37 m (41/2 ft) in length and weighing up to 120 kg (264 lb).

Lathraea A genus of plants of the Broomrape family, *Orobanchaceae.* e.g. L. squamaria - The Toothwort, a perennial plant, 5-35 cm high, with scale-like leaves and white or pale-pink inflorescences of one-sided racemes. Widespread and parasitic on roots of woody plants. Place of origin: Europe; native in Britain.

Lathyrus A genus of plants of the Pea family, *Leguminosae.* e.g. L. niger - The Black Pea, a perennial plant, 40-80 cm high; the leaflets are often micronate or abruptly tipped with a short point, and the stem is angled. The ingestion of the seeds results in a morbid condition known as lathyrism, characterized by spastic paraplegia, pain, hyperesthesia and paresthesia. Place of origin: southeast Europe and Russia. Native in Britain.

Laticauda A genus of Reptiles of the Sea-snake family, *Hydrophiidae.* e.g. L. colubrina - The Amphibious Sea-snake, an Asiatic species distributed from India to Japan and Australia. It has enlarged ventral scales to aid movement on land, when it leaves the water in order to reproduce. The females are oviparous and reach a length of up to 1.5 m (5 ft); they are twice the size of males.

Latimeria A genus of Fishes of the Coelacanth family, *Latimeriidae.* e.g. L. chalumnae - The Coelacanth fish, about 90 cm (3 ft) long; first discovered off the coast of South Africa in 1938. It was previously known only in Fossil or Carboniferous forms of the Devonian to the Cretaceous periods (350-120 million years ago). A deep-bodied, greyish-blue fish, it has a 3-lobed, diphycercal tail (i.e. the tail fin is equally developed on both the dorsal and the ventral sides of the vertebral column).

Latrodectes See **Latrodectus.**

Latrodectus A genus of Arachnids, also known as *Lactodectes,* of the Spider family, *Therididae.* e.g. L. mactans (= *Latrodectes* m.) - The Black Widow, the most notorious of all venomous spiders, widely distributed throughout the warmer regions of the world, although it is the North American subspecies which has acquired the worst reputation. It is black with a red, hourglass-shaped mark on its underside, and produces a neurotoxin which causes severe pain, muscular cramps, paralysis and hypertension. Distribution: United States.

Laurencia A genus of plants of the Red Algae family, *Gigartinaceae*. e.g. L. pinnatifida - The Pepper Dulse, a species of marine Red Algal plants, up to 10 cm high, cartilaginous, flat and branched. Place of origin: the North Sea; native around Britain.

Laurus A genus of Trees of the Laurel family, *Lauraceae*. e.g. L. nobilis - The Bay Laurel or Sweet Bay, an evergreen shrub or tree, growing up to 8 m (26 ft); the leaves are lanceolate, wavy-edged and with an aromatic odor, and the scented yellow flowers are dioecious i.e. the male and female flowers are borne on separate plants. The berries, Laurel Berries (= Bay-laurel Berries; Lauri Fructus; Lorbeer-Frucht; Sweet Bay Berries) were formerly used as a carminative, emmenagogue and diuretic. The leaves are used to flavor food and medicine. In ancient times the Romans used to fashion the leaves into wreaths for crowning heroes and poets. Place of origin: probably in Asia Minor or southeast Asia, but is plentiful in the mountainous wooded areas of the Mediterranean region. Introduced into Britain in 1562.

Lavandula A genus of plants of the Mint or Thyme family, *Labiatae*. e.g. L. stoechas - French Lavender, a dwarf shrub, 30 cm (1 ft) high, with small flowers and inflorescences with violet bracts. The leaves are small and leathery. Grows in poor soils on sunny hillocks and stony slopes. The flower, Flos Lavendulae yields an oil, Lavender Oil, used chiefly in perfumery; also formerly employed as a carminative in the treatment of flatulence and colic. Place of origin: the Mediterranean region.

Lavatera A genus of plants of the Mallow family, *Malvaceae*. e.g. L. thuringiaca - The Thuringian Mallow, a hairy perennial herb, 50-120 cm high, with 3 to 5 lobed leaves and pink flowers. Uncommon, growing in meadows and scrubland. Place of origin: Armenia and the Caucasus; found in southeast and central Europe.

Laverania A genus name formerly applied to the malarial parasite, *Plasmodium*, and of the family *Plasmodidae*. e.g. L. vivax (= *Plasmodium* v.) (see **Plasmodium**).

Lawsonia A genus of plants of the Loosestrife family, *Lythrac* to the Old World. From the dried and powdered leaves is prepared henna, a reddish-orange, cosmetic dye used for tinting the hair, and in Arab countries, the hands and feet; also used as an astringent. Henna contains Lawsone, a quinine compound (= 2-hydroxy-1,4-naphthaquinone).

Lebistes A genus of Fishes of the suborder of Guppies, *Cyprinodontei*, belonging to the family *Poeciliidae*. Formerly known by the generic name, *Acanthophacetus*. e.g. L. reticulatus (= *Acanthophacetus* r.) - The Guppy fish, a small, ornamental, viviparous freshwater species of top-feeding minnows popular in home aquariums and commonly known as "millions". The adults are cannibalistic and often eat their young. They are native to and cultivated in, the south West Indies, including the Barbados and north South America, where they feed on and thus eliminate mosquito larvae.

Lecanium A genus of Scale Insects of the order *Homoptera*. e.g. L. corni - A species of Scale insects, found on the branches of untended fruit trees.

Ledum A genus of plants of the Heath family, *Ericaceae*. e.g. L. palustre - The Labrador Tea, Marsh Tea or Wild Rosemary plant, an evergreen shrub, 50-100 cm high; with dark green leaves rusty down beneath, and a dense terminal cluster of cream-colored flowers. The petals are not joined, and the oblong capsule contains flat, narrow seeds. Grows in bogs and woods. Place of origin: the northern parts of Europe and Asia; native in Britain, although restricted to certain parts of Scotland.

Leersia A genus of plants of the Grass family, *Gramineae*. e.g. L. oryzoides - Cut-grass, a creeping perennial plant, 30-120 cm high; the culm is hairy at the nodes and the leaf-blades are flat, pale yellowish-green, and spiny on the margins. Grows locally on river-banks, in ditches and wet places. Place of origin: Europe; native in Britain.

Leeuwenhoekia A genus of Arachnids of the order of Mites, *Acarina*. e.g. L. australiensis - A mite found in New South Wales, which causes great irritation by burrowing into the skin.

Legionella A genus of gram-negative microorganisms. e.g. L. pneumophila - The causitive bacterial organism of Legionnaires' Disease.

Legousia A genus of plants of the Bellflower or Campanula family, *Campanulaceae*. e.g. L. speculum-veneris - The Common Venus's Looking-glass, an annual plant 10-50 cm high, with violet flowers and calyx-teeth equal ovary. Grows locally on arable and waste land. Place of origin: South and Central Europe.

Lehmannia A genus of Gastropod Molluscs of the Slug family, *Limacidae*. e.g. L. macroflagellata - Length about 50 mm, brownish colored; inhabits mountain areas, mostly broadleaved (beech) forests. Distribution: Central Europe (Carpathian and Sudeten mountains).

Leiognathus A genus of Arachnids of the order of Mites, *Acarina*. Also known by the generic name, *Liponyssus*. e.g. L. bacoti (= *Liponyssus* b.) - A species transmitting typhus.

Leiolepis A genus of Reptiles of the Agamid Lizard family, *Agamidae*. e.g. L. belliana - The Butterfly Agama, a vividly-colored ground lizard, blackish-blue, yellow and a vivid orange hue. The male may be as long as 90 cm (3 ft). Distribution: the tropical parts of southeast Asia.

Leiopa A genus of Birds of the Megapode family, *Megapodiidae*. Also known as Brush-turkeys, Incubator birds or Moundbuilders. e.g. L. ocellata - The Mallee fowl; this species of Brush turkeys builds huge incubation mounds of dry sandy earth, measuring as much as 4.57 m (15 ft) in diameter and 60-90 cm (2-3 ft) high. Distribution: Australia, New Guinea and the neighboring islands.

Leiopus A genus of Insects of the Long-horned or Longicorn Beetle family, *Cerambycidae*. e.g. L. nebulosus - Length 6-10 mm; an inconspicuous and common cerambycid of broad-leaved forests. The larva develops in dry slender trunks and stronger branches of broadleaved trees, chiefly hornbeam, oak and beech. It bores zig-zag tunnels under the bark, which become filled with dust. Distribution; central and northern Europe.

Leishmania A genus of flagellate protozoan parasites, also called by other generic names such as *Helcosoma*, *Ovoplasma*, etc. e.g. L. donovani - A species causing visceral leishmaniasis or kala-azar. L. tropica (= *Ovoplasma* orientale; *Helcosoma* tropicum). The cause of cutaneous leishmaniasis.

Leiurus A genus of Arachnids of the order of Scorpions, *Scorpiones* or *Scorpionida*. e.g L. quinquestriatus - The most venomous scorpion in the world, ranging from the eastern part of North Africa through the Middle East to the shores of the Red Sea. The poison is a powerful neurotoxic venom which, especially in cases of young children, can cause death.

Lelaps A genus of Arachnids of the order of Mites, *Acarina*. e.g. L. echidninus - A mite species parasitic in rats, and acting as the intermediate host of *Hepatozoon perniciosum*.

Lemmus A genus of Mammals of the Common Rat and Mouse family, *Muridae* (= *Cricetidae*) - Lemming, muskrat and vole subfamily, *Microtinae*. e.g. L. lemmus - The Norway or Norwegian Lemming, a close rodent relative of the vole, 15 cm (6 in) long with a very short tail; the upperside is brightly colored, yellow and black. It digs a complex system of corridors at a shallow depth in the ground and builds a spherical nest of shredded vegetation. It is given to extraordinary mass migration at times due to cyclic increases in population, and in their frantic search for food these lemmings pour down the valleys, some reaching the sea and drowning. Distribution: the mountainous wastes of northern Norway and Lapland, and northern Russia.

Lemna A genus of plants of the Duckweed family, *Lemnaceae*. e.g. L. polyrhiza - The Great Duckweed, a small, floating, perennial plant consisting of flattened green leaves or thalli and stems some 5 mm high, from which several long roots hang down into the water. Place of origin: in still waters of most of Europe; native in Britain.

Lemonia A genus of Insects of the Butterfly and Moth family, *Lemoniidae*. e.g. L. dumi - Size 25-29 mm; the caterpillars are brown and hairy with black, elongate patches on the sides of the body, and grow to a length of 60-70 mm, feeding on various Daisy plants (family *Compositae*). Distribution: Europe, from the Balkans to southern Scandinavia, the Urals.

Lemur A genus of Mammals of the Lemur family, *Lemuridae*. e.g. L. catta - The Ring-tailed Lemur, 1.2 m (4 ft) long including its long, ringed tail; it is greyish, with dark rings round the eyes. This mammal lives in thinly wooded, dry and rocky country, whereas most other lemur species live in thick forests. It is

also kept as a domestic animal in its normal habitat, in Madagascar and Mauritius.

Lens A genus of plants of the Pea family, *Leguminosae*. e.g. L. culinaris - The Lentil, an annual plant, 15-50 cm high, with broad, flat pods containing 1 to 2 seeds. Widely cultivated and naturalized. Place of origin: Europe, being a native of the east Mediterranean region.

Leo A genus of Mammals, also known by the generic name, *Felis*, of the Cat family, *Felidae*. e.g. L. leo (= *Felis* l.) - The Lion, one of the largest among the Cats, reaching a length of 195 cm (61/2 ft), with a tail of 90 cm (3 ft). The coat is sandy yellow or reddish yellow. Main habitat: East, Central and West Africa.

Leonotis A genus of plants of the Mint or Thyme family, *Labiatae*. e.g. L. leonurus - The Lion's Ear, a shrub 1-2 m (3 ft 3 in = 61/2 ft) tall with squarish stems, opposite, oblong, lanceolate, dentate leaves and whorls of bright orange-red, nettle-like flowers, each 5-7 cm (2-21/2 in) long. Place of origin: South Africa; introduced into Europe in 1712.

Leonotodon A genus of plants, now better known by the generic name, *Taraxacum*, of the Daisy family, *Compositae*. e.g. L. officinale (= *Taraxacum* o.) - The Common Dandelion (see **Taraxacum**).

Leonotopodium A genus of plants of the Daisy family, *Compositae*. e.g. L. alpinum - Lion's-foot, a flowering stem has 5-6 heads in a terminal cyme; the bracts are large, white and woolly. Grows on scree and rock-crevices up to 3,400 m. Place of origin: South and Central Europe, the Pyrenees and the Balkan mountains.

Leontocebus A genus of Mammals of the Marmoset and Tamarin family, *Callithricidae*. e.g. L. leonicus - The Lion Marmoset, an anthropoid more primitive than the monkey (family *Cebidae*), with the presence of claws on most of the fingers, a non-prehensile tail and litters of more than two. About 40 cm (16 in) long, of which half the length is its tail; it has a yellowish mane. Habitat: the Amazon basin.

Leonurus A genus of plants of the Mint or Thyme family, *Labiatae*. e.g. L. cardiaca - The Common Motherwort or Herba Lemuri, a perennial plant 30-100 cm high; the lower palmate leaves have 3-7 irregularly toothed lobes. Grows mainly in cultivated fields, hedgerows, waste places and grassland. From the dry flowering tops is extracted a tincture with similar properties to Valerian; used in the treatment of hysteria and other nervous conditions. Place of origin: Europe; introduced and naturalized in Britain.

Leopardus A genus of Mammals, also known by the generic name, *Felis*, of the Cat family, *Felidae*. e.g. L. pardalis (= *Felis* p.) - The Ocelot, 1 m (3 ft 4 in) long with a tail about 40 cm (16 in). It is a tawny reddish-brown with dark markings. A forest predator, it is found in Central and South America, and its range extends as far north as Texas, U.S.A. There are a number of sub-species.

Lepadogaster A genus of Fishes of the order of Clingfishes, Skittlefishes and Cornish suckers, *Xenopterygii* or *Gobiesociformes*. e.g. L. candolli - The Clingfish, a small fish with a small terminal mouth, a projectile upper jaw and strong teeth. The eyes are on top of the head and the anal and dorsal fins are opposite each other at the back of the body. Found mainly in tidal pools in the warmer seas.

Lepas A genus of Crustaceans of the subclass of Barnacles and Cirripedes, *Cirripedia*. e.g. L. anatifera - The Goose Barnacle, a Stalked Barnacle species; the stalk is up to 30 cm (1 ft) long. It is found on rocks and objects floating in the sea. The spiral-shaped appendages, the cirri or feeding limbs agitate the

water and drive food into the mouth. Distribution: Europe; native in Britain.

Lepeophtheirus A genus of Copepod Crustaceans of the order *Caligoida*. e.g. L. salmonis - A parasite of salmon when in the sea. The presence of this copepod on a salmon caught in freshwater is usually taken as an indication that the fish has recently returned from the sea. Although strictly a marine parasite, these copepods can live for about 2 weeks in freshwater, but the eggs are said to drop from the females within 2 days of leaving the sea.

Lepidium A genus of plants of the Cress or Crucifer family, *Cruciferae*. e.g. L. campestre - The Common Pepperwort, an annual to biennial plant 5-80 cm high, with densely hairy upper leaves clasping the stem, and fruits with small white, scale-like vesicles. Grows in dry pastures, on walls, banks, by roadsides, and in cultivated fields and waste places. Place of origin: widespread in most of Europe.

Lepidochelys A genus of Reptiles of the order of Turtles, *Chelonia* (= *Testudinata*). e.g. L. kempii - The Atlantic Ridley, the smallest marine turtle in the world, measuring from 50-70 cm (20-28 in) shell length and weighing up to 36 kg (80 lb).

Lepidochitona A genus of Coat-of-mail Shell Molluscs or *Polyplacophora* of the amphineuran Chiton family, *Chitonidae*. e.g. L. cinereus - The Grey Coat-of-mail Shell or Chiton, length 10-22 mm; with 8 articulated shell-plates, usually colored greyish-brown, greenish or reddish. Can roll up into a ball when danger threatens, but is normally attached to rocks like limpets. Feeds on algae. Distribution: Mediterranean, Atlantic and North Sea shores; native around the coasts of Britain.

Lepidopus A genus of Fishes of the Frostfish, Scabbardfish or Cutlassfish family, *Trichiuridae*. These are the Frostfishes, tropical fishes with a continuous dorsal fin and a small forked tail fin. Often regarded as degenerate mackerels.

Lepidorhombus A genus of Fishes (subphylum *Vertebrata*). e.g. L. whiffiagonis - The Megrim fish, a left-eyed species, up to cm (2 ft) long, with a large oblique mouth cleft, and the lateral line strongly curved above the pectoral fin. Distribution: the seas of Europe; native around Britain.

Lepidoselaga A genus of Insects of the Tabanid Fly family, *Tabanidae*. e.g. L. lepidota - The common "motuca fly" of Brazil.

Lepidosiren A genus of Fishes of the Loalach family, *Lepidosirenidae*. e.g. K. paradoxa - The South American Lungfish or Loalach, a sluggish species living in the marshy regions of the Gran Chaco plains in tributaries of the Amazon River. The gills are small and unable to supply the respiratory needs of the fish; consequently it breathes air into its lungs, and hence the common name.

Lepidurus A genus of Crustaceans (subclass *Branchiopoda*) of the order *Notosraca*. e.g. L. productus - An omnivirous species, about 2 cm long freshwater pools in Central Europe.

Lepiota A genus of plants of the family of Gill Fungi, *Agaricaceae*. e.g. L. procera - The Parasol Mushroom, a large, edible gill fungus species; the convex cap is 10-30 cm (4-12 in) wide, whitish with brown scales; the stem is 25 cm (10 in) long and has brownish, concentric, snake-like markings. The soft, crowded gills are white. Found frequently near trees during the summer and autumn. An excellent edible mushroom. Place of origin: Europe; native in Britain.

Lepisma A genus of Apterygote or primitively wingless Insects of the Silverfish family, *Lepismatidae* (order of Bristletails, *Thysanura*). e.g. L. saccharina - The Domestic Silverfish Insect, 7-10 mm long, a widely distributed species found in cool, damp situations and a well-known nocturnal visitor to the larder in the United States and Europe. It can live for several years, moulting every few weeks throughout its life and damaging paper, bookbindings, etc. The body is covered with fine greyish scales. Found in households, pantries, warehouses and libraries. Distribution: worldwide.

Lepismachilis A genus of Apterygote Insects of the Silverfish and Bristletail order *Thysanura*. e.g. L. notata - An inhabitant of sun-warmed rocks of Central Europe, with similar-appearing relatives in the United States. A small, primitive, wingless insect with an elongated body and three long, tail-like processes at the end of the abdomen.

Lepisosteus A genus of Fishes of the Gar family, *Lepisosteidae*. e.g. L. spatula - The Great or Alligator Gar, up to 4.5 m (15 ft) long, olive-green above and whitish below; found in the streams of the southern United States, the Mississippi River and its tributaries; also in the Gulf of Mexico and Central America; likewise in Cuba. It spawns in freshwater where many fishes congregate together, usually 3 or 4 males to every female. It is the largest freshwater fish in North America.

Lepomis A genus of Fishes of the Sunfish, Freshwater Bass, and allied family, *Centrarchidae*. e.g. L. gibbosus - The Pumpkinseed Sunfish, a centrarchid of North America, growing up to 25 cm (10 in) long, which has been introduced successfully into the Danube River. Found in lakes and river inlets with abundant vegetation. It has a single dorsal fin, the spiny frontpart of which is lower than the rest. It is green in color, sometimes with a golden tinge, with a rainbow-colored sheen and red spots. Native to North America, from Dakota to the Gulf of Mexico.

Leptailurus A genus of Mammals, also known by the generic name, *Felis*, of the Cat family, *Felidae*. e.g. L. serval (= *Felis* s.) - The Serval, also known as the Servaline Cat; it lives in Africa south of the Sahara. It is about 75 cm (2 1/2 ft) long with a tail some 40 cm (16 in), and beautifully marked, with very long legs, enabling it to run at great speed. It has a second color phase in which its stripes are replaced by a fine powdering of specks.

Leptinotarsa A genus of Insects of the Leaf Beetle family, *Chrysomelidae*. e.g. L. decemlineata - The Colorade Potato Beetle, 10 mm long; yellow elytra with 10 black lengthwise stripes, and plump, orange and black larvae. It was originally restricted to wild solanaceous plants in western North America, but with the cultivation of the potato, has spread eastwards across the Atlantic into Europe as far as Russia. It is now a notorious pest of the potato plant.

Leptocladodia A genus of plants, sometimes known by the generic name, *Mammillaria*, of the Cactus family, *Cactaceae*. e.g. L. elongata (= *Mammillaria* e.) - The Golden Lace Cactus, a columnar plant with upright, cylindrical stems forming clusters up to 15 cm (6 in) high and covered with numerous rosettes of short yellowish spines, tipped with brown. The flowers are white or yellowish-white. Place of origin: eastern Mexico (see **Mammillaria**).

Leptoconops A genus of Insects of the Bloodsucking Fly family, *Heleidae*.

Leptodactylus A genus of Amphibians of the Horned Toad family, *Leptodactylidae*. e.g. L. labialis - A species found in Central and South America, Mexico and Texas. It is a vividly-colored frog with horn-like protuberances above the eyes. It is usually hidden in sand or mud, with only the head showing.

Leptodera A genus of minute nematode worms, better known by the generic name, *Rhabditis* of the superfamily *Rhabditoidea*. e.g. L. pellio (= *Rhabditis* genitalis) - See **Rhabditis**.

Leptodora A genus of Crustaceans of the order of Water-fleas, *Cladocera*. The species are predators, living on the plankton of lakes in the northern hemisphere, and feeding on smaller crustaceans. It is the only cladoceran with a nauphilius larva which hatches from a fertilized egg.

Leptomitus A genus of Molds or Mycetic Fungi of the phylum *Eumycophyta*. e.g. L. vaginae - A species found in chronic vaginitis.

Leptonychota A genus of Mammals of the Earless Seal family, *Phocidae*. e.g. L. weddelli - Weddell's Seal, up to 3 m (10 ft) long. A carnivorous animal which looks very much like a whale, and leaves the water to breed on land. Distribution: Antarctica.

Leptophtergius A genus of extinct Reptiles, the Ichthyosaurs of the order of Mesozoicfossil reptiles, *Ichthyosauria*. e.g. L. acutirostris - A marine ichthyosaur which swam in the seas round what is now Europe, about 140 million years ago. It had a 2.13 m (7 ft) long skull, 4 paddle-like flippers, a long and powerful tail, dorsal and caudal fins, conical teeth in grooves, which were adapted for catching fish, and measured about 12.19 m (40 ft) in total length. It occurred chiefly in the Lias or Blue Limestone, the lower division of the Jurassic period of the Mesozoic era in geological time, that is about 140 million years ago.

Leptophyes A genus of Insects of the Bush-cricket or Katydid family, *Tettigoniidae*. e.g. L. punctatissima - A European species of Bush-crickets with auditory organs in the front tibiae. The males stridulate by rubbing their wings together.

Leptopsylla A genus of Insects of the order of Fleas, *Siphonaptera*, better known by the generic names, *Ctenopsylla* or *Ctenopsyllus*. e.g. L.

282

musculi (= *Ctenopsylla* segnis; *Ctenopsyllus* s.) - See **Ctenopsyllus**.

Leptoptilos A genus of Birds, also known by the generic name, *Leptoptilus*, of the Stork and Jabiru family, *Ciconiidae*. e.g. L. crumeniferus (= *Leptoptilus* c.) - The Marabou or Adjutant Stork of Africa south of the Sahara, an ugly vulture-like bird, standing about 1.5 m (5 ft) high, with slate-grey and white feathers. It has a large wingspread of anything up to 3.65 m (12 ft). Its stiff strutt has earned it its military name. It has a pouch hanging on the front of its featherless neck. It is a scavenger bird, feeding on refuse and carrion and is frequently found in towns, especially near slaughterhouses.

Leptoptilus See **Leptoptilos**.

Leptospira A genus of microorganisms of the family *Treponemataceae*. These are tightly coiled spirochetes frequently appearing (in spirals), with one bent and forming a hook. There are approximately 60 recognized types, placed in "saprophytic" and "pathogenic" groups. e.g. L. biflexa (= Lept biflexa) - A saprophytic species, commonly present in stagnant water, streams and lakes. L. icterohaemorrhagiae (= Lept i.) - The causative organism of the most severe leptospiral infection, Weil's disease. This is an infectious jaundice with a worldwide distribution, and the most common sources of infection are rat urine and contaminated water.

Leptothrix A genus of microorganisms of the family *Chlamydobacteriaceae*. Unbranched, gram-negative, algae-like, filamentous trichomes or rods usually found in fresh water. e.g. L. buccalis - Now known to be any of two different organisms, and renamed *Leptotrichia* buccalis and *Bacterionema* matruchotii (see **Leptotrichia**; **Bacterionema**). L. placoides - A species found in tooth canals in humans.

Leptotila A genus of Birds of the Dove and Pigeon family, *Columbidae*. e.g. L. wellsi - The Grenada Dove, one of the world's rare species of birds, believed to have a total population of less than 100.

Leptotrichia A former generic name of certain microorganisms, and frequently confused with the genus *Leptothrix* though now known to be a different entity. No longer recognized. e.g. L. buccalis - Part of the normal flora of the mouth and urogenital tracts, and a common non-pathogenic inhabitant of these areas (see **Leptothrix**).

Leptotrombidium A genus of Arachnids, better known by the generic name, *Trombicula*, of the family of Acarine Mites, *Trombiculidae*. e.g. L. akamushi (= *Trombicula* a.) - The Kedani Mite (see **Trombicula**).

Leptotyphlops A genus of Reptiles of the family of Blind Burrowing Snakes, the Thread or Worm Snakes, *Leptotyphlopidae*. These are blind reptiles, sometimes mistaken for large earthworms, harmless snakes which burrow in the ground, feeding on ants and termites in whose nests they are often found. They have teeth only in the lower jaw. Distribution: the warmer parts of the earth, especially Africa. e.g. L. bilineata - The very rare Thread Snake, the shortest snake in the world, 10.8 cm (41/2 in) long. Distribution: the islands of Martinique, Barbados and St. Lucia in the West Indies.

Leptura A genus of Insects of the Longhorned or Longicorn Beetle family, *Cerambycidae*. e.g. L. rubra - Length 10-19 mm; the sexes differ in shape, size, and color. The male is more slender, smaller, with dingy-yellow elytra and black scutum. The female is larger, stouter and colored orange-red. A very common species, settling on flowers, stumps, felled logs and stacked wood. The larvae generally develop in old stumps, in which they bore tunnels and later pupate. Distribution: Europe, Siberia, North Africa.

Leptus A name for the larval form of mites of the genus *Trombicula* and

Eutrombicula. e.g. L. akamushi (=
Trombicula a.) (see **Trombicula**).

Lepus A genus of Mammals of the Hare
and Rabbit family, *Leporidae.* e.g. L.
europaeus - The Common, Brown or
European Hare, about 60 cm (2 ft) long
without the tail, with a ruddy brown fur
above and white beneath, and a great
capacity for running and jumping.
Distribution: worldwide. Originally
from Europe, Asia Minor and East
Africa; it has been introduced into the
Americas and Australasia. Native in
Britain.

Lernaea A genus of Copepod
Crustaceans of the order *Lernaeoida.*
e.g. L. ranae - A North American species
of parasitic copepods, found in tadpoles
of the Green Frog in freshwater, unlike
other, especially African, species of this
genus which are parasites of fish.

Lernaeocera A genus of Copepod
Crustaceans of the order *Lernaeoida.*
e.g. L. branchialis - A common parasite
of the Cod and Whiting fish. The head is
inserted into a gill bar of this fish host,
while the swollen twisted body lies in
the gill chamber.

Lestes A genus of Insects of the order of
Damselflies and Dragonflies, *Odonata.*
e.g. l. viridis - A Damselfly species living in freshwater; up to 4.5 cm long; the
wings are light and of uniform color,
with distinct stalks, and can be folded
over the back when at rest.

Leto A genus of Insects of the Ghost
Moth or Swift Butterfly family,
Hepialidae. e.g. L. stayci - An
Australian hepialid species, with a
wingspan of up to 22.5 cm (9 in), and
very large larvae which feed in the
trunks of eucalyptus trees.

Leucaspius A genus of Fishes of the
Carp, Minnow, Tench, and allied family,
Cyprinidae. e.g. L. delineatus - The
Moderlieschen, one of the smallest
European Cyprinid species, no more
than 5-10 cm (2-4 in) long, with a slen-
der, laterally compressed body; the

mouth is tilted upwards and the lateral
line is short. It lives in shoals in small
ponds or gently flowing streams.
Distribution: central and eastern Europe,
as far north as southern Sweden (see
Carassius).

Leuchtenbergia A genus of plants of the
Cactus family, *Cactaceae.* e.g. L. prin-
cipis - The Agave Cactus plant, a native
of Mexico. It has a short woody trunk
bearing spreading fleshy tubercles 5-10
cm (2-4 in) long. The widely expanded
flowers have numerous petals, and the
fruit contains many dark brown seeds.

Leucifer A genus of Decapod
Crustaceans of the Shrimp family,
Leuciferidae. The species live on plankton of the warmer oceans. They have a
very long head, no gills, and lack the
fourth and fifth pairs of legs.

Leuciscus A genus of Fishes of the Carp,
Minnow, Tench, and allied family,
Cyprinidae. e.g. L. leuciscus (= L.
rutilis) - The Dace, length 30 cm (1 ft),
a European freshwater fish rarely
exceeding 450 gm (1 lb) in weight. The
body is slender, almost circular in cross-
section, with a small mouth and notched
or forked caudal (anal) fin. It is found in
clear foothill and lowland rivers and
streams. It feeds on insects, insect larvae
and other invertebrates and transmits the
Siberian Liver Fluke, *Opisthorcis*
felineus to cats, dogs, pigs and man.
Distribution: Europe; native in Britain
(see **Carassius**; **Opisthorcis**).

Leuckartiara A genus of Celenterate animals of the class of Jellyfishes,
Scyphozoa. e.g. L. octona - The Medusa,
a True Jellyfish marine species, with a
15 cm (6 in) bell which has a spherical
bulge. Found in the Atlantic, North Sea
and the west Baltic. Native around
Britain.

Leucobryum A genus of plants of the
class of Liverworts and Mosses,
Bryophyta. e.g. L. glaucum - The White
Fork-moss, a perennial plant, 5-14 in)
high, forming very compact, roundish,

284

bluish-green cushions (whitish-green when dry). Grows on acid soils in woods and on wet moorland. Place of origin: Europe; native in Britain.

Leucocytozoon A genus of sporozoon parasites found in the blood cells of birds. e.g. L. danilewskyi - A species found in owls.

Leucoium A genus of plants, also called *Leucojum*, of the Daffodil family, *Amaryllidaceae* (see **Leucojum**).

Leucojum A genus of plants, sometimes also called *Leucoium*, of the Daffodil family, *Amaryllidaceae*. e.g. L. vernum (= *Leucoium* v.) - The Spring Snowflake, a poisonous perennial plant, 8-30 cm high, found in garden culture. It has a large bulb and bright green leaves up to 30 cm (1 ft) long. The nodding, bell-shaped flowers are borne singly on erect stems, while the 6 perianth segments are white and tipped with green. The pear-shaped capsule contains pale seeds. The Snowflake grows on the hills of South and Central Europe, and in damp scrub and on hedgebanks of northwestern Europe. Native in southwest Britain.

Leucoma A genus of Insects of the Tussock Moth family, *Lymantriidae*. e.g. L. salicis - The White Satin Moth, length 22-26 mm; found in poplar avenues, parks and along streams bordered by poplars and willows on which the caterpillars feed. Distribution: Europe, Asia Minor, Central Asia, Siberia, North Africa.

Leuconostoc A genus of microorganisms of the order *Eubacteriales*. These are slime-forming saprophytic bacteria found in milk and fruit juices. e.g. L. mesenteroides (= NCIB 8710) - A certain strain of this organism is used in the fermentation of sucrose, producing degraded dextrans (glucose polymers), from which is prepared dextran injections.

Leucophoyx A genus of Birds of the Bittern, Egret, and Heron family, *Ardeidae*. e.g. L. thula thula - The Snowy Egret, of southern North America and South America; snowy white in color with a black beak, up to 25 cm (10 in) long; tragically very popular as an ornamental bird, and has been slaughtered in large numbers for this nefarious purpose.

Leucopsis A genus of Insects of the order of Ants, Bees and Wasps, *Hymenoptera*. e.g. L. dorsigera - A species of parasitic Bees.

Leucorchis A genus of plants of the Orchid family, *Orchidaceae*. e.g. L. albida - The Small White Orchid, a perennial plant, 10-25 cm high, with whitish flowers 5 mm long, and the central lobe of the lip longer than the lateral lobes. Distribution: widespread in Europe; native in Britain.

Leucosolenia A genus of Ascon Sponges of the suborder *Homocoela*. These are parozoan animals and members of the phylum *Porifera*. e.g. L. botryoides - A calcareous sponge growing on seaweed. The whole central cavity, the spongocoel is lined with collared cells or choanocytes (see **Clathrina**).

Leucothea A genus of Celenterate animals of the class of Jellyfishes, *Scyphozoa*. e.g. L. multicornis - A yellowish brown comb-jelly or ctenophore, 20-40 cm long, with gelatinous lobes. A marine species found in the Mediterranean and further north. Native around British seacoasts.

Leucothoe A genus of plants of the Heath family, *Ericaceae*. e.g. L. catesbaei - The Dog-hobble or Fetterbush, an evergreen shrub up to 1.8 m (6 ft) high. It has slender arching stems and leathery, lanceolate, sharply-toothed leaves and small, white, bell-shaped flowers borne in dense axillary racemes. The fruit is a dry capsule. Place of origin: the southeastern United States.

Leucothrix A genus of microorganisms of the family of schizomycetes, *Leucotrichaceae*. These resemble the blue-green algae, but do not contain photosynthetic pigments.

Leukocytozoon A genus of microorganisms, the Hemosporidia, the species of which are parasitic in the blood corpuscles of birds.

Leuresthes A genus of Fishes (subphylum *Vertebrata*). e.g. L. tenuis - The California Grunion fish.

Lewisia A genus of plants of the Purslane family, *Portulacaceae*. e.g. L. brachycalyx - A species growing in the Rocky Mountains; an annual herb with succulent leaves and attractive flowers. Frequently grown in gardens.

Leymus A genus of plants, better known by the generic name, *Elymus*, of the Grass family, *Gramineae*. e.g. L. arienarius (= *Elymus* a.) - Lyme Grass (see **Elymus**).

Liatris A genus of plants of the Daisy family, *Compositae*. e.g. L. spicata - The Spike Gayfeather, Button-Snakeroot or Blazing-Star, a stout, herbaceous perennial plant with bright green leaves, and spikes of lilac to purplish-red flowers. It attains a height of 90 cm (3 ft). Place of origin: east and south United States; introduced into Europe in 1732.

Libellula A genus of Insects of the family of Damselflies and Dragonflies, *Libellulidae*. e.g. L. quadrimaculata - The Four-spotted Libellula, size 4-7.5 cm; the bases of the four wings (2 pairs) are brownish-yellow, each with a center and end spot; 4 spots in all, hence its scientific name. A migratory species and a swift flier, with a well-defined beat over the pools and ponds where it hunts other insects. It spreads its wings out horizontally when at rest. Distribution: Europe, Asia, North America. Native in Britain.

Libocedrus A genus of Trees of the Cypress family, *Cupressaceae*. The species are catalogued more specifically under other genera, such as *Austrocedrus*, *Calocedrus*, etc. e.g. L. chilensis (= *Austrocedrus* c.) - The

Chilean Incence Cedar (see **Austrocedrus**). L. decurrens (= *Calocedrus* d.) - The Incence Cedar (see **Calocedrus**).

Libythea A genus of Insects of the Tortoiseshell family, *Nymphalidae*. e.g. L. celtis - The Short Butterfly, size up to 4 cm; the wings are brown with orange-red spots, and have a zigzag edge. Distribution: the Mediterranean littoral of Europe.

Lichanura A genus of Reptiles of the Boa Snake family, *Boidae*. e.g. L. roseofusca - The Rosy Boa, a slowly-moving desert snake found in California, U.S.A.

Lichen A genus of plants composed of symbiotic algae and fungi, of the Lichen family, *Parmeliaceae*. It has other more specific generic names, such as *Cetraria* and *Lobaria*. e.g. L. islandicus (= *Cetraria* islandica; *Lobaria* islandica) (see **Cetraria; Lobaria**).

Licheterodon A genus of Reptiles of the Colubrid Snake family, *Colubridae*. e.g. L. madagascariensis - The Madagascan colubrine snake, an egg-eating species which, though no thicker than a man's finger, can swallow an entire egg 3 or 4 times the size of its own head without breaking it, the egg first being held between coils of the body so that the jaws can seize it. The whole egg passes into the stomach and both shell and contents are digested.

Lichtheimia A genus of Fungi or Molds of the Fungus family, *Mucoraceae*. e.g. L. corymbifera - A species pathogenic in rabbits. Found in man mycosis of the ear and the respiratory tract.

Licuala A genus of Trees of the Palm family, *Palmae* (= *Arecaceae*). e.g. L. grandis - An attractive palm from the islands north of New Guinea. The leaves have long stalks and orbicular, fan-like blades.

Ligia A genus of Isopod Crustaceans of the Woodlice Sea Slater family, *Ligiidae*. e.g. L. oceanica - The Great Sea slater, up to 3 cm long; a semi-aquatic species of woodlice or large sea slaters, found scavenging on the upper seashores all around the world. The body is dotted with grey, the antennae are long, and the uropods have two long pointed processes. They are one of the most successful group of terrestrial crustaceans, and live along the rocky coasts of the Mediterranean; native around Britain.

Lignognathoides A genus of Insects of the order of Lice, *Anoplura*. e.g. L. montanus - The common sucking louse of the California ground squirrel, *Citellus* beecheyi.

Ligula A genus of Tapeworms (class *Cestoda*). e.g. L. intestinalis - A large, fleshy tapeworm found in the alimentary tract of diving and wading birds. One of the shortest-living species of such worms, living only a few days.

Ligularia A genus of plants, also known by the generic name, *Senecio*, of the Daisy family, *Compositae*. e.g. L. sibirica (= *Senecio* sibiricus) - A short perennial plant, 90-120 cm (3-4 ft) high; it has triangular or kidney-shaped, toothed leaves up to 30 cm (1 ft) across. Handsome spikes of golden-yellow capitula develop, each having a few ray florets. Grows in wet meadows. Distribution: Europe, Asia - from France to Japan (see **Senecio**).

Ligusticum A genus of plants of the Carrot family, *Umbelliferae*. e.g. L. mutellina - The Alpine Lovage, a perennial plant, 10-50 cm high, with only 1 or 2 stem leaves, and purple petals. Grows in stony grasslands, on screes, rocks and stony mountain slopes. Place of origin: Europe, in the Alps and Carpathian mountains.

Ligustrum A genus of Trees of the Olive family, *Oleaceae*. e.g. L. vulgare - The Common Privet, a deciduous shrub, up to 5 m (16 1/2 ft) high; with smooth, lanceolate leaves, opposite or in whorls of three. Small white flowers grow in terminal panicles. The fruit is a shiny,

spherical, black berry containing 2-4 seeds. Widespread on calcareous soils in woods and scrubs. Place of origin: England and Wales, most of Europe and North Africa.

Lilioceris A genus of Insects of the Leaf Beetle family, *Chrysomelidae*. e.g. L. lilii - Length 6-8 mm; it has a black head and brown elytror carapace. Found in gardens, on cultivated plants of the Lily family, *Liliaceae*, usually on the leaves of white lilies. Distribution: Europe, Asia, North Africa.

Lilium A genus of plants of the Lily family, *Liliaceae*. e.g. L. candidum - The Bourbon Lily, Madonna Lily or White Lily, a native of the eastern Mediterranean, the oldest Lily plant in cultivation in the world, stretching back thousands of years. The stems are 60-150 cm (2-5 ft) high, and the flowers are pure white and trumpet-shaped. Place of origin: not positively known.

Lima A genus of Bivalve Molluscs of the File Shell Pterioid family, *Limidae*. e.g. L. lima - The File Shell; length 35 mm; height 50 mm; thickness 18-22 mm; a beautiful, thick-wall shell species, porcelain white both outside and inside. The surface is ornamented with 19-24 ribs radiating from the apex, bearing projecting scales arranged in concentric rows coincidental with the growth lines. A plentiful and common species found on stony bottoms, frequently amongst marine sponges and coral. Distribution: the Mediterranean Sea, coast of southwestern Europe, Canary and Cape Verde Islands, and North America.

Limacina A genus of Molluscs of the order of Sea Butterflies, *Thecosomata*. These are shelled planktonic pteropods with a mantle cavity and a transparent shell. They are extremely abundant.

Limanda A genus of Fishes of the Flounder family, *Hippoglossidae*. e.g. L. limanda - The Lemon Sole or Dab Flatfish, up to 40 cm long, a European species with a large, asymmetrical mouth and a deflected (in front) lateral line; the scales are ctenoid or spiny and the skin is rough. Distribution: Europe; native around Britain. Found also in the North Atlantic.

Limax A genus of Molluscs of the order of Land Snails, *Stylommatophora*. e.g. L. maximus (= L. maximum) - The Great Grey Slug, 12-15 cm long land snail and one of the largest of slugs; light grey with black markings; it spends the day sheltering in the shade in gardens and fields, and emerges at night or on a rainy day to search for food. Distribution: Europe; native in Britain. Also found in the Mediterranean region and North Africa.

Limenitis A genus of Insects of the Butterfly and Moth family, *Nymphalidae*. e.g. L. sibilla - The White Admiral Butterfly, blackish-brown in with white stripes and spots on the wings, which are orche and green on the underside. Found in forest paths and fringes of damp deciduous forest patches. Distribution: Europe.

Limicola A genus of Birds of the Sandpiper and Wader family, *Scolopacidae*. e.g. L. falcinellus - The Broad-billed Sandpiper, size 16.5 cm migrant bird nesting on moors and heaths. The upper parts are dark with creamy white stripes. Distribution: the far north and eastern Europe, wintering further south.

Limnadia A genus of Crustaceans of the order *Conchostraca*. e.g. L. lenticularis - A species widespread in North America and Europe, living mainly in lakes bordering the Arctic Ocean.

Limnaea A genus of Gastropod Molluscs of the Snail family, *Helicidae*. These are the invertebrate, intermediate hosts of the trematode parasite, *Schistosoma* pathlocopticum, a species pathogenic to mice (see **Schistosoma**).

Limnatis A genus of Annelids or Segmented Worms of the Leech family, *Hirudinidae* e.g. L. granulosa - A genus

of Horse Leeches, found in the nasal passages of animals, especially horses.

Limnephilus A genus of Insects of the Caddis Fly family, *Leptoceridae*. e.g. L. rhombicus - A species of Freshwater Caddis flies, 2 cm long; the wings are roof-like with few hairs. Found on still or slow-flowing water. Distribution: Europe; native in Britain.

Limnodromus A genus of Birds of the Curlew, Sandpiper, Snipe, and Woodcock family, *Scolopacidae*. e.g. L. griseus - The Short-billed Dowitcher of North America, the only Snipe of that country to be found on open shores. Readily identified by the white lower back, rump and tail, and by the straight snipe-like bill. The method of feeding is characteristic; the bird jabs its bill vertically into the mud with rapid movements rather like the action of a sewing-machine needle.

Limnodorum A genus of Saprophytic plants of the Orchid family, *Orchidaceae*. e.g. L. abortivum - The Violet Limodor, a perennial plant, 30-60 cm (1-2 ft) high; the stem is violet and the violet flowers grow in loose racemes of more than four. Distribution: the Mediterranean region, south and western Europe.

Limonium A genus of plants of the Sea-lavender family, *Plumbaginaceae*. e.g. L. vulgare - The Sea-lavender, an erect, perennial plant, 15-45 cm high; the inflorescence is corymbose and the spikelets are crowded into spikes. Grows in muddy salt marshes. Distribution: eastern and southern Mediterranean regions of Europe; native in Britain.

Limosa A genus of Birds of the Curlew, Sandpiper, Snipe, and Woodcock family, *Scolopacidae*. e.g. L. limosa - The Black-tailed Godwit, closely related to the Sandpiper, one of the finest European waders; a migrant bird, about 40 cm (16 in) long, blackish-brown, with long legs and an unusually long, straight bill. It favors marshy ground in the vicinity of swamps and ponds. Distribution: the eastern parts of Central Europe; eastern Europe; and western Asia. Native in Britain.

Limosella A genus of plants of the Figwort or Snapdragon family, *Scrophulariaceae*. e.g. L. aquatica - The Mudwort, an annual plant, 1-8 cm high; radical, spathulate, long-stalked leaves with white or lavender, sometimes purple-tinged flowers. Grows in wet mud at the edges of pools, rivers, and ditches. Place of origin: Europe; native in Britain.

Limulus A genus of Horseshoe Crabs, sometimes erroneously called King Crabs, of the class of arthropods, *Merostomata*. Sometimes also called by the generic name, *Xiphosura*. e.g. L. polyphemus (= *Xiphosura* p.) - The so-called King Crab, about 50 cm (20 in) long, commonly found along the eastern seaboard of the United States and around the shores of the Gulf of Mexico. It lives in the mud, and the sharp pointed spine at one end of its body helps it to turn over, while swimming on its back the "right way up", in order to crawl on to the seabed for feeding.

Linaria A genus of plants of the Figwort or Snapdragon family, *Scrophulariaceae*. e.g. L. vulgaris - The Toadflax, a perennial plant, 15-90 the yellow flowers have an orange-yellow palate or narrow basal spur. It has erect stems up to 60 cm high and a creeping rhizome. Common in grassy and waste places. Place of origin: Europe and Asia; native in Britain. Has been introduced into North America.

Linckia A genus of Starfishes of the family *Ophidiasteridae* (class *Asteroidea*). e.g. L. columbiae - An echinoderm species, with a diameter (4 in), capable of regenerating even from a piece of arm, and while the new disc and arm are regrowing it looks like a "star" with a long tail, what is called the Comet form, the "tail" being the arm from which regeneration is proceeding. Habitat: the seas off the coast of California, and the eastern Pacific Ocean.

Lindernia A genus of plants of the Figwort or Snapdragon family, *Scrophulariaceae*. e.g. l. procumbens - The Lindernia, an annual plant, 3-10 cm high; entire, stalkless leaves and stalked, solitary, white or pinkish flowers. Occurs in damp, sandy, and muddy habitats. Place of origin: Central Europe.

Lineus A genus of Ribbon Worms (phylum *Nemertina* or *Rhynchocoela*). e.g. L. longissimus - The "Bootlace" Ribbon Worm, some 30 m (100 ft) long; a carnivorous, soft-bodied, marine invertebrate which breathes by means of gills. It produces a free-swimming larval form or pilidium known as Desor's larva, which remains within the egg. Distribution: the shallow coastal waters of the North Sea.

Linguatula A genus of Arthropods, also known as the Tongue Worms, found in animals and man. e.g. L. rhinaria (serrata) - A species found in man. The larvae are known as *Porocephalus* denticulatus (see **Porocephalus**).

Lingula A genus of Lamp-shells of the phylum *Brachiopoda*. These are the horny-shelled brachiopods, marine invertebrate animals with a superficial resemblance to mussels and oysters. They have a pair of arms with tentacles protruding through the bivalve shell, which cause a current of water to bring microscopic food to the mouth. Habitat: the shallow waters off the shores of Japan, the Indo-Pacific islands and Queensland, Australia.

Linnaea A genus of plants of the Honeysuckle family, *Caprifoliaceae*. e.g. L. borealis - Linnaea, a perennial plant, up to 20 cm high; the prostrate stems are glandular, hairy and slender and the flowers are usually borne in pairs on long stalks. Place of origin: northern Europe; native in Britain.

Linognathus A genus of Insects of the order of Lice, *Anoplura*. e.g. L. piliferus - A species of sucking lice, infesting dogs.

Linuche A genus of Jellyfishes of the order *Coronatae*. These are thimble-shaped species inhabiting the shallow waters off the Bahamas and Florida.

Linum A genus of plants of the Flax family, *Linaceae*. e.g. L. flavum - Yellow Flax, a perennial plant, 20-60 cm high; the stem is angled in the upper part, the flowers are glabrous and yellow, and grow in cymose inflorescences. Occurs in meadows and open scrubland. Place of origin: east, southeast and central Europe. L. usitatissimum - Cultivated Flax, an annual plant from which linen fibers and linseed oil are obtained. Flax for fiber is typically grown in a cool, moist, summer climate and good, porous soil. The stems are up to 1.2 m (4 ft) high. The fiber crop is used for the manufacture of linen, for textile thread, twine, writing paper, sailcloth, etc. Flax for seed is grown, mainly in tropical countries, to yield linseed oil.

Lionotus A genus of Protozoan Animals of the suborder *Rhabdophorina*. These are voracious, carnivorous, microscopic creatures with an extended trichocyst and cilia-bearing anterior processes giving a comb-like appearance to the region in front of the mouth.

Liopelma A genus of Amphibians of a family of primitive Frogs, *Liopelmidae*. There are two species living in the mountainous uplands of New Zealand, where they are the only native frogs. In their haunts there is no standing water and the eggs are laid on the damp ground, the froglets developing wholly within the egg.

Liopsetta A genus of Fishes of the Flounder family, *Hippoglossidae*. e.g. L. glacialis - The Arctic Flounder; it has a large mouth and a straight lateral line.

The males have ctenoid or spiny scales (i.e. scales with rough edges), and the females cycloid scales (i.e. scales with evenly curved edges). Some of these fishes swim quite a way up some northerly rivers (North Dvina and Vyga Rivers in Russia).

Liparis A genus of Fishes of the Sea-snail Fish or Snail-fish family, *Liparidae* (suborder of Scorpionfishes, *Scorpaenoidea*). e.g. L. liparis - The Common Sea-snail, a species of fish 15 cm long; the pelvic fins are fused to form a sucker disc on its belly. It has 2 pairs of nostrils. Occurs chiefly amid growths of seaweed in shallow coastal waters. Distribution: the Atlantic coasts, the North and Baltic Seas. Native around Britain.

Liparus A genus of Insects of the Weevil Beetle family, *Curculionidae*. e.g. L. glabrirostris - Length 14-21 mm; one of the largest of the European weevils. Found in piedmont and mountainous regions, generally on the leaves of butterburs, around streams and in forest margins. Distribution: central and southwestern Europe.

Liphistius A genus of Arachnids of the Trapdoor Spider family, *Ctenizidae*, found only in southeast Asia. They are a most elusive species, living in silken tubes buried in the ground, and are extremely difficult to detect because the hinged circular door which secures the entrance to the lair is coated on the outside with plant matter and is beautifully camouflaged.

Liponyssus A genus of Arachnids of the order of Mites, *Acari* or *Acarina*; also known by the generic name, *Leiognathus*. e.g. L. bacoti (= *Leiognathus* b.) - A species transmitting typhus (see **Leiognathus**).

Lipoptena A genus of Insects of the Dipterous Louse-fly family, *Hippoboscidae*. e.g. L. cervi - The Deer Louse-fly or Deer-fly, length 5.2-5.8 mm; a flattish, firmly built insect difficult to squash. Early in spring it flies and sucks the blood of birds. About June, it seeks out animals such as stags and roe-deer, shedding its wings and living as a parasite in the fur of its host, and sucking its blood. At this period it may also attack humans by a lighting on the hair or beard and biting painfully. Distribution: the Palaearctic region.

Lipotes A genus of Mammals of the River Dolphin family, *Platanistidae*. e.g. L. vexillifer - The Chinese River Dolphin, a species discovered this century in Tung Ting Lake on the Yangtze-Kiang river.

Lippia A genus of plants, now more correctly named *Aloysia*, of the Verbena or Vervain family, *Verbenaceae*. e.g. L. citriodora (= *Aloysia* c.) - The Lemon Verbena (see **Aloysia**).

Liquidambar A genus of Trees of the Witch-hazel family, *Hamamelidaceae*. e.g. L. orientalis - The Oriental Sweet Gum, a deciduous tree, native to Asia Minor. Grows up to 30 m (100 ft) in the wild. The leaves are hairless and rounded at the lobe tips. From the trunk is extracted a purified balsam, liquid Storax or prepared Storax, formerly used as an ointment in the treatment of scabies and other parasitic skin diseases.

Liriodendron A genus of Trees of the Magnolia family, *Magnoliaceae*. e.g. L. tulipifera - The Tulip or Whitewood Tree, a deciduous tree up to 30 m (100 ft) tall; with alternate, truncate leaves and 6-petalled greenish-white flowers with orange blotches. The cone-like fruit consists of a group of elongated archenes each containing a single seed. The timber, called "white wood" is used in North America for house interiors, and a heart stimulant has been extracted from the bark. Place of origin: the eastern United States.

Liriope A genus of Celenterate Animals of the Medusa order, *Narcomedusae* (= *Trachymedusae*). These hydrozoans are a small group with no polyp phase.

They have broad flat bells and lack a manubrium; the mouth opens directly into the stomach region. Their life-history is considered by some to be ancestral. They live in the warmer waters of the tropics (see **Geryonia**).

Lispa A genus of Insects of the Housefly family, *Muscidae*. e.g. L. dentaculata - A species of Houseflies, 6.5 mm long; it resembles the Housefly (see **Musca** domestica); the mouth-feelers are spoon-shaped. It is associated with still or slow-flowing freshwater, laying its eggs there where the larvae develop. Distribution: Europe; native in Britain.

Lissodelphis A genus of Mammals of the Dolphin and Grampus family, *Delphinidae*. e.g. L. borealis - The Right Whale Dolphin, a species found coast of California, U.S.A.

Lissotis A genus of Birds of the Bustard family, *Otididae*. e.g. L. melanogaster - The Black-bellied Bustard, a species living in high grass and preferring burnt areas. It feeds mainly on insects and flower buds. Distribution: Africa.

Listera A genus of plants of the Orchid family, *Orchidaceae*. e.g. L. cordata - The Lesser Twayblade, a perennial plant with slender creeping rhizomes and erect stems, from 5-20 cm (2-8 in) high. Near the base are two ovate leaves and at the top is a loose cylindrical spike of tiny reddish-green flowers. Grows on peaty moors and in damp mountain woods. Distribution: Europe, Asia and North America. Native in Britain.

Listerella A genus of microorganisms, better known as *Listeria*, of the order *Eubacteriales*. e.g. L. monocytogenes (= *Listeria* m.) - See **Listeria**.

Listeria A genus of microorganisms, also known by other generic names, such as *Bacterium* and *Listerella* of the family *Corynebacteriaceae* (order *Eubacteriales*). e.g. L. monocytogenes (= *Bacterium* m.; *Listerella* m.) - The single species of this genus. A gram-positive diphtheroid-like bacillus, often present in pairs and resembling diplococci. Produces monocytosis in laboratory animals such as rabbits and guinea pigs. May cause an influenza-like illness, septicemia, skin lesions, meningitis, upper respiratory-tract infections, etc. in man. Many of the infected pregnant women have a history of ingesting raw cow or goat milk.

Lithobius A genus of Arthropods of the family of Centipedes, *Lithobiidae*. e.g. L. mutabilis - A very common European Chilopod or Centipe which lives under stones and in damp places, feeding mainly on small insects. It has one pair of walking legs on each body segment. Distribution: Europe; native in Britain.

Lithocarpus A genus of Trees of the Beech family, *Fagaceae*. e.g. L. densiflorus - The Tanbark Oak, an evergreen tree, native to California and Oregon, where its bark is used in tanning leather. Height to 21 m (70 ft), sometimes twice as much. The fruit is the acorn.

Lithocolletis A genus of Insects of the Butterfly and Moth family, *Gracillariidae*. The species are widely distributed; from them develop leaf-mining larvae.

Lithocranius See **Litocranius**.

Lithodes A genus of Crustaceans of the Stone Crab family, *Lithodidae*. e.g. L. maia - The North American and European Stone Crab; this species reaches its southern limit in Britain. The carapace is triangular and heavily beset with spines, while the small last leg is normally carried between the thorax and abdomen.

Lithophaga A genus of Molluscs of the subclass of Date Mussels, *Lamellibranchia*. These animals are confined to calcareous rocks, and penetrate by chemical means.

Lithophyllum A genus of plants of the family of Red Algae, *Gigartinaceae*.

Lithops A genus of plants of the family *Aizoaceae* (sometimes known as the family *Mesembryanthemaceae*). e.g. L. gracilidelineata - A succulent species found growing in stony places and producing showy flowers and two, mottled, fleshy leaves. Cultivated as ornamentals. Place of origin: South Africa.

Lithospermum A genus of plants of the Borage family, *Boraginaceae*. e.g. L. officinale - The Common Cromwell, a perennial plant, 30-100 cm high; the leaves are prominently veined and the fruit is glabrous. Grows in hedges and bushy places. Place of origin: Europe; native in Britain.

Litocranius A genus of Mammals, also called *Lithocranius*, of the Cattle family, *Bovidae* (Antelope subfamily, *Antilopinae*). e.g. L. walleri (= *Lithocranius* w.) - The Gerenuk, Giraffe-necked Gazelle or Waller's Gazelle; it has an elongated neck and stands erect on its hind legs to browse on bushes. It is reddish brown in color with a dark band down the back. The males have long, ringed, sublyrate horns. Distribution: East Africa, from southern Ethiopia to Tanzania.

Litomosoides A genus of Nematode or Roundworms, of the order of Filarial worms, *Filaroidea*. e.g. L. carinii - A filarial worm found in the cotton rat, *Sigmodonhispidus*.

Littorella A genus of plants of the Plantain family, *Plantaginaceae*. e.g. L. uniflora - The Shore-weed, a perennial plant, 2-10 cm high; the male flowers are solitary, and are borne on short scapes; the female flowers are solitary or few, and are borne at the base of the male scape or axis. Grows in lakes and ponds. Place of origin: central and northern Europe; native in Britain.

Littorina A genus of Gastropod Molluscs of the Periwinkle Sea Snail family, *Littorinidae*. e.g. L. littorea - The Common or Edible Periwinkle; the shell is 15-40 mm high, greyish with brown lines and composed of 6-7 whorls, with pointed apex, and an operculum with which the animal closes the aperture at low tide. A widely gathered, edible species living on fucoid weeds on rocky shores the world over. Distribution: very common in Europe, from the North and Baltic Seas and Scandinavia to the Mediterranean and Atlantic coasts. Also (introduced to) the eastern coasts of the United States. Native around Britain.

Llama A genus of Mammals, better known by the generic name, *Lama*, of the Camel family, *Camelidae*. e.g. L. guanacos (= *Lama* g.) - see **Lama**.

Lloydia A genus of plants of the Lily family, *Liliaceae*. e.g. L. serotina - The Snowdon Lily, a perennial plant, 7-10 cm high; with two, very narrow, basal leaves and solitary flowers. Grows in alpine meadows, pastures, on ridges and crevices of rocks, above 2,200 m. Place of origin: southcentral and northwestern Europe and the Balkans.

Loa A genus of Nematodes or Roundworms of the order of Filarial Worms, *Filaroidea*. e.g. L. loa - A species, formerly known as *Filaria* loa; this is the "eyeworm" causing the disease known as loiasis or loaiasis due to infection of the skin and eye tissue, with itching and edematous "Calabar" or "Fugitive" swellings. The intermediate host is the Tabanid fly, *Chrysops* silacea. In man these worms have been known to live up to 15 years.

Lobaria A genus of plants, also known by the general generic name, *Lichen*, of the family of Lichens, *Parmeliaceae*, a group of plants composed of symbiotic algae and fungi, where the mycelium of the fungus forms a matrix in which are distributed algae living in a close (symbiotic) relation with the fungus. One species, L. islandica is better known by the generic name, *Cetraria*. e.g. L. islandica (= *Cetraria* i.; *Lichen* islandicus) - The Iceland Moss (see **Cetraria**). L. pulmonaria - The Tree Lungwort, a perennial plant; the lobes are 10 cm (4 in) or more in length; the thallus is leaflike and strongly lobed. It grows on rocks and old deciduous trees, in mountainous areas with high rainfall. Place of origin: Europe; native in Britain.

Lobelia A genus of plants of the Lobelia family, *Lobeliaceae*. e.g. L. dortmanna - The Water Lobelia, a freshwater perennial, 20-60 cm high; the leaves grow in radical rosettes and the pale, lilac flowers are borne in loose racemes. Occurs in lakes and ponds. Place of origin: western and northern Europe; native in Britain. L. inflata - A herbaceous species, characterized by a deeply split corolla and blue, red or white flowers. The dried leaves and aerial parts (Lobelia Herb; Indian Tobacco) yield poisonous alkaloids, containing not less than 0.3% calculated as Lobeline. Used as an antispasmodic in the treatment of bronchial asthma and chronic bronchitis, and as an emetic to induce vomiting. Was once claimed to be of value as a smoking deterrent, but with disappointing results. Distribution: North America, India, and other parts of the world.

Lobodon A genus of Mammals of the Earless Seal family, *Phocidae*. e.g. L. carcinophagus - The Crab-eating Seal, a large Antarctic species, up to 2.5 m (8 1/2 ft) long.

Lobularia A genus of plants, also known by the generic name, *Alyssum*, of the Cressor Crucifer family, *Cruciferae*. e.g. L. maritima (= *Alyssum* maritimum) - Sweet Alison or Sweet Alyssum, an annual to perennial plant, 9-30 cm high; the stems are branched, the petals white and entire, and the fruits are obovate and slightly hairy. Place of origin: a native of the Mediterranean region; introduced in Britain; also West Africa (see **Alyssum**).

Lochnera A genus of plants, better known by other generic names such as *Catharanthus* or *Vinca*, of the Periwinkle family, *Apocinaceae* or *Apocynaceae*. e.g. L. rosea (= *Catharanthus* roseus; *Vinca* rosea) - The

Madagascar Periwinkle (see
Catharanthus; **Vinca**).

Locusta A genus of Orthopterous Insects
of the Grasshopper and Locust family,
Acrididae. e.g. L. migratoria - The
Migratory Locust, 33-65 mm long; a
widely distributed pest with several subspecies and two ecological phases in the
warmer parts of Asia, in Africa, and in
southeast Europe. The hoppers of the
solitary or sedentary phase hatch in
sandy reed-beds, while from time to
time individuals of the migratory phase
come into being. Distribution: Europe
(except the north), much of Asia, Africa,
Madagascar.

Locustella A genus of Birds of the Old
World Warbler family, *Sylviidae*. e.g. L.
fluviatilis - The River Warbler, length 15
cm; a migrant birth; upper parts are dark
olive-brown while the breast has pale
streaks. Found along rivers and backwaters, in damp overgrown meadows, etc.
Distribution: East and South Europe;
western Siberia.

Loddigesia A genus of Birds of the
Humming-bird family, *Trochilidae*. e.g.
L. mirabilis - The Loddige's Racket-
tailed Humming-bird, found in a single
valley in Peru, 230-300 m (7,000-9,000
ft) above sealevel. It is a strange bird,
with only 4 tail feathers instead of the
usual 10, of which the outer, elongated
pair overlap and end in purple, racket
shapes.

Lodoicea A genus of Trees of the Palm
family, *Palmae* (= *Arecaceae*). e.g. L.
maldivica - The Double Coconut, a tree
from the Seychelle Islands which reaches a height of 30 m (100 ft). The leaves
have stalks from 2.4-3 m (8-10 ft) long
and blades up to 1.8 m (6 ft) across. The
plants are unisexual. The fruit is a nut
containing a single seed, weighing up to
22 kg (50 lb). The seed takes 10 years to
develop on the tree and about 3 years to
germinate.

Loeschia A genus of parasitic amebae,
now known by the generic name,
Entameba, causing the disease loeschia-
sis or amebiasis. e.g. L. histolytica (=
Entameba h.; *Entamoeba* h.) - see
Entameba.

Loiseleuria A genus of plants of the
Heath family, *Ericaceae*. e.g. L.
procumbens - The Mountain Azalea, an
evergreen shrub up to 30 cm (12 in)
high, decumbent branches ascending up
to 2.5-7.5 cm (1-3 in). Pinkish flowers
grow in few-flowered clusters. Found on
heaths and moors. Place of origin: North
and Central Europe, the Pyrenees, south
Alps and north Balkans. Native in
Britain.

Loligo A genus of Cephalopod Molluscs
of the family of Squids, *Loliginidae*. e.g.
L. vulgaris - The Long-finned or
Common European Squid, up cm long;
it has an elongated, cigar-shaped, light
colored body with 8 short and 2 long
tentacles, and large rhombic, swimming
fins. It also possesses an internal shell
like the cuttlefish (see **Sepia**).
Distribution: Atlantic Ocean,
Mediterranean Sea, occasionally in the
North and Baltic Seas.

Lolium A genus of plants of the Grass
family, *Gramineae*. e.g. L. temulentum -
The Common or Poisonous Darnel, an
annual plant, 30-80 cm high; the
spikelets have one edge against the axis.
Grows in waste places. A poisonous
species, the seeds cause poisoning
known as lolism or loliism. Place of origin: the Mediterranean region; naturalized in Britain.

Lomaspilis A genus of Insects of the
Geometer or Looper Moth family,
Geometridae. e.g. L. marginata - The
Clouded Border Moth or Butterfly, mm;
it has a variable patterning on the wings.
Found in forest margins; valley mead-
ows, along rivers, etc. Distribution:
Europe, Central Asia, southeastern
Siberia.

Lomatogonium A genus of plants of the
Gentian family, *Gentianaceae*. e.g. L.
carinthiacum - The Lomatogonium, an

annual plant, 1-13 cm high, with pale blue or white, terminal, long-stalked, solitary flowers. Grows on pastureland. Place of origin: Europe, in the Alps and Carpathian mountains.

Lonchocarpus A genus of Trees of the Pea family, *Leguminosae*. e.g. L. utilis - The Cubé Root, Timbo or Barbasco- trees and shrubs containing Rotenone, a poisonous compound used as an insecti- cide.

Lonchura A genus of Birds of the Grassfinch, Mannikin, Java Sparrow and Waxbill family, *Estrildidae*. e.g. L. molucca - The Moluccan Mannikin, found in Indonesia, living near human habitation or on the edge of the forest, and commonly seen in gardens or plagu- ing the paddy fields, eating the seeds.

Lonicera A genus of plants of the Honeysuckle family, *Caprifoliaceae*. e.g. L. periclymenum - The Common Honeysuckle or Woodbine, a twining shrub, up to 6 m long; the leaves are elliptic or ovate, and the two-lipped flowers are yellow. Grows abundantly in woods and in scrub. Place of origin: Europe, from Scandinavia and Germany to the Mediterranean. Native in Britain.

Lonomia A genus of Insects of the Emperor Butterfly or Giant Silkworm Moth family, *Saturniidae*. e.g. L. cynira - An Emperor Moth species; the poison of this moth caterpillar is said to cause a fatal type of hemophilia in humans. Distribution: Venezuela.

Loosia A genus of Trematode Worms or Flatworms (class *Trematoda*), better known as *Metagonimus*. e.g. L. dobro- giensis (= *Metagonimus* yokogawai) - see **Metagonimus**.

Lophiomys A genus of Mammals of the Crested Rat or Rodent family, *Lophiomyidae*. e.g. L. imhausi - The Crested Rat, a stout-bodied rodent, 30 cm (1 ft) long with a tail slightly less than this, covered with grizzled, black and white hair, which can be raised along the midline of the back to form a crest, flanked on either side by naked skin. It is a nocturnal, slow-moving, vegetarian animal, living in rocky ravines. Habitat: Ethiopia and northern Kenya.

Lophius A genus of Fishes of the family of Angler and Goosefishes, *Lophiidae*. e.g. L. americanus - The American Angler, also known as the Monkfish, Goosefish or Great Fishing Frog Fish of the North Atlantic. It grows to a maxi- mum of 1.2 m (4 ft), the body is flat- tened from above, the mouth is very wide with strong jaw-muscles, and it has very sharp, strong teeth. This sea fish preys upon smaller fish, attracting them by filaments on its head, hence its common names.

Lophoceros A genus of Birds of the Hornbill family, *Bucerotidae*. e.g. L. flavirostris - The Yellow-billed Hornbill of Botswana and the Transvaal. It has a huge bill, but is light in weight, being no more than a horny covering to a beak formed of a bony mesh containing numerous air-chambers.

Lophocolea A genus of plants of the phy- lum of Liverworts and Mosses, *Bryophyta*. e.g. L. bidentata - A leafy liverwort plant, growing on the grass of lawns and pastures. The stem is slender and bears two lateral rows of thin, toothed leaves.

Lophodytes A genus of Birds of the Duck family, *Anatidae*. e.g. L. cuculla- tus - The Hooded Merganser, Sawbills or Diving Ducks found only in North America. A sea bird with marked diving ability.

Lopholatilus A genus of Fishes (subphy- lum *Vertebrata*). e.g. L. chamaeleonti- ceps - The Tile-fish.

Lophophora A genus of plants, also known by the generic names, *Anhalonium* and *Mammillaria*, of the Cactus family, *Cactaceae*. e.g. L. williamsii (= *Anhalonium* w.; A. lewinii; *Mammillaria* lewinii) - The Mescal Button or Peyote plant, a globular cactus plant with eight shallow lobes and small white or pink flowers. A native of Texas and north Mexico, it is highly valued by various Mexican tribes, and contains several narcotics which produce a feeling of exhileration. In Mexico it is known by the Aztec name "peyote" or "peyotl". The tops are dried and the Aztecs called them by the name: Teonancatyl. Dried slices of the cactus are known as "mescal buttons", from which is extracted the alkaloid Mescaline (with properties similar to LSD - Lysergide). Used in psychiatry as a hallucinogen, and is capable of producing trance-like states.

Lophophorus. A genus of Birds of the Pheasant, Quail and game bird allied family, *Phasianidae*. e.g. L. impejanus - The Impeyan Pheasant, a short-tailed pheasant ormonal, with an iridescent, glittering plumage of blues, greens and bronze, like burnished metal.

Lophophyton A genus of Achorion fungi. e.g. L. gallinarum - A species of fungus causing comb disease in fowls (see **Achorion**).

Lophortyx A genus of Birds of the Pheasant and Quail family, *Phasianidae*. e.g. L. californica (= L. californicus) - The Californian Quail, a brightly colored bird, about 25 cm (10 in) long, living in the United States, from the east coast across to California.

Lophura A genus of Game Birds of the Pheasant and Quail family, *Phasianidae*. e.g. L. igniti igniti - The Borneo Fireback Pheasant (see **Plasmodium** lophurae).

Lophyrus A genus of Hymenopterous Insects of the Sawfly superfamily, *Tenthredinoidea*. e.g. L. pini - The Pine Sawfly, a well known forest pest in Europe. The male which is only 7.5 mm (3/10 in) long, is black in color and has comb-like feelers. The female is just over 1 cm (2/5 in) in size, and is brown in color; it has an ovipositor with saw-like teeth for cutting slits in leaves,

within which the eggs are laid. The larvae have curled tails, the body often having the shape of an "S", with an elevated tail.

Lora A genus of Gastropod Molluscs of the order of Dog Whelk Sea-snails, *Neogastropoda*. e.g. L. turricula - The Turreted Conelet, a carnivorous Dog-Whelk, 2 cm long, with whorls set back like steps. It feeds on mussels and barnacles. Distribution: the coasts of Europe; native around Britain.

Loranthus A genus of plants of the Mistletoe family, *Loranthaceae*. e.g. L. europaeus - The Yellow-berried Mistletoe, a deciduous shrub 30-120 cm high; the leaves are oblong-ovate, the dioecious flowers grow in terminal racemes, and the berries are yellow. Parasitic on branches of deciduous trees, mainly oaks. Place of origin: southeast and central Europe.

Loricaria A genus of Fishes of the Loricariid Armoured Catfish family, *Loricariidae*. e.g. L. typus - The Brazilian Catfish, a common fish in the tropical freshwaters of South America. The lower or ventral lip of the male parent becomes enlarged to form a brood-pouch where the eggs are incubated; the thick lips in both male and female of the species are used in the torrents of rivers and streams to hang on to stones.

Loris A genus of Mammals of the Loris family, *Lorisidae*. e.g. L. tardigradus - The Slender Loris, 30 cm (10 in) long, brownish yellow, with extremely thin lips. Lorises are also known as "slow lemurs"; they creep about the trees at night, approaching their insect prey very slowly and deliberately. Distribution: southern India and Sri Lanka (Ceylon).

Lota A genus of Fishes of the Cod family, *Gadidae* (Burbot, Cusk and Ling subfamily, *Lotinae*). e.g. L. lota - The Burbot, 30-60 cm (1-2 ft) long. A predator, witha barbelled lower jaw, it will eat even large fish, frogs and crayfish. The only freshwater member of the Cod family, it is found in rivers and lakes, as well as the sea. Distribution: Europe and Central Asia, ranging as far as India and northern North America. Native in Britain.

Lottia A genus of Gastropod Molluscs (class *Gastropoda*). e.g. L. gigantea - A large Limpet or Snail species, with a lifespan of at least 15 years.

Lotus A genus of plants of the Pea family, *Leguminosae* (subfamily *Papilionaceae*). e.g. L. corniculatus - The Common Birdsfoot-trefoil, a perennial plant 10-45 cm high; the stem is brightly angled and the yellow flowers are borne 2 to 6 in each head. Grows in meadows and pastureland. Place of origin: widespread in Europe.

Loxia A genus of Birds of the Finch family, *Fringillidae*. e.g. L. curvirostra - The Red Crossbill, 17 cm long; it has red plumage and the tips of its mandibles cross. The female is yellowish-green. It is found in coniferous woods and forests such as spruce, pine and larch. Distribution: North Europe and North America. Native in Britain.

Loxodonta A genus of Mammals of the Elephant order, *Proboscidea*. e.g. L. africana - The African Elephant; there are three races: 1. L. africana africana - The African Bush or Savannah Elephant, up to 3.3 m (11 ft) high at the shoulders, and weighing nearly 6 tons; larger and heavier than its Indian cousin, and with much larger ears (see **Elephas**). An exclusively vegetarian mammal. As in India, the elephant in Africa can be trained and used for work. 2. L. africana cyclotis - The African Forest Elephant, found in the rain forests of Guinea, French Equatorial Africa and the Congo. A shorter animal than the Bush Elephant, it measures up to 2.3 m (71/2 ft) at the shoulder, and is stockier and proportionately heavier, the average bull weighing about 21/2 tons. 3. L. africana pumilio - The African Pygmy Elephant, found in the swampy

forests of Gabon and the Congo, with a maximum shoulder height of 2 m (6 ft 8 in). Sometimes considered a race of undersized Forest Elephants.

Loxosceles A genus of Arachnoids of the order of True Spiders, *Araneae*. e.g. L. reclusa - The Brown Recluse Spider, an extremely poisonous species; the venom is hemolytic and produces widespread ulceration. Distribution: central and southern U.S.A., South America, and more recently Australia (accidentally introduced).

Loxosoma A genus of Entoproctal or Kamptozoan animals of the family *Loxosomatidae*. These are small aquatic, marine creatures, less than 1 mm (1/25 in) in length, bearing a superficial resemblance to the hydroids. Each consists of a jointed calcareous stalk which supports a rounded or bell-shaped body known as a polyp. The sexes are separate, but reproduction also takes place asexually by budding.

Loxotrema A genus of Flukes or Trematode flatworms, better known by the generic name, *Metagonimus* (class *Trematoda*). e.g. L. ovatum (= *Metagonimus* yokogawai) - see **Metagonimus**.

Lucaena A genus of plants of the Pea family, *Leguminosae*. e.g. L. glauca - The Wild Tamarind or Jambul, a South American plant; its seed is said to reduce libido and cause obesity. The bark and roots are used as emmenagogues.

Lucanus A genus of Insects of the Stag Beetle family, *Lucanidae*. e.g. L. cervus - The Common Stag Beetle, one of the largest European beetles, 2.5-7.5 cm (1-3 in) long; the male has long branched mandibles resembling antlers. The large larvae and pupae develop in the rotting wood of old oak trees. Distribution: Europe; native in Britain.

Lucernaria A genus of Jellyfishes of the order of Stalked Jellyfishes, *Stauromedusae*. These stauromedusae are sessile and live fastened to seaweeds or stones by the exumbrella surface, feeding on small crustaceans.

Lucilia A genus of Insects of the Blowfly family, *Calliphoridae*. e.g. L. caesar - The Common Sheep Blowfly or Green Bottle fly, also known as the "Goldfly" or "Sheep maggot", 12 mm long; the body shines a golden green. A scavenger species, the larvae feed on the flesh of dead animals such as the sheep, boring into the hindquarters and sometimes succeed in killing them. The adult feeds on nectar. Distribution: Europe; native in Britain.

Lucioharax A genus of Fishes of the Characin family, *Characinidae*. The species are predators found in the rivers of Africa and South America.

Lucioperca A genus of Fishes of the Perch family, *Percidae*. e.g. L. lucioperca - The Volga Pikeperch, 40-50 cm long; the dorsal fins have dark spots and the pelvic fins are set far forward. A freshwater fish, it inhabits the river basins of the Black Sea, and at the mouth of the Volga River it may enter the sea. An important food fish in Europe and Russia (and other countries of the former U.S.S.R.).

Ludwigia A genus of plants of the Willow-herb family, *Oenotheraceae*. e.g. L. palustris - The Marsh Ludwigia, an annual to perennial plant, 5-30 cm high; with prostrate, ovate to elliptical leaves and solitary flowers in leaf-axils. Grows in freshwater in shallow pools in fens. Place of origin: Europe; native in Britain.

Luidia A genus of Echinoderm animals of the class of Starfishes, *Asteroidea*. e.g. L. magnifica - A marine species from Hawaii, with a major of 40 cm (20 in) and a span of approximately 75 cm (30 in).

Lullula A genus of Birds of the Lark family, *Alaudidae*. e.g. L. arborea - The Woodlark, 15 cm long, a species inhabiting waste ground, lightly tree-clad slopes and woods. The tail is very short and the wings have black and white edge-marks. A partial migrant bird. Distribution: Europe; native in Britain.

Lumbricus A genus of Annelids or Segmented Worms of the Earthworm family, *Lumbricidae*. e.g. L. terrestris - The Common Earthworm, length 9-30 cm, width 6-9 mm; the body is composed of 110-180 segments and the hind end is flattened. Found chiefly in loamy soils in gardens and forests. It burrows in the soil, aerating and fertilizing it. Reproduction is by means of eggs laid in a cocoon. Distribution: Europe; introduced worldwide.

Lumpenus A genus of Fishes of the Band Fish family. e.g. L. lampretaeformis - The Band Fish, 41 cm long; there are longish spots along the center of each side. A marine species inhabiting northern waters. Distribution: seas of Europe; native around Britain.

Lunaria A genus of plants of the Cress or Crucifer family, *Cruciferae*. e.g. L. rediviva - The Common Moonwort, a perennial plant, 50-100 cm high; the flowers are pale purple to violet, though rarely white; the pods are small. Grows in shady woods in most of Europe. Introduced and naturalized in Britain.

Lupinus A genus of plants of the Pea family, *Leguminosae*. e.g. L. luteus - The Yellow Lupin or Lupine, an annual herb 15-70 cm high; it has palmate leaves with 5 to 9 leaflets, and the yellow flowers are borne in whorls. Place of origin: the western part of the Mediterranean region of Europe.

Luronium A genus of plants of the Water Plantain family, *Alismataceae*. e.g. L. natans - The Floating Water-plantain, a perennial plant, up to 50 cm high; with long-stalked, ovate or elliptic leaves and submerged linear leaves. Grows in ponds, lakes and marshes. Place of origin: western, north and southeast Europe. Native in Britain.

Luscinia A genus of Birds of the Babbler,

Nightingale, Robin and Thrush family, *Turdidae* (Blackbird, Nightingale, Robin and Thrush subfamily, *Turdinae*). e.g. L. megarhynchos - The Nightingale, 17 cm long; a reddish brown, migrant bird, darker on the back than on the underparts. Found in trees and bushes of woodlands, parks and heaths, singing day and night. Distribution: western and central Europe (native in Britain), northwest Africa, the Middle East, and Central Asia.

Lusciniola A genus of Birds of the Old World Warbler family, *Sylviidae*. e.g. L. melanopogon - The Moustached Warbler, 13 cm long; a partial migrant. The crown is almost black. It usually nests in bushes overhanging freshwater areas and in reed-beds. Distribution: southern Europe.

Lutjanus A genus of Fishes of the Snapper family, *Lutjanidae*. e.g. L. analis - The Mutton Snapper of the West Indies and a much favored food fish. L. bohar - A snapper species from Samoa; highly poisonous to eat. L. cyanopterus - The Cubera Snapper of Cuba; poisonous to eat.

Lutra A genus of Mammals of the Badger, Otter, Skunk and Weasel family, *Mustelidae* (Otter subfamily, *Lutrinae*). e.g. L. lutra - The Common or Old World Otter; 75 cm (21/2 ft) long, with a tail 45 cm (11/2 ft) in length, and weighing up to 13.5 kg (30 lb). It has a glossy brown coat and inhabits thickly overgrown banks of streams, in which it makes its den with an underwater entrance. A rare fur animal. Distribution: widespread in Europe (including Britain), North Africa, northern Asia and parts of India. Occurs as ten geographic races throughout Eurasia.

Lutraria A genus of Bivalve Molluscs of the Gaper Shell family, *Myidae*. e.g. L. lutraria - The Common Otter Shell, up to 13 cm long; valve is longish and the external skin is brown. The "gapers" are known as such because their shells can-

not close completely; they burrow in sandy mud off the seashore. Distribution: the coasts of Europe; native in Britain.

Lutreola A genus of Mammals of the Badger, Otter, Skunk and Weasel family, *Mustelidae*. e.g. L. lutreola - The European Mink; length of body 32-40 cm, tail 12-19 cm; found in the vicinity of rivers and streams with thick herbaceous vegetation along the banks. A nocturnal animal. Distribution: occurs as several geographical races in Europe, Russia, and western Siberia.

Lutrogale A genus of Mammals of the Badger, Otter and allied family, *Mustelidae* (Otter subfamily, *Lutrinae*). e.g. L. perspicillata - The Smooth-coated Otter, found in India and Malaya (see **Lutra**).

Luzula A genus of plants of the Rush family, *Juncaceae*. e.g. L. pilosa - The Hairy Woodrush, a grass-like, hairy, perennial plant, 15-30 cm high; the single, dark-brown flowers are borne in loose cymes, and the capsules are acuminate (i.e. they are gradually narrowed to a point). There are only three seeds in the fruit. Grows mainly in woods and scrub. Place of origin: Europe (native in Britain) and northwest Asia.

Lybia A genus of Decapod Crustaceans of the Crab family, *Xanthidae*. These crabs are non-swimmers, heavily armored and particularly abundant on tropical reefs and rocky shores. The species carry small sea anemones in their pincers.

Lycaena A genus of Insects of the Blues, Coppers and Hairstreaks Moth family, *Lycaenidae*. e.g. L. arion - The Large Blue Butterfly, 2.5 cm (1 in) long; the upper surface of the wings is blue, the under surface is speckled brown. The eggs are laid on wild thyme. The caterpillars produce honey-dew which is very attractive to ants, with whom they are always associated. Distribution: Europe; native in Britain.

Lycaon A genus of Mammals of the Dog family, *Canidae*. e.g. L. pictus - The Cape Hunting Dog, up to 105 cm (31/2 ft) long, with a tail of up to 37.5 cm (15 in). It is irregularly spotted black, white and yellow-brown, and has conspicuously large, round, erect ears. It travels in packs and is exclusively carnivorous. Habitat: open savannah country south of the Sahara, as in East Africa.

Lychnis A genus of plants of the Pink family, *Caryophyllaceae*. Some species are now commonly included under the generic name, *Agrostemma*. e.g. L. alpina - The Alpine Catchfly, a perennial species 5-15 cm high; the plant is not sticky, and the inflorescences are dense and head-like. Grows widely in stony mountain pastureland, on rocks and scree, and in bushy places. Place of origin: Europe; native in Britain. L. githago (= *Agrostemma* g.) - The Corn Cockle (see **Agrostemma**).

Lycodon A genus of Reptiles of the Colubrid Snake family, *Colubridae*. e.g. L. aulicus - The Wolf Snake, 60 cm (2 ft) long; found in India and Sri Lanka (Ceylon), where it enters houses and other habitations. A ratter, it catches mice infesting granaries.

Lycogala A genus of Slime Molds or Fungal plants. e.g. L. epidendrum - A slime mold species found widely on rotten wood. The small, round reproductive bodies are soft and red when young, but at maturity they are brittle and violet. Eventually they rupture and the minute spores are blown away.

Lycoperdon A genus of Fungal plants of the Puffball family, *Lycoperdaceae*. e.g. L. perlatum - The Warted Puffball, an Earth Ball fungus species with a cup 3-5 cm wide; the fruit-body is cone-shaped and spiny, with an elongate, cylindrical, stem-like base. When mature the upper cavity contains a mass of loose, dry, olive-brown spores. Edible when young.

It grows in coniferous or deciduous woods and in pastures. Place of origin: Europe; native in Britain.

Lycopersicum A genus of plants of the Nightshade family, *Solanaceae*. e.g. L. esculentum - The Tomato, a widely growing plant, cultivated worldwide. A native of South America from which a large number of cultivated varieties have been developed. It has dissected leaves and loose axillary clusters of yellow flowers. The fruit is a berry rich in vitamin C.

Lycopodiella A genus of plants of the Clubmoss family, *Lycopodiaceae*. e.g. L. inundata - The Marsh Clubmoss, a slender, bright green, rather moss-like species; the stems are slightly creeping, up to 20 cm (8 in) long, little-branched and only half-evergreen, and the leaves are untoothed. The cones are solitary, terminal and unstalked, with leaf-like toothed scales. Grows on moors, wet heaths, dune slacks, especially on acid soils. Distribution: Britain and northern Europe.

Lycopodium A genus of plants of the Clubmoss family, *Lycopodiaceae*. e.g. L. clavatum - The Stag's-horn Clubmoss, a perennial plant up to 20 cm high; the stems creep on the ground up to 1 m long, with 2 to 3 long-stalked cones. The bright-green leaves are spirally arranged, and end in long, whitish hairs. Grows on moors, heaths, and in coniferous forests. Place of origin: Europe; native in Britain.

Lycopsis A genus of plants of the Borage family, *Boraginaceae*. e.g. L. arvensis - The Small Bugloss, a biennial to perennial plant, 20-60 cm high; the plant is bristly, the leaves have undulate margins and the flowers have white throat-scales. Grows in arable and wasteplaces, short grasslands and on roadsides. Place of origin: Europe; native in Britain.

Lycopus A genus of plants of the Mint or Thyme family, *Labiatae*. e.g. L. europaeus - The Gipsy-wort, a non-aromatic, perennial, mint plant and the bitter bugle-weed of Europe, formerly used in medicine, 30-100 cm high; the leaves are pinnately shallowly lobed, the white flowers have purple dots on their lower lip, and there are two stamens. Grows around lakes and ponds, in ditches and marshy places. Place oforigin: Europe; native in Britain.

Lycoris A genus of plants of the Daffodil family, *Amaryllidaceae*. e.g. L. radiata - The Spider Lily, a bulbous plant, 45 cm (18 in).

Lycoris A genus of plants of the Daffodil family, *Amaryllidaceae*. e.g. L. radiata - The Spider Lily, a bulbous plant, 45 cm (18 in) high; with funnel-shaped, bright red or deep pink flowers growing in rounded umbels and linear, strap-shaped, glaucous leaves. Place of origin: China.

Lycosa A genus of Reptiles of the Wolf Spider family, *Lycosidae*. e.g. L. Tarantula - The European Tarantula, a venomous spider, whose effect is hemolytic and produces widespread ulceration.

Lygus A genus of Insects of the Capsid True Bug family, *Miridae*. e.g. L. pratensis - A heteropterous species, 5.8-6.7 mm long; a very common, destructive pest of fruit-tree crops. It has a light reddish-brown carapace. Distribution: the entire Palaearctic region.

Lymanopoda A genus of Insects of the order of Moths and Butterflies, *Lepidoptera*. A migrating butterfly, the species are known to fly at great heights, and they have been seen crossing the Andes, in South America at heights up to 4,700 m (15,419 ft).

Lymantria A genus of Insects, also known by the generic name, *Porthetria*, of the Tussock Moth family, *Lymantriidae*. e.g. L. dispar (= *Porthetria* d.) - The Gipsy (Gypsy) Moth, size 18-36 mm, a brownish or white European moth, now also common in the United States, whose larvae are destructive to the leaves, sometimes also in gardens. The female covers the eggs with a layer of hairs, the whole slightly resembling a tree fungus. Distribution: much of Europe (now extinct in Britain); North Africa; North America (introduced in 1868 and now a pest in certain places) (see **Calosoma**).

Lymnaea A genus of Gastropod Molluscs of the Pond Snail family, *Lymnaeidae*. e.g. L. truncatula - The Dwarfed Pond Snail, a common species found in ponds, pools, ditches and irrigation works. The shell is 7-14 mm high and 3.5-6.3 mm wide, small, conical, elongate and composed of 5 to 51/2 whorls, separated by a deep suture. It feeds on algal plants. This snail serves as an invertebrate, intermediate host for the Liver fluke (*Fasciola* hepatica), for *Trichobilharzia* ocellata, etc., which cannot complete their life cycle without first parasitizing the snail. Distribution: the Palaearctic and Nearctic regions.

Lymnocryptes A genus of Birds of the Curlew, Sandpiper, Snipe and Woodcock family, *Scolopacidae*. e.g. L. minima (= L. minimus - The European Jacksnipe, 19 cm long, a small, slender, migrant bird with a delicate beak, featherless legs, longitudinally striped plumage, and preference for open terrain. It has a dark, center stripe on its crown. Distribution: Britain and much of Europe.

Lymphocytozoon A genus name for ameboid bodies found in leukocytes. e.g. L. cobayae (Kurloff's bodies). L. pallidum (Ross's bodies).

Lynchia A genus of Insects of the order of Two-winged flies, *Diptera*, better known by the generic name, *Pseudolynchia*. e.g. L. canariensis (= *Pseudolynchia* c.) - A species of Pigeon flies, transmitting malaria to pigeons (see **Pseudolynchia**).

Lynx A genus of Mammals, also known by the generic name, *Felis* of the Cat family, *Felidae*. e.g. L. lynx (= *Felis* l.)

- The European Lynx, a carnivorous predator, about 90 cm (3 ft) long with a short tail of only about 12.5 cm (5 in). Found in the larger submontane and montane forests, the den being located in thickets, among rocks or in tree cavities, where sufficient cover is provided from hunters and trappers. A rare species, it now survives in some countries of northern Europe, in the Carpathians, the Balkans, and eastwards as far as China.

Lyonetia A genus of Insects of the Butterfly and Moth family, *Lyonetiidae*. e.g. L. clerkella - Length 4-6 mm; a very common, little moth. The caterpillars live in the leaves of various broadleaved trees where they make serpentine tunnels or mines in which they deposit their dark excrement. Distribution: much of the Palaearctic region.

Lyrurus A genus of Birds of the Grouse or Ptarmigan family, *Tetraonidae*. e.g. L. tetrix - The Black Grouse; the male is about the size of a fowl cock, 53 cm long and its plumage is a bluish black, with a metallic sheen, and its tail is lyre-shaped; the female is 41 cm long, its tail is forked and the plumage brown. A resident bird, the Black Grouse is found in wooded regions and mountainous areas. Distribution: northern Asia and Europe. Native in Britain.

Lysichiton A genus of plants of the Arum family, *Araceae*. e.g. L. americanum - The Skunk-cabbage, a species growing in Alaska, Oregon and California. It has oblong leaves and the spadix, bearing many bisexual flowers, is enclosed by a hooked yellow spathe or sheathlike bract (modified leaf). The fruit consists of a two-seeded berry embedded in the spadix.

Lysimachia A genus of plants of the Primrose family, *Primulaceae*. e.g. L. nemorum - The Wood Pimpernel or Yellow Pimpernel; a slender, creeping, perennial plant, 10-30 cm high; with

solitary, axillary flowers. It grows in woods, hedgebanks and scrub. Place of origin: West and Central Europe, and western Asia. Native in Britain.

Lysmata A genus of Crustaceans of the suborder of Shrimps, *Natantia*. e.g. L. grabhami - The Cleaner Shrimp, an edible marine crustacean with ten legs and a long slender body. It feeds on parasites, removing them from the fish which submit themselves to a delicate cleaning process. Distribution: the Red Sea and the Gulf of Filat (Akaba).

Lythrum A genus of plants of the Loosestrife family, *Lythraceae*. e.g. L. salicaria - The Purple Loosestrife, a perennial herb, 60-120 cm (2-4 ft) high; with lanceolate leaves and handsome spikes of reddish-lilac or purple flowers. The fruit is an ovoid capsule. Grows on banks of ditches, lakes and ponds. Place of origin: North Europe; native in Britain.

Lytta A genus of Insects, better known by the generic name, *Cantharis*, of the family of Oil and Blister Beetles, *Meloidae*. e.g. L. vesicatoria (= *Cantharis* v.) - The Spanish Fly; the body contains the pharmacologically active substance Cantharidin (see **Cantharis**).

M

Macaca A genus of Mammals, also known by the generic name, *Macacus*, of the Old World or Guenon-like Catarrhine Monkey family, *Cercopithecidae*. e.g. M. mulatta (= M. mulata; *Macacus* m.) - The Rhesus Monkey, about 60 cm (2 ft) long with a 30 cm (1 ft) tail; a widespread species inhabiting northern India and southeast Asia. A gregarious, agile animal with a greyish, long-haired coat. Used in laboratories all over the world for research.

Macacus See **Macaca**.

Machilis A genus of Insects (class *Insecta*). e.g. M. tirolensis - A primitive insect, 12-14 mm long; the antennae are as long as the body. Found up to a height of almost 4,000 m. Distribution: Europe.

Maclura A genus of Trees and Shrubs of the Mulberry family, *Moraceae*. e.g. M. pomifera - The Osage-orange (= Bow-wood or Hedge), a deciduous tree up to 15 m (50 ft) tall, with spiny branches and lanceolate leaves. The small greenish male and female flowers are borne in stalked, spherical clusters on separate trees. The fruit formed after fertilization is rough, greenish-yellow and fleshy, and reaches about 10 cm (4 in) across. Known in the United States as the Bow-wood tree because its silky yellow wood was used for bows by the Osage tribe of Indians; also as the Hedge tree from its usefulness along boundaries. The bark has been used for tanning leather. Place of origin: the eastern United States; introduced into Europe, where it is sometimes grown in gardens for its ornamental value.

Macoma A genus of Bivalve Molluscs of the deposit-feeding, lamellibranch, Tellin Shell superfamily, *Tellinacea*. These are the rock-borers, burrowing and nestling in crevices, with long, separate, inhalant and exhalant siphons. The species are very mobile and each possesses a large foot. e.g. M. balthica - The Baltic Tellin, size 2.2 cm. The valve is broadly triangular, yellow, green, red or brown in color. Distribution: Europe; native around Britain.

Macracanthorhynchus A genus of parasites of the class *Acanthocephala*. e.g. M. hirudinaceus - A species parasitic in swine in the U.S.

Macraspis A genus of Flukes (class *Trematoda*) of the family of Aspidogastrid trematodes, *Aspidogastridae*. These are endoparasites in the gall-bladder of chimaerid fishes, and have one row of alveoli (see **Callorhynchus**).

Macrobdella A genus of Segmented Worms of the Hirudenoid jawed leech family, *Gnathobdellae*. e.g. M. decora - A common North American species, this jawed leech sucks the blood of various invertebrates or vertebrate animals, piercing the skin with its lancet-like jaws and covering the wound with the anterior sucker.

Macrobiotus A genus of Water Bear Animalcules (phylum *Tardigrada*). These Tardigrades are minute organisms, widely distributed throughout the world and less than 0.5 mm long. They are minute arthropods with four pairs of stumpy legs and no definite respiratory system. They live in the water film surrounding terrestrial mosses and lichens.

Macrocheira A genus of Crustaceans of the Spider Crab family, *Majidae*. e.g. M. kaempferi - The Giant Spider Crab, also called the "Stilt Crab", the longest species of the family. It has a heart-shaped or triangular carapace measuring about 30 cm (1 ft) by 30 cm (1 ft) and its legs or claws span over 2.4 m (8 ft). Distribution: the deep waters off the southeast coast of Japan.

Macrochelys A genus of Reptiles, better known by the generic name, *Macroclemys*, of the Snapping Turtle

family, *Chelydridae*. e.g. M. temmincki
(= *Macroclemys* t.) - The Alligator
Snapping Turtle (see **Macroclemys**).

Macroclemys A genus of Reptiles, also
known by the generic name,
Macrochelys, of the Snapping Turtle
family, *Chelydridae*. e.g. M. temmincki
(= *Macrochelys* t.) - The Alligator
Snapping Turtle, about 75 cm (2 1/2 ft)
long, with a reduced shell, lateral spines
and a long tail. Found in rivers and
marshes, often covered with green algae
which render it almost invisible among
the vegetation. It is very aggressive and
possesses hooked jaws which can inflict
severe injuries. A long-lived species,
some may live more than 50 years.
Distribution: North America, east of the
Rockies, and southwards to northern
South America.

Macrocyclops A genus of Crustaceans of
the Cyclops family, *Cyclopidae*. e.g. M.
fuscus - A Copepod crustacean, length
1.8-4 mm, a species of freshwater
plankton. Found in water with dense
vegetation. Distribution: Europe.

Macrodontia A genus of Insects of the
Long-horned or Longicorn Beetle family, *Cerambycidae*. e.g. M. cervicornis -
The Common Rhinoceros Beetle, about
12.5 cm (5 in) long, a black and brown
species and one of the largest beetles.
Distribution: southern Brazil.

Macrogastra A genus of Gastropod
Molluscs of the family *Clausiliidae*. e.g.
M. plicatula - The light-brown shell is
10-15 mm high and 2.8-3 mm wide;
slender, ribbed and covered with fine
spiral ridges. Found in forests from lowlands to mountainous elevations amidst
fallen leaves, on tree trunks, in stumps,
on rocks and on walls. Distribution:
Europe, from northern Italy to southern
Scandinavia.

Macroglossum A genus of Insects of the
Hawk Moth family, *Sphingidae*. e.g. M.
stellatorum - The Humming-bird Hawk
Moth, 4.5 cm long. The hind wings are
rust-red. It flies by day, looking like a
humming-bird hovering in front of
plants. Distribution: temperate parts of
Europe; native in Britain.

Macroglossus A genus of Mammals of
the suborder of Fruit Bats or Fruit-sucking Bats, *Megachiroptera*. These are the
long-tongued fruit bats found in India,
New Guinea, Australia and Africa.

Macromonas A genus of microorganisms
of the order *Pseudomonadales*.
Cylindrical to bean-shaped bacteria. e.g.
M. mobilis.

Macroperipatus A genus of Segmented
Worms of the class of Velvet Worms,
Onychophora. The central American
type species reaches 12.5 cm (5 in) in
length. Found in moist habitats, under
stones, or the bark and rotten wood of
fallen trees.

Macropipus A genus of Dacapod
Crustaceans of the suborder of Crabs,
Brachyura. e.g. M. puber - The Velvet
Crab; the carapace is up to 12 cm (4 3/4
in) long. A swimming crab, it is an edible, marine species with swimming
plates on the last pair of walking legs. A
most aggressive rock-pool creature; if
disturbed it will hold its claws out in an
attacking posture. Distribution: the
Mediterranean and North Seas; the
Atlantic Ocean.

Macroplectrum A genus of plants of the
Orchid family, *Orchidaceae*. e.g. M.
sesquipedale - An orchid species native
to Madagascar, with broadly-lanceolate
leaves and white or creamy-white flowers, each having a long, thin, curved
spur up to 30 cm (1 ft) long.

Macropodia A genus of Crustaceans of
the Spider Crab family, *Majidae*. e.g. M.
rostrata - The Slender-legged Crab, a
short-tailed, marine Crustacean, up to 2
cm long, with a triangular carapace,
long, pointed rostrum and long legs. It
camouflages itself with seaweed.
Distribution: the North Sea and western
Baltic; native around Britain.

Macropus A genus of Mammals of the
Kangaroo family, *Macropodidae*. e.g.

M. rufus (= M. rufa) - The Red Kangaroo, the largest of all the marsupials, the males of which weigh more than 100 kg (220 lb) and are larger than 1.5 m (5 ft) from the muzzle to the root of the tail. Distribution: the highlands of central, southern and eastern Australia.

Macroscelides A genus of Mammals of the family of Jumping or Elephant Shrews, *Macroscelididae*. These are rat-sized, jumping animals with mobile, trunk-like snouts. They have long, scaly tails and large eyes and ears. They are diurnal and may be seen in bright sunshine hopping after insects. At night they take refuge from the cold in holes in the ground or in burrows of other animals. The 5 genera are all from Africa and include, besides *Macroscelides*, also *Elephantulus, Nasilio, Petrodomus* and *Rhynchocyon*. e.g. M. proboscideus - The Short-eared Elephant Shrew of South Africa, a rat-sized creature with a long, sensitive snout. It walks on all fours with a plantigrade gait, using the soles of the feet besides its toes.

Macrosternodesnus A genus of Arthropods of the class of Millipedes, *Diplopoda*. e.g. M. palicola - A minute species found in Britain, measuring only 3.5 mm (0.137 in) in length.

Macrostoma A genus of Flagellate Protozoans of the class *Zoomastigophorea*. The species are included under other generic names, such as *Chilomastix* and *Trichomonas*. e.g. M. mesnili (= *Chilomastix* m.; *Trichomonas* hominis, etc.) - an intestinal flagellate (see **Chilomastix**; **Trichomonas**).

Macrostomum A genus of Turbellarian Flatworms of the order *Archoophora*.

Macrostylis A genus of Crustaceans of the order of Woodlice, Pill Bugs, and other Isopods, *Isopoda*. e.g. M. galathae - A marine species of isopods recovered from the deep seas of the Philippine Trench at a depth of 9,790 m (32,119 ft).

Macrotermes A genus of Insects of the order of Termites, *Isoptera*. Also known as the "white ant", it is in fact closely related to the cockroach. The species subsist on fructification of specially cultivated "fungus gardens" within their nests. A colony of these termites can gradually grow in size to number ultimately over a million individuals and to endure for scores of years. In these colonies the queen lies inert in a special royal cell attended by crowds of workers, she becoming virtually an egg-laying machine. e.g. M. bellicosus - The African Termite, the world's largest termite, reaching a length of 12.7 cm (5 in), and building huge mounds up to 12.8 m (42 ft) tall.

Macrotus A genus of Mammals of the American Leaf-nosed Bat family, *Phyllostomatidae*; known erroneously as the Vampire Bats, because they were at one time thought to be blood-suckers. e.g. M. californicus - The Californian Leaf-nosed Bat, an insectivorous and occasionally fruit-eating bat. Distribution: North America.

Macrourus A genus of Fishes of the Grenadier or Rat-tail family, *Macrouridae*. e.g. M. cinereus - The Pop-eyed Grenadier, a deep-water oceanic fish found commonly in the Bering Sea. It has a peculiar truncated snout, often with a rostrum.

Macrozoarces A genus of Fishes of the Eelpout family, *Zoarcidae*. e.g. M. americanus - The Eelpout or Ocean Pout fish, a marine species resembling the Scaleless Blenny (family *Blenniidae*) (see **Zoarces**).

Mactra A genus of Bivalve Molluscs of the Trough Shell family, *Mactridae*. e.g. M. corallina - The Rayed Trough Shell; length 40-60 mm, height 35-45 mm, thickness 16-28 mm; the shell is whitish with concentric lines, sometimes also radiating bands. Found in sandy bottoms at depths of 5-30 m. An edible species. Distribution: Black Sea, west coast of

Europe and the Mediterranean. Native around British coasts.

Madhuca A genus of Trees of the Sapodilla family, *Sapotaceae*. Also known by the generic name, *Bassia*. e.g. M. indica (= *Bassia* latifolia) (see **Bassia**).

Madoqua A genus of Mammals of the Horned Ungulate family, *Bovidae*. e.g. M. swaynei - Swayne's Dik-dik, a slender antelope species from Somalia, East Africa. The adult weighs only 2.26-2.72 kg (5-6 lb), and has a height of about 33 cm (13 in) at the withers. It is one of the world's smallest antelopes.

Madrepora A genus of True or Stony Corals (order *Scleractina*) of the Staghorn Coral variety (see **Acropora**).

Madtsoia A genus of Prehistoric, now extinct Reptiles of the Snake family. This fossil snake from Patagonia in South America, had an estimated length of 10.05 m (33 ft), and is one of the longest prehistoric snakes that ever existed.

Madurella A genus of Granular Fungi. e.g. M. madurae - The black melanoid variety of granular fungi seen in Mycetoma or Madura Foot (as opposed to the "pale" variety, *Actinomyces* madurae) (see **Actinomyces**).

Maena A genus of Fishes of the Sea Bream family, *Sparidae*. e.g. M. vulgaris - The Blotched Picarel, an edible, sea bream fish, up to 25 cm (10 cm) long. The body is lead-colored, with dark, length-wise stripes and it has a black spot behind the pectoral fin, hence the common name. Distribution: the Mediterranean Sea.

Magnolia A genus of Trees and Shrubs native to Asia and North America, of the Magnolia family, *Magnoliaceae*. e.g. M. acuminata - The Cucumber Tree, up to 18-27 m (60-90 ft) tall, a species of deciduous, forest trees with large alternate leaves and large terminal white, pink or yellowish waxy flowers. The 7.5 cm (3 in) long fruits, when young and green, resemble a cucumber, giving this tree its common name. The bitter, aromatic bark of the magnolia tree is used for the preparation of diaphoretics and antipyretics. Place of origin: eastern North America; cultivated in eastern states (of the U.S.) and in Europe as an ornamental.

Mahonia A genus of plants of the Barberry family, *Berberidaceae*. e.g. M. aquifolium - The Mahonia, Oregon Grape or Oregon Holly Grape, an evergreen shrub, up to 1.2 m (4 ft) high, with tough pinnate leaves having 5-9 stalkless leaflets, which are rich glossy green and turn purplish in winter. The yellow flowers grow in dense terminal racemes followed by decorative bunches of grape-like, blue-black berries. Place of origin: western North America; introduced into Britain in 1823.

Maianthemum A genus of plants of the Lily family, *Liliaceae*. e.g. M. bifolium - The May Lily, a perennial plant, 10-25 cm (4-10 in) high; it has two heart-shaped leaves and the tetramerous (4-portioned) flowers grow in terminal racemes. Found commonly in deciduous woods and in scrub. Place of origin: Europe; native in Britain.

Majorana A genus of plants, also known by the generic name, *Origanum*, of the Mint and Thyme family, *Labiatae*. e.g. M. hortensis (= *Origanum* majorana) - The Sweet Majoram plant, a shrubby, aromatic, perennial plant, from 30-60 cm (1-2 ft) high, with pairs of hairy ovate leaves, small white or purplish flowers borne in dense ovoid heads in clusters, and small oval, dark brown nutlets. Place of origin: the Mediterranean region.

Makaira A genus of Fishes of the Marlin family, *Istiohoridae*. e.g. M. nigricans marlina - The Black Marlin fish, a big-game sea fish.

Malachius A genus of Insects of the Beetle family, *Melyridae*. e.g. M. bipustulatus - A green-colored beetle, 5-6 mm long, common on grasses and flowering plants in summer. When danger threatens it produces peculiar reddish pouches or vesicles from the sides of the body. Distribution: Europe, Asia Minor, Siberia.

Malaclemys A genus of Reptiles of the Freshwater Tortoise or Turtle family, *Emydidae*. e.g. M. terrapin - The Diamond-back Terrapin, a popular food in the United States. Its carapace is greenish-brown with concentric dark lines, while the sides and plastron are yellow. It lives in coastal brackish and marine waters of the southern and eastern United States.

Malacocincla A genus of Birds of the Babbler, Thrush and Warbler family, *Muscicapidae* (Babbler subfamily, *Timaliinae*). e.g. M. abbotti - Abbott's Jungle Babbler, a widespread Asiatic species, a common lowland bird particularly along the sea coast. It is olive brown above and pale below. The ball-like nest is built close to the ground and holds spotted eggs.

Malacosarcus A genus of Fishes of the Pricklefish family, *Stephanoberycidae* (see **Stephanoberyx**).

Malacosoma A genus of Insects of the Eggar and Lappet Moth or Butterfly family, *Lasiocampidae*. e.g. M. neustria - The Lackey Moth, size 2-3 cm; the forewings have two reddish or yellowish cross-bands. It lives gregariously in large, silken, tent-like nests in broadleaved woods and fruit orchards. The caterpillars are garden and orchard pests, feeding on fruit trees, on blackthorn and hawthorn bushes, and on oak. Distribution: Europe and North America; native in Britain.

Malania A genus of Fishes of the Coelacanth order, *Actinistia*. These are the so-called Lungfishes, members of which order first existed about 300 million years ago (see **Latimeria**).

Malapterurus A genus of Fishes of the suborder *Siluroidea*. e.g. M. electricus -

The Electric Catfish, a freshwater fish found in the rivers and lakes of tropical and North Africa. It can produce a powerful electrical discharge, and measurements up to 350 Volts at 1 A have been recorded. It is a voracious eater, with barbels and a scaleless, bony-plated body, and a fleshy fin near the tail.

Malassezia A genus of Fungi, also called by other generic names, such as *Microsporon* and *Microsporum*. e.g. M. furfur (= M. macfadyani; M. tropica; *Microsporon* f.; *Microsporum* f.) - the causative fungus of pityriasis versicolor.

Malaxis A genus of plants of the Orchid family, *Orchidaceae*. e.g. M. monophyllos - The One-leaved Malaxis, a perennial plant, 7-45 cm high; it has one or sometimes two leaves, and yellow flowers in which the lip points upwards. A rare species, it is found on moorlands and in peat bogs. Place of origin: North and Central Europe.

Malcomia A genus of plants of the Cress or Crucifer family, *Cruciferae*. e.g. M. maritima - The Virginia Stock plant, a cultivated species.

Malleomycea A name formerly given to a genus of microorganisms, *Schizomycetes*. These are straight or slightly curved bacilli with rounded ends. e.g. M. mallei (= *Actinobacillus* m.; *Bacillus* m.; *Pfeifferella* m.) - the causative organism of glanders. M. pseudomallei (= M. whitmori; *Bacillus* w.; *Pfeifferella* w.; *Pseudomonas* p.) - the causative organism of melioidosis.

Malletia A genus of Bivalve Molluscs (class *Bivalvia*). e.g. M. obtusa - Length of shell up to 13 mm; the shell is oval, thin and transparent with fine, concentric stripes; the siphons are long and grow together. These animals live in deeper water, up to 600 m. Distribution: the coasts of Europe.

Mallomonas A genus of Protozoan animals of the Chrysomonad order, *Chrysomonadina*. This is a solitary species of unicellular, elongate microorganisms, about 40-801/2 in length. It has siliceous spines covering the body. It appears in 2 stages, a flagellar stage (with one flagellum and one chloroplast) and an ameboid stage. e.g. M. schwemmlei - A species with overlapping silicified scales, some or all of which bear long, needle-like projections.

Mallotus 1. Animal Kingdom. A genus of Fishes (subphylum *Vertebrata*). e.g. M. villotus - The Atlantic Capelin fish. 2. Vegetable Kingdom. A genus of plants of the Spurge family, *Euphorbiaceae*. e.g. M. philippinensis - The Kamala plant (= Camala; Glandulae Rottlerae; Rottlera plant), from whose fruits is produced an odorless, tasteless, reddish-brown powder, used as an anthelmintic for tapeworm infestation in veterinary medicine.

Malpighia A genus of plants of the family *Malpighiaceae*. e.g. M. punicifolia - The Acerola or Puerto Rican Cherry tree, the juice of which contains 100 times the ascorbic acid content of orange juice. A valuable source of vitamin C, especially in infants allergic to orange juice.

Malus A genus of Trees of the Rose family, *Rosaceae*. e.g. M. sylvestris - The Crab Apple or Wild Crab, a small deciduous shrub or tree, up to 10 m (33 ft) high, with a fissured brownish-grey bark. An ancestor of the apple tree. Its petals have a reddish tinge and the anthers (the part of the stamen containing the pollen) are yellow. The rounded, yellowish-green fruits are often tinged with red but the flesh is usually sour, even when the fruit is ripe, and they make delicious conserves, especially when mixed with other fruits. Place of origin: Europe (native in Britain), Western Asia.

Malva A genus of plants of the Mallow family, *Malvaceae*. e.g. M. sylvestris - The Common Mallow, a robust, herbaceous, perennial plant, 30-60 cm (1-2 ft) high, with large red or rosy-purple flowers and upright fruitstalks. A decoction of the flowers and leaves has long been used in Europe as an antitussive, emollient, laxative, sedative, etc. Place of origin: Europe (native in Britain), Asia. Introduced into North and South America, and Australia.

Mamestra A genus of Insects of the Noctuid or Owlet Moth family, *Noctuidae*. e.g. M. brassicae - The Cabbage Moth, size 19-23 mm; very common in fields and gardens. The caterpillars feed on vegetables including peas, lettuce and cabbage, and may cause damage. Distribution: Europe, Asia.

Mammillaria A genus of plants now more correctly called *Leptocladodia*, of the Cactus family, *Cactaceae*. Other species are also catalogued under the generic names, *Anhalonium* and *Lophophora* (see **Anhalonium**; **Leptocladodia**; **Lophophora**).

Mammuthus A genus of extinct mammals of the mammoth or prehistoric elephant family, *Mammutidae*. e.g. M. primigenius - The Woolly Mammoth, a prehistoric elephant species which lived about one million years ago in Central Europe and North America. Estimated at having a height of 4.2 m (14 ft) at the shoulders, with a tusk of the same length.

Manayunkia A genus of Segmented Worms of the Serpulid family, *Serpulidae*. There are 2 species, found in North America in freshwater in Pennsylvania, New Jersey and the Great Lakes. They secrete hard limey tubes which are attached to stones.

Mandragora A genus of plants of the Nightshade family, *Solanaceae*. e.g. M. autumnalis (= M. officinalis; M. officinarum) - The oriental Mandrake plant, having properties similar to belladonna. Its forked root was formerly used for its narcotic, sedative, and emetic effects. Place of origin: the Mediterranean littoral.

Mandrillus A genus of Mammals of the Old World Monkey family, *Cercopithecidae.* e.g. M. sphinx - The Mandrill, over 90 cm (3 ft) long, one of the ugliest and mosy brutal of monkeys as well as the largest. The muzzle is purple and the enormous cheek swellings are pale blue. It lives in very dense forests on the ground. Distribution: South Cameroons, Gabon, Guinea and the Congo, in Africa.

Mangifera A genus of plants of the Cashew family, *Anacardiaceae.* e.g. M. indica - The Mango tree, a large tree cultivated widely for its fruit; with opposite, compound leaves and small flowers borne in dense clusters. The fruit is a large, succulent drupe with aromatic pulp, eaten ripe or used green in jam, pickles, etc. It is of great economic importance in tropical and subtropical countries. Place of origin: India and southeast Asia.

Manica A genus of Insects of the Ant family, *Myrmicidae.* e.g. M. rubida - Length 5-9 mm (worker ant); found in foothills and mountains, nesting under stones and in the ground. The sting is quite painful. Distribution: southern and central Europe.

Manihot A genus of plants of the Spurge family, *Euphorbiaceae.* e.g. M. utilissima (= M. esculenta) - The Cassava or Tapioca plant, a stout herbaceous species, up to 2.7 m (9 ft) high, with large palmate leaves. The basal clusters of thick cylindrical roots contain abundant starch; they are more or less poisonous owing to the presence of hydrocyanic acid, but this is destroyed in heating during the manufacture of tapioca and other substitutes for starch and arrowroot. Cassava Starch (= Brazilian Arrowroot; Mandioca; Manihot Starch; Manioc Starch; Rio Arrowroot; Tapioca Starch; etc.), are starch granules prepared from the rhizomes of bitter cassava, the staple food of many South American Indians. Place of origin: Brazil, where it has been cultivated since early times, and is now widely planted in other tropical regions.

Maniola A genus of Insects of the Browns Butterfly or Moth family, *Satyridae.* e.g. M. jurtina - The Meadow Brown Butterfly, 3-6 cm long, probably the commonest British butterfly. Found in meadows, fields, forest margins and clearings. In the female the forewings have a white-centered eye-spot. Distribution: Europe (except the far north); native in Britain, Asia Minor, the Middle East, North Africa.

Manis A genus of Mammals of the order of Pangolins or Scaly Anteaters, *Pholidota.* e.g. M. tricuspis - The Tree Pangolin, a small-scaled, arboreal species found in Africa. The body is covered with a coat of chain mail made up of large, overlapping, sharp-edged pointed scales that are used for attack as well as body protection. The tail is strongly prehensile, gripping branches by scaleless callous parts towards the end; the claws are sharpened for gripping.

Manocirrhus A genus of Fishes (subphylum *Vertebrata*). e.g. M. polyacanthus - The Amazon Leaf-fish.

Mansonella A genus of unsegmented worms or Nematodes (phylum *Nematoda*). e.g. M. ozzardi - A filarial nematode parasite found in the mesentery and body cavities of man in Panama, Yucatan, and neighboring islands.

Mansonia A genus of Insects of the Mosquito family, *Culicidae*, several species of which transmit the parasitic worm, *Wuchereria* bancrofti. Formerly called by the generic name, *Taeniorhynchus*. Some species of this genus may also transmit viruses such as those causing equine encephalomyelitis.

Mansonoides A subgenus of *Mansonia* (see **Mansonia**).

Manta A genus of Fishes of the order of Rays or Ray-fishes, *Hypotremata.* e.g.

M. birostris - The Manta Ray or Devil-fish (Devilfish), a gigantic ray species, with a flattened, cartilaginous body up to 6 m (20 ft) wide, and inhabiting American waters.

Mantichora A genus of Insects of the Tiger Beetle family, *Cicendelidae*. These are very large, black, carnivorous species from Africa, with unusually fearsome mandibles in the male sex, and are active predators. The larvae are ground-dwellers, living in burrows.

Mantis A genus of Orthopterous Carnivorous Insects of the Praying Mantid family, *Manteidae*. e.g. M. religiosa - The Praying Mantis, a slender creature up to 7 cm (2 3/4 in) long, pale green in color, with prehensile forelegs and a long prothorax, found in western Europe and also in the eastern United States, to which it was introduced early this century. The voracious habits of mantids sometimes makes mating a hazardous operation for the male, because if not recognized in time, it is liable to be treated as a legitimate prey by the female, devouring her partner (starting at the head end) even while copulating is in progress. Found in stony fields and among bushes. Distribution: Southern Europe and the Mediterranean region. Also in the New World.

Maranta A genus of plants of the Arrowroot family, *Marantaceae*. e.g. M. arundinacea - The Arrowroot plant. The starch granules of the rhizome, Amylum Marantae (= Araruta or Maranta), have the general properties of starch and are used in the treatment of diarrhea. Place of origin: Brazil.

Marasmius A genus of plants of the Gill Fungus family, *Agaricaceae*. e.g. M. scorodonius - A gill fungus species with a slender, smooth, reddish-brown stalk and a thin pale cap, with radiating gills on its undersurface. The sporophore smells strongly of garlic. Grows mainly in coniferous woods.

Marchantia A genus of plants of the class of Liverworts, primitive green landplants related to the mosses, *Hepaticae* (phylum *Bryophyta*). e.g. M. polymorpha - The Common Liverwort, a perennial, freshwater plant 8 cm (3 1/8 in) high, with a flattened thallus which has airpores on the upper surface. It is dioecious, with male and female flowers growing on separate plants. Occurs on damp ground. Place of origin: Europe; native in Britain.

Margaritana A genus of Bivalve Molluscs, better known by the generic name, *Margaritifera*, of the Pearl Mussel family, *Margaritiferidae*. e.g. M. margaritifera (= *Margaritifera* m.) - The Pearl Oyster Mussel (see **Margaritifera**).

Margaritifera A genus of Bivalve Molluscs (class *Lamellibranchia*), also known by the generic name, *Margaritana*, of the Freshwater Pearl Mussel family, *Margaritiferidae*. e.g. M. margaritifera (= *Margaritana* m.) - The Pearl Oyster or Pearl Mussel; ovate to kidney-shaped shell, length 12-15 cm, height 5-7 cm, thickness 3-4.5 cm, dull, blackish-brown in color. Found only in clean streams and on river beds in lowlands and foothills. Lives to an age of 60-80 years. Because of water pollution this species is rapidly decreasing in numbers. It forms genuine pearls which were once collected for the jewellery trade. Distribution: Europe, Siberia, North America.

Margaropus A genus of Arachnids, better known by the generic name, *Boophilus*, of the Mite and Tick order, *Acari*. e.g. M. annulatus (= *Boophilus* a.) - The Cattle Tick (see **Boophilus**).

Marginaster A genus of echinoderm animals of the class of Starfishes, *Asteroidea*. These are the smallest starfish species, less than 1.27 cm (1/2 in) in diameter when fully grown. They have five arms and no head and are marine creatures. e.g. M. capreensis - A Mediterranean deep-sea species, not exceeding 20 mm (0.78 in) in diameter (see **Marthasterias**).

Marmosa A genus of Mammals of the family of American Opossums, *Didelphidae*. These are the murine or rodent arboreal opossums, ranging from Mexico to Brazil. These American marsupials have no pouch at all, unlike the species of other genera of the family. They are about 37.5 cm (15 in) long, not including the long, prehensile tail. When threatened with danger or caught, they pretend to be dead.

Marmota A genus of Mammals of the Marmot rodent and Tree squirrel family, *Sciuridae*. e.g. M. monax - The Woodchuck or Ground-hog, a familiar rodent or marmot, living in woodlands and on farms, where its depredations on crops make it a pest at times. A somewhat heavy, short-legged, thick-bodied animal, its fur is thick but coarse and dull-colored, with a short, furry tail. It hibernates deep underground throughout the winter, and emerges at the first sign of spring. Distribution: North America.

Marrubium A genus of plants of the Thyme or Mint family, *Labiatae*. e.g. M. vulgare - The White Horehound (= Hoarhound; Marrubii Herba), a perennial plant, 25-60 cm high; the leaves are orbicular and crenate, the flowers are white and hairy, and the calyx has 10 hooked teeth. Grows at the edge of cultivated fields, on village greens, by waysides and in hedgerows. The dried leaves and flowering tops are expectorants and in large doses, laxatives. Place of origin: in scattered localities in most of Europe. Native in Britain.

Marsdenia A genus of plants of the Milkweed family, *Asclepiadaceae*. Better known by the generic name, *Gonolobus*. e.g. M. condurango (= *Gonolobus* c.) - The Condurango or Eagle-vine (see **Gonolobus**).

Marsilea A genus of freshwater plants of the Clover Fern family, *Marsileaceae*. e.g. M. quadrifolia - The Marsilea or Clover Fern, a semi-aquatic, perennial plant 10-50 cm high, with a slender, creeping rhizome and long-stalked leaves like 4-leaved cloves. Grows at the margins of stillwaters, muddy ponds, marshes and ditches. Place of origin: most of Europe.

Martes A genus of Mammals of the Badger, Marten, Otter, Skunk, and Weasel family, *Mustelidae*. e.g. M. martes - The European Pine Marten, a small, long-bodied, short-legged animal about 48 cm (19 in) long, with a 23 cm (9 in) tail. A predatory, arboreal, mustelid carnivore, it inhabits the thick pinewoods of the northern hemisphere. It has dark brown fur with a cream-colored throat patch and a long, bushy tail. Distribution: the northern half of Europe, extending into western Siberia, the Caucasus, and most of Asia.

Marthasterias A genus of echinoderm animals of the class of Starfishes, *Asteroidea*. These are the largest starfish species, most being 10-12.5 cm (4-5 in) across, but some are known to reach up to 90 cm (3 ft). e.g. M. glacialis - A long-living, large starfish with an average lifespan of 7-10 years (see **Marginaster**).

Marumba A genus of Insects of the Hawk Moth or Butterfly family, *Sphingidae*. e.g. M. quercus - The Oak Hawk Butterfly, a light ochre-colored insect. Its green caterpillar lives on oaks in southern Europe and southeast Asia, as well as in the warmer parts of central and eastern Europe.

Masticophis A genus of Reptiles of the Colubrid Snake family, *Colubridae*. e.g. M. flagellum - The Coachwhip Snake, a long and slender reptile, particularly in the region of its lash-like tail. Distribution: the agricultural areas of the United States.

Mastigamoeba A genus of Protozoans of the order *Rhizomastigina*. These are large rhizopods, 150-200 μ in length, possessing pseudopodia and a free flagellum. Some species of the genus are free-living, others are parasitic, especially in the gut of amphibians.

Mastigoproctus A genus of Arachnids of the order of Whip Scorpions, *Uropygi*. The species is found in South America, the southern United States and Central America, in India and throughout East Asia, and much of Indonesia. The prosoma is covered by a single large carapace, and the abdomen terminates in a long flagellum or tail.

Mastodon A genus of Extinct Mammals of the prehistoric Elephant or Mammoth family, *Mammutidae*. e.g. M. giganteus (= M. americanus) - A mammoth species first discovered in 1845 on the Hudson River, New York, U.S.A. The skeleton was beautifully preserved; it had an estimated shoulder height of 4.2 m (14 ft).

Mastomys A genus of Mammals of the Rodent or Rat family, *Muridae*. e.g. M. natalensis - A rat species transmitting the arenavirus, the cause of Lassa fever in West Africa.

Mastotermes A genus of Insects of the primitive Termite family, *Mastotermitidae*. e.g. M. darwiniensis - A primitive termite from Australia, part of the termite order, *Isoptera*.

Matricaria A genus of plants, the species of which are sometimes called by other generic names such as *Tripleorospermum*, etc. of the Daisy family, *Compositae*. e.g. M. chamomilla (= M. recutita) - The Matricaria or Wild Camomile (Chamomile), an aromatic, annual plant, 30-50 cm (12-20 in) high, with rather downy bipinnate leaves, deeply cut into segments and solitary disc-shaped yellow flowers, with white rays which have a tendency to curl back. The flowerheads produce a volatile oil, which is used as an aromatic tea, an aromatic bitter and a febrifuge. It has mild sedative properties, acting on the nervous system, and is also an antispasmodic, digestive, and emmenagogue.

Grows on waste land and by roadsides. This species is not the true Chamomile, and has been confused with the True Chamomile plant, *Anthemis* nobilis (or *Chamaemelum* nobile), although it possesses many of the properties of that plant and is often used as a substitute. Matricaria flowers are less bitter and nauseating than those of the true chamomile. Place of origin: widespread in most of Europe; western Asia to India. It has also been introduced into North America and Australia (see **Anthemis; Chamaemelum**). M. indora (= M. maritimum; *Tripleorosporum* inodorum) - The German Camomile or Scentless Mayweed, an erect, branched, bushy annual or biennial plant, 30-60 cm (1-2 ft) high, with smooth, deeply cut leaves and white flowers with yellow disc florets, growing in terminal heads. Place of origin: Europe (including Britain).

Matteucia A genus of plants of the Lady Fern family, *Athyriaceae*. e.g. M. struthiopteris - The Ostrich Fern, a perennial plant, 1.5 m (5 ft) high. It has one-pinnate, fertile fronds with rolled-up pinnae, surrounded by sterile fronds. Grows in damp woodlands. Distribution: Europe; native in Britain.

Matthiola A genus of plants of the Cress or Crucifer family, *Cruciferae*. e.g. M. tricuspidata - The Three-horned Stock, an annual plant, 10-40 cm high, with purple flowers and 4-10 cm long fruits with three horns. Grows on seashores. Place of origin: southern parts of the Mediterranean littoral.

Mauremys A genus of Reptiles of the Freshwater Turtle family, *Emydidae*. e.g. M. caspica - The Striped-neck Terrapin, length 30 cm; commonly found in fast-flowing streams as well as quiet pools. Distribution: from North Africa through southern Europe to western Asia.

Mayermys A genus of Mammals of the order of Rodents, *Rodenta*. This is the New Guinea Mouse; it has 10 teeth.

Mayetiola A genus of Insects of the Gall Midge family, *Cecidomyiidae*. e.g. M. destructor - The Hessian Fly, a pest of wheat which was carried from Europe to North America in the 18th century. The phytophagus larvae produce characteristic galls on their host-plants; the adults are minute, delicate midges - hence the common family name.

Mazocraës A genus of Tapeworms of the subclass of Merozoic cestodes, *Eucestoda* or *Cestoda* merozoa. e.g. M. alosae - A cestode species parasitic in the Shad fish (genus *Alosa*).

Meandrina A genus of True Corals of the order of True or Stony Corals, *Scleractina*. These are the reef-building Brain Corals; in these animals the polyps share a common trough-like theca. Symbiotic algae in the endoderm are believed to facilitate the laying down of the skeleton, which resembles a human brain. The colony may reach as much as 2.4 m (8 ft) in diameter.

Mecistocirrhus A genus of nematode parasites found in the fourth stomach of ruminants.

Medicago A genus of plants of the Pea family, *Leguminosae*. e.g. M. sativa - The Alfalfa or Lucerne plant, a perennial species, 30-90 cm (1-3 ft) high, with trifoliate leaves and dense clusters of purple flowers which are pollinated by bees. The twisted pod contains up to 20 seeds. Widely grown for green fodder and has become naturalized the world over. Place of origin: western Asia and the Mediterranean region.

Medinilla A genus of plants of the family *Melastomataceae*. e.g. M. magnifica - The Medinilla plant, an evergreen shrub about 2 m (6 1/2 ft) tall, with rich green, oval, opposite leaves prominently veined and stalkless. The flowers grow in long, one foot (30 cm), showy, drooping inflorescences consisting of many rosy-red flowers with yellow stamens and purple anthers, surrounded by large pink bracts. Place of origin: the

Philippine islands; introduced into Europe in 1888.

Megaceryle A genus of Birds of the Kingfisher family, *Alcedinidae* (the fishing Kingfisher subfamily, *Cerylinae*). e.g. M. alcyon - The Belted Kingfisher, the commonest of the fishing kingfishers found in the New World; it has a blue-grey back and white underparts, with a blue-grey band across the breast of both sexes, and a penetrating rattle of a call.

Megachile A genus of Insects of the Bee family, *Apidae*. e.g. M. centuncularis - The Leaf-cutter Bee, up to 12 mm long; the thorax is rusty-red, the abdomen black and with rust-red pollen brushes on the underside. The bee lines its nest-cells with rose leaves. Distribution: Europe; native in Britain.

Megaderma A genus of Mammals of the False Vampire family, *Megadermatidae*. e.g. M. gigas - One of the largest species of false bats or vampires; it has both nose leaves and ear tragi, but no tail, and a wingspan of some 75 cm (30 in). It is extremely bold and fierce, with a vicious bite. Distribution: Australia.

Megadyptes A genus of Birds of the order of Penguins, *Sphenisciformes*. e.g. M. antipodes - The Yellow-eyed Penguin, the only species of this genus.

Megalobatrachus A genus of Amphibians of the Salamander family, *Cryptobranchidae*. e.g. M. japonicus (= M. maximus) - The Japanese Giant Salamander, the largest of the amphibia, up to 1.8 m (6 ft) long; a species found in deep mountain streams, where it hides in holes. The gill slits are all closed in the adult, and there are no external gills. Respiration is accomplished by a capillary network of blood vessels beneath the skin. It is a sluggish creature which comes up only now and again for a breath of air. It is long-lived, reaching and even passing the age of 50 years. Distribution: China and Japan.

Megaloceros A genus of Mammals of the True Deer family, *Cervidae*. e.g. M.

giganteus - The Irish Elk or Giant Deer, now extinct. It was a gigantic fallow deer which lived in northern Europe (especially Ireland) and northern Asia, in Neolithic times (between 6,000 and 3,000 B.C.). The antlers were enormous, measuring nearly 3.3 m (11 ft) from tip to tip, and its weight varied between 27.2-45.4 kg (60-100 lb). Its fossils are found chiefly in Irish peat, in which it probably sank when attempting to graze.

Megaloglossus A genus of Mammals of the suborder of Fruit Bats or Fruit-sucking Bats, *Megachiroptera*. The species are the long-tongued fruit-bats (see **Macroglossus**).

Megalomma A genus of Segmented Worms or Annelids of the Sabellid Fan or Feather-duster worm family, *Sabellidae*. e.g. M. vesiculosum - This elegant annelid binds particles of shell and sand to form a tube in which it lives. The upper end forms a widely spreading, delicate crown, with a separate eye on each ray of the crown. Chlorocruorin pigmentation gives the blood of these sabellids a greenish color.

Megaloprepus A genus of Insects of the order of Damsel flies and Dragon flies, *Odonata*. e.g. M. caerulatus - A large damsel-fly species, one of the largest in the world, only second to *Tetracanthagyna*. It has been measured up to 19 cm (7.48 in) across the wings and 12.7 cm (5 in) in overall length. Distribution: Central and South America (see **Tetracanthagyna**).

Megalops A genus of Fishes, more commonly known by the generic name, *Tarpon* (subphylum *Vertebrata*). e.g. M. atlantica (= M. atlanticus; *Tarpon atlanticus*) - The Tarponfish (see **Tarpon**).

Megalopyge A genus of Insects of the order of Butterflies and Moths, *Lepidoptera*. e.g. M. opercularis - A flannel moth species, known as the Puss Moth. The larvae, the so-called Puss Caterpillars, may cause a contact dermatitis with or without pyrexia and nervous symptoms.

Megalorhina A genus of Insects of the Dung Beetle or Scarab Beetle family, *Scarabaeidae* (Rose Chafer Beetle subfamily, *Cetonidae*). e.g. M. harrisi - A most decoratively colored species found in west equatorial Africa.

Megalosaurus A genus of extinct prehistoric Reptiles (class *Reptilia*), known as the "Large Lizards", the first dinosaur to be scientifically decribed in 1824. This 6 m (20 ft) bipedal theropod carnivore, the bones of which were first found in 1818, stalked across what is now southern England about 130 million years ago.

Megalosporon A large-spored genus of the Trychophyton fungi. e.g. M. ectothrix - The form found on the surface of the hair shaft. M. endothrix - The form found inside the hair shaft.

Meganeura A genus of prehistoric Insects of the order of Dragon flies, *Odonata*. e.g. M. monyi - The largest prehistoric insect, an extinct Dragon-fly species, fossils of which have been found in Central France. It lived between 280 to 325 million years ago, and is estimated as having a wing-expanse of up to 70 cm (271/2 in).

Megaptera A genus of Mammals of the Rorqual Whale family, *Balaenopteridae*. e.g. M. novaeangliae - The Humpback Whale, up to 15 m (50 ft) long; it differs from the true rorquals in the low hump on the back, the bosses on the snout and the very long pectoral fins or flippers. It is black above and white below, sometimes with black spots. Distribution: worldwide; in summer they may reach European coastal waters in schools of 3-20 or more.

Megarhinus A genus of Insects of the two-winged Fly or Mosquito family, *Culicidae*. The species are harmless large mosquitoes of tropical and subtropical countries.

Megarhyssa A genus of Insects of the parasitic family, *Ichneumonidae* (order of Ants, Bees and Wasps, *Hymenoptera*). These insects lay their eggs in the wood-boring larvae of the Sawfly family, *Siricidae* (large wood-wasps or horntails), and are able to drill through solid wood in order to reach their hosts.

Megascolides A genus of Segmented Worms, the Annelids of the class of Earthworms, *Oligochaeta*. e.g. M. australis - A giant earthworm species from Australia, measuring some 1.2 m (4 ft) in length (when contracted), and nearly 2.1 m (7 ft) when naturally extended.

Megasea A genus of plants, better known by the generic name, *Bergenia*, of the Saxifrage family, *Saxifragaceae*. e.g. M. crassifolia (= *Bergenia* c.) - The Leather Bergenia or Saxifrage (see **Bergenia**).

Megaselia A genus of Insects of the Two-winged Fly order, *Diptera*. Their larvae may cause intestinal myiasis.

Megasoma A genus of Insects of the order of Beetles, *Coleoptera*. This genus belongs to the Elephant Beetle family. e.g. M. elephas - The Elephant Beetle, found in tropical America, extremely bulky and measuring up to 12.5 cm (5 in) in length.

Megastoma A genus of Flagellate Protozoans, better known by the generic name, *Giardia*. e.g. M. entericum (= *Giardia* lamblia) (see **Giardia**).

Megathura A genus of primitive Gastropod Molluscs of the order of Topshells or Limpets, *Archaeogastropoda*. These are the Keyhole Limpets from California, a relatively large, marine species with a solitary apical opening, and a black mantle covering the shell.

Megatrichophyton A genus of parasitic fungi, the species of which infest man and domestic animals.

Megicica A genus of Insects of the order of Locusts and short-horned Grasshoppers, *Caelifera*. e.g. M. septemdecim - The North American Grasshopper, also known as the

Seventeen-year Locust because it lives underground in a larval state for about 17 years before digging to the surface and living as an adult for only a few weeks.

Megophrys A genus of Amphibians of the Horned Frog and Spadefoot Frog family, *Pelobatidae*. These are the Horned Frogs, distributed from the Himalayas through southeast Asia to the Philippines. While some species occur at sea level others are found only in mountainous areas above 900 m (3,000 ft).

Melaleuca A genus of Trees of the Myrtle family, *Myrtaceae*. e.g. M. alternifolia - The Australia Tea tree, from the leaves of which is extracted Melaleuca Oil (= Tea Tree Oil; Ti-tree Oil), containing terpenes, cineole and terpineol. Has been used as an antiseptic in medicine and surgery. M. leucadendron - The leaves and twigs yield Cajaput Oil, containing cineole. Used as a mild counter-irritant and a carminative.

Melampyrum A genus of plants of the Grass family, *Gramineae*. The Field Cow-wheat, an annual plant, 15-50 cm high. The bracts are purple, lanceolate, pinnatifed and with long, slender teeth. Place of origin: Europe; native in Britain.

Melanargia A genus of Insects of the Browns Moth and Butterfly family, *Satyridae*. e.g. M. galathea - The Marbled White Butterfly, size 23-28 mm; it has whitish or yellowish wings with variable dark bands and patches. There are many color forms and geographical races. Distribution: Europe, the Caucasus, northern Iran, North Africa.

Melanerpes A genus of Birds of the Woodpecker family, *Picidae*. e.g. M. formicivorus - The Acorn Woodpecker, a species found in North America. It gathers acorns and stores them for the winter in individual holes dug into the boles of trees, telephone poles, agave and yucca stems, etc.

Melanitis A genus of Insects of the Satyrine Nymphalid Butterfly family, *Nymphalidae*. e.g. M. leda - A satyrine butterfly, ranging from South Africa to Japan (where its larva is a pest of bamboo and sugar-cane). A brown or greyish species with a number of distinctive spots, often arranged in a row, near the margin of the wings.

Melanitta A genus of Birds of the Duck, Goose and Swan family, *Anatidae*. e.g. M. nigra - The Common Scoter or Sea Duck, size 48 cm; in the male the plumage is black and the bill has a yellow spot; in the female the plumage is brownish and the crown is dark. A migrant bird, it nests in the tundra, by lakes or on moors in the north, and in the remoter parts of North Scotland. In winter it moves further south. Distribution: Europe; native in Britain.

Melanocorypha A genus of Birds of the Lark family, *Alaudidae*. e.g. M. calandra - The Calandra Lark, 19 cm long, a shy bird with a powerful bill and a large black patch on either side of its neck. A predominantly resident bird, it prefers to stay on bare, dry ground. Distribution: southern Europe.

Melanogrammus A genus of Fishes of the Cod, Haddock, Hake and Whiting family, *Gadidae* (subfamily *Gadinae*). e.g. M. aeglefinus - The Haddock, 50-80 cm long, an important, commercial, marine fish. The lateral line is dark, and there is a dark spot above the pectoral fin. Distribution: the North Atlantic.

Melanoides A genus of Snails, species of which are intermediate hosts of *Paragonimus*.

Melanolestes A genus of Insects. e.g. M. picipes - The "black corsair" or "kissing bug"; its bite resembles a wasp-sting, but is often much more severe.

Melanoplus A genus of Insects of the Migratory Locust or Short-horned Grasshopper family, *Acrididae*. These are migratory locust species found in the United States.

Melanosuchus A genus of Reptiles of the order of Crocodiles, *Crocodilia*. e.g. M. niger - The Black Caiman, a crocodile species found in the Amazon Basin, reaching a length of 5-6 m (16 1/2-20 ft).

Meleagris A genus of Birds of the Turkey family, *Meleagrididae*. e.g. M. gallopavo - The Common Domestic Turkey, extending from southern Canada by way of the United States, south to Mexico. A commercial bird, the most valuable of the breeds being the bronze-colored variety, in which the cock may weigh over 18 kg (40 lb) and the hen up to 9 kg (20 lb). It was introduced into Britain via Spain from Mexico in 1549.

Meles A genus of Mammals of the Badger, Otter, Skunk and Weasel family, *Mustelidae* (Badger subfamily, *Melinae*). e.g. M. meles - The Common or European Badger, up to 90 cm (3 ft) long, including 15 cm (6 in) of tail. A squat animal with a flattened head, small ears and eyes, and a grizzled coat. It has a broad, black stripe on each side of the head. The badger inhabits forests from lowlands to mountain elevations, and digs long, deep tunnels in concealed places. Distribution: Europe, western Asia. Native in Britain.

Melia A genus of plants of the family *Meliaceae*. Better known by the generic name, *Azadirachta*. e.g. M. azadirachta (= *Azadirachta* indica) - The Margosa or Neem plant (see **Azadirachta**).

Melica A genus of plants of the Grass family, *Gramineae*. e.g. M. uniflora - The Wood Melick, a creeping perennial plant, 20-60 cm high; the panicles are very loose and the branches and spikelets usually upright. Grows in dry woods, especially beechwoods on rocky slopes, wood margins and shady banks. Place of origin: Europe; native in Britain.

Meligethes A genus of Insects of the family of Flower Beetles, *Nitidulidae*.

e.g. M. aeneus - Length 1.5-2.7 mm; found very early in spring on various flowers. The eggs are laid in the rape flower (*Brassica* napus, family *Cruciferae*). Distribution: the Palaearctic and Nearctic regions (see **Brassica**).

Melilotus A genus of plants of the Pea family, *Leguminosae*. e.g. M. officinalis - The Common Melilot, a biennial plant, 30-100 cm high; the pods are hairy and blunt, and brown when ripe. Grows commonly on embankments, roadsides, edges of fields and in waste places. Place of origin: Europe; introduced and naturalized in Britain.

Melinna A genus of Segmented Worms of the Ampharetid family, *Ampharetidae*. e.g. M. palmata - An ampharetid species found in long, slender tubes built vertically in the mud, which usually project slightly as chimneys from the surface. The blood pigment is red, unlike the greenish hue of other ampharetids.

Melipona A genus of Insects of the Bee family, *Apidae*. These are very small, tropical, stingless bees. They were domesticated in the past as a source of honey by the Maya Indians of Mexico.

Melissa A genus of plants of the Mint and Thyme family, *Labiatae*. e.g. M. officinalis - The Melissa or Honey Plant. From the leaves and flowering tops is prepared a fragrant, lemon-scented herb, containing tannin and an essential oil, Melissa Balm (= Balm Gentle; Balmmint; Honey Plant; Lemon Balm). Used as a carminative and diaphoretic.

Melitta A genus of Insects of the Bee family, *Apidae*. e.g. M. leporina - Length 11-13 mm; common on flowering clover and lucerne in July and August. The nest consists of a small number of underground cells filled with pollen in which the larvae feed. Distribution: northern and central Europe.

Melittangium A genus of bacteria occurring in manure.

Mellisuga A genus of Birds of the Humming-bird family, *Trochilidae*. e.g. M. helenae - The Bee Humming-bird, the smallest in the family, and no longer than 5 cm (2 in), including the tail and long bill. The world's smallest bird, its diminutive eggs measure about 11.4 x 8 mm (0.45 x 0.31 in) and weigh 0.5 g (0.176 oz). Distribution: Cuba and the Isle of Pines (Isla de Pinos), hugging the southern edge of Cuba.

Mellivora A genus of Mammals of the Badger, Otter, Skunk, and Weasel family, *Mestelidae*. e.g. M. capensis - The Ratel or Honey Badger, one of the largest of the Mustelidae, ranging from India to Arabia, and over all Africa south of the Sahara. It is 75 cm (21/2 ft) long, excluding the 25 cm (10 in) tail; the underside of the body is black and sharply contrasts with the grey back. It is both a burrower and arboreal, nocturnal inhabit, and feeds on small mammals, birds, reptiles and insects.

Meloe A genus of Insects of the Oil Beetle or Blister Beetle family, *Meloidae*. e.g. M. proscarabacus - The Oil Beetle; female size 36 mm, males 10 mm; the body is plump and the elytra short. It is parasitic on bees. It contains the vesicant cantharadin and can raise blisters on the skin if handled. Far more cantharadin however, is contained in the body of the Spanish Fly (*Lytta* vesicatoria), another member of this family. Distribution: Europe; native in Britain (see **Lytta**).

Melogale A genus of Mammals of the Badger, Otter, Skunk, and Weasel family, *Mustelidae*. e.g. M. moschata - The Ferret Badger from Assam, Burma and South China. A partly arboreal creature. In China its fur is called Pahmi.

Melolontha A genus of Insects of the Lamellicorn Beetle family, *Scarabaeidae*. e.g. M. melolontha - The Common Cockchafer Beetle or May

Bug, length 2.5 cm (1 in); the abdomen has black and white jagged bands. A colored species found in pastures and woodlands, burrowing into the soil and feeding on roots of grasses and other plants, to whom they are serious pests. Distribution: much of Europe, except Spain and southern Italy.

Melophagus A genus of Insects of the wingless Hippoboscid or Louse-fly family, *Hippoboscidae*. e.g. M. ovinus - The Sheep Ked, Sheep Keel or Sheep Tick, a wingless fly and a minor pest of sheep, spending its entire life-cycle in the fleece. It is viviparous and gives birth to fully grown larvae which pupate immediately, the new adult generations emerging after about three weeks.

Melopsittacus A genus of Birds of the Parrot and Parakeet family, *Psittacidae*. e.g. M. undulatus - The Budgerigar, a species of parakeets (small parrots with long tails) found in Australia. In the wild state they are only 17.8 cm (7 in) long, the plumage is predominantly green with some yellow, but several varieties have been produced by selective breeding, the most popular having a sky-blue shade. In breeding, they may reach up to 30 cm (1 ft) in length. They were introduced into England from Australia in 1840.

Melursus A genus of Mammals of the Bear family, *Ursidae*. e.g. M. ursinus - The Sloth Bear, an aberrant species with a narrow snout and long, mobile lips. It is nearly 1.8 m (6 ft) high and weighs over 135 kg (300 lb). It has a rough, coarse coat, naked face and a long tongue. It is an ugly animal, yet comic in its movements and can be taught tricks. Habitat: the forested hills of India and Sri Lanka.

Membranipora A genus of aquatic animals of the phylum of Sea Mats, *Bryozoa*. e.g. M. membranacea - A species of Sea Mats 1 mm thick; the body is tubular with a crown of ciliated tentacles and it aggregates in colonies, which form a small-meshed covering possessing numerous fine bristles. These colonies form a crusty covering to seaweeds or shells. Distribution: from the Arctic Ocean to the Adriatic Sea.

Membranoptera A genus of plants of the Red Algae family, *Gigartinaceae*. e.g. M. alata - The Membranoptera plant, a Red Algae species, up to 10 cm (4 in) high; the terminal segments are 2-lobed. Place of origin: the North Sea; also the West Baltic.

Menacanthus A genus of Insects of the family of parasitic Biting Lice, *Menoponidae*. e.g. M. stramineus - A biting louse species of bird-parasites, and a pest of domestic poultry. It is a rather flat, wingless insect with a bristly body and without compound eyes.

Menidia A genus of Fishes (subphylum *Vertebrata*). e.g. M. beryllina - The Whitebait fish or Tide-water Silverside fish.

Meningococcus A generic name for gram-positive cocci, commonly used for the species *Neisseria* meningitidis, and one of the main causes of meningitis. e.g. M. neisseris (= *Neisseria* meningitidis) (see **Neisseria**).

Menispermum A genus of plants. e.g. M. canadense - The Moonseed or Yellow Parilla plant; the rhizome and roots are tonic and alternative.

Menopon A genus of Insects of the family of Biting Lice, *Menoponidae* (order *Mallophaga*). e.g. M. gallinae - A bird parasite and a pest of domestic poultry (see **Menacanthus**).

Mentha A genus of plants of the Mint or Thyme family, *Labiatae*. e.g. M. piperita - The Peppermint plant; from the leaves and flowering tops is extracted a volatile oil, Peppermint Oil (= Oleum Menthae Piperitae; Ol Menth Pip) - Used as an aromatic carminative.

Menura A genus of Birds of the Lyrebird family, *Menuridae*. e.g. M. superba - The Superb Lyrebird, the largest of the perching birds or passerines. It lives

318

alone in the eucalyptus forests of eastern Australia. It is terrestrial and rarely takes wing. The inner tail feathers resemble the strings of a lyre stretched between the S-shaped outer pair.

Menyanthes A genus of plants of the Bogbean family, *Menyanthaceae*. e.g. M. trifoliata - The Bogbean, Buckbean or Marsh-trefoil, a perennial plant 15-30 cm (6-12 in) high, with a creeping rhizome and large trifoliate leaves borne on long stalks; the S-lobed corolla is colored pink. Grows commonly in damp places such as fens and bogs. The liquid extract of this plant is used as a bitter tonic, and in large doses as a purgative. Place of origin: Europe (native in Britain), northern and central Asia, North America, Greenland and Iceland.

Meoma A genus of Echinoderm animals of the Sea Urchin family, *Echinidae*. e.g. M. ventricosa - A marine species found in the Bahamas; size 12.5 cm (5 in).

Mephitis A genus of Mammals of the Badger, Otter, Skunk and Weasel family, *Mustelidae* (skunk subfamily, *Mephitinae*). e.g. M. mephitis - The Striped Skunk, about 40 cm (15 3/4 in) long with a slightly shorter tail. Its odor is even more offensive than other species of the skunk family. It has two musk glands situated at the base of the tail, from which it can squirt an irritating and nauseous liquid to a distance of 3 m (10 ft). In contrast, it has a beautiful coat of glossy black fur with a white stripe on each side running into the tail. Distribution: southern Canada and most of the United States, all the way south to Mexico.

Mercenaria A genus of Bivalve Molluscs, better known by the generic name, *Venus*, of the Venus and Carpet Shell family, *Veneridae*. e.g. M. mercenaria (= *Venus* m.) - The American Hard-shell Clam or Quahaug (see **Venus**).

Mercurialis A genus of plants of the Spurge family, *Euphorbiaceae*. e.g. M. annua - The Annual Mercury, an annual plant, 10-80 cm high; the branched stem is bluntly 4-angled, and the dioecious flowers are borne in many-flowered spikes (i.e. male and female flowers on separate plants). Grows in waste places and as a garden weed. Has alternative properties and was formerly used as a tonic. Place of origin: Europe; native in Britain.

Mergus A genus of Birds of the Duck, Goose and Swan family, *Anatidae*. e.g. M. merganser - The Goosander, Common Merganser or Diving Duck, also known as the Saw-bill, Goldeneye or Surf Scoter. A sea bird of the northern hemisphere; it is about 60 cm (2 ft) long, has a green head, black beak, and white breast when in breeding plumage, and breeds in northern latitudes. A partial migrant, it flies south only during extreme cold. It has marked diving ability, hence its common name. Distribution: Europe; native in Britain.

Meriones A genus of Mammals of the Mouse and Rat family, *Muridae* (Sand rat subfamily, *Gerbellinae*). These are the gerbils or sand rats; the species are also known as jirds, and inhabit the drier regions of North Africa and southwest Asia. They are sandy or pale buff colored creatures, and leap on their hind legs kangaroo fashion, using their long tails as a counterpoise.

Merluccius A genus of Fishes of the Cod, Haddock, Hake and Whiting family, *Gadidae* (= Hake family, *Merlucciidae* or Hake subfamily, *Merluccinae*). e.g. M. merluccius - The European Hake; 125 cm long, weighing 10 kg; it has two dorsal fins and no barbel on the lower lip. Found in moderately deep water along the coast of Europe as far north as the North Sea, and southwards into the Mediterranean and North Africa. Native around Britain.

Mermis A genus of Nematodes or Roundworms of the order *Enoplida*.

These are parasitic worms found in arthropods.

Merodon A genus of Dipterous Insects of the Hover-fly family, *Syrphidae*. These are the Bulb flies, feeding out of the center of narcissus bulbs. They have been introduced from Europe into North America.

Merops A genus of Birds of the Bee-eater family, *Meropidae*. e.g. M. apiaster - The Common Bee-eater, 28 cm long; a colorful bird in its lovely blues, yellows, orange, reds and black. A migrant species, it usually nests in large colonies, burrowing into sandy cliffs or river banks, or in road cuttings - the horizontal passage may be 90-180 cm (3-6 ft) long, and ends in a chamber. Its favorite food is wasps and bees (especially honey bees). Distribution: from southern Europe and Central Asia to South Africa, Madagascar, and Australia.

Merulius A genus of Fungi or Molds. e.g. M. lacrimans - The fungus of the dry-rot of wood; inhaled in dust it causes a persistant, sometimes fatal bronchitis.

Mesamphisopus A genus of Crustaceans of the order *Isopoda* (suborder *Phreatoicidea*). e.g. M. capensis - A species found under moss on stones in swift mountain streams in South Africa. It has a cylindrical or laterally compressed body with a well-marked downward curve of the tail. The abdominal appendages are flat.

Mesembryanthemum A genus of plants, now known by the generic name, *Dorotheanthus*, of the Mesembryanthemum family, *Aizoaceae*. e.g. M. criniflorum (= *Dorotheanthus* bellidiflorus; D. bellidiformis) - The Livingstone Daisy (see **Dorotheanthus**).

Mesidotea A genus of Crustaceans, better known by the generic name, *Saduria*, of the order of Isopods, Sowbugs and Woodlice, *Isopoda*. e.g. M. entomon (= *Saduria* e.) (see **Saduria**).

Mesoacidalia A genus of Insects of the Tortoiseshell Butterfly or Moth family, *Nymphalidae*. e.g. M. charlotta - The Dark Green Fritillary moth; size 5.5 cm; the wings are reddish-yellow with black spots above, the hind wing is greenish with iridescent spots beneath. It lays its eggs on dog violets (*Viola* canina). Place of origin: Europe; native in Britain (see **Viola**).

Mesocerus A genus of Insects of the Squash Bug family, *Coreidae*. e.g. M. marginatus - Length 12-14 mm; found in damp places, chiefly on sorrel, dock, blackberry, groundsel, etc. Distribution: Europe, Asia Minor, Central Asia.

Mesocricetus A genus of Mammals of the Mouse and Rat family, *Muridae* (Hamster subfamily, *Cricetinae*). e.g. M. auratus - The Golden Hamster, length 18 cm with a 2 cm tail; a species with 14-22 mammae, as opposed to the Common Hamster which has only 8 (*Cricetus* cricetus). Popular as a pet and laboratory subject, the latter partly because of its high rate of breeding, the gestation age being only 15 days and the young reaching full adulthood at 11 weeks. The coat is brightly colored and with black spots, the belly is light-colored, the ears large, and the (2 cm) tail is very short. Distribution: Europe; Syria and Asia Minor.

Mesocypris A genus of Crustaceans of the subclass of Ostracods or Seed Shrimps, *Ostracoda*. e.g. M. terrestris - A terrestrial species of ostracods, which though usually found in water environments, in this case live in the leaf litter of forest floors. Distribution: Africa.

Mesogonimus A genus of Flukes, better known by the generic name, *Heterophyes* of the order *Digenea*. e.g. M. heterophyes (= *Heterophyes* h.) (see **Paragonimus**).

Mesonychoteuthis A genus of Molluscs of the class *Cephalopoda*. e.g. M. hamiltoni - A gelatinous cranchid mollusc, an exceptionally large squid, found in the rich feeding grounds of the Antarctic Ocean. This cephalopod may reach a length of over 3 m (10 ft) excluding the tentacles.

Mesoplodon A genus of Mammals of the Beaked and Bottle-nosed Toothed Whale family, *Ziphiidae*. e.g. M. bidens - Sowerby's Whale, a beaked-whale species, under 5 m (17 ft) long, with all its teeth in the lower jaw, and the snout elongated into a beak. Found in the North Atlantic; also in the Mediterranean.

Mespilus A genus of plants, better known by the generic name, *Eriobotrya*, of the Rose family, *Rosaceae*. e.g. M. germanica - The Medlar, a deciduous tree or shrub, up to 6 m (20 ft) tall; it is a thorny plant with large, 5-styled flowers, red anthers, and subglobose fruit. It is grown for its fruit and naturalized in many countries, including Britain. Place of origin: southeastern Europe and southwest Asia. M. japonica (= *Eriobotrya* j.) - The Japanese Plum Tree (see **Eriobotrya**).

Messor A genus of Insects of the order of Ants, *Hymenoptera*. e.g. M. barbarus - The Harvester Ant; the species develops in two sterile castes, one the small-headed minor workers and the other, the large-headed major workers or soldiers. They make large nests in the soil. Towards the end of the season the soldiers are killed off by the minor workers, their bodies eaten, while their large heads are piled up in a chamber of the nest. Distribution: widely, in Africa.

Metacrinus A genus of Echinoderm animals of the class of Crinoids or Sea-lilies, *Crinoidea*. These are very large stalked Sea-lilies, with a stem height of up to 60 cm (2 ft) excluding the 15 cm (6 in) arms. The species are distributed throughout the marine waters of the Japan-Malay-Australian region.

Metagonimus A genus of trematode parasites, also known by other generic names such as *Loosia* and *Loxotrema*, of the class of Flatworms or Flukes, *Trematoda*. e.g. M. yokogawai (= M. ovatus; *Loosia* dobrogiensis; *Loxotrema* ovatum; Yokogawa's Fluke) - An intestinal fluke causing diarrhea; found in the Far East, Middle East, and the Balkans.

Metasequoia A genus of Trees of the family *Taxodiaceae*. e.g. M. glyptostroboides - The Dawn-redwood, first discovered in a small area of Central China in 1945. Until that time the species was known only from fossil remains. Now known as the "living fossil", it flourishes in many parts of the world.

Metastrongylus A genus of nematodes or roundworms (class *Nematoda*). e.g. M. apri (= M. elongatus) - The porcine lung worm; it can cause human infestation.

Methanobacterium A genus of microorganisms of the family *Spirillaceae* - Long slender rods with rounded ends, containing deeply staining granules. e.g. M. vulgaris.

Methanococcus A genus of anaerobic, gram-variable microorganisms of the family *Micrococcaceae* - Chemoheterotrophic and saprophytic cocci of soil and sewage, producing methane.

Methanomonas A genus of microorganisms of the family *Methanomonadaceae* - Monotrichous cells, obtaining energy from the oxidation of methane to CO_2 and water. e.g. M. methanica.

Metridium A genus of Sea Anemones of the order *Actiniaria*. e.g. M. senile - The Plumose Sea Anemone, up to 13 cm long, widely distributed in the temperate waters around Britain and North America. It has numerous delicate, white or many other-colored tentacles, giving it a feathery appearance. It feeds on plankton which are swept towards the mouth by beating cilia on the tentacles and disc. Distribution: Europe and North America. Native around Britain.

Metrosideros A genus of Trees of the Myrtle family, *Myrtaceae*. e.g. M.

robusta - A tree found in New Zealand, yielding a hard, red wood known as Rata.

Meum A genus of plants of the Carrot family, *Umbelliferae*. e.g. M. athamanticum - The Spignelmeu, a perennial plant, 15-60 cm high; the leaves are 3-4-pinnate, the segments capillary and usually whorled, and the flowers are yellowish-white. Grows in grassy places in mountain districts. Place of origin: north, central and western Europe. Native in Britain.

Mibora A genus of plants of the Grass family, *Gramineae*. e.g. M. minima - Early Sand-grass, an annual plant, 3-9 cm high, one of the smallest European grasses; the racemes are spike-like and one-sided. A rare species, growing on bare, sandy ground, often by the sea. Place of origin: southwest and southern Europe; native in Britain.

Micraspides A genus of Crustaceans of the order, *Anaspidacea*. This is a colorless subterranean species, lacking eyes and living in freshwater. The small body is elongated, less than 5 cm (2 in) long, with all the segments distinct, except the first which joins with the head. Distribution: South Australia.

Microbacterium A genus of microorganisms of the family *Corynebacteriaceae*. Gram-positive rods found in dairy products, with high resistance to heat.

Microbatrachella A genus of Amphibians of the order of Frogs and Toads, *Salientia*. e.g. M. capensis - The smallest frog species in Africa, restricted to the Cape Flats, near Cape Town, South Africa. It has a maximum snout-vent length of 1.6 cm (0.62 in).

Microbuthus A genus of Arachnids of the order of Scorpions, *Scorpionida*. e.g. M. pusillus - The smallest scorpion species in the world, measuring only 1.3 cm (0.51 in) in total length. Distribution: the Red Sea coast.

Microcebus A genus of Mammals of the Lemur family, *Lemuridae*. e.g. M. murinus - The Lesser Mouse Lemur, a mouse-like species of dwarf lemurs, measuring as little as 25 cm (10 in) in length and weighing about 60 g (2.1 oz). It dwells in the plains, lodging in holes in the ground and in rock crevices. Distribution: Madagascar.

Microchaetus A genus of Segmented Worms or Annelids of the order of Earthworms, *Oligochaeta*. e.g. M. rappi (= M. microchaetus) - A giant earthworm species found in South Africa. The average-sized specimen measures 136 cm (41/2 ft) in length, or 65 cm (251/2 in) when contracted, but the longest individual ever reported was said to have measured just under 6.6 m (22 ft)!

Micrococcus A genus of microorganisms of the family *Micrococcaceae*. These are gram-positive, saprophytic cocci widely found in nature. The species are also classified under other generic names. e.g. M. catarrhalis (= *Diplococcus* c.; *Neisseria* c.) - A frequent commensal in the nose and throat. M. tetragenus (= *Gaffkya* tetragena) (see **Gaffkya**).

Microcoelum A genus of Trees of the Palm family, *Palmae*, some of whose species were formerly included in the genera, *Cocos* or *Syagrus*. e.g. M. weddeliana (= *Cocos* w.; *Syagrus* w.) - The Cocos Weddeliana or Weddel Palm, a tree-like palm with black fiber-netted trunk, and arching sprays of fern-like leaves which bend over and almost reach the ground. Mostly used as an elegant pot plant. Place of origin: Brazil.

Microcyclus A genus of microorganisms of the family *Spirillaceae* - Non-motile curved rods, forming a closed ring during growth. e.g. M. aquaticus.

Microdon A genus of Dipterous Insects of the family of Hover-flies, *Syrphidae*. The adult fly is a bee-like insect, found near the bases of grasses and other plants. The eggs are laid in nests of various species of ants. The larvae are scavengers and are so slug-like in appearance that they were originally described as a new species of molluscs.

Microfilaria A genus name, sometimes used for the prelarval stage of *Filarioidea* in the blood of man and in the tissues of the vector. e.g. M. bancrofti - The microfilaria of *Wuchereria* bancrofti (see **Wuchereria**). M. streptocerca - The larval form of *Onchocerca* volvulus, found in skin lesions in West Africa (see **Onchocerca**).

Microgaster A genus of Hymenopterous Insects of the Braconid Wasp family. The species are found in the United States.

Microglossus A genus of Birds of the Parrot and Parakeet family, *Psittacidae*. The genus is represented by a single black-colored, cockatoo species found in Australia and New Guinea, where its native name is "kasmalos". The bird is crested and has bare cheek patches like the macaws. It has a long and cylindrical tongue, which ends in a spoon-shaped swelling, characteristically featured for the act of swallowing.

Microhierax A genus of Birds of the order of Falcons and other birds of prey, *Falconiformes*. e.g. M. latifrons - The Bornean Falconet, a sparrow-sized bird of prey, one of the smallest, only 15 cm (6 in) long, found in the forest country of northwest Borneo. It weighs about 35 g (11/4 oz).

Microhydra A generic name formerly given to the minute polyp of what is now known to be actually *Craspedacusta*, a genus of Hydroids (class *Hydrozoa*) of the order *Limnomedusae*. The polyp buds off other small medusae (see **Craspedacusta**).

Microhyla A genus of Amphibians of the Frog family, *Microhylidae*. e.g. M. carolinensis - A Toad species with adhesive properties found during amplexus (close embrace during mating), when the male and female adhere to one another by means of their sticky skin secretions,

which are also highly irritant. Distribution: Florida, U.S.A.

Micromesistius A genus of Fishes of the Cod family, *Gadidae*. e.g. M. poutassou - The Blue Whiting, length up to 50 cm (20 in), a Cod species, with a deeply forked caudal fin, a bluish-grey back and a brownish lateral line. A commercial fish. Distribution: the coasts of western Europe, from the Arctic to the Mediterranean, but not in the North Sea.

Micrometrus A genus of Fishes of the Perch family, *Percidae*. e.g. M. minimus - The Dwarf Perch of California; it gives birth to living young which are capable of taking care of themselves from birth.

Micrommata A genus of Arachnids of the True Spider family, *Sparassidae*. e.g. M. rosea - A True Spider (order *Araneae*), length 9-13 mm; the abdomen is green, with a red stripe in the male. It does not spin a web, but captures its prey with its front legs. Found among fallen leaves and on shrubs. Distribution: the Palaearctic region.

Micromonospora A genus of microorganisms of the family *Streptomycetaceae*. Occurs in soil and water. e.g. M. purpurea - From cultures of this species is produced the antibiotic Gentamycin.

Micromyces A former name given to a genus of microorganisms, the species of which are now classified under other genera such as *Actinomyces*, *Nocardia*, *Streptomyces*, and *Streptothrix* (see **Actinomyces**, **Nocardia**, **Streptomyces** and **Streptothrix**).

Micromys A genus of Mammals of the Old World Mouse and Rat family, *Muridae* (subfamily *Murinae*). e.g. M. minutus - The Old World European or Eurasian Harvest Mouse, the smallest known rodent, about half the size of the common housemouse (*Mus* musculus), measuring up to 13.5 cm (5.3 in) in length, including the long (5 cm - 2 in), prehensile tail, and weighing 4.2-10.2 g

(0.14-0.36 oz). It is a bright red-brown above and colored white below, living in wheat fields, wet meadows, on the edges of swamps, or on the shores of streams and ponds. Distribution: Europe, native in Britain; Asia, east to Japan and North Vietnam; Africa.

Micropterus A genus of Fishes of the Sunfish and Freshwater Bass family, *Centrarchidae* (a member of the Perch family, *Percidae*). e.g. M. salmoides - The Large-mouth Black Bass, 35-40 cm long and weighing over 2 kg which has been successfully introduced into Russia and Europe. It has an attractive, olive-green color with dark specks, the scales are large and the huge mouth extends to behind the eyes. There is a single dorsal fin, the spiny front part of which is lower than the rest. It is a carnivorous species, eating crayfishes and frogs, and is found in slow-moving or stagnant waters with dense vegetation. Distribution: a native of the United States and southern Canada.

Micropteryx A genus of Insects of the primitive Moth family, *Micropterygidae*. These are small moths with well-developed mandibles, feeding on pollen as adults, and found on buttercups in the spring.

Micropus A genus of plants of the Daisy family, *Compositae*. e.g. M. erectus - The Upright False Cudweed, an annual plant, 5-20 cm high; the heads grow in spherical clusters. Rare, found in sandy fields and sunny hillocks. Place of origin: South and Central Europe, and the Balkans.

Micropyrum A genus of plants of the Grass family, *Gramineae*. e.g. M. tenellum - The Gravel Fescue, a short species, 5-20 cm (2-8 in) high, growing in two-rowed, gravel spikes in dry places, particularly on sand, gravel or shingle. Place of origin: Europe, especially in the north of France.

Microscilla A genus of saprophytic microorganisms, the Schizomycetes, of the family *Vitreoscillaceae*. The species resemble those of the genus *Vitreoscilla* (see **Vitreoscilla**).

Microsorex A genus of Mammals of the Shrew family, *Soricidae*. e.g. M. hoyi - The Pigmy Shrew, with a head and body length of 6.25 cm (21/2 in), a red-toothed North American species, and like most shrews, solitary in habit and remarkably savage, always ready to bite.

Microspira A genus name formerly used for a group of small spiral-shaped microorganisms.

Microspironema A former generic name for organisms now included in the genus *Treponema*. e.g. M. pallidum (= *Treponema* p.) (see **Treponema**).

Microsporon See **Microsporum** (see also **Malassezia**).

Microsporum A genus of small-spored ringworm fungi, formerly also called by other generic names such as *Cercosphaera, Malassezia, Microsporon*, etc., causing diseases of the skin and hair (tinea or dermatophytosis). e.g. M. audouini (= *Cercosphaera* addisoni; *Microsporon* audouini) - The common fungus in ringworm of the scalp in children. M. furfur (= *Malassezia* f.; *Microsporon* f.) - Fungus of pityriasis versicolor (see **Malassezia**).

Microstomus A genus of Fishes of the Sole family, *Soleidae*. e.g. M. kitt - The Lemon Sole, up to 50 cm (20 in) long; the head is small, it is right-eyed (lying on the left side of the body), the lateral line is straight and the tail short. A commercially valuable flat-fish. Distribution: Europe, in the North Sea and Icelandic waters. Native around Britain.

Microtrombidium A genus of Insects of the Acarine Mite family, *Trombiculidae*, better known by the generic name, *Trombicula*. e.g. M. akamushi (= *Trombicula* a.) - The Kedani mite (see **Trombicula**).

Microtus A genus of Mammals of the Cricetid Mouse, Rat and Vole family,

Cricetidae. e.g. M. montebelli (= M. montebelloi) - The Field Mouse or Field Vole, a mouse-like animal frequently harboring the *Leptospira* hebdomadis, an organism causing the "Seven-day" fever of the East.

Micrurus A genus of Reptiles, also known by the generic name, *Elaps*, of the Cobra, Krait and Mamba family, *Elapidae.* e.g. M. fulvius (= *Elaps* f.) - The Harlequin or Eastern Coral Snake, up to 90 cm (3 ft) long, and the most poisonous of American snakes, though not the most dangerous. It is front-fanged and bites slowly and deliberately, gradually sinking its fangs into the victim and not letting go until late after biting (see **Elaps**).

Midgardia A genus of Echinoderm animals of the class of Starfishes, *Asteroidea.* e.g. M. xandaros - A fragile, brisingid starfish species, one of the largest known, measuring some 1.4 m (41/2 ft) from arm tip to arm tip, though the disc diameter is only about 2.5 cm (1 in)! It has 11 or 12 arms. Distribution: the Gulf of Mexico.

Miescheria A genus of Sporozoan parasites of the order *Sarcosporidia*, found in the muscles of warm-blooded animals and producing anemia and cachexia. e.g. M. muris - A sarcosporidium species infesting the muscles and livers of animals. In humans it produces what are called Miescher's tubules, long elongated cysts containing the parasites, in the muscles of infected subjects. The disease is known as Sarcosporidiosis or Sarcosporidiasis.

Mikiola A genus of Insects of the Gall-gnat or Gall-midge family, *Cecidomyidae.* e.g. M. fagi - A Gall-midge, 4-5 mm long, an inconspicuous, small fly, best known for its green, yellowish and reddish galls which are ovate, pointed and smooth, and are produced by the larvae and located on the upper side of beech leaves. These galls are about 4-12 mm long and are shaped

like small skittles. Distribution: Europe; native in Britain.

Milium A genus of plants of the Grass family, *Gramineae.* e.g. M. effusum - The Wood Millet, a perennial plant, 45-180 cm high; the panicles are very loose, nodding, with spreading or deflexed branches and the spikelets are one-flowered. Widespread in deciduous woods, especially in damp and calcareous soils. Place of origin: Europe; native in Britain.

Millepora A genus of Sea firs or Hydroids, Hydras and Syphonophores (class *Hydrozoa*) of the order *Hydrocorallinae* (suborder *Milleporina*). This suborder consists of the above genus, whose species are found in tropical shallow seas, where they often lay down their calcareous skeletons, forming extensive coral reefs, with upright white or whitish-yellow, leaf-like or branching growths. The surface of the calcareous skeleton is pitted with a ring of small pores surrounding each of the larger pores.

Milvus A genus of Birds of the Eagle, Harrier, Hawk, and Old World Vulture family, *Accipitridae* (Old World Vulture subfamily, *Aegypiinae*). e.g. M. milvus - The Red Kite, length 60 cm (2 ft), a partially migrant bird living in plains, meadows and marshes, and less commonly in hilly or mountainous country. The plumage is reddish, and the long tail deeply forked. Distribution: Europe; native in Britain; resident in Wales. Also found in the Middle East and northwest Africa.

Mima A genus of gram-negative, coccobacillary microorganisms of the family *Brucellaceae*, now better known by the generic names, *Acinetobacter* or *Moraxella.* e.g. M. polymorpha (= *Acinetobacter* lwoffi; *Moraxella* l.) (see **Acinetobacter; Moraxella**).

Mimas A genus of Insects of the Hawk Moth family, *Sphingidae.* e.g. M. tiliae - The Lime Hawk Moth, size 3-6 cm; the forewings are ochre to grey, clouded with brown, and the hindwings are very small. They lay their eggs on the foliage of lime, birch, oak, ash and other deciduous trees on which the larvae feed. Distribution: Europe; native in Britain.

Mimosa A genus of plants of the family *Mimosaceae* (closely related to the Pea family, *Leguminosae*). e.g. M. pudica - An American shrub species, frequently cultivated in hothouses as a novelty. The bipinnate leaves are very sensitive and when touched fold up and droop. Place of origin: Tropical America.

Mimulus A genus of plants of the Figwort or Snapdragon family, *Scrophulariaceae* e.g. M. guttatus - The Monkey-flower, a perennial plant, about 30 cm (1 ft) high, with ovate irregularly toothed leaves. The calyx is 5-toothed and the 2-lipped yellow corolla is marked with small red spots. Place of origin: North America, from Alaska and Montana to northwestern Mexico. Extensively naturalized in Europe and throughout Britain, where it is found on the banks of streams.

Mimus A genus of Birds of the Catbird, Mockingbird and Thresher family, *Mimidae*. e.g. M. polyglottos - The Common Mockingbird, a slender and long-tailed species with a well-developed song of its own, as well as its capability of mimicing or mocking nearly any other bird song or call. It will sing all day and even at night, without seeming to take time out to hunt for worms or insects. Distribution: from southern Canada southwards to all but the extreme south of South America.

Miniopterus A genus of Mammals of the Bat order, *Chiroptera*. e.g. M. schreibersi - Shreiber's or Long-winged Bat, 6 cm (2 3/8 in) long with a similar wing size, with short ears. It lives in groups, often in large colonies, in open countrysides and hibernates in caves and cellars. Distribution: South Europe.

Minuartia A genus of plants of the Pink family, *Caryophyllaceae*. e.g. M. verna - The Spring Sandwort, a perennial plant, 3-15 cm high; the inflorescences are usually forked and the white petals are as long as the calyx. Grows on stony pastureland, on scree, rocky slopes and in rock-crevices. Distribution: widespread in Europe; native in Britain.

Mirabilis A genus of plants of the family *Nyctaginaceae*. e.g. M. jalapa - The Four-o-clock plant or Marvel of Peru, a herbaceous perennial, about 60 cm (2 ft) high, with tuberous, thickened roots; opposite, entire, smooth, ovate-lanceolate leaves, the lower ones stalked, the upper leaves sessile. The rosy-purple or red to white and yellow, fragrant flowers grow in clusters with long perianth tubes. Place of origin: Tropical America, Mexico and Peru; introduced into Europe about 1525.

Mirafra A genus of Birds of the Lark family, *Alaudidae*. These are the Bush Larks, higher perching song birds, found in North Africa.

Mirapinna A genus of Fishes of the family *Mirapinnidae*, found in the deeper waters of the Atlantic Ocean. They have very large pelvic fins situated in front and below the pectoral fins.

Mirounga A genus of Mammals of the Earless Seal family, *Phocidae* (Sea Elephant subfamily, *Cystophorinae*). e.g. M. leonina - The Southern Sea Elephant Seal, up to 6 m (20 ft) long and weighing up to 3 tons, the largest of the 32 known species of pinnipeds. The male has an inflatable snout. These pinnipeds inhabit the Arctic waters from Spitsbergen to Canada, but are also seen along the eastern and western sides of the northern North Atlantic, off the shores of Britain and France, and New England respectively.

Misgurnus A genus of Fishes of the Loach family, *Cobitidae*. e.g. M. fossilis - The European Weatherfish or Pond Loach, length 30 cm (1 ft), weight 150 g (5 1/3 oz); an elongated species. The

326

upper jaw has 6 long barbels and the lower 4 short ones; there is a wide brown band along the side. A freshwater fish, it lives in brackish waters, ponds and river inlets. In oxygen-deficient water it swallows air, absorbing oxygen from it as it passes through the digestion tract. Distribution: in European rivers, from the Seine to the Neva and from the Danube to the Volga. Absent in Britain, Scandinavia and southern Europe.

Misopates A genus of plants of the Figwort or Snapdragon family, *Scrophulariaceae*. e.g. M. orontium - The Lesser Snapdragon, an annual plant, 20-80 cm high; the leaves are lanceolate and the flowers pale red. Grows in cultivated fields, by roadsides and on waste ground. Place of origin: Europe; native in Britain.

Mistichthys A genus of Fishes of the Goby family, *Gobiidae*. e.g. M. luzonensis - The Sinarapan, a Goby species found in Lake Buhi in southern Luzon in the Philippines. A diminutive fish, it measures only 10-14 mm (0.39-0.55 in) in length. It is so tiny, it has to be studied under a microscope. Although so small it is in demand as food and has considerable local commercial importance. These fish are caught by the natives with large close-web nets, they are then packed tightly into woven baskets until the water drains out, and then sold in dried cake form, each 1 lb (0.45 kg) "cake" containing about 70,000 fish.

Misumena A genus of Arachnids of the family of Crab Spiders, *Thomisidae*. e.g. M. vatia - The Flower-haunting Crab Spider, length of male 4 mm, female 10 mm. Pale yellow in color, the female is able to adapt its color to that of the object on which she is resting. It hides in flowers, feeding on honey bees and other insects. Very common on flowering daisies and other flowers of the Daisy family, *Compositae*. Place of origin: Europe; native in Britain.

Mitra A genus of Gastropod Molluscs of

the order *Stylommatophora*. e.g. M. papalis - A beautiful species from the Indian Ocean, with a white shell decorated with reddish-brown specks. A marine animal.

Mitu A genus of Birds of the Curassow family, *Cracidae*. e.g. M. mitu - The Great Razor-billed Curassow; it has a blade-like casque on the beak. Distribution: America, from Texas in the north to the Argentine in the south.

Miyagawanella A genus of microorganisms of the family *Chlamydiaceae* (order *Rickettsiales*) - the psittacosis-lymphogranuloma venereum group of virus-like agents. e.g. M. lymphogranulomatosis - The etiological agent of the human disease known as lymphogranuloma venereum (= L. inguinale; climatic bubo; esthiomene). M. psittaci - Cause of psittacosis (parrot fever) in man and psittacine birds.

Mnemiopsis A genus of Comb Jellies or Sea Walnuts of the Tentacular order, *Lobata*. These ctenophores or comb jellies are found in coastal waters of America from Cape Cod southwards to Carolina.

Mniotilta A genus of Birds of the Woodwarbler family, *Parulidae*. e.g. M. varia - The Black-and-white Warbler, named for its black and white, striped plumage; about 12.5 cm (5 in) long. It has a slender, pointed beak and has become adapted to a tree-climbing habit, though it builds its nest on the ground. Distribution: most of Canada and the U.S.

Mnium A genus of Moss plants of the order of True Mosses, *Eubryales*. e.g. M. undulatum - The Palm-tree Moss, a perennial plant, up to 15 cm high; the large leaves often form a rosette at the end of the stem. Grows in shady woods. Place of origin: Europe; native in Britain.

Moehringia A genus of plants of the Pink family, *Caryophyllaceae*. e.g. M. mucosa - The Moss Sandwort, a perennial plant 5-20 cm high; the leaves are thread-like and the white flowers usually tetramerous (4-petalled). Grows in stony pastureland, on scree, rocky slopes and in rock-crevices. Place of origin: the mountains of South and Central Europe.

Moenchia A genus of plants of the Pink family, *Caryophyllaceae*. e.g. M. erecta - The Dwarf Chickweed, an annual plant, 3-10 cm high; the white flowers are usually 4-petalled. Grows in pastures, sandy and gravelly turf, and on cliffs and dunes near the sea. Place of origin: an uncommon species found in South and Central Europe; also in Britain.

Moho A genus of Birds (class *Aves*). e.g. M. braccatus - The Hawaiian O-o, a species doomed to extinction, whose last count in 1975 was just one pair!

Mola A genus of Fishes of the Headfish, Mola or Ocean Sunfish family, *Molidae*. e.g. M. mola - The Common Ocean Sunfish, a large marine fish seen in all seas, up to 2.5 m (81/4 ft) long and weighing up to 550 kg (1,200 lb). It has a flat, short, deep body, a strong leathery skin, and high dorsal and anal fins. It is often seen on the open sea with the pointed dorsal fin projecting out of the water. One of the most fertile fishes, laying as many as 300 million eggs. Distribution: found mainly in tropical, subtropical and temperate waters.

Molinia A genus of plants of the Grass family, *Gramineae*. e.g. M. coerulea (= M. caerulea) - Purple Moor grass, a tussock-forming perennial plant, 30-150 cm (1-5 ft) high; the culm is one-noded and the lugule is a dense fringe or ring of short hairs. Grows in damp, peaty moorland, heaths, and in fens and marshes on acid soils. Place of origin: Europe; native in Britain.

Mollienesia A genus of Fishes of the order of Foureyed Fishes, Guppies, Killfishes, and Swordtails, *Microcyprini* (suborder *Cyprinodontei*). e.g. M. formosa - The Amazon Molly fish; this

species of aquatic fishes has apparently no males, fertilization being carried out by males of other species. All the progeny are female.

Moloch A genus of Reptiles of the Agamid Lizard family, *Agamidae*. e.g. M. horridus - The Spiny Lizard, Moloch or Thorny Devil, an agamid lizard found in the deserts of Australia, an inoffensive creature despite its fearsome armor; it is completely covered with spines, the largest of which are on the head and neck, above the eyes, behind the nostrils and in front of the ears. It feeds mainly on ants caught with its tongue.

Molorchus A genus of Insects of the Long-horned or Longicorn Beetle family, *Cerambycidae*. e.g. M. minor - Length 6-16 mm; in this species of cerambycids the elytra cover only part of the abdomen. The larvae develop in dead trunks and in branches of spruce and pine trees, in which they bore zigzag tunnels. Distribution: Europe, the Caucasus, Asia Minor, the Middle East, Siberia and Japan.

Molothrus A genus of Birds of the American Blackbird and Oriole family, *Icteridae*. e.g. M. ater - The Brownheaded Cowbird of North America; the males are black with a brownish head, while the females are a uniform grey; the bill is short and conical like that of a sparrow. They frequent cattle pens for food, hence the common name. The species is credited with a ten-day incubation period. Distribution: southern Canada and most of the United States.

Molucella A genus of plants of the Mint or Thyme family, *Labiatae*. e.g. M. laevis - A species with ovate, round-toothed leaves and conspicuous spikes of curious flowers, in which the calyx forms a large expanded green funnel wih a small purple corolla at the bottom. Place of origin: Asia Minor, Syria and Iraq. Occasionally found in Iran.

Molva A genus of Fishes of the Cod, Haddock, Hake and Whiting family,

Gadidae (Burbot, Cusk and Ling subfamily, *Lotinae*). e.g. M. molva - The Common Ling fish; up to 2 m long and weighing 30 kg; a long, slender, codlike fish of northern Europe, and the largest member of the genus; used for food either salted or dried. It has two dorsal fins but no barbels. A very fertile species, found at depths of 100-600 m, the Common Ling is credited with bearing more than 28 million eggs in its ovaries. Distribution: Europe, in the northeastern Atlantic, extending to the Bay of Biscay. Native around Britain.

Momordica A genus of plants of the Cucumber, Gourd and Melon family, *Cucurbitaceae*. e.g. M. charmantia - The Balsam Pear, a variable climbing member of the Cucumber family, grown for its warty fruits. It has simple tendrils, conspicuously lobed leaves, yellow 5-petalled flowers and oblong, slender and pointed warted fruits. These are yellowish or copper-colored when ripe and open to disclose white or brown seeds with scarlet arils. Place of origin: southeast Asia, Tropical Africa; introduced into Europe in 1710.

Monacanthus A genus of Fishes of the Triggerfish family, *Balistidae*. e.g. M. ciliatus - The Fringed Filefish, a pink colored, marine species inhabiting warm seas; it has a laterally compressed body with strong movable spines in the dorsal, pelvic and anal fins.

Monacha A genus of Gastropod Molluscs of the family of Land Snails, *Helicidae*. e.g. M. incarnata - A European forest species of land snails; a hermaphrodite in which sexual union does take place, the eggs being laid by the individual acting as the female, usually in the earth or under stones.

Monachus A genus of Mammals of the Earless Seal family, *Phocidae*. e.g. M. monachus - The Monk Seal, up to 2.4 m (8 ft) long; plain grey above and white below. This pinniped is now rare, having been almost exterminated by

hunters. Distribution: the Black Sea, the Mediterranean, particularly in the Adriatic and round the Greek islands, and in the Atlantic Ocean near Madeira and the Canary Islands.

Monadenium A genus of plants of the Spurge family, *Euphorbiaceae*. e.g. M. guentheri - An erect species with spreading spatulate leaves. Place of origin: tropical Africa.

Monas A genus of minute, solitary, free-swimming, protozoan organisms. e.g. M. lens - A species found in sputum.

Monbretia A genus of plants, better known by the generic name, *Crocosmia*, of the Iris family, *Iridaceae* - Also called *Montbretia*. e.g. M. crocosmiiflora (= *Crocosmia* c.; *Montbretia* c.) - The Garden Montbretia (see **Crocosmia**; **Montbretia**).

Moneses A genus of plants of the Wintergreen family, *Pyrolaceae*. e.g. M. uniflora - The One-flowered Wintergreen, a perennial plant, up to 7 cm high, with large, white, solitary, rotate and wide open flowers. Grows in damp mossy coniferous or deciduous woods. Place of origin: Europe; native in Britain.

Monias A genus of Birds of the Mesite, Monia or Roatelo family, *Mesitornithidae*. e.g. M. benschi - Bensch's Monia or Rail, a polyandrous species, just under 30 cm (1 ft) long, a rail-like, near flightless bird found only in Madagascar; the male builds the nest and incubates the eggs.

Moniezia A genus of Tapeworms (class *Cestoda*), of the family *Anoplocephalidae*. e.g. M. expansa - The Sheep Tapeworm; in this species each segment has two sets of male and female reproductive organs, marginal in position.

Monilia The former name for a genus of fungi, now better known as *Candida*. e.g. M. albicans (= *Candida* a.) - A yeast-like fungus found in thrush (moniliasis), now called candidiasis (see **Candida**).

Moniliformis A genus of acantho-cephalous nematodes or roundworms (class *Nematoda*). e.g. M. moniliformis - A parasite of dogs and rats, sometimes infesting man.

Monocelis A genus of Turbellarian Flatworms of the Proticlad order, *Proticladida* (suborder *Crossocoela*). These are marine animals with a triclad type of pharynx and a simple undivided intestine.

Monocystis A genus of Gregarine Spore-formers or Sporozoa of the Acephaline suborder, *Acephalina*, the species of which are parasitic in the seminal vesi-cles of the earthworm. The mature trophozoite or sporont is large, but sel-dom exceeds 200 μ in length. e.g. M. epithelialis - Protozoan bodies (named by Pfeiffer), found in skin cells.

Monodella A genus of Crustaceans of the order *Thermosbaenacea*; the species are less than 5 mm in length; four of them have been found in subterranean water, both fresh and brackish, around the Mediterranean Sea, and a fifth species has been discovered in freshwater in a cave in Texas. The eyes are reduced or absent, but the antennules are large and well equipped with sensory chitinous bristles or setae.

Monodon A genus of Mammals of the Narwhal family, *Monodontidae*. e.g. M. monoceros - The Narwhal, a toothed whale, the only species of the family, and a close relative of the White Whale or Beluga (see **Delphinapterus**). It reaches a length of 3.6-4.5 m (12-15 ft), excluding the (male) tusk, one of which may be as long as 2.4 m (8 ft). The nar-whal tusk contributed to the legend of the unicorn, and the animal itself has consequently been dubbed the "Sea-uni-corn Fish". It has black spots on its white skin, and inhabits Arctic waters. It also regularly visits European coastal waters.

Monodonta A genus of Gastropod Molluscs of the family *Trochidae*. e.g. M. articulata - It has a large strong shell, 30-50 mm high and 24-42 mm wide, and is composed of 6-7 whorls; it is very varied in coloration; and is closed by a calcareous operculum. Found in abundance on rocky coasts and often, on stone harbor constructions. Distribution: Mediterranean Sea, Portugal.

Monodontus A genus of Tapeworms of cattle and sheep (class *Cestoda*).

Monochus A genus of Nematode para-sites or Roundworms (phylum *Nematoda*), found in the urine of inhabi-tants in the Canal Zone (Panama).

Monopsyllus A genus of wingless insects of the order of Fleas, *Siphonaptera*. e.g. M. anisus - The Common Rat Flea of Japan and North China.

Monoraphis A genus of Sponges (phy-lum *Parazoa*) of the suborder *Amphidiscophora*. e.g. M. chuni - The Single-rod Sponge, a deep-sea hexactin-nelid or glass sponge species found in the Indian Ocean. It has a single siliceous spicule or rod almost 90 cm (3 ft) long, based on a hexaradiate or six-rayed plan, an amphidisc with the glass sponge itself situated near the upper end, and the lower end of the rod being embedded in the mud of the sea floor.

Monochotrema A genus of heterophyid flukes (class *Trematoda*) found in Japan and Taiwan, with an operculate snail (*Melania*) as the invertebrate host, and an edible fish as first vertebrate host.

Monosporium A genus of Fungi or Molds. e.g. M. apiospermum - One of the causative fungi of maduromycosis (e.g. Madura foot).

Monostoma A genus of Trematode Worms. One species occurs in the crys-talline lens, causing the disease mono-stomidosis. Also known by the generic name, *Monostomum*.

Monostomum See **Monostoma**.

Monotropa A genus of plants of the Birdsnest family, *Monotropaceae*. e.g.

M. hypopitys - The Yellow Birdsnest or Pinesap, a non-green, perennial herb, 8-30 cm high; the leaves are scale-like, the yellow flowers grow in short racemes and droop when in bloom. A saprophyte, this species is found in coniferous and deciduous (beech) woods. Place of origin: Europe; native in Britain.

Monsonia A genus of African and Asian geraniaceous plants of the Cranesbill and Geranium family, *Geraniaceae*. The species are used in medicine as astringents.

Monstera A genus of plants of the Arum family, *Araceae*. e.g. M. deliciosa - The Monstera or Swiss Cheese plant, mistakenly called the Split Leaf Philodendron, a climbing evergreen, up to 3.6 m (12 ft) high; with huge, tough, dark green leaves which have deeply incised margins and large holes in the blades. Thick aerial roots hang down, which may penetrate into most soil under favorable conditions. There are creamy arum flowers followed by juicy, cone-shaped fruits with a delicious pear-pineapple flavor. In the tropics the fruit pulp is used in ices and drinks. Place of origin: Mexico and Guatemala.

Montacuta A genus of Bivalve Molluscs of the Cockle family, *Cardiidae*. e.g. M. ferruginosa - The Rusty Montacute Shell, a cockle species with a shell up to 9 mm long; the valves are twice as long as the height, bluish to violet in color, and fragile. Found only in the burrows of the Sea Potato (*Echinocardium* cordatum) or sometimes between its spines. Distribution: the coasts of Europe; native around Britain.

Montastrea A genus of Stony Corals of the order of True or Stony Corals, *Scleractinia*. These are the Star Corals, in which the skeleton shows individual thecae with ridges or septa radiating from the outer margins. These coral animals grow into boulder-like masses, yellow-brown in color, in the West Indies and Florida.

Montbretia A genus of plants, better known by the generic name, *Crocosmia*, of the Iris family, *Iridaceae*. Also known as *Monbretia*. e.g. M. crocosmiiflora (= *Crocosmia* X c.; *Monbretia* c.; *Tritonia* c.) - The Garden Montbretiia (see **Crocosmia**).

Montia A genus of plants of the Purslane family, *Portulacaceae*. e.g. M. fontana - The Common Blinks, an annual to perennial, freshwater plant, 3-10 cm high; the stem is often erect and forked. Grows by streamsides, springs, in wet places among rocks and in moist places. Place of origin: Europe; native in Britain.

Monticola A genus of Birds of the Babbler, Thrush, Old World Warbler and allied family, *Muscicapidae* (Nightingale, Old World Blackbird, Robin and Thrush subfamily, *Turdinae*). e.g. M. saxatilis - The Rock Thrush, length 19 cm; a solitary, partially migrant, multicolored bird living in open rocky areas and nesting in rock-crevices and holes in mountain sides. In the male the head is blue and the rump white; the female is brownish with an orange-red tail. Distribution: the Alps and other mountainous areas of southern Europe; northwest Africa, the Middle East, central and eastern Asia.

Montifringilla A genus of Birds of the Finch family, *Fringillidae*. e.g. M. nivalis - The Snow Finch, length 18.5 cm; a resident mountain bird with a grey head, black throat, brown back and a cream-colored belly with white in the wings. It builds its nest in rock crevices, on and above the tree-line. Distribution: the mountains of Europe and Asia.

Moraxella A genus of gram-negative diplobacilli of the family of microorganisms, *Brucellaceae*. The species are also classified under other generic names such as *Acinetobacter*, *Bacterium*, *Hemophilus*, *Mima*, etc. Found as parasites and pathogens in warm-blooded animals. e.g. M. bovis (= *Hemophilus* b.) - causing acute conjunctivitis in cattle M. lacunata (= *Bacterium* duplex; *Hemophilus* duplex; *Hemophilus* of Morax-Axenfeld; Morax-Axenfeld bacillus; *Moraxella* d.; *Moraxella* liquefaciens) - causing blepharo-conjunctivitis in humans. M. lwoffi (= *Acinetobacter* l.; *Mima* polymorpha) (see **Acinetobacter**).

Morchella A genus of plants of the order of Cup Fungi (subclass *Ascomycetes*). e.g. M. esculenta - The Edible Morel, a cup fungus species found growing in clearings in woods and pastures. The rounded, yellow-brown, sponge-like cap is 3-7 cm wide, and has firm, branched ribs forming deep, wide, irregular pits. It is edible when boiled, and is often used for flavoring soups and stews. Place of origin: Europe; native in Britain.

Morimus A genus of Insects of the Longicorn or Long-horned Beetle family, *Cerambycidae*. e.g. M. funereus - A beautiful Longicorn beetle, light greyish-blue in color with black velvety spots on the carapace. The upper surfaces of the forewings are rough grained and wrinkled. Distribution: southeast Europe.

Moringa A genus of plants. e.g. M. pterygosperma - An East Indian plant, called Sajina, which yields benoil.

Mormyrops A genus of Fishes of the Mormyrid suborder, *Mormyroidea*. e.g. M. boulengeri - A freshwater fish found in Africa, with a very long snout, rather like an elephant's trunk.

Morpho A genus of Insects of the Tortoiseshell Butterfly family, *Nymphalidae*. These large moths or butterflies are brilliantly metallic in color, often an iridescent blue, and they soar among the trees of tropical forests in Central and South America.

Morus 1. Animal Kingdom. A genus of Birds of the Booby and Gannet family, *Sulidae*. e.g. M. bassana - The Northern Gannet or Common Booby, a common species and a colonial cliff-nester; it congregates on crowded coastal cliff-sites to nest and rear its young. The adult plumage is white with a black tip to the wings. Distribution: on temperate seacoasts. 2. Vegetable Kingdom. A genus of Trees of the Mulberry family, *Moraceae*. e.g. M. nitra - The Black Mulberry, Common Mulberry, Amoras or Mûre, a deciduous tree, about 9 m (30 ft) tall, with dark green leaves above and hairy below. The fruit is dark red to purple, and is used as a mild laxative and expectorant; also in jam and wine making. Place of origin: the Far East; cultivated in Europe and Asia.

Moschus A genus of Mammals of the True Deer family, *Cervidae* (Musk subfamily, *Moschinae*). e.g. M. moschiferus - The Musk Deer, up to 50 cm (20 in) high. Its thick, spring coat fits it for a life in exposed situations in the mountains. The male has a musk gland in front of the navel which is highly odoriferous, and is sought after in the manufacture of perfume. Musk (Almiscar; Deer Musk; Moschus) is the dried secretion from the preputial follicles of the musk deer. It has a powerful, penetrating and persistant odor due to the presence of the ketone, Muskone. Used as a fixative in perfumery. Distribution: the natural habitat is in East Asia.

Motacilla A genus of Birds of the Pipit and Wagtail family, *Motacillidae*. e.g. M. alba - The Pied or White Wagtail, length 17.5 cm (7 in); it possesses a black and white plumage, slender legs and a long tail. A field bird, it spends most of its time on the ground. When walking it wags its tail up and down. It is a partial migrant, found in open country, and makes its nest at the water's edge. Distribution: Europe; native in Britain; almost all of Asia; North Africa.

Mucor A genus of Fungi forming delicate, white tubular filaments and spherical, black sporangia. e.g. M. mucedo - A non-pathogenic species, made up of non-septate hyphae or phycomycetes,

332

found in feces and other nitrogenous
substances. A frequent contaminant of
food and of culture media in the laboratory. It produces a fatal disease in bees.

Mucuna A genus of plants of the Pea
family, *Leguminosae*. e.g. M. pruriens
(= M. prurita) - The Cowage, Cowhage
or Cowitch plant - has been used as a
vermifuge.

Muellerius A genus of Lung-worms. e.g.
M. capillarius - A species parasitic in
sheep and goats.

Muggiaea A genus of Celenterate animals of the order of Siphonophores,
Siphonophora. They have a single bell,
forming swimming or floating colonies
and are often found off the southwest
coast of Britain.

Mugil A genus of Fishes of the Mullet
family, *Mugilidae*. These are the Grey
Mullets, in shore or coastal marine fishes, living in river estuaries or harbours
in shallow water, and prowling around
the muddy bed in large shoals. They are
good food fishes, with thick lips and
broad serrated teeth. e.g. M. cephalus -
The Striped Mullet, length 50 cm.
Distribution: sea and lower reaches of
rivers from the Baltic along the entire
coast of Europe, including the
Mediterranean and the Black Sea.

Mullus A genus of Fishes of the Mullet
family, *Mugilidae*. e.g. M. surmuletus -
The Red Mullet, length up to 45 cm; a
pink colored fish, usually with 3-5 yellow stripes along each side; the chin has
2 long barbels. A predator species, preying on other fishes. Distribution: the
Mediterranean Sea and along the
Atlantic coasts of Europe, as far as the
North Sea. Native around Britain.

Multicalyx A genus of Flukes (class
Trematoda) of the family of
Aspidogastrid Trematodes,
Aspidogastridae. These are endopara-
sites found in the gallbladder of selachi-
an fishes, and have one row of alveoli.

Multiceps A genus of Tapeworms (class
Cestoda) of the Cyclophyllidean order,

Cyclophyllidae, the bladder worms of
which are found in herbivorous animals,
and the adult forms in carnivorous ones.
e.g. M. multiceps - The "Gid" Worm. Its
larval stage (*Coenurus* cerebralis) develops in the central nervous system of
goats and sheep, and occasionally in
man. It causes giddiness and staggering
in sheep, hence the name "gid" worm.
The adult stage is parasitic in dogs. In
humans infection is transmitted from
eggs in dog feces.

Multicotyle A genus of Flukes (class
Trematoda) of the family of
Aspidogastrid trematode worms,
Aspidogastridae. These are endopara-
sites of molluscs and cold-blooded ver-
tebrates.

Multidictus A genus of Tapeworms (class
Cestoda) of the order *Diphyllidea*. e.g.
M. physeteris - A giant diphyllobothriid
cestode, measuring some 20 m (22
yards), found in the bile ducts of the
sperm whale. It has a diameter of up to
3 cm (1.18 in).

Mungos A genus of Mammals of the
Civet, Genet and Mongoose family,
Viverridae. e.g. M. mungo - The Banded
Mongoose, a dun-colored animal with
dark bands across the back.
Distribution: from tropical Africa south-
wards.

Muntiacus A genus of Mammals of the
True Deer family, *Cervidae*. These are
the Muntjacs or Barking Deers, with
short, 2-tined antlers at the ends of long,
bony, skin-colored pedicles or burrs.
They also have tusk-like upper canines.
e.g. M. reevesi - Reeves' Muntjac, 1 m
tall with a 15 cm tail; the antlers are
borne on long hairy pedicles, simply
forked. The female has no antlers.
Distribution: the jungles of India; introduced from there into England and
France, and has become naturalized in
other countries.

Muraena A genus of Teleostean Fishes of
the Moray Eel family, *Muraenidae*. e.g.
M. helena - The Moray Eel, Greek

Moray or Murry, up to 1.8 m (6 ft) long; a ferocious fish with powerful jaws and a savage bite. The sharp teeth have poisonous glands at the base, and the species is highly dangerous. Like the other moray eels, the body is slender, smooth and serpentine, devoid of ventral fins and with a continuous dorsal fin. Distribution: the warm seas of Europe, especially the Mediterranean; rarely as far north as Britain.

Murex A genus of Gastropod Molluscs of the Carnivorous Sea-snail family, *Muricidae*. e.g. M. purpurea - A gastropodous mollusc of the Mediterranean, from which an ancient purple dye known as Tyrian purple, was extracted. Used in homeopathic medicine in uterine diseases. The active substance is a neurotoxic substance, Murexin, identified as β-[imidazoyl-(4)]-acrylcholine, and believed to be the same as purpurin. The species lives on muddy sea bottoms, feeding on other shells or barnacles in the Mediterranean Sea.

Murgantia A genus of Insects of the Shield Bug family, *Pentatomidae*. These are the Terrestrial Bugs or the Geocorisae. e.g. M. histrionica - The Harlequin Bug, a conspicuous species attacking cabbages in North America. It has a large, triangular, shield-like, thoracic plate known as the scutellum.

Murimyces A genus name proposed for the pleuropneumonia-like organisms isolated from rats.

Mus A genus of Mammals of the Common Rat and Mouse family, *Muridae* (Old World Rat and Mouse subfamily, *Murinae*) (L) = "Mouse". e.g. M. alexandrinus - The Egyptian or Roof Rat. M. decumanus (= M. norvegicus). M. musculus - The Common House Mouse, one of the oldest known species of domestic rodents and a formidable pest. Distribution: worldwide. M. norvegicus - The Brown or Barn Rat. M. rattus - The English Black Rat. This genus has a very short gestation period,

particularly in the case of the species M. musculus, which is only 17 days (see **Sorex**).

Musa A genus of plants of the Banana family *Musaceae*. e.g. M. sapientum - A Banana species, a large tree-like herb of extreme economic importance, up to 6-9 m (20-30 ft) tall with oblong leaves up to 3 m (10 ft) in length. The inflorescence bears a number of half-whorls of unisexual flowers, male at the top and female towards the base. The male flowers are shed, and bunches of yellow curved fruits, the banana, develop on the pendulous axis. Place of origin: the tropics - Africa, South America, India and southeast Asia.

Musca A genus of Dipterous Insects of the Fly family, *Muscida* (L) = "Fly". e.g. M. domestica - The Common House Fly, whose larvae develop in all kinds of rotting plant and animal matter, including animal and human excrement. Size 10 mm, with a black body and hairless eyes. A very short-lived insect with an average lifespan of some 18 days. The adults may transmit a variety of pathogenic and saprophytic microorganisms, infecting animals and humans. Distribution: worldwide.

Muscardinus A genus of Mammals of the Doormouse or Dormouse family, *Muscardinidae*. e.g. M. avellanarius - The Common Doormouse or Dormouse, the only dormouse native to Britain. It is no bigger than a House Mouse, size 8.5 cm with a 6 cm tail (see **Mus**); it is tawny yellow with white below. A nocturnal creature, it lives in small colonies, and is found in broadleaved and mixed woodlands, less frequently in coniferous forests. Distribution: widely distributed from Britain to Turkey and from Scandinavia to Tuscany.

Muscari A genus of plants of the Lily family, *Liliaceae*. e.g. M. tenuiflorum - The Slender Hyacinth, a perennial plant, 25-50 cm high; the flowers are greenish-white with blackish teeth. Grows in scrub and on sunny slopes. Place of origin: southeast and western Europe.

Muscicapa A genus of Birds of the Babbler, Baldcrow, Old World Flycatcher and Warbler, and Thrush family, *Muscicapidae* (Old World Flycatcher subfamily, *Muscicapinae*). e.g. M. striata - The Spotted Flycatcher, length 14 cm; a dull grey-brown, migrant bird with brown-black streaking on its throat; found in trees and bushes in reasonably open country, particularly parklands. Has the habit of hunting after flies and other insects. Winters in tropical and South Africa. Distribution: all Europe, native in Britain; Asia; northwest Africa.

Muscivora A genus of Birds of the Tyrant Flycatcher or Kingbird family, *Tyrannidae*. e.g. M. forficata - The Scissor-tailed Flycatcher; the species owes its name to the habit of opening and closing its long outer tail feathers while in flight. The total length of the adult male is 35 cm (14 in), of which the tail is 25 cm (10 in) long. The head and back are pearly grey and the crown and wing-linings salmon pink. The tail-feathers are white, with the exception of the black-tipped central pair. Usually found in open, sparsely wooded country. Distribution: throughout the New World.

Musculus A genus of Bivalve Molluscs of the subclass of Mussels, *Lamellibranchia*. e.g. M. marmoratus - The Marbled Crenella, a marine species; length of shell 1.7 cm; there are about 16 ribs in front of the valve and 25 ribs at the back. Often found in Sea-squirts or sheltering in other shells. Distribution: widespread around the coasts of Europe; native around Britain.

Mustela A genus of Mammals of the Badger, Mink, Otter, Skunk, Stoat, and Weasel family, *Mustelidae*. e.g. M. vison - The North American Mink, a species bred for the important mink fur trade from both North America and Europe. It is a larger animal than its European cousin (M. lutreola), and its coat is thicker and softer. It is 50 cm long and has a 16 cm long tail. The coat is uniformly dark brown, wooly-haired and soft, but there are many different-colored varieties.

Mustelus A genus of Fishes of the Soft-mouthed Hound Shark, Smooth hound or Smooth Dogfish family, *Triakidae*. e.g. M. californicus - The Smooth Dogfish of the eastern Pacific, a harmless Smooth hound Shark with flat teeth. An edible species, often eaten by humans.

Mutilla A genus of Hymenopterous Insects of the Velvet Ant family, *Mutillidae*. e.g. M. europaea - The Large Velvet Ant, 10-15 mm long; the front abdominal segments are bluish and unspotted. The male is winged, the female wingless. The larvae are parasites in the nests of bumblebees. Distribution: the Palaearctic region.

Mutinus A genus of plants of the class of Fungi or Molds, *Gasteromycetes*. e.g. M. caninus - A species of fungus with a strong and characteristic smell that can be detected some distance away. At the top is a loose, conical cap with a coarse, honey-combed surface, covered with a shiny mass of black spores. Owing to the strong fetid smell, flies are attracted to it and within a few hours all the spores are carried away.

Mutisia A genus of plants of the Daisy family, *Compositae*. e.g. M. viciaefolia - A composite-flowered plant of South America; used there as a sedative.

Mya A genus of Bivalve Molluscs of the Gaper family, *Myidae*. e.g. M. arenaria - The American Soft-shell Clam or Sand Gaper; shell length 120 mm, height 70 mm, thickness 40-45 mm; a lamellibranch animal with an oval shell rounded at the front end and tainted brown. It lives buried about 20 cm deep in sand or mud on the sea bottom, extending only its long, massive siphon sometimes as much as four times the length of the shell. An edible species harvested for food. Distribution: coasts of northern Europe, North America, Japan, the Black Sea. Native around Britain.

Mycelis A genus of plants of the Daisy family, *Compositae*. e.g. M. muralis - The Wall Lettuce, a perennial plant 20-150 cm high, with thin leaves. Widespread and commonly found on walls, rocks, and occasionally in woodlands. Place of origin: Europe; native in Britain.

Mycobacterium A genus of microorganisms of the family *Mycobacteriaceae*. Gram-positive slender, rod-shaped cells distinguished by acid-fast staining (i.e. resisting acid-alcohol decolorization). Also abbreviated: Myco, as in Myco tuberc. e.g. M. tuberculosis (= *Bacillus* t.; *Bacillus* t. var. hominis; Koch's bacillus; M. tuberc; Myco tuberc) - the causative agent of tuberculosis (TB), most commonly pulmonary, in man. M. tuberculosis var. avium. - TB in birds. M. tuberculosis var. bovis - TB in mammals (e.g. cattle), and extra-pulmonary tuberculosis in humans.

Mycocandida A genus of yeast-like Fungi or Molds of the family *Cryptococcaceae*, better known by the genus name, *Candida*. e.g. M. albicans (= *Candida* a.) (see **Candida**).

Mycococcus A genus of microorganisms of the family *Mycobacteriaceae*. Saprophytic rod-shaped cells found in soil and water.

Mycoderma A genus of Fungi or Molds, the species of which are included under various generic names. e.g. M. aceti (see **Acetobacter**). M. dermatitidis (see **Blastomyces**). M. inmite (see **Blastomyces**; **Coccidioides**).

Myconostoc A genus of microorganisms of the family *Pseudomonadaceae*. Branching cells found in the soil. e.g. M. bullata.

Mycoplasma A taxonomic name given to a genus of highly pleomorphic, very small, gram-negative, wall-defective microorganisms including the pleuropneumonia-like organisms (PPLO). e.g. M. hominis - Causes non-specific urethritis, pyelonephritis, pelvic abscess, postpartum and postabortal fever, and septicemia. M. pneumoniae (= Eaton agent) - A species causing a typical viral pneumonia.

Mycotoruloides A genus of yeast-like Fungi or Molds of the family *Cryptococcaceae*, better known by the generic name, *Candida*. e.g. M. albicans (= *Candida* a.) (see **Candida**).

Mycteria A genus of Birds of the Jabiru and Stork family, *Ciconiidae*. e.g. M. americana - The Wood Stork, a common American species and the only true stork in the United States. It is found from the swamps of Florida south to the Argentine, and was once misleadingly called the Wood Ibis (see **Threshkiornis**). It has a white coloring, with a distinctive black tail and pink legs.

Mycteroperca A genus of Fishes of the Sea Bass family, *Serranidae*. e.g. M. bonaci - The Jewfish of Florida, a species of Sea Basses, also erroneously called the Black Grouper (the true Grouper fishes belong to the genus *Epinephelus*), being dark in color, rough-scaled and sluggish. A marine species, found off the coast of Florida, U.S.A.

Myctophum A genus of Bony Fishes of the Lanternfish family, *Myctophidae*. e.g. M. punctatum - The Spotted Lanternfish, 10 cm long. These are deep-sea fishes with transparent larvae, which remain in groups near the water surface until they attain a length of about 2 cm. Only then, as they attain their adult form, do they move downward into deeper water. They are important as food for other sea-fish of commercial value. Distribution: the middle depths of the Atlantic Ocean and the Mediterranean Sea.

Mydas A genus of Insects of the family

of Robber Flies, *Asilidae.* e.g. M. heros - The Robber Fly of tropical South America, the largest known fly with a body length of up to 6 cm (2 3/8 in), and similar wing-expanse.

Mygale A genus of Arachnid Spiders of the family *Avicularoideae* (= *Mygalomorphae*). Better known by the genus name, *Avicularia.* e.g. M. avicularia (= *Avicularia* a.) (see **Avicularia**).

Myiarchus A genus of Birds of the Tyrant Flycatcher or Kingbird family, *Tyrannidae.* e.g. M. crinitus - The Great Crested Flycatcher. It has a prominent crest, grey throat and breast, yellow underparts and a rufous tail. It nests in cavities in rocks where it lays its eggs. Distribution: mainly in eastern North America, but also in Texas. A migrant bird, it winters in South America.

Myiatropa A genus of Dipterous Insects of the Hover Fly family, *Syrphidae.* e.g. M. florea - Length 12-16 mm; it has various types of markings on the thorax and is commonly found on flowers. Distribution: much of the Palaearctic region.

Myiochanes A genus of Birds of the New World Flycatcher family, *Tyrannidae.* e.g. M. richardsoni - The Western Pewee, the flycatcher bird of North America, olive-green in color. It has the highest recorded temperature for a bird, being as much as 44.8° C (112.7° F).

Myletes A genus of Fishes of the Characin family, *Characinidae.* These are South American species with strong molar-like teeth which are used for grinding molluscs.

Myleus A genus of Fishes of the Characin family, *Characinidae*, found in South America, and with similar features as the genus *Myletes* (see **Myletes**).

Myocastor A genus of Mammals of the Coypu or Nutria family, *Myocastoridae.* e.g. M. coypus - The Coypu or Nutria, a large, rat-like animal, 40-80 cm long with a 40-45 cm long tail, and webbed toes on the hind legs. A fur-bearing relative of the porcupine, it has a fine, thick, long coat chestnut-brown in color; the fur is known as Nutria. Found along flowing or calm bodies of water with plenty of aquatic vegetation. Distribution: native to South America, from where it was introduced into Europe after World War I.

Myopus A genus of Mammals of the Rodent family, *Muridae.* e.g. M. schisticolor - The Wood Lemming, 9.5 cm long with a 1.9 cm tail. Dark grey; lighter in winter, with a reddish-brown back. A forest dweller of the mouse family. Distribution: northern Europe.

Myosciurus A genus of Mammals of the Marmot, Prairie Dog and Squirrel family, *Sciuridae.* These are the tiny pygmy squirrels of Cameroun in West Africa, extremely small tree squirrels, no larger than mice. Distribution: the tropical forests of both Old and New Worlds.

Myosotis A genus of plants of the Borage family, *Boraginaceae.* e.g. M. alpestris - The Alpine Forget-me-not, a perennial plant, 2-10 cm high, with densely hairy stems and blue flowers. Grows in alpine grassland and on rocks. Place of origin: Europe; native in Britain.

Myosoton A genus of plants of the Pink family, *Caryophyllaceae.* e.g. M. aquaticum - The Water Chickweed, an annual to perennial plant, 8-80 cm high; a fragile, often trailing stem carries 5 styles or prolongations of ovarian segments (carpels) supporting the stigma - the tip of the female part of the flower which receives the pollen. Grows in damp places. Place of origin: most of Europe, except the north. Native in Britain.

Myosurus A genus of plants of the Buttercup family, *Ranunculaceae.* e.g. M. minimus - The Mousetail, an annual to biennial plant, 1-12 cm high; it has a basal rosette of narrowly linear leaves, and the fruits grow in long spikes resembling a mouse's tail. Found in damp habitats and in places which have been trampled by cattle, and in orchards. Place of origin: Europe; native in Britain.

Myotis A genus of Mammals of the Long-eared Vespertilionid Bat family, *Vespertilionidae*. e.g. M. myotis - The European Common Mouse-eared or Brown Bat, a large bat 6-8 cm long, with a wingspan of 40 cm; abundant in Europe, but seldom found in Britain. The ear tragi are long and pointed, hence the common name, and it is a high and fast flyer, found over fields, gardens, woods and rivers, hunting for insects. Distribution: widespread in Europe, particularly in the south; Asia Minor; North Africa.

Myoxocephalus A genus of Bony Fishes of the Sculpin family, *Cottidae*. e.g. M. scorpius - The Father lasher or Shorthorn Sculpin, length 30-50 cm; a marine fish inhabiting the littoral zone and sometimes entering river estuaries. Distribution: the northeastern Atlantic from the Bay of Biscay to Iceland and Greenland; the Arctic Ocean.

Myrica A genus of plants of the Bayberry, Bog Myrtle, Buckbean or Sweet Gale family, *Myricaceae*. e.g. M. gale - The Sweet Gale or Bog Myrtle, a bog shrub up to 1.5 m high; with lance-olate leaves, serrated near the tip and dotted with bitter-tasting, fragrant, resinous glands. Widely distributed in bogs, on wet heaths and in fens. Place of origin: Europe, in the north temperate zone; native in Britain; North America.

Myricaria A genus of Shrubs of the Tamarisk family, *Tamaricaceae*. e.g. M. germanica - The German Tamarisk, a rare evergreen shrub up to 2.4 m (8 ft) tall; the grey leaves are small and scale-like and the small flowers are pale pink or red. Grows locally on stony banks of mountain streams. Place of origin: Europe.

Myriophyllum A genus of plants of the Water-milfoil family. e.g. M. spicatum - The Spiked Water-milfoil, a perennial plant, 50-250 cm high; the leaves are usually 4 in a whorl and the inflorescences are many-flowered. Grows in still or slow-flowing freshwater. Place of origin: widespread in Europe; native in Britain.

Myriotrochus A genus of Echinoderm animals, spiny-skinned, marine invertebrates (phylum *Echinodermata*). e.g. M. brunni - An echinoderm marine species found at great depths in the Western Pacific, as much as 10,000 m (11,000 yards).

Myristica A genus of plants of the family *Myristicaceae*. e.g. M. fragrans - The Nutmeg Tree; the dried kernels of the seeds, known as Mace, yield a volatile oil, Nutmeg Oil or Oleum Myristicae, consisting of myristic acid (macene), myristicene, myristicol, myristin and smaller amounts of palmitic, oleic, linoleic and lauric acids. Used as an aromatic, carminative, and flavoring agent.

Myrmecia A genus of Insects of the Ant family, *Myrmicidae*. e.g. M. brevinoda (= M. gigas) - The Bull Ant, a rare species found in Queensland and northern New South Wales, Australia, measuring up to 37 mm (1.44 in) in length. M. forficata - The Black Bulldog Ant, of the coastal regions of Australia and Tasmania, and the most dangerous ant in the world, which uses its sting and jaws simultaneously when attacking. Its sting can be fatal to man.

Myrmecobius A genus of Mammals of the Numbat family, *Myrmecobiidae*. e.g. M. fasciatus - The Banded Anteater or Numbat, a rat-sized marsupial; it has a pointed snout and a long, extensile, worm-like tongue which it uses to catch ants and termites. Its red-brown coat is marked on the back with white transverse stripes. The tail is long and velvety and it has no pouch. Distribution: Australia.

Myrmecocystus A genus of Insects of the Honeypot Ant family, *Formicidae*. The species store their food by gorging themselves with honeydew, and the grossly distended bodies of these worker-ants hang almost immobile from the roof of the nest. Two or three hundred of these "repletes" or gorged creatures may be found in one nest. Distribution: southwest U.S.A.

Myrmecodia A genus of Shrubs of the Bedstraw or Madder family, *Rubiaceae.* e.g. M. echinata - A parasitic species growing on other plants and attached by adventitious roots, its base forming a swollen corky tuber. This base is penetrated by a network of tunnels and galleries which are always inhabited by ants. Place of origin: southeast Asia.

Myrmecophaga A genus of Mammals of the Anteater family, *Myrmecophagidae.* e.g. M. jubata - The Great Anteater, the largest of all the edentates or Toothless Anteaters (order *Edentata*), more than 2.4 m (8 ft) long. It is a weird-looking creature with an awkward gait. From the end of its long snout, projects the worm-like tongue, about 20 cm (8 in) long, which it uses to suck in hundreds of ants and other insects sticking to it, and each time its tongue emerges from the mouth, it is bathed in fresh viscous saliva. Distribution: the tropical grasslands of Central and South America.

Myrmeleon A genus of Insects of the Ant Lion family, *Myrmeleontidae* (order of Lacewings and Ant Lions, *Planipennia*). e.g. M. formicarius - The Ant Lion, 35 mm long with a wingspan of 65 mm; it slightly resembles a dragonfly. It has 4 similar net-veined wings. The larva lies hidden in a pit which it digs in the sand, feeding on ants which fall in. Distribution: most of Europe, north to Scandinavia. Absent in Britain.

Myrmica A genus of Insects of the Ant family, *Myrmicidae.* e.g. M. rubra laevinoides - The Red Ant, length 4-5 mm; reddish yellow; it has a poisonous sting on the end of the abdomen. Distribution: common in Europe; native in Britain.

Myroxylon A genus of Trees of the Pea family, *Leguminosae.* e.g. M. pereirae - The Peru Balsam tree; the trunk yields a balsam, Peru Balsam (= Balsam of Peru; Bals Peruv; Peruvian Balsam). It is a dark brown, viscid liquid with a bitter, acrid, burning taste, containing 50-60% balsamic esters. It is used in the treatment of eczema and pruritus, and in suppositories, for the symptomatic relief of hemorrhoids.

Myrrhis A genus of plants of the Carrot family, *Umbelliferae.* e.g. M. odorata - The Sweet Cicely, a perennial plant, 50-100 cm high with a strong aromatic smell, leaves bristly-woolly beneath and fruits up to 2.5 cm long. Grows widespread in woody meadows and grassland. Place of origin: Europe; native in Britain. The Sweet Cicely is known in North America as an entirely different genus, *Osmorhiza* (see **Osmorhiza**).

Myrtophyllum A genus of Protozoan organisms. e.g. M. hepatis - A species found in hepatic abscess.

Myrtus A genus of Trees of the Myrtle family, *Myrtaceae.* e.g. M. communis - The Common Myrtle, the Old World Myrtle; an aromatic, dense, evergreen shrub, 3-4.5 m (9-14 ft) tall; a strongly scented species; the leaves have translucent glandular dots, the flowers are white and the berries blue-black. Grows in the Maquis, a type of vegetation in the Mediterranean region composed of a large number of evergreen woody plants. Famous since classical times - wreaths of myrtle were worn by victors in the Olympic Games. It is still used with orange blossom as a traditional bridal flower or wreath. The leaves are antiseptic and astringent. Place of origin; southern Europe (the Mediterranean region) to West Asia. Cultivated in Britain since 1597.

Mysia A genus of Bivalve Molluscs of the family of Venus and Carpet Shells, *Veneridae.* e.g. M. undata - The Wavy Venus; size of shell 3 cm; the mantle cavity is large. Found commonly in fine sand. Distribution: the coasts of Europe; native around Britain.

Mysis A genus of Crustaceans of the Opossum Shrimp order, *Mysidacea* (suborder *Mysida*). e.g. M. relicta - A notable freshwater species found in lakes in North America and Europe. Considered to have been isolated in these lakes when they were cut off from the sea by rising land at the end of the Ice Ages. Because of its glacial origin it breeds in water in deeper areas of the lake in summer. It has no gills, the gill function being taken over by the carapace, which is very well supplied with blood vessels.

Mystax A genus of Mammals of the Marmoset and Tamarin family, *Callithricidae*. These tamarin species are expert climbers and can drop long distances without harm. They have no ear tufts or tail rings and live on fruit, eggs and insects.

Mytilicola A genus of Copepod Crustaceans of the Cyclopoid order, *Cyclopoida*. The species look like maggots and live in the intestine of the Common Mussel (see **Mytilus**).

Mytilus A genus of Bivalve Molluscs of the Mussel Shell family, *Mytilidae* (subclass of Mussels, *Lamellibranchia*). Some of the species are edible, others poisonous. e.g. M. edulis - The Common Mussel; length of shell 60-80 mm, height 40 mm, thickness 35 mm; an edible, bivalve, marine mollusc, found in shallow water, on rocks and breakwaters, where it is firmly secured to the bottom by means of horny threads, the byssus, which is formed by a gland in the foot. Of high food value. Distribution: coasts of Europe; native around Britain; off California and Japan. Other species of *Mytilus* are poisonous, and contain a toxic substance, Saxitoxin. This poison is also present in the clam, *Saxidomus* and the plankton, *Gonyaulaux*.

Myxilla A genus of Gelatinous Sponges of the order *Tetraxonida* (suborder *Sigmatosclerophora*). The microscleres or spicules in this suborder are based on C- or S-shapes.

Myxine A genus of Hagfishes and Lampreys of the Hagfish family, *Myxinidae*. e.g. M. glutinosa - The Common Atlantic Hag or Hagfish, a marine fish 25-50 cm long, with a well-developed nostril in the form of a tube-like cylinder which opens on the palate. It has poorly developed eyes, which are covered with pigmented skin. There are 6 gill pouches, and it is parasitic on other, larger fishes. A nocturnal species, it is found at depths of 20-800 m on muddy sea bottoms. Distribution: both coasts of the North Atlantic.

Myxobolus A genus of Protozoan parasites infecting fish. e.g. M. cyprini - A species causing pox disease in the carp. M. pfeifferi - A species parasitic to the fish, *Barbus* fluviatilis.

Myxococcidium A genus of Sporozoan organisms. e.g. M. stegomyiae - A sporozoan species found in the body of the mosquito, *Stegomyia* fasciata.

Myxococcus A genus of microorganisms of the family *Myxococcaceae*, found in decaying organic matter.

Myxovirus A general generic name for various groups of viruses, such as the influenza, mumps and Newcastle disease viruses. Has been given formal generic status by some workers. e.g. M. influenzae.

Myzomyia A subgenus of anopheline mosquitoes, several species of which act as the carriers of malarial parasites.

Myzorhynchus A subgenus of anopheline mosquitoes, several species of which are the carriers of malarial parasites. e.g. M. paludis - A species found in Africa. M. sinensis - A Japanese species.

Myzus A genus of Homopterous Insects of the Aphid family, *Aphididae*. e.g. M. cerasi - Length 2 mm; found on cherry leaves, parasitic, sucking the sap and

causing the leaves to curl. Distribution:
worldwide.

N

Nabis A genus of Insects of the Damsel Bug family, *Nabidae*. e.g. N. rugosus - A damsel bug species, 6-smaller plant-eating insects, and commonly found on expanses of grass. Distribution: the Palaearctic region.

Naegleria A genus of coprozoic, free-living Amoebae, of the order *Rhizomastigina*, found particularly in warm springs or in freshwater warmed by hot effluents, and capable of causing disease in humans. Better known by the generic name, *Dimastigamoeba*. e.g. N. fowleri (= *Dimastigamoeba* f.) - This species enters the body by way of the nose and the olfactory nerves, producing a primary meningo-encephalitis. N. gruberi (= N. punctata; *Dimastigamoeba* g.) - A small limax-type ameba, living in organically rich soils, in dry conditions or in high osmotic pressure liquids (see **Dimastigamoeba**).

Naja A genus of Reptiles of the Cobra, Krait, and Mamba Snake family, *Elapidae*. These are front-fanged, extremely poisonous snakes. The ejected venom flows down a groove at the back of each fang. e.g. N. naja - The Asiatic, Indian, or Spectacled Cobra, up to 1.8 m (6 ft) long; a poisonous species responsible for about a quarter of snake-bite deaths recorded in the Indian subcontinent. N. tripudians - The "Cobra di capello", a venomous snake from India. From its venom is prepared Naja, a homeopathic medicine.

Najas A genus of plants of the *Naiad* family. e.g. N. marina - The Holly-leaved Naiad, a dioecious annual plant, 20 cm high; the stem has occasional teeth near the top and the leaves are strongly spinous-toothed. It grows in brackish freshwater. Place of origin: widespread in Europe.

Nannizia A genus of soil-living dermato-phytic Fungi or Molds, botanically related to the genus *Microsporum*.

Nannosciurus A genus of Mammals of the Squirrel family, *Sciuridae*. These are the Tufted-ear Squirrels found in Borneo, the species of which are no larger than mice.

Nanophyes A genus of Flukes or Trematode Worms of the class *Trematoda*, also called *Nanophyetus*, now known by the generic name *Troglotrema*. e.g. N. salmincola (= *Troglotrema* s.) - An intestinal parasitic species of flatworms (see **Troglotrema**).

Nanophyetus A genus of Flukes or parasitic flatworms, better known by the generic name, *Nanophyes*, and now, *Troglotrema*. e.g. N. salmincola (= *Nanophyes* s.; *Troglotrema* s.).

Napaeozapus A genus of Mammals of the Jumping Mouse family, *Zapodidae*. These are forest species found in North America, chiefly in Canada and the northern United States.

Narcissus A genus of plants of the Daffodil family, *Amaryllidaceae*. e.g. N. stellaris - The Narrow-leaved Daffodil, a perennial plant, 20-30 cm high; the leaves are 2-5 mm wide, and the peri-anth-segments do not overlap. Place of origin: Europe; mainly in the Alps.

Nardostachys A genus of plants of the Valerian family, *Valerianaceae*. e.g. N. jatamansi - The Indian Nard, Spikenard or Jatamansi plant; the dried rhizome and root are used as a substitute for valerian, chiefly in oriental medicine.

Nardurus A genus of plants of the Grass family, *Gramineae*. e.g. N. tenellus - The Delicate Nardurus, an annual plant, 20-40 cm high; the spikelets are shortly stalked and grow in simple racemes. A rare species, found on cultivated and waste land and in dry places. Place of origin: South and Central Europe.

Nardus A genus of plants of the Grass family, *Gramineae*. e.g. N. stricta - Mat-grass, a perennial plant, 15-20 cm high; the leaves are bristle-like and stiff, the spikes are one-sided and slender and the spikelets are one-flowered, narrow and finely pointed. Common in grasslands and sometimes on high ground. Place of origin: widespread in Europe.

Narke A genus of Marine Fishes of the family of Electric Rays, *Torpedinoidae*. The species have 2 dorsal fins; they are sluggish, poor swimmers and most of the time they lie on the bottom partly buried in sand. They have a flattened body, with highly developed electric organs in the front part, capable of releasing an electric discharge, hence the common name.

Narthecium A genus of plants of the Lily family, *Liliaceae*. e.g. N. ossifragum - The Bogasphodel, a perennial plant, 10-40 cm high; it grows in racemose inflorescences with long-stalked flowers. Found in bogs and on wet heaths and moors. Place of origin: Europe; absent in the southern parts of the continent.

Nasalis A genus of Mammals of the Old World Monkey family, *Cercopithecidae*. e.g. N. larvatus - The Proboscis Monkey; the male has a swollen, pendent nose about 7.5 cm (3 in) long, hanging down from the mouth. The females and young have snub noses. Habitat: the humid forests of Borneo near rivers, where it is frequently found swimming.

Nasilio A genus of Mammals of the Jumping or Elephant Shrew family, *Macroscelididae* (see **Macroscelides**).

Nasonella A genus of Insects of the Dwarf Beetle family, *Ptiliidae* or *Trichopterygidae*. e.g. N. fungi - A hairy-winged, tropical dwarf beetle, which measures only 0.25 mm (0.0098 in), one of the smallest insects recorded.

Nasonia A genus of Insects of the order of Wasps, *Hymenoptera*. e.g. N. vitripennis - A tiny parasitic species of wasps.

Nassarius A genus of Gastropod Molluscs of the Sea-snail Dog Whelk family, *Nassariidae*. e.g. N. reticulatus -

The Netted Dog Whelk; the shell is about 30 mm high, conical, pointed and with a spiral pattern of broad flat ribs; its 8 whorls have ornamental transverse ridges. A carnivorous animal, it feeds mainly on mussels and barnacles. A common species, it is found in abundance on soft or sandy sea bottoms. Distribution: the Mediterranean Sea, European coast of the Atlantic, the North Sea and the West Baltic.

Nasturtium A genus of plants of the Cress or Crucifer family, *Cruciferae*. e.g. N. officinale - The Common Watercress, a perennial plant, 10-60 cm high; the flowers are borne in terminal clusters, the anthers are yellow and the leaves have a peppery taste. Occurs mainly in running freshwater, in rivers and on lakes. Place of origin: Europe; native in Britain.

Nasua A genus of Mammals of the Racoon family, *Procyonidae*. e.g. N. rufa - The Ring-tailed Coati, a red-brown species of American racoons or procyonids; about 60 cm (2 ft) long with a tail almost as long; the fur is reddish-brown and the tail has blackish rings. It has a very long snout, which it uses in grubbing for insects and larvae. It is an unfriendly and refractory animal as are the other members of this genus. It inhabits forests, feeding on fruit, insects, molluscs, and small vertebrates. Distribution: South and Central America.

Nasutitermes A genus of Insects of the order of Termites, *Isoptera*. e.g. N. triodiae - An Australian species, extremely long-lived; the queen termites may live 100 years or more, though this has not been proven; some have been known to lay eggs for up to 50 years. An ant-like species, they have very small, white, soft bodies. Each colony has a large winged king and queen, with multitudes of wingless, sterile, workers and soldiers. They do great damage to wood.

Natica A genus of Molluscs of the order

of Periwinkles and Slipper Limpets, *Mesogastropoda*. These are the Moonshells, rounded mollusc, marine animals, living under sand, where they seize and bore into bivalves. e.g. N. nitida - The Common Necklace Shell, 1.5 cm long, with 5-7 whorls, and highly polished. A burrowing carnivore. Distribution: Europe; native around Britain.

Natrix A genus of Reptiles of the Colubrid Snake family, *Colubridae*. e.g. N. natrix - The British Snake, European Grass, or Ringed Snake, a non-venomous species, about 90 cm (3 ft) long, with a yellowish-white collar round the nape, hence the name Ringed Snake. Because it is found near water it is also called the Water Snake, especially in the United States. Distribution: central and southern Europe (including England and Wales). Also found in the United States.

Naucrates A genus of Fishes of the Horse-mackerel, Jack, Pompano and Scad family, *Carangidae*. e.g. N. ductor - The Pilotfish, a well-known species in which the scales of the lateral line have comb-like extensions. These fishes make long migrations, accompanying sharks or ships, hence the common name.

Naumanniella A genus of microorganisms of the family *Siderocapsaceae*. These are ellipsoidal or rod-shaped cells with rounded ends, found in iron-containing water. e.g. N. elliptica.

Naumburgia A genus of plants of the Primrose family, *Primulaceae*. e.g. N. thyrsiflora - The Tufted Loosestrife, a perennial plant, 30-60 cm high; the leaves are opposite and the flowers are borne in dense bracteate racemes. Grows widespread in still and slow-flowing freshwaters, by ponds and lakes, and in marshes. Place of origin: Europe; native in Britain.

Nausithoe A genus of Jellyfishes of the order *Coronotae* (class *Scyphozoa*). The species inhabit the shallow waters around the Bahamas and Florida. They are characterized by the presence of a coronal groove midway between the center of the bell and the margin.

Nautilius A genus of Cephalopod Molluscs of the subclass *Nautiloidea*. These are tetrabranchiate squids with two ctenidia or tentacular appendages. They are carnivorous molluscs inhabiting the West Pacific and eastern Indian oceans at depths down to 200 m (650 ft).

Neanthe A genus of plants, better known by the generic name, *Chamaedorea*, of the Palm family, *Palmae*. e.g. N. bella (= *Chamaedorea* elegans; *Collinia* e.) - The Camedorea or Feathers Palm (see **Chamaedorea**).

Neanthes A genus of Annelids or Segmented Worms of the Ragworm family, *Nereidae*. e.g. N. diversicolor - A Ragworm species; length 60-120 mm width 2-6mm; the long, elongated body is composed of 70-120 segments. The color is variable, ranging from yellow, orange and green to reddish. It lives in muddy sand sea bottoms off shallow coastlines, where it digs a tube about 30 cm long. Distribution: European Atlantic coast, North Baltic, and Mediterranean Seas.

Nebalia A genus of Crustaceans of the order *Leptostraca*. e.g. N. bipes - A species with large limbs lying below the carapace, forming an elaborate feeding mechanism. The eggs are bright red and are borne on short stalks. Found on the lower parts of the seashore, under stones and among weeds. Distribution: from Europe to China and Japan.

Nebaliopsis A genus of Crustaceans of the order *Leptostraca*. A planktonic species which appears to lay its eggs freely into the sea, where they float as they develop.

Necator A genus of Nematodes or Roundworms of the order *Strongylida*. e.g. N. americanus (= *Ancylostoma* americanum; *Uncinaria* americana) - The American Hookworm, a parasite of man, the adults being parasitic and the larvae free-living. The third-stage larvae are picked up on the bare feet of the host and bore through the skin until they reach a blood vessel and are carried to the lungs. From the lungs they make their way to the esophagus, moult once, and are then swallowed, entering the intesine, where they moult again before growing into adults. These may live 15-16 years. Also known as the "American murderer", the females can produce between 25,000 to 30,000 eggs every day for 5 years.

Necrophorus A genus of Insects, also called *Nicrophorus*, of the Carrion Beetle family, *Silphidae*. e.g. N. vespillo (= *Nicrophorus* v.) - The Common Burying Beetle, length 10-24 mm; the elytra have two orange-red bands. Like the other species of this genus, it seeks out small dead (vertebrate) animals and buries them, thus providing a food store for the offspring. Distribution: Europe; native in Britain.

Nectarinia A genus of Birds of the Sunbird family, *Nectariniidae*. e.g. N. taccaze - The Taccaze, a very small bird, resembling the humming-bird, with a tubular tongue and a long, finely pointed bill which it uses for puncturing flowers to feed on nectar. The plumage is a brilliant bronze in the male; in the female it is less ornate, but with a distinctive eyestripe. Distribution: East Africa.

Nectonema A genus of Nematodes or Roundworms of the order *Nectonematoidea*. These are marine species of roundworms, the Nematomorphs or Pseudocelomates. They have an armed retractile proboscis with a superficially segmented trunk. The larvae are parasitic in various crustaceans such as hermit crabs and true crabs of various species.

Necturus A genus of Amphibians of the order of Salamanders, *Caudata*. They superficially resemble lizards but possess a soft, moist, brightly colored skin, no scales, and large external gills. The species are employed in physiological research.

Negaprion A genus of Fishes of the order of Sharks, *Selachii*. e.g. N. brevirostris - The Lemon Shark.

Neisseria A genus of microorganisms of the family *Neisseriaceae*. These are gram-negative cocci, usually appearing in pairs. e.g. N. gonorrhoeae (= *Gonococcus* neisseriae) - the causative agent of gonorrhoea. Found intracellularly in capsulated pairs.

Nelumbium A genus of plants, better known by the generic name, *Nelumbo*, of the Waterlily family, *Nelumbonaceae*. e.g. N. speciosum (= *Nelumbo* nucifera) - The Sacred Lotus (see **Nelumbo**).

Nelumbo A genus of plants, also known by the generic name, *Nelumbium*, of the Waterlily family, *Nelumbonaceae* or *Nymphaeaceae*. e.g. N. nucifera (= *Nelumbium* speciosum) - The Sacred, East Indian, or Hindoo Lotus, a freshwater plant with large leaves and rose-colored, fragrant flowers which are carried above the surface of the water on 90-120 cm (3-6 ft) stems. When ripe the curious fruit consists of hard nuts contained in separate pits on the upper surface. Place of origin: India; introduced into Europe in 1787.

Nemacheilus A genus of Fishes of the Loach family, *Cobitidae*. e.g. N. barbatulus - The Loach, a 15 cm (6 in) long, edible, freshwater fish living in brackish or running water; its upper jaw has four short and two long barbels, and the end of the caudal fin is straight. A valuable commercial fish. Distribution: most of Europe; native in Britain.

Nematodirus A genus of Nematode parasites or Roundworms found in the duodenum of ruminants.

Nematus A genus of Insects of the Sawfly family, *Tenthredinidae*. e.g. N. ribesii - The Gooseberry Sawfly, a species injurious to gooseberries and currants in Europe and North America.

The larvae devour the leaves and defoliate much of the plant they attack. When fully grown they fall to the ground and pupate within a cocoon in the soil.

Nemertes A genus of Ribbon-worms (phylum *Nemertina*). e.g. N. borlasi - A Mediterranean species, up to 12.19 m (40 ft) long. A marine creature which can shrink to less than one third of its normal length. A carnivorous, soft-bellied invertebrate, which breathes by means of gills.

Nemobius A genus of Insects of the True Cricket family, *Gryllidae*. e.g. N. silvestris - The Ground or Wood Cricket, 7-10 mm long; the body is dark brown with light spots; the forewings are very short. It lives amidst fallen leaves in woodlands. Distribution: Europe, North Africa. Native in Britain.

Nemophora A genus of Insects of the Butterfly and Moth family, *Incurvariidae*. e.g. N. degeerella - Size 7-8 mm; the flight is jerky and slow. It is found in forest undergrowth. The male has remarkably long antennae. Distribution: Europe.

Neobalaena A genus of Mammals of the Right Whale family, *Balaenidae*. Grooves on the throat and chest, and the absence of a dorsal fin distinguishes right whales from all other cetaceans. The color varies from slate-grey to black according to the species (see **Balaena**; **Eubalaena**).

Neoceratodus A genus of Fishes of the order of Lungfishes, *Dipnoi* or *Dipteriformes* (suborder *Ceratodei*). e.g. N. forsteri (= *Ceratodus* f.) - The Burnett Salmon or Australian Lungfish (see **Ceratodus**).

Neodiprion A genus of Insects of the Sawfly family, *Dipronidae*. These are serious defoliators in the coniferous forests of Canada, whose larvae feed externally on the needles.

Neodrepanis A genus of Birds of the Asity and False Sunbird family, *Philepittidae*. e.g. N. coruscans - The Wattled False Sunbird, a species closely resembling the true Sunbirds (genus *Nectarinia*), but with a more slender build, shorter legs and a long down-curving bill with which it sips nectar and swallows small insects. Distribution: the humid forests of the island of Madagascar (see **Nectarinia**).

Neofiber A genus of Mammals of the Common Rat and Mouse family, *Muridae* (Lemming, Muskrat and Vole subfamily, *Microtinae*). e.g. N. alleni - The Florida Water Rat, sometimes called the Round-tailed Muskrat. It spends its life in ponds, marshes and rivers, feeding on aquatic vegetation. It has an odorous smell due to a substance secreted by its inguinal glands. Its fur resembles a beaver's, thick, soft and shiny, and so does its behavior.

Neomeris A genus of Mammals of the Porpoise family, *Phocaenidae*. e.g. N. phocaenoides - The Black Porpoise, a cetacean found in the Indian Ocean and the Far East; a finless mammal, it has no dorsal fin. It often ascends rivers and has been known to swim for a thousand miles up the Yellow River or the Hwang Ho River.

Neomys A genus of Mammals of the Shrew family, *Soricidae*. e.g. N. fodiens - The European or Eurasian Water Shrew, 9.5 cm (3 3/4in) long with a 7.5 cm (3 in) tail. It lives by the water and is an excellent swimmer. It has a "keel" of bristles along the ventral side of the tail. Distribution: the greater part of Europe (including Britain), Asia Minor, and Siberia.

Neophilina A genus of Molluscs (phylum *Mollusca*), also known by the name *Neopilina* of the class monoplacophora. e.g. N. galathea (= *Neopilina* galatheae) - The earliest known mollusc species, which lived about 500 million years ago. It was discovered alive off Costa Rica in 1952, and off the Pacific coast of Mexico, deep in the ocean at a depth of 3,000 m or more. A monoplacophoran limpet, it has a flattened body, about 35 mm (1 1/3 in) long at the apex and bears a spirally coiled embryonic shell near the anterior end.

Neophoca A genus of Mammals of the Eared Seal family, *Otariidae*. e.g. N. cinerea - The Australian Sea-Lion, an eared seal species with external ears and a coarse coat of hairs, sometimes called a Hair Seal, and found in Australia.

Neophron A genus of Birds of the Old World Harrier, Hawk, and Vulture family, *Accipitridae* (Old World Vulture subfamily, *Aegypiinae*). e.g N. percnopterus - The Egyptian White Vulture or Pharaoh's Chicken, a scavenger and partial migrant, with white plumage tinged with brown on the underparts, a wing span of up to 1.65 m (51/2 ft), and a wedge-shaped, white tail. It was among the sacred birds of ancient Egypt and appears in stylised form in Egyptian sculpture and reliefs. Distribution: North Africa, the Mediterranean littoral, India, and southwest Asia.

Neopilina A genus of Molluscs, better known by the generic name, *Neophilina*, of the class *Monoplacophora*. e.g. N. galatheae (= *Neophilina* galathea) - A monoplacophoran limpet species (see **Neophilina**).

Neorickettsia A genus of microorganisms of the family *Rickettsiaceae*. e.g. N. helminthoeca - Causing disease in dogs, and found (as a reservoir of infection) in the intestinal trematode, *Nanophyetus* salmincola.

Neoschöngastia A genus of Arachnids of the Tick superfamily, *Ixodoidae*. These are related to mites but larger. e.g. N. americana - A tick species attacking chickens in the southern United States.

Neotoma A genus of Mammals of the Common Mouse and Rat family, *Muridae* (Hamster and New World Mouse and Rat subfamily, *Cricetinae*). e.g. N. albigula - The White-fronted Wood Rat, also known as the Pack Rat or Trade Rat, one of the native species

of North American rodents, a nocturnal animal living in colonies, building large, globe-shaped nests of grass, sticks and other materials. A large soft-furred, rat-like animal, it carries off every small article from deserted log cabins and outhouses, and replaces them with pebbles, hence the alternative common name, Trade Rat. Distribution: the western United States.

Neotragus A genus of Mammals of the Cattle family, *Bovidae* (Antelope subfamily, *Antilopinae*). e.g. N. pygmaeus - The Pygmy or Royal Antelope, the so-called neotragines, the smallest of all ruminants, only 30 cm (1 ft) at the shoulder, and living in the dense forests of West Africa.

Neottia A genus of plants of the Orchid family, *Orchidaceae*. e.g. N. nidus-avis - The Bird's nest Orchid, a saprophytic perennial plant, 15-60 cm high, with creeping rhizomes and a mass of short, fleshy, blunt roots; the stems have dense brownish scales. Place of origin: widespread in Europe; native in Britain.

Nepa A genus of Insects of the Water Scorpion family, *Nepidae*. e.g. N. cinerea (= N. rubra) - The Water Scorpion, an actively predacious species, 26 mm long, with a grey, flat, oval body; the forelegs are modified to serve as grasping organs. It breathes through a long tube at the rear end of the abdomen. It is found among plants at the edge of the water in lakes, ponds, and streams. Distribution: Europe; native in Britain, especially in the south.

Nepenthes A genus of plants of the Old World Pitcher-plant family, *Nepenthaceae*. There are about 60 species, occurring mainly in Borneo, although some are found in South China and Australia. They are mostly herbaceous climbing plants with leaves modified into tubular pitchers holding secretions, for trapping and digesting their insect prey.

Nepeta A genus of plants of the Mint or

Thyme family, *Labiatae*. e.g. N. cataria - The Wild Cat-mint, a perennial plant, 10-100 cm high; the leaves are ovate and long-stalked, and the lower lips of the flowers have red spots. A fragrant species, it grows in hedgerows and by waysides, at the edges of cultivated fields and on village greens, in scattered localities. Place of origin: most of Europe.

Nephila A genus of Arachnids of the order of True Spiders, *Araneae*. e.g. N. eremiana - A species of tropical orb weaver spiders, inhabiting the lower steppes of Central Australia. Like other members of the genus, it is a large spider which spins some of the largest webs, some cobwebs measuring up to 3.6-4.5 m (12-15 ft) across. The silk produced in these webs is incredibly strong and possesses great elasticity, and in them live commensally the smaller Silver Spiders, such as *Argyrodes* (see **Argyrodes**).

Nephrolepis A genus of plants of the Fern family, *Oleandraceae*. e.g. N. exaltata - The Swordfern, a rapidly growing, tropical plant, 45-75 cm (11/2-21/2 ft) high, with pale green pinnate fronds. It is one of the most beautiful ferns and is used as a pot plant. Place of origin: the Tropics; introduced into Europe in 1793.

Nephrops A genus of Crustaceans of the Lobster family, *Homaridae*. e.g. N. norvegicus - The Norwegian Lobster, Scampi, or Dublin Bay Prawn, a light reddish-brown, decapod crustacean, 14-18 cm in length, with long angled, pincer-limbs, and ranging from the Norwegian coast to the Adriatic. It is much smaller than the Common Lobster, and is found in deep water (see **Homarus**).

Nepthys A genus of Annelids or Segmented Worms of the family *Nephthyidae*. These are the polchaetes and are found on "clean" sandy beaches. They may be recognized by the creamy color and rather iridescent sheen of the body cuticle. There are no eyes. These are carnivorous and scavenging worms.

Nepticula A genus of Insects of the order of Butterflies and Moths, *Lepidoptera*. e.g. N. microtheriella - One of the smallest moths found in Britain, with a wingspan of 3-4 mm (0.11-0.15 in) and a body length of 2 mm (0.78 in) (see **Johanssonia**; **Stigmella**).

Neptunea A genus of marine Gastropod Molluscs of the carnivorous Sea-snail family. e.g. N. antiqua - The Spindle Shell, up to 20 cm long; the last whorl of the shell is rounded and has a wide aperture. A carnivorous animal, it feeds on other shells or barnacles. It searches its prey using its siphon, which acts as a scent organ, killing its victim and eating it by boring through the shell into the flesh with its rasp-like tongue. Distribution: the seas of Europe; native around Britain.

Neptunus A genus of Crustaceans of the order of Crabs, *Decapoda*. e.g. N. pelagines - One of the slowest-moving crabs. One specimen tagged in the Red Sea took 29 years to travel the 162.4 km (1011/2 miles) to the Mediterranean via the Suez Canal, at an average speed of 5.5 km (31/2 miles) a year.

Nereis A genus of Segmented Worms of the Nereid family, *Nereidae*. Also known as ragworms or clam worms. e.g. N. diversicolor - One of the commonest species of ragworms or nereids in western Europe, up to 10 cm (4 in) long; colored reddish, orange or green, etc.; hence the name. There are well-developed parapodia down each side of the body, giving the worm the appearance of a tattered strip of rag. The species has adapted itself to living in river estuaries, digging into the mud, and has penetrated the Baltic Sea as far Finnish waters.

Nerine A genus of plants of the Daffodil family, *Amaryllidaceae*. e.g. N. bowdenii - The Cape Colony Nerine, a handsome bulbous plant, with a stout, smooth, fleshy stem up to 60 cm (2 ft) high, terminating in large, round umbels of pale pink flowers. Place of origin: South Africa; introduced into Europe in 1889.

Nerita A genus of Molluscs of the order *Neritacea*. The species are formed only on rocky seasides on both oceanic shores of the United States. Absent around Britain. These are primitive molluscs, similar to the limpets and topshells (order *Archaeogastropoda*) (see **Haliotis**).

Neritina A genus of Molluscs of the order *Neritinacea*, similar to the genus *Nerita* (see **Nerita**).

Nerium A genus of plants of the Periwinkle family, *Apocynaceae*. e.g. N. oleander - The Rose Bay or Oleander, an evergreen shrub, up to 5 m (161/2 ft) tall; with long slender, upright branches, leathery grey-green leaves, large red, pink, peach, yellow or white flowers borne in terminal corymbs, and a poisonous milk-like sap from which is prepared oleandrin (= neriolin), a cardiac glycoside with action similar to that of strophanthin. The plant grows widespread at the edges of ponds and in the beds of streams. Place of origin: the Mediterranean littoral; introduced into Britain in 1596.

Nerophis A genus of Fishes of the family of Pipefishes and Seahorses, *Syngnathidae*. e.g. N. ophidion - The Straight-nosed Pipefish, about 15 cm long. The body is covered in bony rings, and the male has a brood-pouch on its belly. There is no caudal fin. Distribution: the seas of Europe: the Atlantic coast, the North and Baltic Seas. Native around Britain.

Neslia A genus of plants of the Cress or Crucifer family, *Cruciferae*. e.g. N. paniculata - The Ball Mustard, an annual plant, 15-80 cm high; the leaves clasp the stem and the fruits are spherical. Place of origin: southern Europe; introduced and naturalized in Britain.

Nesoryctes A genus of Mammals of the Tenrec family, *Tenrecidae*. These are the Rice Tenrecs, mole-like animals related to the tenrecs, and also living in Madagascar. The species cause damage to crops as they burrow for insects (see **Centetes**).

Nestor A genus of Birds of the Parrot family, *Psittacidae*. e.g. N. notabilis - The Kea, a distinct and aberrant parrot species, brownish-green in color and variously marked with reds and yellows. A large bird, the size of a crow, about 56 cm (22 in) long. It lives in low-level forests usually in flocks, and feeds on fruits, nectar and grubs dug out of rotten wood with its sharp, sickle-shaped beak. It also feeds on blow fly larvae living in the wool of sheep, occasionally inflicting fatal injuries on these animals by its incessant or constant pecking. It nests in tree cavities, where it lays four white eggs. Distribution: the mountainous regions of the South Island of New Zealand.

Netta A genus of Birds of the Duck, Goose and Swan family, *Anatidae*. e.g. N. rufina - The Red-crested Pochard, length 51-57 cm; in the male the bill is red and the head light-brown, in the female the cheeks and bill are white and the crown dark brown. A partial migrant and a diving bird, it is found in inland and coastal waters, such as lakes and ponds, as well as shallow and brackish waters. Distribution: South Europe; sporadically in central and eastern Europe; Central Asia.

Neuroterus A genus of Insects of the Gall Wasp family, *Cynipidae*. e.g. N. quercusbaccarum - The Spangle Gall Fly, size 2-3 mm, with sexual and asexual generations; found in oaks, where the females of the sexual generation lay eggs on the underside of leaves, inducing the formation of spangle galls, which in autumn fall to the ground, all producing asexual females. These lay eggs in the male catkin flowerbuds of the oak, inducing the formation of little

Spherical "currant" galls in which the sexual generation develops. Distribution: Europe, Asia Minor, North Africa.

Neurotrichus A genus of Mammals of the Mole family, *Talpidae*. e.g. N. gibbsii - The Shrew Mole, one of the smallest of the mole species.

Nevskia A genus of microorganisms of the family *Caulobacteriaceae*. These are stalked, rod-shaped cells, the long axis of the cell being at right angles to the axis of the stalk. e.g. N. ramosa.

Nezera A genus of Hemipterous Insects of the Shield-bug family, *Pentatomidae*. e.g. N. viridula - A bright-green terrestrial bug (section *Geocorisae*) and a widely distributed phytophagous or plant-eating pest.

Niadvena A genus of plants of the Waterlily family, *Nymphaeaceae*. These are the common American Spatterdock plants.

Nicotiana A genus of plants of the Nightshade family, *Solanaceae*. e.g. N. tabacum - The Tobacco plant, a stout viscid annual or biennial plant, up to 1.8 m (6 ft) high. The large, thin, alternate leaves are lanceolate to ovate, and their bases clasp the stem. The small pinkish flowers are borne in dense, stalked, drooping clusters. The plant contains the alkaloid Nicotine, and has been cultivated from earliest times for the tobacco prepared from its dried leaves, making it an extremely important plant in commerce. The Nicotine itself has no therapeutic use but is employed in veterinary medicine as a parasiticide and anthelmintic, and in agriculture as an insecticide. Place of origin: Tropical America; now cultivated worldwide.

Nicrophorus A genus of Insects, also called *Necrophorus* of the Carrion Beetle family, *Silphidae*. e.g. N. vespillo (= *Necrophorus* v.) - The Common Burying Beetle (see **Necrophorus**).

Nigella A genus of plants of the Buttercup family, *Ranunculaceae*. e.g.

N. arvensis - The Field Fennel-flower, an annual plant, 7-30 cm high; the flowers are pale blue and the carpels are united only below the middle. Place of origin: the Mediterranean region; naturalized elsewhere in Europe.

Nigritella A genus of plants of the Orchid family, *Orchidaceae*. e.g. N. nigra - The Common Black-orchid, a perennial plant, 5-25 cm high; the flowers are brownish black and vanilla-scented. Found in dry turf and damp alpine meadows up to 2,300 m. A widespread species. Place of origin: Europe.

Niphargus A genus of Crustaceans of the Freshwater Shrimp family, *Gammaridae*. These are the blind Well Shrimps, consisting of about 85 different species living in wells and caves. e.g. N. puteanus - Size of male 30 mm, female 18 mm; the body is transparent and it has a long forked tail. Lives in caves and wells.

Nipponia A genus of Birds of the Ibis and Spoonbill family, *Threskiornithidae*. e.g. N. nippon - The Nippon Ibis, a rare Ibis species from Japan, believed to have a total population of less than 100. A wading bird, allied to the stork, about 60 cm (2 ft) tall with a long, downward-curving bill.

Nippostrongylus A genus of Nematodes or Roundworms of the Hookworm family, *Ancylostomatidae*. e.g. N. muris - A hookworm species found in rats.

Niptus A genus of Insects of the Spider Beetle family, *Ptinidae*. e.g. N. hololeucus - The Golden Spider Beetle or Shining Niptus, up to 4 mm long; a general scavenger species, with a narrow head and thorax, and a very wide abdomen. The body is covered with dense, brass-colored hairs. Distribution: Europe; native in Britain.

Nitrobacter A genus of microorganisms of the family *Nitrobacteraceae*. Rod-shaped cells which oxidize nitrites to nitrates. e.g. N. agilis.

Nitrocystis A genus of microorganisms of

the family *Nitrobacteraceae*, which oxidize nitrites to nitrates. e.g. N. micropunctata.

Nitrosocystis A genus of microorganisms of the family *Nitrobacteraceae*, which oxidize ammonia to nitrite. e.g. N. coccoides.

Nitrosogloea A genus of microorganisms of the family *Nitrobacteraceae*, which oxidize ammonia to nitrite. e.g. N. membranacea.

Nitrosomonas A genus of microorganisms of the family *Nitrobacteraceae*, which oxidize ammonia to nitrite more rapidly than other genera of the same family. e.g. N. europaea.

Nitrosospira A genus of microorganisms of the family *Nitrobacteraceae*, which slowly oxidize ammonia to nitrite. e.g. N. antartica.

Nocardia A genus of microorganisms of the family *Actinomycetaceae*, formerly classified under the generic names *Micromyces* or *Streptothrix*. These are gram-positive, true bacteria, superficially resembling fungi, with branching filaments. e.g. N. gardneri - A microorganism producing an antimicrobial substance, proactinomycin, acting strongly against gram-positive bacilli. N. lurida - A filamentous microorganism producing two antimicrobial substances, ristocetin A and B. Other species, such as N. asteroides cause pulmonary lesions, brain abscesses, and lesions in other organs in man.

Nochtia A genus of Nematodes or Roundworms (class *Nematoda*). e.g. N. nochti - A species of nematode worms found in the stomach of Javanese monkeys.

Noctiluca A genus of Protozoan animals of the order of Dinoflagellates, *Dinoflagellata*. These are marine organisms with sculptured cellulose cell walls. They have two flagella, the longitudinal flagellum which is used for propulsion and the equatorial flagellum which rotates the animal, serving possibly also in orientation. *Noctiluca* is well known as an organism causing phosphorescence of the sea at night. It is a large and aberrant dinoflagellate, up to 2 mm in diameter and with a permanent tentacle formed for the hypocone. Phosphorescent granules in rows form a close mesh in the cytoplasm. The discharge of light occurs mainly when the water is disturbed.

Noctua A genus of Insects of the Noctuid or Owlet-moth Butterfly family, *Noctuidae*. e.g. N. pronuba - The Large Yellow Underwing, size 3-5 cm; the hindwings are yellow with a transverse black band at the edges, and the forewings are mottled brown. The caterpillars feed on sorrel, cabbage, auricula and other plants. Distribution: Europe; native in Britain.

Noemacheilus A genus of Bony Fishes of the Loach family, *Cobitidae*. e.g. N. barbatulus - The Stone Loach, length 10-18 cm; primarily a nocturnal fish with a dark-marbled, cylindrical body and 6 barbels around the mouth. Lives on the bottom in flowing waters, ponds and lakes. Distribution: Europe.

Noguchia A genus of microorganisms of the family *Brucellaceae*. These are small gram-negative rods with peritrichous flagellae, found in the conjunctiva of man and animals. e.g. N. granulosis - A species at one time thought to be the etiological agent of trachoma.

Nomeus A genus of Fishes (subphylum *Vertebrata*). e.g. N. gronovii - The Portuguese Man o'War Fish, a small marine species about 7.5 cm (3 in) long, common in the Gulf of Mexico where it hides amongst, and is unaffected by, the poisonous tentacles of such jellyfishes as the Portuguese Man o'War (*Physalia physalis*) (see **Physalia**).

Nonnea A genus of plants of the Borage family, *Boraginaceae*. e.g. N. pulla - The Wrinklenut, a perennial plant, 12-40 cm high; with soft-hairy entire leaves and dark reddish-brown flowers, also rarely red, yellow or white. Grows at the edges of fields or by roadsides. Place of origin: southern Europe. Naturalized or a casual in Central Europe.

Nosema A genus of Protozoan animals of the order *Microsporidia*. e.g. N. apis - A species causing the nosema disease (Gr. nosema = a sickness) of bees, known as nosematosis. They have non-motile spores, each containing two or more polar filaments, small sacs with curled, spring-like structures within. They are parasitic to honey bees, where a debilitating and often fatal disease (nosematosis) results from infection.

Nosopsyllus A genus of Insects of the order of Fleas, *Siphonaptera*. e.g. N. fasciatus - The Common Rat Flea of North America and Europe, being a vector of murine typhus and probably of plague, formerly known as *Ceratophyllus* fasciatus. Also a vector of *Trypanosoma* lewisi, a species of Trypanosomal sporozoa, parasitic in the blood of the rat.

Notechis A genus of Reptiles of the suborder of Snakes, *Ophidia*. e.g. N. scutatus - The Tiger Snake, the most venomous land snake, found in southern Australia and Tasmania; up to 1.5 m (5 ft) long; its bite causes vomiting and excessive sweating followed by respiratory failure, usually resulting in death in 2 to 3 hours.

Notemigonus A genus of Fishes (subphylum *Vertebrata*). e.g. N. crysoleucas - The Golden-shiner fish.

Notharchus A genus of Birds of the Puffbird family, *Bucconidae*. e.g. N. macrorhynchus - The White-necked Puffbird, about 25 cm (10 in) long; the head is disproportionately large, the plumage is a drab grey or brown, and the tail is short. The overall effect is that of a squatt, puffy bird, hence the common name. Habitat: the tropical forests of Central and South America.

Nothobranchius A genus of Fishes of the Killifish suborder, *Cyprinodontei*. e.g.

N. guentheri - A species found in temporary ponds, drainage ditches and even in water-filled footprints of large animals. The eggs are laid in mud at the bottom of the water. When the pool or ditch dries up, the eggs aestivate or fall into a state of torpor, until the next wet season, when they hatch, grow at great speed and spawn, dying in the next drought. These are the shortest-lived fishes known, living about 8 months in the wild state.

Nothocrax A genus of Birds of the Curassow family, *Cracidae*. e.g. N. urumutum - The Urumutum Curassow, a glossy black colored bird, with ovoid red, bare patches on its head, and head-feathers curled at the ends. A gregarious species, it nests in trees. Distribution: British Guiana in South America.

Nothofagus 1. A genus of Homopterous Insects of the family *Peloridiidae*. These are the *Coleorrhyncha*, homopterous insects found mainly in forests of the southern beech in Australasia and South America. 2. A genus of Trees of the Beech family, *Fagaceae*. e.g. N. antarctica - The Antarctic Beech, Southern Beech or Nirre, a deciduous, evergreen tree, up to 15 m (50 ft) tall, but often low and straggly on exposed mountain sites; with simple alternate leaves; the minute male flowers are crowded in slender catkins, and the female flowers have a tiny nut shape with a red tassel, growing singly or in small groups. The fruit is a one-seeded nut, surrounded by a capsule. Place of origin: southern Chile and the southern hemisphere.

Notiophilus A genus of Insects of the Ground Beetle family, *Carabidae*. e.g. N. biguttatus - Length 5 mm; a very common small ground beetle species of European forests; it has strikingly large eyes and is brownish-black in color. Distribution: Europe.

Notodelphys A genus of Copepod Crustaceans of the Cyclopod family, *Notodelphyidae*. All members of this family are parasites of sea-squirts (class *Ascidiacea*), and the females have an internal brood-pouch, distended into a globular shape. They cling to the host by means of a hook at the end of the antenna (see **Botryllus**).

Notodonta A genus of Insects of the Puss Moth or Prominent Butterfly family, *Notodontidae*. e.g. N. zicsac (= N. ziczac) - A Puss Moth species; its wings are serrated at the edges; the red caterpillars have tooth-like projections on their backs, and adopt curious postures. They can be found in young poplar trees.

Notoedres A genus of Arachnids of the order of Mites, *Acarina*. e.g. N. cati - An itch mite causing a fatal mange or itch disease in cats; may temporarily infest man.

Notonecta A genus of Heteropterous Insects of the Water Boatman or Backswimmer family, *Notodectidae*. e.g. N. glauca - The Water Boatman, size 13-16 mm; also known as the Backswimmer, an insect which moves upside-down with powerful strokes of its very long and hairy hindlegs, and carries a store of air on the underside of the abdomen and beneath the wings. Found in calm freshwater, in ponds, pools, paddles and amongst vegetation. It lays its eggs in or on aquatic plants. Distribution: worldwide; common in Europe, the Caucasus, North Africa. Native in Britain.

Notornis A genus of Birds of the Coot, Moorhen and Rail family, *Rallidae*. e.g. N. mantelli - The Takahe or Notornis, a large flightless species, related to the Moorhen or Common Gallinule and believed to be extinct since 1855, until in 1948 a small group of these birds was found living in a remote valley in South Island, New Zealand. It is the size of a large domestic hen and has an almost parrot-like beak. Its plumage is purple-blue with a green sheen, and the frontal shield is a brilliant red.

Notoryctes A genus of Mammals of the Marsupial Mole family, *Notoryctidae*. e.g. N. typhlops - The Marsupial Mole, a mole-like animal with a pointed snout, short close fur and powerfully clawed front feet. Its eyes are buried beneath closed eyelids, and it feeds on insects. It moves underground, digging a passage with its muzzle and forepaws and closing it behind with its hind paws; thus no permanent tunnel is left.

Nototrema A genus of Amphibians of the Tree Frog family, *Hylidae*. e.g. N. marsupiatum - The Pouched Frog, a tree-frog species found in the forests of Ecuador and Peru in South America. The female has a pouch in its back which is filled with eggs at breeding time, inserted by the male during mating. The eggs develop in this receptacle and are finally discharged as newly-hatched tadpoles into the water.

Notropis A genus of Fishes of the Cyprinid Carp, Minnow, Tench and allied family, *Cyprinidae*. e.g. N. cornutus - The Common Shiner, a North American freshwater fish found in almost every brook east of the River Missouri.

Noturus A genus of Fishes of the North American freshwater Catfish family, *Ameiuridae* (= *Ictaluridae*). These are the Mad-toms, small North American catfishes, with their pectoral spines armed at their bases with a poison sac; the effect of a sting is painful but not dangerous to man.

Nucella A genus of Molluscs of the order of Conches, Oysterdrills and Whelks, *Neogastropoda*. These are the Dogwhelks, of the family *Nassariidae*, small scavenging, carnivorous, Sea-snail molluscs, intertidal marine animals living on acorn barnacles, mussels, and so on. They have an eversible proboscis. e.g. N. lapillus - The Dog Whelk, size of shell 3-4 cm; shell ovate, often with black or yellow bands; the lip is thick and dentate. Distribution: Europe; native around Britain.

Nucifraga A genus of Birds of the Crow, Jay and Magpie family, *Corvidae*. e.g. N. caryocatactes - The Nutcracker, a species of Crows, 32 cm long; with a brown plumage heavily spotted with white, a square tail, and prominent white under the tail-coverts. A predominantly resident bird found in mountain forests, and nesting in conifers. Distribution: Europe: the Alps, Balkan and Carpathian mountains; Siberia; Central Asia.

Nucula A genus of Molluscs of the Nut Shell family, *Nuculidae*. e.g. N. nucleus - The Common Nut Shell; length 12 mm, height 9 mm, thickness 5.6 mm; surface of the shell is yellowish-white, the lower margin is finely toothed, the ventral shell edge is crenate, the inside walls have a nacreous gloss and the valve is concentrically striped. Found in abundance in coarse sediments on sea bottoms down to depths of 150 m (500 ft). Distribution: European coast of the Atlantic Ocean from Portugal to Norway, Mediterranean Sea, Black Sea, west and southeast coast of Africa. Native around Britain.

Nuculana A genus of Molluscs of the Nut Shell family, *Nuculanidae*. e.g. N. minuta - The Beaked Leda or Beaked Nuculana; length of shell 14 mm, broadly rounded at the front and narrow at the rear. The valve has 2 edges. A filter feeder, the Beaked Leda feeds through a small mouth as it burrows, the toothed edge of the foot grasping fragments of sand and pulling the shell along. Found on sea bottoms of muddy sand and gravel at depths of 10-180 m, burrowing well below the low-water mark. Distribution: widely distributed in the northern hemisphere on the Atlantic coast of Europe, south to the English Channel. Reaches California, Japan, and Nova Scotia.

Numenius A genus of Birds of the Curlew, Sandpiper, Snipe, and Woodcock family, *Scolopacidae*. e.g. N. americanus - The Long-billed Curlew of North America, the largest of the curlews, with a length of 60 cm (2 ft). It is a migrant bird and breeds at high latitudes, migrating to the far north in summer.

Numida A genus of Birds of the Guinea-fowl family, *Numididae*. e.g. N. meleagris - The Common, Domestic or Helmet Guinea-fowl, about 55 cm (1 ft 10 in) long, a pearly-green plumaged bird with red wattles, which first originated from West Africa and the Cape Verde Islands. These birds were later domesticated in Greek and Roman times, when they were known as African fowls, but were later allowed to revert to the wild state. The same species exists in Central America, where it became established from domesticated stock originally brought from Europe. Guinea-fowls are easy to rear; the eggs make good eating and the flesh has the same taste as pheasant meat.

Nuphar A genus of plants of the Waterlily family, *Nymphaeaceae*. e.g. N. lutea - The Yellow Waterlily, Brandy-bottle, European Cow-lily or Nenuphar, a perennial plant, up to 3 m tall; the leafblade is 10-30 cm long and the cup-shaped, yellow flowers are 4-6 cm wide. The fruit has an unpleasant, alcoholic smell and matures above the waterline. Grows widely in still and slow-flowing freshwater. Place of origin: Europe, northern Asia, North America. Native in Britain.

Nuttallia A genus of small protozoan parasites found in the red blood corpuscles of dogs and horses. e.g. N. equi (also called *Babesia* caballi; B. equi) - A species causing hemoglobinuric fever of horses in South Africa; probably transmitted by the tick, *Rhinicephalus* everti (see **Boophilus**, also **Babesia**).

Nyctalus A genus of Mammals of the Long-eared, Vespertilionid Bat family,

Vespertilionidae. e.g. N. noctula - The Common Noctule or Great Bat, the largest and fastest flying bat, commonly found in Britain; 8 cm long; with a wingspan of 38 cm (15 in). The tragi are short and swollen; it is a fast and high flier, and has been clocked at up to 50 km/h as it sweeps over fields, rivers and woods, hunting for insects. Distribution: Europe; former Asiatic U.S.S.R., northern China and northwest Africa. Native in Britain.

Nyctanassa A genus of Birds of the Bittern, Egret and Heron family, *Ardeidae*. e.g. N. violacea - The Yellow-crowned Night Heron, a species of Night-herons of North and South America. It lives in colonies in nests set up high in trees.

Nyctea A genus of Birds of the Typical Owl family, *Strigidae*. e.g. N. scandiaca - The Snowy Owl, 50-60 cm long, a rare Arctic bird with pure white plumage and a varying number of dark, barred markings. It lives and nests in the Arctic tundra, where it feeds on lemmings, voles and Arctic hares. A migrant bird, it inhabits the Arctic region, sometimes wintering in Scandinavia, Canada, the United States, and Russia.

Nyctereutes A genus of Mammals of the Dog, Fox and allied family, *Canidae*. e.g. N. procyonoides - The Racoon-like Dog or Racoon-Dog; length 60 cm (2 ft) with a 15 cm (6 in) long tail. Lives in broadleaved woodlands, scrub country and in cultivated steppes. Usually active at night. Also known as the "Japanese Fox". Distribution: native to the Amur and Ussuri region of Russia; China and Japan. In the 1930s it was introduced as a fur animal into Central Europe.

Nyctibius A genus of Birds of the Potoo or Wood-nightjar family, *Nyctibiidae*. The five species of this family are distributed from southern Mexico to Brazil. They are solitary birds and lay a single white egg, both parents taking turns at incubating.

Nycticebus A genus of Mammals of the Loris family, *Lorisidae*. e.g. N. coucang - The Slow Loris, a loris species, the size of a small cat, with short, thick limbs, no tail and vestigial index fingers. A very slow creature, creeping about the trees at night, hunting for insects. Distribution: India, Sri Lanka and southeast Asia.

Nycticorax A genus of Birds of the Bittern, Egret and Heron family, *Ardeidae*. e.g. N. nycticorax - The Black-crowned Night Heron, 60 cm (2 ft) long; a widely distributed species of night herons breeding in both Americas, Eurasia, Africa and the East Indies. It has a black headcap and back, with a white forehead and yellowish legs. A migrant bird, nesting is colonial, in stick nests set high up in trees.

Nyctidromus A genus of Birds of the Nightjar family, *Caprimulgidae*. e.g. N. albicollis - The White-necked Nighthawk, a nightjar species inhabiting Central and South America. A typical nightjar or goatsucker, it belongs to the subfamily, *Caprimulginae*.

Nyctotherus A genus of infusorian microparasites, so-called because they are found in infusions after exposure to air. e.g. N. gigantus - A large species found in the human intestine in Germany.

Nymphaea A genus of plants of the Waterlily family, *Nymphaeaceae*. e.g. N. alba - The White Waterlily, a perennial freshwater plant, up to 3 m tall; grows in still and slow-flowing water. The leaves have a deep and wide basal cleft. Place of origin: widespread in Europe; native in Britain.

Nymphalis A genus of Insects of the Tortoiseshell Butterfly or Moth family, *Nymphalidae*. e.g. N. antiopa - The Mourning Cloak or Camberwell Beauty, 5 cm (2 in) long; a species of nymphalid butterflies familiar to Europe and North America. The wings are a dark velvet-brown with yellow, more rarely white, margins. It lays its eggs on birches, willows and poplars.

Nymphicus A genus of Birds of the Cockatoo family, *Kakatoidae* (order of Parrots, *Psittaciformes*). e.g. N. hollandicus - The Australian Cockatiel, a slender bird, about 33 cm (13 in) long; light greyish-brown in color, with an orange spot behind the eye and a pale yellow crest. A gregarious species, it lives in large flocks, nesting in hollows. Can be tamed, and taught to repeat words like other species of the order. Distribution: Australia and the East Indies.

Nymphoides A genus of plants of the Bogbean family, *Menyanthaceae*. e.g. N. peltata - The Fringed Waterlily or Yellow Floating-heart, a perennial plant, up to 1 m (39 in) high; the leaves are simple, circular and deeply heart-shaped at the base. The long stalks float above the water. A local plant, it is found in ponds and slow rivers. Place of origin: central and southern Europe (native in Britain); northern and western Asia, the Himalayas and Japan. Has become naturalized in North America.

Nymphon A genus of Arthropods of the class of Sea Spiders, *Pycnogonida*. These are the pycnogonids, bizarre marine organisms. e.g. N. charcoti - A pycnogonid species with long legs, compressed body and minute abdomen. It has a large proboscis, through which it feeds on sedentary celenterates. Distribution: the deep waters of the South Atlantic around South Georgia.

Nymphula A genus of Insects of the Butterfly and Moth family, *Pyraustidae*. The aquatic larvae of these moths are equipped with well-developed gills.

Nyssa A genus of Trees of the family *Nyssaceae*. e.g. N. sylvatica - The Sourgum, Tupelo, Black Gum, or Pepperidge, a deciduous tree up to 30 m (100 ft) tall, with a tapering trunk; growing in swamps or damp ground. The male and female flowers are borne on separate trees; the fruits are about 1.2 cm (1/2 in) long and ripen to dark bluish-black in autumn. Place of origin: eastern North America. Cultivated as an ornamental in the eastern states of the U.S. and in Europe.

Nyssorhynchus A genus of Insects of the Mosquito family, *Culicidae*. These are anopheline mosquitoes, several species of which act as carriers of the malarial parasite in Central and South America.

O

Obelia A genus of Sea-firs, Thecate Hydroid animals, or Hydrozoan Polyps of the family *Campanulariidae*. e.g. O. geniculata - A Sea Fir celenterate, up to 4 cm high, and found commonly on oar-weeds, with single polyps in open goblet-shaped thecae arranged alternately, giving a zigzag appearance and releasing free medusae, each of which has 16-20 tentacles. Distribution: the Atlantic coasts and the North Sea. Occasionally also in the Mediterranean.

Oberea A genus of Insects of the Long-horned or Longicorn Beetle family, *Cerambycidae*. e.g. O. oculata - Length 15-21 mm; found on various flowering plants; the larvae develop in the twigs of roses, hawthorn and fruit trees. The infected twigs die back from the tips and the bark falls off. Distribution: Europe, Transcaucasia, Asia Minor, North Africa.

Oceanites A genus of Birds of the Fulmar, Petrel, Sea-Bird, and Tubenose family, *Procellariidae*. e.g. O. oceanicus - Wilson's Petrel, an abundant, migrant sea-bird which flies to the North Atlantic every summer from its breeding-grounds at the edge of the Antarctic. It breeds in crevices and holes in rocks, laying a single egg.

Oceanodroma A genus of Birds of the Fulmar, Petrel, Sea-bird, and Tubenose family, *Procellariidae*. e.g. O. leucorhoa - Leach's Petrel, 20 cm long, a migrant, marine bird, with a forked tail and a white rump. Like other members of the family, the nostrils are enclosed in tubes along the top of the bill. It breeds along the coasts of Britain and Iceland.

Ocenebra A genus of Gastropod Molluscs of the family *Muricidae*. e.g. O. erinacea - The Sting Winkle or Oyster Drill; the shell is about 60 mm high, thick-walled, yellowish white to dark brown, and the oval mouth terminates in a siphonal canal. The surface is covered with large spines, hence its Latin name "erinaceosus", meaning hedgehog. Found on hard sea bottoms at shallow depths. It is collected for its edible, fleshy foot, and the ornamental shells are sold as souvenirs. Distribution: Mediterranean Sea, Atlantic coast from western Scotland to the Azores and Madeira.

Ochlerotatus A genus of Insects of the two-winged Fly or Mosquito family, *Culicidae*. The species are said to transmit African horse sickness.

Ochlodes A genus of Insects of the Skipper Moth or Butterfly family, *Hesperiidae*. e.g. O. veneta - The Large Skipper Moth, size 14-17 mm; a brown-colored species which hibernates. The caterpillars feed on grasses. Distribution: Eurasia.

Ochotona A genus of Mammals of the Pika, Mouse-hair or Piping Hare family, *Ochotonidae*. These are the pikas, mouse-hairs, calling hairs, or rock rabbits. They resemble rabbits but have short ears and are no larger then guinea-pigs. e.g. O. pusilla - The Steppe Pika, 15 cm long, with a rudimentary tail. It lives in colonies, digging burrows about 1 m long, which terminate in a living chamber. Distribution: chiefly Asia (eastern Russia and China); also in eastern Europe.

Ochrobium A genus of microorganisms of the family *Siderocapsaceae* - Ellipsoidal to rod-shaped cells partially surrounded by a thickening impregnated with iron. e.g. O. tectum.

Ochromonas A genus of Protozoan animals of the Chrysomonad order, *Chrysomonadina*. This organism cannot synthesize vitamin B12 and also cannot grow without it; it is used therefore as a guage in estimating the amount of this vitamin present in various foods. Thus the extent of growth of a culture on food media affords a measure of the vitamin

B12 content in the food. The organisms have the appearance of a small round cell with two stubby flagella, one slightly longer than the other.

Ochromyia A genus of Insects of the Two-winged order of Flies, *Diptera*. e.g. O. anthropophaga - A species of flies found in Senegal, whose larva, the cayor worm attacks humans.

Ocimum A genus of plants of the Mint or Thyme family, *Labiatae*. e.g. O. basilicum - The Sweet Basil, an erect, herbaceous, aromatic, annual plant about 45 cm (11/2 ft) high. The leaves are bright green, lanceolate and short-stemmed; the flowers are white, pink, or lilac and are borne in terminal spikes. Grows in fresh, light and fertile soil. The plant is sacred to the Hindu religion and is used in homeopathic medicine. Place of origin: tropical America and the Pacific islands; introduced into Britain in 1548.

Ocinebra A genus of Molluscs of the order of Oysterdrills, *Neogastropoda*. These are pests on European oyster beds, boring through the shells of bivalve molluscs.

Ocotea A genus of trees of the Laurel family, *Lauraceae*. From the wood is prepared Brazilian Oil or Sassafras Oil (see **Sassafras**).

Octolasium A genus of Annelids or Segmented Worms of the Earthworm family, *Lumbricidae*. e.g. O. lacteum - An earthworm species, 30-180 mm long and 2-8 mm wide; generally colored milky-grey, less often reddish-brown. Found in all types of soil, but not in sand. Distribution: native to Europe; introduced to other continents.

Octomitus A genus of minute flagellate protozoa. e.g. O. hominis - A species found in the human intestine.

Octomyces A genus of Fungi or Molds. e.g. O. etiennei - A species isolated from a severe pleuropulmonary infection.

Octopus A genus of marine Cephalopod Molluscs of the Octopus family, *Octopodidae*. Also known by other generic names, such as *Hapalochlaena* and *Paroctopus*. e.g. O. punctatus (= *Paroctopus* apollyon) - The largest of the 150 recorded species of octopods. Found in the North Pacific, it regularly exceeds 4 m (13 ft 1/2 in) in radial spread and 18 kg (39.68 lb) in weight. O. rugosus (= *Hapalochlaena* lunulata) - A blue-ringed octopod of the Indo-Pacific region. The most venomous cephalopod in the world. The fast-acting neurotoxic venom carried by these small molluscs (with a radial spread of 101-152 mm or 4-6 in) is so potent, the amount injected through the horny beak in one bite is said to be enough to kills even people. Distribution: The Indo-Pacific; especially around these as off Australia. O. vulgaris - The Common Octopus; size of head 20-30 cm, with 8 long arms up to 90 cm (3 ft) long, each with two rows of suckers. The body color can change rapidly. A common species, found along the coast of the Mediterranean Sea, and considered a favorite food throughout the region. It extends north as far as the coast of Britain, especially after a mild winter.

Ocyphaps A genus of Birds of the Dove and Pigeon family, *Columbidae*. e.g. O. lophotes - The Crested Pigeon, an Australian species, with a pointed crest like that of a lapwing.

Ocypode A genus of Crustaceans of the Crab family, *Oxypodidae*. These are the Ghost Crabs, also called the Racing Crabs, because they can run sideways at high speeds across the tropical beaches.

Odobaenus See **Odobenus**.

Odobenus A genus of Mammals, also known by the generic name, *Odobaenus*, of the Walrus family, *Odobenidae* or *Odobaenidae*. e.g. O. rosmarus (= *Odobaenus* r.) - The Atlantic Walrus; it resembles a true seal in having no external ears. About 3-3.6 m (10-12 ft) long, it weighs up to 1,200 kg (3,000 lb). The female is much smaller. This species is found in Arctic waters from the Kara Sea westwards as far as Labrador and Hudson Bay.

Odocoileus A genus of Mammals of the Deer family, *Cervidae*. e.g. O. virginianus - The White-tailed or Virginian Deer, length 180 cm with a 30 cm long tail; the summer coat is reddish-brown, the winter coat is grey, and the antlers (in the male) are many-tined and bend upwards and forwards. The female has no antlers. Distribution: the New World; introduced from America it has become naturalized in Finland.

Odonestis A genus of Insects of the Emperor or Giant Silkworm Moth family, *Saturniidae*. e.g. O. pruni - A rare European Silkworm moth, orange in color with white specks in the middle of its head. Found in summer on alder, birch, blackthorn and lime trees, as well as other deciduous trees and bushes.

Odontaspis A genus of Fishes, also known by the generic name, *Carcharias*, of the Sand Shark family, *Carchariidae*. e.g. O. littoralis (= *Carcharias* l.) - The Common Atlantic Sand Shark (see **Carcharias**).

Odontaster A genus of Echinoderm animals of the class of Starfishes, *Asteroidea*. e.g. O. validus - An Antarctic Starfish species with great longevity (up to 100 years).

Odontites A genus of plants of the Figwort or Snapdragon family, *Scrophulariaceae*. e.g. O. lutea - The Yellow Bartsia, an annual plant, 15-50 cm high with yellow flowers. Grows in grassy places and in scrub. Place of origin: South and Central Europe.

Odontogadus A genus of Fishes of the Cod family, *Gadidae*. e.g. O. merlangus - The Whiting, a commercially valuable fish, 50 cm long; the lateral line is bent below the second dorsal fin, and it has a dark spot at the base of the pectoral fins. A marine species, it is widely distributed in the Baltic Sea and North Atlantic Ocean.

Odontoglossum A genus of plants of the Orchid family, *Orchidaceae*. e.g. O. grande - A species from Guatemala; it has oblong, dark green leaves and from 4-7 orange-yellow, banded flowers on each erect stem. There are also cultivated forms with different-colored flowers.

Odontomyia A genus of Insects of the Soldier-Fly family, *Stratiomyiidae*. Found in damp areas. The larvae are aquatic and are provided with a long respiratory tube by which they hang suspended from the water surface.

Odynerus A genus of Insects of the Wasp family, *Eumenidae*. e.g. O. spinipes - A wasp species, 9-12 mm long. The nest is a small chamber with a tube cemented of bits of earth round the entrance, which the female excavates by a slender thread from the chamber ceiling. Distribution: all Europe, Asia Minor, Siberia, North Africa.

Oecanthus A genus of Insects of the True Cricket family, *Gryllidae*. These are the True Crickets, delicate, nocturnal, arboreal insects that lay their eggs in the pith of plant stems, unlike most gryllids which deposit them in the ground.

Oeciacus A genus of Insects of the order of True Bugs, *Hemiptera* or *Hetoroptera*. e.g. O. hirundinis - The Swallow Bug, a pale yellow, hairy species found in birds' nests. Does not infest man. Distribution: North America.

Oecophylla A genus of Insects of the order of Ants, Bees and Wasps, *Hymenoptera*. e.g. O. smaragdina - An Indian Ant species; it builds its nest in trees. Several worker-ants first pull the edges of two leaves together; after this other workers bring up larvae which produce silken threads used to sew the leaves together.

Oedemera A genus of plants of the Beetle family, *Oedemeridae*. e.g. O. podagrariae - A species of beetles, 8-13 mm long; the elytra diverge at the back to reveal the folded membranous wings beneath. The males have stout hind femora. Found in summer on flowers. Distribution: much of Europe, Asia Minor, Siberia.

Oedipoda A genus of Insects of the Short-horned Grasshopper or Locust family, *Acrididae*. e.g. O. caerulescens (= O. coerulescens) - The Blue-winged Grasshopper; length 15-28 mm; the head is yellowish-brown and the hind wings are usually blue with a broad, black transverse band before the apex. Strongly resembles other species, such as those of the genus *Sphingonotus*. Found in dry places, hillsides, meadows, steppes and pastureland. Distribution: Central Europe (absent in Britain), Asia Minor, the Middle East and North Africa (see **Sphingonotus**).

Oenanthe A genus of Birds of the Babbler, Thrush and Old World Warbler family, *Muscicapidae* (Nightingale, Robin and Thrush subfamily, *Turdinae*). e.g. O. oenanthe - The Wheatear, a mountain bird, 15 cm (6 in) long; an insectivorous species which lives in stony, rocky or sandy areas such as quarries and sandpits, making its nest in a hole in the ground. It has a white rump, with white sides to its black tail. A migrant, it winters in tropical Africa. Distribution: all Europe, Norht America, Greenland.

Oeneis A genus of Insects of the Browns family, *Satyridae*. e.g. O. aëllo - The Glacier Butterfly, 4.8 cm long; the forewings usually have two eye-spots, the hindwings one. Found abundantly in alternate years on edges of snow fields. Distribution: Europe, mainly in the Alps.

Oenothera A genus of plants of the Wilow-herb family, *Oenotheraceae* (= *Onagraceae*). Some species are also classified under the generic name, *Hartmannia*. e.g. O. biennis - The Common Evening Primrose, a biennial herb with erect stems, up to 90 cm (3 ft) high, and hairy, lanceolate, radical leaves and ovate, upper leaves. The fra-

grant yellow flowers are long and funnel-shaped. Place of origin: Eastern and North America. It has become widely naturalized in Europe, including Britain, and also in New Zealand. Introduced into Europe in 1612. O. speciosa (= *Hartmannia* s.) - The Showy Evening Primrose or White Sundrop, a perennial plant with horizontal, leafy stems 10-15 cm (4-20 in) long, smooth leaves and pendulous, white flowers which become pink with age. Place of origin: south Central United States to Mexico; introduced into Europe in 1821.

Oesophagostomum A genus of nematode worms (class *Nematoda*), of the family *Strongylidae*, found in the intestines of various animals. e.g. O. apiostomum (= Oe. a.) - A species found in monkeys and occasionally in man in Africa and Java.

Oestrus A genus of Insects of the Botfly or Warble Fly family, *Oestridae*. e.g. O. hominis - A bot-fly species whose larvae occasionally infest man. O. ovis - The Sheep Bot-fly, 10-12 mm long; the body is colored brown, the end of the abdomen is hairy and checkered, and it pupates in the ground. It deposits its larvae in the nostrils of the sheep, sometimes causing obstruction in the nasal and frontal sinuses (see **Cephenomyia**).

Ogocephalus A genus of Fishes of the Batfish family, *Ogocephalidae*. e.g. O. vespertilio - The Longnose Seabat or Batfish of the West Indies; it has a small, bony mouth and an angling spine, hidden under along snout. It grows to a length of 20-30 cm (8-12 in).

Oiceoptoma A genus of Insects of the Carrion Beetle family, *Silphidae*. e.g. O. thoracicum - Length 12-16 mm; easily identified by its red scutum; very common on dead animals, decaying plant matter such as old mushrooms, and in excrement. Distribution: Europe, Siberia, Japan.

Oidium The former generic name for the

fungi, now called *Candida*. e.g. O. albicans (= *Candida* a.) (see **Candida**).

Okapia A genus of Mammals of thanimal, standing 1.5 m (5 ft) at the shoulders, with a sloping back and shorter neck and limbs than the latter, suggesting what would seem a half-formed Giraffe. It has horizontal stripes on the hind quarters and front legs, contrasting with the rest of its hazel coat, but has no relationship with the zebra. Only the males have horns. It is regarded as a primitive member of the Giraffe family. Distribution: the dense rain forests of the eastern Congo.

Olae A genus of Trees of the Olive family, *Oleaceae*. e.g. O. europaea - The Olive tree, an evergreen tree up to 12 m (40 ft) tall; the leaves are leathery and silver-hairy beneath, and the flowers are white and scented. From the ripe fruit is expressed Olive Oil (see below). A woody plant, it grows wild on mountains, and is widely cultivated for its food value. Used in medicine. Place of origin: the Mediterranean region. O. fragrans (= *Osmanthus* f.) - The Sweet Olive tree (see **Osmanthus**). The Olive is of great economic importance. The green, unripe fruit is pickled in salt or vinegar and eaten as a savory. Olive Oil (= Oleum Olivae; Ol Oliv; Ol Olivea; Azeite) is obtained by pressing the ripe fruit. The best edible oil, the almost colorless virgin oil is obtained by lightly pressing the fruit in the cold. A second, rather stronger pressing produces a clear yellow, edible oil which is also used in medicine as a nutrient, demulcent and a mild laxative. A third pressing under heat, produces the so-called tree oil which is used as a fuel and in the manufacture of soap, and externally in medicine, as an emollient in the preparation of linaments, ointments and plasters.

Olibrus A genus of Insects of the Shining Flower Beetle family, *Phalacridae*. e.g. O. aeneus - Length 2.5-2.8 mm; shining dark-blue to black; common on flower-

ing camomile plants (genus *Matricaria*), the food plant of its larvae. Distribution: Europe, Siberia.

Oligomyrmex A genus of Insects of the Ant family, *Formicidae*. e.g. O. bruni - The worker minor of Sri Lanka, the world's smallest ant; only 0.8-0.9 mm (0.031-0.035 in) in length.

Olindias A genus of Hydrozoan animals of the order *Limnomedusae*. e.g. O. phosphorica - A free-swimming medusal celenterate form, with a characteristic radial canal system and bell formation, and numerous tentacles in sets round the bell margin. Some of these tentacles bear suckers, enabling the medusae to walk on seaweeds. Distribution: shallow tropical waters.

Oliva A genus of Molluscs of the predacious, carnivorous order of Oysterdrills, *Neogastropoda*. These are the smooth-shelled Olive and Pencil shells; they move just below the surface of the sand.

Ollulanthus A genus of Nematodes or Roundworms (class *Nematoda*). e.g. O. tricuspis - The smallest known species of parasitic roundworms, found in the stomach walls of cats. Adult females measure only 1 mm (0.039 in) in length.

Ommastrephes A genus of Cephalopod Molluscs, better known by the generic name, *Dosidicus*, of the Squid family, *Loliginidae*. e.g. O. gigas (= *Dosidicus* g.) (see **Dosidicus**).

Omphalodes A genus of plants of the Borage family, *Boraginaceae*. e.g. O. scorpioides - The Small-flowered Navelwort, an annual plant, 10-30 cm high; the flowers are blue, with yellow throat-scales. Grows in woods, scrubland and waste land. Place of origin: South and Central Europe.

Onchidium A genus of plants of the Orchid family, *Orchidaceae*. e.g. O. varicosum - An epiphytic orchid from Brazil, with conical, ribbed pseudobulbs, broadly lanceolate leaves from 15-23 cm (6-9 in) long, and yellowish-green and brown flowers borne on inflorescences. The large, lobed labellum is bright yellow.

Onchocerca A genus of Microfilarian Worms, also known by the generic names, *Onchocercus*, *Oncocerca* or *Oncocercus*, of the class of Roundworms, *Nematoda*. e.g. O. volvulus (= *Onchocercus* v.; *Oncocerca* v.; *Oncocercus* v.) - A species causing subcutaneous nodules on the heads of patients in Africa and tropical America. It also causes punctate keratitis and loss of vision (blinding filarial disease). It is transmitted by the bite of flies of the genus *Simulium* in Africa, and *Eusimulium* in Mexico and Central America, both causing the disease known as coast erysipelas (see **Microfilaria**; **Simulium**).

Onchocercus See **Onchocerca**.

Onchorhyncus A genus of Fishes, also known by the generic name, *Oncorhyncus*, of the Char and Salmon family, *Salmonidae*. e.g. O. gorbuscha - The Pink Salmon or Onchorhyncus; like all members of the family, it has a naked head, complete opercula, a forked caudal fin, a lateral line, and a dorsal fin of moderate length. From the testes is obtained protamine sulfate, which is used intravenously to check hemorrhage caused by heparin overdosage.

Onciola A genus of Acanthocephalous parasites. e.g. O. canis - A species found in the intestines of dogs in Texas and Nebraska.

Oncocerca or **Oncocercus** See **Onchocerca**.

Oncomelania A genus of Molluscs of the class of Snails, *Gastropoda*. The species transmit Japanese Schistosomiasis in China (see **Schistosoma**).

Oncorhyncus See **Onchorhyncus**.

Ondatra A genus of Mammals of the Common Rat and Mouse family, *Muridae* (Lemming, Muskrat and Vole subfamily, *Microtinae*). e.g. O. zibethica (= O. zibethicus) - The Muskrat or Musquash, a sort of vole found in Canada and the United States. It is the size of a rabbit, about 30 cm (1 ft) long, and weighs around 1.2 kg (3 lb). The body is plump and the tail is flattened laterally. Its fur resembles a beaver's, thick, soft and shiny, brown above and grey below. Essentially a water rat, the Muskrat spends its life in ponds and rivers. Like the beaver it digs deep burrows in riverbanks, dykes and weirs with two openings, one under the water and the other above. Its popular name is derived from the odorous substance comparable to a beaver's castorcum secreted by its inguinal glands. Distribution: Besides Canada and the United States, it is now widespread throughout much of Europe, eastern Russia, Mongolia and China.

Oniscus A genus of Isopod Crustaceans of the Sow Bug or Woodlouse family, *Oniscidae*. e.g. O. asellus - Length up to 18 mm; the body is dark grey, and is marked with light spots down the middle. A variable species, it is found in large numbers in damp places in forests, in cellars and old buildings, under stones, amidst damp leaves, etc. Distribution: much of Europe and North America.

Onobrychis A genus of plants of the Pea family, *Leguminosae*. e.g. O. viciifolia (= O. viciaefolia) - The Sainfoin, a perennial plant, 30-60 cm high; the flowers are bright pink or red, and the one-seeded pods are tubercled on the lower margin. Grows locally in meadows, pastures and fields. Place of origin: southeastern Europe; native in Britain.

Onoclea A genus of plants of the Lady Fern family, *Athyriaceae*. e.g. O. sensibilis - The Sensitive Fern, a Lady Fern species with leaves up to 50 cm long, growing singly along a creeping rootstock; there are triangular, coarsely-toothed, barren leaves and leaves which are fertile, with the pinnae rolled up into overwintering blackish, berry-like globules, which contain the sporecases, which crack open to release the spores. Found in damp, shady places. Place of origin: North America. Naturalized in Britain and northern Europe.

Ononis A genus of plants of the Pea family, *Leguminosae*. e.g. O. spinosa - The Spiny Restharrow, a perennial plant, 30-60 cm high; the stems are usually spiny, with two lines of hairs and the podis as long or longer than the calyx. The root, Restharrow Root or Radix Ononidis contains saponins, and has been used as a diuretic. It grows widespread in dry, rough, grassy places. Place of origin: Europe; native in Britain.

Onopordum A genus of plants of the Daisy family, *Compositae*. e.g. O. acanthium - The Scotch Thistle, a biennial plant, up to 1.5 m (5 ft) high; the stem has very broad, spiny wings and is covered with woolly hairs; the heads are large and solitary, and the toothed, elliptical leaves are spiny. Grows on waste land. Place of origin: mainly in South and Central Europe (native in Britain but rare in Scotland, despite its common name!); western Asia. Introduced and naturalized in North America.

Onychogalea See **Onychogales**.

Onychogales A genus of Mammals, also called *Onychogalea*, of the Kangaroo family, *Macropodidae*. e.g. O. unguifera (= *Onychogalea* u.) - The Nail-tailed Wallaby. This species of kangaroo has a nail at the end of its tail. Distribution: Australia.

Onychoteuthis A genus of Molluscs of the class of Squids, *Cephalopoda*. e.g. O. banksii - A small, surface-dwelling, marine squid, one of the swiftest creatures in the sea with a rapid trajectory flight above the water level, especially when endangered.

Oospora A genus of Fungi or Molds. e.g. O. lactis - A species found on the surface of cheese, milk, etc., and forming a white mold.

Opalina A genus of Protozoan animals of the Mastigophoran order of parasitic infusoria, *Opalina*. e.g. O. ranarum - A large species of parasites, generally ranging from 100-800 μ in length, and found in the rectum of frogs and toads.

Opercularia A genus of Protozoan animals of the class *Ciliata*. These are ciliates, microscopic animals which form branching colonies on various freshwater organisms.

Operculina A genus of plants of the Bindweed or Convolvulus family, *Convolvulaceae*. e.g. O. macrocarpa (= O. tuberosa) - The Brazilian Jalap plant, from the root of which is prepared a resin. Used as a powerful purgative.

Operophtera A genus of Insects of the Geometer or Looper Moth family, *Geometridae*. e.g. O. brumata - The European Winter Moth, a slender, large-winged species and a serious plant defoliator. The females are wingless. The larvae progress with characteristic looping movements and are variously known as Measuring Worms, Inchworms, Loopers or Earth-measurers.

Ophiacantha A genus of Echinoderm animals of the family of Brittle-stars, *Ophiurae*. These marine species live at great depths in the sea, sometimes exceeding 4 miles (6,432 m or 21,120 ft).

Ophiactis A genus of Echinoderm animals of the class of Brittle-stars and Basket-stars, *Ophiuroidea*. e.g. O. savigni - A self-dividing Brittle-star found on the coral reefs of the West Indies. These animals reproduce with asexual regeneration, tearing themselves into two, and each half regenerating the part lost.

Ophiarachna A genus of Echinoderm animals of the family of Brittle-stars, *Ophiurae*. e.g. O. increassata - A species of Brittle-stars, the largest found in the Indo-Pacific Ocean, with a disc diameter of over 51 mm (2 in) and a span of about 508-609 mm (20-24 in).

Ophidion A genus of Bony Fishes of the Cusk-eel family, *Ophidiidae*. These are the deep-sea cusk-eels; they have elongated bodies with pelves reduced to bifid rays, which look like barbels.

Ophidium A genus of Bony Fishes of the Cusk-eel family, *Ophidiidae*. e.g. O. barbatum - The Bearded Ophidium, length 25 cm; a seafish dwelling on sandy bottoms, often buried up to the head in the sand. Active by night, when it feeds on invertebrates such as marine crayfish, polyps and molluscs. Distribution: the Mediterranean Sea, the Atlantic European coast and the Black Sea. Native around Britain.

Ophiocephalus A genus of Fishes of the Snakehead family, *Ophiocephalidae*. e.g. O. striatus - The Striped Snakehead fish; a freshwater species, it cannot live on atmospheric oxygen alone, although it possesses an extra respiratory organ which it uses for breathing air, a structure formed from parts of the first gill arch; to survive it must keep its body in the water. Distribution: the freshwaters of India, Malaya, China and Africa.

Ophiodaphne A genus of Echinoderm animals of the class of Brittle-stars and Basket-stars, *Ophiuroidea*. e.g. O. materna - A species of Brittle-star found in Indo-Pacific waters. Regeneration is by sexual reproduction, where something very much like coupling takes place, the male lying across the female with arms entwined, and the eggs and sperms being shed in the usual way.

Ophioderma A genus of Echinoderm animals of the Brittle-star or Serpent Star family, *Ophiodermatidae*. e.g. O. longicauda - A brittle-star species, beautifully colored, with a body disc about 25 mm (1 in) in diameter, found on rocky shores under stones, at depths of up to 70 m (230 ft). Distribution: Mediterranean Sea and the Atlantic north to Biscay.

Ophioglossum A genus of plants of the Adderstongue or Adder's Tongue family, *Ophioglossaceae*. e.g. O. vulgatum - The Adder's Tongue Fern, or Adderstongue Fern, a perennial plant, 5-

25 cm high; the leaf is entire. Grows widespread in meadows, heath and woodland. Place of origin: Europe; native in Britain.

Ophiomisidium A genus of Echinoderm animals of the Brittle-star family, *Ophiurae*. The species are the smallest known ophiuroids or brittle-stars, measuring only 3-5 mm (0.11-0.19 in) in diameter.

Ophion A genus of Insects of the Ichneumon Fly family, *Ichneumonidae*. e.g. O. luteus - The Yellow Ophion, a fly species, 15-20 mm long; the abdomen is laterally compressed. Found commonly in gardens, where the eggs are laid in the caterpillars of various Moths and Butterflies (order *Lepidoptera*). Distribution: the entire Palaearctic region.

Ophionotus A genus of Echinoderm animals of the class of Brittle-stars and Basket-stars, *Ophiuroidea*. e.g. O. hexactis - An Antarctic Brittle-star, an ovoviviparous species in which the eggs are hatched in the ovaries, which themselves are attached to the inner walls of the bursae or pouches, at the bases of the arms which serve for respiration. The young brittle-star develops from this egg within the ovary, and eventually makes its way to the exterior through the slit-like opening of the bursa.

Ophiophagus A genus of Reptiles of the Cobra Snake family, *Elapidae*. e.g. O. hannah - The King Cobra or Hamadryad, the largest of the venomous snakes, up to 3.6 m (12 ft) long, yellowish-green in color with black bands. It feeds chiefly on other snakes. A very aggressive species. Distribution: South China, Vietnam, Burma, Malaya and the Philippines.

Ophiosaurus A genus of Reptiles, also known as *Ophisaurus*, of the Glass Snake, Lateral Fold Lizard and Slowworm family, *Anguidae*. e.g. O. apodus (= *Ophisaurus* a.) - The Glass Snake or European Glass Lizard, also known as the Scheltopusik; related to the Slow-worm, it has a brownish, cylindrical body with a lateral fold of skin, a deep furrow from behind the head to the vent, and there are rudimentary limbs at the sides of the cloaca. It is the largest of the family of legless or Anguid lizards, reaching a length of almost 1.2 m (4 ft); it has a brown back and light underparts. It is very tame and when in contact with humans does not usually even attempt to bite. Found in sunny meadows, fields, scrub country, as well as around rock formations in the foothills. Distribution: southeast Europe, the Balkans, Asia Minor, and southwest and central Asia.

Ophiosphalma A genus of Echinoderm animals of the Brittle-star family, *Ophiurae*. The species live at great depths in the sea, sometimes reaching excesses of up to 7,210 m (23,359 ft).

Ophiothrix A genus of Echinoderm animals of the class of Brittle-stars and Basket-stars, *Ophiuroidea*. e.g. O. fragilis - A common, shallow, water species of Brittle-stars; the disc is 2.5 cm (1 in) across and the five 15 cm (6 in) long arms are very fragile and have long, glassy spines which seem to fall to pieces, almost as soon as they are touched. Regeneration is associated with asexual reproduction. It lives on muddy sand. Distribution: Europe; native around Britain.

Ophisaurus See **Ophiosaurus**.

Ophiura A genus of Echinoderm animals, also called *Opiura*, of the class of Brittle-stars and Serpent-stars, *Ophiuroidea*. e.g. O. albida (= *Opiura* a.) - The Small Sandstar; size up to 10 cm, with a disc 15 mm across and five arms, each 5 cm long, possessing thorny teeth along the sides. Distribution: the seas of Europe; native around Britain.

Ophrys A genus of plants of the Orchid family, *Orchidaceae*. e.g. O. insectifera - The Fly Orchid, a perennial plant, 15-30 cm high; the flower lips are 3-lobed, the middle lobe is deeply bifid, downy,

dark purplish-brown with bluish markings, resembling a fly. Grows in woods, copses and in shrubs, especially in mountainous areas up to 1,600 m. Place of origin: Europe; native in Britain.

Opilio A genus of Arachnids or Spiders of the Harvestman family, *Phalangiidae*. e.g. O. parietinus - A Harvestman spider species, size 8-10 mm; the ovate body is lightish brown, with a greyish-white belly and dark markings on the back. It lives on walls, in cellars or sheds, or under stones. Like other Harvestmen, it does not spin webs. Distribution: Europe (native in Britain); Asia.

Opisocrostis A genus of Insects of the order of Fleas, *Siphonaptera*. e.g. O. bruneri - A squirrel flea, said to be a vector of sylvatic plague.

Opisthocomus A genus of Birds of the Hoatzin family, *Opisthocomidae*. e.g. O. hoazin - The Hoatzin of northeastern South America, and the sole species of this family. Sometimes known as the Reptile-bird; it has a musky odor like a crocodile, and its call sounds are more like a reptile's than those of a bird. It is the size of a crow, and is brown in color with pale streaks on the breast. The tail is long, consisting of ten loosely arranged feathers, and it has an untidy crest on a very small head.

Opisthognathus A genus of Fishes (subphylum *Vertebrata*). e.g. O. aurifrons - The Yellow-headed Jaw-fish.

Opisthonema A genus of Fishes of the Herring family, *Clupeidae*. e.g. O. oglinum - The Thread Herring fish.

Opisthorcis A genus of Flukes or Flatworms (class *Trematoda*), having testicles near the posterior end of the body. e.g. O. felineus - The Siberian Liver Fluke, found in the livers of cats, dogs, pigs and man. Infection recurs through eating fish, such as *Leuciscus* rutilis and *Idus* melanotus (see **Leuciscus**; **Idus**). O. sinensis (= *Clonorchis* s.) (see **Clonorchis**).

Opiura See **Ophiura**.

Opsana A genus of Fishes of the order of Midshipmen and Toadfishes, *Haplodocior batrachoidiformes*. e.g. O. tau - The Common Toadfish or Oysterfish, a predatory and voracious fish, with an undivided post-temporal bone in the skull. A bone connects the shoulder girdle to the vertebral column. When young it clings to rocks by means of a sucker on its belly; this sucker is lost in adulthood. Distribution: the Atlantic coast of North America.

Opuntia A genus of plants of the Cactus family, *Cactaceae*. e.g. O. vulgaris (= O. monacantha) - The Prickly Pear, a strong-growing cactus tree, up to 1.8 m (6 ft) high, with flattened, ovoid, jointed stems, growing in dry regions. It bears raised areoles, each of which has a small pointed leaf, one or several spines, numerous barbed bristles and many large, golden-yellow or reddish flowers, about 7.5 cm (3 in) across. Used in homeopathic medicine. Place of origin: from Canada to Brazil, Uruguay and the Argentine.

Orcaella A genus of Mammals of the River Dolphin family, *Platanistidae*. e.g. O. brevirostris - The Irrawadi Dolphin, a true dolphin found far up the Irrawadi River in India.

Orchestia A genus of Crustaceans of the Sandhopper family, *Orchestiidae*. e.g. O. gammarella - A Sandhopper species, a shrimp-like crustacean, up to 17 mm long; both the first two pairs of thoracic legs have pincers. It lives on shingly or sandy beaches, leaping in great numbers on the seashore after the tide has receded. Distribution: Europe; native around Britain.

Orchis A genus of plants of the Orchid family, *Orchidaceae*. e.g. O. mascula - The Early Purple Orchid or Salep (= Salep Tuber), a perennial plant, 15-60 cm high; the flowers are purple in color, with the lip paler, and deeply 3-lobed. From this species is prepared Mucilage of Salep (= Mucilago Salep), a veg-

etable demulcent having nutritive properties and allaying gastrointestinal irritation. Grows in meadows and occasionally in woods. Place of origin: Europe; native in Britain.

Orcinus A genus of Mammals of the Grampus Whale and Dolphin family, *Delphinidae*. e.g. O. orea - The Grampus or Common Killer Whale, the largest of the *Delphinidae*, sometimes mistakenly classified under the generic or scientific name, *Grampus*, a genus of Dolphins. A swift, predaceous creature with over 50 teeth; it has a dorsal fin and broad flippers with rounded tips. The male reaches a length of 9 m (30 ft), and the female is only half as long. It is black all over except for a white patch over the eye and a white ventral band. A ferocious species, the Grampus attacks seals, porpoises, dolphins, all kinds of fishes, and birds such as penguins. In schools these Killer Whales will even attack other whales, biting great chunks out of them. Distribution: worldwide - off Greenland, in the Bering Strait, in the Antarctic, and off the coasts of Europe. Native around Britain (see **Grampus**).

Orconectes A genus of Crustaceans of the Crayfish subfamily, *Cambarinae*; better known by the generic name, *Cambarus*. e.g. O. limosus (= *Cambarus affinis*) (see **Cambarus**).

Oreamnos A genus of Mammals of the Wild Cattle family, *Bovidae* (Sheep, Goat, and Goat-antelope subfamily, *Caprinae*). e.g. O. americanus - The Rocky Mountain Goat of North America, a goat-antelope with a shaggy coat and short, black horns about 23 cm (9 in) long. It lives on the craggiest and most remote mountain slopes, usually well above the tree-line. Distribution: the Rocky Mountains from the northwestern United States along the Pacific Coast to Alaska.

Orectolobus A genus of Fishes of the Carpet or Nurse Shark family,

Orectolobida. e.g. O. barbatus - A large, warm water, carpet shark found commonly from Japan to Australia. This Nurse Shark species is allied to the Cat Shark, and may reach a length of 3 m (10 ft) or more.

Oreochloa A genus of plants of the Grass family, *Gramineae*. e.g. O. disticha - The Two-rowed Moorgrass, a perennial plant, 10-20 cm high; the leaves are bristle-like and the spikelets grow in two rows. Found in stony mountain pastures, dry turf, scree and on rocks. Place of origin: The Pyrenees, Central Alps, South Carpathians.

Oreopteris A genus of plants of the Marsh Fern family, *Thelypteridaceae*. e.g. O. limbosperma - The Lemon-scented Fern, an often tall, 30-90 cm high, fairly stout, tufted fern plant with a characteristic lemon scent when crushed; its leaves are narrowly tapered at each end, and its spores lack spore-cases and are black when ripe, being borne near the margins instead of near the midribs of the pinnules. Grows in woods, heaths, screes and stream banks on acid soils, usually in hilly districts. Place of origin: Britain and northern Europe.

Oreotyx A genus of Birds of the Partridge, Pheasant and Quail family, *Phasianidae*. e.g. O. pictus - The Mountain Quail of North America; this bird nests at altitudes up to 8,686 m (9,500 ft) in the California mountains, making a short migration trip on foot in single file, in groups of 10-30 individuals into the sheltered valleys in September. In the spring the birds make the return trip, again on foot, to the higher altitudes.

Orester A genus of Echinoderm animals of the class of Starfishes, *Asteroidea*. e.g. O. reticulatus - The five-armed Starfish Echinoderm of the West Indies, a sea-animal massively built and practically all disc, with a span up to 508 mm (20 in).

Orgyia A genus of Insects of the Tussock Moth family, *Lymantriidae*. e.g. O. antiqua - The Vaporer Moth or Vapourer Moth, size 11-15 mm; the male is winged while the female is stout with only stunted wings. The strikingly ornamental caterpillars with their long tufts and brushes, feed on broadleaved trees and shrubs, including oak, beech, sallow, and fruit trees. Distribution: much of Europe, Siberia, northern China, Japan and North America.

Origanum A genus of plants of the Mint or Thyme family, *Labiatae*. e.g. O. majorana (= O. majoranoides; *Majorana hortensis*) - Marjolaine or Sweet Marjoram (see **Majorana**).

Oriolus A genus of Birds of the Old World Oriole family, *Oriolidae*. e.g. O. oriolus - The Golden Oriole, length 24 cm, a lovely black and yellow bird migrating into Europe in May and leaving in August for Africa. It is wild, noisy and bold, with a sonorous, rather lovely song heard in the morning and evening. It nests high up in the trees in parklands, orchards and meadow forests. Distribution: Europe, western and southern Asia; northwest Africa.

Orlaya A genus of plants of the Carrot family, *Umbelliferae*. e.g. O. grandiflora - The Orlaya, an annual plant, up to 1 m high; the leaves are 2-3-pinnate and the outer, white flowers have petals 5-8 mm long. An uncommon species, it grows on calcareous soil. Place of origin: South and Central Europe; the Mediterranean region.

Orneodes A genus of Insects of the Butterfly or Moth family, *Pyralidae* (subfamily *Orneodinae*). e.g. O. hexadactyla - A butterfly species found in the United States.

Ornithodoros A genus of Arachnids of the family of Ticks, *Ixodidae*. These are the argasid ticks, many species of which transmit the spirochetes (*Borrelia*) of relapsing fevers. e.g. O. moubata - The Tampan Tick of South Africa, transmitting *Borrelia* duttoni, the causative organism of West African relapsing fever (African Tick Fever).

Ornithogalum A genus of plants of the Lily family, *Liliaceae*. e.g. O. umbellatum - The Common Star-of-Bethlehem, a perennial plant, 10-30 cm high; the inflorescence is a corymbose raceme. Grows in grassy places. Place of origin: Mediterranean region, northwards to Sweden. Native in Britain. Grows wild in the United States, having escaped from cultivation.

Ornithomyia A genus of Insects of the Louse-fly family, *Hippoboscidae*. e.g. O. avicularia - The Common Louse-fly; a parasitic species inhabiting various species of birds. The larvae develop in the mother's body, pupating immediately after being deposited, and continue infesting the bird host.

Ornithoptera A genus of Insects of the Birdwing Butterfly family. e.g. O. alexandrae - The Queen Alexandra birdwing of New Guinea, the world's largest known butterfly and also the heaviest, a rare tree-dwelling species; the females (the males are smaller) have an average wingspan of 202-228 mm (8-9 in).

Ornithopus A genus of plants of the Pea family, *Leguminosae*. e.g. O. sativus - The Serradella, an annual to perennial plant, 10-60 cm high; the bracts are shorter than the white or pink flowers. Widely cultivated as a fodder plant. Place of origin: southwest Europe.

Ornithorhynchus A genus of Mammals of the order of egg-laying Mammals, *Monotremata*. e.g. O. anatinus - The Duckbill, Duckmole, Duckbilled Platypus or Platypus, about 50 cm (20 in) long and well-adapted for swimming, having a stream-lined body and webbed feet. Its short, velvety fur is reddish-brown above and greenish below. It is the most primitive mammal, and most closely related to the reptile, being an oviparous creature. It is a burrowing marsupial, but has no pouch. Found in the rivers of Tasmania and southeast Australia, where it is also known as the Water Mole.

Orobanche A genus of plants of the Broomrape family, *Orobanchaceae*. e.g. O. hederae - The Ivy Broomrape, a perennial plant up to 60 cm high; the stem is glandular and hairy. It is parasitic on the roots of the Ivy. Place of origin: south, central and western Europe. Native in Britain (see **Hedera**).

Oropsylla A genus of Insects of the order of Fleas, *Siphonaptera*. e.g. O. idahoensis - The Rodent Flea of the western United States, said to transmit sylvatic plague.

Orthezia A genus of Scale Insects of the Ensign coccid family, *Ortheziidae*. e.g. O. urticae - A scale insect generally found on nettles; about 3 mm long; the male has one pair of wings, and the female is wingless, her body covered with symmetrically arranged strips of a chalk-white waxy substance. Distribution: Europe, Asia.

Orthilia A genus of plants of the Wintergreen family, *Pyrolaceae*. e.g. O. secunda - The Serrated Wintergreen, a perennial plant, up to 15 cm high; the racemes are one-sided and the leaves have small serrations. Grows in shady woods, damp rock-ledges, moors and heaths. Place of origin: Europe; native in Britain.

Orthochela A genus of Decapod Crustaceans of the Porcelain Crab family, *Porcellanidae*. e.g. O. pumila - A Porcelain crab species, found in the Gulf of California, where its bright yellow color matches the fan corals (*Gorgonacea*), on which it spends most of its time. It lives in the shallower parts of the ocean or on the seashore.

Orthotomus A genus of Birds of the Babbler, Baldcrow, Thrush and allied family, *Muscicapidae* (True Warbler subfamily, *Sylviinae*). These are the Tailor-birds, warblers of several Asiatic species. They owe their popular name to the way they sew together several leaves of the tree in which they nest, edge to edge, to make the walls of the nest.

Orya A genus of Centipedes (class *Chilopoda*) of the order *Geophilomorpha*. e.g. O. barbarica - A worm-like centipede species, found in North Africa and measuring up to 177 mm (7 in).

Orycteropus A genus of Mammals of the Aardvark or Earth-pig family, *Orycteropidae*. e.g. O. afer - The Aardvark or Earth-pig, a nocturnal burrowing animal which resembles a pig, is about the same size with pig-like bristles, but its ears are donkey-like, its elongated muzzle ends in wide nostrils, its thin tongue extends almost 45 cm (18 in) from the small, tubular mouth, and its tail is long and heavy. Distribution: from Ethiopia to South Africa.

Oryctes A genus of Insects of the Chafer and Dung Beetle family, *Scarabaeidae*. e.g. O. nasicornis - The European Rhinoceros Beetle, 3.7 cm long; the body is a shining chestnut-brown, and the male has a horn 10 mm long, hence the name. The larvae develop in rotten tree-stumps. Distribution: Europe.

Oryctolagus A genus of Mammals of the Hare and Rabbit family, *Leporidae*. e.g. O. cuniculus - The European Wild Rabbit, 40 cm (16 in) long with a 7 cm (2 3/4 in) white tail; the ears are short. A gregarious and burrowing animal, it is found in large numbers living in complex, ramifying burrows, with galleries running from one to another, and with many entrances. Distribution: Native to northwest Africa and Spain. Spread by man throughout central and eastern Europe, Australia and New Zealand.

Oryx A genus of Mammals of the Cattle family, *Bovidae* (Horned Ungulate and Antelope subfamily, *Antilopinae* or *Hippotraginae*). e.g. O. algazel - The White or Scimitar Oryx, a somewhat large antelope, related to the ox and goat, about 2 m (61/2 ft) long and 1 m (31/2 ft) high at the shoulders, with very long horns (up to 112 cm or 3 ft 9 in long) in both sexes, curving gracefully. Found in the deserts of western Sudan; the southern fringe of the Sahara, ranging from Ethiopia to Mauretania. O. nasomaculatus (= *Addax* m.) - The Addax Antelope (see **Addax**).

Oryza A genus of plants of the Grass family, *Gramineae*. e.g. O. sativa - The Rice plant, one of the most important cereal foodcrops, mostly grown and eaten in India, China and Japan. It has been cultivated in these countries for hundreds of years and there are many different strains. More recently fairly extensive cultivation has been undertaken in the southern United States and the Northern Territory of Australia. It is estimated that at least half the world's population subsists wholly or partially on this grain. It is an erect, annual grass with loose inflorescences of one-flowered spikelets. Methods of cultivation vary but usually during the period of vegetative growth the paddy fields are flooded. The water is drained off to facilitate harvesting. It is doubtful whether the species exists today in the wild state. From the rice plant is prepared Amylum or Starch.

Oryzomys A genus of Mammals of the order of Rodents, *Rodentia*. e.g. O. swarthi - The James Island rice rat or Swarth's rice rat, believed to be the rarest rodent in the world; only 4 species were discovered on this island in the Galapagos group in 1906, and were not heard of again until January 1966, when the skull of a dead animal, recently demised, was found.

Oryzoryctes A genus of Mammals of the Tenrec family, *Tenrecidae*. These are the Rice-tenrecs, mole-like animals related to the tenrecs; they cause damage to crops as they burrow for insects. Distribution: Masagascar.

Oscanius A genus of Sea Slug Molluscs, better known by the generic name,

Pleurobranchus, of the order *Notaspidea* or *Pleurobranchomorpha* (see **Pleurobranchus**).

Oscarella A genus of Sponges of the class *Gelatinosa*. The species belong to the suborder *Homoscleropha*. In these animals the differentiation of the spicules into megascleres and microscleres is slight, and in some species there is none at all.

Oscillaria A genus of Algae or Molds. e.g. O. malariae.

Oscinella A genus of Acalyptrate Insects of the Fly family, *Chloropidae*. e.g. O. frit - The Frit Fly; its larvae are injurious to graminaceous plants (grass family, *Gramineae*), especially to oats.

Oscinis A genus of Insects of the order of Flies. e.g. O. pallipes - A fly species transmitting yaws.

Osmanthus A genus of Trees, also known by the generic name, *Olea*, of the Olive family, *Oleaceae*. e.g. O. fragrans (= *Olea* f.) - The Sweet Olive, an evergreen shrub or tree with opposite and entire, or subtly dentate, lanceolate leaves. The white flowers are small but richly fragrant and 4-petalled in the leaf axils. The Chinese use them for scenting tea. Place of origin: China and Japan; introduced into Europe in 1771.

Osmerus A genus of Fishes of the Smelt family, *Osmeridae*. e.g. O. mordax - The American Smelt-fish, a small marine fish which spawns in freshwater (rivers and streams), and is much esteemed as food. It resembles the salmon in structure, is translucent greenish in color above, and silvery on the sides. The average length is 20-25 cm (8-10 in). Found in the northern hemisphere.

Osmia A genus of Insects of the Bee family, *Apidae*. e.g. O. papaveris - The Poppy Mason-bee, 10-15 mm long; the black body has a yellow belly with dense hairs, which collect pollen. It lines its underground nest with petals of poppies or cornflowers. Distribution: Europe.

Osmorhiza A genus of plants of the Carrot family, *Umbelliferae*. The species of this genus found in North America, are called the Sweet Cicely plants as opposed to the European genus, *Myrrhis* to which the same common name applies (see **Myrrhis** odorata).

Osmunda A genus of plants of the Royal Fern family, *Osmundaceae*. e.g. O. regalis - The Royal Fern, a handsome, perennial plant, 50-150 cm (20-60 in) high, with a short erect stem and pinnate leaves 30-240 cm (1-8 ft) long. The spores are produced on the upper pinnae on the inner fronds. Found infrequently in wet heaths and bogs. This is one of the tallest ferns in Europe. Place of origin: Europe (native in Britain); eastern North America.

Osmylus A genus of Insects of the Lacewing and Ant Lion Fly family, *Osmylidae*. e.g. O. chrysops - A Lacewing fly, 25 mm long with a wingspan of 37-52 mm; it flies at night near flowing water. The larva which overwinters, is amphibious, and is a predator of fly larvae. Distribution: Europe.

Osphronemus A genus of Fishes of the Labyrinth family, *Anabantidae*. e.g. O. gourami - The Giant Gourami from Indonesia, a large freshwater fish, 60 cm (2 ft) long, grown especially in the irrigation ditches of rice paddies, and is used as food in southeast Asia.

Osteodontornis A genus of now extinct Birds, related to the Pelicans and Storks. e.g. O. orri - A species which flew over what is now the State of California about 20 million years ago. It had a wing-expanse of about 4.87 m (16 ft), and is estimated to have weighed as much as 27.2 kg (60 lb), allowing it probably to fly, if at all, only for short distances.

Osteoglossum A genus of Bony Fishes of the order of Anchovies, Herrings, Pikes, Salmon, Sardines and Trout, *Clupeiformes*. e.g. O. bicirrhosum - A freshwater fish, up to 50 cm (20 in) long; mainly silver grey, with red on the throat, shining with all the colors of the rainbow. Distribution: from the Guianas and the Amazon region.

Osteolaemus A genus of Reptiles of the order of Crocodiles, *Crocodylia*. e.g. O. tetraspis - The Broad-fronted Crocodile, only up to 1.8 m (6 ft) long; mostly found in small forest streams. It does not attack humans. Distribution: equatorial West Africa.

Ostericum A genus of plants of the Carrot family, *Umbelliferae*. e.g. O. palustre - The Marsh Angelica, a perennial plant, 50-120 cm high; the stem is ridged, the leaves 3-4-pinnate with blunt segments and short cartilaginous tips. Grows mainly in fens and damp meadows. Place of origin: east, southeast and central Europe.

Ostertagia A genus of filiform, nematode, parasitic worms (class *Nematoda*). The species are found mostly in cysts on the walls of the abomasum (the fourth stomach) of cattle.

Ostiolum A genus of Flukes or Flatworms (class *Trematoda*). e.g. O. medioplexus - A distome fluke, parasitic in the lungs of frogs.

Ostracion A genus of Fishes of the suborder of Trunkfishes, *Ostracoidei*. e.g. O. lentigenosum - The Blue Trunkfish from Indonesia; it has a strong case around its body, which is about 25-30 cm (10-12 in) long, like a trunk or luggage-box. Also called the Cowfish because of the skeletal, dermal prolongations growing from the head, rather like cows' horns.

Ostrea A genus of Bivalve Molluscs of the Edible Oyster family, *Ostreidae*. e.g. O. edulis - The Common, Flat or Native Oyster, or European Edible Oyster, a flat species with a skull diameter of 70-150 cm; thick, irregularly rounded, non-identical valves; and of great commercial importance as food. The larvae are incubated. A gregarious species, it lives in oyster colonies, which are cultivated in "oyster beds". Distribution: Atlantic coast of Europe, Arctic Circle, coast of North Africa, the Black and Mediterranean Seas.

Ostrinia A genus of Insects of the Butterfly or Moth family, *Pyraustidae*. e.g. O. nubilalis - The European Corn Borer, a pyraustid moth species which was introduced into the United States in c. 1917, and has since spread over many eastern and central states, causing great harm to maize in the stems of which the larvae live.

Ostrya A genus of Trees of the Ostrya or Birch family, *Betulaceae* or *Carpinaceae*. e.g. O. carpinifolia - The European Hop-Hornbeam, a deciduous tree, 9-18 m tall; the infructescence or fruit-bearing stem is similar to a hop cone, hence the common name, and the leaves are similar to those of the Hornbeam. The mature fruits are enclosed in bracts on ovoid catkins. The wood of this species is very hard and is used for making mallets. Place of origin: southern Europe, Asia Minor and the Caucasus (see **Carpinus**).

Oswaldocruzia A genus of Trichostrongyline parasites found in the lungs and intestines of reptiles and batrachians.

Otaria A genus of Mammals of the Eared Seal family, *Otariidae*. e.g. O. byronia - The Patagonian or Southern Sea-lion, a yellowish-brown giant; the males are over 2.1 m (7 ft) long and have a mane; the females are smaller. It frequents the Galapagos Islands, ranging down the coast of Peru and Chile, and up that of Patagonia as far as the River Plate.

Otiorhynchus A genus of Insects of the Weevil Beetle family, *Curculionidae*. e.g. O. niger - The Black Weevil, a beetle species 6.5-10 mm long, found in spruce woods in piedmont and mountainous country. The eggs are laid in the ground and the larvae feed on various small roots. Distribution: central and southern Europe.

Otis A genus of Birds of the Bustard family, *Otididae*. e.g. O. tarda - The Great Bustard; male length 102 cm female 80 cm; with a normal weight range of 10.9-15.9 kg (24-35 lb). It looks somewhat like a tan, black-spotted turkey, but has longer legs and a more pointed beak. A robust bird, it has been credited with the title of the world's heaviest flying bird. It is found in lowland fields, meadows, steppes and open plains. Distribution: western Europe, North Africa, and Central Asia to the Pacific Ocean.

Otobius A genus of Mite Arachnids of the family of Ticks, *Ixodidae*. These are the Argasid Ticks, the spinous ear ticks. e.g. O. lagophilus - The nymphs of rabbits. O. megnini - The nymphs of cattle; they are sometimes found in the ears of humans.

Otocentor A genus name given to Ticks (Mite order, *Acari*), formerly classified as *Dermacentor*. e.g. O. nitens (= *Dermacentor* n.) - A species of inornate ticks, yellow-brown in color, found on horses and related animals in southern Texas (see **Dermacentor**).

Otocolobus A genus of Mammals, also known by the generic name, *Felis*, of the Cat family, *Felidae*. e.g. O. manul - Pallas's Cat or Manul, up to 52.5 cm (1 ft 9 in) long, with a long, broad tail up to 25 cm (10 in). It has long, pale, greyish-yellow fur and round pupils, and lives chiefly in the mountain steppes of China and Mongolia.

Otodectes A genus of Arachnid Arthropods of the order of Mites, *Acari*. The species cause otoacariasis (infection of the ears or parasitic otitis) in cats, dogs and other domestic animals. e.g. O. cynotis - A parasitic mite, infesting the ears of dogs and cats; about 8.5 mm (1/3 in) in length.

Otomyces A genus of Molds or Fungi of the phylum *Eumycophyta*. e.g. O. pur-pureus - A species found infesting the human ear.

Otus A genus of Birds of the Typical Owl family, *Strigidae*. e.g. O. scops - The Scops Owl, a useful insectivorous species and a night bird of prey. A miniature owl, it averages about 20 cm (8 in) in length and 50 cm (20 in) in wingspan. Also known as the Screech Owl, and a partial migrant, its somewhat melancholy but not unpleasant song may frequently be heard at night. It appears in Europe in spring and migrates to the African interior in autumn.

Oulema A genus of Insects of the Leaf Beetle family, *Chrysomelidae*. e.g. O. melanopa - Length 4-4.8 mm; found chiefly in oats and barley; both beetles and larvae feed on the leaves of the host plants; in some regions it is regarded as a pest of grain fields, when present in large numbers. Distribution: Europe, Siberia, North Africa. Since it was introduced into North America it has become a pest there also.

Ovibos A genus of Mammals of the Horned Ungulate and Wild Cattle family, *Bovidae*. e.g. O. moschatus - The Musk-ox, a ruminant species intermediate between oxen and sheep, and closely related to the Chamois of Switzerland (genus *Antilope*); found in northeast Canada and Greenland. It is 250 cm long, with a 10 cm tail, has a compact neck, an equiline profile, hairy muzzles, thin lips and small pointed ears, and has a musky odor, hence the name. The horns are bent downwards. Recently it has been threatened with extinction because of overhunting by the Eskimo, and has been declared a protected species by the Canadian and Danish Governments in restricted reserves (see **Antilope**).

Ovis A genus of Mammals of the Horned Ungulate family, *Bovidae*. e.g. O. aries - The Domestic Sheep; from the internal fat of the abdomen is prepared Sevum or Suet.

Ovoplasma A genus of Flagellate Protozoans, now known by the generic name, *Leishmania*. e.g. O. orientale (= *Leishmania* tropica) - A species causing cutaneous leishmaniasis or Oriental Sore (see **Leishmania**).

Owenia A genus of Segmented Worms (phylum *Annelida*) of the Owenid family, *Oweniidae*. e.g. O. fusiformis - A polychaete annelid, with separate sexes. The larva is known as the Mitraria, because of its bishop's mitre-shape, which is very unlike the cylindrical, tube-dwelling, adult worm found in fine sand, to which it changes at the end of its free-swimming life.

Oxalis A genus of plants of the Oxalis or Wood-sorrel family, *Oxalidaceae*. e.g. O. acetosella - The Common Wood-sorrel or Wood Sorrel, a low-growing, woodland, perennial plant, 5-15 cm high; the solitary white or pinkish flowers are borne in the axils of trifoliate leaves on a creeping rhizome. The leaves contain an acid sap. Place of origin: Europe (native in Britain), Asia, North America (see **Oxydendrum**).

Oxeamnos A genus of Mammals of the Horned Ungulate and Wild Cattle family, *Bovidae*. e.g. O. americanus - The Rocky Mountain Goat, more closely related to the European Chamois than the True Goats. It is white except for the black horn and hoofs (see **Antilope**).

Oxybelis A genus of Reptiles of the suborder of Snakes or Scaly Reptiles, *Ophidia*. e.g. O. acuminatus - The Vine Snake, a rear-fanged species, greyish-brown in color, well camouflaged with the twigs and aerial roots among which it lives; it preys on small lizards. Distribution: the tropical regions of America.

Oxydendrum A genus of Trees of the Heath family, *Ericaceae*. e.g. O. arboreum - The Sorrel Tree or Sour Wood, a deciduous tree native to eastern North America. It is cultivated in parks and gardens in eastern states and in western Europe. The leaves are rather acid and have been used medicinally as a tonic and diuretic. It grows in the wild to a height of 18 m (60 ft), but only half when cultivated. The small whitish-yellow flowers grow in clusters, while the fruits are hard, woody, bell-shaped capsules about 1.2 cm (1/2 in) long. The leaves are alternate and tapered, coloring a bright red in autumn (see **Oxalis**).

Oxyporus A genus of Insects of the Rove Beetle family, *Staphylinidae*. e.g. O. rufus - The Red Rove Beetle, length 7-12 mm; a carnivorous species hunting various insects in mushrooms, in which it excavates tunnels. Often found perching on the underside of the cap. Abundant in forests, especially in the summer months. Distribution: much of Europe, the Caucasus, Siberia.

Oxyria A genus of plants of the Dock or Knotgrass family, *Polygonaceae*. e.g. O. digyna - The Mountain Sorrel, a perennial plant 5-15 cm high; the stem is leafy only at the base, and the leaves are kidney-shaped and long-stalked. Grows in mountain pastures, on scree, rocky and grassy slopes, up to 3,000 m and more. Place of origin: Europe; native in Britain.

Oxyruncus A genus of Birds of the Flycatcher allied family, *Tyrannidae* (Sharpbill subfamily, *Oxyruncinae*). e.g. O. cristatus - The Sharpbill, a single species of this subfamily; is about the size of a starling, olive-green above, yellowish with dark spots below, and with a yellow and scarlet crest. It has a straight, sharp-pointed bill and feeds on fruit. Distribution: tropical America.

Oxyspirura A genus of Nematode parasitic worms (class *Nematoda*). e.g. O. mansoni - A microscopic species found beneath the nictitating membrane of chickens and other fowl in the southern United States.

Oxytropis A genus of plants of the Pea

family, *Leguminosae*, many of whose
species are poisonous to farm animals,
and are known as "loco" or loco-weeds
(Spanish = insane), because of the selenium they contain. e.g. O. pilosa - The
Hairy Milk-vetch, a perennial plant, 15-
30 cm high; a white-hairy species with
pale yellow flowers, growing in globose
heads, in warm regions on hillsides and
in grassy places. Place of origin: southeast, central, and southern part of North
Europe (see **Astragalus**).

Oxyura A genus of Birds of the Duck
family, *Anatidae*. e.g. O. leucocephala -
The White-headed Duck, 45.5 cm long;
in the male the head is white and the bill
blue, while the female has a cheek-
stripe. A partial migrant bird, it is the
only European stiff-tailed duck.
Distribution: a freshwater species, found
on inland and brackish waters near the
Mediterranean coast.

Oxyuranus A genus of Reptiles of the
order of Lizards and Snakes, *Squamata*.
e.g. O. scutellatus - The Taipan, a snake
from Northern Australia and southeast
New Guinea. A comparatively rare
species.

Oxyuris A genus of Nematode intestinal
worms of the family *Ascaridae*, now
known by the generic name, *Enterobius*.
e.g. O. vermicularis (= *Enterobius* v.) -
The Pinworm or Common Thread-worm
(see **Enterobius**).

Ozaena A genus of Marine Cephalopod
Molluscs of the Octopus family,
Octopodidae and better known by the
generic name, *Eledone*. e.g. O. moschata (= *Eledone* m.) - The Musk Octopus
(see **Eledone**).

P

Pachymerium A genus of Centipedes of the order *Geophilomorpha*. e.g. P. ferrugineum - A geophilid species with a life-span of at least four years.

Paecilomyces A genus of Fungi or Molds. e.g. P. varioti banier var. antibioticus - A species producing the antifungal substance pecilocin.

Pagellus A genus of Fishes of the family of Sea Breams, *Sparidae*. e.g. P. centrodontus - The Common Sea Bream or Axillary Bream a marine food fish, 40-50 cm long, occurring in large shoals in deeper waters of the eastern Atlantic.

Pagophilus A genus of Seals of the order *Pinnipedia*. e.g. P. groenlandicus - The Harp Seal or Greenland Seal; a spectacular diver, once reported to have been caught in a net dropped to a depth of 275 m (902 ft).

Paguma A genus of Mammals of the Civet, Genet, and Mongoose family, *Viverridae*. e.g. P. larvata - The Masked Palm Civet, an oriental viverrid, as big as a cat, found east and southeast of the Himalayas in northern India.

Paguristes A genus of Crabs of the order *Decapoda*. e.g. P. oculatus - A species of crab inhabiting the sea-orange sponge, *Suberites* domuncula (see **Suberites**).

Pagurus A genus of Hermit Crabs of the family *Paguridae*. e.g. P. bernhardus - The Common Hermit Crab, length about 35 mm. It inhabits the abandoned shells of marine gastropods. Found on the seashore, or sandy or stony sea bottom along the Atlantic coast of Europe and the Mediterranean.

Palaeopithecus A genus of prehistoric Mammals of the Anthroid Ape family, *Pongidae*. The species lived in the Cainozoic era.

Palaeostenzia A genus of prehistoric Spiders of the class *Arachnida*. e.g. P. crassipes - A species belonging to the earliest spiders, believed to have lived about 370 million years ago.

Palaquium A genus of evergreen trees and plants of the family *Sapotaceae*, native to Malaya and the East Indies. e.g. P. gutta.- The Gutta Percha or Gummi Plasticum plant, from which the milky juice or sap is distilled and purified, to produce a latex known as guttapercha. This rubber-like substance is quite tough and strong at room temperature, becoming thermoplastic at about 70° C. It is used as a substitute for collodion, a temporary filling material in dentistry, in electric insulation, and in the production of golf-ball covers.

Paleosuchus A genus of Caiman Crocodiles of the family *Crocodylidae*. e.g. P. palpebrosus - Cuvier's smooth-fronted caiman crocodile of northeastern South America, one of the smallest known crocodiles, with a maximum recorded length of 1.45 m (4 ft 9 in).

Palomena A genus of Bugs of the family *Pentatomidae*. e.g. P. viridissima - The Green Bug, 12-14 mm long, found in shrubs and trees, mostly in Europe.

Pan A genus of Anthropoid Apes of the Chimpanzee family. e.g. P. troglodytes - A long-living chimpanzee species, with a potential lifespan of 60 years.

Panax A genus of aromatic plants of the family *Araliaceae*. e.g. P. schinseng (= P. ginsing) - The Ginseng plant; from the dried root is prepared ginseng, reputed to have a sedative effect on the cerebrum and a mildly stimulating action on the vital centers. It has been given in fatigue, hypotonia, and neurasthenia. Place of Origin: China.

Pandaka A genus of very tiny Fishes. P. pygmaea - The Dwarf Pygmy Goby fish, the shortest known fish and the shortest of all vertebrates, adult males reaching a maximum length of 1 cm. It is so small that it has to be studied under a microscope, rather than with a strong magnifying glass.

Pandinus A genus of African Scorpions,

having together with the genus *Heterometrus*, the largest sting of any other scorpions; it can inject a massive dose of venom, though the potency is low and human fatalaties are extremely rare. e.g. P. imperator - One of the world's largest scorpions, and found in West Africa. One specimen received at the London Zoo from Ghana in 1931 measured 228 mm (9 in) in total length, and was described as the biggest in living memory.

Pandion A genus of Osprey Birds of the family *Pandionidae*. e.g. P. haliaetus - The Osprey, a bird of prey measuring 55 cm in length. It feeds chiefly on fishes, and builds a huge nest in the tops of trees.

Pangasianodon A genus of large Fishes. e.g. P. gigas - The rare Pa beuk or Pla buk, a giant catfish, attaining a length of up to 3 m (9 ft 10 in). Found in the deep waters of the Mekong River in Laos and Thailand.

Panicum A genus of plants. e.g. P. miliare - Grain, from which is obtained Moriyo Starch, commonly used as a dietary food in India. Also used as a disintegrating agent in various tablets such as calcium carbonate, calcium gluconate, etc.

Panopea A genus of Bivalve Molluscs of the class *Bivalvia*. e.g. P. norvegica - The Norwegian Rock Borer, size 8 cm; thick elongated skull, with upper and lower edges almost parallel. Found along the Danish, British, and Icelandic coasts.

Panorpa A genus of Scorpion Flies of the family *Panorpidae*. e.g. P. communis - The Scorpion Fly, length 20 mm, wingspan 25-30 mm; the male has a pincers-like, clasping organ at the tip of the abdomen. Found in waterside vegetation and forest undergrowth; in Europe.

Panstrongylus A genus of cone-nosed Bugs of the family *Reduviidae*, species of which are vectors of Trypanosomes.

e.g. P. megistus - Formerly called *Conorhinus* megistus, *Lamus* megistus or *Triatoma* megista; a species in Brazil which is an important insect vector of *Trypanosoma* cruzi, the cause of human trypanosomiasis (see **Triatoma**).

Panthera A genus of Mammals, also known by the generic name *Felis*, of the Cat family, *Felidae*. e.g. P. tigris altaica (= *Felis* t. a.) - The rare, long-furred, Siberian tiger, also called the "Amur" or "Manchurian" tiger, the largest member of the cat family. Adult males average 3.12 m (10 ft 3 in) in length (nose to tip of extended tail), stand 99-107 cm (39-42 in) at the shoulder, and weigh about 265 kg (585 lb). P. onca (= *Felis* o.; *Jaguarius* o.) - The Jaguar or Leopard of the New World. Its coat is yellow, with whitish underparts, and is marked with black spots arranged in rosettes of 4-5 around a central spot.

Panulirus A genus of Lobsters of the family *Astacidae*. e.g. P. argus - The Spiny Lobster, characterized by its behavior of marching in parallel, single-file columns for mutual defence (the vulnerable abdomen of each lobster is protected by the hard thorax of the individual behind).

Panurgus A genus of Insects of the Bee family, *Apidae*. e.g. P. calcaratus - The Spurred Panurgus; 8-9 mm long; a generally common little black bee found on flowering dandelion and other yellow composites. Distribution: Europe.

Panurus A genus of Perching Birds of the family *Paradoxornithidae*. e.g. P. biarmicus - The Bearded Tit, inhabiting large reed beds bordering rivers, lakes, and brackish waters.

Papaver A genus of plants of the Poppy family, *Papaveraceae*. e.g. P. somniferum - The Opium poppy plant; from the dried latex is obtained opium, a potent narcotic containing the alkaloids morphine, codeine, papaverine, narceine, noscapine, and theibane. Its narcotic action is due mainly to its morphine content. Opium has other names, such as: Army disease; Black pill; Black stuff; Block; Brown stuff; Can (morphine; C&M (cocaine and morphine mixture); Dilaudid (dihydromorphinone hydrochloride); Ex-10-029 (dihydromorphine derivative); God's medicine (morphine); Gow; Gum (raw opium); HMC (heroin, morphine, and cocaine); Hop; Mary (morphine); Mash Allah (opium); M&C (morphine and cocaine); Metopon (methyldihydromorphinone); Mojo (morphine, heroin, or cocaine); Morph; MS (morphine sulfate); O; O&B (opium and belladonna); Op; Opium Crudum (crude opium / opium crude); Raw opium (= Gum); Yerli (a fine quality of Turkish opium).

Papilio A genus of Insects of the Swallow-tail Butterfly family, *Papilionidae*. e.g. P. antimachus - The African giant swallow-tail butterfly; females of the species measure up to 254-279 mm (10-11 in) across the wings. They are virtually impossible to catch with a net, and collectors have to resort instead to "shooting" them down with sporting guns loaded with dust or water.

Papio A genus of Mammals of the Monkey family. e.g. P. anubis - The Anubis baboon of Equatorial Africa, one of the largest members of the monkey family. Adult males of this species have a head and body length of 76-101 cm (30-40 in) and weigh 29-41 kg (65-90 lb).

Paracaudina A genus of Sea-cucumbers of the class *Holothuroidea*. e.g. P. chilensis - The Pacific sea-cucumber, a relatively long-living holothurian echinoderm. This species lives at least four years.

Parachordodes A genus of Hairworms or "Horsehair" Worms of the family *Chordodidae* (order *Gordiacea*). The adults are free living and the larvae are parasitic to insects. Human infection is

accidental. e.g. P. pustilosus - A species from Italy.

Paracoccidioides A genus of yeast-like fungi, morphologically similar to *Saccharomyces*, and better known by the generic name *Blastomyces*. e.g. P. brasiliensis (= *Blastomyces* b.) - The fungus causing South American blastomycosis (see **Blastomyces**).

Paracolobactrum A genus of microorganisms of the family *Enterobacteriaceae*. Short rods found in the intestines of man and animals, and in surface waters, soil and grains. e.g. P. aerogenoides.

Paradoxurus A genus of Mammals of the Civet, Genet and Mongoose family, *Viverridae*. e.g. P. hermaphroditus - The Palm Civet, an oriental viverrid about the size of a cat, with a tail as long as the body. Distribution: India and Sri Lanka (Ceylon).

Parafossularus A genus of Snails (phylum *Mollusca*) - A vector transmitter of the larval stage of the Chinese liver fluke, *Clonorchis* sinensis.

Paragnathia A genus of Isopod Crustaceans of the suborder *Gnathiidea*. e.g. P. formica - This species lives as an adult in holes and banks at the edge of salt marshes on European estuaries. The larvae suck the blood of fishes, taking enormous meals.

Paragonimus A genus of Trematode Flukes of the order of Digenetic trematodes, *Digenea*. Formerly known by the generic name *Distoma*. e.g. P. westermani (= *Distoma* pulmonale; D. ringeri; D. westermani; P. ringeri) - The species is found in the Far East, producing the disease known as Paragonimiasis (= Pulmonary distomiasis or Endemic hemoptysis).

Paragordius A genus of Hairworms or "Horsehair" Worms of the family *Chordodidae* (order *Gordiacea*). As in *Parachordodes*, human infection is accidental. e.g. P. varius - A species causing worm infection in North Africa (see **Parachordodes**).

Paralichthys A genus of Fishes of the Turbot or Lefteye Flounder family, *Bothidae*. e.g. P. californicus - The California Halibut, a large flounder species of economic importance, living off the Pacific coast, and weighing as much as 27 kg (60 lb).

Paralithodes A genus of Crustaceans of the Stone Crab family, *Lithodidae*. e.g. P. camtschatica - The King Crab, a large species with a legspan of over 90 cm (3 ft), and weighing up to 71/4 kg (16 lb). The carapace is triangular and heavily beset with spines, and the legs are relatively long. Distribution: off the coast of Alaska, across the North Pacific from the Bering Sea to the Sea of Japan.

Parameba A former name for the genus *Craigia*. Also known as *Paramoeba* (see **Craigia**).

Paramecium A genus of elongated holotrichous Ciliate Protozoans (subclass *Holotricha*). Certain strains have been employed in the protozoan test. e.g. P. coli (= *Balantidium* c.) - see **Balantidium**.

Paramoeba See **Parameba**.

Paramphistomum A genus of Flatworms of the class of Flukes, *Trematoda*. e.g. P. cervi - A species found in the stomach of cattle and sheep. Distribution: Egypt; United States.

Pararge A genus of Insects of the Brown Moth or Butterfly family, *Satyridae*. e.g. P. aegeria - The Speckled Wood Butterfly, 22-25 mm long; it flies along forest margins, forest rides, and in clearings from lowlands to mountainous areas. Distribution: Europe, Asia and North Africa.

Pararutilus A genus of Fishes, better known by the generic name *Rutilus*, of the Carp and Minnow family, *Cyprinidae*. e.g. P. frisii (= *Rutilus* f.) - The Black Sea Roach carp (see **Rutilus**; **Carassius**).

Parasaccharomyces A genus of yeast-like Fungi, various species of which have been isolated from human lesions. e.g. P. ashfordi (= *Monilia* psilosis) (see **Candida**).

Parasitus A genus of Mites (order *Acari*), of the family *Parasitidae*. e.g. P. coleoptratorum - An orange-colored mite, 1-1.2 mm long; the larvae live chiefly on the underside of the bodies of dung beetles (*Geotrupes*), which carry them to piles of horse dung where they find food - the larvae of flies and small aschelminths.

Paraspirillum A genus of microorganisms of the family *Spirillaceae* - Spiral or S-shaped cells, thick at the middle and tapering at the ends. e.g. P. vejdovskii.

Parateuthis A genus of Squids of the class *Cephalopoda*. e.g. P. tunicata - The smallest squid so far recorded, with a total length of only 12.7 mm (1/2 in).

Paravespula A genus of Insects of the Wasp family, *Vespidae*. e.g. P. vulgaris - The Common Wasp, 11-20 mm long; nests underground, often in the abandoned nest of a mouse, mole, etc. The complete nest contains as many as 3,000 individuals. It hunts flies, caterpillars, etc. as food for its larvae, and also sucks nectar; in autumn it feeds on ripe fruits. Distribution: the Palaearctic region and North America.

Pardanthus A genus of plants of the Iris family, *Iridaceae*. e.g. P. chinensis - An iridaceous plant of Asia, of high repute as an aperient or laxative.

Parelephas A genus of extinct land Mammals. e.g. P. trogontherii - A species of giant proboscideans or elephants, which lived about 1 million years ago in Central Europe and North America, and believed to have had a shoulder height of at least 4.5 m (14 ft 9 in), and to have weighed about 40,000 lb (17.8 ton).

Parendomyces A genus of yeast-like fungi, species of which have been isolated from various lesions in man.

Parietaria A genus of plants of the Nettle family, *Urticacaea*.

Parnassius A genus of Insects of the Apollo Butterfly family, *Papilionidae*. e.g. P. apollo - The Apollo butterfly, 34-50 mm from wing to wing; found in mountainous regions, usually resting on thistles of mountain slopes and meadows.

Paroctopus A genus of Octopods of the class *Cephalopoda*. e.g. P. apollyon (= *Octopus* punctatus) (see **Octopus**).

Parus A genus of Birds of the Typical Tit family, *Paridae*. e.g. P. caeruleus - The Blue Tit, 12 cm long; a perching bird found in groves, parks, gardens, and open mixed woods. The hen lays from 7-24 eggs in a "single clutch". A resident bird found in Europe and the Middle East.

Passer A genus of Birds of the Weaver family, *Ploceidae*. e.g. P. domesticus - The House Sparrow, a perching bird and an associate of man, living in small villages as well as large cities. A resident bird with worldwide distribution.

Passiflora A genus of plants of the Passionflower family, *Passifloraceae*. e.g. P. incarnata - The Passion Flower plant (= Grenadille; May-pop; Pasionaria); has been used as a nerve sedative in various neuralgias.

Pasteurella A genus of bacillary microorganisms of the family *Brucellaceae*. Motile or non-motile, small coccoid to rod-shaped, gram-negative cells. Sometimes also known by the generic names *Francisella* and *Yersinia*. e.g. P. pestis (= *Bacillus* p.; *Yersinia* p.) - The bacterium of the dreaded Oriental or Bubonic Plague, transmitted from rat to rat and from man to man by the rat flea, and from man to man by the human body louse. P. tularensis (= *Bacterium* tularense; *Francisella* tularensis) - A species causing tularemia (see **Francisella**).

Pasteuria A genus of microorganisms of the family *Pasteuriaceae* - Pear-shaped, non-motile cells attached in cauliflower-like masses, which are parasitic on freshwater crustacea. e.g. P. ramosa.

Patella A genus of Molluscs of the family *Patellidae*. e.g. P. vulgata - The Common Limpet, an edible mollusc; the shell is 50-60 mm long, with rough ridges on its green or brown outer surface. Distribution: along the Atlantic and Mediterranean coasts.

Patu A genus of Arachnids of the Spider family, *Symphytognathidae*. e.g. P. marplesi - A species of the smallest spider known; found in Western Samoa, in the southwest Pacific, a typical male specimen measuring 0.43 mm (0.016 in) in length.

Paullinia A genus of plants of the Soapberry family, *Sapindaceae*. e.g. P. cupana - The Guarana or Brazilian Cocoa plant; from the roasted seeds is prepared a paste, Guarana Paste, containing caffeine, catechu tannic acid, and starch. Has been used in the treatment of migraine and as an astringent in diarrhea.

Pausinystalia A genus of plants of the Madder family, *Rubiaceae*. e.g. P. yohimbe (= *Corynanthe* yohimbi) - see **Corynanthe**.

Pecten A genus of Molluscs of the Scallop family, *Pectinidae*. e.g. P. jacobaeus - The Fan Shell, length 130-150 mm; shaped like a fan. An edible mollusc found on sandy sea bottoms at depths of more than 25 m. Distribution: along the Mediterranean coast, and off the Canary and Cape Verde islands.

Pedaliodes A genus of Butterflies, species of which migrate at great heights; some have been observed crossing the Andes at heights up to 4,700 m (15,419 ft).

Pediculus A genus of Insects of the order of Sucking Lice, *Anoplura*. e.g. P. humanus var. capitis - The Head Louse, which may carry typhus fever, favus, and impetigo. P. humanus var. corporis - The Body Louse, which transmits typhus fever, trench fever, and relapsing fever, also provoking urticaria and melanoderma. P. pubis (= P. inguinalis; *Phthirus* pubis) - Infests the hair of the pubic region; sometimes found in the eye brows and eye lashes. Causes severe local itching.

Pediococcus A genus of microorganisms of the family *Lactobacillaceae*. e.g. P. acidilactici - A saprophytic form found in fermenting fruit juices.

Peganum A genus of plants of the Rue family, *Rutaceae*. e.g. P. harmala - Armel; Harmel; Hurmal; Syrian Rice; Wild Rice plant - Contains the alkaloid harmaline in association with harmine, harmalol and peganine (vasicine); used in India as an anthelmintic and narcotic agent (see **Banisteria**; **Haemadictyon**; **Tetrapteris**).

Pelagothuria A genus of Echninoderm animals of the class of Sea-cucumbers, *Holothuroidea*. A species of deepsea holothurians, having a central body with a ring of long tentacles round the mouth, and ornamented with lobes and finger-like processes.

Pelamis A genus of Sea snakes of the family *Hydrophidae*. e.g. P. platurus - The Black and Yellow sea snake of the Pacific, and the swiftest member of the family, widely distributed throughout the Pacific. It is 914 mm (3 ft) long; and its velocity may reach as high as 16 km/h (10 miles/h) for short bursts, when pursuing prey.

Pelargonium A genus of plants of the Geranium family, *Geraniaceae*. From the plant is extracted Geranium Oil (= Aetheroleum Pelargonii; Geran Oil; Pelargonium Oil; Rose Geranium Oil) - Used for perfuming tooth powders, ointments, talcum powders and various cosmetics.

Pelecanus A genus of Birds of the Pelican family, *Pelecanidae*. e.g. P. onocrotalus - The White feeds exclusively on fish, and nests in colonies. A migratory bird. Distribution: Eurasia.

Pelecus A genus of Fishes of the Carp and Minnow family, *Cyprinidae*. e.g. P. cultratus - The Ziege fish, length up to 60 cm; weight 1 kg; typical fish of the upper layers; found in the brackish waters of the Baltic and Black Seas, the Caspian Sea and Lake Aral.

Pelmatohydra A genus of Hydras or Polyps, of the family *Hydridae*. e.g. P. oligactis - A brownish-colored hydra, 30 mm long, with a single body opening which serves both as a mouth and vent, and surrounded by several tentacles furnished with special sting cells, containing a coiled thread-like structure, enabling it to whip out and numb the victim upon which the hydra feeds.

Pelobates A genus of Spadefoot Toads of the family *Pelobatidae*. e.g. P. fuscus - The Spadefoot Toad, length 5-8 cm; burrows in the ground and hibernates deep underground, often at depths of more than 2 m. Found in sandy areas in Central Europe.

Peloploca A genus of microorganisms of the family *Peloplocaceae* - colorless, filamentous, cylindrical cells containing false vacuoles which emit a reddish gleam of light. e.g. P. taeniata.

Penicillium A genus of common food molds and contaminants of culture media. e.g. P. expansum - The common green mold found on food. P. notatum - From this species is extracted the antimicrobial substance penicillin.

Pennatula A genus of Corals of the order *Pennatularia*. e.g. P. phosphorea - The Phosphorescent Sea-pen, length 20 cm; forms colonies resembling a branching bird feather. The skeleton is red, the polyps white. Found on sandy bottoms. Luminescent; when brought to the surface, the slightest touch will cause the whole colony to glow. Distribution: worldwide.

Pentastoma A genus of endoparasitic wormlike Arthropods. e.g. P. taenioides - A species found in man.

Pentatrichomonas A genus of intestinal Trichomonads marked by having 5 anterior flagella. e.g. P. ardin delteili - A species resembling *Trichomonas hominis*; could be pathogenic to man. P. hominis - A commensal species found in man.

Pepsis A genus of Insects of the Spider-wasp family, *Pompilidae*. These are the Spider-hunting Wasps, the largest wasps in the world. Also known as the "Tarantula Hawks". Found in South America. e.g. P. formosa - A species 3.75 cm (11/2 in) long of the southern U.S. and Central America. P. frivaldskii - A species of which a specimen, preserved in the British Museum, has a wingspan of 114 mm (41/2 in).

Peptococcus A genus of microorganisms of the family *Micrococcaceae* - Spherical, gram-positive, anaerobic forms, some species being saprophytic and others behaving as non-pathogenic parasites or commensals. Found in bedsores and peripheral venous (varicose) ulcers.

Peptostreptococcus A genus of microorganisms of the family *Lactobacillaceae* - Occurring as parasites of the intestinal tract; occasionally found in gangrenous and necrotic lesions as secondary invaders.

Perameles A genus of Mammals of the Bandicoot family, *Peramelidae*. These are the true bandicoots. e.g. P. gunni - The Barred Bandicoot, one of the most common species found in Tasmania; its hind quarters are marked with a few black stripes arranged symmetrically on either side of a longitudinal black band (see **Choeropus**; **Isodon**).

Perca A genus of Fishes of the Perch family, *Percidae*. e.g. P. fluviatilis - The European Perch fish, length 50 cm, weight 4 kg. One of the commonest of freshwater fishes, found in the middle and lower reaches of rivers, in backwaters, inlets, lakes, ponds and valley reservoirs.

Percina A genus of Fishes. e.g. P. peltata - The Shield Darter fish.

Perdix A genus of Birds of the Partridge, Pheasant, and Quail family, *Phasianidae*. e.g. P. perdix - The Grey Partridge, length 29 cm; found in fields and meadows in lowland as well as hilly country. A resident species and important game bird, laying 12-20 eggs in a "single clutch". Found in Europe and Central Asia.

Perforatella A genus of Molluscs of the family *Helicidae*. e.g. P. bidentata - This species of snail has a shell 7 mm high, 10 mm wide, is glossy brown, and is composed of 7 whorls. Found in lowlands in moist areas, including riverine forests and alder stands. Distribution: central and eastern Europe, and the Balkans.

Periophthalmus A genus of Fishes (subphylum *Vertebrata*). e.g. P. koelreuteri - The Mud-skipper fish.

Periplanea A genus of Insects of the Cockroach family. e.g. P. orientalis - The Oriental cockroach.

Periplaneta A genus of Insects of the Cockroach family. e.g. P. americana - The American cockroach.

Periploca A genus of plants of the Milkweed family, *Asclepiadaceae*. e.g. P. graeca - The Silk-vine, a deciduous shrub found in the Black Sea region, from which is isolated the cardiac glycoside periplocin (= Glucoperiplocymarin; Periplocoside). Has similar actions and toxic effects as digitalis and is used for this purpose (by slow IV injection) in Russia.

Perla A genus of Insects of the Stonefly family, *Perlidae*. e.g. P. burmeisteriana - Length 17-28 mm, wingspan 55-58 mm; has a dark thorax and light abdomen. The adults fly near water. Distribution: Europe.

Pernis A genus of Birds of the Buzzard, Eagle, and Hawk family, *Accipitridae*. e.g. P. apivorus - The Honey Buzzard, length 50-57 cm; found in woodlands at elevations up to 1,000 m. Feeds chiefly on honeybees, bumblebees, and especially wasps. A migrant bird, found in almost all Europe and Asia. It winters in Africa, often travelling in large flocks.

Petricola A genus of Molluscs of the family of False Angel Wings, *Petricolidae*. e.g. P. pholadiformis - The American Piddock; the shell is 60-80 mm long, whitish and thin. It lives in shallow depths where it bores into hard clay, chalk, limestone, etc. Distributed along the North Atlantic coast.

Petrodomus A genus of Mammals of the family of Jumping or Elephant Shrews, *Macroscelididae* (see **Macroscelides**).

Petrognatha A genus of Insects of the Long-horned or Longicorn Beetle family, *Cerambycidae*. e.g. P. gigas - A large tropical species from West Africa.

Petromyzon A genus of Eel-like water animals or Lampreys of the family, *Petromyzonidae*. e.g. P. marinus - The Sea Lamprey, a blood-sucking marine species, up to 1 m long, parasitic on fish, sucking up their blood and muscle tissues. Swims up rivers to spawn and breed in their shallow waters. Common along the Atlantic coast of North America and Europe.

Petroselinum A genus of plants of the Carrot and Parsley family, *Umbelliferae*. e.g. P. crispum (= P. sativum; *Apium* petroselinum; *Carum* p.) - Both the fruit and the root have been used medicinally as diuretics and emmenagogues. From the fruit is distilled Parsley Oil (= Oleum Petroselini; Parsley Fruit Oil), containing apiole. This has been used as an emmmenagogue but is of doubtful therapeutic value. Severe toxic effects, including nephrosis, have resulted from its use.

Peucedanum A genus of plants, better known by the generic name *Anethum*, of the Carrot family, *Umbelliferae*. e.g. P. graveolens (= *Anethum* g.) - The Dill plant (see **Anethum**).

Peumus A genus of plants of the family *Monimiaceae*. e.g. P. boldus - The Boldo plant; from the dried leaves is extracted the alkaloid boldine and the glycoside boldin or boldoglucin. Has been used as a diuretic in the treatment of hepatic congestion.

Pfeifferella A genus name formerly given to certain bacillary microorganisms, now included in the family *Brucellaceae*. e.g. Pf. anapestifer (= Pf. anapestifer; *Pasteurella* a.). P. mallei (= *Actinomyces* m.; Pf. mallei; *Bacillus* m.; *Malleomyces* m.) - The causative organism of glanders.

Phacochoerus A genus of Mammals of the family of Wild Pigs, *Suidae*. e.g. P. aethiopicus - The Wart Hog, an ugly animal with a grey, naked skin, misshapen head with enormous, sickle-shaped, upper canines, and large, wart-like excrescences on the cheeks. Distribution: Africa.

Phalacrocorax A genus of Birds of the Cormorant or Pelican family, *Phalacrocoracidae*. e.g. P. carbo - The Black Cormorant, length 80-90 cm; lives on freshwaters as well as at sea. A deep-diving flying bird, it feeds on fish which it pursues under water. A migrant bird with worldwide distribution.

Phalaenoptilus A genus of Birds which enter a torpid or hibernating state during which the body temperature is lowered to a level near that of the surrounding air. e.g. P. nuttallii - The Torpid Nuttall's Poor-will, found in California. Has extremely low body temperature (for birds), ranging from 18-19.2° C (64.4-67.7° F).

Phalangium A genus of Arachnids (order *Opiliones*), of the Harvestmen Spider family *Phalangidae*. e.g. P. opilio - Length 6-9 mm; black-greyish to rufous-brown; a common species found on the ground, on trees, rocks, telegraph poles, etc. Distribution: worldwide.

Phalaropus A genus of Birds of the Auk, Gull, and Shorebird family, *Phalaropidae*. e.g. P. lobatus - The Red-necked Phalarope, length 17 cm; a migrating bird of the north, found in wet grassy places on the shores of lakes and rivers, coasts, and in marshes. Distribution: northern Europe, far northerly parts of Asia, and North America.

Phalera A genus of Insects of the Prominent Moth family, *Notodontidae*. e.g. P. bucephala - The Buff-tip Moth, size 24-32 mm; has a large yellow spot at the apex of each forewing. Found in Europe, Asia Minor, and northeast Africa.

Pharbitis A genus of plants of the Morning-glory family, *Convolvulaceae*. e.g. P. nil (= *Ipomoea* hederacea) - The Ivy Leaf morning-glory plant, the dried seeds of which (Kaladana or Pharbitis Seeds) yield an extract with purgative properties similar to jalap. Used for this purpose in India and the Far East.

Pharnacia A genus of Insects of the order of Leaf and Stick Insects, *Phasmida* (suborder of the large Stick Insects, *Phasmatodea*). e.g. P. serratipes - The Tropical Stick Insect, a creature having a long, thin, stick-like body, the females of which have been measured up to 33 cm (13 in) in head and body length.

Phascolosoma A genus of Sipunculoid animals of the phylum *Sipunculoidea*. Worm-like marine animals with a long, cylindrical, unsegmented body. Sedentary inhabit, living in permanent or semi-permanent burrows in mud or sand, in rocky fissures, under stones or amongst algal holdfasts (see **Dendrostoma**; **Goldfingia**).

Phaseolus A genus of plants of the Bean family, *Leguminosae*. e.g. P. vulgaris - A species of beans from the seeds of which acted the mucoprotein, phytohemagglutinin (= phytohemagglutin) which causes hemagglutination, serum-protein precipitation, and stimulation of mitosis in leukocytes. It has been used in the treatment of aplasticanemia, with equivocal results.

Phasianus A genus of Birds of the Pheasant family, *Phasianidae*. e.g. P. colchicus - The Ring-necked Pheasant; length of male 79 cm, of female 60 cm; marked sexual dimorphism. A resident species and an important game bird, naturalized in Europe; original home in central and eastern Asia.

Phialophora A genus of hyphomycetous Fungi. e.g. P. verrucosa - A species causing a skin lesion resembling blastomycosis.

Philanthus A genus of Insects of the Digger Wasp family, *Sphecidae*. e.g. P. triangulum - The Bee-killer Wasp, length 12-18 mm; found chiefly in warm places, where the female excavates a main tunnel 20-100 cm long into her nest, with several lateral tunnels, each ending in a cellin which the larvae develop. The larvae feed solely on honeybees, which the female wasp captures on flowers, first paralysing them and then carrying them off to the nest. Distribution: the Palaearctic region, except the far north.

Philomachus A genus of Shorebirds, Auks and Gulls of the Snipe and Sandpiper family and their allies, *Scolopacidae*. e.g. P. pugnax - The Ruff, length of male 25 cm, of female 23.5 cm; marked difference between the sexes. A migrant bird, it inhabits damp lowland meadows, marshes, or in the northern regions the tundra, wintering in tropical and South Africa, and very occasionally in western and central Europe.

Phlebotomus A genus of Insects of the Biting Sandfly family, *Psychodidae*; about 60 species, all the females of which are blood-sucking. e.g. P. argentipes - The species transmitting kala-azar in India. P. chinensis - The species transmitting kala-azar in China. P. intermedius - A species suspected of transmitting leishmaniasis in South America. P. macedonicum - An Italian species. P. noguchi - A species transmitting *Bartonella* bacilliformis, the organism

of Oroya fever. P. papatasii - The sandfly, a dipterous insect of India and the
Mediterranean region, which transmits
pappatasi fever and sandfly fever, dermal leishmaniasis and tropical leishmaniasis. P. sergenti - A species causing the
Oriental sore. P. verrucarum - A fly
found in Peru, regarded as the transmitter of the infection of Verruga peruana
or Oroya fever. P. vexator - A species of
the U.S.; does not transmit any disease.

Phobaeticus A genus of large Insects of
the order of Stick Insects, *Phasmida*.
e.g. P. fruhstorferi - A species found in
Burma, the female measuring 300 mm
(11.81 in) (see **Pharnacia**).

Phobosuchus A genus of now Extinct
Reptiles of the Crocodile family. e.g. P.
hatcheri (see **Deinosuchus**).

Phoca A genus of Pinnipeds, also known
by the generic name *Pusa*, of the family
of True or Earless Seals, *Phocidae*. e.g.
P. vitulina (= *Pusa* v.) - The Common
Seal, length 165-180 cm. Spends most
of its life in water, although spending
some time resting on sandy and clay
shores. Also sleeps in water, and feeds
chiefly on fish, marine crustaceans, molluscs and other invertebrates. Found off
the coasts of the North Atlantic.

Phoenetria A genus of migrating Birds,
related to the Petrels. e.g. P. fusca - The
Sooty Albatross - Between breeding seasons it flies round the world at 40° south
latitude, covering a distance of 30,400
km (19,000 miles) in about 80 days.

Phoenicurus A genus of Perching Birds
of the Thrush family, *Turdidae*. e.g. P.
phoenicurus - The Redstart, length 14
cm; found in broadleaved and mixed
woods, parks and gardens; continually
twitches its tail and is hardly ever still.
A migrant bird, found in Europe, northwest Africa and Asia. Winters in tropical
Africa and southern Arabia.

Pholas A genus of Bivalve Molluscs of
the Piddock family, *Pholadidae*. e.g. P.
dactylus - The Common Piddock, 90-
150 mm long; has a whiteshell with yel-

lowish epidermis; drills corridors in soft
rock and wood. An edible species, found
along the Mediterranean Sea, Black Sea,
and Atlantic coasts of Europe.

Pholcus A genus of Arachnids of the
Daddy Longlegs Spider family,
Pholcidae. e.g. P. opilionides - The
Daddy Longlegs spider, length 4.5-5
mm; commonly found in buildings, stables, under stones, etc. The female carries cocoon-containing eggs in the chelicerae. Distribution: central and southern Europe and China.

Phoneutria A genus of Arachnids of the
order of True Spiders, *Araneae*. e.g. P.
nigriventer - The Aranha Armedeira, the
world's most dangerous spider, the
largest true spider of South America. It
frequently enters human dwellings and
hides in clothing or in shoes. Apart from
being the most aggressive of all the
highly venomous spiders, it also has the
largest venom glands (up to 10.4 x 2.70
mm; 0.4 x 0.1 in) of any living spider.
The venom can be lethal to humans.

Phormia A genus of Flies. e.g. P. regina -
A Blow fly causing a cutaneous myiasis
of sheep in the U.S. and Canada.

Phororhacos A genus of now extinct
gigantic flightless Birds. e.g. P. longissimus - The carnivorous ratite, which
stalked across what is now Patagonia,
South America about 10 million years
ago. It stood about 3.04 m (10 ft) tall
and had a skull as large as that of a
horse.

Photobacterium A genus of microorganisms of the family *Pseudomonadaceae* -
Coccoid or rod-shaped cells producing
luminescent substances, and found on
dead fish in seawater. e.g. P. phospho-
reum.

Phoxinus A genus of Fishes of the
Minnow family, *Cyprinidae*. e.g. P. laevis - A minnow fish used to demonstrate
the presence of the chromatophore-stim-
ulating hormone (CSH). When an
extract of posterior pituitary is injected
into a minnow, a red coloration appears

at the point of attachment of the thoracic, abdominal, and anal fins (see **Carassius**).

Phragmidiothrix A genus of microorganisms of the family *Crenotrichaceae* - Small disk-shaped cells occurring in attached trichomes. e.g. P. multiseptata.

Phreatoicopsis A genus of Isopod Crustaceans of the suborder *Phreatoicidea* (order of Isopods, Sowbugs, Woodlice and allies, *Isopoda*). e.g. P. terricola - A terrestrial species of isopods or gammarid amphipods, burrowing in the deep earth in the forests of Victoria, southeast Australia.

Phronima A genus of Crustaceans of the order of Amphipods, Scuds, and Sideswimmers, *Amphipoda*. e.g. P. sedentaria - A species usually found enclosed in a transparent barrel-shaped house, made from the empty test or outer coat of a salp.

Phryganea A genus of Insects of the Caddis Fly family, *Phryganeidae*. e.g. P. grandis - A trichopterous species, 15-21 mm long, with a wingspan of 40-60 mm; found in summer from lowlands to mountainous elevations round calm bodies of water. Distribution: the Palaearctic and Nearctic regions.

Phrynobatrachus A genus of extremely small Amphibians of the Frog family. e.g. P. chitialaensis - Found in Malawi; adult males measure 14-18 mm (0.55-0.70 in) in snout-vent length.

Phrynomerus A genus of Amphibians of the Frog family, *Phrynomeridae*. e.g. P. bifasciatus - A species found in the southern half of Africa. It hides in burrows, and its skin secretions are highly irritant.

Phryxe A genus of Insects (order of Two-winged Flies, *Diptera*), of the family, *Tachinidae*. e.g. P. vulgaris - A beneficial species, as are most tachinid flies; length 5-8 mm; adults found on flowers. Distribution: all over Europe.

Phthirus A genus of Insects of the order *Anoplura*. e.g. P. pubis - A species infesting the hair of the pubic region; also found in the eyebrows and eyelashes (see **Pediculus**).

Phyllanthus A genus of plants. e.g. P. engleri - A plant of northern Zimbabwe (formerly, Rhodesia), known as the suicide plant. When the bark or root is smoked it causes death.

Phyllobates A genus of Amphibians of the Venomous Frog family, *Ranidae*. e.g. P. bicolor - The Koko's Arrow-poison frog, found near the headwaters of the Rio San Juan river and its tributaries, N.W. Columbia, South America. Its skin secretions contain a powerful poison, batrachotoxin, of which 0.0001 g (0.0000004 oz) is sufficient to kill an average-sized man, which means that 28.3 g (1 oz) of this poison would be enough to wipe out 2,500,000 people.

Phyllobrotica A genus of Insects of the Leaf Beetle family, *Chrysomelidae*. e.g. P. quadrimaculata - Length 5-7 mm; has 4 black spots on a brown-colored elytra. Found in damp places, on the edges of bogs, etc. Distribution: Europe.

Phyllodecta A genus of Insects of the Leaf Beetle family, *Chrysomelidae*. e.g. P. vulgatissima - Length 4-5 mm; emerges early in spring, and is abundant until autumn on the leaves of the willow, where the larval stage is passed. Distribution: Europe and North America.

Phyllomedusa A genus of Amphibians of the Typical Tree-frog family, *Hylidae*. These are tree-nesting species from Mexico and Central America.

Phyllopertha A genus of Insects of the Chafer and Dung Beetle family, *Scarabaeidae*. e.g. P. horticola - The Garden Chafer Beetle, length 8-11 mm; very common in gardens, forest margins, hedgerows, etc.; fond of flowering roses. Distribution: Europe, Siberia, and Mongolia.

Phyllopora A genus of Crickets. e.g. P. grandis - A species found in New Guinea, the largest bush-cricket in the world, with a wingspan of almost 266 mm (101/2 in).

Phylloscopus A genus of Birds of the Old World Warbler family, *Sylviidae*. e.g. P. sibilatrix - The Wood Warbler, length 13 cm; found in open broadleaved and mixed woodlands in lowland regions, particularly in beech woods. Nests on the ground. Distribution; most of Europe and western Siberia.

Physalia A genus of siphonophores (class *Hydrozoa*), closely related to the Jell-yfish, of the order *Siphonophora*. e.g. P. physalis - The Portuguese Man-of-war, the most dangerous of all the coelenter-ates found in British waters; its sting is poisonous but seldom fatal to man. It measures up to 304 mm (12 in) in length and 152 mm (6 in) across, and its tentacles can extend nearly 12.19 m (40 ft). Most people stung by this highly colored float, experience a burning pain followed by a large wheal which may last for a week before fading. Others however, especially children, or those who are sensitive to the venom can be seriously ill, and a severe shock could lead to drowning.

Physalis A genus of plants of the Nightshade family, *Solanaceae*. e.g. P. alkekengi - Alkekengi; Bladder Cherry; Chinese Lantern; Ground Cherry; Strawberry Tomato; Winter Cherry plant. The Physalis berry is reported to have diuretic properties.

Physaloptera A genus of Nematode Worms of the family *Strongylidae*, found in the stomach and intestines of man and other vertebrates. e.g. P. mordens - A round worm found in Negroes in East Africa.

Physeter A genus of Sperm Whales (order *Cetacea*), of the family *Physeteridae*. e.g. P. catodon (= P. macrocephalus) - The Sperm Whale, the largest toothed mammal ever recorded; the average bull measures 14.32 m (47 ft) in length and weighs 33 tons. It is one of the most highly valued of

cetaceans, practically all of it being processed for commercial purposes. Very much in demand is a solid wax, known as ambergris or spermaceti (= Cetaceum or Walrat), found in the gut and used in the preparation of cold creams and suppositories. The Sperm Whale often visits the shores of Europe, and occasionally also enters the Mediterranean (see also **Hyperoodon**).

Physocephalus A genus of Worms. e.g. P. sexalatus - A species of thick, non-bursate worms found in the stomach of pigs.

Physopsis A genus of Molluscs or Snails, a carrier of the miracidial stage in schistosomiasis, especially of S. hematobium (see **Schistosoma**).

Physostegia A genus of plants, also known by the generic name *Drachocephalum*, of the Thyme family, *Labiatae*. e.g. P. virginiana (= *Drachocephalum* virginianum) - The False Dragon-head or Obedient plant (see **Drachocephalum**).

Physostigma A genus of Tropical plants of the Pea family, *Leguminosae*. e.g. P. venenosum - The Calabar Bean or Chopnut plant; the poisonous seed of this climbing plant of Africa contains the alkaloids calabarine and physostigmine or eserine. Physostigmine is an anticholinesterase, used chiefly as a miotic, and to decrease intra-ocular pressure in glaucoma.

Phytocoris A genus of Bugs of the Capsid family, *Miridae*. e.g. P. tiliae - Length 6.1-6.9 mm; brownish color; lives on various trees such as oak, lime, ash, etc., and feeds on larvae of insects and mites. Distribution: Europe and North Africa.

Phytolacca A genus of plants of the family *Phytolaccaceae*. e.g. P. decandra (= P. americana) - The Poke plant; it is an emetic, purgative, and mild narcotic but is rarely used in medicine.

Phytomonas 1. A genus of flagellate worms of the family *Trypanosomatidae*

- Short leishmanial and leptomonad forms; they have invertebrate and plant hosts, and do not form cysts. 2. A genus of bacteria that produce pathogenic necrosis in plants.

Pica A genus of Birds of the Crow and other allied perching (bird) family, *Corvidae*. e.g. P. pica - The Magpie, length 46 cm (head to tip of long tail); found in parklands, fruit orchards and hedgerows with tall trees. A resident bird. Distribution: worldwide.

Picea A genus of Trees of the Pine family, *Pinaceae*. e.g. P. abies (= P. excelsa) - The Norway spruce tree, yielding a purified resinous exudate, Burgundy Pitch (= White Pitch; Poix de Bourgogne) - Used in plasters as a mild counter-irritant.

Picraena A genus of Trees, also known by the generic name *Picrasma*. e.g. P. excelsa (= *Picrasma* e.) - The Jamaica quassia tree (see **Aeschrion**).

Picrasma A genus of Trees (see **Picraena**).

Picromerus A genus of Heteropterous Insects of the True Bug family, *Pentatomidae*. e.g. P. bidens - Length 11-14 mm; found in broadleaved forests. Distribution: the Palaearctic region.

Picrorhiza A genus genus of plants, also known by the generic name *Picrorrhiza*, of the Figwort family, *Scrophulariaceae*. e.g. P. kurroa (= *Picrorrhiza* k.) - The Picrorhiza or Kutki plant; the dried rhizome is a bitter. Used in India as a tonic and antiperiodic.

Picrorrhiza A genus of plants (see **Picrorhiza**).

Picus A genus of Birds of the Woodpecker family, *Picidae*. e.g. P. viridis - The Green Woodpecker, length 32 cm; found in broadleaved woods, parks, etc., excavating a nesting hole in the trunk of the tree, in which the female lays 5 to 7 eggs. A resident bird, found througout Europe and the Middle East.

Piedraia A genus of Fungi or Molds of the family *Endomycetales*, several species of which are parasitic on hair, forming small, black, adherent nodules. e.g. P. hortai - A fungus species causing black piedra, a disease of the scalp hairs, especially in tropical regions.

Pierus A genus of Insects of the White and Yellow Butterfly family, *Pieridae*. e.g. P. brassicae - The Large White Cabbage butterfly, length 29-34 mm; a migrating butterfly found in gardens, fields, and meadows. The female lays 200-300 eggs on the underside of the host plants, mainly cabbages (in which it can be a pest), and other crucifers.

Pilocarpus A genus of plants of the Rue family, *Rutaceae*. e.g. P. microphyllus - The Jaborandi shrub of tropical America; the dried leaflets yield the principal alkaloid pilocarpine, a parasympathetic cholinergic drug. Used as a miotic (to constrict the pupil), to decrease the intra-ocular pressure in glaucoma and detachment of the retina, and to antagonize the effects of atropine on the eye.

Pimenta A genus of Trees and Shrubs of the Myrtle family, *Myrtaceae*. e.g. P. acris (= P. racemosa) - Yields Bay Oil or Myrcia Oil; used in the preparation of Bay rum, a hair lotion and astringent. P. officinalis (= *Eugenia* pimenta) - The Pimento tree (= Pimenta; Allspice; Jamaica Pepper), a tree of tropical America; the dried full-grown, unripe fruit yields Pimento Oil (= Allspice Oil; Pimenta Oil), a volatile oil, used as an aromatic, carminative, and stimulant.

Pimpinella A genus of plants of the Carrot family, *Umbelliferae*. e.g. P. anisum - The Anise or Aniseed plant; the dried ripe fruit contains a volatile oil, Anise Oil or Anethol (= Aniseed Oil; Esencia de anis; Essence d'Anis; Ol Anis; Oleum Anisi); used as an aromatic carminative, expectorant, and flavoring agent. P. saxifraga - The Burnet Saxifrage; the dried rhizome and roots (Pimpinella Root; Racine de Boucage;

Bibernellwurzel) contain a bitter principle, pimpinellin; used as a tonic, diuretic, emmenagogue, and an aromatic carminative.

Pimpla A genus of Insects of the Ichneumon Fly family, *Ichneumonidae*. e.g. P. instigator - Length 10-24 mm. Its larvae are parasites of the caterpillars of various *Lepidoptera*. Distribution: Europe and North Africa.

Pinguinus A genus of Shorebirds, Gulls and Auks (order *Charadriiformes*) of the Auk family, *Alcidae*. e.g. P. impennis - The Great Auk, a now extinct bird, the last member of this species having been seen in 1844 in Eldey, Iceland.

Pinna A genus of Bivalve Molluscs of the family *Pinnidae*. e.g. P. nobilis - The Fan-mussel, an edible species, length 70-90 cm; the shell is shaped like a cudgel, and is found at depths of more than 3 m on sandy coasts, with the pointed end embedded in the sand. Distribution: the Mediterranean and Atlantic coasts.

Pinnotheres A genus of Pea Crabs, the smallest crabs in the world, of the family *Pinnotheridae*, living in the mantle cavities of bivalve molluscs such as oysters, mussels and scallops. e.g. P. pisum - A tiny Pea Crab, with a shell diameter of only 6.35 mm (0.25 in), found in British waters.

Pinoyella A genus of Fungi or Molds. e.g. P. simii - A fungus of the trichophyton group, which produces a transmissible disease of the glabrous skin of monkeys.

Pinus A genus of Trees of the Pine family, *Pinaceae*. e.g. P. palustris - The Pine Tree; from the wood is extracted secondary and tertiary terpene alcohols, mainly terpineol. Used as disinfectants and deodorants. P. pinaster (= P. maritima) - From the wood is extracted an oleoresin, from which is distilled terpentine (Oleum Terebinthinae). Used in linaments externally and in evacuant enemas.

Piophila A genus of Flies. e.g. P. casei - The fly whose larvae are the "cheese skippers", and a common cause of intestinal myiasis.

Piper A genus of plants of the Pepper family, *Piperaceae* - The species produce betel, cubeb, kava-kava, matico, and pepper. e.g. P. nigrum - The Black Pepper plant (= Pepper; Piper; Pimenta); the fruit contains a pungent resin, chavicine, with piperine, piperidine, and a volatile oil. Black pepper stimulates the tastebuds, increasing reflex gastric secretion. It has diaphoretic and diuretic properties.

Pipistrellus A genus of Mammals of the Vespertilionid Bat family, *Vespertilionidae*. e.g. P. pipistrellus - The Common Pipistrelle; body length 33-45 mm; wingspan 180-230 mm. Originally a tree-dweller, but now often found in areas of human habitation such as wall crevices and attics. Forms large colonies in summer. Distribution: Europe, Asia, and northwest Africa.

Piptadenia A genus of plants of the family *Mimosaceae*. e.g. P. peregrina - From the seeds and leaves is isolated an indole alkaloid, bufotenine (= mappine; N,N-dimethylserotonin), and an active principle, dimethyltryptamine. Both of these are hallucinogenic agents, producing effects on the mental state which are similar in some respects to those of lysergide. In South America a narcotic snuff (Cohoba snuff) is prepared from the seeds and leaves by some Indian tribes. Elsewhere, the hallucinogenic snuff goes by the name yopo (see **Amanitamappa**; **Bufo**).

Piratinga A genus of Fishes of the Catfish family. e.g. P. piraiba - The Pirahyba catfish of the Amazon River has been called the Goliath of catfishes. The maximum length and weight of this species has been put at 2.13 m (7 ft) and 181 kg (400 lb), respectively.

Pironella A genus of Snails, whose species are intermediate carriers in the lifecycle of the fluke, *Heterophyes* heterophyes.

Piroplasma A name formerly given to a genus of protozoan parasites found in the red blood cells of animals, now known as *Babesia*; also known as *Pyroplasma* or *Pyrosoma*. e.g. P. bigemina - The causative organism of Texas fever, transmitted by the tick *Margaropus* annulatus (= *Boophilus* annulatus) (see **Babesia**; **Nuttallia**; **Theileria**).

Pisaster A genus of Starfishes (class *Asteroidea*). e.g. P. brevispinus - The Five-armed Starfish, a very bulky asteroid; has been weighed at 4.53 kg (10 lb).

Pisaura A genus of Arachnids of the family of True Spiders, *Pisauridae*. e.g. P. mirabilis - Length 11-13 mm; found chiefly in sunny places at lowland elevations, on herbaceous plants and shrubs, in forests and forest steppes. Distribution: the Palaearctic region.

Piscidia A genus of Trees of the Pea family, *Leguminosae*. e.g. P. piscipula (= P. erythrina) - The Jamaica Dogwood or Fishpoison Bark tree; the bark is used as an anodyne to relieve neuralgia and in the treatment of dysmenorrhea.

Pisidium A genus of Bivalve Molluscs of the Freshwater Cockle family, *Sphaeriidae*. e.g. P. amnicum - The River Peashell, length 8-10 mm; one of the largest species of this genus, found in fine mud on the bottoms of rivers and larger streams. Distribution: Europe, northern Asia, and North Africa.

Pistacia A genus of plants of the Cashew family, *Anacardiaceae*. e.g. P. lentiseus - The Pistacia plant; yields a resinous exudate, Mastic (= Mastiche; Mastix; Almáciga). Used as temporary fillings for carious teeth, and as a protective covering for wounds.

Pithecophaga A genus of Birds of the Eagle family, *Accipitridae*. e.g. P. jefferyi - The Monkey-eating Eagle, one of the world's rarest birds; believed to have a total population of less than 100.

Pitymys A genus of Mammals of the rodent Vole family, *Cricetidae*. e.g. P. subterraneus - The Pine Vole, length 80-105 mm, tail 26-40 mm. Found in forest margins, among dense vegetation and damp areas. Distribution: throughout most of Europe.

Pityophagus A genus of Insects of the Flower Beetle family, *Nitidulidae*. e.g. P. ferrugineus - Length 4-6.5 mm; brownish in color; lives under the bark of trees where it hunts bark beetles. Found in the coniferous forests of Central Europe.

Pityrosporon A genus of Fungi or Molds, also known by the generic name *Pityrosporum*, which are yeast-like and produce no mycelium. e.g. P. orbiculare (= *Pityrosporum* o.) - A resident of the normal skin, capable of causing disease (tinea versicolor).

Pityrosporum See **Pityrosporon**.

Placentonema A genus of Roundworms (class *Nematoda*). e.g. P. gigantissima - The largest known roundworm, a marine nematode found in the Pacific. Adult females measure 6.75-8.40 m (22 ft - 27 ft 6 in) in length, and adult males 2.04-3.75 m (6 ft 10 in - 12 ft 3 in).

Placobdella A genus of Annelids or Leeches.

Plagiodera A genus of Insects of the Leaf Beetle family, *Chrysomelidae*. e.g. P. versicolora - Length 2.5-4.5 mm; usually colored blue, sometimes blue-green or coppery. Beetles and larvae found on leaves of the willow and poplar trees; the larvae strip the leaves. Distribution: the Palaearctic region and North America.

Plagiodontia A genus of Mammals of the Coypu family, *Echimyidae*. A very primitive rodent species of the New World, found in San Domingo.

Plagionotus A genus of Insects of the Long-horned or Longicorn Beetle family, *Cerambycidae*. e.g. P. arcuatus - Length 6-20 mm; coloration resembles that of a wasp, with yellow patches and bands on the elytra. Found in oak woods, resting on felled logs, where the female lays 30 eggs which she inserts with her ovipositer into cracks in the bark. Distribution: Europe, Asia Minor, and North Africa.

Plagiostomum A genus of Flatworms of the order *Eulecithophora*. The species have a simple intestine.

Planaria A genus of Turbellarian Flatworms of the Planarian order, *Tricladida*, nowadays classified as the genus *Dugesia*. e.g. P. gonocephala (= *Dugesia* g.) (see **Dugesia**).

Planigale A genus of Mammals of the Marsupial family. e.g. P. subtilissima - The world's smallest marsupial, the very rare Kimberley planigale, a flat-skulled marsupial "mouse" found only in the Kimberley district of Western Australia. Adult males have a head and body length of about 44 mm (1.73 in), a 50 mm (1.96 in) long tail, and weigh about 4 g (0.141 oz). Females are smaller.

Planorbarius A genus of Gastropod Molluscs or Snails of the family *Planorbidae*. e.g. P. corneus - The Ram's-horn Snail; shell 11-13 mm high, 25-33 mm wide, and composed of 41/2-5 whorls. It inhabits calm or gently flowing waters with rich vegetation. Distribution: much of Europe, and Asia Minor.

Planorbis A genus of Gastropod Molluscs or Snails of the family *Planorbidae*. Several species act as intermediate hosts for *Schistosoma* mansoni and S. haematobium (see **Schistosoma**). e.g. P. boissyi - In Egypt. P. guadelupensis - In Venezuela. P. olivaceus - In Brazil. P. planorbis - The Margined Trumpet-snail, found in the Palaearctic and Nearctic regions; shell 3.5-4 mm high, 12-17 mm wide, composed of 51/2-6 whorls. Situated at lower elevations in calm waters with vegetation.

Plantago A genus of plants of the Psyll family, *Plantaginaceae*. e.g. P. psyllium - The Plantago plant; the dried ripe seed, Psyllium (= Flea Seed; Flohsame; Psyll; Psylli Semen; Psyllium Seed), contains mucilage and is used as a demulcent, absorbing and retaining water, and producing a bulk-providing medium in the treatment of chronic constipation.

Plasmodiophora A genus of one-celled animals, the *Rhizopods*. e.g. P. brassicae - A rhizopod organism which causes a disease of cabbages and other cruciferous plants, called "fingers and toes" or "stump root".

Plasmodium A genus of unicellular malarial parasites (class *Sporozoa*) of the order *Hemosporiidea*, and the family *Plasmodidae*, parasitic in red blood cells and formerly classified under the generic name, *Laverania*. e.g. P. bovis - A species found in cattle. P. brazilianum - A species found in monkeys in South America, similar to P. malariae. P. canis - A species found in dogs in India. P. capistrani - A species causing malaria in birds. P. cathemerium - A species causing malaria in birds. P. cynomolgi - A species causing malaria in monkeys, of the genus *Macaca* (see **Macaca**). P. danilewskyi - A species found in birds in Italy, India, and Africa. Ross first traced the development of the Plasmodium in the mosquito with this species. P. durae - A species pathogenic for turkeys. P. equi - A species found in the horse. P. falciparum - The species causing "malignant" estivo-autumnal malaria in man. P. gallinaceum - One of the many species causing avian malaria. P. inui - A species causing malaria in monkeys (see **Macaca**). P. knowlesi - A species causing malaria in monkeys. P. kochi - A species pathogenic for chimpanzees and for monkeys. P. lophurae - A species pathogenic for the domestic fowl. It has been recovered from a Borneo fireback pheasant, *Lophura* igniti igniti. P. malariae - The species causing quartan malaria in man. P. ovale - A species causing tertian malaria, found in the Congo. P. pitheci - A species found in the orangutan and in chimpanzees. Resembles P. vivax, but non-pathogenic to man. P. pleurodyniae - The name given to certain inclusion bodies found in red blood cells in cases of epidemic diaphragmatic pleurodynia. P. relictum - A species causing malaria in birds. P. relictum var. matutinum - A species isolated from the robin, morphologically similar to P. reticulum, but biologically very different, with a strict quotidian periodicity. P. richenowi - A species found in chimpanzees. P. tenue - A species causing malaria in India, distinguished by its tenuity and ameboid activity. P. vassali - A species found in the squirrel. P. vivax - The species causing benign tertian malaria in man; it has 12-24 merozoites. P. vivax minuta - A species similar to P. vivax, but is smaller and has only 4-10 merozoites.

Platanista A genus of Mammals of the River Dolphin family, *Platanistidae*. e.g. P. gangetica - The Gangetic Dolphin, a blind species, the eyes being mere vestiges. Found in the Brahmaputra, Indus and Ganges rivers.

Platichthys A genus of Fishes of the Flounder family, *Hippoglossidae*. e.g. P. stellatus - The Starry Flounder; it has no scales and is covered with star-like platelets; the eyes are on the left side of the body. Distribution: off the Asiatic and American coasts of the North Pacific, often entering freshwater river estuaries.

Platypoccilus A genus of Fishes of the order of Guppies, Killifishes, Swordtails, and Foureyed fishes, *Microcyprine* or *Cyprinodontiformes*. These are the Platy fishes, brightly colored species kept in aquariums.

Plecotus A genus of Mammals of the Vespertilionid Bat family, *Vespertilionidae*. e.g. P. auritus - The Long-eared Bat, length of body 47-51 mm, wingspan 220-260 mm. Found near human habitations, emerging after dark and flying slowly, close to the ground. It hunts prey in parks, gardens, forests and forest margins. Distribution: the foothills and mountains of northern Europe and Asia.

Plectrophenax A genus of Birds of the Bunting family, *Emberizidae*. e.g. P. nivalis - The Snow Bunting, length 17 cm; found in tundra, rocky country near the seashore and mountains of northern Europe, arctic regions of Asia, and North America. A migrating perching bird.

Pleodorina A genus of Flagellate Protozoans of the order, *Phytomonadina*; the arrangement of cells, in the species, is spherical.

Plethodon A genus of Amphibians of the Newt or Salamander family, *Plethodontidae*. e.g. P. cinereus - The Red-backed Salamander, a nocturnal creature, found by day under or inside old logs, or rock crevices. One of the most abundant species in the eastern United States.

Pleurobranchus A genus of Sea Slug Molluscs, also known by the generic name, *Oscanius*, of the order *Notaspidea* or *Pleurobranchomorpha*. These are small flattened animals with a reduced internal shell, but without a mantle cavity; they are carnivorous, feeding on large squirts and swim with the aid of parapodia or muscular projections, segmentally arranged, from the side of the body.

Pleurodeles A genus of Amphibians of the Newt or Salamander family, *Salamandridae*. These are the western Mediterranean salamanders. In these species the males clasp the females during mating, using their tails as nooses to hold their mates, so that their cloacae are very close together.

Pleuronectes A genus of Fishes of the Right-eyed Flounder family, *Pleuronectidae*. e.g. P. platessa - The European Plaice, length 25-90 cm; often found in brackish waters, tolerating various degrees of salinity; it also enters rivers and spawns in water at great depths. Distribution: northeast Atlantic and western Mediterranean.

Pleuronichthys A genus of Fishes of the Turbot family, *Bothidae*. e.g. P. ritteri - The Spotted Turbot fish.

Plexaura A genus of Horny Corals of the order *Gorgonacea*. e.g. P. homomalla - The Sea Whip or Sea Fan coral, found in coral reefs off the coast of Florida and the Caribbean islands. An important source of prostaglandins, this animal is used in the treatment of articular rheumatism.

Plocamium A genus of plants of the Red Algae family, *Gigartinaceae*. e.g. P. vulgare - The Cockscomb, up to 25 cm high; a somewhat cartilagenous, rosered plant with comblike branchlets. Found on rocky shallows in the North Sea.

Plumaria A genus of plants of the Red Algae family, *Gigartinaceae*. e.g. P. elegans - The Plumaria, up to 10 cm high; a brownish-red to blackish plant in which the branches grow in one plane, and the plumose segments are in two rows. Grows chiefly on rocky shallows in the North Sea.

Plumbago A genus of climbing plants of the family, *Plumbaginaceae*. e.g. P. capensis - Leadwort or Plumbago, a tender, climbing, perennial plant with terminal spikes of pale-blue, 5-lobed flowers. Once believed to be an antidote to lead poisoning. Place of origin: South Africa; introduced into Europe in 1818. P. larpentae (= *Ceratostigma* plumbaginoides) - False Plumbago (see **Ceratostigma**).

Pluvialis A genus of Birds of the Plover family, *Charadriidae*. e.g. P. apricaria - The European Golden Plover, 28 cm long; it inhabits tundras, northern marshes and moorland in Britain, Iceland, Scandinavia and northern Russia. A migrating bird, it winters in

the Mediterranean and in southwest Asia.

Pneumocystis A genus of Protozoan parasites. e.g. P. carinii (= PC) - A species normally commensal in the respiratory tract, which becomes an opportunist pathogen, causing interstitial Plasma Cell Pneumonia (PCP), especially in patients receiving large doses of corticosteroids or immunosuppressive drugs in the treatment of lymphomas and leukemias, or after organ transplantation; also in diseases where the reticuloendothelial system has been severely compromised, such as Acquired immunodeficiency syndrome (AIDS).

Pocadius A genus of Insects of the Flower Beetle family, *Nitidulidae*. e.g. P. ferrugineus - Length 2.8-4.5 mm; develops in puffball fungi. Distribution: Europe.

Podangium A genus of saprophytic microorganisms of the family of schizomycetes, *Polyangiaceae*, found in soil and decaying organic matter.

Podiceps A genus of Birds, also known by the generic name *Colymbus*, of the family *Podicepedidae* or *Colymbidae*. e.g. P. cristatus (= *Colymbus* c.) - The Great Crested Grebe, length 48 cm; prefers stagnant waters; common inhabitant of ponds and lakes. About the size of a seagull, it has a long white neck and white front in contrast to its brown upper parts. Two stiff tufts of black feathers project backwards from the head. Distribution: almost worldwide (almost everywhere in Europe except the far North; Asia; Northwest Africa) (see **Colymbus**).

Podilymbus A genus of Birds of the family *Podicepedidae* or *Colymbidae*. e.g. P. gigas - The Great Pied-billed Grebe, one of the world's rarest birds, believed to have a total population of less than 100 (see **Colymbus**).

Podophyllum A genus of plants of the Barberry family, *Berberidaceae*. e.g. P. peltatum - The Podophyllum plant (=

American Mandrake; May Apple Root); the dried rhizome and roots yield podophyllum resin, a drastic purgative with slow action. Used also as a caustic agent for certain skin tumors.

Podura A genus of Insects (order *Collembola*), of the Springtail family, *Poduridae*. e.g. P. aquatica - Length 1.1-1.5 mm; occurs in large numbers on the surface of puddles. Very common in Europe, Asia, and North America.

Poecilia A genus of Bony Freshwater Fishes (subphylum *Vertebrata*). e.g. P. reticulatus - The Guppy, a tiny freshwater fish producing the least number of eggs. The average yield is 40-50; one female measuring 30 mm (1.2 in) in length was found to have only 4 eggs in her ovaries. The species is often kept in aquariums because of its brilliant coloring. Distribution: Barbados, Trinidad, and Venezuela.

Poephila A genus of Birds of the Grassfinch, Mannikin, and Waxbill family, *Estrildidae*. e.g. P. castanotis - The Common Zebra Finch of Australia, a colorful and popular cage bird, with its black and white barred tail, chestnut ear patches and pink bill. It builds a bottle-shaped nest of dried grasses in a bush or low tree.

Polemaetus A genus of Birds of the Falcon family, the Eagles. These have a visual acuity 8 to 10 times stronger than that of humans. e.g. P. bellicosus - The Martial eagle, a bird of prey.

Polemonium A genus of herbs known as Greek Valerian. Some species are said to have medicinal properties, and to be expectorant and diaphoretic.

Polistes A genus of Insects of the Wasp family, *Vespidae*. e.g. P. nimpha - Length 12 mm; makes a small nest attached to a plant by a stalk. Distribution: the Palaearctic region.

Pollachius A genus of Fishes (subphylum *Vertebrata*). e.g. P. virens - The Pollack fish.

Polyangium A genus of saprophytic microorganisms of the family of Schizomycetes, *Polyangiaceae*, found in soil and decaying organic matter.

Polycelis A genus of Planarian Flatworms of the order *Tricladida*. e.g. P. cornuta - Length 18 mm. Has greatly elongated "ears"; it is furnished with many eyes arranged parallel to the front edge of the body. Distribution: most of Europe, except the north.

Polydesmus A genus of wormlike Arthropods (class *Diplopoda*), of the Millepede (= Millipede; Milliped) family, *Polydesmidae*. The species possess 2 pairs of legs on each of most of their segments (L. mille = thousand; pes/pedis = foot). e.g. P. complanatus - Length 15-23 mm, width 2.3-3.2 mm. A common arthropod species found in damp places such as alder groves and broadleaved forests, under the bark of old tree-stumps. Distribution: Central Europe.

Polydrusus A genus of Insects of the Weevil Beetle family, *Curculionidae*. e.g. P. sericeus - This beetle species is 6-8 mm long and beautifully colored, with gleaming, greenish-gold scales. Found on various broadleaved trees. Distribution: Europe.

Polygala A genus of plants of the Milkwort family, *Polygalaceae*. The species have showy flowers of various colors; they were formerly believed to increase lactation in nursing women, hence their common family name. e.g. P. senega - Seneca Snakeroot, a plant of North America, containing polygalin (= polygalic acid) and a saponin, senegin. It was formerly used in the treatment of asthma, and sometimes in dropsy as a cathartic or hydragogue.

Polygonatum A genus of plants of the Lily family, *Liliaceae*, called Solomon's seal. Several of the species are tonic, vulnerary, diuretic and purgative.

Polygonia A genus of Insects of the Butterfly or Moth family, *Nymphalidae*. e.g. P. c-album - The Comma Butterfly, size 22-25 mm; white pattern on the underside of the hindwings resembles the letter "C". Distribution: Eurasia and North Africa.

Polygonoporus A genus of Tapeworms (class *Cestoda*). e.g. P. giganteus - The world's largest known tapeworm, individuals measuring up to 5 m (16 ft 5 in) in length and 45 mm (1.77 in) in diameter, have been found in the intestine of a sperm whale, indicating a total length of c. 30 m (98 ft 5 in)!

Polygonum A genus of plants of the Knotgrass family, *Polygonaceae*. e.g. P. bistorta - Bistort Rhizome, Snake-weed, or Easter-ledge. The plant root contains tannin, and has been used as an astringent in diarrhea.

Polymnia A genus of flowered plants of the Daisy family, *Compositae*. e.g. P. uvedalia - Leafcup or Bearsfoot; it is an anthelmintic, alterative, and antispasmodic.

Polyodon A genus of Fishes (subphylum *Vertebrata*). e.g. P. spathula - The Paddle-fish.

Polyommatus A genus of Blue Butterflies or Moths of the family, *Lycaenidae*. e.g. P. icarus - The Common Blue Butterfly, size 14-18 mm; found from lowlands to mountainous elevations. Very abundant in the Palaearctic region.

Polyplax A genus of Sucking Lice of Rats and Mice. e.g. P. spinulosa - The common louse of rats.

Polyporus A genus of Mushrooms or Fungi of the family *Polyporaceae*. e.g. P. officinalis (= *Fomes* o.) - Surgeon's Agaric, growing on beech and oak trees; used as a hemostatic (see **Fomes**).

Polypterus A genus of Fishes (subphylum *Vertebrata*). e.g. P. weeksi - The Nile Bichir fish.

Polystoma A genus of worms (class of Flukes, *Trematoda*), also known by the generic name, *Polystomum*, of the order of Monogenetic Trematodes, *Monogenea*. e.g. P. integerrimum (=

Polystomum i.) - A species of monogenetic flukes, occurring in the bladder of frogs, and notable for synchronization between the young stages of the parasite and the host.

Polystomum See **Polystoma**.

Polytoma A genus of Flagellate Protozoan animals of the phytomonad order, *Phytomonadina*. The species are saprozoic in their nutrition, with no cytostome or cell mouth, food intake appearing to take place by diffusion.

Polytrichum A genus of Mosses. e.g. P. juniperinum - Haircap or juniper moss; a diuretic.

Polyxenus A genus of Arthropods of the class of Millipedes, *Diplopoda* (see **Polydesmus**). e.g. P. lagurus - The world's shortest millipede; a British species, measuring 2-3 mm (0.07-0.11 in) in length and 0.5 mm (0.019 in) in breadth.

Pomacanthus A genus of Fishes (subphylum *Vertebrata*), of the Angelfish family. e.g. P. semicirculatus - A striking, tropical species of angelfishes, with striped patterns of alternating colors across, from head to tail.

Pomacentrus A genus of Fishes of the Damsel-fish family. e.g. P. trichourus - The White-tailed Damsel Fish, found in the coral reefs of the Red Sea (see **Dascyllus**).

Pomatoceros A genus of Segmented Worms or Annelids of the Serpulid family, *Serpulidae*. e.g. P. triqueter - A common species, secreting hard, limey tubes, and usually attached to stones; found off the seas of western Europe. There are several color varieties, the crown appearing orange, brown or blue according to different combinations of several distinct pigments.

Pomatomus A genus of marine Fishes of the Bluefish family, *Pomatomidae*. e.g. P. saltatrix - The Bluefish, the world's most ferocious marine fish, weighing on the average about 2 1/4 kg (5 lb), and found in the warmer parts of the seas and oceans. It has been described as "an animated chopping machine", cutting to pieces and destroying as many as ten times the number of fish it can actually eat.

Pomatoschistus A genus of Fishes of the Goby family, *Gobiidae*. e.g. P. minutus - The Sand Goby, length 10 cm; found on sandy bottoms along the Atlantic and Mediterranean coasts.

Pomolobus A genus of Fishes of the Herring family, *Clupeidae*. e.g. P. pseudoharengus - The Alewife, about 25 cm (10 in) long; a marine fish which spawns in freshwater but spends most of its time in the sea. A commercial species.

Pomoxis A genus of Fishes (subphylum *Vertebrata*). e.g. P. nigromaculatus - The Black Crappie fish.

Poncirus A genus of Shrubs of the Rue family, *Rutaceae*; some species are classified under the generic name *Citrus*. e.g. P. trifoliata (= *Aegle* sepiaria; *Citrus* t.) (see **Aegle**; **Citrus**).

Pongamia A genus of East Indian trees of the Pea family, *Leguminosae*. e.g. P. glabra - This species yields a fixed oil.

Pongo A genus of Mammals of the Anthropoid Ape family, *Pongidae*. e.g. P. pigmaeus - The Orangutan or Orang-utan, the sole species of this genus, strictly confined to the low-lying forests of Sumatra and Borneo. The arms are longer than those of the gorillas and chimpanzees, the legs shorter, and the canines more fang-like. Its cranial capacity is midway between gorillas and chimpanzees. The Orang-utan is the only anthropoid whose head has developed upwards, producing a high, arched forehead, and its facial resemblance to man is remarkable.

Pontederia A genus of plants of the family *Pontederiaceae*, named after Guilo Pontedera (1688-1757), a botanist at Padua. e.g. P. cordata - The Pickerel Weed or Wampee, a robust, herbaceous, perennial plant, 45-90 cm (11/2-3 ft)

high, with smooth, ovate-cordate, dark green leaves growing on erect stems, and sky-blue flowers in terminal spikes. Place of origin: North America; introduced into Europe in 1597.

Pontoporia A genus of Mammals of the River Dolphin family, *Platanistidae*. e.g. P. blainvillei - The La Plata Dolphin, a very short animal, about 1.5 m (5 ft) long; a river dolphin with a long, almost bird-like beak, but furnished with teeth. It has lost all contact with the sea and must not be confused with the Marine Dolphins (family *Delphinidae*).

Popillia A genus of Insects of the Beetle family. e.g. P. japonica - The Japanese beetle (see **Rickettsiella**).

Populus A genus of Salicaceous Trees and Shrubs, the Aspens, Cottonwoods, Poplars and Willows, of the Willow family, *Salicaceae*. The bark is tonic, containing the glucosides, populin and salicin. e.g. P. candicans (= P. balsamifera) - The Poplar Tree. The air-dried, closed, winter leaf buds (Popular Bud; Populi Gemma), yield an extract called Balm of Gilead Bud (of commerce), affording a variety of tacamahae - stimulant, expectorant, and tonic internally, while being counter-irritant and vulnerary externally. P. tacamahacca - A species yielding the "Balsam Poplar Bud". Uses as above.

Porcellana A genus of Crustaceans of the Porcelain Crab family, *Porcellanidae*. e.g. P. platycheles - An Anomura species, a crab-like crustacean possessing a tail fin, and resembling the encrusting growths of animals and plants found on the undersides of stones on the seashore, thus resulting in a very effective camouflage. Distribution: Europe.

Porcellanaster A genus of Echinoderm animals of the class of Starfishes, *Asteroidea*. e.g. P. ivanovi - A species recovered from the greatest depth (7,584 m = 24,872 ft) in the Mariana Trench in the western Pacific, by a Russian research vessel.

Porcellio A genus of Insects of the Sow bug or Woodlouse family, *Oniscidae*. e.g. P. scaber - Length up to 16 mm; ground color yellowish with grey markings; common in damp places both in the open and in buildings. Distribution: worldwide.

Porocephalus A genus of wormlike Arthropods, formerly known by the generic name *Armillifer*, of the order *Linguatulida*, parasitic in man and animals. e.g. P. armillatus (= *Armillifer* a.) - The adult is found in the lungs and trachea of the python (*Python* sebae); the larval forms are found in monkeys, and occasionally in man. P. denticulatus - The larva of *Linguatula* rhinaria (serrata) (see **Linguatula**).

Porthetria A genus of Insects, better known by the generic name *Lymantria*, of the Tussock Moth family, *Lymantriidae*. e.g. P. dispar (= *Lymantria* d.) - The Gipsy Moth (see **Lymantria**).

Portunus A genus of Swimming Crabs of the family *Portunidae*. e.g. P. holsatus - Length about 40 mm; has a smooth, grey-green carapace; found by the seashore as well as at greater depths; locally abundant. Distribution: the Atlantic coast of Europe.

Porzana A genus of Birds of the Coot, Gallinule, and Rail family, *Rallidae*. e.g. P. porzana - The Spotted Crake, length 23 cm; lives a concealed life in marshes and along the banks of calm bodies of water with dense vegetation. A migrant bird, found in Europe and Asia.

Potamobius A genus of Crayfishes of the family *Astacidae*. e.g. P. astacus - The Common European Crayfish (see **Astacus**).

Potamochoerus A genus of Mammals of the family of Wild Pigs, *Suidae*. e.g. P. porcus - The Red River Hog, with a white mane and a grey face; it has pointed ears ending in tufts of hair, and lives in reed beds and thickets, causing great damage to crops. Spread over Africa, this Bush pig has a related form in Masagascar.

Potamogale A genus of insectivore rodents, the Shrews. e.g. P. velox - The Giant African Water Shrew of Central Africa; looks remarkably like a small otter; has been credited with the title of "largest insectivore", having a body weight of up to 1,000 g (2.2 lb) and a head and body length of 290-350 mm (11.4-13.7 in).

Potentilla A genus of plants of the Rose family, *Rosaceae*. e.g. P. erecta - The Common Tormentil plant (= Erect Cinquefoil; Consolda Vermelha). The dried rhizome contains a red coloring principle and tannin. Has been used as an astringent.

Poterolebias A genus of Fishes of the suborder *Cyprinodentei*. e.g. P. longipinnis - A species of shortest-living killing fish of South America, living about 8 months in the wild state. They are found in temporary ponds and drainage ditches, the eggs being laid in the mud at the water bottom. When this dries up the fish die and the eggs aestivate (i.e. fall into a state of torpor) until the next wet season, when they hatch, grow at great speed and spawn, dying in the next drought.

Potosia A genus of Ants. e.g. P. cuprea - The Wood ant.

Priapulus A genus of Worms of the phylum *Priapulida*. e.g. P. caudatus - Length 30-80 mm; flesh-red in color; elongated, cylindrical body, furnished with numerous cuticular hooks. Lives in the littoral zone along the Atlantic coast, where it burrows in the mud.

Primula A genus of plants of the Primrose family, *Primulaceae*. e.g. P. veris (= P. officinalis) - The Cowslip plant; the dried rhizome and roots contain saponins, and have been used as expectorants.

Prinos A genus or subgenus of aquifolaceous shrubs, commonly assigned to the genus *Ilex* (holly) (see **Ilex**). e.g. P. verticillatus - The black alder, or winterberry of North America; has a tonic and astringent bark.

Prionace A genus of Fishes of the Requiem Shark family, *Carcharhinidae*. e.g. P. glauca - The Great Blue Shark, length up to 4 m (13 ft); an inhabitant of the open seas, travelling swiftly along great distances at speeds of more than 21.3 knots (39.2 km/h = 24.5 miles/h). It is a viviparous fish, dangerous to man; the newly born measure 50-60 cm (20-24 in). In Japan it is hunted in large numbers as food.

Prionotus A genus of Fishes (subphylum *Vertebrata*). e.g. P. carolinus - The Common Sea-robin fish.

Prionus A genus of Insects of the Longhorned or Longicorn Beetle family, *Cerambycidae*. e.g. P. coriarius - Length 18-45 mm; found in old broadleaved and coniferous forests; they are on the wing at dusk, being most abundant in August. They emit rasping sounds by rubbing the margins of the elytra against their hind legs. Distribution: Europe, Asia and North Africa.

Pristinotherium A genus of one of the largest prehistoric Mammals ever recorded, and like the genus *Indricotherium*, the fossil remains have been discovered in Kazakhstan in Central Asia (see **Baluchitherium**; **Indricotherium**).

Pristis A genus of Fishes (subphylum *Vertebrata*). e.g. P. pectinaus - The Sawfish; the saw teeth are much larger than those of the Saw sharks (family *Squalidae*).

Proactinomyces A name formerly given to a genus of microorganisms.

Procambarus A genus of Crustaceans of the North American Crayfish family, *Astacidae*. e.g. P. clarkii - An eastern North American crayfish found in Louisiana, maturing to about 45 mm (1

3/4 in) in length. It lives in caves and ponds.

Procavia A genus of Mammals of the Hyrax family, *Procaviidae*. e.g. P. capensis - The Rocky Hyrax or Dassy, a small herbivorous mammal found throughout Africa, as well as in Syria and southern Arabia. It lives in colonies on rocky cliffs or hill-tops. Each foot has 4 toes ending in blunt claws, and the soles have footpads that assist in climbing.

Procerodes A genus of Flatworms of the order of Triclads, *Tricladida*. These are marine forms which have been catalogued in a suborder, *Maricola*.

Profelis A genus of Mammals, also known by the generic name *Felis*, of the Cat family, *Felidae*. e.g. P. aurata (= *Felis* a.) - The African Golden Cat, a slim creature, about 75 cm (21/2 ft) long, with a small head and a tail of some 40 cm (1 ft 4 in). Variously colored, it is most frequently a reddish-brown and pale underneath, but it may be greyish, and the coat is often heavily spotted. It lives in the forests of equatorial Africa, feeding on birds and small rodents.

Promachocrinus A genus of Echinoderm animals of the class of Sea-lilies, *Crinoidea*. e.g. P. kerguelensis - The Atlantic feather-star, a crinoid echinoderm having a greater age potential than most sea-lilies, because it does not reach maturity until the tenth year. Most of the small feather-stars have a lifespan of only 2 to 3 years.

Promicrops A genus of Fishes (subphylum *Vertebrata*). e.g. P. itaiara - The Spotted Jew-fish.

Propeomyces A genus of yeast-like Fungi of the family *Eremascaceae*, various species of which cause dermal lesions.

Propionibacterium A genus of microorganisms of the family *Propionibacteriaceae*, better known by the generic name *Corynebacterium*. Non-spore-forming, anaerobic, gram-positive bacilli found as saprophytes in dairy products. e.g. P. acne - A species causing skin acne (see **Corynebacterium**).

Prosthagonia A genus of Trematode parasites. e.g. P. macrorchis - A species occurring in chickens, turkeys, pheasants and other birds.

Prostherapis A genus of Amphibians of the Frog family, *Ranidae*. The males of the species carry the tadpoles on their backs.

Protaminobacter A genus of microorganisms of the family *Pseudomonadaceae*. Motile or nonmotile, frequently pigmented cells in soil or water. e.g. P. alboflavus.

Protea A genus of Trees of many species from various wet and warm regions, several of which are medicinal.

Proteoglypha A genus of poisonous Snakes that have small, grooved, stationary fangs that must be held in the wound if the poison is to reach the deeper tissues. e.g. The Harlequin snake; The Sonoran coral snake.

Proteosoma A genus of parasitic microzoa from the blood of birds.

Proterorhinus A genus of Fishes of the Goby family, *Gobiidae*. e.g. P. marmoratus - The Mottled Black Sea Goby, length 7-11 cm; locally abundant in irrigation canals with dense vegetation, dwelling near the bottom in brackish waters and also in rivers. Distribution: rivers flowing into the Black Sea.

Proteus A genus of microorganisms of the family *Enterobacteriaceae*. Gram-negative, generally active, motile, rod-shaped bacteria of limited pathogenicity, usually found in fecal and other putrefying material, and appearing as concomitants with the dysentery bacilli. e.g. P. mirabilis - A species which is usually saprophytic; occasionally acts as a human pathogen, causing urinary tract and other infections. P. morgani (= Morgan's bacillus; P. morganii; *Salmonella* morgani) - A non-sporing,

gelatin-nonliquefying, aerobic bacillus, serologically related to some strains of *Bacillus* proteus. Found in the intestine and associated with summer diarrhea of infants.

Protocalliphora A genus of Flies whose larvae feed on nesting birds.

Protoglossus A genus of Chordate animals of the class of Acorn Worms, *Enteropneusta* (see **Harrimania**).

Protominobacter A genus of bacteria containing several species found in soil and water.

Protospirura A genus of Nematode parasites. e.g. P. gracilis - A species found in cats.

Protostrongylus A genus of Lung Worms. e.g. P. rufescens - A species infecting sheep, goats, and deer.

Protula A genus of Segmented Worms (class *Annelida*) of the family, *Serpulidae*. e.g. P. tubularia - Length up to 50 mm; body composed of 100-125 segments; lives concealed in a tube attached to rocks, shells or stones, with only the numerous brightly colored tentacles extending. Distribution: Atlantic coast and the Mediterranean.

Prowazekella See **Prowazekia**.

Prowazekia A genus of Flagellate organisms, also known by the generic name *Prowazekella*, having two nuclei and two flagellae, found in human feces and urine; non-pathogenic.

Prunella 1. A genus of Perching Birds of the Accentor family, *Prunellidae*. e.g. P. modularis - The Dunnock, length 18 cm; inhabits broadleaved and mixed woods, and parks with thick undergrowth. Distribution: almost all Europe. 2. A genus of aromatic plants of Thyme family, *Labiatae*. e.g. P. vulgaris - The "Heal-all", an astringent and tonic species.

Prunus A genus of Trees and Shrubs of the Rose family, *Rosaceae*. e.g. P. americana - The Plum tree. P. amygdala - The Almond tree. P. amygdalus var. amara (= *Amygdalus* communis var. a.) -

The Bitter Almond tree (see **Amygdalus**). P. amygdalus var. dulcis (= *Amygdalus* communis var. d.) - The Sweet Almond tree (see **Amygdalus**). P. armeniaca - The Apricot tree. P. avium - The Wild Cherry tree (black cherry). P. cerasus - The Red or Sour Cherry tree. P. domestica - The Prune tree. P. lauro-cerasus - The Cherry-laurel tree. P. per-sica - The Peach tree. P. serotina - The Wild or Black Cherry tree (see P. avium). P. spinosa - The Sloe or Blackthorn tree (a species of plum tree) P. virginiana - The Choke-cherry tree of North America.

Psammechinus A genus of Echinoderms or spiny-skinned, invertebrate, marine animals. e.g. P. miliaris - A species known to live up to eight years under laboratory conditions.

Psammodromus A genus of Reptiles of the Lizard family, *Lacertidae*. e.g. P. hispanicus - The Spanish Psammodromus, length 10-20 cm. Very swift and shy, the species are found in dry rocky regions, and in coastal sand dunes with sparse vegetation. Distribution: the Iberian Peninsula and southern France.

Psammothuria A genus of Echinoderm animals of the class of Sea-cucumbers, *Holothuroidea*. e.g. P. ganapatii - The smallest known sea-cucumber from southern India, which does not exceed 4 mm (0.15 in) in length.

Psetta A genus of Fishes of the Turbot Flatfish family, *Bothidae*. e.g. P. maxima - A European flatfish or Turbot fish, weighing up to 13.6-18 kg (20-30 lb) and highly valued as food.

Psettodes A genus of Fishes of the family *Psettodidae*. e.g. P. erumeni - It has one dorsal fin and spiny rays in the fins. Found in the Indian Ocean, the Red Sea and the Malaysian archipelago.

Pseudacris A genus of Amphibians of the Typical Treefrog family, *Hylidae*. These are the Chorus frogs found in the United States.

Pseudamphistomum A genus of Trematode Worms. e.g. P. truncatum - A species found in the bile ducts of cats and dogs; occasionally in man.

Pseudis A genus of Amphibians of the Frog family, *Pseudidae*. e.g. P. para-doxus - An aquatic species found in high-altitude lakes in South America. The contrast between the very large tad-poles and the much smaller adults is very marked, hence the name.

Pseudocarcinus A genus of Crustaceans (section *Brachyura*) of the Crab family, *Portunidae*. e.g. P. gigas - The Great Xanthid crab, found offshore in the Bass Strait area separating Australia from Tasmania. This edible crab has been accurately measured up to 456 mm (18 in) across the carapace, and may reach a weight of 13.6 kg (30 lb).

Pseudodiscus A genus of Flukes. e.g. P. watsoni - A fluke found in the small intestine of man in Africa.

Pseudogorilla A genus of Primates of the family of Gorillas. e.g. P. mayema - The "Pygmy gorilla" of the Gabon coastal region, West Africa; some female speci-mens are known to have weighed only 14 kg (31 lb) and measured only just over 61 cm (2 ft) in height, at the age of about four years.

Pseudois A genus of Rodents of the fami-ly of Hares, living at high altitudes. e.g. P. nayaur - The Bharal hare; has been found above 5,486 m (= 18,000 ft) at very cold temperatures (-17° C or +1.4° F).

Pseudolynchia A genus of Insects, also known by the generic name *Lynchia*, of the Pigeon Fly family. e.g. P. canariensis (= *Lynchia* c.) - A species of Pigeon flies that transmits *Plasmodium* capistrani, cathemerium, etc., the organisms caus-ing malaria in pigeons (see **Plasmodium**).

Pseudomonas A genus of microorgan-isms, also called by the genus name *Actinobacillus* of the family *Pseudomonadaceae*. Gram-negative, aerobic and facultative anaerobic, monotrichous, lophotrichous or non-motile straight rods. e.g. P. aeruginosa (= Ps. aeruginosa; *Bacillus* pyocyaneus) - A species pathogenic to man, producing a blue-green pigment, pyocyanin and giving a blue color to the pus of certain suppurative lesions. Usually found in mixed infections, with pyogenic cocci. May occur as a commensal in the bowel of man and animals, and may cause enteritis and even a general infection, such as septicemia. P. mallei (= *Actinobacillus* m.) - A species causing glanders.

Pseudomonilia A genus of yeast-like Fungi, many species of which have been isolated from blastomycotic lesions in man.

Pseudomydura A genus of rare Tortoises. e.g. P. umbrina - The protected Short-necked tortoise, the rarest living chelonian in the world, found in a swamp area known as the Ellen Brook Reserve, some 32 km (20 miles) northeast of Perth, Western Australia. In 1973 the total world population of this species was estimated at 150 animals.

Pseudophyllanax A genus of Crickets. e.g. P. imperialis - A large cricket of New Caledonia, S.W. Pacific; body length 105 mm (4.13 in), and wingspan 240 mm (9.44 in).

Pseudopleuronectes A genus of Fishes (subphylum *Vertebrata*). e.g. P. americanus - The Winter Flounder fish.

Psidium A genus of Trees of the Myrtle family, *Myrtaceae*. e.g. P. guajava - The Guava tree - The rind from the fruit contains tannin, and has been used as an adjunct in the treatment of infantile diarrhea.

Psilocybe A genus of Gill Fungi of the family *Agaricaceae*. e.g. P. mexicana - A species of gill fungi or mushrooms grown in Mexico, from which are obtained the alkaloids, Psilocybin (= 4-Phosphoryloxy-N,N-dimethyltryptamine); 3-[2-Dimethylaminoethyl]indol-

4-yl dihydrogen phosphate) and Psilocin (= 4-Hydroxydimethyltryptamine). Both are hallucinogens. Psilocybin is converted in the body to psilocin and is also derived from the mushrooms *Stropharia* cubensis and *Conocybe* spp.

Psithyrus A genus of Insects of the Bee family. e.g. P. rupestris - The parasitic Cuckoo-bee, which does not build its own nest, but invades the nest of other bees, such as the Bumblebee (*Bombus* lapidarius). Distribution; Eurasia.

Psittacus A genus of Birds of the Parrot family. e.g. P. erythacus - The African grey parrot. Reportedly said to be a longlived bird.

Psophus A genus of Insects of the Locust and Short-horned Grasshopper family, *Acrididae*. e.g. P. stridulus - Length 23-32 mm; the name is derived from the grating sounds so characteristic of the species. Found in mountainous elevations in meadows and clearings. Distribution: Europe (absent in Scandinavia and the British Isles).

Psoralea A genus of plants of the Pea family, *Leguminosae*. e.g. P. corylifolia - Psoralea Fruits (= Psoralea Seeds; Babchi). Contain the oleoresins psoralen and isopsoralen; used in the treatment of leucoderma (vitiligo) and other skin diseases.

Psoroptes A genus of Itch Mites. e.g. P. equi - A species causing skin lesions in horses (psoroptic mange).

Psylla A genus of Insects of the Jumping Plant Louse family, *Psyllidae*. e.g. P. mali - The Apple Sucker Louse, length 3.5 mm; found on apple trees. The flat nymphs, which are injurious to the leaves, excrete large amounts of sweet honeydew which causes the leaves and flowerbuds to become sticky. Distribution: almost worldwide.

Pteranodon A genus of now extinct large flying Reptiles, the Pterosaurs, that lived about 70 million years ago. e.g. P. ingens - A species of pterosaurs, estimated at having had a wingspan of up to 8 m (26 ft 3 in).

Pterocarpus A genus of plants of the Pea family, *Leguminosae*. e.g. P. santalinus - Red Sanders Wood (= Heartwood; Lignum Santali Rubrum; Red Sandal Wood; Red Saunders; Ruby Wood; C. I. Natural Red 22). Has been used as a red coloring agent for tinctures and similar alcoholic preparations.

Pterocles A genus of rapidly flying Birds. e.g. P. alchata - The Pintailed Sandgrouse; found in deserts. The maximum speed of this bird through still air in level flight has been estimated at 99 km/h (62 miles/h).

Pterois A genus of Fishes (subphylum *Vertebrata*). e.g. P. volans - The Lion-fish or Zebra Fish.

Pteromys A genus of Rodents of the Tree Squirrel family, *Sciuridae*. e.g. P. volans - The European Flying Squirrel, length of body 135-205 mm; tail 90-140 mm. Glides to the ground by means of an extensible skin fold which can be spread out like an aerofoil between the fore and hind feet. Found in mixed woods and aspen groves. Mainly an inhabitant of northern Asia and Europe.

Pteronura A genus of Otters. e.g. P. brasiliensis - The Giant Brazilian otter, capable of feeding on the ferocious freshwater Piranha fish (of South American rivers) with relative impunity.

Pterophyllum A genus of Fishes (subphylum *Vertebrata*). e.g. P. eimekei - The Freshwater Angel-fish (Scalare).

Pteropus A genus of Mammals of the family of large Bats. e.g. P. vampyrus - The Kalong fruit bat or Flying Fox; found in Malaysia and Indonesia, and the world's largest known bat in terms of wing expanse. It has a wingspan measurement of up to 1.70 m (5 ft 7 in), a head and body length of about 400 mm (15 3/4 in), and weighs up to 900 g (31.8 oz).

Pterostichus A genus of Insects of the Ground Beetle family, *Carabidae*. e.g. P.

niger - A species of black ground beetles, 16-21 mm long; found in fields, forests, and gardens. Distribution: Europe.

Ptinus A genus of Insects of the Spider Beetle family, *Ptinidae*. e.g. P. fur - Length 2-4.3 mm; usually found in households, stables, attics, and bird nests. Distribution: the Palaearctic region, and North America.

Ptiolcercus A genus of primate Mammals, the Shrews. e.g. P. lowii - The Feather-tailed shrew; the smallest known primate, with a total length of 230-330 mm (9-13 in): head and body 100-140 mm (3.93-5.51 in), tail 130-190 mm (5.1-7.5 in), and weight of 35-50 g (1.23-1.76 oz).

Ptychocheilus A genus of Fishes (subphylum *Vertebrata*). e.g. P. lucius - The Colorado Squaw-fish.

Ptychodera A genus of Chordate animals of the class of Acorn Worms, *Enteropneusta* (see **Harrimania**).

Pudu A genus of Mammals of the Deer family, *Cervidae*. e.g. P. mephistophiles - The Pudu of northern South America; the smallest true deer, adult males measuring 330-381 mm (13-15 in) at the withers and weighing 8.16-9.07 kg (18-20 lb).

Pueraria A genus of plants of the Pea family, *Leguminosae*. e.g. P. hirsuta - The Pueraria plant; from the root is obtained Pueraria Starch.

Pulex A genus of Insects of the Flea family, *Pulicidae*, which are parasitic on man, and on badgers, cats, and dogs. e.g. P. cheopis (= *Xenopsylla* c.) - A rat flea transmitting plague and murine typhus. P. irritans - The Common Flea or the Human Flea, length 2-3.5 mm; attacks man as well as domestic and wild animals, sucking their blood and causing intense itching by its bite. It is an excellent jumper, able to leap 130 times its own height (at least 196 mm or 7 3/4 in), subjecting itself to a force of 200 g, the energy coming from the elas-

ticity of resilin protein situated above
the jumping legs. In contrast, a human
being would black out at 14 g! The
female lays about 400 eggs.
Distribution: worldwide. P. penetrans (=
Tunga p.) - The Chigo, Jigger, or Sand
flea of Tropical America and the southern United States. Causes intense itching in humans. P. serraticeps (=
Ctenocephalides canis; *Ctenocephalus*
c.) - Flea parasite on dogs (see
Ctenocephalides).

Pullularia A genus of Fungi, species of
which have been isolated from pigmented lesions resembling mycetoma.

Pulsatilla A genus of plants, also known
by the generic name *Anemone*, of the
Buttercup or Crowsfoot family,
Ranunculaceae. e.g. P. vulgaris (=
Anemone pulsatilla) - see **Anemone**.

Punica A genus of Trees of the
Pomegranate family, *Punicaceae*. e.g. P.
granatum - The Pomegranate tree. The
bark (Pomegranate Root Bark; Granado;
Granati Cortex; Granatrinde; Granatum;
Grenadier; Melograno; Romeira) contains alkaloids which have been used in
the expulsion of tapeworms. The rind
contains gallotannic acid, and has been
used as an astringent in the treatment of
diarrhea and dysentery.

Puntius A genus of Freshwater Fish (subphylum *Vertebrata*). e.g. P. javanicus - A
species placed in freshwater ponds in
order to eliminate the weeds necessary
for the propogation of mosquitoes.

Pupilla A genus of Gastropod Molluscs
of the family *Pupillidae*. e.g. P. musco-
rum - Shell 3-4 mm high; 1.8-2 mm
wide; small, thick, and reddish-brown.
A common species living on dry grassy
slopes, particularly on chalky-loamy
substrates. Distribution; Eurasia and
North America.

Pusa A genus of Mammals, also known
by the generic name *Phoca*, of the Seal
family, *Phocidae*. e.g. P. vitulina (=
Phoca v.) - The Common Seal (see
Phoca).

Putorius A genus of Mammals of the
family of Weasels and their allies,
Mustelidae. e.g. P. putorius - The
European Polecat, length of body 400-
440 mm, tail 130-190 mm; partial to
human dwellings; may be encountered
in fields as well as forests. When threatened or in danger it discharges a fetid
liquid from the anal gland at the base of
its tail. Distribution: Eurasia and northwest Africa.

Pycnanthemum A genus of American
plants of the family *Labiata*, called
"basil" and "mountain mint" - aromatic
and carminative.

Pycnopodia A genus of Echinoderms of
the class of Starfishes, *Asteroidea*. e.g.
P. helianthoides - The Sunflower or
Twenty-rayed starfish of the North
Pacific. One of the bulkiest of starfish
echinoderms, spanning up to 121.9 cm
(4 ft).

Pygocentrus A genus of ferocious freshwater fishes, the Piranhas, which live in
the sluggish waters of the large rivers of
South America. They are utterly fearless
cannibals, having razor-sharp teeth, and
their jaw muscles are so powerful they
can bite off a man's finger or toe like a
carrot. Other genera of the Piranhas are
Pygopristis and *Serrasalinus*.

Pygopristis A genus of ferocious freshwater fishes, the Piranhas (see
Pygocentrus).

Pygoscelis A genus of fast-swimming
Birds of the Penguin family. e.g. P.
papua - The Gentoo penguin of
Antarctica, the fastest swimmer among
birds, reaching speeds of 40-43 km/h
(25-27 miles/h).

Pyracantha A genus of plants, also
known by the generic name *Crataegus*,
of the Rose family, *Rosaceae*. e.g. P.
coccinea (= *Crataegus* pyracantha) -
The Firethorn plant (see **Crataegus**).

Pyralis A genus of Insects of the
Butterfly or Moth family, *Pyralidae*. e.g.
P. farinalis - The Meal Moth, size 8-12
mm. The larvae are found in mills, grain

stores, bakeries, etc., where they feed on flour, grain and straw. Locally still abundant, but no longer considered as important a pest as before. Distribution: almost worldwide.

Pyretophorus A genus of Mosquitoes. e.g. P. costalis - Transmits malaria and filariasis in Africa.

Pyrgus A genus of Insects of the Skipper Moth family, *Hesperiidae*. e.g. P. malvae - The Grizzled Skipper Moth, size 10-11 mm; found in forest clearings and forest margins; the larvae are concealed in rolled leaves. Distribution: the Balkans to southern Scandinavia; Asia.

Pyrochroa A genus of Insects of the Cardinal Beetle family, *Pyrochroidae*. e.g. P. coccinea - Length 14-15 mm; head and scutellum are black, the elytra are red. Found on flowers, stumps, and felled trees. Distribution: Europe.

Pyroplasma or **Pyrosoma** See **Piroplasma**.

Pyrrhidium A genus of genus of Insects of the Long-horned or Longicorn Beetle family, *Cerambycidae*. e.g. P. sanguineum - Length 8-12 mm; velvety-red hairs cover the whole body. Found on felled oak logs in which eggs are laid from April onwards. Distribution: Eurasia, the Middle East, and North Africa.

Pyrrhocorax A genus of Birds. e.g. P. graculus - The Alpine Chough; has the highest acceptable altitude recorded for a bird, namely 8,200 m (26,902 ft), in the May 1924 British Everest expedition to Camp V.

Pyrrhocoris A genus of Insects of the Cotton-stainer Bug family, *Pyrrhocoridae*. e.g. P. apterus - The Firebug, length 7-12 mm; found at the base of trees in parks, tree avenues, etc., usually in great numbers. Red-colored elytra with black spots. Distribution; almost worldwide.

Pyrrhula A genus of Perching Birds of the Finch family, *Fringillidae*. e.g. P. pyrrhula - The Bullfinch, length 15 cm;

the male has a striking red breast; the female is greyish-brown. Found in coniferous and mixed forests, in parks, and occasionally also in gardens. Distribution: Eurasia.

Pyrrosoma A genus of Insects of the Damselfly and Dragonfly family, *Coenagriidae*. e.g. P. nymphula - Length 35 mm; wingspan 45 mm; body reddish-black. Found near calm or very slow-flowing waterways. Distribution: Europe and Asia Minor.

Pyrus A genus of Trees of the Rose family, *Rosaceae*. e.g. P. malus - The Apple tree.

Python A genus of very large Snakes. e.g. P. sebae - The African rock python; some specimens have exceeded 10.66 m (35 ft) in length.

Pyxicephalus A genus of long-living Frogs. e.g. P. adspersus - A large Bullfrog, known to live up to 20 years in captivity.

Q

Quassia A genus of Trees of the
Ailanthus family, *Simarubaceae*. e.g. Q.
amara - Surinam quassia, a tropical tree,
the dried stem-wood of which has been
used as a bitter stomachic; also by
enema as an anthelmintic for the expulsion of threadworms (see **Aeschrion**).

Quelea A genus of Birds, class *Aves*. e.g.
Q. quelea - The Red-billed quelea, the
most abundant species of birds, and the
most destructive, estimated as having a
world population of up to 10 billion.
Distributed throughout the drier parts of
Africa south of the Sahara, it is popularly known as the "feathered locust", pos-
ing a serious threat to cereal crops in 25
developing African countries.

Quercus A genus of Oak Trees of the
family *Fugaceae*. e.g. Q. infectoria -
Excrescences of the twigs of this tree,
by development in them of the larva of
the gall wasp, *Adleria* gallae tinctoria,
results in the production of gall (=
Allepo galls; Blue galls), containing gallotannic acid. Gall is used as an astringent in ointments and suppositories for
the treatment of hemorrhoids. It is the
main source of tannic acid (see
Adleria). Q. robur (= Q. pediculata) -
The common oak tree. The dried bark
yields the astringent, quercitannic acid.

Quillaja A genus of Trees of the Rose
family, *Rosaceae*. e.g. Q. saponaria -
From the dried bark of this tree is prepared Quillaia (= Quill; Panama Bark;
Soap Bark). It contains toxic saponins
(quillaic acid), and is used chiefly as an
emulsifying agent in creosote and tar
preparations, a foam stabilizer in contra-
ceptive tablets, and a frothing agent for
various technical purposes.

R

Raillietina A genus of Tapeworms (class *Cestoda*), of the family *Davainiidae*, many species of which infest chickens, guinea-fowl, and turkeys. e.g. R. mada-gascariensis - The Madagascar tape-worm, which occasionally infests man; it has a scolex with four deeply excavated suckers with minute rim spines.

Raja A genus of Fishes of Skate family, *Rajidae*. e.g. R. clavata - The Thornback Ray or Skate-fish, length 70-120 cm; has many large spines on the back. An important fish on the market, for its meat and liver. Distribution: the sea-coasts of Europe.

Rallus A genus of Birds of the Rail family, *Rallidae*. e.g. R. aquaticus - The Water Rail, length 28 cm; found on the banks of calm bodies of water among dense growth of reeds, in marshes, wet meadows, and moorland. Distribution: Europe, Asia, and northwest Africa.

Ramibacterium A genus of microorganisms of the family *Lactobacillaceae*. Non-sporulating, anaerobic, gram-positive bacilli found in the intestinal tract; occasionally pathogenic.

Rana A genus of Frogs of the True Frog family, *Ranidae*. e.g. R. esculenta - The Edible Frog, length 10-12.5 cm; has a dark-spotted belly; lives around quiet and flowing waters, and hibernates in mud on the water bottom. Lives as much as 16 years in captivity. Distribution: all of Europe.

Rangifer A genus of Horned Ungulates of the Deer family, *Cervidae*. e.g. R. tarandus - The Reindeer, length of body 130-220 cm; height at shoulder 110-120 cm. Both male and female have antlers. Found in northern tundras as well as in forests, in the northern parts of Europe, Asia, and North America.

Raphidia A genus of Insects of the Alder and Snake Fly family, *Raphidiidae*. e.g.

R. notata - The Snake-fly, length 15 mm; span of forewings 25-29 mm. Found in forests in shrubs and ground vegetation. Distribution: Northern and Central Europe.

Rattus A genus of Old World Rodents, formerly known by the generic name *Epimys* of the Rat family, *Muridae*. e.g. R. rattus (= *Epimys* r.) - The Black Rat or Indian Plague Rat; length of body 158-252 mm, tail 190-240 mm; found in dry places and on the higher floors of buildings. Often mistaken for the Brown Rat (R. norvegicus), but is smaller and more slender, feeding chiefly on plant food, seeds, cereal grains and fruit. Distribution: worldwide.

Rauwolfia A genus of tropical Trees and Shrubs of the Dogbane family, *Apocynaceae*, many species of which have been used in South America and Africa as a source of arrow poisons. Also found widely in India, Burma, Java and Siam. e.g. R. serpentina - From the dried root is obtained Rauwolfia (= Rauwol; Rauwolfia Serpentina; Chotachand), containing numerous alkaloids, the most active as hypotensive agents being the ester alkaloids, reserpine and rescinnamine. Reserpine is widely used in the treatment of essential hypertension, anxiety states and neuropsychiatric disorders. Other, non-esterified alkaloids of rauwolfia, related to reserpic acid, include ajmaline (rauwolfine) - closely resembling the actions of quinidine and used in the treatment of cardiac arrhythmias; also ajmalinine, ajmalicine, isoajmaline (isorauwolfine), serpentine, rauwolfinine, and sarpagine.

Recurvirostra A genus of Birds of the Avocet and Stilt family, *Recurvirostridae*. e.g. R. avosetta - The Avocet, length 43 cm; a bird found on coasts near brackish waters and river estuaries. Distribution: Western and southern Europe.

Regalecus A genus of Fishes (subphylum *Vertebrata*). e.g. R. glesne - The Oar or

Ribbon fish, one of the world's largest fishes, measuring as much as 7.62 m (25 ft) in length.

Regulus A genus of Birds of the Old World Warbler family, *Sylviidae*. e.g. R. regulus - The Goldcrest, length 9 cm, the smallest European bird. Found chiefly in spruce and pine woods. A resident species, distributed in almost all Europe and North America.

Remijia A genus of Shrubs of the Madder family, *Rubiaceae*. e.g. R. pedunculata - A species furnishing cuprea bark and the derivatives of cupreine.

Remiz A genus of Birds of the Penduline Tit family, *Remizidae*. e.g. R. pendulinus - The Penduline Tit, length 11 cm. Found along calm and flowing waters bordered with dense reeds and thickets. A resident bird. Distribution: Europe, Central and Far East Asia.

Reticulitermes A genus of Insects (Termite order, *Isoptera*) of the family *Rhinotermitidae*. e.g. R. flavipes - A destructive subterranean species of termites from the northeastern United States.

Retortamonas A genus of intestinal Flagellates, formerly known as *Embadomonas*. Several species have been found in lower animals. e.g. R. intestinalis (= *Embadomonas* i.) - This species has been found in humans.

Rhabditis A genus of minute Nematodes or True Roundworms of the superfamily, *Rhabditoidea*, living mostly in damp earth, but occasionally infesting man. Also known by the generic names *Rhabdonema* and *Leptodera*. e.g. R. intestinalis (= *Rhabdonema* i.) - A species infesting humans. R. genitalis (= *Leptodera* pellio) - A minute roundworm species.

Rhaa A genus of microorganisms of the family *Thiorhodaceae* - Irregular long rods or filaments, occurring singly. e.g. R. gracilis.

Rhabdonema See **Rhabditis**.

Rhacophorus A genus of Amphibians of the Tropical Treefrog family, *Rhacophoridae*. Some species are known as the "Flying Frogs" because of their long slanting glides from tall trees. Found in Africa and eastern Asia.

Rhagio A genus of Insects of the Biting Snipe Fly family, *Rhagionidae*. e.g. R. scolopacea - Length 13-18 mm; often rests on logs and treetrunks with its head turned towards the ground and front part of the body raised. An annoying biter. Distribution: Europe.

Rhagium A genus of Insects of the Longhorned or Longicorn Beetle family, *Cerambycidae*. e.g. R. inquisitor - Length 10-21 mm. Found in coniferous forests. Distribution: Europe and North America.

Rhagoletis A genus of Two-winged Flies (order *Diptera*), of the family *Tephritidae*. e.g. R. cerasi - Length 3.5-5 mm; found in cherry orchards and groves. The larva develops in the cherry fruit and gradually tunnels its way to the pip. One method of avoiding damage is timely harvesting of the fruit. Distribution: Europe, except the U.K.

Rhagonycha A genus of Insects of the Soldier Beetle family, *Cantharidae*. e.g. R. fulva - Length 7-10 mm. Found on flowering umbelliferous plants in meadows. Distribution: Europe.

Rhamnus A genus of Trees and Shrubs of the Buckthorn family, *Rhamnaceae*. e.g. R. purshiana (= *Rham* p.) - The dried bark yields Cascara (= Casc; Cascara Sagrada; Sacred Bark). Used as a mild purgative in the treatment of constipation due to colonic stasis.

Rhamphosuchus A genus of extinct Saurian Crocodiles that existed about 7 million years ago. The huge Gharial species whose remains have been found in the Suoalik Hills of North India, reached a length of 15.24 m (50 ft).

Rhaphidophora A genus of plants of the Arum family, *Araceae*. e.g. R. aurea (formerly called *Scindapsus* aureus) -

The Ivy Arum or Pothos plant, a perennial climber with heart-shaped, dark green, shiny leaves flecked with yellow. Used as an indoor plant because of the foliage and rapid growth. Place of origin: warm regions of Asia and Australia.

Rheum A genus of Cathartic plants of the family *Polygonaceae*. e.g. R. palmatum - The rhizome known as Rhubarb (= Rheum; Chinese Rhubarb; Rhabarber; Ruibarbo) is a mild anthraquinone purgative; it also contains tannin which acts as an astringent after purgation.

Rhincodon A genus of Sharks, better known by the generic name *Rhineodon*, of the class *Chondrichthyes*. e.g. R. typus - The Whale Shark (see **Rhineodon**).

Rhinencephalus See **Boophilus**.

Rhineodon A genus of Sharks, also known by the generic name *Rhincodon*, of the class *Chondrichthyes*. e.g. R. typus (= *Rhincodon* t.) - The Whale Shark, a rare plankton-feeding fish, the largest in the world, found in the warmer areas of the Atlantic, Pacific, and Indian Oceans. Some have been found to measure more than 13.71 m (45 ft) in length. The largest whale shark on record, 15.24 m (50 ft) long, was found off the Koh Chik (Chick Island) on the east coast of the Gulf of Siam in early 1919. This species may live 70 years or more; it lays the largest egg of any living mammal. In one case, in 1955, the egg was found to measure 304 x 139 x 88 mm (12 x 5.5 x 3.5 in), and the embryo was 350 mm (13.78 in) long.

Rhinicephalus A genus of Ticks. e.g. R. everti - A species transmitting the protozoan parasite *Nuttalia* equi, and causing hemoglobinuric fever of horses in South Africa (see **Nuttalia**).

Rhinichthys A genus of Fishes (subphylum *Vertebrata*). e.g. R. atratulus - The Black-nosed Dace fish.

Rhinobatos A genus of Fishes (subphy-

lum *Vertebrata*). e.g. R. lentiginosus - The Spotted Guitar-fish.

Rhinocephalus A genus of cattle Ticks. e.g. R. annulatus (see **Boophilus**).

Rhinocladium A genus of Fungi, formerly included under the generic name, *Sporotrichum* (see **Sporotrichum**).

Rhinocricus A genus of Millipedes of the class *Diplopoda*. e.g. R. lethifer - A millipede found in Haiti which can squirt a very strong repugnatorial fluid, containing hydrocyanic acid, causing blindness in chickens and other small animals.

Rhinodrilus A genus of giant Earthworms (class *Oligochaeta*). e.g. R. fafner - A species weighing as much as 453 g (1 lb).

Rhinoestrus A genus of Flies whose larvae are found in the nasal passages of horses; may be deposited in the human eye. Distribution: Europe, Asia, and Africa.

Rhinolophus A genus of Mammals of the Horseshoe Bat family, *Rhinolophidae*. e.g. R. ferrumequinum - The Greater Horseshoe Bat, length of body 58-70 mm; wingspan 330-400 mm; lives in caves, rock cavities, and ruins in warm areas of Europe and Asia. Can live as long as 26 to 27 years.

Rhinoptera A genus of Fishes (subphylum *Vertebrata*). e.g. R. bonasus - The Cow-nosed Ray fish.

Rhinosporidium A genus of Fungi related to the *Phycomycetes*. e.g. R. seeberi (= R. kinealyi; R. sieberi) - a species causing rhinosporidiosis, a polypoid disease of the nose and tumor-like lesions of the conjunctiva, uvula, and mouth. Occurs in India and South America.

Rhipicephalus A genus of cattle Ticks, species of which are the agents in transmitting *Babesia, Borrelia, Rickettsia, Theileria* and other diseases in animals and man. e.g. R. appendicularis (= Rh. a.) - The Brown Tick, transmitting the *Theileria* parva of East African Coast fever in cattle, and *Rickettsia* rickettsii (typhus) in man. R. sanguineus - The

Common Dog Tick, transmitting *Rickettsia* connori, the causal agent of Spotted Fever.

Rhipsalidopsis A genus of plants of the Cactus family, *Cactaceae*. e.g. R. gaertneri (= *Epiphyllum* g.; *Schlumbergera* g.) - The Easter Cactus, an epiphytic plant with starry scarlet flowers at the tips of the branches. An indoor or hothouse species. Place of origin: Tropical America, Brazil.

Rhizobium A genus of microorganisms of the family *Rhizobiaceae*. Rod-shaped, gram-negative, symbiotic, nitrogen-fixing bacteria, producing nodules on the roots of leguminous plants and fixing free nitrogen in this symbiosis. e.g. R. japonicum (= *Rhiz* j.).

Rhizoglyphus A genus of Mites. e.g. R. parasiticus - The "Coolie-itch" Mite, which lives on the ground and causes sore feet; it infests especially the barefooted coolies or porters of India; hence the name.

Rhizopus A genus of Fungi of the family *Mucoraceae*. e.g. R. equinus - Causes mycosis in domestic animals.

Rhizostoma A genus of Jellyfish (class *Scyphozoa*) of the order *Rhizostomeae*. e.g. R. pulmo - The European Jellyfish, 200-800 mm in diameter, whitish with a bluish-violet margin. Distribution: The Mediterranean, and Atlantic coast of Europe.

Rhodactis A genus of Coelenterates of the class of Sea-anemones, *Anthozoa*. e.g. R. howesii - The most toxic sea-anemone, found on the reefs of American Samoa, S.W. Pacific, which when cooked, forms part of the native diet, locally known as "Matamula". This coelenterate, when eaten raw, accidentally or by intent, has been the cause of death in humans due to respiratory failure.

Rhodeus A genus of Fishes of the Carp and Minnow family, *Cyprinidae*. e.g. R. sericeus - The European Bitterling fish, length 9 cm; deep-bodied and laterally compressed; found in calm waters of the lower reaches of rivers and in pools. Distribution: Central Europe (see **Carassius**).

Rhodnius A genus of Bugs. e.g. R. prolixus - A South American bug, capable of transmitting *Trypanosoma* cruzi, the organism causing Chagas' disease.

Rhododendron A genus of evergreen shrubs of the family *Ericaceae*, bearing large pink, red, purplish or white flowers. Distribution: cool areas of the northern hemisphere. e.g. R. hymenanthes - The leaves contain a poisonous compound, rhodotoxin.

Rhodomicrobium A genus of microorganisms of the family *Hyphomicrobiaceae*. Ovoid cells connected by filaments, growing in salmon-pink to orange-red colonies. e.g. R. vannielii.

Rhodopechys A genus of Mountain Birds. e.g. R. sanguinea - The Crimson-winged Finch.

Rhodopseudomonas A genus of microorganisms of the family *Athiorhodaceae*. Motile, spherical or rod-shaped cells. e.g. R. capsulata.

Rhodospirillum A genus of microorganisms of the family *Athiorhodaceae*. Motile, spiral-shaped cells. e.g. R. fulvum.

Rhodospiza A genus of Desert Birds (class *Aves*). e.g. R. obsoleta - The Desert Finch.

Rhodotheca A genus of microorganisms of the family *Thiorhodaceae* cells, each having a wide capsule. e.g. R. pendens.

Rhogogaster A genus of Insects of the Typical Sawfly family, *Tenthredinidae*. e.g. R. viridis - Length 10-13 mm; bright-green, with a dark stripe on the abdomen. Found in fields, forests, and in gardens. Distribution: Europe and temperate regions of Asia.

Rhopalopsyllus A genus of Fleas. e.g. R. cavicola - The South American cavy flea which transmits *Pasteurella* pestis.

Rhus A genus of Trees and Shrubs, many

of them poisonous, of the Cashew family, *Anacardiaceae*. Contact with certain species produces a severe dermatitis. e.g. R. glabra - The Smooth or Pennsylvanian Sumach. The dried fruit (Sumach Berries) are astringent and have diuretic properties. R. toxicodendron (= R. radicans) - The Poison Ivy plant, an irritant poisonous species, causing severe contact dermatitis. Extracts ofthe leaves and twigs (Rhus extract; Poison ivy extract) have been used in the prophylaxis and treatment of poison ivy dermatitis. R. venenata - The Swamp Sumac plant.

Rhyacophila A genus of Insects of the Caddis Fly family, *Rhyacophilidae*. e.g. R. vulgaris - Length 7-9 mm; wingspan 24-32 mm. Larval stage in fast-flowing streams. Distribution: Europe.

Rhyncocyon A genus of Mammals of the family of Jumping or Elephant Shrews, *Macroscelididae* (see **Macroscelides**).

Rhyniella A genus of prehistoric insects. e.g. R. proecursor - This species, the earliest known insect, lived about 370 million years ago. Its remains have been found in Aberdeenshire, Scotland.

Rhyparia A genus of Insects of the Tiger Moth family, *Arctiidae*. e.g. R. purpurata - Length 20-26 mm; there are several types of different coloring of the hind wings (red, yellow, or orange), with dark spots, and a patterning of the forewings. Distribution: Europe, Asia Minor and Japan.

Rhyssa A genus of Insects of the Ichneumon Fly family, *Ichneumonidae*. e.g. R. persuasoria - Length 18-35 mm; found in forests. Female has an ovipositor 30-35 mm long. Distribution: Europe and North America.

Ribes A genus of Trees of the family *Grossulariaceae*. e.g. R. rubrum - The fresh ripe fruit, Red Currant (= Ribes Rubrum; Groseille; Groselhas) is used as a coloring and flavoring agent.

Ricinus A genus of plants of the Spurge family, *Euphorbiaceae*. e.g. R. communis - The Castor Oil plant; from the seeds is expressed Castor Oil (= Oleum Ricini; Ol Ricin; Huile de Ricin), used internally as a mild purgative (due to its ricinoleic acid content); externally used as an emollient and an ingredient of hair lotions.

Rickettsia A genus of microorganisms of the family *Rickettsiaceae* - Small rod-shaped to coccoid, often pleomorphic cells, about 1 μ or less in their long diameter, known as *Rickettsia* or Rickettsial bodies; non-motile, they stain reddish-purple with Giemsa's stain, occurring intracytoplasmically or free in the lumen of the gut in fleas, lice, mites, and ticks, by which they are transmitted to man and other animals. e.g. R. diaporica (= *Coxiella* burnettii) - Causes Q fever in man (see **Coxiella**). R. prowazekii - The causative agent of epidemic, louse-borne, typhus fever. R. tsutsugamushi (= *Theileria* t.) - The etiologic agent of Scrub typhus (= Rural typhus) and Japanese Flood Fever (Tsutsugamushi fever) (see **Theileria**).

Rickettsiella A genus of microorganisms of the family *Rickettsiaceae* - Minute intracellular rickettsia-like organisms, parasitic on the Japanese beetle (*Popillia* japonica), and non-pathogenic to mammals. e.g. R. popilliae.

Ricolesia A genus of microorganisms of the family *Chlamydiaceae*. Coccoid organisms causing a keratoconjunctivitis in domestic animals. e.g. R. bovis - A species producing keratoconjunctivitis in cattle.

Riparia A genus of Birds of the Martin and Swallow family, *Hirundinidae*. e.g. R. riparia - The Sand Martin or Bank Swallow, 12 cm long; inhabits open country with calm or flowing waters. A migrating bird, it winters in Africa and Asia. Distribution: almost all Europe.

Rissa A genus of Birds of the Gull and Tern family, *Laridae*. e.g. R. tridactyla - The Kittiwake bird, length 40 cm; it nests on rocky coasts but outside the breeding season it keeps to the open seas. Distribution: Northern coast of Europe, Asia, and North America.

Rivea A genus of climbing plants of the Morning-glory family, *Convolvulaceae*. e.g. R. corymbosa - From the seeds is expressed an hallucogenic drug called by the Aztecs of Mexico, by the name of "ololiuqui", and by the Zapotec Indian tribes of Southern Mexico as "badoh". The drug contains about 0.01% alkaloids with the characteristics of the hallucinogen, lysergic acid, viz. D-isolysergic acid amide, D-lysergic acid amide, chanoclavine, elymoclavine, and lysergol (see **Ipomoea**).

Rivulogammarus A genus of Crustaceans, formerly known by the generic name *Gammarus*, of the freshwater Shrimp family, *Gammaridae*. e.g. R. pulex (= *Gammarus* p.; R. pulex fossarum) - The Common Freshwater Shrimp, length 12-15 mm; body laterally flattened, arched, and colored greyish-yellow. Found in flowing clean water, and in springs. Distribution: Europe and Asia Minor. Lake Baikal in Siberia is remarkable for its gammarids, and about 300 species are endemic to this ancient lake.

Roccella A genus of Lichens of the family *Roccellaceae*. e.g. R. tinctoria - A species producing Litmus, a blue pigment used as a test for acidity and alkalinity, with a pH range of 4.5 to 8.3, and Cudbear (= Persio; C.I. Natural Red 28) - a red coloring agent used especially in the preparation of syrups having an acid reaction. Also contains an alcohol, Erythritol, from which is prepared erythrityl tetranitrate, a vasodilator used in the prophylaxis of angina pectoris.

Romericus A genus of prehistoric Reptiles, the species of which are one of the earliest known, having lived about 290 million years ago. Their remains have been found in Nova Scotia.

Rosa A genus of plants of the Rose family, *Rosaceae*. e.g. R. damascena - The Damask Rose plant; from the fresh flowers is distilled Rose Oil (= Oleum Rosae; Ol Ros; Attar of Rose; Otto of Rose) - used in perfumery, lozenges, ointments, toothpastes, and toilet preparations.

Rosmarinus A genus of plants of the Sage and Thyme family, *Labiatae*. e.g. R. officinalis - The Rosemary plant; from the flowering tops and leafy twigs is distilled Rosemary Oil (= Esencia de romaro; Essence de Romarin; Essencia de Alecrim; Ol Rosmarin; Oleum Rosmarini; Oleum Roris Marini; Rosemarinol). It contains an ester, bornyl acetate and free alcohols, including borneol, and is used as a carminative; also externally in hair lotions and linaments.

Rossiella A genus of piroplasma-like organisms, parasitic in the blood of certain animals. e.g. R. rossi - A species found in the jackal in East Africa.

Rousettus A genus of long-living Bats. e.g. R. leachii - The Rousette fruit bat, a member of which died in the Giza Zoo, Cairo, Egypt in 1918, having reached the age of 22 years 11 months.

Rubia A genus of plants of the Madder family, *Rubiaceae*. e.g. R. tinctoria - The Madder plant, reputed to be medicinally active.

Rubus A genus of plants of the Rose family, *Rosaceae*, including the blackberry, bramble, cloudberry, dewberry, and raspberry trees. e.g. R. idacus - The Raspberry tree; the fresh ripe fruit, the Raspberry (= Framboise; Fructus Rubi Idaci; Himbear; Rubus Idacus), is used as a coloring and flavoring agent; also as an antispasmodic "tea" infusion in dysmenorrhea.

Rudbeckia A genus of plants of the Daisy family, *Compositae*, now more correctly called by the generic name, *Echinacea*. e.g. R. purpurea. (= *Echinacea* p.) - The Rudbeckia (see **Echinacea**).

Rupicapra A genus of Horned Ungulates

of the family *Bovidae*. e.g. R. rupicapra - The Chamois, length of body 110-113 cm; height at shoulder 70-80 cm. Found on high mountains up to the region of permanent ice and snow. Distribution: the high mountains of Europe, the Caucasus and Asia Minor.

Ruta A genus of plants of the Rue family, *Rutaceae*. e.g. R. graveolens - The dried herb, Rue (= Herbygrass; Ruta; Rutae Herba) contains about 0.1% of volatile oil; formerly used as an antispasmodic and emmenagogue.

Ruthenica A genus of Gastropod Molluscs of the family *Clausiliidae*. e.g. R. filograna - The shell is 7.5-9 mm high, 2-2.2 mm wide, and composed of 9-10 convex whorls. Found in damp woods with abundance of decaying leaves. Distribution: the mountainous woodlands of Europe.

Rutilus A genus of Fishes of the Carp and Minnow family, *Cyprinidae*. e.g. R. rutilus - The Roach, length 40 cm; weight 1 kg; one of the commonest fish in all types of fresh waters except mountain trout streams. Distribution: fresh waters of most of Europe (see **Carassius**).

S

Sabatinca A genus of Insects of the family of Butterflies and Moths, *Micropterygidae* (order *Lepidoptera*). A primitive species from New Zealand, in which the larvae feed on mosses and liverworts.

Sabella A genus of segmented worms (phylum *Annelida*), of the class of Bristleworms, *Polychaeta*. e.g. S. pavonina - The Peacock Bristleworm, a relatively long-lived worm, not sexually mature until it reaches ten years of age.

Sabellastarte A genus of Tube Worms or Tubicolous worms found in the waters off the Red Sea coast. e.g. S. indica - A tube worm living in tubes composed of a soft cementing compound mixed with mud and extending cilia covered arms to sieve small food particles from the water. When disturbed it instantly retracts into the tube. Other examples of tube worms are those of the genus *Spirobranchus*.

Sabouraudites A genus name for some species of *Microsporon*, *Trichophyton*, etc.

Saccharomyces A genus of ascomycetous unicellular Fungi or Yeasts, belonging to the family, *Saccharomycetaceae*. e.g. S. albicans (= *Candida* a.) (see **Candida**). S. cantlici - A genus causing a tropical blastomycosis. S. cerevisiae - One of the chief sources of Dried Yeast (= Cerevisiae Fermentum Siccatum; Faex Siccata; Fermento de Cerveja; Levedura Sêca; Levure de Bière; Saccharomyces Siccum) - used for the prevention and treatment of vitamin B deficiency; also in the treatment of skin diseases such as acne and furunculosis. S. epidermica (= *Cryptococcus* epidermidis) (see **Cryptococcus**). S. galacticolus - A species producing fermentation in milk. S. granulomatosus - A species producing granulomatous tumors in pigs. S.

hominis (= *Schizosaccharomyces* h.) - a species causing fungus infection in man.

Saccoglossus A genus of Chordate animals of the Acorn worm class, *Enteropneusta* (see **Harrimania**).

Saduria A genus of Crustaceans, also known by the generic name *Mesidotea*, of the order of Isopods, Sowbugs and Woodlice, *Isopoda*. e.g. S. entomon (= *Mesidotea* e.) - A glacial relict found in brackish water and in freshwater lakes in North America and Europe. It has large uropods which can close like a pair of doors over the respiratory limbs.

Sagittaria A genus of plants of the Water Plantain family, *Alismataceae*. e.g. S. sagittifolia - The Old-World Arrowhead, a herbaceous plant with an almost round tuberous rootstock. Leaves are slender and arrow-shaped. The tubers are eaten, particularly by the Chinese, and the green parts are fed to pigs. Found in Europe and Asia.

Saiga A genus of Mammals of the Horned Ungulate family, *Bovidae*. e.g. S. tartarica - The Saiga Antelope. A steppe-dweller living in herds in the lowlands and in mountains on upland plateaus. Length of body 130-135 cm; height at shoulder, up to 70 cm. Distribution: Ukraine and northern Asia.

Saintpaulia A genus of plants of the family *Gesneriaceae*. e.g. S. ionantha - The African Violet, an evergreen plant producing loose sprays of violet flowers. Found in tropical East Africa.

Salamandra A genus of Reptiles of the Salamander family, *Salamandridae*. e.g. S. salamandra - The European or Fire salamander, length 15-20 cm. Found in mountainous regions in broadleaved forests around clear streams. The female bears up to 60 larvae, who live in water for 2 to 3 months, then metamorphizing to live on dry land. This species lives up to 24 years of age. Distribution: Central and southern Europe.

Salix A genus of plants of the Willow family, *Salicaceae*. e.g. S. reticulata -

The Neatleaf or Dwarf Willow, a low or prostrate deciduous shrub with dark green, wrinkled and prominently veined leaves growing in rock or alpine gardens. The bark yields a crystalline β-glucoside, salicin, which was once used in the treatment of rheumatic fever. Found in the mountains of Europe and Labrador.

Salmo A genus of Fishes of the Salmon and Trout family, *Salmonidae*. e.g. S. trutta fario - The Brown Trout, a common fish of Europe's mountain streams, rivers and lakes. Also planted into cool ponds, and has been cultured in almost every part of Europe.

Salmonella A genus of microorganisms of the family *Enterobacteriaceae*. Rod-shaped, gram-negative, usually but not invariably motile bacteria, which do not ferment lactose. Formerly known by the genus name *Eberthella*, and includes the typhoid-paratyphoid bacilli. e.g. S. typhosa (= S. typhi; Sal. typhi; Salm. typhi; *Bacillus* typhosus; *Eberthella* typhosa; E. typhi) - the etiological agent of typhoid fever occurring only in man. S. morgani (= *Proteus* morgani; P. morganii; Sal. morgani; Salm. morgani (see **Proteus**).

Salticus A genus of Arachnids of the Jumping Spider family, *Salticidae*. e.g. S. scenicus - The Zebra Spider, length 4.5-7.5 mm. Abdomen marked with dark stripes. Found on sunny walls and rocks. Distribution: Europe, North Africa, and North America.

Salvelinus A genus of Fishes, formerly known by the generic name *Cristivomer*, of the Salmon and Trout family, *Salmonidae*. e.g. S. alpinus - Alpine Trout; length 50-70 cm; weight 2-4 kg; an ocean fish spawning in autumn in streams, rivers and freshwater lakes. Many mountain and northern lakes contain permanent, non-migrant freshwater forms. Distribution: Northern seas, and lakes of Europe and the British Isles. e.g. S. namaycush (= *Cristivomer*

n.) - The Great Lake Char or "Trout" of North America. These are large, sluggish, game fishes reaching a weight of 80 lb.

Salvia A genus of plants of the Mint, Sage, and Thyme family, *Labiatae*. e.g. S. officinalis - The Red or Golden Sage, a perennial, strongly aromatic plant with thick, velvety, greyish leaves and violet-blue, tubular flowers. The dried sage leaves have been reported to have spasmolytic and bactericidal properties, and to diminish salivation. Place of origin: the Mediterranean region.

Sambucus A genus of Trees and Shrubs of the Honeysuckle family, *Caprifoliaceae*. e.g. S. nigra - The Common European Elder, a small deciduous tree up to 32 ft (10 m) high, with creamy-white fragrant flowers. The tree is grown ornamentally to make loose hedges and thickets. The soft pith is employed for botanical section cutting in microscopy. The flowers were once considered to be diuretic and diaphoretic, and the berries were thought to possess some laxative principles. Found in Europe, North Africa and western Asia.

Sanguisuga A genus of Leeches. e.g. S. medicinalis - The Swedish or Official leech.

Sanninoidea A genus of Insects of the Butterfly and Moth order *Lepidoptera* (suborder *Ditrysia*). e.g. S. exititiosa - The American Peach Tree Borer, an injurious species; the larvae infest and destroy peach trees.

Sansevieria A genus of plants of the Lily family, *Liliaceae*. e.g S. trifasciata "laurentii" - The Bowstring Hemp (= Mother-in-law's tongue; Snake plant), a plant with flat-concave, thick and hard, sword-shaped leaves, with fragrant, creamy flowers in long racemes. Used as a tough indoor plant even under unfavorable conditions. The leaves are a source of fiber used for fishing lines, nets, hats, mats, and shoes. Place of origin: West Africa; introduced into Europe in 1904.

Santalum A genus of plants, also known by the generic name *Eucarya*, of the Heart-wood or Sandalwood family, *Santalaceae*. e.g. S. album - The Sandalwood plant, from which is prepared Sandalwood oil (= Oil of Sandal Wood). Was formerly used as a urinary antiseptic. Oils of sandalwood are also used in perfumery. S. spicatum (= *Encarya* spicata) - see **Encarya**.

Santolina A genus of Shrubs of the Daisy family, *Compositae*. e.g S. chamaecyparissus - The Lavender Cotton plant, a bushy aromatic, evergreen shrub, with globular long-stalked yellow flowerheads. Normally grown for its aromatic foliage. Used as an insecticide and vermifuge. Also as an ingredient of potpourri. Distribution: Southern Europe.

Saperda A genus of Insects of the Longicorn or Long-horned Beetle family, *Cerambycidae*. e.g. S. populnea - Length 9-15 mm; found in aspen and willow trees; the larvae pupate inside the tree branches and the adults emerge through a circular hole. Distribution: Europe, Asia Minor, North Africa, and North America.

Saponaria A genus of plants of the Honeysuckle family, *Caprifoliaceae*. e.g. S. officinalis - The Soapwort (= Bouncing Bet; Fuller's Herb), a perennial herb with creeping rhizome and pink flowers. The plant contains saponins which lather with water. The chopped roots were once used as a substitute for soap (L. sapo), and as an expectorant and diuretic. Now currently used for cleaning precious fabrics and wall tapestries.

Saprolegnia A genus of partially saprophytic phycomycetous Fungi. e.g. S. ferax - A species destructive to salmon and to various water animals.

Saprospira A genus of microorganisms of the family *Spirochaetaceae*; actively motile cells, containing spiral protoplasm without evident axial filament, and with a definite periplast membrane, found free-living in marine ooze. e.g. S. grandis.

Saraca A genus of plants of the Pea family, *Leguminosae*. e.g. S. indica - The Ashok or Asoka plant; the dried stem bark contains tannins with small amounts of an essential oil, and an organic compound. Used as an astringent and uterine sedative.

Sarcina A genus of microorganisms of the family *Micrococcaceae*; spherical, gram-positive cells in cubical packets of 8 cells, found in soil and wateras non-pathogenic saprophytes.

Sarcocystis A genus of protozoan parasites of the family *Sarcosporidia*, found in animals; rarely man. e.g. S. tenella - A species found in sheep and cattle (in the muscles and internal organs), and sometimes in man.

Sarcophaga A genus of Two-winged Flies of the family, *Sarcophagidae*. e.g. S. carnaria - The Flesh-fly, length 13-15 mm; has striking chequer-board markings. The eggs are dropped into meat or decaying flesh on which the larvae hatch and feed. The larvae are also found in wounds, ulcers, and the nasal passages. This species also causes genito-urinary myiasis. Distribution: Europe; Africa.

Sarcophyton A genus of Octocorals, a group of coral animals that do not usually deposit calcareous skeletons. Each living individual (polyp) in an octocoral has 8 feathered tentacles surrounding the mouth. Found on the Red Sea coast and the Gulf of Eilat (Akaba).

Sarcoptes A genus of Ascarids or Mites, also known by the generic name *Acarus*, of the order *Acari* or *Acarina*. e.g. S. scabei (= *Acarus* s.; A. scabiei; A. siro; S. scabiei) - The Itch Mite which bores beneath the skin, forming canaculi or burrows, and causes scabies in man (see **Acarus**).

Sarda A genus of Fishes of the Mackerel family, *Scombridae*. e.g. S. sarda - The

Atlantic Bonito fish, length 80 cm. Occurs in large shoals near the surface by the coast; an important fish of commerce. Distribution: warmer parts of the Atlantic and the Mediterranean.

Sardinia A genus of small fishes, better known by the generic name *Sardinus*, of the Herring family, *Clupeidae*. e.g. S. pilcharda (= *Sardinus* pilchardus) - The Pilchard or Sardinefish (see **Sardinus**).

Sardinops A genus of Fishes (subphylum *Vertebrata*). e.g. S. sagax - The Californian Sardine.

Sardinus A genus of Fishes, also known by the generic name *Sardinia*, of the Herring family, *Clupeidae*. e.g. S. pilchardus (= *Sardinia* pilcharda) - The Sardine or Pilchard, length 25-26 cm. A pelagic, shoaling fish, spawning in the open sea. An important source of food. Distribution: the seas around Europe.

Sarothamnus A genus of Shrubs of the Pea family, *Leguminosae*. e.g. S. scoparius (= *Cytisus* s.) - The Scoparium plant (Brown Tops; Planta Genista; Scoparii Cacumina), from which is obtained the dibasic alkaloid sparteine and the principle scoparin. Scoparium lessens the irritability and conductivity of cardiac muscle, and was formerly used in the treatment of tachycardia; also as an oxytocic agent to induce labor at term, as a useful alternative to pituitary oxytocics, and as a mild diuretic, purgative, and emetic (see **Cytisus**).

Sarracenia A genus of plants of the Polypetalous plant order, *Sarraceniaceae*. e.g. S. purpurea - The Side-saddle flower or Pitcher plant of North America. The secretion of the pitcher of this plant is said to contain digestive enzymes. It is a stimulant, diuretic, and a perient.

Sassafras A genus of Trees of the Laurel family, *Lauraceae*. e.g. S. albidum - From the root or root bark is distilled Sassafras Oil (= American Sassafras Oil; Ol Sassaf; Oleum Sassafras), used externally as a pedicide. Formerly used internally as a flavoring agent, but can cause fatty changes in the liver and kidneys. This volatile oil contains safrene and safrol, and is an antinarcotic (see **Ocotea**).

Saturnia A genus of Insects of the Emperor Moth family, *Saturniidae*. e.g. S. pyri - The Great Peacock Moth, length 60-72 mm; the largest European moth. The caterpillar is found on forest trees and on ash. Distribution: southern Europe, Asia Minor, the Middle East.

Saurophagus A genus of now extinct animals, the Carnosaurs. e.g. S. maximus - The Carnosaur or lizard eater.

Saussurea A genus of plants of the Daisy family, *Compositae*. e.g. S. lappa - The Saussurea plant (= Costus; Kut; Pachak); the dried roots are said to have antiseptic, antispasmodic, carminative, expectorant, and diuretic properties. The plant has been used in India in the treatment of bronchial asthma.

Saxicola A genus of Perching Birds of the Thrush family, *Turdidae*. e.g. S. rubetra - The Whinchat, length 13 cm. Inhabits rolling, open hill country with isolated bushes and streams. A migrating bird. Distribution: Europe and southwest Asia.

Saxidomus A genus of poisonous Clams, containing the toxic substance saxitoxin (see **Mytilus**).

Saxifraga A genus of plants of the Saxifrage or Geranium family, *Saxifragaceae*. e.g. S. stolonifera (= S. sarmentosa) - The Strawberry Geranium (= Mother of Thousands); a tufted perennial plant with many thin, trailing strawberry-like runners; grown for its foliage and used for hanging baskets, greenhouse or container decoration. Place of origin: China and Japan; introduced into Europe in 1815.

Scabiosa A genus of plants of the family *Dipsacaceae*. e.g. S. caucasica - The Caucasian Scabious (= Pincushion Flower; Scabiosa); a perennial border plant with showy pale-blue flowers coming in round, pin-cushion like heads. Popularly regarded as a blood depurative or purifying substance. Distribution: Caucasus; in Europe since 1803.

Scaeva A genus of Insects of the Hover Fly family, *Syrphidae*. e.g. S. pyrastri - Length 14-19 mm; partial to flowering plants of the carrot family; found from lowlands to mountainous areas. Distribution: Eurasia, North Africa, and North America.

Scalaria A genus of Gastropod Molluscs, also known by the generic name *Clathrus* (see **Clathrus**).

Scaphistostreptus A genus of Millipedes (class *Diplopoda*). e.g. S. seychellarum - The millipede of the Seychelles in the Indian Ocean, one of the longest known species of millipedes, measuring up to 280 mm (11.02 in) in length and 20 mm (0.78 in) in breadth.

Scardinius A genus of Fishes of the Carp and Minnow family, *Cyprinidae*. e.g. S. erythrophthalmus - The Rudd fish, length more than 30 cm; weight 1 kg. Common in coves and inlets on the lower reaches of rivers, and in backwaters and pools. Distribution: Europe (see **Carassius**).

Scarus A genus of Fishes (subphylum *Vertebrata*). e.g. S. gibbus - The Parrot Fish, a many-colored species occurring in coral seas, especially in the Red Sea and the Gulf of Eilat (Akaba). Its teeth are fused into a "beak" used for breaking off pieces of stony coral which it eats.

Scatophaga A genus of Insects of the Dung-fly family, *Scatophagidae*. e.g. S. sterocorarium - The Yellow Dung-fly; length 9-11 mm. Found on dung heaps and excrement, from lowlands to mountainous areas. Distribution: the Palaearctic regions.

Schindleria A genus of Fishes (subphylum *Vertebrata*). e.g. S. praematurus - Found off the island of Samoa; one of the shortest recorded marine fishes,

measuring only 12-19 mm (0.47-0.74 in). The lightest of all vertebrates, mature specimens have been known to weigh only 2 mg (0.0007 oz).

Schinopsis A genus of plants. e.g. S. lorenzii - The Red Quebracho plant of Argentina, the leaves of which cause a dermatitis called "paaj" or "mal de quebracho".

Schinus A genus of Trees of the Cashew family, *Anacardiaceae*. e.g. S. molle - The Pepper tree of tropical America; yields a resinous substance which is mildly purgative and aromatic.

Schistocera A genus of Insects of the order of Locusts, *Caelifera*. e.g. S. gregaria - The Desert Locust; the locust of the Bible; the most destructive insect in the world, whose habitat is the dry and semi-arid regions of Africa, the Middle East, and the Indian subcontinent. This species of locust, which is in fact a short-horn grasshopper, can eat its own weight in food in one day, and during long migratory flights of up to 3200 km (2000 miles), a large swarm will consume 3054 tonne (3000 tons) of green stuff per day, causing famine to whole communities.

Schistosoma A genus of Trematode parasites or blood Flukes (class *Trematoda*). e.g. S. bovis - A species found in the portal system of sheep and oxen in Africa, the Mediterranean region, and the Middle East. S. haematobium - A species found in tropical countries, especially in Egypt. It enters the body by the alimentary tract in drinking water or through the skin while bathing or wading, the invertebrate (intermediate) host being the small snails of the genera *Bulinus*, *Planorbis*, and *Physopsis*. It enters the cystic vein, causing irritability of the urinary bladder and hematuria, and the mesenteric venous system causing dysentery. This species was formerly called *Distoma* haematobium or *Billharzia* haematobia. S. indicum - A species found in cattle, sheep, and goats in India and Rhodesia. S. intercalatum - A species causing intestinal schistosomiasis in the Belgian Congo (Zaire). S. japonicum - A species causing infection of the liver and spleen, with ascites by penetrating the skin of persons in infested waters. The transmitting hosts are snails of the genera *Katayama* and *Oncomelania*. Found in Japan, China, and the Philippines. S. mansoni - A species very similar to S. haematobium, but found in feces instead of the urine, and causing intestinal or visceral schistosomiasis. The intermediate host is the freshwater snail, *Planorbis*. S. mattheei - A species found in the portal mesenteric vein of sheep in South America. S. pathlocopticum - A species pathogenic to mice. The intermediate host is the snail, *Limnaea*. S. spindale - A species parasitic in the buffalo and goat in India.

Schistosomatium A genus of blood Flukes allied to *Schistosoma*. e.g. S. douthitti - A species found in the hepatic portal veins of the meadow mouse.

Schizoblastosporion A genus of yeast-like Fungi of the family *Eremascaceae*, various species of which have been isolated from onychomycotic lesions.

Schizocardium A genus of Chordate animals of the Acorn worm class, *Enteropneusta* (see **Harrimania**).

Schizophyllum A genus of Millipedes (class *Diplopoda*), of the family *Iulidae*. e.g. S. sabulosum - Length 15-47 mm; width 1.6-4 mm. A common species, identified by the two longitudinal bands composed of yellow-red spots. Found in forests, on tree branches, and in chalky and sandy places under stones, etc. Distribution: Europe.

Schizosaccharomyces A genus of Fungi or Yeasts. e.g. S. hominis (= *Saccharomyces* h.) (see **Saccharomyces**).

Schizosiphon A genus of nematogenous Schizomycetes with flagelliform filaments, slender toward the extremity.

Schizotrypanum A genus of Sporozoan parasites, now known by the generic name, *Trypanosoma*. e.g. S. cruzi (= *Trypanosoma* c.) (see **Trypanosoma**).

Schlumbergera A genus of plants of the Cactus family, *Cactaceae*. e.g. S. gaertneri - The Easter Cactus plant (see **Epiphyllum**; **Rhipsalidopsis**).

Schoenocaulon A genus of plants of the Lily family, *Liliaceae*. e.g. S. officinale - The Sabadilla plant (= Caustic Barley; Cevadilha; Cevadilla; Sabadillsamen); the dried ripe seeds contain several alkaloids, the most important group of which is cevadine or crystalline veratrine, used externally as a parasiticide, especially for pediculosis capitis.

Scilla A genus of plants, now more correctly called *Endymion* or *Urginea*, of the Lily family, *Liliaceae*. e.g. S. hispanica (= S. campanulatus) - The Spanish Bluebell or Spanish Squill (see **Endymion**).

Scindapsus A genus of plants, now more correctly called *Rhaphidophora*, of the Arum family, *Araceae*. e.g. S. aureus (= *Raphidophora* aurea) - The Ivy Arum or Pothos plant (see **Rhaphidophora**).

Sciurus A genus of Rodents of the Tree Squirrel family, *Sciuridae*. e.g. S. vulgaris - The Red Squirrel, length of body 20-24 cm. A rodent often found in areas of human habitation, in woodlands, parks, and gardens. It builds a nest in trees, where the female bears 3-8 young twice a year. Distribution: the Palaearctic region.

Sclerostoma A genus of Nematode worms, the species of which appear under other generic names. e.g. S. duodenale (= *Ancylostoma* d.) (see **Ancylostoma**). S. syngamus (= *Syngamus* trachealis) (see **Syngamus**).

Scolia A genus of Insects of the family *Scoliidae*. e.g. S. quadripunctata - Length 10-18 mm; has a variable number of yellow patches on the abdomen, usually four. Distribution: Eurasia, North Africa.

Scolopax A genus of Birds of the

Sandpiper and Snipe family, *Scolopacidae*. e.g. S. rusticola - The Woodcock, length 34 cm; common in broadleaved and mixed woodlands. A migratory bird, it feeds on invertebrates, hunting its prey on the surface and in the ground, probing the soil and pulling up worms with its long beak. The nest is a depression in the ground, lined with old leaves. Distribution: Eurasia.

Scolopendra A genus of poisonous Centipedes (class *Chilopoda*). e.g. S. gigantea - The 46-legged giant scolopender, the largest centipede species, reaching 250-265 mm (9.84-10.43 in) in length when fully extended and 25 mm (1 in) in breadth. Measurements of up to 381 mm (15 in) in length have been reported. Distribution: the rain forests of Central and South America.

Scomber A genus of Fishes of the Mackerel family, *Scombridae*. e.g. S. scombrus - The Atlantic Mackerel, 50 cm long; one of the most valuable fish of commerce. Distribution: the North Atlantic and Mediterranean; also the Black Sea.

Scomberomorus A genus of Fishes of the Mackerel family, *Scombridae*. e.g. S. cavalla - The King Mackerel fish.

Scophthalmus A genus of Marine Fishes (order *Heterosomata* or *Pleuronectiformes*), of the Flatfish family. e.g. S. maximus - The Turbot Fish; this species has a very high fertility level, laying an enormous number of eggs. It has been estimated to contain as much as 9 million eggs in its ovaries.

Scopolia A genus of plants of the Nightshade family, *Solanaceae*. e.g. S. carniolica (= S. atropoides) - The Scopolia or Scopola plant of Europe; the dried rhizome contains about 0.5% of alkaloids, chiefly hyoscyamine. Closely resembles belladonna root in its action, as a sedative and narcotic, but is rarely used medicinally.

Scopula A genus of Insects of the

Butterfly and Moth family, *Geometridae*. e.g. S. ornata - The Lace Border butterfly; size 11-12 mm; yellowish-brown in color. Distribution: Europe; North Africa.

Scopulariopsis A genus of penicillin-like Molds or Fungi, causing primary damage of the nail plates, and other non-dermatophyte mold infections (scopulariopsis) in man. e.g. S. americana - A species causing a cutaneous condition resembling blastomycosis. S. blochi - A species causing a condition resembling sporotrichosis. S. brevicaulis var. hominis - A species causing onychomycosis. S. koningi - A species causing oculomycosis.

Scopus A genus of Birds (class *Aves*). e.g. S. umbretta - The Humming bird.

Scorpaena A genus of Fishes (subphylum *Vertebrata*), also known by the genus name, *Scorpaenopsis*. e.g. S. gibbosa - The Humpbacked Scorpion Fish, dangerous to other fish and to man. Has one of the most efficient camouflage in the animal kingdom. Capable of causing excruciating pain and death to humans. Distribution: the Red Sea coast.

Scorpaenopsis See **Scorpaena**.

Scotoplanes A genus of Echinoderms of the class of Sea-cucumbers, *Holothuroidea*. e.g. S. globosa - A deepsea holothurian; it has a rounded, oval body, with a ring of short tentacles around the mouth, a dozen or so finger-like processes on the ventral surface and four long ones on the dorsal surface.

Scrobicularia A genus of Bivalve Molluscs of the family *Scrobiculariidae*. e.g. S. plana - The Peppery Flower Shell, length 50-55 mm, height 40 mm, thickness 15 mm. Whitish to yellowish-white shells, found on the seashore. An edible species harvested and sold commercially. Distribution: Mediterranean and Atlantic coasts.

Scutiger A genus of Centipedes (class *Chilopoda*). e.g. S. coleoptrata - The fastest-moving centipede, found in southern Europe. It can outpace a fast-walking human travelling at a rate of 50 cm (19.68 in) a second (= 7.15 km/h; 4.47 miles/h), over short distances.

Scyliorhinus A genus of Fishes of the Cat Shark family, *Scyliorhinidae*. e.g. S. caniculus - The Common Spotted Dogfish, a nocturnal cat shark almost 1 m long; hunts small fish and crustaceans, etc. Found in sandy areas at depths of 10-85 m. Distribution: along the coasts of Europe and North Africa.

Scytonema A genus of Algae with cylindrical branching filaments.

Sebastes A genus of Fishes of the Scorpion and Rock family, *Scorpaenidae*. e.g. S. marinus - The Red fish, 1 m long; found in the open sea and along the coast at depths of 100-400 m. An important game fish. Distribution: North Atlantic.

Sebastodes A genus of Fishes (subphylum *Vertebrata*). e.g. S. paucispinis - The Bocaccio, or Pacific Rock-fish.

Secale A genus of plants of the Grass family, *Gramineae*. e.g. S. cereale - The Common Rye plant, in whose ovary develops the fungus *Claviceps* purpurea, from which is produced ergot (see **Claviceps**).

Sedum A genus of plants of the family *Crassulaceae*. e.g. S. acre - Gold Moss (= Stone Crop; Wallpepper); a small evergreen mat-forming plant, with green to yellow leaves which have a peppery taste. Bright yellow 5-petalled flowers. Grows on dry walls and crevices, and used in rock gardens. Distribution: Europe, North Africa, western Asia.

Segmentina A genus of Gastropod Molluscs of the family, *Planorbidae*. e.g. S. nitida - Has a shell 5-7 mm wide, tiny, flat, and composed of 41/2 to 5 whorls. A common species found in pools, ditches, and ponds. Distribution: the Palaearctic region.

Selaginella A genus of plants of the Fern family, *Selaginellaceae*. e.g. S. martensii - Club-moss, an evergreen plant, resem-bling moss with much branched stems and bright green leaves, grown in bottle gardens or shallow pans indoors, or in damp woodlands. Place of origin: Mexico.

Selasphorus A genus of birds (class *Aves*), of the Hummingbird family. e.g. S. rufus - The Rufous hummingbird, a diminutive, migrating bird, which every autumn leaves Alaska and flies along the entire west coast of North America to its winter quarters in Mexico, returning home the following spring, a total odyssey of 6400 km (4000 miles).

Selatosomus A genus of Insects of the Click Beetle family, *Elateridae*. e.g. S. aeneus - Length 10-17 mm. Has a metallic sheen of variable color, usually greenish, sometimes blue, blue-violet, or black. Found on plants in meadows, field paths, etc. Can cause damage to cultivated areas. Distribution: Europe; Siberia.

Selenomonas A genus of microorganisms of the family *Spirallaceae*; motile kidney- to crescent-shaped cells, with blunt ends. e.g. S. palpitans.

Semnopithecus A genus of the highest order of animals, the *Primates*, of the Monkey family. e.g. S. priam - The Langur monkey.

Sempervivum A genus of plants, better known by the generic name *Aeonium*, of the Cactus family, *Cactaceae*. e.g. S. canariense - The Aeonium cactus (see **Aeonium**).

Senecio A genus of plants of the Daisy family, *Compositae*. Some of the species have other generic names, such as *Kleinia*, *Ligularia*, etc. e.g. S. cruentus - The Cineraria plant; has palmate and pubescent leaves, and large massed heads of white, pink, red, blue or violet daisy flowers. Grown indoors or in greenhouses. Place of origin: Canary Islands. S. cuneatus (= *Kleinia* cuneifolia) - A species from South Africa (see **Kleinia**). S. jacobaea (= S. jacobea) - The Ragwort plant; contains a poisonous alkaloid, jacobine, which may cause necrosis of the liver, and is poisonous to livestock. Has been used externally as a vulnerary. Place of origin: the British Isles. S. sibiricus (= *Ligularia* sibirica) - An Eurasian species (see **Ligularia**).

Sepia A genus of Cephalopod Molluscs of the Cuttle family, *Sepiidae*. e.g. S. officinalis - The Common Cuttle, length 30-40 cm. Around the mouth are two long tentacles, each with 5-6 large suckers; also 8 short arms furnished with suction pads. The Cuttle is a predator; when threatened it ejects a dark pigment. Distribution: the Mediterranean and the North Sea coasts.

Sepsis A genus of Flies. e.g. S. violacea - The Common Dung fly; found also in human habitations.

Serenoa A genus of Trees of the Palm family, *Palmae*. e.g. S. serrulata - The Saw palmetto or Sabal palm tree of the southern U.S. A fluid extract of the berries is diuretic, expectorant, and an aphrodisiac, and is used in diseases of the prostrate and urinary bladder.

Serica A genus of Insects of the Chafer and Dung Beetle family, *Scarabaeidae*. e.g. S. brunnea - Length 8-10 mm; yellowish-brown in color, it conceals itself during the day in moss, under tree barks and stones, emerging at dusk. Distribution: much of Europe, and North Africa.

Sericopelma A genus of Spiders (order *Araneae*). e.g. S. communis - The Black Tarantula spider of Panama, a venomous species which has been credited with human fatalities.

Serinus A genus of Perching Birds, the Finches, of the family *Fringillidae*. e.g. S. canaria - The Canary, the longest-lived small cagebird, which was imported into Europe from the Canary Islands in the 16th century, and has an average lifespan of 12-15 years. S. serinus - The Serin Finch, length 12 cm, a migrating bird, native to the Mediterranean region.

Found near human habitations, in gardens, parks and woodlands. Distribution: Europe, Asia Minor and northwest Africa.

Serranelus A genus of Bony Fishes of the Grouper and Sea Bass family, *Serranidae*. e.g. S. scriba - The Sea Bass, length 25 cm; lives in shallow waters with rocky or sandy bottoms. Distribution: northeastern Atlantic, the Mediterranean and Black Seas.

Serrasalmus A genus of ferocious freshwater Fishes, the Piranhas (subphylum *Vertebrata*). e.g. S. nattereri - An ever-hungry Piranha fish, also known as the "river shark", living in the sluggish waters of the large rivers of South America (see **Pygocentrus**).

Serratia A genus of microorganisms of the family *Enterobacteriaceae*; small gram-negative rods producing red pigments, and saprophytic on decaying plant or animal matter. Former generic name, *Erythrobacillus*. e.g. S. indica (= *Erythrobacillus* i.).

Sertularella A genus of Thecate Hydroids or Sea Firs, of the family *Sertulariidae*. e.g. S. polyzonias - A species of Sea Fir, up to 5 cm in height; the colonies have few branches, and the polyps are barrel-shaped immediately below the jount. Distribution: oceans of Europe; native off British coasts.

Sertularia A genus of Thecate Hydroids or Sea Firs, also known as *Dyamena* or *Dynamena*, of the family *Sertulariidae*. e.g. S. pumila (= *Dyamena* p.; *Dynamena* p.) - The Sea Oak; size up to 5 cm (2 in); a thecate hydroid, a colonial relative of the Sea Anemone. These animals live in a protective case, usually in unbranched creeping colonies, and are common on Wracks or Brown Algae such as *Fucus* (see **Fucus**), with polyps growing in pairs opposite each other in thecae with lids. They are joined to the other members of the colony by an extension of the polyp body. Known commercially as "white weed", they

grow offshore and are dried, stained and sold for decorative purposes.
Distribution: the Atlantic coasts, in the North Sea and the Baltic; native on British seacoasts.

Sesamum A genus of plants of the family *Pedaliaceae*. e.g. S. indicum - The Sesame plant; the seeds yield a fixed oil, Sesame Oil (= Benne Oil; Gingelly Oil; Ol Sesam; Oleum Sesami; Teel Oil); used like olive oil in the preparation of linaments, plasters, ointments, and soaps, and as a solvent for steroids and other insoluble drugs, in capsules and injections.

Sesia A genus of Insects of the Butterfly and Moth family, *Sesiidae* (order *Lepidoptera*). e.g. S. apiformis - The Hornet Clearwing, size 15-20 mm; often found on poplar trees fringing streams. Distribution: the Palaearctic region; North America. S. tipuliformis - A species damaging current and gooseberry bushes in Europe, North America and New Zealand.

Sessinia A genus of Blistering Beetles of certain Pacific islands.

Setaria A genus of filarial Nematodes. e.g. S. equina - A species found in the abdominal cavity of the horse.

Shigella A genus of microorganisms of the family *Enterobacteriaceae*; non-motile, rod-shaped, gram-negative bacteria, causing dysentery, and separated into non-mannitol fermenting (A), and mannitol-fermenting (B, C, and D) groups. Some species are now classified with the coliform bacilli under the generic name, *Escherichia*. e.g. S. dysenteriae (= Sh. d.; Shig. d.) - The species name given to group A dysentery bacilli, and separated into numbered serotypes. S. dispar (= *Escherichia* d.) - A slow lactose-fermenting dysentery bacillus of limited or doubtful pathogenicity.

Shorea A genus of resinous plants of the family *Dipterocarpaceae*. Used in the manufacture of varnishes and as a

mountant in microscopy. An important constituent of East Indian or Singapore Dammar, a resin compound (see **Balanocarpus**).

Sialis A genus of Insects of the Alder Fly family, *Sialidae*. e.g. S. lutaria - The Alder Fly, length 10-15 mm; wingspan 25 mm. Found on vegetation near water. Distribution: Europe.

Sicista A genus of Rodents of the Jumping Mice family, *Zapodidae*. e.g. S. betulina - The Northern Birch Mouse, body length 50-75 mm; tail 76-108 mm. Found in mountain meadows and fields. Feeds on grass seeds, fruits, and insects. Distribution: Northern Europe, Siberia, and the Caucasus.

Siderobacter A genus of microorganisms of the family *Siderocapsaceae*; bacilliform cells with rounded ends, and iron or manganese compounds on the membrane or on the surface of the cells. e.g. S. brevis.

Siderocapsa A genus of microorganisms of the family *Siderocapsaceae*; ellipsoid cells embedded in a primary capsule, with iron stored on the surface. e.g. S. coronata.

Siderococcus A genus of microorganisms of the family *Siderocapsaceae*; small coccoid cells without a gelatinous capsule. e.g. S. communis.

Sideromonas A genus of microorganisms of the family *Siderocapsaceae*; short coccoid to rod-shaped cells embedded in a large capsule, impregnated or completely encrusted with iron or manganese compounds. e.g. S. duplex.

Sideronema A genus of microorganisms of the family *Siderocapsaceae*; coccoid cells in short chains, embedded in a gelatinous sheath. e.g. S. globuliferum.

Siderophacus A genus of microorganisms of the family *Caulobacteraceae*; biconcave or rod-like cells on horn-shaped stalks, in which ferric hydroxide accumulates. e.g. S. corneolus.

Siderosphaera A genus of microorganisms of the family *Siderocapsaceae*;

small, coccoid cells, in pairs and embedded in a primary capsule. e.g. S. conglomerata.

Sigmodon A genus of Rodents of the Rat and Mice family. e.g. S. hispidus - The cotton rat.

Silene A genus of plants of the Pink family, *Caryophyllaceae*. e.g. S. pendula - The Drooping Catchfly plant, a showy annual plant with drooping sprays of pink, 5-petalled flowers. Also white, purple and deep rose colors occur. Grown in rockeries and gardens. Distribution: Southern Europe; introduced into Britain in 1731.

Silurus A genus of Fishes of the Eurasian Catfish family, *Siluridae*. e.g. S. glanis - The Wels or European Catfish, length up to 3 m. One of Europe's largest freshwater fishes. One specimen on record has been measured at 4.57 m (15 ft). Inhabits large rivers, dam reservoirs and lakes. Distribution: the rivers of Europe east of the upper reaches of the Rhine.

Simaruba A genus of plants of the Ailanthus family, *Simarubaceae*. e.g. S. amara (= S. officinalis) - The Simaruba plant; the dried rootbark, Simaruba Bark (= Orinoco Simaruba; Surinam Simaruba), has been used as a bitter, and as an astringent in the treatment of diarrhea.

Simulium A genus of Insects of the Fly family, *Simuliidae*; widely distributed, and a great pest. e.g. S. arcticum - A species found in Alaska. S. callidum - A species found in Guatemala. S. columbaczense - A species found in southern Europe; sometimes fatal to animals. S. damnosum - A species found in Africa, known as the Black Fly or Black Gnat, and especially acting as vectors or intermediate hosts of the microfilarial nematode worm, *Onchocerca* volvulus (see **Onchocerca**). S. decorum katmai - A species transmitting tularemia. S. exiguum - A species found in Mexico. S. griseicollis - The nimetti fvolvulus. S. neavi - A species found in Africa. S.

ochraceum - A species found in Guatemala. S. pecuarum - The buffalo gnat, a scourge.

Sinningia A genus of plants of the family *Gesneriaceae*, formerly mistakenly called by the generic name *Gloxinia*. e.g. S. speciosa - The Gloxinia plant, an upright herbaceous perennial with 5-lobed, tubular flowers flowing outwards at the margins. The horticultural varieties are of various colors, ranging from red to pink, white and purple, plain or bicolor flowers. Used as an indoor or greenhouse plant. Place of origin: Brazil; introduced into Europe in 1815 (see **Gloxinia**).

Sinodendron A genus of Insects of the Stag Beetle family, *Lucanidae*. e.g. S. cylindricum - Length 12-16 mm. Inhabits broadleaved forests in foothills and mountainous regions. Distribution: Europe; Siberia.

Siphonochalina A genus of Marine Sponges (phylum *Porifera*), one species of which is known as the Tube Sponge, and is found along coral reefs, especially in the Red Sea. As in other sponges, these are animals with a single body structure, somewhere between the one-celled protozoans, and the more complex organ-composed higher animals. They lack a defined nervous system and muscular tissue, and feed on microscopic animals, the plankton that float in the water, by pumping seawater in and out of their bodies. The brightly colored sponges add much beauty to the coral reef, especially in its shadier regions.

Siphonophanes A genus of Crustacean Arthropods of the family *Chirocephalidae*. e.g. S. grubii - Length 12-28 mm; including appendages which measure 3 mm. Found locally in early spring in snow melt waters. Distribution: Central Europe, eastern France, Denmark.

Siphonophora A genus of Millipedes (class *Diplopoda*). e.g. S. panamensis - A species found in Panama; it has 175 segments, but not every segment has two pairs of legs. Like other millipedes, it starts life with even fewer legs than centipedes, the usual number being three pairs.

Siphunculina A genus of Insects of the order *Diptera*. e.g. S. funicola - The common "eye-fly" of India, spreading conjunctivitis, the Naga sore, and trachoma.

Siren A genus of Amphibian animals (class *Amphibia*). e.g. S. lacertina - The Great Siren, a long-lived amphibian, living as much as 25 years.

Sirex A genus of Insects of the Horntail or Wood Wasp family, *Siricidae*. e.g. S. juvencus - Length 14-30 mm. Larvae develop in the trunks of pine or spruce trees. Distribution: all Europe; Japan; North Africa; Australia.

Sistrurus A genus of Rattlesnakes.

Sisyra A genus of Insects.

Sisyrinchium A genus of South American indaceous plants of the family, *Indaceae*. e.g. S. striatum - The Spring Bell, an erect growing plant, up to 30 cm (1 ft) high, with sword-shaped foliage and creamy yellow flowers. Used in gardens for flower and foliage effect. Its bulbs are purgative and diuretic. Place of origin: Chile; introduced into Europe in 1788.

Sitta A genus of Perching Birds of the Nuthatch family, *Sittidae*. e.g. S. europaea - The Nuthatch, length 14 cm. Found in open woodlands, parks, and large gardens. Nests in tree holes; a resident bird. Distribution: Eurasia; northwest Africa.

Smaragdina A genus of Insects of the Leaf Beetle family, *Chrysomelidae*. e.g. S. cyanea - Length 4.5-6.5 mm; found chiefly on sorrel. Distribution: Europe.

Smerinthus A genus of Insects of the Hawk Moth family, *Sphingidae*. e.g. S. ocellatus - The Eyed Hawk Moth; 33-44 mm. Has a stiking dark-pupilled and dark-bordered eyespot on each hind-

wing. Found in woods and groves. Distribution: Asia Minor.

Smilacina A genus of aromatic plants. e.g. S. racemosa - The False Spikenard, a North American plant, used as an aromatic and stimulant.

Smilax A genus of climbing smilaceous plants of the Lily family, *Liliaceae*. From the roots of various species is prepared Sarsaparilla (= Sarsa; Sarsaparilla Root; Salsepareille; Salsaparrilha); used chiefly as a vehicle and flavoring agent for medicaments. Formerly used in the treatment of chronic rheumatism and skin diseases. e.g. S. aristolochiaefolia - A species yielding Sarsaparilla. S. ornata - A species growing in Morocco, and traditionally used in that country to treat leprosy. There have been recent reports of its effectiveness as an adjunct to dapsone therapy.

Sminthillus A genus of Amphibians of the order *Salientia*. e.g. S. limbatus - The Arrow-poison frog, the smallest-known amphibian having a snout-vent length ranging from 8.5 to 12.4 mm (0.33 to 0.48 in). Found in Cuba.

Sminthopsis A genus of Marsupial animals (class *Mammalia*). e.g. S. psammophia - The Sandhill Dunnart, also called the Narrow-footed marsupial "mouse"; found in the Northern Territory of Australia, near Lake Amadeus.

Soja A genus of plants, better known by the generic name *Glycine*, of the Peafamily, *Leguminosae*. e.g. S. hispida (= *Glycine* soja) - The Soya Bean (see **Glycine**).

Solanum A genus of Herbal plants and Shrubs of the Nightshade family, *Solanaceae*, which includes the potato, several of the nightshades, and many poisonous and medicinal species. e.g. S. caroliense - A species found in the U.S. The fluid extracts of the root and berries have been used in the treatment of epilepsy. S. dulcamara - The Dulcamara plant (= Bittersweet; Douce-Amere;

Dulcamarae Caulis; Woody
Nightshade); formerly a popular remedy
for chronic rheumatism and dermatitis.
All parts of the plant are poisonous due
to the presence of solanine. The berries
have caused poisoning in children. S.
mammosum - The so-called apple of
Sodom, a plant of North America. Used
in homeopathic remedies. S. oleaceum -
The Jaquerioba, a herb of tropical
America, used in homeopathy. S.
tuberosum - The Common Potato plant,
grown in most temperate regions for its
edible tubers. A rich source of starch
(amylum), the tuber is eaten cooked as a
vegetable, and used for stock food, and
also to produce alcohol. Starch is
absorbent and is widely used in dusting
powders and protective ointments, as
mucilage in enemas, and as an antidote
in the treatment of iodine poisoning, etc.
S. wendlandii - The Giant Potato Vine
or Costa Rican Nightshade, a vigorous,
prickly, climbing plant, reaching 6 m
(20 ft), under good conditions. Produces
bright green leaves and bluish-lilac
flowers, growing in trusses at the ends
of drooping branches. Grown on walls
and to cover arbors, etc. Place of origin:
Costa Rica; introduced into Europe in
1882.

Solea A genus of Bony Fishes of the Sole
family, *Soleidae*. e.g. S. solea - The
Common European Sole, length 30-60
cm. Found in shallow coastal waters. An
important fish of commerce.
Distribution: the European coasts of the
northeast Atlantic and the
Mediterranean.

Solenoglypha A genus of Reptiles of the
poisonous Snake family, *Crotalidae*,
with fangs that are hollow like a hypo-
dermic needle, which can be erected for
striking and piercing. Common species
are the massausauga and the rattlesnakes.

Solenopotes A genus of parasitic Insects
of the order of Lice, *Anoplura*. e.g. S.

capillatus - A species of sucking lice,
occasionally parasitic on cattle.

Solenostomus A genus of Fishes of the
Pipefish family, *Syngnathidae*. The
species, about 4 cm long, inhabit black
feather stars, camouflaging themselves
among their arms. Distribution: the Red
Sea.

Solidago A genus of plants of the goldenrod Daisy family, *Compositae*. e.g. S.
canadensis - The Canada Goldenrod, a
common late summer perennial with
stiff leafy stems, terminating in broad
pannicles of many small yellow flowers.
It is an aromatic and a diuretic. Found in
open woodland, shrub borders, slopes,
and banks. Place of origin: eastern
North America; originally introduced
into Europe in 1648.

Somateria A genus of Birds of the Duck
and Goose family, *Anatidae*. e.g. S. mol-
lissima - The Eider, length 56-62 cm. A
migrating bird. Distribution: northern
Europe, Greenland, North America.

Somniosus A genus of large Sharks (class
Chondrichthyes). e.g. S. microcephalus -
The Greenland or Sleeper Shark, a carnivorous fish found in the arctic and
northern seas. It rarely exceeds 4.26 m
(14 ft) in length, but one specimen was
seen from a research submarine in 1966
at the 1219 m (4000 ft) bottom of the
San Diego Trough, whose length was
calculated at approximately 9.14 m (30
ft)!

Sophora A genus of Trees and Shrubs of
the Pea family, *Leguminosae*. e.g. S.
japonica - The Chinese Pagoda-tree; its
flower buds yield Rutin (= Rutoside),
used in the treatment of capillary bleeding, especially in retinal hemorrhages.
Many such species are poisonous to
horses, cattle and sheep in certain arid
regions, and are known as loco (Spanish
= insane), because of the selenium they
contain (see **Astragalus**). e.g. S.
lupinoides (= *Thermopsis* lanceolata)
(see **Thermopsis**).

Sorangium A genus of saprophytic soil

microorganisms of the family *Soriangiaceae* (class *Schizomycetes*); unicellular organisms, considered plants.

Sorbus A genus of Trees of the Rose family, *Rosaceae*. e.g. S. aucuparia - Mountain Ash or Rowan tree, a deciduous tree, 9-15m (30-50 ft) high, with variously pinnate leaves coloring to fine autumnal tints of red and crimson before falling. Flowers grow in dense inflorescences of white, followed by bunches of bright red berries. Distribution: Europe; Asia.

Sorex A genus of Insectivores of the Shrew family, *Soricidae*. e.g. S. araneus - The Common Shrew, length of body 65-85 mm; tail 32-47 mm. Found in damp places where it lives in burrows, more often dug by other small mammals. Like other shrews it is a very short-lived mammal, usually dying before the age of one year. It has the shortest of all mammalian gestation periods, the average being 16 days, compared to 17 days for the House Mouse (*Mus* musculus), within which period the young are fully developed at birth. Distribution: Eurasia.

Spalax A genus of Rodents of the Palaearctic Mole Rat family, *Spalacidae*. e.g. S. leucodon - The Lesser Mole Rat, length of body 150-240 mm; very short tail. Totally blind, the atrophied eyes are covered with skin. Found on dry flatlands as well as in mountains. Lives underground where it tunnels long corridors with the aid of its strong head and large incisors. Distribution: Eastern Europe and Asia Minor.

Spartium A genus of Shrubs of the Pea family, *Leguminosae*. e.g. S. junceum - The Spanish Broom (= Weaver's Broom); a vigorous, deciduous shrub with large, pea-shaped, fragrant yellow flowers, used for garden decoration. A good seaside scrub. Distribution: Mediterranean region; introduced into Britain in 1548.

Sparus A genus of Fishes of the Porgy, Sea bream and Snapper family, *Sparidae*. e.g. S. auratus - The Gilt Head Fish, length up to 70 cm. Found in the upper sea layers. An edible fish. Distribution: the east Atlantic and the Mediterranean.

Spathiphyllum A genus of plants of the Arum family, *Araceae*. e.g. S. wallisii - White Sails, a perennial evergreen plant, with long-stalked shiny lanceolate and arum-like flowers. Used as an indoor plant, also as an outdoor shade plant. The variety "Mauna Loa" is highly scented. Place of origin: Columbia.

Spenceriella A genus of Earthworms (class *Oligochaeta*). e.g. S. gigantea - The giant earthworm of New Zealand, measuring up to 1300 m (41/2 ft).

Spermophilus A genus of Marmots, a fur-bearing rodent of Manchuria, several species of which may harbor rat fleas.

Sperosoma A genus of Echinoderms of the class of Sea-urchins, *Echinoidea*. e.g. S. giganteum - The giant sea-urchin of Japan, the largest of the 800 known species of this animal. It has a horizontal shell diameter of 320 mm. (12.59 in).

Sphaeranthus A genus of plants. e.g. S. indicus - A species yielding an aphrodisiac oil.

Sphaeria A genus of Fungi. e.g. S. sinensis - A fungus species found in China, used for medicinal preparations.

Sphaeridium A genus of Insects of the Water-scavenger Beetle family, *Hydrophilidae*. e.g. S. scarabaeoides - Length 5.7 mm. Despite its name this insect is not aquatic, but lives on dry land, where it feeds on fresh cow-dung. Distribution: the Palaearctic and Nearctic regions.

Sphaerium A genus of Bivalve Molluscs of the Freshwater Cockle family, *Sphaeriidae*. e.g. S. lacustre - The Lake Orb-shell, length 8-9.3 mm; a plentiful species found in muddy, calm and slow-flowing waters. Distribution: the Palaearctic region.

Sphaerodactylus A genus of insect-eating Lizards (class *Reptilia*), of the Gecko family. e.g. S. parthenopion - The tiny Gecko, the smallest known species of reptile; found only on the island of Virgin Gorda, one of the British Virgin Islands in the West Indies. One of the largest specimens was found to measure only 18 mm (0.71 in) from snout to vent, with a tail of approximately the same length.

Sphaerophorus A genus of minute, gram-negative, non-sporulating, obligate, anaerobic, pleomorphic bacteria, also called by the generic name, *Spherophorus*, found in necrotic tissues as a secondary invader. e.g. S. necrophorus - A species producing multiple sclerotic abscesses in cattle, hogs, and other animals.

Sphaerophysa A genus of Trees of the Pea family, *Leguminosae*. e.g. S. salsula - A species yielding the alkaloid sphaerophysine, a ganglion-blocking agent used in Russia, in the treatment of hypertension and uterine atony.

Sphaerotilus A genus of microorganisms of the family *Chlamydobacteriaceae*; attached or free-floating filaments, frequently with false branches. e.g. S. dichotomus.

Sphagnum A genus of Bog Moss or Peat Moss plants of the family *Sphagnaceae*; it absorbs about 25 times its weight of water, and is used as an absorbent dressing and as a bedding for patients with urinary incontinence.

Sphenodon A genus of Reptiles of the order of Tuatara Reptiles, *Rynchocephalia*. e.g. S. punctatus - The Tuatara of New Zealand, a green-colored reptile, the only surviving member of tuataras which flourished 150 million years ago.

Spherophorus See **Sphaerophorus**.

Sphingonotus A genus of Insects of the Locust family, *Acrididae*. e.g. S. caerulans - Length 15-28 mm; hindwings usually blue. Resembles the genus *Oedipoda*. Distribution: Europe, Asia Minor, Middle East and North Africa (see **Oedipoda**).

Sphinx A genus of Insects of the Hawk Moth family, *Sphingidae*. e.g. S. ligustri - The Privet Hawk Moth, size 44-50 mm. Flies around flowering lilac where it sucks the nectar. It rests with the front part of the body raised, resembling a sphinx, hence its name. Distribution: Europe; North Asia.

Sphyraena A genus of Bony Fishes of the Barracuda family, *Sphyraenidae*. e.g. S. sphyraena - The Barracuda, length 1 m. A slender, very aggressive ocean fish known to attack man. Found near the shore where it hunts small fish. Distribution: the Atlantic Ocean, Mediterranean and Black Seas.

Sphyrna A genus of Fishes (class *Chondrichthyes*), of the Shark family, *Squalidae*. e.g. S. zygaena - The Common or Smooth Hammerhead Shark, a large carnivorous fish found in the Atlantic; over 4 m long, and weighing as much as 450 kg. From the liver is extracted Shark-liver Oil (Oleum Selachoidei).

Spigelia A genus of plants of the family *Loganiaceae*. e.g. S. marilandica - The rhizome and roots of this species contain a bitter volatile principle spigelin, which is purgative and has been used as an anthelmintic.

Spilanthes A genus of plants of the Daisy family, *Compositae*. e.g. S. acmella - The Para Cress of tropical America and Asia, formerly used as a remedy for toothache.

Spiraea A genus of plants, now known by the generic name *Aruncus*, of the Rose family, *Rosaceae*. e.g. S. aruncus (= *Aruncus* dioicus) - The Goat's Beard or Wood Goatsbeard plant (see **Aruncus**).

Spiranthes A genus of plants of the Orchid family, *Orchidaceae*. e.g. S. autumnalis - A species reputed to be an aphrodisiac.

Spirillum A genus of microorganisms of the family *Spirillaceae*; cells which form long spirals or portions of a spiral. e.g. S. minus (= Sp. minus; *Spirochaeta morsus muris*) - The cause of rat-bite fever (sodoku) in man; pathogenic for guinea pigs, mice, rats, and monkeys.

Spirobolus A genus of African Millipedes (class *Diplopoda*), of the order *Spirobolida*. Most species are locally venomous, causing skin necrosis with scarring.

Spirobranchus A genus of Worms. e.g. S. giganteus - The Giant Tube Worm (see **Sabellastarte**).

Spirocerca A genus of Nematode Worms of the family *Spiruridae*. e.g. S. sanguinolenta - A species found in the walls of the aorta, esophagus, and stomach of dogs.

Spirochaeta A genus of microorganisms of the family *Spirochaetaceae*; flexible, undulating, spiral-shaped rods, found in fresh- or seawater slime, especially when hydrogen sulfide is present. e.g. S. marina. S. morsus muris (now known as *Spirillum* minus) (see **Spirillum**).

Spironema A genus of spiral-shaped microorganisms, most of which are now included in the genus *Borrelia*. e.g. S. duttoni (= *Borrelia* d.) (see **Borrelia**).

Spiroschaudinnia A genus name for a group of spiral-shaped microorganisms found in the blood, most of which are now included in the genus *Borrelia* (see **Borrelia**).

Spirostreptus A genus of Millipedes (class *Diplopoda*), of the order *Spirostreptida*. A large species, found in the Sunda Isles, Malay Archipelago, is reported to have an exceptionally venomous secretion.

Spondylis A genus of Insects of the Long-horned or Longicorn Beetle family, *Cerambycidae*. e.g. S. buprestoides - Length 12-24 mm; black in color with large mandibles; during the day it is found under the bark of stumps and logs, flying out in the evening. Distribution: Eurasia.

Spongia A genus of Calcareous Sponges, also known by the generic name *Euspongia*, of the family *Spongiidae*. e.g. S. officinalis (= *Euspongia* o.) - The Bath Sponge, a very common animal, diameter 15-20 cm or more; soft, composed of spongin fibers, colored red, brown, green, grey, and black. Found anchored to one spot near the coast at shallow depths. Due to its soft skeleton, it is collected commercially, dried, and used in the household. Distribution: the Mediterranean (see **Euspongia**).

Spongilla A genus of Freshwater Sponges, also known by the generic name, *Euspongilla*, of the family *Spongillidae*. e.g. S. lacustris (= *Euspongilla* l.) - The Pond Sponge; soft and slippery body, colored yellowish-grey or brown. This species attains a greenish hue in sunlit waters due to the presence of green algae (*Pleurococcus*). It coats various underwater objects such as stones, submerged logs, etc., and is found in freshwater lakes. Distribution: Palaearctic and Nearctic regions (see **Euspongilla**).

Sporothrix See **Sporotrichum**.

Sporotrichon See **Sporotrichum**.

Sporotrichum A genus of Fungi, formerly called *Sporothrix* or *Sporotrichon*, some of whose species are now included under the generic name, *Rhinocladium*. e.g. S. schenckii (= S. beurmanni; S. councilmani; S. equi; S. jeanselmei; *Sporothrix* schenckii; *Sporotrichon* s.) - The species causing sporotrichosis.

Sprattus A genus of Bony Fishes of the Herring family, *Clupeidae*. e.g. S. sprattus - The Sprat fish, length up to 16.5 cm; a small seafish found in shoals, which tolerates marked fluctuations in water salinity; it is also found in river mouths and brackish waters. A very important fish of commerce. Distribution: the seas around Europe.

Sprekelia A genus of plants of the

Amaryllis family, *Amaryllidaceae*. e.g.
S. formosissima - The Aztec Lily (=
Jacobean Lily; St. James' Lily); a monotypic genus with one or occasionally
two, deep crimson flowers. Place of origin: Mexico; intrduced into Europe in
1658.

Squaliolus A genus of small Sharks
(class *Hondrichthyes*). e.g. S. laticaudus
- The smallest known species of sharks,
not exceeding 152 mm (6 in), and found
in the Gulf of Mexico.

Squalus A genus of Fishes of the Dogfish
Shark family, *Squalidae*. e.g. S. acan-
thias - The Common Spiny Dogfish
Shark, length up to 1.2 m; viviparous;
has a sharp poisonous spine in front of
each of its two dorsal fins. Used in commerce. Distribution: most plentiful shark
of the North Sea.

Squatina A genus of Fishes of the Angel
Shark family, *Squatinidae*. e.g. S.
squatina - The Monkfish, length 1-2.5
m; a shark with a ray-like body. Feeds
chiefly on fish and crustaceans.
Distribution: along the European
Atlantic coast and the Mediterranean.

Stagmomantis A genus of Insects of the
suborder of Praying Mantids, *Mantodea*
(order of Cockroach and Praying
Mantids, *Dictyoptera*). e.g. S. carolina -
A southern species of Praying mantids
found in North America. An exceedingly voracious insect (see **Tenodera**).

Staphylinus A genus of Insects of the
Rove Beetle family, *Staphylinidae*. e.g.
S. caesareus - Length 17-25 mm; feeds
on various larvae and dead animals, as
well as decaying plant matter.
Distribution: Europe; North America.

Staphylococcus A genus of microorganisms of the family *Micrococcaceae*. e.g.
S. aureus (= Staph. aureus) - A species
of pigmented, coagulase, gram-positive
organisms pathogenic to man and animals, particularly affecting the skin.

Stegomyia A subgenus of Insects of the
Mosquito family, *Culicidae*. e.g. S.
argenteus (= S. Calopus; S. fasciata; S.

fasciatus) - These are former names for
Aedes aegypti (see **Aedes**).

Stegosaurus A genus of Extinct Reptiles,
known as the "plated reptile", a heavily
built herbivore which roamed across the
Northern Hemisphere about 150 million
years ago. It has been dubbed the most
brainless of all the dinosaurs, because it
had a poorly organized walnut-sized
brain weighing only 70.7 g (2.5 oz), rep-
resenting 0.004% of its body weight,
compared with 0.074% for an elephant,
and 1.88% for an adult human being.

Stellaria A genus of plants of the
Chickweed or Pink family,
Caryophyllaceae. e.g. S. media - A
species of chickweed, formerly used as
a demulcent.

Stena A genus of Insects of the Ant order,
Hymenoptera. e.g. S. westwoodi -
Westwoodi's ant, a long-lived species of
ants, attaining as much as 18 years of
age.

Stenella A genus of Mammals of the
Dolphin family. e.g. S. attenuata - The
Spotted Dolphin, a rapid swimmer,
reputedly clocking at 39.36 km/h (21.4
knots; 24.62 miles/h).

Stenobothrus A genus of Insects of the
Short-horned Grasshopper family,
Acrididae. e.g. S. lineatus - The Stripe-
winged Grasshopper, length 16-25 mm.
Found in meadows and forest paths.
Variable in color. Distribution: Eurasia.

Stenopus A genus of Crustacean
Arthropods of the Shrimp family,
Crangonidae. e.g. S. hispidus - The
Cleaner Shrimp, a species feeding on
parasites they remove from their clients,
the fish, that submit themselves to a delicate cleaning process. Found among the
coral branches in the Red Sea region.

Stenoteuthis A genus of Squids of the
class *Cephalopoda*. e.g. S. bartrami - A
small surface-dwelling decapod, said to
be one of the swiftest creatures of the
sea, generating speeds up to 55 km/h
(34.35 miles/h).

Stenotomus A genus of Fishes (subphylum *Vertebrata*). e.g. S. chrysops - The Northern (Porgy) Scup fish.

Stephanoberyx A genus of Fishes of the Pricklefish family, *Stephanoberycidae*. These are elongated fishes with large eyes, the head being covered with spines. The body is covered with perpendicular, spiny scales, and the airbladder remains open. There are three main genera, the other two being *Acanthochaenus* and *Malacosarcus*.

Stephanofilaria A genus of Filarian Nematodes. e.g. S. stilesi - A species causing skin infection in domestic animals.

Stephanurus A genus of Nematode parasites. e.g. S. dentatus - A species parasitic in the urinary tract of swine.

Stercorarius A genus of Shorebirds of the Skua family, *Stercorariidae*. e.g. S. skua - The Great Skua, length 59 cm; feeds on fish, robbing other sea birds of their catch. Nests either singly or in small colonies close to the sea. Distribution: the North Atlantic coast.

Sterculia A genus of Trees and Shrubs of the family *Sterculiaceae*; especially found in the tropics; some species have edible seeds and others are medicinal, while still others yield a gummy exudate, sterculia (= Stercul; Goma Caraia, Indian Tragacanth; Karaya Gum; Sterculia Gum), resembling tragacanth. e.g. S. urens - A species yielding sterculia gum, used internally to stimulate peristalsis in the treatment of constipation; it has adhesive properties and is used in dental fixative powders. It has been used as a tragacanth, to prepare bases for medicinal jellies and pastes but is a less efficient suspending agent.

Sterigmatocystis See **Sterigmocystis**.

Sterigmocystis A genus of Molds, also known as *Sterigmatocystis*, resembling *Aspergillus*, except that secondary phialides project from each primary phialide. e.g. S. nidulans - A species found in otomycosis and in the white granules of mycetoma (see **Actinomyces** madurae).

Sterna A genus of Shorebirds of the Tern family, *Laridae*. e.g. S. hirundo - The Common Tern, length 35 cm. Found on the shores of fresh waters and on muddy islets, as well as on seacoasts. A migrant bird, wintering in the Mediterranean and Africa. Distribution: most of Europe, Asia, and North America.

Sternbergia A genus of plants of the Amaryllis family, *Amaryllidaceae*. e.g. S. lutea - Lily of the Field or Winter Daffodil, a crocus-like plant with rich golden, solitary flowers. Considered by some to be the "lilies of the field" of the Bible. Place of origin: Mediterranean region; introduced into Britain in 1596.

Sternotherus A genus of Turtles of the order *Chelonia*. e.g. S. odoratus - The Stinkpot, a long-lived turtle, attaining more than 50 years of age. This reptile is found in the southeastern U.S.

Stichopus A genus of Sea-cucumbers of the class *Holothuroidea*. e.g. S. variegatus - This bulky echinoderm from the Philippines, is one of the largest of the approximately 1000 known species of sea-cucumber. It has been measured up to 1000 mm (39.37 in) in length when fully extended, and 210 mm (8.26 in) in diameter.

Stigmella A genus of Moths of the order *Lepidoptera*. e.g. S. ridiculosa - One of the smallest species, having a wingspan of only 2 mm (0.08 in). Found in the Canary Islands.

Stillingia A genus of Trees, Shrubs and Herbs of the family *Euphorbiaceae*. e.g. S. sylvatica - A plant of North America. The root is a sialagogue and diuretic.

Stipa A genus of plants of the Grass family, *Gramineae*. e.g. S. viridula - A species of grass found in the southwestern Unit States, called sleepy grass. It is poisonous to cattle and horses, and is said to be a powerful narcotic, diuretic, sudorific, and cardiac poison.

Stizostedion A genus of Fishes of the

Perch family, *Percidae*. e.g. S. lucioperca - The Zander or Pike-perch fish, length more than 1 m; weight 12 kg or more; lives near the bottom in deeper sections of rivers with sandy or soil beds. Bred in ponds and planted into dam reservoirs, lakes, and rivers. An important freshwater fish of commerce. Distribution: the rivers and lakes of Europe.

Stomoxys A genus of Two-winged Flies (*Diptera*), of the family, *Muscidae*. e.g. S. calcitrans - The Common Stable-fly, also called the Leg-sticker fly, length 5.5-7 mm. A country fly, biting severely and feeding on the blood of mammals, including man. Capable of transmitting anthrax, tetanus, and infectious anemia of horses. Distribution: worldwide.

Storeria A genus of Snakes (order *Squamata*). e.g. S. dekayi - De Kay's snake.

Strangalia A genus of Insects of the Long-horned or Longicorn Beetle family, *Cerambycidae*. e.g. S. masculata - Length 14-20 mm; the elytra are yellowish, with various large dark spots. Found on flowers in forest meadows. Distribution: Europe, western Asia.

Stratiomys A genus of Insects of the Soldier Fly family, *Stratiomyidae*. e.g. S. chamaeleon - Length 14-15.5 mm. Found in marshes, damp meadows and hedgerows. Distribution: central and southern Europe.

Strelitzia A genus of plants of the Banana family, *Musaceae*. e.g. S. reginae - Bird of Paradise Flower or Strelitzia, an exotic evergreen plant, named after Queen Charlotte, wife of King George III of England, who died in 1818. Purple and orange flowers are large and showy, with free sepals and blue protruding petals, bearing a resemblance to the plumaged head of tropical birds. Place of origin: South Africa; introduced into Europe since 1773.

Streptobacillus A genus of microorganisms, also known by the generic name

Haverhillia, of the family *Bacteroidaceae*; pleomorphic bacteria, varying from short rods to long, interwoven filaments. e.g. S. moniliformis (= *Haverhillia* multiformis) - The etiological agent of ratbite fever and Haverhill fever (see **Haverhillia**).

Streptococcus A genus of microorganisms of the family *Lactobacillaceae*; separated into the *Pyogenic, Vividans, Enterococcus*, and *Lactic* groups. e.g. S. pyogenes (= Str. p.; Strep. p.) - β-hemolytic, toxigenic, pyogenic streptococci of group A; causes septic tonsillopharyngitis, scarlet-fever, rheumatic fever, acute glomerulonephritis, etc. S. zymogenes (= *Enterococcus z.; Zymogenes z.*) - Isolated from human feces and the vagina.

Streptomyces A genus of microorganisms of the family *Streptomycetaceae*, formerly called by the generic name *Micromyces* or *Streptothrix*, and separable into 150 different species, usually soil forms, and notable as the source of various antibiotics. e.g.:
S. alanosinicus - A species producing the antibiotic Alanosine, which has antiviral and antitumor activity.
S. albo-niger - A species producing the amebicide Puromycin; especially effective against *Entameba* histolytica and *Trypanosoma* infections.
S. ambofaciens - A species producing the narrow-spectrum antibiotics Spiramycin I, II and III; especially effective against gram-positive bacteria.
S. antibioticus - A species producing:
1. the antibiotic Oleandomycin, active against a wide range of gram-positive bacteria, including cocci (especially resistant staphylococci), *Bacillus* anthrasis, B. subtilis, *Clostridia*, etc.;
2. the antimicrobial chromopeptide Actinomycin D (= Dactinomycin; Meractinomycin), a cytotoxic and antineoplastic agent; also produced by growth of *Streptomyces* chrysomallus.
S. ardus - A species producing the

430

antibiotic Porfiromycin (= Methylmitomycin); has antitumor activity.

S. argillaceus - A species producing the antimicrobial acid Mithramycin. Has antibacterial activity against gram-positive organisms, and moderate antineoplastic activity. Mithramycin is also produced by S. tanashiensis.

S. aureofaciens - A species producing the antibiotic Dimethylchlortetracycline, a broad-spectrum antibiotic with range and action similar to tetracycline.

S. caespitosus - A species producing the antibiotic Mitomycin C (= Mitomycin X), and also two other mitomycins, viz. Mitomycin A and B. Mitomycin C has some antibacterial activity, as well as cytotoxic and antineoplastic properties.

S. canus - A species producing the antimicrobial substance Amphomycin; acts against gram-positive cocci as well as *Trypanosoma* gambiense and T. rhodesiense.

S. capreolus - A species producing the antimicrobial polypeptide Capreomycin (= Capromycin; Compound 29275); active against *Mycobacterium* tuberculosis.

S. chrestomyceticus - A species producing the antibiotic Aminosidin, an effective amebicide.

S. chrysomallus - A species producing the antimicrobial chromopeptides, which have cystostatic and antineoplastic properties. These are:
1. Actinomycin C (= Cactinomycin; HBF 386) - a mixture of Actinomycin C2 and C3;
2. Actinomycin D (= Dactinomycin; Meractinomycin) (see **Streptomyces** antibioticus);
3. Actinomycin F1;
4. Actinomycin K.

S. decaris - A species producing the antibiotic Framycetin (consisting of Neomycin B with a small amount of Neomycin C); particularly useful in the treatment of staphylococcal infections (see **Streptomyces** fradiae).

S. erythreus - A species producing the antibiotic Erythromycin (= Eritromicina; Erythromyc); a broad-spectrum antibiotic, particularly useful in patients sensitized to penicillin, or in infections due to penicillin-resistant strains of bacteria.

S. flocculus - A species producing the antineoplastic substance Streptonigrin (= Flomycin; NSC 45383).

S. fradiae - A species producing the antibiotic Neomycin (= Fradiomycin; Neomycin Sulfas). Neomycin consists of two isomers, Neomycin B and Neomycin C, and is a multipurpose, poorly absorbed antibiotic for local use, and orally as an intestinal antiseptic (see **Streptomyces** decaris).

S. garyphalus - A species producing the antibiotic Cycloserine (= Cycloserinum), used in the treatment of severe pulmonary tuberculosis. Cycloserine is also produced by the growth of *Streptomyces* orchidaceus.

S. griseoflavus - A species producing the iron-containing antibiotic Ferrimycin A, which resembles Albomycin (produced by the growth of *Actinomyces* subtropicus) and is chiefly active against gram-positive organisms.

S. griseus - A species producing one of the first well-known antibiotics Streptomycin, widely used in the treatment of tuberculosis. A bactericide like penicillin, with which its action is synergistic, streptomycin is especially effective in subacute bacterial endocarditis caused either by *Streptococcus* mitis or Str. fecalis.

S. griseus var. purpureus - A species producing the antibiotic Viomycin, used in the treatment of resistant cases of tuberculosis.

S. hachijoensis - A species producing the antimycotic agent Trichomycin, active against *Candida* albicans, *Trichomonas* vaginalis, *Trichophyton* and *Aspergillus* spp.

S. halstedii - A species producing the antibiotic Carbomycin (consisting of Carbomycin A and Carbomycin B), a broad-spectrum antibiotic with action similar to that of erythromycin (see **Streptomyces** erythreus).

S. hygroscopicus var. augustmyceticus - A species producing three antimicrobial substances Augustmycins A, B and C, which have antineoplastic activity.

S. kanamyceticus - A species producing the antibiotic Kanamycin, used in many severe systemic infections, including pulmonary tuberculosis.

S. kitasatoensis - A species producing the macrolide antibiotic Leucomycin, used in coccal infections, with actions similar to those of carbomycin and erythromycin (see **Streptomyces** erythreus). Especially useful in the local treatment of trachoma.

S. lincolnensis var. lincolnensis - A species producing a monobasic antibiotic Lincomycin, a bacteriostatic agent with a range of antibacterial action similar to that of erythromycin (see **Streptomyces** erythreus).

S. mediterranei - A species producing a group of antibiotics, the Rifamycins (= Rifomycins). Rifamycins A, B, C, D, and E have been isolated naturally; Rifamycins O, S, and SV have been derived in the laboratory from Rifamycin B, while Rifamycins AG and X are further derivatives of Rifamycin O. Rifamycin SV is the most useful and least toxic of the rifamycins, and inhibits the growth of gram-positive and gram-negative bacteria. Its action against *Mycobacterium* tuberculosis is equal to that of isoniazid and much greater than that of streptomycin.

S. natalensis - A species producing the amphoteric antibiotic Natamycin (= Pimaricin), a powerful antifungal agent acting against *Aspergillus*, *Candida*, *Coccidiodes*, *Cryptococcus*, *Histoplasma*, and *Trichophyton* spp.

S. niveus - A species producing the antibiotic Novobiocin (= Streptonivicin), useful in cases sensitized to penicillin, or for the more resistant coccal infections. Novobiocin is incompatible with the common basic antibiotics such as streptomycin, kanamycin, neomycin, spiramycin, and viomycin, forming water-insoluble salts.

S. nodosus - A species producing the broad-spectrum antifungal agent Amphotericin B, used in the treatment of deep-seated mycotic infections, particularly histoplasmosis and North American blastomycosis.

S. noursei - A species producing the antimycotic substance, Nystatin (= Fungicidin; Mycostatin; Nystatinum), used especially in the local treatment of moniliasis, especially that caused *Candida* albicans.

S. orchidaceus - A species producing the antibiotic Cycloserine (see **Streptomyces** garyphalus).

S. orientalis - A species producing the amphoteric antimicrobial substance Vancomycin, highly active against *Staphylococcus* aureus, *Streptococcus* pyogenes, and Str. pneumoniae.

S. pimprina - A species producing the antimicrobial substance Hamycin, which has antifungal, antitrichomonal, and anti-inflammatory properties.

S. pristina spiralis - A species producing the bacteriostatic agent Pristinamycin (= RP 7293), acting particularly against gram-positive bacteria.

S. rimosus - A species producing the antibiotic Oxytetracycline, a broad-spectrum bacteriostatic agent, with actions similar to those of tetracycline.

S. rimosus forma paromomycinus - A species producing the antibiotic Paromomycin, a broad-spectrum antibiotic, especially useful in the treatment of intestinal amebiasis, *Shigella* infections, and gastroenteritis due to *Escherichia* coli.

S. spectabilis - A species producing the antibiotic Spectinomycin (=

Actinospectacin), active against some bacteria, especially species of *Brucella*, *Proteus*, *Pseudomonas*, and *Staphylococcus*.

S. tanashiensis - A species producing the antimicrobial acid Mithramycin (see **Streptomyces** argillaceus).

S. variabilis - A species producing a mixture of antimicrobial substances, Streptovarycin, which has been tried in tuberculosis and in the lepromatous form of leprosy.

S. venezuelae - A species producing the well-known bacteriostatic agent Chloramphenicol (= Choramphen; Laevomycetin), which is now mainly prepared synthetically. This is a broad-spectrum antibiotic of first choice used only in typhoid and paratyphoid fever, because of the possible occurrence of serious toxic effects, such as severe depression of bone-marrow activity, leading to aplastic anemia, agranulocytosis, and thrombocytopenic purpura.

S. virginiae - A species producing a mixture of antimicrobial substances, Virginiamycin (= Virgimycin), which has been tried in infections caused by *Bordetella* pertussis, and in staphylococcal infections resistant to other antibiotics.

Streptopelia A genus of Birds of the Dove and Pigeon family, *Columbidae*. e.g. S. turtur - The Turtle Dove, length 27 cm; the tail has a narrow, white, terminal band which is clearly visible in flight. Found in open, mixed, and broadleaved woodlands, as well as in thickets. A migrant bird. Distribution: Eurasia and North Africa.

Streptosporangium A genus of microorganisms of the family *Actinoplanaceae*; mycelium-forming saprophytes found in soil and water.

Streptothrix A name formerly given to a genus of microorganisms, species of which are now classified under *Actinomyces*, *Nocardia*, *Streptomyces*, and other genera. Formerly also known as *Micromyces* (see **Actinomyces**, **Nocardia**, and **Streptomyces**).

Strigops A genus of Birds of the Parrot family, *Psittacidae*. e.g. S. habroptilus - The Kakapo, a curious nocturnal parrot found in New Zealand, and the sole species of this genus. It was formerly classified among the owls (order *Strigiformes*), since its eyes are partly turned forwards and its beak, like that of a typical owl, is hidden in its feathers. A ground bird, it returns during the day into a burrow made among the tree roots.

Strix A genus of Birds of the Typical Owl family, *Strigidae*. e.g. S. aluco - The Tawny Owl, length 38 cm; a night bird of prey, feeding chiefly on rodents and other small mammals, birds, and larger insects. It inhabits woodlands with old broadleaved trees, parks, and large gardens. A resident bird. Distribution: Eurasia and North Africa.

Strombus A genus of Marine Snails of the class *Gastropoda*. e.g. S. gigas - The Queen Conch; this mollusc species is found in the Florida Keys and the West Indies, and weighs up to 2.27 kg (5 lb); its shell measures 304-609 mm (12-24 in) in length.

Strongylocentrotus A genus of Sea-urchins of the class *Echinoidea*. e.g. S. drobachiensis - This echinoderm, collected off the coast of Norway, has been reported to have a maximum age of 6 years, while most shallow-water echinoderms live less than 4 years.

Strongyloides A genus of Nematodes or Roundworms, widely distributed in the intestines of mammals. e.g. S. stercoralis (= S. intestinalis) - A roundworm found in man and widely occurring in tropical and subtropical countries. The female adult and her larvae inhabit the mucosa and submucosa of the small intestine where they cause diarrhea and ulceration (intestinal strongyloidiasis). The larvae, expelled from an infected person with his feces, develop in the soil and penetrate the human skin on contact. They eventually reach the lungs where they cause hemorrhage (pulmonary strongyloidiasis).

Strongylus A genus of parasitic Nematode Worms, also known as *Eustrongylus*, some of whose species are now included under the generic names Dioctophyma and Trichostrongylus. e.g. S. equinus - The palisade worm found in the intestines of horses. S. gibsoni - A species found in cattle and sheep, and occasionally in man. S. gigas (= S. renalis; *Dioctophyma* renale) - The giant kidneyworm (see **Dioctophyma**). S. subtilis (= *Trichostrongylus* instabilis) - A species found in sheep and goats, and occasionally in man (see **Trichostrongylus**).

Strophanthus A genus of plants of the Dogbane family, *Apocyanaceae*. e.g. S. gratus - A species, the seeds of which yield the glycoside, Oubain (= Oubainum; G-strophanthin; strophanthin-g; Strophanthinum; Strophanthoside-G; Uabaina; Ubaina); a cardiac glycoside with an action similar to that of digitalis. Also obtained from the wood of *Acokanthera* schimperi.

Stropharia A genus of Fungi, Molds, or Mushrooms. e.g. S. cubensis - A mushroom species yielding the Mexican Indian drug Teonanacatl, containing the indole alkaloid Psilocybin, a hallucinogen (see **Psilocybe**).

Struthio A genus of Birds of the order of Ostriches, *Struthioniformes*. e.g. S. camelus - The African Ostrich, the world's largest living bird. There are five subspecies. S. camelus camelus - The Northern African ostrich, the largest of these subspecies, found south of the Atlas Mountains of Upper Senegal and Niger, to the Sudan and Central Ethiopia. Adult cock examples of this ratite or flightless bird can reach heights up to 2.74 m (9 ft), and weigh as much as 157 kg (345 lb).

Strychnos A genus of Tropical Trees of the family *Loganiaceae*. e.g. S. nux-vomica - The dried ripe seeds of this species yield Nux Vomica (= Nux Vom; Strychni Semen; Noix Vomique; Neuz vómica; Brechnuss), which contains the alkaloids strychnine and brucine, and traces of strychninine, vomexine, and the glycoside loganin. Strychnine has been used in the past as an analeptic, but it stimulates the respiratory and cardiovascular centers only if used in convulsive doses. Nux Vomica is used as a bitter, and as an ingredient of purgative pills and tablets.

Sturnus A genus of Birds of the Starling family, *Sturnidae*. e.g. S. vulgaris - The Starling, length 21 cm; originally living in broadleaved forests, it is now found also in tree-less regions. Sometimes it migrates in winter to western Europe and the Mediterranean region. Distribution: Eurasia.

Stylaria A genus of Segmented Worms (phylum *Annelida*), of the family *Naididae*. e.g. S. lacustris - Length 5.5-18 mm; transparent body; found in freshwater amidst vegetation. Distribution: Eurasia and North America.

Stylosanthes A genus of Herbs of the Pea family, *Leguminosae*. e.g. S. elatior - The Pencil-flower of North America; acts as a uterine sedative.

Styrax A genus of Trees and Shrubs of the Storax family, *Styracaceae*. e.g. S. benzoin - A shrub or small tree of Sumatra, from the incised stem of which is expressed a balsamic resin, Benzoin (= Benjoim; Benzoe; Gum Benjamin; Gum Benzoin; Sumatra Benzoin), containing various esters of benzoic and cinnamic acids together with the free acids. Used as a reflex expectorant and inhalant in catarrh of the upper respiratory tract, and externally as a tincture for its local antiseptic, astringent, and protective properties.

Suberites A genus of Siliceous Sponges

of the family *Suberitidae*. e.g. S. domuncula - The Sea-Orange, diameter 40-60 mm; a globular smooth-surfaced, orange-colored animal covering the shells of marine gastropods, inhabited by the crab *Paguristes* oculatus. This sponge is abundant on the sandy bottom of seacoasts. Distribution: Mediterranean Sea and Atlantic Coasts; the English Channel and the North Sea.

Succinea A genus of Gastropod Molluscs of the family *Succineidae*. e.g. S. putris - The Large Amber Snail; the shell is 16-22 mm high, 8-11 mm wide, and composed of 3 to 4 yellowish, thin, semi-transparent whorls. Lives in riverine forests, on reeds and water plants, chiefly in the lowlands. Distribution: Eurasia.

Sufflamen A genus of Fishes (subphylum *Vertebrata*). e.g. S. chrysoptera - The Bluethroat fish; belongs to the trigger fish family, and is found along coral reefs. It has a small spine on its back that locks or releases one or more of the larger back spines. This locking mechanism is utilized by the fish for wedging itself into small crevices when danger approaches.

Sula A genus of Birds of the Pelican Gannet family, *Sulidae*. e.g. S. bassana - The Gannet, length 94 cm; a seabird found on the open sea, only rarely appearing inland. Nests in large colonies, located on coastal cliffs. Distribution: the shores and islands of the North Atlantic, including the British Isles.

Suncus A genus of Insectivore Mammals of the Shrew family, *Soricidae*. e.g. S. etruscus - Savi's White-toothed Pygmy Shrew, also called the "Etruscan Shrew", the world's smallest mammal, having a head and bodylength of 36-52 mm (1.32-2.04 in), a tail length of 24-29 mm (0.94-1.14 in), and weighing 1.5-2.5 g (0.052-0.09 oz). Found in damp places near water. Distribution: along the coasts of the northern Mediterranean and southwards to Cape Province, South Africa.

Sus A genus of Ungulate Mammals of the Old World Pig family, *Suidae*. e.g. S. scrofa - The Wild Boar, an even-toed ungulate, length of body 110-180 cm; height at shoulder 85-115 cm. Generally found in herds in oak and beech woods next to fields and meadows, mainly in foothills and mountains. Distribution: Eurasia and North Africa. From the fresh stomach is prepared a coarse granular powder, Desiccated Stomach (BPC) (= Dried Stomach; Powdered Stomach; Ventriculus Desiccatus), formerly used in the treatment of pernicious anemia before the advent of vitamin B12.

Swertia A genus of plants of the Gentian family, *Gentianaceae*. e.g. S. chirata (= Chirat; Chirayta; Chiretta; East Indian Balmony) - The Chirata plant, from which is extracted a simple bitter, administrated as a tincture or a concentrated infusion.

Syagrus A genus of Trees, now classified under the generic name *Microcoelum*, of the Palm family, *Palmae*. e.g. S. weddeliana - The Weddel Palm (= *Cocos* w.; *Microcoelum* w.) (see **Microcoelum**).

Sycon A genus of Calcareous Sponges of the family, *Sycettidae*. e.g. S. raphanus - Length 5-20 mm; ovate or globular, yellowish or brownish body, covered with bristles. The skeleton of this animal is composed of calcareous spicules. Lives in salt water along the Mediterranean coast.

Syctalidium A genus of non-dermatophyte Molds causing both nail infections and skin changes resembling tinea.

Sylvia A genus of Perching Birds of the Old World Warbler family, *Sylviidae*. e.g. S. nisoria - The Barred Warbler, length 15 cm; found on banks, in thickets, forest margins, parks and gardens, and along streams. A migrating bird. Distribution: Eurasia.

Symbiotes A genus of microorganisms of the family *Rickettsiaceae*. e.g. S. lectularius - Only species of this genus; it is parasitic on the bedbug, *Cimex* lectularius (see **Cimex**).

Sympetrum A genus of Insects of the Dragonfly and Damselfly family, *Libellulidae*. e.g. S. sanguineum - Length 35-40 mm, wingspan 50-60 mm; found on banks by the waterside, on shrubs, etc. Distribution: Eurasia, North Africa.

Symphoricarpus A genus of Shrubs of the family *Caprifoliaceae*. e.g. S. albus (= S. racemosus) - The Snowberry, a slender, deciduous shrub, about 3 ft (1 m) high, with small pink flowers on slender spikes. These are succeeded by large, round, white berries. Used in hedges and shrubberies, and to decorate slopes and rocky places. Place of origin: N.E. America; introduced into Europe in 1879.

Symphysodon A genus of Fishes (subphylum *Vertebrata*). e.g. S. discus - The Discus Fish.

Symphytum A genus of plants of the Borage family, *Boraginaceae*. e.g. S. officinale - The Comfrey plant of Europe and North America.

Symplocarpus A genus of Tapeworms. e.g. S. foetidus (= *Dracontium* foetidum) (see **Dracontium**).

Synanceja A genus of Fishes (phylum *Vertebrata*), a member of the Scorpion Skinfish family and with the same characteristics. e.g. S. verrucosa - The Stone Fish (see **Scorpaena**).

Synangium A genus of saprophytic microorganisms of the family of schizomycetes, *Polyangiaceae*. Found in soil and decaying organic matter.

Synapta A genus of Sea-cucumbers (class *Holothuroidea*); some species of these worm-like echinoderms can stretch themselves out to lengths of 1000 mm (39.37 in), or even 2000 mm (78.74 in), but they only measure about 12 mm (0.5 in) in diameter.

Syngamus A genus of Nematode Worms that are parasitic in fowl and other birds.

It may occur in man. Also known as *Sclerostoma*. Found in Tropical America. e.g. S. trachea (= S. laryngeus; S. trachealis; *Sclerostoma* syngamus) - A species of worms which are parasitic to chickens, pheasants, turkeys, and various wild birds, inhabiting the upper respiratory tract and causing hemoptysis and asthma. They have also been reported in man.

Syngnathus A genus of Fishes of the Pipefish and Seahorse family, *Syngnathidae*. e.g. S. acus - The Great Pipefish, length 35 cm; usually found at depths of less than 15 m in places along the coast where there is abundant seaweed; also in brackish waters and in the estuaries of large rivers. Distribution: northeastern Atlantic.

Synodus A genus of Fishes (subphylum *Vertebrata*). e.g. S. variegatus - The Lizard fish, a coral fish usually found on the sandy floor on coral reefs, lying motionless and camouflaged, blending into the surrounding background. Distribution: Red Sea coast.

Synsepalum A genus of plants. e.g. S. dulcificum - Miracle fruit of West Africa, and a natural sweetener; is tasteless but causes sour fruit to taste sweet.

Syphacia A genus of Nematode oxyurid parasites found in the intestines of rodents. e.g. S. obvelata - A common cecal parasite of laboratory rats; occasionally reported in man.

Syringa A genus of plants of the Olive family, *Oleaceae*. e.g. S. vulgaris - The Lilac plant, cultivated in numerous forms, with richly scented flowers in various shades of mauve or violet, red, white and primose-yellow, in dense, pyramidal panicles. Used in shrubberies and hedges in temperate climates. Place of origin: mountains of eastern Europe.

Syringospora Former generic name of the Fungus, now called *Candida* (see **Candida**).

Syrinx A genus of Marine Snails of the class *Gastropoda*. e.g. S. auranus - The

436

Trumpet or Baler conch of Australia. It weighs up to 2.27 kg (5 lb) and has a shell measuring 304-609 mm (12-24 in) in length.

Syrnium A genus of Birds of the Owl family (class *Aves*). e.g. S. aluco - The Wood Owl.

Systenocerus A genus of Insects of the Stag Beetle family, *Lucanidae*. e.g. S. caraboides - Length 10-14 mm; colored metallic green, blue-green, blue-violet to blue-black. Found chiefly in oak and beech woods. Distribution: western and central Europe.

Syzygium A genus of Tropical Trees of the Myrtle family, *Myrtaceae*. e.g. S. jambolanum - The Jambul tree of India; is an astringent.

T

Tabanus A genus of Insects of the biting Horse Fly family, *Tabanidae*. They transmit trypanosomiasis and anthrax to lower animals. e.g. T. bovinae - The fast-flying Gadfly of cattle and horses. The female sucks the blood of animals and man; the male feeds on plant juices. Abundant in fields and meadows. Distribution: Asia, Africa, and South America.

Tachina A genus of Insects of the Two-winged Fly family, *Tachinidae*. e.g. T. fera - Length 11-14 mm; larvae are internal parasites of the caterpillars of the Black Arches Moth and Gypsy Moth. Distribution: the Palaearctic region.

Tadorna A genus of Birds of the Duck family, *Anatidae*. e.g. T. tadorna - The Shelduck, length 62-66 cm. Found on flat, sandy, or muddy seashores with dunes or river mouths, and occasionally also in salty inland waters. Distribution: coasts of the North Baltic, Black, Caspian, and Mediterranean Seas.

Taenia A genus of Common Tapeworms (class *Cestoda*), of the family *Taeniidae*. e.g. T. saginata (= T. mediocanellata; *Taeniarhynchus* s.) - The Beef Tapeworm of Fat Tapeworm, the common tapeworm of man; the largest species of human infesting tapeworms, usually 10-12 m (32 ft 91/2 in - 39 ft 41/2 in) long, but females have reached up to 25 m (82 ft) in length. Found in the adult form in the human intestine, and in the larval stage (*Cysticercis* bovis) in the muscles and other organs of the ox. Man acquires the disease by eating infected meat (see **Taeniarhynchus; Taeniarhynus**).

Taeniarhynchus A genus name formerly given to Beef Tapeworms, now called *Taenia*. e.g. T. saginata (= *Taenia* s.) - The Beef Tapeworm (see **Taenia**).

Taeniarhynus A subgenus of Tapeworms, a name formerly given to the Beef *Taenia* saginata (= *Taeniarhynchus* s.); as in the adult worm, the scolex has no rostellum or rostellar hooks (see **Taenia**).

Taeniorhynchus A genus of Mosquitoes now called *Mansonia* (see **Mansonia**).

Tagetes A genus of plants of the Daisy family, *Compositae*. e.g. T. erecta - The African Marigold, a strongly aromatic half-hardy annual plant with pinnate, pungent, deeply-cut leaves and large single, yellow or orange daisy-like flowers. Used in gardens for bedding and in containers, etc. Place of origin: Mexico; in Europe since 1596.

Talauma A genus of plants. e.g. T. elegans - A plant of Java, with stomachic, antispasmodic, and sedative properties.

Talpa A genus of Insectivore Mammals of the Mole family, *Talpidae*. e.g. T. europaea - The European Mole; length of body 130-170 mm; tail 23-38 mm. Found in meadows, parks and forest margins. Lives underground using its shovel-like forefeet to dig a series of tunnels, resulting in molehills on the surface. Distribution: Europe, Asia.

Tamarindus A genus of Trees of the Pea family, *Leguminosae*. e.g. T. indica - The Tamarind tree, from the seeds of which is extracted a polysaccharide, Tamarind Polyose, used as a binding, emulsifying, and suspending agent. Tamarind is also a mild laxative and an ingredient of Senna Confection.

Tamias A genus of Mammals of the Squirrel family, *Sciuridae*. The species are the American Chipmunks, friendly little animals; these rodents are brightly colored, ground Squirrels, familiar figures throughout the United States and Canada (see **Eutamias**).

Tamus A genus of plants of the family *Dioscoreaceae*. e.g. T. communis - The Black Bryony, a species used in homeopathic medicine.

Tanacetum A genus of plants of the

Daisy family, *Compositae*, now known by the generic name *Chrysanthemum*. e.g. T. vulgare (= *Chrysanthemum* v.) - Buttons; The Common Tansy; Tansyplant (see **Chrysanthemum**).

Taniopygis A genus of Birds of the Finch family. e.g. T. castanotis - The Zebra Finch, a species laying very small eggs. Some measure only 9.5 x 7 mm (0.37 x 0.27 in).

Taraktogenos A genus of tropical plants of the family *Flacourtiaceae*, also known by the generic name *Hydnocarpus*. e.g. T. kurzii (= *Hydnocarpus* k.) - From the seed is extracted the fixed oil, Hydnocarpus Oil, employed in the treatment of leprosy (see **Hydnocarpus**).

Taraxacum A genus of plants of the Daisy family, *Compositae*, formerly known as *Leontodon*. e.g. T. officinale (= *Leontodon* o.) - The Common Dandelion, a perennial herb, with bright yellow flower heads 3 to 6 cm (11/2 in) across, borne on smooth hollow stalks up to 50 cm (20 in) in height. Widespread and common in grassy and waste places. It contains the principle taraxacin, and is used as a cholagogue, diuretic, and mild laxative. Distribution: worldwide; native in Britain.

Tarbosaurus A genus of extinct animals, the Dinosaurs, the "meat reptile" therapods or carnosaurs. These were huge animals, measuring up to 14.32 m (47 ft) in overall length. Their remains were found in the Nemegetu Basin, Mongolia.

Tarentola A genus of Lizards of the Gecko family, *Gekkonidae*. e.g. T. mauritanica - The Moorish Gecko, length 16 cm; brownish color; nocturnal. The suction pads on its feet enable it to move swiftly and safely even on smooth vertical surfaces. Eggs are laid under stones and cracks in walls. Distribution: the Mediterranean region.

Taricha A genus of Amphibians of the Salamander family, *Salamandridae*. e.g.

T. torosus - The California newt; this amphibian produces the powerful poison tetrodotoxin, so virulent that in experiments, 9 mg (0.0003 oz) have been found sufficient to kill 7,000 mice. Distribution: the United States.

Tarpon A genus of Fishes (subphylum *Vertebrata*), also known by the generic name *Megalops*. e.g. T. atlanticus (= *Megalops* a.) - The Tarpon fish, a large, marine, game fish, up to 91 kg (200 lb) in weight, common in the Gulf of Mexico off the coast of Florida.

Tatusia A genus of Trypanosomes. e.g. T. novemcincta - A trypanosome found in the armadillo, transmitted by the insect *Triatoma* geniculata (see **Triatoma**).

Taurotragus A genus of Antelopes (class *Mammalia*). e.g. T. derbianus derbianus - The Central African Giant Derby Eland, the largest of all antelopes. A rare species; adult bulls average 1.65 m (5 ft 9 in) at the withers and weigh about 714 kg (1575 lb).

Tautoga A genus of Fishes (subphylum *Vertebrata*). e.g. T. onitis - The Tautog Fish.

Taxodium A genus of Trees of the family *Taxodiaceae*. e.g. T. distichum - The Swamp Cypress or Bald Cypress tree, a deciduous, coniferous, timber tree of North America; up to 45 m (150 ft) tall. The resin was formerly used in the treatment of rheumatism.

Taxotes A genus of Fishes (subphylum *Vertebrata*). e.g. T. jaculatrix - The Archer-fish.

Taxus A genus of Trees. e.g. T. brevifolia - The Pacific Yew tree, a slow-growing, evergreen, prickly species found in the Pacific northwest, U.S.A. The bark yields Taxol, used especially in the treatment of ovarian and breast cancers.

Tayassu A genus of Ungulate Mammals of Tropical America, the Peccaries of the family *Tayassuidae*. e.g. T. angulatus - The Collared Peccary, a wild, gregarious species; they look like small, tusked pigs. Their tanned skins are used for gloves, etc.

Tedania A genus of Sponges of the family *Tedaniidae*. e.g. T. anhelans - Length about 70 mm; colored brownish, greenish, or bluish. Found on muddy-sandy bottoms at depths of less than 2 m. Distribution: the Mediterranean Sea coast.

Tegenaria A genus of Arachnids of the True Spider family, *Agelenidae*. e.g. T. ferruginea - The House Spider, length 9-14 mm. One of the largest of "domestic" spiders; found in attics; cellars, sheds, etc. in the neighborhood of man. Brownish in color. Distribution: Europe.

Teinopalpus A genus of Insects of the Butterfly or Moth family. e.g. T. imperialis - A rare species found in the mountains of S.E. China.

Telescopus A genus of Reptiles of the Colubrid Snake family, *Colubridae*. e.g. T. fallax - The European Cat Snake, 1 m long; brownish in color, with "stripes"; it has poisonous teeth at the back of its jaws. Found in dry stony regions with low shrubs. Distribution: the Balkans, Greece, and the Middle East.

Tellina A genus of Bivalve Molluscs of the family *Tellinidae*. e.g. T. tenuis - The Thin Tellin; length up to 25 mm; height 15-17 mm, thickness 5-6 mm. Thin, yellowish-white or rose colored shell, sculptured with concentric bands. Found on sandy coasts. Distribution: Mediterranean Sea coast and the coasts of Europe.

Tenebrio A genus of Insects of the Darkling Beetle family, *Tenebrionidae*. e.g. T. molitor - The Mealworm Beetle, length 15 mm. Used as food for birds and pet animals. Black in color; entire development takes place in flour. Needs Vitamin Bt for its nutrition. Distribution: worldwide.

Tenodera A genus of Insects of the order of Cockroaches and Praying Mantids, *Dictyoptera* (Mantid suborder, *Mantodea*). e.g. T. aridifolia - A large

species of praying mantids found in North America, which were introduced from China. These are predacious insects with voracious appetites. If not recognized in time, the male may be devoured by the female even while copulating (starting at the head end!).

Tenthredo A genus of Insects of the Typical Sawfly family, *Tenthredinidae*. e.g. T. zonula - Length 9-10 mm; found on flowering umbelliferous plants in early summer. Distribution: Europe, Asia Minor, North Africa.

Tenuibranchiurus A genus of exceedingly small crustaceans. Some species of these freshwater crayfish are the smallest known, and do not exceed 25 mm (1 in) in total length. They are found in Queensland, Australia.

Teratornis A genus of extinct prehistoric Birds (class *Aves*). e.g. T. incredibilis - The condor-like species which flew over what is now Nevada, U.S.A., between 50,000 and 1,750,000 years ago. It was a gigantic flying bird, and remains discovered in 1952 in Smith Creek Cave, Nevada, indicate a wingspan of 5 m (16 ft 41/2 in), and a weight of nearly 22.2 kg (50 lb).

Teredo A genus of Bivalve Molluscs of the family of Ship Worms, *Teredinidae*. e.g. T. navalis - The Common Ship Worm; the body measures 100-200 mm in length; the shell is reduced and looks like a helmet. The animal bores into the wood with the aid of the helmet, causing great damage, particularly to harbour constructions and ships, where these molluscs occur in large numbers. Distribution: worldwide.

Terminalia A genus of Trees of the family, *Combretaceae*. e.g. T. chebula - The Myrobalan tree or Jangi Harara; the fruit contains tannin and a greenish oleoresin. Used in India and the Far East as an equivalent of gall, but with a basis of paraffin instead of lard.

Ternidens A genus of Nematodes or Roundworms of the superfamily,

Strongyloidea. e.g. T. diminutus - A species common in simians but has also been found in man. It has three complex teeth arising from a lobe of the esophagus. Found in East Africa.

Testudo A genus of Reptiles of the Tortoise family, *Testudinidae*. e.g. T. graeca - The Common European Tortoise or Mediterranean Spur-thighed Tortoise, length 30 cm; herbivorous, inhabiting shrubby regions; female lays eggs in self-dug holes in the ground. A long-living animal, sometimes reaching the age of 120 years. Distribution: southern Europe, North Africa, southwest Asia.

Tethys A genus of Snails of the class *Gastropoda*. e.g. T. californicus - The Sea Hare or Sea Slug, a seaweed-eating mollusc found in the shallow waters off the coast of California, U.S.A. This is the largest known species of snail, the average weight being 3.18-3.63 kg (7-8 lb).

Tetracanthagyna A genus of Insects of the order of Dragon-flies, *Odonata*. e.g. T. plagiata - The largest dragon-fly in the world, found in Neborneo. One specimen preserved in the British Museum of National History, has a wingspan of 194 mm (7.63 in) and an overall length of 108 mm (41/2 in).

Tetrachilomastix A genus of Flagellate Protozoa, sometimes found in the human intestine. e.g. T. intestinalis - A pyriform-shaped, coprozoic, flagellate organism, with four anterior flagellae, sometimes found in human feces.

Tetraclinis A genus of Trees of the family *Cupressaceae*. e.g. T. articulata - The Sandaraca tree; from the incision of the stem is obtained a pale yellow resin, Sandarac (= Gum Juniper; Sandaraca), used as a temporary filling for carious teeth, and in pill and industrial varnishes.

Tetracoccus A genus of microorganisms of the family *Micrococcaceae*; spherical cells appearing in plates of four (or mul-tiples of four), division occurring in two planes at right angles.

Tetragnatha A genus of Arachnids of the True Spider family, *Tetragnathidae*. e.g. T. extensa - Length 8.5-12 mm; found in the characteristic outstretched position, on the leaves of aquatic plants by the banks of ponds. Distribution: worldwide.

Tetrameres A genus of worms parasitic in the alimentary tract of chickens and other fowl. e.g. T. americana - A filaroid parasite of the proventriculus of chickens and other birds.

Tetramitus A genus of protozoa, now known by the generic name *Chilomastix*. e.g. T. mesnili (= *Chilomastix* m.) (see **Chilomastix**; **Trichomonas**).

Tetranychus A genus of Mites. e.g. T. autumnalis (= *Trombicula* a.) (see **Trombicula**). T. telarius - The Spider mite, which sometimes infests man.

Tetrao A genus of Fowl-like Birds of the Grouse family, *Tetraonidae*. e.g. T. urogallus - The Capercaillie, length 65-95 cm; multicolored; inhabits forests, nesting on the ground. Distribution: the mountain forests of central and northern Europe and northern Asia.

Tetraodon A genus of Fishes (subphylum *Vertebrata*). e.g. T. hispidus - The Japanese Puffer fish, a poisonous fish of the deadly puffer family (see **Arothron**), producing the venom tetrodotoxin, which causes rapid respiratory paralysis in humans and has no known antidote.

Tetrapteris A genus of plants of the family *Malpighiaceae*. e.g. T. methystica - From this species, together with those of the genus *Banisteria*, is produced a narcotic drink known as "caapi" in Brazil, "Yagé" in Columbia, and "Ayahuasca" in Bolivia, Ecuador, and Peru. The main active principles are the alkaloids harmine and harmaline (see **Banisteria**; **Haemadictyon**; **Peganum**).

Tetrapturus A genus of Fishes, the

Marlins (subphylum *Vertebrata*), many species of which are among the fastest fishes in the world, over a sustained distance.

Tetrasomus A genus of Fishes (subphylum *Vertebrata*). e.g. T. gibbosus - The Triangular Boxfish; has a hard, thorny covering as an excellent defence against predators. Lives on sandy bottoms especially on coastal reefs, such as those of the Red Sea coasts.

Tetrastes A genus of Fowl-like Birds of the Grouse family, *Tetraonidae*. e.g. T. bonasia - The Hazelhen, length 34-37 cm; a resident bird found in mixed woodlands, nesting on the ground. Distribution: northern and central Europe, and central and northern Asia.

Tetrastoma A genus of Trematode Worms, sometimes found in the urine.

Tetratrichomonas A genus of parasitic Flagellate Protozoa, now known as *Trichomonas*. e.g. T. buccalis (= *Trichomonas* b.) (see **Trichomonas**).

Tetrix A genus of Insects of the Grouse-locust family, *Tetrigidae*. e.g. T. subulata - Length 7-10 mm; found in lowlands to mountain elevations. Distribution: Europe, North Africa, North America.

Tetropium A genus of Insects of the Longicorn or Long-horned Beetle family, *Cerambycidae*. e.g. T. praeusta - Length 3-5 mm; found in various flowering plants and in fruit trees. Infested twigs die back from the tips and the bark falls off. Distribution: Europe, Asia Minor, North Africa.

Tettigonia A genus of Insects of the Bush Cricket family, *Tettigoniidae*. e.g. T. viridissima - The Great Green Bush Cricket, length 28-42 mm. Found on shrubs, meadow and field plants. The males stridulate day and night. Distribution: Europe, Asia Minor, North Africa.

Teucrium A genus of plants of the family *Labiatae*. e.g. T. chamaedrys - The Wall Germander, a shrubby, branching, perennial with sage-like, pinkish-purple flowers clustered in irregular whorls. Has been used in the past as a febrifuge, and for stomach complaints, dropsy, and gout. Place of origin: Europe.

Thalarctos A genus of Mammals of the Bear family, *Ursidae*. e.g. T. maritimus - The Polar Bear; body length over 2.5 m, weight 400-600 kg; inhabits regions of permanent ice and snow around the North Pole. Distribution: Greenland, and the arctic coasts of North America and Asia.

Thalassema A genus of Echiuroid Worms of the phylum *Echiuroidea*. These are found in mud or under stones, and fissures in rocks. They have a gutter-like proboscis, short, funnel-like or shovel-like, which they use for feeding.

Thalassoma A genus of Fishes (subphylum *Vertebrata*). e.g. T. ruppelli - The Rainbow Wrasse fish, one of the most common fishes of the wrasse family. Found on coastal reefs in the Red Sea.

Thalazia A genus of worms allied to Filaria (see **Thelazia**).

Thalia A genus of plants of the family *Marantaceae*. e.g. T. dealbata - Powdery Thalia or Water Canna; a fine-foliaged aquatic plant up to 2 m (7 ft) tall, with broad basal leaves and purplish 6-petalled flowers on leafless waxy stems. Place of origin: southeastern U.S.A.

Thamnidium A genus of Molds, resembling Mucor (see **Mucor**), often found growing on meat in cold storage. It can grow at 28° F (-2.2° C) and forms a profuse hairy growth. e.g. T. elegans.

Thamnophis A genus of Snakes. e.g. T. sauritus - The Ribbon snake, a slender snake, about 70 cm in length.

Thanasimus A genus of Insects of the Checkered Beetle family, *Cleridae*. e.g. T. formicarius - Length 7-10 mm. Found on stacked wood and felled logs, where it hunts bark beetles and kills great numbers of potentially harmful insects. Distribution: Eurasia, North Africa, North America.

Thapsia A genus of plants of the Carrot

family, *Umbelliferae*. e.g. T. garganica - A species of North Africa, affording an irritant resin in plasters; the plant is locally used as a polychrest remedy.

Thaumatococcus A genus of tropical plants. e.g. T. daniellii - A West African plant, from which is extracted a natural sweetener, talin which is about 5000 times as sweet as sucrose.

Thaumetopoea A genus of Butterflies or Moths of the family *Thaumetopoeidae*. e.g. T. processionea - The European Processionary Moth, length 13-18 mm; found in oak woods. Distribution: Central and southern Europe.

Thea A genus of Insects of the Ladybird Beetle family, *Coccinellidae*. e.g. T. vigintiduopunctata - Length 3-4.5 mm. This species has bright yellow elytra with 22 black spots. Found in mountain regions. Distribution: southern part of the Palaearctic region.

Theileria A genus of minute intraglobular protozoan parasites. e.g. T. dispar - A species causing heavy mortality in cattle in North Africa, and transmitted by the tick *Hyalomma* mauritanicum. T. parva (= *Babesia* p.) - A species causing East African Coast Fever in cattle (see **Babesia**). T. tsutsugamushi (= *Rickettsia* t.) - A species causing scrub typhus (see **Rickettsia**).

Thelazia A genus of Nematode Roundworms allied to *Filaria*, also known by the generic name *Thalazia*, of the superfamily *Spuroidea*. e.g. T. callipaeda (= *Thalazia* c.) - The Oriental Eye Worm, a species parasitic in the eyes of animals; the worm is creamy colored and thread-like; it passes across the eye, causing lacrymation, conjunctivitis, and scarring of the cornea.

Thelohania A genus of microsporidium parasites. e.g. T. magna - A species parasitic in the larvae of the mosquito, *Culex* pipiens.

Thelyphonus A genus of Arachnids of the order of Whip Scorpions, *Uropygi* (see **Mastigoproctus**).

Theobaldia A genus of Mosquitoes of the temperate regions of the Old and New Worlds.

Theobroma A genus of plants of the family *Sterculiaceae*. e.g. T. cacao - The Cacao or Cocoa plant (called theobroma and cacao). Its seed, the Theobroma Seed (= Cacao or Cocoa Seed; Cacau; Theobromatis Semen) contains the alkaloid theobromine (= santheose; 3,7-dimethylxanthine), and is used in the preparation of cacao and chocolate. Theobromine is used as as smooth muscle relaxant to dilate the coronary or peripheral arteries.

Theragra A genus of Fishes (subphylum *Vertebrata*). e.g. T. chalcogrammus - The Wall-eye Pollack Whiting fish.

Theraphosa A genus of Arachnids of the northern South America, having a legspan of 254 mm (10 in) when fully extended, a body length reaching 89 mm (31/2 in), and weighing almost 57 g (2 oz).

Therioplectes A genus of Tabanid Flies.

Therizinosaurus A genus of extinct animals, the Dinosaurs, similar to present day ant-eaters. These animals lived in forest and savannah regions some distance from lakes or inland seas about 70 million years ago. They had a feeble skull partly or entirely lacking teeth, and claws measuring as much as 700 mm (27.55 in) along the outside curve.

Thermoactinomyces A genus of microorganisms of the family *Streptomycetaceae*; saprophytic soil and water microorganisms distinguished by their ability to grow at high (50°-65° C) temperatures.

Thermopsis A genus of Shrubs and Trees of the Pea family, *Leguminosae*, also known by the generic name *Sophora*. e.g. T. lanceolata (= *Sophora* lupinoides) - It yields the alkaloid Pachycarpine (= d-Sparteine), a ganglion-blocking agent with oxytocic properties similar to those of l-Sparteine (see **Sophora**).

Thevetia A genus of plants of the family

Apocynaceae. e.g. T. neriifolia - This plant yields a glycoside Thevetin with action similar to that of digitalis. Used in Europe (on the Continent) in the treatment of mild myocardial insufficiency, as a substitute for digitalis.

Thiobacillus A genus of microorganisms of the family *Thiobacteriaceae*; small gram-negative, rod-shaped cells. e.g. T. concretivorus.

Thiobacterium A genus of microorganisms of the family *Thiobacteriaceae*; rod-shaped sulfur bacteria, found in fresh or salt water, or in the soil. e.g. T. bovista.

Thiocapsa A genus of microorganisms of the family *Thiorhodaceae*; spherical cells within a slime capsule. e.g. T. floridana.

Thiocystis A genus of microorganisms of the family *Thiorhodaceae*; spherical to ovoid cells embedded in a common gelatinous capsule. e.g. T. violacea. *Thioderma*. A name formerly given to a genus of sulfur bacteria.

Thiodictyon A genus of microorganisms of the family *Thiorhodaceae*; rod-shaped cells arranged end to end in a net-like structure. e.g. T. elegans.

Thiopedia A genus of microorganisms of the family *Thiorhodaceae*; spherical to short rod-shaped cells. e.g. T. rosea.

Thiopolycoccus A genus of microorganisms of the family *Thiorhodaceae*; spherical cells occurring in irregularly shaped dense aggregates, held together by mucus. e.g. T. ruber.

Thiosarcina A genus of microorganisms of the family *Thiorhodaceae*; spherical cells in cubical packets. e.g. T. rosea.

Thiospira A genus of microorganisms of the family *Thiobacteriaceae*; colorless, slightly bent, large rods with a small number of polar flagellae and containing sulfur granules within the cells. e.g. T. bipunctata.

Thiospirillum A genus of microorganisms of the family *Thiorhodaceae*; occurring singly as spirally wound cells. e.g. T. sanguineum.

Thiothece A genus of microorganisms of the family *Thiorhodaceae*; spherical to elongated cells embedded in a gelatinous capsule. e.g. T. gelatinosa.

Thiovulum A genus of microorganisms of the family *Thiobacteriaceae*; round to ovoid unicellular organisms, normally containing sulfur in the cytoplasm. e.g. T. majus.

Thracia A genus of Bivalve Molluscs of the class *Bivalvia*. e.g. T. phaseolina - The Papery Lantern Shell, length up to 3 cm; unequal brittle valves. Common in the North Sea, the Atlantic, and the Mediterranean. Native around Britain.

Threskiornis A genus of Birds of the Ibis and Spoonbill family, *Threskiornithidae*. e.g. T. aethiopica - The Sacred or Tantalus Ibis, venerated by the ancient Egyptians and familiar among the hieroglyphics on ancient Pharonic monuments. About 60 cm (2 ft) tall, it has a long, downward-curving beak, and white plumage, except for the wings which are partly black. The head and neck are naked and covered with black skin. It is a wading bird, allied to the storks (see **Mycteria**). Found especially in tropical regions.

Thuja A genus of Coniferous Trees called arbor vitae. e.g. T. occidentalis - The White Cedar tree; the fresh tops yield thuja, an antipyretic, diuretic, emmenagogue, and sudorific.

Thunbergia A genus of plants of the family *Acanthaceae*. e.g. T. alata - Black-eyed Susan or Thunbergia, a slender plant with 5-lobed flat, orange yellow flowers with chocolate-brown centers and curved tubes behind. A climbing and spreading plant used in houses and conservatories in pots and containers. Place of origin: Tropical Africa.

Thunnus A genus of Perciform Bony Fishes of the Thunnic family, *Thunnidae*. e.g. T. thynnus - The Great Bluefin (= Atlantic Tuna; Tunny) fish,

444

up to 3 m in length; spawns in summer; the sperm yields a protamine, thynnin. An important fish of commerce. Distribution: The Atlantic Ocean and the Mediterranean Sea.

Thyatira A genus of Insects of the Butterfly and Moth family, *Thyatiridae*. e.g. T. batis - The Peach Blossom Moth, length 18-19 mm; has pink patches on the wings, and is found in clearings, forest margins, gardens and parks. Caterpillars feed on raspberry and blackberry. Distribution: Eurasia.

Thylacinus A genus of rare carnivorous Marsupials, found in the remoter parts of S.W. Tasmania. e.g. T. cynocephalus - A species known as the "Tasmanian tiger" or the "Tasmanian wolf", the largest of the carnivorous marsupials, now almost extinct.

Thylacomys A genus of Mammals of the Bandicoot family, *Peramelidae*. These are the Bilbies Rabbit Bandicoots, extremely swift animals and lively in their movements (see **Choeropus**; **Isodon**).

Thymallus A genus of Salmoniform Bony Fishes of the Grayling family, *Thymallidae*. e.g. T. thymallus - The Common European Grayling, length 50 cm; weight more than 1 kg. Lives in shoals in the submountain sections of rivers with sandy and gravelly bottoms. Distribution: the rivers of northern and central Europe.

Thymus A genus of plants of the family *Labiatae*; native to South Central Europe and grown extensively in other countries. e.g. T. vulgaris - Garden Thyme (= Common Thyme; Rubbed Thyme; Thym; Thymi Herba); the dried leaves and flowering tops yield a volatile oil which is aromatic and carminative. Thyme also contains thymol, thymene, and cumene. T. serpyllum - Wild Thyme (= Herba Serpylli; Feldthymian; Mother of Thyme; Quendel; Serpolet); contains a volatile oil with similar properties to thyme.

Thyphlops A genus of Reptiles, better known by the generic name *Typhlops*, of the diminutive blind Snake family, *Typhlopidae* (see **Typhlops**).

Thysania A genus of Insects of the order of Butterflies and Moths, *Lepidoptera*. e.g. T. aggrippina - The Great Owlet Moth of Central America and northern South America. It has the greatest wingspan of all the *Lepidoptera*, sometimes reaching as much as 360 mm (14-17 in).

Thysanosoma A genus of Tapeworms. e.g. T. actinioides - The Fringed Tapeworm, found in the bile ducts and small intestine of sheep, antelope, and deer in western and southwestern U.S.

Tibicen A genus of Homopterous Insects of the Cicada family, *Cicadidae*. e.g. T. haematodes - Length 26-38 mm; span of forewings 75-85 mm. One of the largest of the European cicadas. Found in warm places such as vineyards. Distribution: southern and central Europe.

Tibouchina A genus of plants of the family *Melastomataceae*. e.g. T. urvilliana (= T. semidecandra) - The Brazilian Glorybush or Spider Flower plant; a beautiful large shrub with longitudinal, veining large, vivid purple, 5-petalled flowers with long spidery, purple stamens. Used in greenhouses and flower beds. Place of origin: Brazil; introduced into Europe in 1864.

Tichodectes A genus of Lice. e.g. T. canis - The biting Dog Louse which is an intermediate host of the double-poured dog tapeworm, *Dipylidium* caninum.

Tichodroma A genus of Birds (class *Aves*). e.g. T. muraria - The Wall creeper, a high altitude bird; has been seen at heights of 6,400 m (21,000 ft) in the Karakorum Range of the Himalaya Mountains of North India.

Tilapia A genus of Fishes (subphylum *Vertebrata*). e.g. T. macrocephala - The West African Mouth-breeder fish.

Tilia A genus of Trees of the family

Tiliaceae. e.g. T. x europaea - The
Common Lime Tree; the dried inflorescences contain tannin and a volatile oil,
and are mildly astringent, with spasmolytic and diaphoretic properties.
Lime-flower "tea" is a traditional
domestic remedy.

Tilletia A genus of ustilageneous fungi
causing smut on cereals.

Tinca A genus of Fishes of the Carp and
Minnow family, *Cyprinidae*. e.g. T.
tinca - The Golden Tench fish, length
50-60 cm; weight 5-6 kg, a robust fish
with small scales and small eyes. Lives
close to the bottom of slow-flowing
rivers and backwaters with muddy bottoms. Tolerates marked lack of oxygen
in water; planted as a valuable supplementary fish in carp ponds. Distribution:
Europe (see **Carassius**).

Tindaria A genus of long-living Molluscs
(phylum *Mollusca*). e.g. T. callistiformis
- The Deep-sea Clam, the longest-lived
of all molluscs, taking an estimated 100
years to reach a length of 8 mm (0.31
in), the slowest rate of growth in the
animal kingdom. The greatest age
recorded for a clam in the largest size
category is 98 years.

Tineola A genus of Insects of the
Butterfly and Moth family, *Tineidae*.
e.g. T. bisselliella - The Common
Clothes Moth, 4-8 mm long; found in
houses chiefly in spring, often in great
numbers. The larva eats into fabrics,
furs, etc., hence its name. Distribution:
worldwide.

Tinospora A genus of plants of the Vine
family, *Menispermaceae*.

Tipula A genus of Two-winged Insects of
the Crane Fly family, *Tipulidae*. e.g. T.
oleracea - The Daddy Longlegs Fly,
length 15-23 mm; the larvae feed on
plant roots, especially of grasses and
pupate in the ground. Distribution:
Europe, North Africa, North America.

Titanus A genus of Insects of the order of
large Beetles, *Coleoptera*. e.g. T. giganteus - The rare Longhorn Beetle of the

Amazon basin, exceeding 120 mm (4.72
in) in length (excluding the antennae).

Tityus A genus of venomous Scorpions
of the order *Scorpionida*. e.g. T. serrula-
tus - The highly poisonous scorpion of
Brazil whose sting has proved fatal in
many cases.

Toddalia A genus of Shrubs of the Rue
family, *Rutaceae*. e.g. T. aculeata - A
species found in the East Indies; the root
is used as an aromatic stomachic.

Tolypocladium A genus of Fungi or
Molds. e.g. T. inflatum - A species of
Fungi, producing a metabolite, cyclic
polypeptide, also known as Cyclosporin
A or Ciclosporin, an immunosuppres-
sive agent consisting of 11 amino acids.

Tomicus A genus of Insects of the Bark
Beetle family, *Scolytidae*. e.g. T.
piniperda - The Large Pine Short Beetle,
length 3.5-4.8 mm; found under the bark
of pine trees. Distribution: the
Palaearctic region.

Torpedo A genus of Fishes (class
Chondrichthyes), of the Electric Ray
family, *Torpedinidae*. e.g. T. marmorata
- The Marbled Electric Ray Fish, length
up to 1.5 m. Viviparous; has a large
electric organ on either side of the head
with which it stuns its prey, producing a
charge of 45 to 220 volts. Distribution:
the shallow waters along the coast of
Europe, to the Mediterranean and northwest Africa.

Tortrix A genus of Insects of the
Butterfly and Moth family, *Tortricidae*.
e.g. T. viridana - The Green Oak Tortrix
Moth, length 9-11 mm. Lives among
host trees such as the oak, the female
laying eggs on the young oak twigs.
Distribution: Europe, Asia Minor.

Torula A genus of Molds or Yeast-like
Fungi of the family *Cryptococcaceae*,
now named *Cryptococcus*. e.g. T.
hominis (= T. capsulata; T. histolytica;
Cryptococcus neoformans, etc.) (see
Candida; **Cryptococcus**; **Torulopsis**).

Torulopsis A genus of Molds or Yeast-
like Fungi of the family

Cryptococcaceae, now named *Cryptococcus*, and also included in the generic name, *Candida*. e.g. T. hominis (= T. histolytica; *Torula* hominis; *Torula* histolytica; *Cryptococcus* neoformans, etc.) (see **Cryptococcus**; **Torula**). T. utilis (= *Candida* u.; *Torula* u.) - Torula Yeast; Dried Torula Yeast (see **Candida**).

Toxascaris A genus of parasitic nematodes of the family *Ascaridae*. e.g. T. leonina - A species of nematode worms occurring in lions, tigers, and other large *Felidae*. May occur in older dogs and cats, but its larvae do not pass through the lungs, as they do in the case of the genus *Toxocara* (see **Toxocara**).

Toxocara A genus of Nematode Roundworms of the family *Ascaridae*. e.g. T. canis - The dog ascarid, a species found in the intestine of dogs; sometimes occurs in man, especially in children. The larvae pass through the lungs (see **Toxascaris**). T. cati - The cat ascarid, a species occurring in cats; closely related to T. canis. Has been seen in children who had eaten cat feces (see **Toxascaris**).

Toxoplasma A genus of parasitic sporozoa found in the epithelial cells of many mammals and birds, and also occurring in humans. The parasite divides by binary fission instead of schizogony, although it resembles a Plasmodium in appearance. Furthermore it has no specific alternative host and lacks a sexual stage. e.g. T. gondii -A protozoan parasite causing a chronic infection in a North American rodent, *Ctenodactylus* gondi, and also in man, causing the disease toxoplasmosis.

Toxopneustes A genus of Echinoderms of the class of Sea-urchins, *Echinoidea*. e.g. T. pileolus - The Sea-urchin of the Indo-Pacific region, the world's most venomous echinoderm. Its sting can produce intense pain, followed by a curare-like paralysis and sometimes, though exceptionally, death.

Toxostoma A genus of Birds of the Catbird, Mockingbird, and Thrasher family, *Mimidae*. e.g. T. rufum - The Brown Thrasher; a slim bird, larger than a thrush, with a longer tail and yellow eyes. Found mainly in thickets. Its song resembles that of the Catbird, *Dumetella* carolinensis, although it is more musical and is repeated once. Distribution: as in the Catbird, in the New World (see **Dumetella**).

Toxothrix A genus of microorganisms of the family *Chlamydobacteriaceae*; made up of colorless, filamentous, cylindrical cells with a sheath originally thin but later thickened with impregnated iron oxide. e.g. T. gelatinosa.

Trachelospermum A genus of plants of the family *Apocynaceae*. e.g. T. jasminoides - The Chinese Star-Jasmine plant, an evergreen climber of up to 6 m (20 ft), with very fragrant white flowers in axillary and terminal sprays. Used as a wall shrub. Prized for its scented flowers. Place of origin: China and Japan; introduced into Europe in 1846.

Tracheophilus A genus of trematode parasites. e.g. T. cymbius - A species parasitic in the trachea and esophagus of ducks in Europe.

Trachinus A genus of Fishes of the Weever family, *Trachinidae*. e.g. T. draco - The Greater Weever Fish, length 40 cm. Found on sandy bottoms. On the gill covers and on the first dorsal fin are sharp spines with grooves containing venom. They are a menace to swimmers, causing severe wounds. Distribution: the coast of the northeast Atlantic, the Mediterranean, and the Black Sea.

Trachurus A genus of Bony Fishes of the Crevalle, Jack, Pompano, and Scad family, *Carangidae*. e.g. T. trachurus - The Horse Mackerel or Saurel Fish, length up to 40 cm. A fish of commerce. Distribution: the Atlantic, Baltic, Black, Mediterranean and North Seas.

Trachybdella A genus of Leeches. e.g. T.

bistriata - A leech found in Brazil which attacks man and other animals.

Trachylobium A genus of plants of the Pea family, *Leguminosae*. e.g. T. hornemannianum - The Copal plant or Zanzibar Copal; this species yields a fossil resin, Copal or Gum Animi, used for covering cement fillings in tooth cavities to protect them from saliva, and as an ingredient in some varnishes.

Trachys A genus of Insects of the Metallic Wood-borer Beetle family, *Buprestidae*. e.g. T. minutus - Length 3-3.5 mm; dark-brown to black elytra; locally abundant on flowers and willow leaves. Distribution: Europe and Asia Minor.

Trachyspermum A genus of plants of the Carrot family, *Umbelliferae*. e.g. T. ammi (= *Carum* copticum) - The Ajowan plant (= Tachyspermum; Ptychotis), yielding Ajowan Oil, a commercial source of Thymol, an antiseptic, antioxidant, fungicide, and parasiticide. The dried, ripe fruits are antispasmodic and carminative (see **Carum**).

Tradescantia A genus of plants of the family *Commelinaceae*. e.g. T. virginiana - "Isis"; the Spiderwort; an erect, branching, herbaceous perennial, with deep blue 3-petalled flowers with fluffy stamens. Used as a house or pot plant. Place of origin: eastern North America; introduced into Europe in 1629.

Tragelaphus A genus of vertebrate animals, the antelopes (class *Mammalia*). e.g. T. spekii - The Antelope.

Tragia A genus of poisonous plants of the family *Euphorbiaceae*. e.g. T. urens - A species of poisonous weed of the southern U.S.

Tragulus A genus of Mammals of the family of Ruminants. e.g. T. javanicus - The Mouse Deer or Chevrotain of S.E. Asia, including Java, Borneo and Sumatra. A very small ruminant with only three compartments to its stomach, and thus not a true deer (*Cervidae*), which has four. A very small animal,

measuring 203-254 mm (8-10 in) at the withers, and weighing only 2.72-3.17 kg (6-7 lb).

Trechona A genus of Arachnids of the order of Spiders, *Araneae*. e.g. T. venosa - The Funnel-web "tarantula" of South America, one of the most venomous species of spiders in terms of neurotoxic potency, killing off white rats within seconds.

Treponema A genus of microorganisms of the family *Treponemataceae*, formerly called *Microspironema* or *Spirochaeta*; made up of cells 3 μ to 18 μ long, in acute, regular, or irregular spirals. e.g. T. pallidum (= Trep. pallidum; *Microspironema* p.; *Spirochaeta pallida*); the causative agent of syphilis in man.

Triactis A genus of Coelenterate animals of the class of Sea-anemones and Corals, *Anthozoa*. e.g. T. producta - The poisonous Sea Anemone of the Red Sea coast; has a soft hollow body with a mouth surrounded by tentacles capable of a potent sting. This is due to thousands of stinging cells that erupt when touched, paralysing the prey, usually fish and crustaceans, which is then brought into the mouth to be digested.

Triakis A genus of Fishes of the Shark family (subphylum *Vertebrata*). e.g. T. semifasciata - The Leopard Shark.

Trianthema A genus of plants of the family *Ficoideae*. e.g. T. portulacastrum (= T. monogyna) - From the white variety of this species is prepared a liquid extract, trianthema, containing an alkaloid which has been reported to possess similar diuretic properties of punaravine, the alkaloid present in Punarnava. Trianthema is used in India as a diuretic, and the name Punarnava has also been applied to it (as well as to the plant *Boerhaavia* (see **Boerhaavia**).

Triatoma A genus of Insects of the Bug family, *Reduviidae*, now included under the generic name *Panstrongylus*; called the cone-nosed bugs; important in medicine as vectors of *Trypanosoma* cruzi. Formerly also known as *Conorhinus*, *Entriatoma*, and *Lamus*. e.g. T. megista (= Conorhinus/ Lamus/ Panstrongylus/ Triatoma megistus) (see **Conorhinus**; **Panstrongylus**). T. sordida (= *Entriatoma* s.) - a species of bugs of Sao Paulo.

Tribolium A genus of small Beetles that live in and are very destructive to flour and other cereal products. e.g. T. confusum - 3.5 mm long; reddish-brown color.

Tricercomonas A genus of human parasites, now identified as the same as *Enteromonas*. e.g. T. hominis (= *Enteromonas* h.) (see **Enteromonas**).

Trichechus A genus of small sirenian Mammals of the family *Sirenia*. e.g. T. inunguis - The freshwater Amazonian Manatee, an extremely small sirenian species, reaching a maximum size of 2.5 m (8 ft 3 in) and 140 kg (308 lb).

Trichia A genus of Gastropod Molluscs of the family *Helicidae*. e.g. T. hispida - The Bristly Snail; shell 4.5-5 mm high, 7-8.5 mm wide, composed of 5-6 whorls, greyish-brown to reddish-brown, and covered with hairs up to about 0.3 mm in length. Found in moist areas in meadows and gardens at lower elevations, concealed in moss, old walls, amongst leaves, and under stones. Distribution: Europe.

Trichina See **Trichinella**.

Trichinella A genus of Nematode Roundworms, formerly called *Trichina*, of the family *Trichinellidae*. e.g. T. spiralis (= *Trichina* s.) - The Pork-muscle worm or Trichina worm, one of the smallest of the parasitic nematodes, measuring 1.5-4 mm (0.05-0.15 in); found coiled in a cyst in the muscles of the rat, pig, and man. The disease is known as trichinelliasis, trichinellosis, trichiniasis, or simply, trichinosis - infestation with the parasitic cysts or trichinae.

Trichobilharzia A genus of Flukes of the superfamily *Schistosomatoidea*. e.g. T. oscellata - A blood fluke parasitic in European ducks; it cannot penetrate human skin but does produce a dermatitis known as Swimmers's Itch.

Trichocephalus A former genus of Nematodes, now called *Trichuris*. e.g. T. trichiurus (= T. dispar; *Trichuris* trichiura) - The Whipworm (see **Trichuris**).

Trichodectes A genus of parasitic insects. e.g. T. equii - A biting louse found on horses.

Trichodina A genus of ciliate infusoria, species of which live on hydras and the gills of fish and salamanders.

Trichoglossus A genus of Birds of the Parrot family, *Psittacidae*. These are the Lories, small parrots found in Australia, New Guinea, Malaya and Polynesia. They have a feebly developed beak and the tongue usually has a papillose or brush-like tip. The species include the Black Lory, the Red Lory, the Green Lory, etc. They do not thrive in captivity.

Trichomastix A genus of parasitic microorganisms. e.g. T. caniculi - A parasitic species resembling a trichomonad, and found in rabbits.

Trichomonas A genus of Flagellate Parasitic Protozoans of the class *Zoomastigophorea* (the Flagellates), formerly known by the generic name *Tetratrichomonas*. Other names for this genus are: *Cercomonas*, *Cheilomastix*, *Chilomastix*, *Macrostoma*, and *Tetramitus*. The species have three flagellae in front, an undulating membrane, and a (fourth) trailing flagellum, and cause a rather serious disease Trichomoniasis in animals and birds; they may cause diarrhea in humans. e.g. T. buccalis (= T. elongata; *Tetratrichomonas* b.) - A species found in the mouth, and especially around the tartar of the teeth. T. hominis (= *Cercomonas* h.; *Cheilomastix* h.; *Chilomastix* h.; *Chilomastix* mesnili; *Macrostoma* h.; *Tetramitus* h.;

Trichomonas intestinalis, etc.) - A common parasite in the human intestine, frequently found in diarrheal stools. T. vaginalis - A species found in the vagina, producing a refractory vaginal discharge. Can cause urethritis in males.

Trichophaga A genus of Insects of the Butterfly and Moth family, *Tineidae*. e.g. T. tapetzella - The Tapestry Moth, 6-8 mm long. Caterpillars are troublesome pests of tapestries, felt, woollen fabrics, etc. Distribution: worldwide.

Trichophyton A genus of Fungi, formerly called *Endodermophyton*, consisting of flat, branched filaments and chains of spores. The species attack the skin, nails, and hair, and are placed in five groups: I. *Gypseum* Group: T. mentagrophytes. II. *Rubrum* Group: T. rubrum. III. *Crateriform* Group: T. tonsurans, etc. IV. *Faviform* Group: T. ferrugineum, etc. V. *Rosaceum* Group: T. rosaceum.

Trichosoma A genus of Roundworms (class *Nematoda*). e.g. T. contortum - A roundworm parasite in domestic fowls.

Trichosomoides A genus of Roundworms (class *Nematoda*). e.g. T. crassicauda - A nematode parasite in rats.

Trichosporon A genus of Fungi or Molds infecting the hair. e.g. T. beigelii (= T. bigelii; T. giganteum) - A species causing white piedra, a fungous disease of the hair, affecting the beard and mustache, in which the shafts are marked by the presence of hard white gritty nodules.

Trichostrongylus A genus of Nematode Worms of the family *Strongylidae*, comprising some of the species formerly included in the genus *Strongylus*; small worms with unarmed heads, parasitic in herbivorous animals, and only rarely affecting man. The heads are buried in the intestinal epithelium of the host. Infection is by the mouth. e.g. T. instabilis (= *Strongylus* subtilis) (see **Strongylus**). T. orientalis - A species found in Japan.

Trichothecium A genus of Mold Fungi. e.g. T. roseum - A species found in the human ear.

Trichuris A genus of intestinal nematode parasites (class *Aphasmidia*), formerly called *Trichocephalus*. e.g. T. trichiura (= *Trichocephalus* dispar; *Trichocephalus* trichiurus) - The Whipworm, a species infesting man; 2 in long, it inhabits the large intestine, and may cause diarrhea and vomiting.

Tridacna A genus of Bivalve Molluscs of the class *Lamellibranchia* (= *Bivalvia*). e.g. T. derasa - The Giant Clam, the largest of all existing bivalve molluscs, found on the Indo-Pacific coral reefs and off the east coast of Africa. It has been known to reach weight levels of up to 454.4 kg (1,000 lb). T. elongata - The Giant Clam of the Red Sea coastal reefs, a much smaller relative, but itself reaching a length of 70 cm (28 in).

Tridontophorus A genus of Nematode Worms of the family, *Stongylidae*. e.g. T. diminutus - A parasite frequently present in monkeys and occasionally in man.

Trigonella A genus of plants of the subfamily *Papilionaceae* (Pea family, *Leguminosae*). e.g. T. foenumgraecum - The Fenugreek, an Asiatic plant cultivated for making curry and for veterinary medicine (L. foenumgraecum = Greek hay).

Trillium A genus of plants of the Lily family, *Liliaceae*. e.g. T. erectum - The Wake-robin, a North American plant; the rhizome contains trilliin.

Trimastigamoeba A genus of Ameba having three equal flagella in the flagellate stage. e.g. T. philippensis - A species obtained in culture from city water.

Tringa A genus of Shorebirds of the family of Sandpipers and Snipes, *Scolopacidae*. e.g. T. hypoleucos - The Common Sandpiper, length 20 cm; inhabits flat, shrub-covered, muddy and sandy shores of lakes and ponds. A migratory bird, it winters mainly in Africa and southern Asia, as far as Australia. Distribution: all of Europe, Asia. T. macularia - The Spotted Sandpiper of North America. Closely related to the Common Sandpiper and with similar characteristics.

Triops A genus of Crustacean Arthropods of the family *Triopsidae*. e.g. T. crancriformis - Length up to 10 cm (including the caudal spines which have the same length as the body). Body composed of 32 to 35 segments, the last 4 to 7 without legs, and largely covered by the broadly rounded shield-like carapace. Found in puddles and damp muddy areas. Distribution: the southern Palaearctic and Nearctic regions.

Tripheustes A genus of Sea-urchins of the class *Echinoidea*. e.g. T. gratilla - This echinoderm is found among other sea-urchins in large numbers on sandy bottoms and coral reefs. Especially abundant in the Red Sea region; they are active night scavengers.

Triphleps A genus of Bugs. e.g. T. insidiosus - The Flower Bug; it causes a skin eruption in man.

Tripleorospermum A genus of plants of the family, *Compositae*, now better known by the generic name *Matricaria*. e.g. T. inodorum (= *Matricaria* inodora; M. maritimum) - German Camomile; Scentless Mayweed) (see **Matricaria**).

Triteleia See **Tritelia**.

Tritelia A genus of plants, also called *Triteleia*, of the family *Alliaceae*, now better known by the genus name *Brodiaea*. e.g. T. laxa (*Brodiaea* l.) - Brodiaea; Grassnut; Ithuriel's Spear plant (see **Brodiaea**).

Triticum A genus of plants of the Grass family, *Gramineae*. e.g. T. aestivum (= T. sativum; T. vulgare; and sometimes, *Agropyron* aestivum) - the bread wheat plant. The origin of modern wheat, known to have grown in the Nile valley in 5,000 B.C. and in England 3,000 years later, is complex and probably involved natural hybridization between members of this genus and species of *Agropyron* (= *Agropyrum*), with which it is freqently identified, and those of *Aegilops*. The main bread wheat cultivated today is T. aestivum; it is unknown in the wild state and its origin is obscure but it probably arose during the Iron Age, about 1,000 B.C. There are many varieties in cultivation, and new ones with desirable features such as short stems, large ears, and resistance to disease are continually being produced. An annual plant, with a fibrous root system, several erect shoots, and inflorescences of terminal spikes with two rows of laterally compressed spikelets containing florets. Besides its use as a staple food, and as an important source of Starch (Amylum), from the bread wheat embryo is extracted Wheat-Germ Oil (= Oleum Tritici Germinis), containing α-tocopherol (Vitamin E) and other tocopherols (see **Agropyron**). T. compactum - The Club Wheat; this species probably had an origin similar to that of the bread wheat. It is grown mostly in Chile, the western United States, and India. T. dicoccum - The Emmer Wheat, an ancient species, still occasionally grown in the mountains of Europe. T. monococcum - The Einkorn Wheat, one of the ancestors of our modern wheats. This species produces only one grain per spikelet and grows in poor, rocky ground. It is still cultivated in mountain regions of southern Europe. One of the earliest cultivated wheat plants, for it has been found on the site of Swiss lake dwellings of the Stone Age. T. repens (= *Agropyron* r.; *Agropyrum* r.) - Couch grass; a diuretic (see **Agropyron**).

Tritoma A genus of plants, better known by the generic name *Kniphofia*, of the Lily family, *Liliaceae*. e.g. T. uvaria (= *Kniphofia* u.) - The Red Hot Poker or Torch Lily plant (see **Kniphofia**).

Tritonia A genus of plants, also called *Montbretia*, and now known better by

451

the generic name *Crocosmia*, of the Iris family, *Iridaceae*. e.g. T. crocosmiiflora - Garden Montbretiia (see **Crocosmia**).

Triturus A genus of Tailed Amphibians of the Newt and Salamander family, *Salamandridae*. e.g. T. vulgaris - The Smooth Newt, length 10-11 cm, the commonest European newt; found at lower elevations. The female lays 200-300 eggs in aquatic vegetation. The larvae hatch 2 to 3 weeks later. Distribution: Europe.

Trochosa A genus of Arthropods, better known by the generic name, *Hogna*, of the order of True Spiders, *Araneae*. e.g. T. singoriensis (= *Hogna* s.) (see **Hogna**).

Trogium A genus of Insects of the order of Booklice, *Psocoptera*. e.g. T. pulsatorium - A species of small, inconspicuous, soft-bodied insects, the booklice, found in houses, barns or gardens, in dusty places or among old papers and other debris. They have modified biting mouth-parts.

Troglodytella A genus of ciliates, species of which occur in the gorilla and chimpanzee.

Troglodytes A genus of Perching Birds of the Wren family, *Troglodytidae*. e.g. T. troglodytes - The European Wren, an extremely small bird, only 9.5 cm long; found in woods, parks, and gardens with thick undergrowth. Nests in thickets, along brooks, in piles of underbrush, etc. Has a short incubation period of only 14 days. Distribution: about 38 races, throughout the northern hemisphere.

Troglotrema A genus of Flukes, known formerly by the name *Nanophyes*. e.g. T. salmincola - A fluke transmitted to dogs and foxes by the ingestion of infected trout or salmon.

Troides A genus of rare Butterflies or Moths, of the order *Lepidoptera*. e.g. T. paradisea - The rare "Butterfly of Paradise", found in New Guinea.

Trollius A genus of plants of the Buttercup family, *Ranunculaceae*. e.g. T. europaeus - The Common Globe Flower plant, an erect little-branched perennial, with globular, lemon yellow, solitary flowers, 3-5 cm (1-2 in) across. A poisonous plant. Place of origin: Europe.

Trombicula A genus of Ascarids of the Mite family, *Trombiculidae*, also variously known by the generic names *Eutrombicula*; *Leptotrombidium*; *Leptus*; *Microtrombidium*, and *Trombidium*. e.g. T. akamushi (= *Leptotrombidium* a.; *Microtrombidium* a.; *Trombidium* a.) - The kedani mite, whose larvae transmit *Rickettsia* tsutsugamushi, the causative organism of Scrub Typhus. T. alfreddugesi (= *Eutrombicula* a.; T. irritans) - The Harvest mite or Red Bug; the 6-legged red larva is known as the chigger, the bites of which produce a wheal on the skin accompanied by intense itching. T. autumnalis (= *Leptus* a.) - The autumnal chigger of Europe. T. holosericeum (= *Trombidium* h.) - The common harvest mite of Europe; length 4 mm; colored a striking bright red; body covered with fine velvety hairs. Larvae are parasitic on insects.

Trombidium A name formerly given a genus of Mites, now included in the genus, *Trombicula*. e.g. T. holosericeum (= *Trombicula* h.) (see **Trombicula**).

Tropaeolum A genus of plants of the family, *Tropaeolaceae*. e.g. T. majus - Nasturtium, a climbing or trailing annual plant, with simple or double flowers, large, long-stalked, irregular, yellow, orange or red in color. Used as a balcony or terrace plant; the leaves can be used in salads and the green fruit pickled as a substitute for capers. Place of origin: Peru; introduced into Europe around 1686.

Tropinota A genus of Insects of the Chafer and Dung Beetle family, *Scarabaeidae*. e.g. T. hirta - Length 8-11 mm; found on composite plants; black in color; larvae live in the ground and feed on old grass roots. Harmless. Distribution: Europe, Middle East, North America.

Truncatellina A genus of Gastropod Molluscs of the family *Vertiginidae*. e.g. T. cylindrica - Cylindrical shell, 1.8-2 mm high, about 1 mm wide, and composed of 5 to 6 finely grooved whorls, yellow-brown in color. Lives in dry areas and on rocks. Distribution: Europe, Asia Minor, North Africa.

Trutta A genus of Fishes (subphylum *Vertebrata*). The sperm or mature testes of these species yield Protamine sulfate, used intravenously to neutralize the anticoagulant action of heparin. Used similarly in the treatment of overdosage of dextran sulfate.

Trypanophis A genus of parasites resembling *Trypanoplasma*, but with a very small kinetonucleus (see **Trypanoplasma**).

Trypanoplasma A genus of sporozoan parasites resembling *Trypanosoma*, but having a posterior, as well as an anterior, flagellum. e.g. T. abramdis - A species found in the Bream, *Abramis* brama. T. borelli - A species found in the blood of fish. T. intestinalis - A species found in a saltwater fish, *Bax* boops. T. truttae - A species found in the Trout, *Salmo* fario. T. ventriculi - A species found in the Lumpfish or Sea hen, *Cyclopterus* lumpus.

Trypanosoma A genus of sporozoan parasites of the family *Trypanosomatidae*, found in the blood plasma of man and animals, characterized by the delicate, undulatory membrane attached to the body, and the whip-like flagellum. Most species live part of their life cycle in insects or other invertebrate hosts, where they undergo remarkable transformations. Also called by the generic names, *Castellanella, Duttonella, Schizotrypanum, Trypanosomonas, Trypanozoon*, and *Undulina*. e.g. T. cazalboui - A species parasitic in the goat in French Guinea and transmitted by the biting fly, *Stomoxys* bouffardi. T. cruzi (= *Schizotrypanum* c.) - A species causing the American form of trypanosomiasis called Chagas' disease. T. gambiense (= T. castellani; T. hominis; T. ugandense) - A species found in the cerebrospinal fluid and the blood of man in cases of tropical splenomegaly and sleeping sickness. It is transmitted by the tsetse fly, *Glossina* palpalis. T. granulosum - A species parasitic in the eel. Its intermediate host is the leech, *Hemiclepsis* marginata. T. inopinatum - A species parasitic in the frog and transmitted by a leech, *Helobdella* algira. T. lewisi - A species found in the blood of the rat, and transmitted by a second host, the rat flea, *Nosopsyllus* fasciatus. T. luis (= *Treponema* pallidum) - the causative organism of syphilis. T. nigeriense - The species causing sleeping sickness in Nigeria. It is transmitted by the biting tsetse fly, *Glossina* tachinoides, and is probably the same as *Trypanosoma* gambiense. T. noctuae - A species found in the blood of the little owl, being disseminated by the gnat, *Culex* pipiens. T. rhodesiense - A species found in the antelope in Nyassaland in South Africa. It may be transmitted to man by the bite of *Glossina* morsitans. It causes a form of sleeping sickness (kaodzera).

Trypanosomonas See **Trypanosoma**.

Trypanozoon See **Trypanosoma**.

Trypocastellanelleae A type of trypanosomes embracing the genera *Castellanella, Duttonella*, and *Schizotrypanum* (see **Trypanosoma**).

Tsuga A genus of Coniferous Trees of the Pine family, *Pinaceae*, also known by the genus name, *Abies*. e.g. T. canadensis (= *Abies* c.) - The Hemlock Spruce tree or Pinus Canadensis. The dried inner bark (Hemlock Spruce Bark or Pine Bark) yields Canada pitch containing tannin, and has been used for its astringent properties (see **Abies**).

Tubifex A genus of Segmented Worms

(phylum *Annelida*) of the family *Tubificidae*. e.g. T. tubifex - Length 25-30 mm; colored red; lives in tubes of cemented mud, particularly in polluted waters. Used as food for aquarium fish. Distribution: Palaearctic, Nearctic regions and Australasia.

Tubipora A genus of Octocorals of the class *Anthozoa*. e.g. T. musica - The Pipe Organ Coral, is an octocoral jelly-fish. This coelenterate lives off the Red Sea coast and the Gulf of Eilat (Akaba), and like the other octocorals does not deposit a calcareous skeleton.

Tulipa A genus of plants of the Lily fami-ly, *Liliaceae*; it has some 100 species of varying heights and seasons of flower-ing. e.g. T. gesneriana - The Tulip; its bulbs are large, and with rich brown skins. Flowers are borne single (or in twos and threes rarely), erect, large, shapely, with 6 perianth segments in various colors or combinations of shades. Used as cut flowers or for for-mal bedding. Place of origin: S.E. Europe and Central Asia; introduced into N. Europe from 1554 onwards.

Tunga A genus of Insects of the Flea family, *Hectopsyllidae*. e.g. T. penetrans - A species widely distributed in the tropical regions of America and Africa. The fertilized female burrows into the skin of the feet, often beneath the nail, causing intense irritation and may lead to spontaneous amputation of the digit (tungiasis).

Tunnus A genus of Fishes (subphylum *Vertebrata*). e.g. T. thynnus - The Bluefin Tuna fish, one of the fastest fish in the sea, with speeds up to 104 km/h (65 miles/h).

Turbatrix A genus of Nematode Roundworms of the superfamily *Rhabditoidea*. e.g. T. aceti - The "Vinegar Eel", a free living rhabditoid, sometimes parasitic in man.

Turbinaria A genus of Stony Corals of the class *Anthozoa*. These species, known as Yellow Coral, because of their bright yellow color, inhabit the Red Sea and other coral reefs and are responsi-ble, among other stony corals, for the depositing of most of the calcareous material that builds the gigantic coral reefs of all tropical seas.

Turdoides A genus of Birds (class *Aves*). e.g. T. squamiceps - The Arabian Babbler, a desert bird.

Turdus A genus of Perching Birds (order *Passeriformes*), of the Thrush family, *Turdidae*. e.g. T. merula - The Blackbird, length 26 cm. A very com-mon resident (as well as migrating) bird, found in large cities as well as in the countryside. It nests in trees, thickets, niches of buildings, etc. Distribution: Europe, northwest Africa, the Middle East, southeast Asia.

Turnera A genus of plants of the family *Turneraceae*. e.g. T. diffusa var. aphro-disiaca - The Damiana or Turnera plant; it is mildly purgative and is reputed to have aphrodisiac properties.

Tursiops A genus of Mammals of the Dolphin family, *Delphinidae*. e.g. T. truncatus - The Bottlenosed Dolphin, length 2.5 m; travels in schools, feeding chiefly on marine fish, and generally producing a single offspring or very occasionally two. The gestation period is about 12 months. This species of cetaceans (order *Cetacea*), whales, dol-phins, and porpoises, has been the sub-ject of intense research, and is the most commonly kept of the dolphin family in oceanariums. Adult specimens have a fairly large brain, weighing 1,500 to 1,800 g; they are extremely intelligent animals. Distribution: the Atlantic Ocean and the Mediterranean Sea.

Tussilago A genus of plants of the family *Compositae*. e.g. T. farfara - The Coltsfoot plant; used as a demulcent to relieve chronic cough.

Tydeus A genus of small Mites. e.g. T. mollestus - A very small mite, found in Belgium, which attacks humans.

Tylenchus A genus of Roundworms of

the class *Nematoda*. e.g. T. polyphyprus - A long-lived plant parasite, this species infests plants such as rhy, and has been found alive after more than 35 years.

Tylophora A genus of plants of the family *Ascelpiadaceae*. e.g. T. asthmatica - A plant species of South Africa; it is emetic, and is useful in dysentery and asthma.

Tylosurus A genus of Fishes (subphylum *Vertebrata*). Species of this genus are predators commonly found in the Red Sea. They float almost motionless just below the water surface, waiting to attack with lightning speed any approaching school of small fish.

Typhlocoelum A genus of Trematode parasites. e.g. T. cucumerinum - A species found in the trachea, esophagus, and thoracic cavity of chicks in Brazil.

Typhlomolge A genus of Amphibians of the family of Newts or Salamanders, *Plethodontidae*. These are the Texas Blind Salamanders, found in deep caves and wells in Texas.

Typhlops A genus of Reptiles, also called *Thyphlops*, of the diminutive blind Snake family, *Typhlopidae*. e.g. T. anchietae - The Angola blind snake, measuring only 101-152 mm (4-6 in). T. vermicularis - The Worm snake, length 20-35 cm (8-14 in); a small, pink, burrowing snake resembling a large earthworm, with two tiny, black eyes and a sharp spine on the tail. Distribution: Europe, Asia.

Typhlotriton A genus of Amphibians of the family of plethodontid Newts or Salamanders, *Plethodontidae*. e.g. T. spelaeus - The Ozark or Grotto Salamander, a lungless, terrestrial species found in grottoes and caves in the United States. They are blind creatures, nocturnal in their habits.

Typhonium A genus of plants. e.g. T. trilobatum - An Asiatic plant highly valued in oriental practice as a polychrest remedy.

Tyrannosaurus A genus of extinct animals, the *Dinosaurs*. e.g. T. rex - The "Tyrant Lizard", one of the larger theropods or carnosaurs ("meat reptiles"), believed to have lived in what are now the states of Montana and Wyoming in the U.S. about 75 million years ago. This dinosaur measured up to 14.32 m (47 ft) in overall length, had a bipedal height of up to 5.63 m (181/2 ft), a stride of 4 m (13 ft) and weighed a calculated 6.78 tonne/ton. It also had a 1.21 m (4 ft) long skull containing serrated teeth measuring up to 184 mm (7.25 in) in length.

Tyria A genus of Insects of the Tiger Moth family, *Arctiidae*. e.g. T. jacobaeae - The Cinnabar Moth, size 18-21 mm. Found in damp areas. Distribution: Eurasia.

Tyroglyphus A genus of pale, soft-bodied Ascarid Mites, related to *Glyciphagus*. e.g. T. farinae - The flour mite, found in flour mills and granaries. T. longior - The species of mites, causing copra itch. T. siro - The cheese mite; can cause gastritis and diarrhea in persons eating infested cheese (see **Glyciphagus**).

Tyrothrix A genus name once given by Duclaux to certain microorganisms, later included in the genera *Baccillus*, *Clostridium*, or *Lactobacillus*.

Tyto A genus of Birds (order *Strigiformes*) of the Barn Owl family, *Tytonidae*. e.g. T. alba pratincola - The Barn Owl, length 34 cm; has a striking heart-shaped facial disc. Found near human habitation. A hunting bird of prey, feeding on voles, birds, insects, etc. A resident or dispersive bird, it has a great sensitivity to low light intensities, that is 50 to 100 times greater than that of human night vision. Distribution: worldwide.

U

Uca A genus of Fiddler Crabs of the family *Oxopodidae*. These species abound in mangrove swamps and mud flats in warm regions. The male has one claw very much longer than the other. This claw is often brightly colored in contrast with the rest of the body, and is used in display to attract females and to defend territory around a burrow.

Uleiota A genus of Beetles of the family *Cucujidae*. e.g. U. planata - Length 4.5-5.5 mm; black-colored with extremely long antennae. Lives under the bark of trees, mostly broadleaved species. Distribution: southern part of the Palaearctic region.

Ulmus A genus of Trees of the Elm family, *Ulmaceae*. e.g. U. fulva - The Slippery Elm; the inner bark is mucilaginous and demulcent.

Umbra A genus of Fishes of the Mud Minnow family, *Umbridae*. e.g. U. krameri - Length 8-10 cm; brownish-red, irregularly spotted fish, found in clear waters with abundant vegetation. Distribution: backwaters of the Danube from Vienna to the mouth of the river, in its tributaries, in the lower reaches of the Dniester and Prut rivers and in Lake Balaton and the Neusiedlersee.

Umbrina A genus of Fishes of the Croaker, Drum, and Grunt family, *Sciaenidae*. Many of them can make noises by using their air bladders, hence their name. e.g. U. roncador - The Yellowfin Croaker Fish; a marine species found in the warmer regions of the Pacific ocean; widely distributed and commercially important.

Uncaria A genus of tropical plants of the Madder family, *Rubiaceae*. e.g. U. gambier - The Catechu plant (= Catech; Gambir; Pale Catechu); a dried aqueous extract of the leaves and young shoots is employed as an astringent in the treat-

ment of diarrhea. The diluted tincture has been used as a gargle.

Uncia A genus of Leopards of the Cat family, *Felidae*. e.g. U. uncia - The Snow Leopard or Ounce; not really a true leopard. Its fur is soft and thick, with handsome black spots on a whitish background. It lives in the mountains of central Asia and the Himalayas, usually above 8,000 ft.

Uncinaria A genus of Nematode worms, the species of which are also classified under other generic names: e.g. U. americana (= *Necator* americanus) (see **Necator**), U. duodenalis (= *Ancylostoma* duodenale) (see **Ancylostoma**), U. stenocephala - The fox hookworm, also affecting dogs.

Undulina A genus of sporozoan parasites, now called *Trypanosoma* (see **Trypanosoma**).

Unio A genus of Bivalve Molluscs of the Swan Mussel family, *Unionidae*. e.g. U. timidus - The Swollen River Mussel; length 70-80 mm, height 36-40 mm, and thickness 23-26 mm. Glossy greenish-brown shell, broadly rounded at the front, pointed at the rear. Found in quiet and slow-flowing waters such as ponds, backwaters of rivers, etc. Distribution: Europe.

Upupa A genus of Birds of the Hoopoe family, *Upupidae*. e.g. U. epops - The Hoopoe, the only species of the family; length 28 cm; has a pinkish-brown plumage, and its wings and tail are black barred with white. On its head is a long crest of chestnut-brown feathers with black tips. Unlike most birds the Hoopoe does not remove the feces from the nest, usually in tree holes, which is identifiable by its filthy condition and foul repellant smell. It also exudes a foul-smelling secretion from the preen gland, which it squirts at the enemy when danger threatens. It lives a solitary existence in open country and is very timid. A migratory bird, it winters in tropical Africa. Distribution: Eurasia and Africa.

Uragoga A genus of plants, also known by the generic name *Cephaelis*, of the Madder family, *Rubiaceae*. e.g. U. ipecacuanha (= *Cephaelis* i.) - Ipecacuanha Root (see **Cephaelis**).

Uranotaenia A genus of Mosquitoes. e.g. U. sapparinus - A species occurring in the eastern United States.

Ureaplasma A genus of microorganisms, the species of which may cause non-specific urethritis, pelvic infections, Reiter's disease, and infertility.

Urechis A genus of Echiuroid Worms of the phylum *Echiuroidea*. e.g. U. caupo - Almost 10 in long and 1 in across; looks like a large uncooked sausage. Found in central California.

Urechites A genus of plants of the Dogbane family, *Apocynaceae*. e.g. U. suberecta - The Savannah flower, an apocynaceous plant of tropical America, with poisonous and antipyretic leaves, containing the poisonous glycosides ure-chitin and urechitoxin.

Urginea A genus of plants, now more correctly called by the generic name, *Endymion*, of the Lily family, *Liliaceae*. e.g. U. maritima (= *Endymion* hispani-cus; *Scilla* campanulatus) - Spanish Squill (see **Endymion**).

Uria A genus of Birds of the Auk, Guillemot, and Puffin family, *Alcidae*. e.g. U. aalge - The Guillemot or Common Murre; length 43 cm; has a black back and white belly, but its bill is larger than that of the auks. Nests on coastal cliffs and islands in large colonies, often together with the Razorbill. Only one egg is laid directly on the bare rock. An inhabitant of arctic waters in Iceland, the Faroes, Greenland, and Labrador.

Uroa A genus of Reptiles of the Gecko Lizard family, *Gekkonidae*. e.g. U. fim-briatus - The Loaf-tailed Gecko of Madagascar. Also called the Flat-tailed or Bark Gecko because of its appear-ance and its ability to change color in imitation of the bark of the trees on which it lives. This species is almost 1 ft long.

Urocerus A genus of Wasps (class *Insecta*). e.g. U. gigas - The Wood Wasp; it lays its eggs in wood.

Urocyon A genus of Foxes of the canine family, *Canidae*. e.g. U. cinereoargen-teus - The Grey Fox, a forest animal; it is sometimes found sheltering among the cacti on desert plains and semi-desert areas. Distribution: the United States.

Urocystis A genus of Fungi or Molds. e.g. U. tritici - A species causing flag smut of wheat in Australia.

Uroglena A genus of free-swimming, fla-gellate, protozoans, also known as *Uroglenopsis*, which sometimes impart a fishy odor to the water supply.

Uroglenopsis See **Uroglena**.

Uromastyx A genus of Mastigure Reptiles of the Lizard family, *Agamidae*. e.g. U. acanthinurus - The Mastigure of the Sahara and North Africa, a species remarkable for its capability to change color according to the temperature. At low temperature its back is blackish-brown, changing to yellowish orange and green in hot weather. In Algeria it is called the Date-palm Lizard, because it feeds on dates.

Uronema A genus ciliate organisms. e.g. U. caudatum - A species found in the feces in dysentery.

Urophycis A genus of Fishes of the Hake subfamily, *Merluccinae* (family *Gadidae*). e.g. U. tenuis - The White Hake; 40 lb in weight and 4 ft long.

Urosalpinx A genus of Molluscs of the order *Neogastropoda*. Its species are the American oyster drills, now also a pest on British oyster beds, which bore through the shells of bivalve molluscs.

Ursus A genus of Mammals of the Bear family, *Ursidae*. e.g. U. arctos arctos - The European Brown Bear; was former-ly very common, but now exists only in

small numbers in the mountainous
regions of Europe, though still plentiful
in North Scandinavia and Russia. It has
poor eyesight and relies chiefly on its
keen sense of smell. Length of body
170-250 cm; tail 6-14 cm; height at
shoulder 100-110 cm; weight 150-300
kg; can reach as much as 400 kg.
Largest European carnivore. Seven full
species and 15 subspecies of brown and
grizzly bears are known in North
America, but their identifications are
largely based on variation in size and
colors of pelts. The modern view is that
they are all races of a single species,
Ursus arctos, and are distributed all over
Europe, Asia, and North America.

Urtica A genus of plants, including the
true and typical nettles. These plants are
covered with stinging hairs and secrete a
poisonous fluid. e.g. U. dioica - A stinging nettle of temperate regions having
stimulant, diuretic, and hemostatic properties. Used in homeopathic medicine.

Usnea A genus of Lichens. e.g. U. barba-
ta - A large lichen growing on forest
trees; used in homeopathic medicine.

Ustilago A genus of moldlike Fungi of
the order *Basidiomycetes*, called smuts,
which are parasitic on other plants. e.g.
U. maydis - The smut of maize, which
when eaten produces a condition resembling ergotism.

V

Vaccaria A genus of plants of the Pink family, *Caryophyllaceae*. e.g. V. pyramidata - The Cow Basil, an annual plant, 20-stem branched, with much-branched inflorescences. Grows on arable and waste land, on walls, sunny hillocks and cliffs. Introduced and naturalized in Britain.

Vaccinium A genus of Dwarf Shrubs of the Heath family, *Ericaceae*. e.g. V. myrtillus - The Common Blueberry (= Baccae Myrtilli; Bilberry; Blaeberry; Huckleberry; Hurtleberry; Myrtilli Fructus; Myrtillus; Whortleberry), a shrub species up to 40 cm high, with solitary, 5-lobed flowers, and black, glaucous, globose, sweet, edible berries. Common on heaths and in lowland and mountain moors. A diuretic and astringent, the extract of the dried fruit or leaves have been used in diarrhea and dysentery. Was formerly used in some European countries as an antiscorbutic. Native in Britain.

Vahlkampfia A genus of Amebae in which there is no flagellate stage of development (see **Dimastigamoeba**).

Valeriana A genus of plants of the Valerian family, *Valerianaceae*. e.g. V. officinalis - The Common Valerian, a stout o 1.5 m (5 ft) high, with opposite pinnate leaves, and dense clusters of small, pale pink, bisexual flowers. The fruit contains a single seed, and bears a terminal pappus of stiff-spreading hairs. It grows mainly on gravelly slopes and in mountain pastureland. This species contains an odorless crystalline principle (methyl 2-pyrrolyl ketone), a volatile alkaloid, and about 1% of volatile oil. Valerian depresses the central nervous system. The dried rhizome and roots, Valerian Rhizome (= Valer; Valerian Root), are used in the form of an extract, infusion, or tincture in conjunc-

tion with bromides, chloral hydrate, and phenobarbitone, in the treatment of hysteria and other nervous conditions. Also as a carminative. Distribution: throughout Britain, temperate Europe and Asia.

Valerianella A genus of plants of the Valerian family, *Valerianaceae*. e.g. V. locusta - The Common Cornsalad, an annual plant, 7-30 cm high, with spathulate basal leaves, oblong stem leaves, and compressed fruits. Grows in grassy places and cultivated fields. Native in Britain.

Vallonia A genus of Gastropod Molluscs of the family *Valloniidae*. e.g. V. costata - Shell 1.3 mm high, 2.1-2.7 mm wide; yellow-grey, with radial ribs, as many as 36 in the last whorl. Found commonly in meadows and on rocks. Distribution: Palaearctic and Nearctic regions.

Valvaster A genus of Starfishes of the class *Asteroidea*. e.g. V. striatus - A starfish from the Pacific. From the underside the five-armed starfish enfolds its prey, seasnails and bivalve molluscs, which it forces open with its tube-feet before devouring the soft parts with its extensile mouth.

Valvata A genus of Molluscs. e.g. V. cristata - The Flat Valve Snail, size 1-2 mm, disc-shaped; the eyes are present inside the bases of the tentacles. A freshwater snail, found more commonly in running water.

Vampyromorpha A genus of Molluscs of the order of Flying Squids, *Vampyromorpha*. e.g. V. infernalis - The Vampire Squid, a small purplish-black animal, formerly classified with the octopods. In addition to eight arms united by a swimming web, it also possesses two long sensory filaments which can be retracted into special pockets. Distribution: It inhabits depths between 1,000 and 2,000 m in the subtropical and tropical oceans.

Vampyrus A genus of Mammals of the South and Central American family of leaf-nosed "vampire" bats,

Phyllostomatidae. e.g. V. spectrum - A species from Guiana and Brazil, 15 cm (6 in) long, with a wingspan of 70 cm (28 in).

Vanda A genus of plants of the Orchid family, *Orchidaceae.* e.g. V. tricolor - A species from Java, with leathery leaves and sprays of fragrant, pale-yellow flowers with brown spots. There are many cultivated varieties of this beautiful orchid.

Vandellia A genus of Fishes (subphylum *Vertebrata*). e.g. V. cirrhosa - The Candiru fish.

Vanellus A genus of Beach and Shorebirds of the Plover family, *Charadriidae.* e.g. V. vanellus - The Lapwing, Green Plover, or Pewit; length 32 cm (13 in); with blackish-green, glossy plumage, and a prominent crest on top of the head. Found in damp meadows, swamps, fields, and on the banks of rivers and ponds. The nest is a depression in the ground, lined with grass. A partial migrant bird. Distribution: Europe (except the far north); temperate parts of Asia.

Vanessa A genus of Insects, also known by the generic name *Cynthia*, of the Tortoiseshell Butterfly family, *Nymphalidae.* e.g. V. cardui (= *Cynthia* c.) - The Painted Lady Butterfly (see **Cynthia**).

Vanilla A genus of plants of the Orchid family, *Orchidaceae.* e.g. V. planifolia - The Vanilla plant; its fruits, the vanilla beans, which contain vanilla, are used as a flavoring agent and a mild stimulant. They are said to be aphrodisiac. Place of origin: Mexico.

Varanus A genus of Reptiles of the Monitor Lizard family, *Varanidae.* e.g. V. komodo - The Komodo Monitor or Ora Lizard, a dragon-like carnivorous reptile found on the Indonesian islands of Rintja, Padar, and Flores. One of the world's largest lizards, the adult males averaging 2.59 m (8 ft 6 in) in total length, and weighing 79-91 kg (175-200 lb). They are as powerful as small crocodiles.

Variimorda A genus of Insects of the Cardinal Beetle family, *Mordellidae.* e.g. V. fasciata - Length 6-9 mm; with grey-brown pattern on the elytra; found on flowers. Distribution: Europe; Asia Minor.

Variocorhinus A genus of Afro-asian Cyprinoid Fishes of the Carp family, *Cyprinidae.* These include the barbels, carps, minnows, etc. (see **Carassius**).

Variola A genus of Fishes (subphylum *Vertebrata*). e.g. V. louti - The Painted Coral Trout, a predator sea fish. Its opened mouth is enormous compared to its body size, and relatively large fish often fall victim to this solitary hunter. Found in the Red Sea, off the coral reefs of the Gulf of Eilat (Akaba).

Variolaria A genus of Lichen plants. e.g. V. amara - A febrifugal and anthelmintic species of the Old World.

Vaspa See **Vespa**.

Vateria A genus of Asian trees of the family *Dipterocarpaceae.* e.g. V. indica - An East Indian tree which affords a resin or gum, Indian copal, piny varnish, white dammar, or Indian anime. Used as a varnish; also as an ingredient in dental medicine for covering cement fillings in tooth cavities to protect them from saliva.

Veillonella A genus of microorganisms of the family *Neisseriaceae.* Made up of minute cocci; found as non-pathogenic parasites in the mouth, intestines, and urogenital and respiratory tracts of man and other animals. e.g. V. parvula.

Velamen A genus of Comb Jellies of the order *Cestida*. The species are restricted to tropical and subtropical waters and the Mediterranean. Some may be seen around Florida. The body is elongated in the sagittal plane to a flattened gelatinous ribbon, which may be more than 1 m long.

Velella A genus of Hydras or Hydrozoans of the order *Athecata.* e.g. V. velella -

The Velella or By-the-wind-sailor, a pelagic animal, 1-5 cm long, inhabiting warmer seas; but shoals are often swept northwards in the Atlantic, blown along by the wind by means of its gas-filled bladder, to become stranded on British and American Atlantic shores. It resembles an elongated, inverted, and solitary version of a polyp, with a round or oval float containing air chambers, and a purple sail. The small medusae released by Velella are known as *Chrysomitra*, which were originally classified as a separate genus. These have no mouth, and die after releasing gametes. The animal often lives in colonies on the surface of the sea. A Mediterranean Siphonophore, it may reach the southwest coast of Britain.

Velia A genus of Insects of the Water Cricket family, *Veliidae*. e.g. V. caprai - The Water Cricket, a stream-dwelling predaceous species, size 6-7 mm; with short legs and thick belly; it runs on six legs on the water surface, and preys on other insects and spiders.

Velifer A genus of Fishes of the Veliferid family, *Veliferidae*. The species have tall anal and dorsal fins. *Velifer* is the single genus of this family, and looks rather like the *Lamprididae*.

Veltheimia A genus of plants of the Lily family, *Liliaceae*. e.g. V. viridiflora - A South African plant with wavy, intense, green leaves from 20-30 cm (8-12 in) long, and crowded heads of tubular, reddish or yellowish, spotted flowers.

Venus A genus of Bivalve Molluscs, also known by the generic name *Mercenaria*, of the Venus and Carpet Shell family, *Veneridae*. e.g. V. verrucosa - Warty Venus, length 40-70 mm, height 39-68 mm, thickness 27-48 mm. Shell very thick, surface marked with rugged concentric ridges. An edible species. Distribution: Mediterranean Sea, European and African coast of the Atlantic, southeast Africa. V. mercenaria (= *Mercenaria* m.) - The American

Hard-shell Clam or Quahaug, a lamellibranch animal living close under the water surface and having very short siphons.

Veratrum A genus of poisonous plants of the Lily family, *Liliaceae*. e.g. V. album - The Common White Hellebore (= European Hellebore; Veratri Rhizoma; Veratrum Album; Weisse Nieswurz; White Hellebore; White Hellebore Rhizome; White Veratrum), a poisonous perennial plant, 60-150 cm high; with broadly elliptical leaves, and white flowers, greenish on the outside. Has an emetic, errhine, and cathartic rhizome. Yields two alkaloids, Protoveratrine A and B, a mixture of which is used the treatment of hypertension and the toxemias of pregnancy. Grows in alpine meadows, pastures, on ridges, and in crevices of rocks, above 2,200 m. Place of origin: the Pyrenees; south, central, and northern Europe. V. viride - Green Hellebore (= American Hellebore; American Veratrum; Green Hellebore Rhizome; Green Veratrum; Veratro Verde; Veratrum Viride); the dried rhizome and roots contain an alkaloid mixture, Alkavervir and other alkaloids, including cryptenamine, protoveratrine, etc. which cause vasodilatation, and have been used in the treatment of hypertension and the toxemias of pregnancy.

Verbascum A genus of plants of the Figwort or Snapdragon family, *Scrophulariaceae*. e.g. V. thapsus - The Common Mullein, a biennial plant, up to 1.8 m high; stem and leaves with soft white wool; lower filaments glabrous or with few hairs. The dried flowers, Verbascum Flowers (= Bouillon Blanc; Mullein Flowers; Wollblumen) have been used extensively in Europe in the treatment of pulmonary complaints, though its sole therapeutic value could be only that of a demulcent. Grows in stony, sandy, and grassy places, by waysides, in enbankments and on waste ground. Distribution: Europe; native in Britain.

Verbena A genus of plants of the Verbena or Vervain family, *Verbenaceae*. e.g. V. officinalis - The Vervain, an annual plant, 30-100 cm high; small flowers in slender terminal spikes. Widespread and local, growing by waysides and in waste places. Distribution: most of Europe; native in Britain.

Vermicella A genus of venomous Australian Serpents. e.g. V. annulata - The Five-ringed Snake, up to 60 cm (2 ft) long; its pale body is marked with 4 to 10 conspicuous bands, olive to reddish in color. Although poisonous it is not dangerous to man. Distribution: the northern parts of Australia.

Vermivora A genus of Birds of the Wood-warbler family, *Parulidae*. e.g. V. peregrina - The Tennessee Warbler, a migrating bird found in the United States.

Veronia A genus of plants. e.g. V. anthelmintica - An anthelmintic plant called Somraj in India.

Veronica A genus of plants of the Figwort or Snapdragon family, *Scrophulariaceae*. e.g. V. chamaedrys - Birdseye Speedwell or Germander Speedwell, a small perennial plant; 5-40 cm high; ovate toothed leaves; each flower has a 4-lobed calyx, and a bright blue 4-lobed corolla; the fruit is a flattened, heart-shaped capsule containing several seeds. Grows in woods, hedges, pastures, and by roadsides. Distribution: Europe, Asia; native in Britain; has become naturalized in North America.

Verticillium A genus of Fungi or Molds. e.g. V. graphii - A mold sometimes causing otitis externa.

Vertigo A genus of Gastropod Molluscs of the family *Vertiginidae*. e.g. V. pygmaea - Shell 1.5-2.2 mm high, 1.2-1.5 mm wide, glossy brown. Mouth furnished with 4 to 6 teeth, generally 5. Found in meadows and hedgerows, also on rocks. Distribution: Europe, the Caucasus, western Asia, and North America.

Vespa A genus of Hymenopterous Insects, also known by the generic name *Vaspa*, of the Wasp family, *Vespidae*. e.g. V. crabro (= *Vaspa* c.) - The Hornet or European Hornet, now introduced into the United States; length 19-35 mm; head much widened behind the eyes. It builds a large nest which in time contains a whole colony of hornets. Only the females have stings. These wasps live only one year, only the young fertilized females surviving the winter. They subsist on nectar, ripe fruit, and sugary liquids, but they are also predacious and feed their larvae with masticated insects. A homeopathic preparation is derived from this species. Distribution: the Palaearctic and Nearctic regions. Native in Britain.

Vespertilio A genus of Mammals of the Smooth-faced or Vespertilionid Bat family, *Vespertilionidae*. e.g. V. murinus - The Particolored Bat, size 6-7 cm; wingspan 28 cm; upper side musty grey, underside very light-colored. A rare, migrant, species found in Europe, particularly in the south.

Vespula A genus of Hymenopterous Insects of the Wasp family, *Vespidae*. This genus includes the black and yellow wasps, known in America as yellow-jackets or hornets. The species are smaller than those of *Vespa* (see **Vespa**).

Vibrio A genus of microorganisms of the family *Spirillaceae* (= *Pseudomonadaceae*). Short, slightly curved, actively motile, gram-negative rods, occurring singly and occasionally found end-to-end. There are 34 species, including the cholera and El Tor vibrios pathogenic for man, agents of specific disease in lower animals, and paracholera or cholera-like vibrios (= NAG/non-agglutinating vibrios). e.g. v. cholerae (= V. cholerae - asiaticae; V. coli; V. comma) - The causitive agent of Asiatic cholera in man. Curved rods with a single flagellum at one end. The hemolytic vibrios of this species are known as the El Tor vibrios. V. fetus (= *Campylobacter* f.) - The etiological agent of abortion in pregnant cattle, sheep, and goats, and of an essentially symptomless infection in the males. Non-pathogenic for man (see **Campylobacter**). V. vulnificus (= *Beneckea* vulnifica) - The L+ (lactose-positive) halophilic vibrio, a marine microorganism, isolated from blood cultures, CSF, and localized infections of the extremities, occurring after their exposure to the sea.

Vibrion A genus of microorganisms of the family *Bacillaceae*. Also classified under the generic name *Clostridium*. e.g. V. septique (= *Clostridium* septicum) - A moderately large bacillus, with rounded ends about 3-10 μ by 0.6-1 μ; mobile, with peritrichous flagellae. Responsible for the disease known as Braxy in sheep, and some cases of Black-leg in cattle and sheep (see **Clostridium**).

Viburnum A genus of Trees and Shrubs of the Honeysuckle family, *Caprifoliaceae*. e.g. V. opulus - The Guelder Rose or Cranberry Tree, a deciduous shrub up to 5.4 m (18 ft) high, with small central fertile flowers and larger white, sterile peripheral ones. The succulent, subglobose, red fruits (drupes) are borne in pendulous clusters. The leaves are 5-lobed and coarsely toothed. The smooth greyish, dried bark, Viburnum Opulus (= Cramp Bark; Guelder Rose Bark; High-bush Cranberry Bark; Snowball Tree Bark - this from the variant, V. opulus roseum), has been used as an antispasmodic and uterine sedative in the treatment of dysmenorrhea and other functional uterine disorders, but evidence of its supposed therapeutic activity on the uterus is conflicting. The species grows in woods and scrub. Place of origin: Eurasia; native in Britain; cultivated in North America. V. opulus var. roseum - The Snowball Tree; it has spherical clusters of large sterile flowers, often grown in gardens. Its bark, Snowball Tree Bark has been used as above.

Vicia A genus of Herbs, including the Vetch and the Broad Bean, of the Pea family, *Leguminosae*. e.g. V. faba (= V. fava) - The Broad Bean plant, also called Fava or Fava Bean; has been cultivated over a long period for its edible seeds. These beans or pollen contain a component that is capable of causing an acute hemolytic anemia, known as favism or the broad-bean syndrome, in susceptible individuals, viz. in Glucose-6-phosphate dehydrogenase or G6PD deficiency states. Distribution: worldwide; a native of southwestern Europe.

Victoria A genus of plants of the Waterlily family, *Nymphaeaceae*. e.g. V. amazonica (= V. regia) - The Royal Waterlily; Royal Water Platter; a huge densely prickly, aquatic plant with short, thick rhizomes, and circular floating leaves up to 2.1 m (7 ft) in diameter, with margins turned up to form a continuous rim which prevent water getting on to the top. Owing to the network of thick veins underneath and the heavily cutinized upper surface, these leaves are so strong that they are able to carry the weight of a child without sinking. The red, pink, or white flowers, more than 30 cm (1 ft) across, are borne on thick, prickly stalks just above the water surface. The pulpy seeds are rich in starch and are used for flour. The rhizomes are used in dyeing and tanning. The plant is grown in botanic gardens in temperate climates, and in large ponds and lakes in the tropics. Place of origin: the backwaters of tropical South America.

Villarsia A genus of plants of the Gentian family, *Gentianaceae*. e.g. V. nymphaeoides - An Old World gentianaceous species; an antiscorbutic.

Vimba A genus of Fishes of the Carp and Marrow family, *Cyprinidae*. e.g. V.

vimba - The Zahrte fish, length 30-40 cm, weight more than 1 kg; has an inferior mouth with a pointed, protruding snout. A partial migrant, inhabiting the lower reaches of larger, slow-flowing rivers. Distribution: The Weser, Elbe, and other rivers as far as the Neva, in southern Finland and in Sweden (see **Carassius**).

Vinca A genus of flowering plants, also known by the generic name *Catharanthus*, of the Periwinkle family, *Apocynaceae*. e.g. V. rosea - The Madagascar Periwinkle (see **Catharanthus**).

Vincetoxicum A genus of plants of the Milkweed family, *Asclepiadaceae*. e.g. V. hirundinaria - The Swallowwort, a perennial plant, 30-120 cm high; short-stalked, opposite leaves; rotate corolla. Widespread in woods and scrub, and on rock-ledges.

Viola A genus of plants of the Viola or Violet family, *Violaceae*. e.g. V. tricolor - The Heartsease or Wild Pansy, an annual to perennial plant, 10-45 cm high; yellow, violet, or pink flowers, or of these colors in combination. Used as an emetic. Distribution: widespread in Britain; Europe and Asia. V. canina - The Dog Violet, a wild scentless species.

Vipera A genus of Reptiles of the Viper Snake family, *Viperidae*. e.g. V. berus - The Common European Viper or Adder, length 1 m; male smaller than the female; with a stout, stumpy appearance and a dark, zigzag band on the back; body coloring is very variable, grey, brown, or almost black. A venomous snake, its bite may be fatal, especially in children. Found in moist and sunny places along forest streams, in clearings and around buildings, and in mountainous areas. Distribution: Europe; central and northern Asia. The only viper found in Britain.

Vireo A genus of New World Birds of the Vireo family, *Vireonidae*. e.g. V. solitar-

ius - The Blue-headed or Solitary Vireo, a small American bird distinguished by bristles covering the nostrils and forehead, many having an eye-ring, usually white in color. Normally a white spot links this ring with the base of the bill. These birds are insectivorous, and live mainly in trees or shrubs. A migrating bird, it returns from the Gulf of Mexico in the spring to breed in the northern United States and southern Canada.

Viscaria A genus of plants of the Pink family, *Caryophyllaceae*. e.g. V. vulgaris - Red German Catchfly, a perennial plant, 20-70 cm high; stem very sticky beneath each node. Native in Britain.

Viscum A genus of plants of the Mistletoe family, *Loranthaceae*. e.g. V. album - The Mistletoe plant, an evergreen shrub, up to 50 cm high; narrow, greenish-yellow leaves in pairs, usually dioecious flowers and white, sticky berries. Parasitic on the branches of deciduous trees, growing chiefly on apple, pine, plum, poplar and spruce trees. The dried semi-parasite, Viscum Album (= Viscum; Visci Caulis; Gui; Tallo de muérdago), has a vasodilator action and is used for lowering blood-pressure. Mistletoe has been used in epilepsy, hysteria, and chorea and for making bird-lime. It has also been suggested for use in the treatment of cancer. Widely used at Christmas-time for decorative purposes. Distribution: most of Europe and northern Asia. Native in Britain.

Vitaliana A genus of plants of the Primrose family, *Primulaceae*. e.g. V. primuliflora - Gold Primrose, a perennial plant, 1-5 cm high; leaves narrowly linear, soft-hairy beneath. Grows on mountains in rock-crevices, on scree, turf, and in snow patches. Distribution: the Alps, at 1,700-3,100 m.

Vitex A genus of plants of the Verbena or Vervain family, *Verbenaceae*. e.g. V. agnus-castus - Monk's Pepper Tree or

Tree of Chastity; a shrub 2-4 m high; leaves white-downy beneath; flowers bluish violet, pink or white. Widespread at the edges of ponds and in the beds of streams in the Mediterranean region.

Vitis A genus of plants of the Grape family, *Vitaceae*. e.g. V. carnosa - An Asiatic species of grape; in India the seeds and roots are used in medicine. V. latifolia - An East Indian grape vine affording detergent, alternative, and soothing medicines. V. vinifera - The Grape Vine or Wine Grape tree, the most important member of this genus with many subspecies. It has been in cultivation since ancient times and its origin is obscure, but it is probably a native of the Caspian area of the Caucasus or western India. A woody climber with thin, deeply lobed, coarsely-toothed leaves, and tendrils which twine round a support. Dioecious flowers. The oval fruits are black when ripe.

Vitrea A genus of Gastropod Molluscs of the family *Zonitidae*. e.g. V. crystallina - Crystal Snail; shell 2 mm high, 3-4 mm wide, and composed of 41/2-5 whorls, the last about 11/2 times broader than the preceding one. Lives in moist locations under old fallen trees, in decayed stumps and moss. Distribution: Europe, northwest Africa.

Vitreoscilla A genus of saprophytic microorganisms, the Schizomycetes of the family *Vitreoscillaceae*; the species are found in decaying organic matter, occurring in colorless trichomes of varying flexibility, which show a gliding motion when attached to a substrate. Other genera of this family are *Bactoscilla* and *Microscilla*.

Vitrina A genus of Gastropod Molluscs of the family *Vitrinidae*. e.g. V. pellucida - The Pellucid Glass-snail; shell about 3.5 mm high, 4.9-6 mm wide; it is wide, glossy and smooth, and composed of 3-31/2 whorls; very delicate, and barely large enough to contain the animal. A very common species, living in both damp as well as dry localities, in forests, steppes, on rocks, in meadows and cultivated fields. Distribution: Eurasia, North America.

Viverra A genus of Mammals of the Civet, Genet and Mongoose family, *Viverridae*. e.g. V. zibetha - The Large Indian Civet, about 1.2 m (4 ft) long; it has a grey coat with black markings, an erectile mane and short tail, and lives in bushy regions. Civet, a musk-like perfume is obtained from these animals, which have been kept in captivity to obtain their scent; this is expressed from the glands about twice a week with a small spatula.

Viverricula A genus of Mammals of the Civet, Genet and Mongoose family, *Viverridae*. e.g. V. indica - The Rasse, or Small Indian Civet. The musk-like perfume, civet is also obtainable from this species, which has been introduced into Madagascar (see **Viverra**).

Viviparus A genus of Gastropod Molluscs of the Freshwater Winkle family, *Viviparidae*. e.g. V. viviparus - The Common River Snail; thin-walled shell, 32-40 mm high, 24-30 mm wide, with sharp-pointed apex and ovate mouth closed by an operculum. Whorls are arched, and the eyes are on short stalks outside the base of the tentacles. A freshwater species, found in calm water with dense vegetation. Herbivorous. Distribution: Europe.

Volucella A genus of Dipterous or Two-winged Insects of the Hover Fly family, *Syrphidae*. e.g. V. pellucens - Length 15-16 mm; distinguished by a false longitudinal vein almost in the center of the wings. Larval development takes place in the nests of the Common Wasp and the German Wasp. Distribution: Europe, Asia.

Volvox A genus of Protozoans of the order of Green Algae, *Phytomonadina*. e.g. V. aureus - This species consists of a hollow sphere with many minute biflagellate cells embedded in a peripheral gelatinous layer, and consisting of green-colored colonies of several thousand of these cells, in which movement is effected by the coordinated lashing of these flagella. Reproduction may be by asexual or sexual means, the daughter colonies being formed inside the sphere. These colonies reach a size of about 0.1 cm (0.04 in), and inhabit freshwater.

Vombatus A genus of Mammals of the Wombat family, *Vombatidae*. e.g. V. hirsutus - The Common Wombat, a badger-like animal, with dentition exactly as that found in rodents. Often called "badger" in Australia. A thick-set, short legged and tailless burrowing species, the burrows sometimes reaching 30 m (100 ft) in length. A nocturnal, omnivorous (though vegetable food predominates), long-lived (up to 30 years of age) animal. Found in Australia.

Vormela A genus of Mammals of the Weasel family, *Mustelidae*. e.g. V. peregusna - The Marbled Polecat, size 38 cm, with a 16 cm tail; back and flanks have distinctive spots. A dusk and nocturnal animal living in cultivated areas. Distribution: the Balkans.

Vorticella A genus of Ciliate Protozoans of the order *Peritrichida*. These Peritrichs are bell-shaped animals mounted with the pointed end on a contractile stalk. They live in clusters with the crown of peristomial cilia expanded, about 500 μ long. At the slightest disturbance, the contraction of the whole cluster is very conspicuous. Reproduction is by binary fission and conjugation. They are found in feces, urine, nasal mucus, etc.

Vriesia A genus of plants of the Pineapple family, *Bromeliaceae*. e.g. V. splendens - Flaming Sword, an epiphytic evergreen plant with leathery rosettes of bluish-green leaves and long flattened spikes of yellow flowers with red bracts. The genus is named after a Dutch botanist W. H. de Vriese. Place of origin: Guyana.

Vulpes A genus of Carnivorous Mammals of the Fox, Jackal, Wild Dog, and Wolf family, *Canidae*. e.g. V. vulpes - The Old World Red Fox, length of body 65-76 cm, tail 35-44 cm; red-brown above and white below. Found chiefly in forests, copses, and field thickets, often in places with rock cavities. Lives singly or in families. Distribution: greater part of Europe and much of northern Asia. Native in Britain.

Vulpia A genus of plants of the Grass family, *Gramineae*. e.g. V. myuros - Rat's tail Fescue, an annual plant, 10-50 cm high; the upper leaf-sheet reaches the panicle, the lower glume is one-third the length of the upper. Place of origin: South and Central Europe. Native in Britain.

Vultur A genus of Birds of Prey of the New World Vulture family, *Cathartidae*. e.g. V. gryphus - The Andean Condor, the heaviest bird of prey, also known as "King of the Andes"; it ranges from western Venezuela to Tierra del Fuego and Patagonia in South America. Adult males weigh 9.09-11.3 kg (20-25 lb), and have the largest wing expanse or "sail area" of any living bird, with an average of 2.81 m (9 ft 3 in). They are more than 90 cm (3 ft) in length. The bird is easily recognized by its shining black plumage embellished with a white band on each wing, and a white collaret at the base of the neck. The males have a high, fleshy wattle which forms a crest extending from forehead to crown. In the Andean Cordilleras, the Condor is said to soar to a height of 4,800-6,000 m (16,000-20,000 ft).

W

Wahlenbergia A genus of plants of the Bellflower family, *Campanulaceae*. e.g. W. hederacea - The Ivy-leaved Bellflower, a creeping, perennial plant, up to 30 cm (1 ft) high, with solitary, long-stalked flowers. Grows on damp moorland, heaths and in open woods. Distribution: West and Central Europe. Native in Britain.

Wasielewskia A genus of Coprozoic Amebae, better known by the generic name *Dimastigamoeba* (see **Dimastigamoeba**).

Weigela A genus of plants, also known by the generic name *Diervilla*, of the Honeysuckle family, *Caprifoliaceae*. e.g. W. florida (= *Diervilla* f.) - The Weigela plant, a deciduous spreading bush, 2 m (61/2 ft) tall, with opposite, ovate-lanceolate, prominently-veined leaves and pink, tubular flowers, growing in clusters. The fruit is a capsule. Place of origin: North China; introduced into Europe in 1845.

Welwitschia A genus of plants of the group *Gymnosperms*, vascular plants containing woody conducting tissue, and producing exposed seeds. e.g. W. mirabilis - A remarkable species growing in the desert regions of tropical West Africa. It may live for 100 years or more, and only has two leaves which continue growing from the base, where they are attached to the top-shaped woody stem. The plant is unisexual, and the naked seeds are dispersed by the wind.

Willemetia A genus of plants of the Daisy family, *Compositae*. e.g. W. stipitata - The Willemetia, a perennial plant, 15-45 cm high, with sinuate-toothed leaves, and golden-yellow flowers in long stalked heads. Grows in damp meadows and on flat moorland in moun-tain regions. Distribution: Europe, in the Alps, East Pyrenees, and the Balkans.

Wisteria A genus of climbing and spreading plants of the Pea family, *Leguminosae*. e.g. W. sinensis - Chinese Wisteria, a woody climbing plant, growing up to 18 m (60 ft) tall; with mauve to deep lilac, fragrant flowers in long, pendulous racemes and pinnate leaves, with oval to oblong leaflets. The flattened pods are about 15 cm (6 in) long. Used to cover walls, arbors, railings, etc. and to decorate balconies and terraces. Place of origin: China; introduced into Europe in 1816.

Withania A genus of plants of the Nightshade family, *Solanaceae*. e.g. W. somnifera - The Aswagandha plant, growing in India. The dried root, Withania Root (= Aagandh; Aswagandha Root) has narcotic and diuretic properties, and is used as a liquid extract in the Indian Pharmacopeia.

Wolffia A genus of plants of the Duckweed family. e.g. W. arrhiza - The Rootless Duckweed, a perennial plant, 1-1.5 mm high; with ovoid to ellipsoid rootless thalli, growing in still water. Distribution: most of Europe; native in Britain.

Wombatula A genus of Mammals of the Wombat family, *Vombatidae*. A hairy-nosed animal, somewhat badger-like, with dentition similar to that of the rodents.

Woodsia A genus of plants of the Polypody or Fern family, *Polypodiaceae* (= *Aspleniaceae*). e.g. W. ilvensis - The Oblong Woodsia, a perennial plant, 10-12 cm high; rachis or main axis and underside of frond-segments are scaly; the industrium or structure enclosing the sporangia has long hair points. Grows in crevices of siliceous rocks. Place of origin: North Europe, the Arctic region and southwest Russia. Native in Britain.

Wuchereria A genus of Filarian Roundworms (class *Nematoda*), of the order *Filaroidea*. e.g. W. bancrofti - A species found in the warm regions of the world, "inoculating" humans with larvae carried by infected mosquitoes which serve as intermediate hosts, and causing the disease known as elephantiasis or filariasis. The species has been known to live as much as 17 years.

Wulfenia A genus of plants of the Figwort or Snapdragon family, *Scrophulariaceae*. e.g. W. carinthiaca - The Wulfenia, a perennial plant, 20-40 cm high; the crenate leaves grow in a rosette, the raceme is one-sided, and the blue corolla has a bearded throat. A rare relic of the Rhododendron zone in the Gaital (Carinthia); found also in Montenegro.

X

Xanthium A genus of plants of the Daisy family, *Compositae*. e.g. X. spinosum - The Spiny Cocklebur, an annual plant, 15-70 cm high; leaf-stalks with 1-2 trifid yellow spines at base. Found growing in rubbish-dumps, waste and arable land, and in gardens. Introduced from America and naturalized in Britain.

Xanthoceras A genus of plants of the Soapberry family, *Sapindaceae*. e.g. X. sorbifolia - A deciduous, woody shrub, up to 6 m (20 ft) tall, a native of China.

Xanthochymus A genus of plants. e.g. X. pictorius - An East Indian plant called Thaikal; it yields a purgative extract resembling cambogia.

Xanthomonas A genus of microorganisms of the family *Pseudomonadaceae*, occurring as monotrichous cells producing a yellow pigment. Pathogenic for plants. e.g. X. campestris - A species yielding a polysaccharide gum, Corn Sugar Gum (= B 1459; Xantham Gum) by bacterial fermentation of dextrose.

Xanthoria A genus of Lichen plants found growing on the bark of trees, particularly the conifers. e.g. X. parietina - The Common Lichen, a perennial plant, up to 10 cm high. Grows on the bark of trees and has a thallus of yellow, leaf-like lobes with orbicular apothecia or fruiting organs. Native in Britain.

Xanthorrhiza A genus of herbs. e.g. X. apiifolia - A North American shrub called yellowroot; the root and wood are bitter and tonic.

Xema A genus of Birds of the Gull or Tern family, *Laridae*. e.g. X. sabini - Sabine's Gull, size 33 cm; a northern gull with a deeply forked tail, black-edged in young; triangular white pattern on wings. A partial migrant.

Xenicus A genus of Birds of the New Zealand Wren family, *Xenicidae* or *Acanthisittidae*. e.g. X. gilviventris - The Rock Wren, a rare wren-like bird, almost extinct today, having been preyed upon by rats, cats, and stoats. Introduced onto the islands of New Zealand (see **Acanthisitta**).

Xenophora A genus of soft-bodied animals, the Molluscs, of the phylum *Mollusca*. e.g. X. solaris - A spiral shell form, growing in coils about an imaginary axis through the apex of the shell. This species is able to cement small pebbles and other shells to itself.

Xenopsylla A genus of Insects, also known by the genus name *Pulex*, of the Fleafamily, *Pulicidae*. e.g. X. cheopis (= *Pulex* c.) - The Oriental or Indian Rat Flea, the most dangerous flea in the world, carrier of the dreaded bubonic plague bacterium, *Pasteurella* pestis. In the 14th century this parasite caused the deaths of some 25 million people in Europe (approximately one quarter of the total population in that continent at the time), and in the great Indian plague of 1867, when over ten million people died. Under 2.5 mm (1/10 in) in size, this deadly insect spreads the infection among rats from whom it is transferred to humans. It is also responsible for spreading murine typhus and has done so along the coast of the United States (see **Pulex**).

Xenopus A genus of Amphibians of the Frog family, *Pipidae* (suborder *Aglossa*). e.g. X. laevis - The Clawed Frog or South African Clawed Toad; the fore feet and flatenned body are well adapted to aquatic life. A species used in the diagnosis of human pregnancy (Xenopus Test): following the injection of the patient's urine, the female frog will respond to the hormones secreted during pregnancy by spawning within a few hours. Natural Habitat: the warm regions of Africa, south of the Sahara.

Xenosaurus A genus of Reptiles of the Lizard family, *Xenosauridae*. The species are little-known and very rare, and are found in Central America, particularly in Central Guatemala and Mexico.

Xenus A genus of Birds of the Sandpiper or Wader family, *Scolopacidae*. e.g. X. cinereus - The Terek Sandpiper, size 23 cm long and thin bill with a slight upward curve. Nests on the waterside. A migrant Waderbird, inhabiting northern Russia.

Xeranthemum A genus of plants of the Daisy family, *Compositae*. e.g. X. annum - An annual plant, about 30-60 cm (1-2 ft) high, with lanceolate, entire leaves and purple capitula. The dried capitula persist for many years, and in cultivation show a wide range of colors. A native of southern Europe.

Xeris A genus of Insects of the Woodwasp order, *Hymenoptera*. e.g. X. spectrum - Size 4 cm; slender black body; the first segment of the thorax has two yellow, lengthwise stripes. It is found in felled logs and trees in coniferous forests, in which it bores holes and lays its eggs. The larvae feed in the wood and develop into adults. Native in Britain.

Xerolycosa A genus of Arthropods of the Wolf Spider family, *Lycosidae*. e.g. X. nemoralis - length 5-7 mm; the cephalothorax is marked with a pale, parallel-sided band covered with fine white hairs. A very plentiful species found in dry locations, such as coniferous woods, clearings and heaths, etc. Distribution: the Palaearctic region.

Xestobium A genus of Insects of the order of Beetles, *Coleoptera*. e.g. X. rufovillosum - The Deathwatch Beetle, a species parasitic on wood and wood housing.

Ximenia A genus of African olacineous trees; the drupes of some species are edible and aromatic.

Xiphias A genus of Fishes of the Swordfish family, *Xiphidae* (= *Xipiidae*). e.g. X. gladius - The Swordfish, the only species of the family, nearly 5 m (16 1/2 ft) long, characterized by an extremely long upper jaw, the so-called sword. A fast-swimming predaceous fish, it lives in the open sea at depths ranging from the surface down to about 600 m. A favorite game fish in North America. Distribution: worldwide; most common in the Mediterranean Sea.

Xiphophorus A genus of Fishes of the order *Microcyprini* (= *Cyprinodontiformes*). e.g. X. helleri - The Swordtail fish, a very popular aquarium fish; the male has an elongated anal fin which is used as an intromittent organ. A freshwater species found in the rivers of Central America.

Xiphorhamphus A genus of Fishes of the Characin family, *Characinidae*. The species are predatory and are found in the rivers of Africa and South America. They are characterized by a scaly body, normal pharyngeal bones, and often an adipose fin. The mouth is not protactile and is usually toothed.

Xiphorhynchus A genus of Birds of the Woodhewer family, *Dendrocolaptidae*. e.g. X. flavigaster - The Ivory-billed Woodhewer, a solitary bird, cryptically colored in brownish tones, and never seen away from the woods. A predator on the large carpenter ants which live in the decaying parts of trees.

Xiphosura A genus of Horseshoe Crabs, better known by the generic name *Limulus* (see **Limulus**).

Xyleborus A genus of Insects of the Bark-beetle family, *Scolytidae*, sometimes included in the Weevil-beetle family, *Curculionidae*. These species, the so-called Ambrosia beetles, bore out elaborate systems of radiating galleries between the bark and wood of many trees and shrubs, and have an organized pattern of polygamy, with one male to about 60 females. They also cultivate special fungi on which the larvae are nourished.

Xyleutes A genus of Insects of the Cossid Moth family, *Cossidae*. e.g. X. boisduvali - An Australian Cossid, with a wingspan of 25 cm (10 in) and an abdomen as large as a small banana.

Xylocopa A genus of Insects of the Bee family, *Apidae*. e.g. X. violacea - The Carpenter Bee, length 21-24 mm. One of Europe's largest bees. The female makes its nest or hive in an old trunk or branch, tunnelling into the wood. Distribution: southern Europe and southern part of Central Europe.

Y

Yersinia A genus of microorganisms. e.g. Y. enterocolitica (= Ye enterocolitica) - A species causing acute infective enterocolitis.

Yoldia A genus of Bivalve Molluscs of the subclass *Protobranchia*. The species live vertically disposed in water, with the inhalant current entering posteriorly through a short siphon ventral to the exhalant current.

Yponomeuta A genus of Insects of the Butterfly and Moth family, *Yponomeutidae*. e.g. Y. padella - size 9-10 mm; a summer butterfly. The larvae live in a communal web where they also pupate. Distribution: the Palaearctic region.

Yucca A genus of plants of the Lily family, *Liliaceae*. e.g. Y. gloriosa - Adam's Needle, Mound Lily, or Spanish Bayonet; an evergreen, shrub-like plant with a short, thick stem, occasionally branched. Glaucous-green, leathery leaves, spined at the tips, up to 60 cm (2 ft) long and 5-8 cm (2-3 in) wide, growing in a rosette. Creamy-white, hooded flowers, tinged with red outside, growing in closely packed, large panicles of 1-2 m (3-6 ft) or more. Place of origin: southeastern United States; introduced into Britain in 1550.

Z

Zabrus A genus of Insects of the Groundbeetle family. e.g. Z. tenebroides - Length 14-16 mm; convex, plump body; upper side brown or black, underside lighter, and short antennae. Zabrus and its larva are plant-eaters and are found in gardens and cultivated fields.

Zaglossus A genus of Oviparous, primitive Mammals of the order *Monotremata*. e.g. Z. bruijni - The Long-beaked Echidna, a monotreme mammal found in New Guinea. The one or two eggs laid by the female are placed in a rudimentary marsupial pouch, into which is discharged the mammary secretion. A burrowing animal, it has strong claws and spines on its back, rolling up like a hedgehog when in danger.

Zalophus A genus of Mammals of the Eared Seal family, *Otariidae* (suborder of Seals, Sea Lions and Walruses, *Pinnipedia*). e.g. Z. californianus - The Californian Sea-lion, over 2.1 m (7 ft) long; the most abundant species today and the most common pinniped seen in zoos and trained in circuses. Its coat is dark brown and its shrill barking or yapping is disagreeable.

Zanclus A genus of Fishes of the Moorish idol family, *Zanclidae*. These are similar and are related to the Surgeon fishes (family *Acanthuridae*). e.g. Z. canescens - The Moorish Idol; the body is laterally compressed and the dorsal and ventral fins extend into long filaments. This species of fish inhabits coral reefs.

Zannichellia A genus of plants of the Horned-pondweed family. e.g. Z. palustris - The Horned-pondweed, a perennial plant, 25-90 cm high; monoecious, with male and female flowers on the same plant; hair-like leaves. Widespread, growing in streams, ditches and pools. Distribution: Europe; native in Britain.

Zantedeschia A genus of plants, also known by the generic name *Calla*, of the Lily family, *Liliaceae*. e.g. Z. aethiopica (= *Calla* a.) - The Arum Lily, Calla Lily, or Lady-of-the-Nile, a herbaceous, moisture-loving, robust, perennial plant up to 60-75 cm (2-2 1/2 ft) high, with plain green, stalked, lanceolate leaves. A beautiful white spathe, 7-25 cm (3-10 in) long, enfolds the base of the spadix and has a recurved margin. The male flowers occupy the upper part of the spadix and the female flowers are lower down. The species is cultivated easily in wet conditions. Place of origin: South Africa; introduced into Europe in 1731.

Zanthoxylum A genus of Trees of the Rue family, *Rutaceae*. e.g. Z. americanum - The Northern Prickly Ash; from the dried bark is prepared Zanthoxylum (= Toothache Bark; Xanthoxylum), used as a carminative. The Prickly Ash Berry or Zanthoxylum Fruit is the dried pericarp of the ripe fruit of the Prickly Ash. Place of origin: North America.

Zapus A genus of Mammals of the Jumping Mouse family, *Zapodidae*. The species are found in North America, chiefly in Canada and the northern United States.

Zea A genus of plants of the Grass family, *Gramineae*. e.g. Z. mays - Corn, Indian Corn or Maize, a coarse, erect, annual plant with prop-roots and large leaves. The male flowers ae produced ina branched terminal inflorescence, the tassel and the female flowers are borne on a short axillary axis protected by large bracts. The maizecob consists of the axis with the fruits attached. There are several kinds: 1. Dent corn - a highly productive variety which has a hard grain with a soft patch. 2. Flint corn - with hard grains. 3. Flour corn - with soft grains. 4. Popcorn - which bursts when heated. 5. Sweet corn - which is sugary. Maize is grown to feed man and animal stock; it yields various oils, such as Maize Oil (= Oleum Maydis; Ol Mayd; Corn Oil), Bourbon whisky, syrups, and building materials. Because of its high content of glycerides of unsaturated acids, maize oil is used in diets intended to reduce high levels of blood cholesterol. Distribution: worldwide. Place of origin: unknown in the wild; may have originated from the Andes region of northwest South America.

Zebrasoma A genus of Fishes of the Surgeonfish family, *Teuthididae*.e.g. *Z. xanthurus* - The Blue Tang, a dark-blue Surgeonfish, distinguished by a horizontal, retractable spine as sharp as a surgeon's scalpel at the base of the bright yellow tail fin, which is used for defence.

Zebrina 1. A genus of Gastropod Molluscs or Land Snails, of the Oleacid family, *Enidae*. e.g. *Z. detrita* - Shell 10 mm wide, 21 mm high, thick, glossy-white with brown lengthwise stripes and 6-7 whorls. Widespread in hedgerows and on calcareous soils, such as walls, stones, etc. Lives in dry regions of the European Mediterranean, Central Germany, and as far east as northwestern Iran. 2. A genus of plants of the Spiderwort family, *Commelinaceae*. e.g. *Z. pendula* - The Wandering Jew, a plant originating in Mexico, with trailing, rooting stems and variegated leaves. A commonly grown house plant; occurs outdoors in Florida.

Zeledonia A genus of Birds of the Babbler and Thrush family *Muscicapidae* (of the Wren-thrush subfamily, *Zeledoniinae*). e.g. *Z. coronata* - The Wren-thrush. It lives in the deep forests of Costa Rica and western Panama, 1,800-3,000 m (6,000-10,000 ft) above sea-level.

Zelkova A genus of Trees of the Elm family, *Ulmaceae*. e.g. *Z. serrata* - An elm tree species from Japan, with serrated leaves and wingless fruits.

Zenkerella A genus of Mammals of the Scaly-tail Squirrel family, *Anomaluridae*. The Flightless Scaly-tail rodent species are very rare, and are known only from a few specimens found in southern Cameroun. They are the size of large Dormice, pure grey in color, and have no gliding membrane.

Zephyranthes A genus of plants of the Daffodil family, *Amaryllidaceae*. e.g. *Z. carinata* (= *Z. grandiflora*) - The Fairy Lily, Flower of the West Wind, Rain Lily, or Zephyr Lily; a pretty little bulbous plant, 12.5-25 cm (5-10 in) high, with narrowly strap-shaped leaves and bell-shaped, solitary, pink flowers with golden anthers (lobed upper part of stamen where pollen is produced), growing in a hollow stem. Place of origin: West Indies; introduced into Europe in 1824.

Zerynthia A genus of Insects of the Butterfly or Moth family, *Papilionidae*. e.g. *Z. hypsipyle* - A beautiful species of butterflies, yellowish-black in color, with black spots on the underside of both pairs of wings and the top of the hind pair. Found in damp woods, dry sandy lowlands, and vineyards. Distribution: southeastern Europe.

Zeuglodon A genus of extinct Mammals, the Whales, better known by the generic name *Basilosaurus* (see **Basilosaurus**).

Zeus A genus of Fishes of the John Dory family, *Zeidae*. e.g. *Z. faber* - The John Dory, 70 cm (28 in) long and weighing about 9 kg (20 lb); the spiny dorsal fin is very well-developed, the mouth is large, and just above the back end of the pectorals is a black spot, which legend purports to be the thumb-print of St. Peter, supposedly made as he handled one of these fishes to withdraw a coin from its mouth! Locally a popular and valuable market fish. Distribution: extremely common in the Mediterranean. Native around Britain and Ireland.

474

Zeuzera A genus of Insects of the Goat Moth family, *Cossidae*. e.g. Z. pyrina - The Leopard Moth, size 18-35 mm; male is smaller than the female, the wings are spotted like the skin of the leopard. Distribution: southern and central Europe, temperate regions of Asia, North Africa.

Zingiber A genus of plants of the Ginger family, *Zingiberaceae*. e.g. Z. ottensii - A rhizomatous herb; the fruit is a capsule or berry.

Zinnia A genus of plants of the Daisy family, *Compositae*. e.g. Z. elegans - Cut and Come Again, Youth and Old Age, or Garden Zinnia; a colorful annual plant, from 7.5-90 cm (3 in - 3 ft) high, with ovate leaves, and flowers in large heads from 5-12.5 cm (2-5 in) across, colored white, red, yellow, violet, orange, crimson, etc., single and in doubles. Place of origin: Mexico; introduced into Europe in 1796.

Ziphius A genus of Mammals of the Beaked and Bottle-nosed Whale family, *Ziphiidae*. e.g. Z. cavirostris - Cuvier's Beaked Whale, length 6-9 m (20-30 ft); the snout is elongated into a beak, and all the teeth are in the lower jaw. Distribution: worldwide.

Zirfaea A genus of Bivalve Molluscs of the Piddock family, *Pholadidae*. e.g. Z. crispata - The Oval Piddock; length of shell up to 8 cm; front of valve shiny, hind part concentrically striped. Found on the Atlantic and West Baltic coasts, boring dwelling-holes or tubes in wood, chalk, peat or firm clay.

Zoarces A genus of Fishes of the Eelpout family, *Zoarcidae* (Blenny suborder, *Blennioidea*). e.g. Z. viviparus - The European Eelpout or Viviparous Blenny, a viviparous species of blenny fishes, length 45 cm (18 in); it has 10-20 oblique bands along the dorsal fin and back. Habitat: the seabottom in seaweed areas, or brackish waters near the coast, from Scandinavia to the Bay of Biscay. Native around Britain.

Zonaginthus A genus of Birds of the Grassfinch, Mannikin, and Waxbill family, *Estrildidae*. e.g. Z. oculatus - The Red-eared Firetail, one of the small seed-eating weaver finches of Australia. The nestlings have conspicuous markings on the palate and tongue, displayed when they beg for food.

Zoogloea A genus of Microorganisms of the family *Pseudomonadaceae*; occurring as rod-shaped cells embedded in a gelatinous matrix. e.g. Z. filipendula.

Zootermopsis A genus of Insects (Termite order, *Isoptera*), of the family *Hodotermitidae* - The rotten-wood termites of the Pacific coast.

Zopfius A former genus of Bacteria, now called *Kurthia*. e.g. Z. zenkeri (= *Kurthia* z.) (see **Kurthia**).

Zorotypus A genus of Insects of the order *Zoraptera*. The species are minute insects living under bark, rotten wood, or associated with termite nests. The adults exist in both winged and apterous (wingless) forms.

Zostera A genus of plants of the Eel-grass family. e.g. Z. marina - The Common Eel-grass, Grass-wrack, or Sea-wrack, a perennial marine plant, 30-40 cm high, with branched flowering stems, leaves 3-7 mm wide, and entire sheaths. Grows on muddy sea-shores. Distribution: Europe; native in Britain.

Zosterops A genus of Birds of the White-eyes family, *Zosteropidae*. e.g. Z. lateralis - The Silvereye or White-eyes bird of Australia; it has a distinctive ring of minute white feathers round the eyes. The bill is short, pointed and slightly decurved. The species is gregarious and arboreal, building its nest of vegetable material in the fork of a tree.

Zuberella A genus of gram-negative Anaerobic bacteria.

Zygadenus A genus of plants. e.g. Z. intermedius - A species yielding the crystalline alkaloid, zygadenine ($C_{39}H_{63}NO_{10}$).

Zygaena A genus of Insects of the Burnet

and Forester Butterfly or Moth family, *Zygaenidae*. e.g. Z. exulans - The European Mountain Burnet-moth, size 3-3.2 cm; color attractive red and black; forewings have five carmine spots. A diurnal insect, flying about on dry hillsides in high summer. Distribution: the palaearctic region; native in Britain.

Zygnema A genus of Green algae plants, the *Chlorophyta*. e.g. Z. circumcarinatum - This species of green filamentous algae has unbranched rows of cells, each containing two stellate, green chromoplasts.

Zygocotyle A genus of Trematode Worms. e.g. Z. ceratosa - A trematode parasite found in the intestine of ducks in North America.

Zygoena A genus of Fishes of the Sand Shark family *Carchariidae*. e.g. Z. tudes - A Sand Shark species; from the fish or carefully preserved livers is obtained Shark-liver Oil (= Oleum Selachoidei), containing not less than 6,000 Units of Vitamin A per gram.

Zymobacterium A genus of microorganisms of the family *Propionibacteriaceae*. Non-spore-forming, anaerobic to microaerophilic, gram-positive bacilli, non-pathogenic but occurring as parasites in the intestinal tract.

Zymogenes A genus of Enterococci, now included under Group D of the β-hemolytic Streptococci (see **Streptococcus**).

Zymonema A genus of Fungi or Molds of the family *Eramascaceae*, also known by other generic names such as *Blastomyces*, *Candida*, *Endomyces*, etc. Infestation with this fungus causes the disease known as Zymonematosis. e.g. Z. albicans (= *Endomyces* a.) (see **Endomyces**). Z. capsulatum (= Z. dermatitidis; Z. gilchristi; *Blastomyces* dermatitidis) (see **Blastomyces**). Z. farciminosum - A species causing epizootic lymphangitis in horses (see **Blastomyces**).

Zymononas A genus of microorganisms of the family *Pseudomonadaceae*; occurring as rod-shaped or ellipsoidal cells in fermenting beverages. e.g. Z. anaerobia.

Glossary

A

Abelia; Abelpharus; Abies; Abietinaria; Abra; Abramis; Abraxas; Abroma; Abrostola; Abrus; Absidia; Abudefduf; Abutilon; Acacia; Acalypha; Acanthamoeba; Acantharus; Acanthaster; Acanthia; Acanthinula; Acanthisittta; Acanthocardia; Acanthocephalus; Acanthochaenus; Acanthocheilonema; Acanthocinus; Acanthocybium; Acanthodactylus; Acantholienon; Acanthophacetus; Acanthophis; Acanthopis; Acanthopsetta; Acanthoptilum; Acanthozostera; Acanthus; Acarapis; Acarus; Accipeter; Accipiter; Acentropus; Acer; Aceraria; Aceras; Acerina; Acestrura; Acetobacter; Achatina; Acherontia; Acheta; Achillea; Achimenes; Achirus; Achnatherum; Achorion; Achras; Achromobacter; Achyranthes; Acilius; Acineta; Acinetobacter; Acinonyx; Acinos; Acipenser; Aciptilia; Aclypea; Acmaea; Acneta; Acocanthera; Acokanthera; Acomys; Aconitum; Acorus; Acrida; Acris; Acrobates; Acrocephalus; Acrocinus; Acronicta; Acronycta; Acropora; Acrosiphonia; Acrotus; Acryllium; Actaea; Actaeon; Actias; Actinia; Actinobacillus; Actinodiscus; Actinomyces; Actinophrys; Actinoptychus; Actinosphaerium; Actitis; Actophilornis; Acuaria; Aculeatus; Adalia; Adamsia; Adansonia; Addax; Addox; Adelges; Adenphora; Adenostyles; Adhatoda; Adiantum; Adioryx; Adleria; Adonis; Adoxa; Adoxophyes; Aechmea; Aechmophorus; Aedes; Aegilops; Aegithalos; Aegle; Aegolius; Aegopodium; Aegosoma; Aegypius; Aegyptianella; Aelia; Aelosoma; Aeonium; Aepyceros; Aepyornis; Aequipecten; Aerobacter; Aeromonas; Aeronautes; Aeschna; Aeschrion; Aesculus; Aeshna; Aethia; Aethionema; Aethopyga; Aethusa; Afrixalus; Afrosubulitermes; Agabus; Agalma; Agama; Agamodistomum; Agamofilaria; Agamomermis; Agamonematodium; Agapanthus; Agapornis; Agaricus; Agathis; Agathosma; Agave; Agelaius; Agelastica; Agenus; Ageratum; Agkistrodon; Aglaiocercus; Aglais; Aglaophenia; Aglivis; Agonum; Agonus; Agrilus; Agrimonia; Agriocharis; Agriocnemis; Agrion; Agriotes; Agriotypus; Agrobacterium; Agromyza; Agropyron; Agropyrum; Agrostemma; Agrostis; Agrotis; Agrumaenia; Agrypnus; Aguila; Agulla; Agyroneta; Ahaetulla; Ailanthus; Ailantus; Ailuropoda; Ailurus; Aiptasia; Aira; Aix; Ajaia; Ajuga; Akbaria; Alabes; Alaria; Alauda; Albizia; Albula; Albulina; Alburnoides; Alburnus; Alea; Alcaligenes; Alcea; Alcedo; Alcephalus; Alces; Alchemilla; Alcippe; Alcyonidium; Alcyonium; Aldrovanda; Alectis; Alectoria; Alectoris; Alectra; Alepisaurus; Aletris; Alextoria; Alginobacter; Alginomonas; Alisma; Alkalescens-Dispar; Alkaligenes; Alkanna; Allactaga; Allenopithecus; Allescheria; Alliaria; Alligator; Allium; Allolobophora; Alloteuthis; Almyracuma; Alnus; Alocimua; Aloe; Aloides; Alonella; Alonopsis; Alopecosa; Alopecurus; Alopercurus; Alopex; Alopias; Alopius; Alopochen; Alosa; Alouatta; Aloysia; Alpinia; Alsophila; Alstoma; Alstroemeria; Althaea; Althea; Alucita; Alyssum; Alytes; Amanita; Amaranthus; Amaroucium; Amaryllis; Amaurobius; Amazona; Amblyglyphidodon; Amblyomma; Amblyopus; Amblyorhynchus; Amblyosomus; Amblystoma; Ambystoma; Ameba; Ameiurus; Amelanchier; Amia; Amidostomum; Amitermes; Ammi; Ammocrypta; Ammodorcas; Ammodytes; Ammomanes; Ammoperdix; Ammophila; Ammospermophilus; Ammotragus; Amoeba; Amoebobacter; Amoebotaenia;

Amoebus; Amomum; Amorpha;
Amorphophallus; Amphibolurus;
Amphileptus; Amphilina; Amphimallon;
Amphimallus; Amphimerus;
Amphiophiura; Amphioxus; Amphipholis;
Amphipnous; Amphiprion; Amphisbaena;
Amphistoma; Amphitirte; Amphitrite;
Amphiuma; Amphiura; Amygdalus;
Anabas; Anableps; Anabrus; Anacamptis;
Anacardium; Anacridium; Anacyclus;
Anagallis; Anagasta; Anagyris; Anaimos;
Anajapx; Analges; Anamirta; Anampses;
Ananas; Anaplasma; Anarhicas;
Anarhynchus; Anarrhinum; Anas; Anasa;
Anaspides; Anastomus; Anatis; Anax;
Anchusa; Ancistrocactus; Ancistrodon;
Ancylostoma; Ancylus; Anda; Andira;
Andrena; Andricus; Androctonus;
Andromeda; Andropogon; Androsace;
Anemone; Anemonia; Anethum; Angelica;
Angiococcus; Angiopteris;
Angiostrongylus; Anguilla; Anguillula;
Anguillulina; Anguina; Anguis;
Anhalonium; Anhima; Anhinga;
Anhydrophryne; Anisomorpha; Anicoplia;
Anisotremus; Ankylostoma; Annarrhicas;
Anoa; Anobium; Anodonta; Anomala;
Anomalurus; Anomia; Anomma;
Anonymus; Anopheles; Anoplius;
Anoplocephala; Anous; Anser; Anseranas;
Anseropoda; Antalis; Antedon;
Antennaria; Antennarius; Antestia;
Anthaxia; Anthemis; Antheraea;
Anthericum; Anthias; Anthobothrium;
Anthocharis; Anthocoris; Anthomyia;
Anthonomus; Anthophora; Anthornis;
Anthoxanthium; Anthrenus; Anthreptes;
Anthriscus; Anthrophora; Anthropoides;
Anthurium; Anthus; Anthyllia; Anthyllis;
Anthyopoides; Antiaris; Antidorcas;
Antilocapra; Antilope; Antipataria;
Antipathes; Antirrhinum; Anurida;
Aoliscus; Aonyx; Aotes; Aotus; Apamea;
Apanteles; Apatele; Apatosaurus; Apatura;
Apeltes; Apera; Aphanes;
Aphanizomenon; Aphelandra;
Aphelenchus; Aphelocheirus; Aphia;
Aphiochaeta; Aphodius; Aphredoderus;
Aphriza; Aphrocallistes; Aphrodite;

Aphrophora; Apis; Apium; Aplocera;
Aplocheilichthys; Aplodinotus;
Aplodontia; Aploppus; Aplysia;
Apodemus; Apoderus; Aporia;
Aporocactus; Aporrectodea; Aporrhais;
Aposeris; Aprocta; Aptenodytes;
Apterona; Apterygiformes; Apteryx; Apus;
Aquarius; Aquila; Aquilegia; Ara;
Arabidopsis; Arabis; Arachis;
Arachnocampa; Arachnoidiscus;
Arachnospermum; Aralia; Aramides;
Aramus; Araneaus; Araneus; Arapaima;
Araschnia; Araucaria; Arbutus; Arca;
Arcella; Archerpeton; Archichauliodes;
Archidiskodon; Archilochus; Architeuthis;
Archulochus; Arctia; Arctica; Arctictis;
Arctium; Arctocebus; Arctocephalus;
Arctogadus; Arctomys; Arctonoe;
Arctonyx; Arctostaphylos; Arctous;
Arcyria; Ardea; Ardeola; Ardeotis;
Ardisia; Arduenna; Areca; Aremonia;
Arenaria; Arenicola; Argas; Argema;
Argiope; Argonauta; Argulus; Argusianus;
Argynnis; Argyrodes; Argyroneta;
Argyropelecus; Argyrosomus; Arianta;
Arion; Ariosoma; Arisaema; Arisarum;
Aristeomorpha; Aristolochia; Arius;
Arixenia; Arizona; Armadillidium;
Armeniaca; Armeria; Armigeres;
Armillaria; Armillifer; Armoracia; Arnica;
Arnoglossus; Arnoseris; Aromia;
Arothron; Arrhenatherum; Artamus;
Artemia; Artemisia; Arthroderma;
Arthrographis; Artocarpus;
Artyfechinostomum; Arum; Aruneus;
Arundo; Arvicola; Asarum; Ascalaphus;
Ascaphus; Ascaris; Ascidiella; Asclepias;
Ascocotyle; Ascophyllum; Asellus; Asilus;
Asimina; Asio; Asopia; Asparagus;
Aspergillus; Asperugo; Asperula;
Asphodeline; Asphodelus; Aspidelaps;
Aspidistra; Aspidogaster; Aspidontus;
Aspidosperma; Aspius; Asplenium; Aspro;
Astacilla; Astacopsis; Astacus; Astarte;
Astasia; Aster; Asteracantha; Asterias;
Asterina; Asterionella; Asterococcus;
Asthenosoma; Astilbe; Astracopsis;
Astraea; Astragalus; Astrangia; Astrantia;
Astroboa; Astrococcus; Astropecten;

Astrophytum; Astroscopus; Astrotoma; Astylosternus; Ateleopus; Ateles; Atelopus; Athena; Athene; Atherestes; Atherina; Atherinopsis; Atherix; Atherosperma; Atherurus; Athous; Athripsodes; Athyrium; Atilax; Atlanta; Atlantisea; Atlantisia; Atolla; Atractaspis; Atrax; Atrichornis; Atriplex; Atropa; Atta; Attacus; Attagenus; Attalea; Attaphila; Attelabus; Atyaephyra; Aubrieta; Aubrietia; Auchmeromyia; Aucuba; Aurelia; Aurinia; Australopithecus; Austrocedrus; Austrofundilis; Austrophlebia; Autographa; Autolytus; Automeris; Avahi; Avena; Avicularia; Axine; Axiolotus; Axis; Aythya; Azadirachta; Azolla; Azotobacter; Azotomonas.

B

Babesia; Bacillus; Bacteriodes; Bacterionema; Bacterium; Bacteroides; Bactoscilla; Balaena; Balaeniceps; Balaenophilus; Balaenoptera; Balaninus; Balanocarpus; Balanoglossus; Balanophyllia; Balantidium; Balanus; Balbiana; Balearica; Balistes; Balistipus; Balsamodendron; Baluchitherium; Bandicota; Banisteria; Baptesia; Barbastella; Barbourula; Barbus; Barnea; Barosma; Bartonella; Bartramia; Basidiobolus; Basilosaurus; Bassariscus; Bassia; Bassogigas; Bathycrinus; Bathyergus; Bathynomus; Bathypterois; Batocera; Batrachoseps; Batrachuperus; Battey Bacillus; Bax; Bdella; Bdellostoma; Bedsonia; Begonia; Belamcanda; Belascaris; Bellis; Belone; Beloperone; Benaratherium; Beneckea; Benthamia; Bentheuphausia; Berardius; Berberis; Bergenia; Beroe; Bertiella; Beryx; Bethesda; Betonica; Betta; Betula; Bibio; Bibos; Bielzia; Bifidobacterium; Bigelovia; Bilharzia; Billbergia; Biorhiza; Birgus; Bison; Bispira; Biston; Bitis; Bitoma; Bittium; Blaniulus; Blarina; Blastocaulis; Blastocerus; Blastocystis;

Blastmyces; Blastophaga; Blatta; Blatella; Blennius; Blepharidopterus; Bletia; Blicca; Blicca; Boa; Boaedon; Bodo; Boerhaavia; Boiga; Bolina; Bolinopsis; Bolitoglossa; Bombina; Bombus; Bombycilla; Bombylius; Bombyx; Bonasa; Bonellia; Boocercus; Boophilus; Borassus; Bordetella; Boreogadus; Boreus; Borophyrne; Borrelia; Bos; Boselaphus; Boswellia; Botaurus; Bothops; Bothriocephalus; Bothrioplana; Bothrops; Botia; Botrychium; Botryllus; Botrytis; Bougainvillea; Bourletiella; Bouvardia; Bovimyces; Brachinus; Brachionus; Brachiosaurus; Brachycephalus; Brachymystax; Brachyplatystoma; Bradicebus; Bradypus; Bragada; Branchinecta; Branchiostoma; Branhamella; Branta; Brassica; Brayera; Brephidium; Breviceps; Brevoortia; Brodiaea; Brontosaurus; Brosme; Brucea; Brucella; Bruchus; Brugmansia; Brunfelsia; Bryobia; Bryonia; Bubalus; Bubo; Buccinum; Bucephala; Bucephalus; Buceros; Buddleia; Buddleja; Budorcas; Bufo; Bugula; Buistia; Bulinus; Bulla; Bulweria; Bungarus; Bupalus; Buphagus; Buprestis; Burhinus; Burretis; Busycon; Butea; Buteo; Buthacus; Buthus; Byctiscus; Byrrhus; Bythnia; Byturus.

C

Cabassous; Cacatua; Cacajao; Cactoblastis; Caenorhabditis; Caesalpinia; Caiman; Cairina; Cakile; Calamagrostis; Calamintha; Calamoichthys; Calamus; Calandrella; Calanus; Calathea; Calcarius; Calceolaria; Caldesia; Calendula; Caleonas; Calepina; Calicotyle; Calidris; Caliroa; Calla; Callaeas; Calliactis; Callianira; Callianthemum; Callicebus; Callichthys; Callidium; Callimorpha; Callinectes; Callionomys; Callionymus; Calliostoma; Callipepla; Calliphora; Calliptamus; Callista; Callista; Callistemon; Callistephus; Callithrix; Callitriche; Callitroga; Callocalia;

Callophyrs; Callorhinus; Callorhynchus; Calluna; Calocedrus; Calopteryx; Calosoma; Calotermes; Calotes; Calothamnus; Calotropis; Caltha; Calures; Calycanthus; Calymmatobacterium; Calyptomena; Calyptraea; Calystegia; Calyx; Cambarellus; Cambarus; Camelina; Camellia; Camelus; Campanula; Campephaga; Campodea; Camponotus; Campostoma; Campsis; Camptoptera; Campylobacter; Cananga; Canarium; Cancer; Candida; Canis; Canella; Canna; Cannabis; Cantharellus; Cantharis; Canthocamptus; Capella; Capillaria; Capparis; Capra; Caprella; Capreolus; Capricornis; Caprimulgus; Capromys; Capsella; Capsicum; Capulus; Carabus; Caracal; Caracara; Caralluma; Caranx; Caraproctus; Carapus; Carassius; Carausius; Carbenia; Carcharhinus; Carcharias; Carcharodon; Carcinus; Cardamine; Cardaminopsis; Cardaria; Cardium; Carduelis; Carduus; Caretta; Carex; Cariama; Carica; Carlina; Carludovica; Carpesium; Carpinus; Carpiodes; Carpocapsa; Carpodacus; Carterocephalus; Carpoglyphus; Carum; Carya; Caryophyllia; Caryophyllus; Caryopteris; Carysomyia; Casarca; Casmerodius; Caspiomyzon; Cassia; Cassida; Cassidaria; Cassiope; Cassiopeia; Cassis; Castanea; Castellanella; Castor; Casuarina; Casuarius; Catabrosa; Catalpa; Catamblyrhynchus; Catasetum; Catastomus; Catha; Catharacta; Catharanthus; Catharsius; Cathartes; Catla; Catocala; Catoptrophorus; Cattleya; Caucalis; Caudamoeba; Caularchus; Caulobacter; Caulophyllum; Causus; Cavia; Cavochalina; Cebuella; Cebus; Cecropia; Cecropis; Cecrops; Cedrus; Ceiba; Celerio; Cellfalcicula; Cellvibrio; Celosio; Cenolophium; Centaurea; Centaurium; Centella; Centetes; Centranthius; Centriscus; Centrocercus; Centrocestus; Centrolenella; Centronotus; Centropelma; Centropristes; Centropus; Centrostephanus; Centruroides; Cenurus; Cepaea; Cephaelis; Cephalanthera;

Cephalodiscus; Cephalopterus; Cephaloscyllium; Cephalosporium; Cephalothrix; Cephenomyia; Cephus; Cepphus; Cerambyx; Ceramium; Cerastes; Cerastium; Cerastoderma; Ceratitis; Ceratium; Ceratodus; Ceratonia; Ceratophrys; Ceratophyllum; Ceratophyllus; Ceratostigma; Ceratotherium; Ceratozamia; Cerbera; Cerceris; Cercis; Cercocebus; Cercomela; Cercomonas; Cercopis; Cercopithecus; Cercosphaera; Cercosporalla; Cercotrichas; Cerdocyon; Cerebratulus; Cereopsis; Cereus; Cerianthus; Cerinthe; Cerithium; Ceropegia; Certhia; Cerura; Cervus; Ceryle; Cestrum; Cestum; Cetarach; Cetengraulis; Ceterach; Cetomimus; Cetonia; Cetorhinus; Cetraria; Cettia; Ceuthophilus; Chaenomeles; Chaenorhinum; Chaerophyllum; Chaetodipterus; Chaetodon; Chaetogaster; Chaetonotus; Chaetophractus; Chaetopterus; Chaetorhynchus; Chaetura; Chalcides; Chalcites; Chalcophora; Chalcosoma; Chamaea; Chamaebuxus; Chamaecyparis; Chamaedaphne; Chamaedorea; Chamaeleo; Chamaemelum; Chamaenerion; Chamaeorchis; Chamaepericlymenum; Chamaerops; Chamaesaura; Chamaespartium; Chaos; Charabdea; Charadrius; Chasmorhynchus; Chauliodus; Chaulmoogra; Chauna; Cheilanthes; Cheilomastix; Cheilotrema; Cheiranthus; Cheirogaleus; Chelidonium; Chelifer; Chelodina; Chelon; Chelonia; Chelydra; Chelys; Chenopodium; Cheyletiella; Chiasmodon; Chiasognathus; Chilocorus; Chilodon; Chilodonella; Chilomastix; Chilomonas; Chilomycterus; Chimbacche; Chimaera; Chimaphila; Chimonanthus; Chinchilla; Chionactis; Chionaspis; Chionis; Chirocephalus; Chirolophis; Chironectes; Chironomus; Chiropotes; Chiropsalmus; Chiropterotriton; Chiroxiphia; Chiton; Chlamydia; Chlamydomonas; Chlamydophrys; Chlamydosaurus; Chlamydoselachus; Chlamydotis; Chlamyphorus; Chlamys;

Chlidonias; Chloephaga; Chloris;
Chlorobacterium; Chlorobium;
Chloroceryle; Chlorochromatium;
Chlorohydra; Chlorophonia; Chloropsis;
Chlorostigma; Choanotaenia; Choeropsis;
Choreopus; Choiromyces; Choisia;
Choleopus; Chologaster; Chondrodendron;
Chondrilla; Chondromyces; Chondrosteus;
Chorisochismus; Choristoneura;
Chorthippus; Chortoicetes;
Chromobacterium; Chromulina;
Chrosomus; Chrysanthemum; Chrysaora;
Chrysemys; Chrysis; Chrysoara;
Chrysochloris; Chrysococcyx;
Chrysocyon; Chrysolophus; Chrysomela;
Chrysomitra; Chrysomyia; Chrysopa;
Chrysopelea; Chrysophanus; Chrysophris;
Chrysopogon; Chrysops; Chrysospalax;
Chrysosplenium; Chrsozona;
Chthonerpeton; Chthonius; Chunga;
Cicada; Cicadella; Cicadetta; Cecendia;
Cicerbita; Cichlasoma; Cichorium;
Cicindela; Ciconia; Cicuta; Ciliata;
Cimbex; Cimex; Cimicifuga; Cinchona;
Cinclus; Cinnamomum; Ciona; Cionus;
Circaea; Circaetus; Circulifer; Circus;
Cirratulus; Cirsium; Cistus; Citellus;
Citrobacter; Citrullus; Citrus; Civettictis;
Cladium; Cladognathus; Cladonia;
Cladorchis; Cladorhynchus;
Cladosporium; Clamator; Clangula;
Clatochloris; Clathrina; Clathrocystis;
Clathrulina; Clathrus; Clausilia; Clava;
Clavaria; Clavelina; Claviceps; Clelia;
Clematis; Clemmys; Cleome;
Clerodendrum; Clethrionomys;
Clevelandella; Climacia; Climacteris;
Clinocoris; Clinophilus; Clinopodium;
Cliona; Clivia; Cloeon; Clonorchis;
Clonothrix; Clossiana; Clostridium;
Clupea; Clymenella; Clytia; Clytocybe;
Clytus; Cnemidocoptes; Cnicus; Cnidium;
Cobaea; Cobitis; Coccidioides;
Coccidium; Coccinella; Coccobacillus;
Coccothraustes; Coccus; Coccytius;
Coccyzus; Cochlearia; Cochlearius;
Cochliomyia; Cochlodina; Cocos;
Codiaeum; Codium; Coelacanthus;
Coelagyne; Coeloglossum; Coeloplana;

Coelosomides; Coenagrion; Coendou;
Coenogonimus; Coenonympha; Coenurus;
Coffea; Cola; Colacium; Colaptes;
Colchicum; Coleonyx; Colesiota;
Colettsia; Coleus; Colias; Colinus;
Colletes; Collinia; Collocalia; Collotheca;
Colobocentrotus; Colobus; Colocasia;
Colopoda; Coluber; Columba;
Columbigallina; Columnea; Colutea;
Colymbetes; Colymbus; Combretum;
Commiphora; Compsomyia; Condylura;
Conger; Conioselinum; Conium;
Conochilus; Conocybe; Conolophus;
Conophytum; Conorhinus; Conringia;
Constrictor; Contopus; Conuropsis;
Conus; Convallaria; Convoluta;
Convolvulus; Conyza; Cooperia;
Copaifera; Copeina; Copernicia; Coprinus;
Copris; Copromastix; Copromonas;
Coptis; Coracia; Coracias; Coracina;
Coragyps; Corallina; Corallium;
Corallorhiza; Corallus; Corchorus;
Corcorax; Cordulegaster; Cordulia;
Cordyline; Cordylobia; Cordylophora;
Cordylus; Coregonus; Coreopsis;
Corethra; Coriandrum; Coris; Corixa;
Cornufcr; Cornus; Coronella; Coronilla;
Coronopus; Coronula; Corophium;
Corrigiola; Cortaderia; Cortusa; Corvus;
Coryaria; Corycaeus; Corydalis;
Corydalus; Corylus; Corymbites;
Corymorpha; Corynactis; Corynanthe;
Coryne; Corynebacterium; Corynephorus;
Coryphaena; Coryphella; Coscinasterias;
Coscinia; Coscinium; Coscinodiscus;
Coscinoscera; Coscoroba; Cosmarium;
Cosmos; Cosmotriche; Cossus; Cothurnia;
Cotoneaster; Cottus; Cotula; Coturnix;
Cotylaspis; Cotyledon; Cotylogaster;
Cotylogasteroides; Coua; Councilmania;
Coura; Cowdria; Coxiella; Crabro;
Cracticus; Craigia; Crambe; Crangon;
Craspedacusta; Craspedella; Crassostrea;
Crassula; Crataegus; Craterellus;
Craterolophus; Crax; Crenilabrus;
Crenothrix; Creophilus; Crepidula; Crepis;
Crex; Cricetomys; Cricetulus; Cricetus;
Crinia; Crinitaria; Crinum; Crisia;
Cristatella; Cristispira; Cristivomer;

Crithidia; Crocethia; Crocidura;
Crocodylus; Crocosmia; Crocus; Crocuta;
Crossandra; Crossomys; Crossoptilon;
Crossoptilum; Crotalus; Crotaphytus;
Croton; Crotophaga; Cruciata; Crynchia;
Cryphalus; Cryptanthus; Cryptobranchus;
Cryptocellus; Cryptocephalus;
Cryptocercus; Cryptochiton;
Cryptococcus; Cryptogamma;
Cryptolucilia; Cryptoprocta; Cryptotis;
Crypturellus; Crystalogobius;
Ctenocephalides; Ctenocephalus;
Ctenodactylus; Ctenolabrus;
Ctenolepisma; Ctenomys;
Ctenophthalmus; Ctenophyryngodon;
Ctenopsylla; Ctenopsyllus; Ctenosaura;
Ctinocephalus; Cucubalus; Cuculia;
Cuculus; Cucumaria; Cucumis; Cucurbita;
Culex; Culicoides; Cultellus; Cuminum;
Cuniculus; Cunila; Cunina;
Cunninghamia; Cuon; Cupressus;
Curculio; Curcuma; Cursorius; Cuscuta;
Cuspidaria; Cyanea; Cyanerpes; Cyaniris;
Cyanocitta; Cyanocorax; Cyanopica;
Cyanopsis; Cyanosylvia; Cyathea;
Cyathura; Cybister; Cycas; Cychrus;
Cyclamen; Cycleptus; Cyclestheria;
Cyclopes; Cyclops; Cycloptera;
Cyclopterus; Cyclothone; Cydia; Cydonia;
Cygnopsis; Cygnus; Cylindrogloea;
Cymbalaria; Cymbidium; Cymbopogon;
Cynara; Cynips; Cynocephalus; Cynodon;
Cynogale; Cynomys; Cynoglossum;
Cynopithecus; Cynops; Cynoscion;
Cynosurus; Cynoxylon; Cynthia; Cyperus;
Cypraea; Cyprinodon; Cyprinus;
Cypripedium; Cypris; Cypselurus;
Cypsilurus; Cyrtodactylus; Cyrtonyx;
Cysticercus; Cystomonas; Cystophora;
Cystopteris; Cytherella; Cytisus;
Cyotospermium.

D

Daboia; Dacelo; Dactylis; Dactylochalina;
Dactylogyrus; Dactylopius; Dactylopterus;
Dactylorchis; Dactylorhiza; Dacus;
Daedalea; Daedolacanthus; Dahlia;

Dalechampia; Dallia; Dama; Dammara;
Danaus; Daphne; Daphnia; Daphnis;
Dascyllus; Dasogale; Dasyatis; Dasychira;
Dasypeltis; Dasyphyllia; Dasypoda;
Dasyprocta; Dasypus; Dasyurops;
Dasyurus; Datura; Daubentonia; Daucus;
Davainea; Davidia; Decticus; Deilephila;
Deinocheirus; Deinosuchus; Delesseria;
Delia; Delichon; Delospermum;
Delphinapterus; Delphinium; Delphinus;
Demodex; Demoixys; Dendroaspis;
Dendroaster; Dendrobates; Dendrobium;
Dendrocalamus; Dendrochilum;
Dendrocoelum; Dendrocopos;
Dendrocygna; Dendrodoa; Dendrogyra;
Dendrohyrax; Dendroica; Dendrolagus;
Dendrolimus; Dendromus;
Dendronephyta; Dendronessa;
Dendrostoma; Dentalium; Dentex;
Deporaus; Deraeocoris; Dermacentor;
Dermacentroxenus; Dermanyssus;
Dermatobia; Dermatocentor;
Dermatophagoides; Dermatophilus;
Dermestes; Dermochelys; Deroceras;
Derocheilocaris; Deschampsia;
Descurainia; Desmacidon; Desmana;
Desmodus; Desmognathus; Desulfovibrio;
Desvoidea; Deutzia; Diadema;
Diadumene; Diaea; Dialister; Diamanus;
Dianthus; Diapensia; Diapheromera;
Diaphus; Diaptomus; Dibothriocephalus;
Dibranchus; Dicamptodon; Dicentra;
Diceromonas; Dicerorhinus; Diceros;
Dichonemertes; Diclodophora;
Dicranocephalus; Dicranum; Dicranura;
Dicrocoelium; Dicrurus; Dictamnus;
Dictyoptera; Didelphis; Didinium;
Didunculus; Dieffenbachia; Dielytra;
Diemictylus; Diendamoeba; Dientamoeba;
Diervilla; Difflugia; Digaster; Digenea;
Digitalis; Digitaria; Digramma; Dileptus;
Diloba; Dimastigamoeba; Dimorphotheca;
Dinomys; Dinoponera; Dinornis;
Dioctophyma; Dioctophyme; Diodon;
Diodora; Diogenes; Diomedea; Dioon;
Dioscorea; Dioscoreophyllum;
Dipetalonema; Diphylla;
Diphyllobothrium; Diplacanthus;
Diplobacillus; Diplobacterium;

Diplococcus; Diplodocus; Diplodus;
Diplogonoporus; Diplolepis; Diploneis;
Diplopylidium; Diploria; Diplotaxis;
Diplozoon; Dipodomys; Diprion;
Dipsacus; Dipteryx; Dipus; Dipylidium;
Diretmus; Dirofilaria; Discoglossus;
Discoma; Disculiceps; Discus;
Dispholidus; Distichopora; Distoma;
Distomum; Ditylenchus; Dizygotheca;
Dochmius; Dodecatheon; Dolichonyx;
Dolichotis; Dolichovespula; Doliolum;
Dolomedes; Dolomys; Dolycoris;
Dombeya; Donax; Donovania; Dorcadion;
Dorcus; Dorema; Dorippe; Dorocidaris;
Doronicum; Dorotheanthus; Dorstenia;
Dorycnium; Doryichthys; Dorylus;
Dosidicus; Dosinia; Draba; Dracaena;
Dracocephalum; Dracontium;
Dracunculus; Drascius; Drawida;
Dreissena; Drepana; Drepanidum;
Dromaius; Dromas; Dromia; Dromiceius;
Drosera; Drosophila; Dryas; Dryocopus;
Dryomys; Dryophis; Dryopithecus;
Dryopteris; Duboisia; Dugesia;
Dugesiella; Dugong; Dulus; Dumetella;
Dunckerocampus; Dunkerocampus;
Duttonella; Dyamena; Dynamena;
Dynastes; Dytiscus.

E

Eberthella; Ecballium; Echeneis;
Echeveria; Echidna; Echidnophaga;
Echinacea; Echinaster; Echinocactus;
Echinocardium; Echinochloa;
Echinococcus; Echinocyamus;
Echinometra; Echinomyia; Echinoprocta;
Echinops; Echinorhinus; Echinorhynchus;
Echinosorex; Echinostoma;
Echinostrephus; Echinus; Echis; Echium;
Echiurus; Eciton; Ectobius; Ectopistes;
Edwardsiella; Egeria; Egretta; Ehrlichia;
Eichhornia; Echornia; Eidolon; Eikenella;
Eimeria; Eisenia; Eiseniella; Elaeagnus;
Elaeis; Elanoides; Elanus; Elaphe;
Elaphurus; Elaps; Elasmosaurus; Elater;
Elatine; Electrophorus; Eledone; Eleginus;
Eleocharis; Elephantulus; Elephas;

Elephe; Elettaria; Eleutherodactylus;
Eliomys; Elodea; Elphidium; Elymus;
Elysia; Emarginula; Embadomonas;
Embelia; Emberiza; Embiotica;
Emericellopsis; Empetrum; Empicoris;
Empusa; Emys; Enantiothamnus;
Encephalitozoon; Encephalocytozoon;
Enchelyopus; Enchytraeus; Encope;
Endamoeba; Endodermophyton;
Endolimax; Endomyces; Endomychus;
Endromis; Endrosa; Endymion;
Engraulicypris; Engraulis; Enhydra;
Enicurus; Enkianthus; Ensatus; Ensis;
Entameba; Entamoeba; Entelurus;
Enterobacter; Enterobius; Enterococcus;
Enteromonas; Enteromorpha;
Entomobrya; Entomospira; Entriatoma;
Eozapus; Ephedra; Ephemera; Ephestia;
Ephippiorhynchus; Ephoron; Ephydatia;
Epicrates; Epidendrum; Epidermophyton;
Epidinium; Epilobium; Epimedium;
Epimys; Epinephelus; Epipactis;
Epiphyllum; Epipogium; Episyrphus;
Epitonium; Epizoanthus; Epomorphorus;
Eptatretus; Eptesicus; Equisetum; Equus;
Eragrostis; Eranis; Eranthis; Erebia;
Erebus; Eremias; Eremicaster;
Eremophila; Eremurus; Eresus;
Erethrizon; Eretmochelys; Erica; Ericulus;
Erigeron; Erignathus; Erinaceus; Erinus;
Eriobotrya; Eriocheir; Eriocheiris;
Eriodictyon; Eriophorum; Eristalis;
Erithacus; Eritrichium; Erodium; Erolia;
Erophila; Erpobdella; Erucastrum;
Erwinia; Eryngium; Erynnis; Erysimum;
Erysipelothrix; Erythraea; Erythrina;
Erythrinus; Erythrobacillus; Erythrocebus;
Erythronium; Erythrophloeum;
Erythroxylon; Erythroxylum; Eryx;
Escherichia; Eschichtius; Eschscholtzia;
Eschscholzia; Esox; Essox; Estrilda;
Etheostoma; Ethusina; Euastacus;
Eubacterium; Eubalaena; Eucalia;
Eucalyptus; Eucarya; Eucharis; Euchlanis;
Euchlora; Eucinostomus; Eucomis;
Euctimsna; Eudendrium; Eudia;
Eudontomyzon; Eudorina; Eudromia;
Eudromias; Eudyptes; Eudyptula;
Eugenia; Euglena; Eulalia; Eulota;

Eumeces; Eumenes; Eumenis; Eumetopias; Eumops; Eunectes; Eunice; Eunicea; Eunicella; Euonymus; Eupagurus; Eupatorium; Eupetes; Euphagus; Eupharynx; Euphorbia; Euphrasia; Euphydryas; Euplagia; Euplectella; Euplotes; Eupomotes; Euproctis; Euproctus; Euptasia; Eurotium; Eurycea; Eurydema; Eurylaimus; Eurypelma; Eurypyga; Eurystomus; Euschemon; Euscorpius; Eusimulium; Euspongia; Euspongilla; Eustrongylus; Eutamias; Eutrombicula; Euxenura; Evasterias; Evernia; Eviota; Excalfactoria; Exochorda; Exocoetus; Exogone; Exogonium; Extastosoma.

F

Fabriciana; Fagopyrum; Fagus; Falcaria; Falco; Fannia; Fasciola; Fasciolaria; Fascioletta; Fascioloides; Fasciolopsis; Fatsia; Faucaria; Favia; Feijoa; Felis; Fennecus; Ferocactus; Ferula; Festuca; Ficedula; Ficus; Filago; Filaria; Filipendula; Fimbriaria; Fissurella; Fistularia; Fittonia; Flavobacterium; Flustra; Fockea; Foeniculum; Fomes; Fontinalis; Forcipomyia; Forficula; Formica; Formicaleon; Forsythia; Fossa; Fragaria; Francisea; Francisella; Francolinus; Frangula; Fratercula; Fraximus; Freesia; Fregata; Fringilla; Fritillaria; Fromia; Frontonia; Fuchsia; Fucus; Fulica; Fulmarus; Fumana; Fumaria; Fundulus; Fungia; Funkia; Furcellaria; Furnarius; Fusarium; Fusidium; Fusiformis; Fusobacterium.

G

Gadus; Gaffkya; Gagea; Gaidropsaurus; Gaillardia; Galaeorhinus; Galago; Galanthus; Galathea; Galaxea; Galega; Galemys; Galeocerdo; Galeodes; Galeopsis; Galeorhinus; Galerida; Galeruca; Galeus; Galinsoga; Galipea;

Galium; Gallinago; Gallinula; Gallionella; Gallirallus; Gallus; Galtonia; Gambusia; Gammarus; Garcinia; Gardenia; Gardonus; Gari; Garrulus; Garypus; Gasteracanthus; Gasteria; Gasterophilus; Gasteroplecus; Gasterosteus; Gastrochaena; Gastrodiscoides; Gastrodiscus; Gastropacha; Gastrophilus; Gastrophyrne; Gastrotheca; Gattyana; Gaultheria; Gavia; Gavialis; Gazania; Gazella; Geaster; Gekko; Gelochelidon; Gempylus; Genetta; Genista; Gentiana; Gentianella; Geocapromys; Geocarcoidea; Geochelone; Geococcyx; Geodia; Geometra; Geomys; Geonemertes; Geophilus; Geotrupes; Geranium; Gerardinus; Gerbera; Gerbillus; Gerris; Geryonia; Geum; Giardia; Gibberella; Gibbula; Gigantocypris; Gigantophis; Gigantornis; Gigartina; Ginglymostoma; Ginglystoma; Ginkgo; Giraffa; Gladiolus; Glandiceps; Glareola; Glaucidium; Glaucium; Glaucomys; Glaux; Glechoma; Glenospora; Gliricola; Glis; Glischrochilus; Globicephala; Globigerina; Globularia; Glomeris; Gloriosa; Glossina; Glossiphonia; Glossobalanus; Glossodoris; Glossoscolex; Glossus; Gloxinia; Glyceria; Glycine; Ggyciphagus; Glycymeris; Glycyrrhiza; Glyphesis; Glyptocephalus; Glyptocranium; Gnaphalium; Gnathoremus; Gnathostoma; Gnathostomum; Gobiesox; Gobio; Gobionellus; Gobius; Godetia; Golfingia; Goliathus; Golofa; Gomophia; Gomphidius; Gomphosus; Gomphus; Gonepteryx; Gongylonema; Goniodes; Gonionemus; Goniopora; Gonium; Gonococcus; Gonodontis; Gonolobus; Gonorhynchus; Gonyaulaux; Gonyaulax; Goodyera; Gopherus; Gordius; Gorgon; Gorgonaria; Gorgonia; Gorgonocephalus; Gorilla; Gorsachius; Gortyna; Gossypium; Goura; Gracilaria; Gracula; Graellsia; Grahamella; Grallina; Grammistes; Grammostola; Grampus; Grantia; Graphidostreptus; Graphiurus; Graphoderus; Grapholitha; Graphosoma;

Graptemys; Graptodytes; Gratiola;
Gregarina; Grevillea; Greyia; Grifolia;
Grindelia; Grison; Groenlandia;
Grubyella; Grus; Grylloblatta; Gryllotalpa;
Gryllus; Grystes; Guaiacum; Guaree;
Gulo; Guzmania; Gymnadenia;
Gymnema; Gymnocalycium;
Gymnocarpium; Gymnocephalus;
Gymnodactylus; Gymnodinium;
Gymnogyps; Gymnothorax; Gynerium;
Gynocardia; Gypaetus; Gyposophila;
Gyps; Gypsophila; Gyrinocheilus;
Gyrinophilus; Gyrinus; Gyrocotyle;
Gyrodactylus; Gyromitra; Gyropus;
Gyrostoma.

H

Haberlea; Habronema; Hacquetia;
Hadogenes; Haemadictyon; Haemadipsa;
Haemagogus; Haemanthus;
Haemaphysalis; Haemapium;
Haematobium; Haematomyzus;
Haematopinus; Haematopota;
Haematopus; Haematoxylon;
Haementaria; Haemobartonella;
Haemodipsus; Haemogregarina;
Haemonchus; Haemophilus;
Haemophoructus; Haemopis;
Haemoproteus; Haemulon; Hagenia;
Haideotriton; Halarachnion; Halcyon;
Halecium; Halesia; Haliaeetus; Haliaetus;
Haliastur; Halichoeres; Halichoerus;
Halichondria; Haliclona ; Haliclystus;
Halictus; Halidrys; Halieutaea; Halimione;
Haliotis; Halistemma; Halobacterium;
Halobates; Haloporphyrus; Hamamelis;
Hammarbya; Hapalochlaena;
Hapalodermes; Haplobothrium;
Haplochilus; Haplopterus;
Haplosporangium; Harpactes; Harpalus;
Harpia; Harpium; Harpodon;
Harpyrhynchus; Harrimania; Hartmanella;
Hartmannia; Haverhillia; Hebeclinium;
Hedera; Hedychium; Hedysarum;
Helarctos; Helcosama; Helenium;
Heliactin; Helianthemum; Helianthus;
Heliaster; Helicella; Helichrysum;

Helicigona; Helicodonta; Helictotrichon;
Helicobacter; Heliometra; Heliopais;
Heliornis; Heliosperma; Heliothrips;
Heliotropium; Helipora; Helix;
Helkesimastix; Helleborus; Helobdella;
Heloderma; Helophilus; Helvella;
Helwingia; Hemaris; Hematopota;
Hementaria; Hemerocallis; Hemerocampa;
Hemicentetes; Hemiclepsis; Hemidactylis;
Hemidactylus; Hemidesmus;
Hemigymnus; Hemimerus; Hemiprocne;
Hemispora; Hemitragus; Hemophilus;
Hemoproteus; Hendersonula; Heniochus;
Heodes; Hepatica; Hepaticola;
Hepatozoon; Heperopteryx; Hepialus;
Heptatretus; Heptranchias; Heracleum;
Herellea; Hermetia; Herminium;
Hermodactylus; Hernanda; Herniaria;
Herpestes; Herpeton; Herpobdella; Herse;
Hesperis; Hesperoloxodon; Heterakis;
Heteralocha; Heterocentrotus;
Heterocephalus; Heterodera; Heterodon;
Heterodontus; Heterodoxus; Heterometra;
Heterometrus; Heterophyes; Heteropoda;
Heteroscelus; Heteroteuthis; Heterotis;
Heurnia; Hevea; Hexabranchia;
Hexactinella; Hexagenia; Hexamita;
Hexanchus; Hibiscus; Hieracium;
Hieraetus; Hierochloe; Hildenbrandia;
Himantarum; Himanthalia;
Himantoglossum; Himantopus;
Hippeastrum; Hippelates; Hippobosca;
Hippocampus; Hippocrepis;
Hippoglossoides; Hippoglossus;
Hippolais; Hippolyte; Hippomane;
Hippophae; Hippopotamus; Hippospongia;
Hippotigris; Hippotragus; Hippuris;
Hirudo; Hirundo; Hister; Histiobranchus;
Histiogaster; Histomonas; Histoplasma;
Histrio; Histrionicus; Histrix; Hogna;
Holacanthella; Holacanthus; Holarrhena;
Holcus; Holocentrus; Hololepta;
Holophrya; Holorusia; Holospora;
Holosteum; Holothuria; Holothyrus;
Homalomyia; Homarus; Homo;
Homogyne; Honkenya; Hopea;
Hoplophorus; Hoplopsyllus; Hordelymus;
Hordeum; Horminium; Hormodendrum;
Hornungia; Hosackia; Hosta; Hottonia;

486

Houttuynia; Hoya; Hucho; Humulus;
Hupertzia; Huso; Hutchinsia;
Hutchinsoniella; Hyacinthus; Hyaena;
Hyaenictis; Hyalinoecia; Hyalomma;
Hyalonema; Hyalophora; Hyas;
Hydatigera; Hydnocarpus; Hydnum;
Hydra; Hydractinia; Hydraena;
Hydrallmania; Hydrangea; Hydrilla;
Hydrobates; Hydrobia; Hydrochara;
Hydrocharis; Hydrochoerus; Hydrocleis;
Hydrocotyle; Hydrocyon; Hydrodamalis;
Hydrogenomonas; Hydrolagus;
Hydromantes; Hydrometra; Hydrophidus;
Hydrophilus; Hydrophis; Hydropogne;
Hydroporus; Hydropotes; Hydrothaea;
Hydrous; Hydrurga; Hygrobia;
Hygrophila; Hyla; Hylaeosaurus;
Hylarnus; Hylecoetus; Hylemyia; Hyles;
Hylobates; Hylobius; Hylochoerus;
Hylocichla; Hyloconium; Hyloicus;
Hylomys; Hylonomus; Hylotrupes;
Hyloxalus; Hymeniacidon; Hymenobolus;
Hymenocallis; Hymenoea; Hymenogorgia;
Hymenolepis; Hymenophyllum;
Hymenopus; Hynobius; Hyocoris;
Hyoscyamus; Hyostrongylus; Hyperchiria;
Hyperia; Hypericum; Hyperoodon;
Hypholomoa; Hyphomicrobium;
Hyphydrus; Hypobosca; Hypochoeris;
Hypocolius; Hypoctonus; Hypoderma;
Hypomesus; Hyponomeuta; Hypopachus;
Hypophthalmichthys; Hypositta;
Hypselosaurus; Hypsignathus;
Hypsopsetta ; Hyssopus; Hysterocrates;
Hystricopsylla; Hystrix.

I

Iberis; Ibis; Ichneumia; Ichthyophthirius;
Icichthys; Icosteus; Ictalurus; Icterus;
Ictinia; Ictobius; Ictonyx; Idaea;
Idiacanthus; Idotea; Idus; Iguana;
Iguanodon; Ijimaia; Ilex; Ilia; Illecebrum;
Illicium; Ilybius; Ilyocoris; Impatiens;
Inachis; Inachus; Incarvillea; Indicator;
Indiella; Indigofera; Indri; Indricotherium;
Ingolfiella; Inia; Inimicus; Inocybe; Inula;
Iodamoeba; Iphiclides; Ipomoea; Ips;

Irena; Iridoprocne; Iris; Isatis;
Ishthyostega; Isidora; Isis; Isodon; Isoetes;
Isognomostoma; Isopyrum; Isospora;
Isotoma; Ispidina; Istiophorus; Isurus;
Iulus; Ixia; Ixobrychus; Ixodes;
Ixodiphagus.

J

Jabiru; Jacana; Jacobinia; Jacquinia;
Jaculus; Jaguarius; Jambosa; Janthella;
Janthinosoma; Jasione; Jasminium;
Jasminum; Jateorhiza; Jatropha;
Johanssonia; Jordanella; Jovibarba;
Juglans; Julus; Juncus; Juniperus; Jurinea;
Jynx.

K

Kakatoe; Kalanchoe; Kallima; Kalmia;
Kalopanax; Karagassiema; Karyamoebina;
Katayama; Katsuwonus; Kentranthus;
Keratella; Kernera; Kerria; Kickxia;
Kinosternon; Kinyxis; Kitaibelia;
Kittacincla; Klebsiella; Kleinia;
Klinophilus; Knautia; Kniphofia;
Kobresia; Kochia; Koeleria; Koelreuteria;
Koenenia; Krameria; Krohnius;
Kronosaurus; Kryptopterus; Kurthia;
Kurtus; Kyphosus.

L

Labea; Labeo; Labia; Labroides; Labrus;
Laburnocytisus; Laburnum; Laccifer;
Lacerta; Lachesis; Lachnus; Lactarius;
Lactobacillus; Lactophrys; Lactrodectus;
Lactuca; Laelaps; Laevicardium; Lafoea;
Lagendelphis; Lagenorhynchus;
Lagerstroemia; Lagidum; Lagochilascaris;
Lagopus; Lagorchestes; Lagostomus;
Lagothrix; Lagria; Lagurus; Lama;
Lamblia; Lamellaria; Lamia; Laminaria;
Lamium; Lamna; Lampetra; Lampranthus;
Lampris; Lamprocystis; Lampropeltis;
Lampyris; Lamus; Lanice; Lanius;

Lankesterella; Lankesteria; Lantana;
Laothoe; Laphria; Laportea; Lappula;
Lapsana; Laptonychotes; Laria; Larix;
Larosterna; Larus; Laser; Laserpitium;
Lasiocampa; Lasiodora; Lasiohelia;
Lasiorhinus; Lasius; Laspeyresia;
Laterallus; Laternaria; Lates; Lathraea;
Lathyrus; Laticauda; Latimeria;
Latrodectes; Latrodectus; Laurencia;
Laurus; Lavandula; Lavatera; Laverania;
Lawsonia; Lebistes; Lecanium; Ledum;
Leersia; Leeuwenhoekia; Legionella;
Legousia; Lehmannia; Leiognathus;
Leiolepis; Leiopa; Leiopus; Leishmania;
Leiurus; Lelaps; Lemmus; Lemna;
Lemonia; Lemur; Lens; Leo; Leonotis;
Leonotodon; Leonotopodium;
Leontocebus; Leonurus; Leopardus;
Lepadogaster; Lepas; Lepeophtheirus;
Lepidium; Lepidochelys; Lepidochitona;
Lepidopus; Lepidorhombus; Lepidoselaga;
Lepidosiren; Lepidurus; Lepiota; Lepisma;
Lepismachilis; Lepisosteus; Lepomis;
Leptailurus; Leptinotarsa; Leptocladodia;
Leptoconops; Leptodactylus; Leptodera;
Leptodora; Leptomitus; Leptonychotes;
Leptophtergius; Leptophyes; Leptopsylla;
Leptoptilos; Leptoptilus; Leptospira;
Leptothrix; Leptotila; Leptotrichia;
Leptotrombidium; Leptotyphlops;
Leptura; Leptus; Lepus; Lernaea;
Lernaeocera; Lestes; Leto; Leucaspius;
Leuchtenbergia; Leucifer; Leuciscus;
Leuckartiara; Leucobryum;
Leucocytozoon; Leucoium; Leucojum;
Leucoma; Leuconostoc; Leucophoyx;
Leucopsis; Leucorchis; Leucosolenia;
Leucothea; Leucothoe; Leucothrix;
Leukocytozoon; Leuresthes; Lewisia;
Leymus; Liatris; Libellula; Libocedrus;
Libythea; Lichanura; Lichen;
Licheterodon; Lichtheimia; Licula; Ligia;
Lignognathoides; Ligula; Ligularia;
Ligusticum; Ligustrum; Lilioceris; Lilium;
Lima; Limacina; Limanda; Limax;
Limenitis; Limicola; Limnadia; Limnaea;
Limnatis; Limnephilus; Limnodromus;
Limnodorum; Limonium; Limosa;
Limosella; Limulus; Linaria; Linckia;

Lindernia; Lineus; Linguatula; Lingula;
Linnaea; Linognathus; Linuche; Linum;
Lionotus; Liopelma; Liopsetta; Liparis;
Liparus; Liphistius; Liponyssus;
Lipoptena; Lipotes; Lippia; Liquidambar;
Liriodendron; Liriope; Lispa;
Lissodelphis; Lissotis; Listera; Listerella;
Listeria; Lithobius; Lithocarpus;
Lithocolletis; Lithocranius; Lithodes;
Lithophaga; Lithophyllum; Lithops;
Lithospermum; Litocranius; Litomosoides;
Littorella; Littorina; Llama; Lloydia; Loa;
Lobaria; Lobelia; Lobodon; Lobularia;
Lochnera; Locusta; Locustella;
Loddigesia; Lodoicea; Loeschia;
Loiseleuria; Loligo; Lolium; Lomaspilis;
Lomatogonium; Lonchocarpus; Lonchura;
Lonicera; Lonomia; Loosia; Lophiomys;
Lophius; Lophoceros; Lophocolea;
Lophodytes; Lopholatilus; Lophophora;
Lophophorus; Lophophyton; Lophortyx;
Lophura; Lophyrus; Lora; Loranthus;
Loricaria; Loris; Lota; Lottia; Lotus;
Loxia; Loxodonta; Loxosceles;
Loxosoma; Loxotrema; Lucaena;
Lucanus; Lucernaria; Lucilia; Lucioharax;
Lucioperca; Ludwigia; Luidia; Lullula;
Lumbricus; Lumpenus; Lunaria; Lupinus;
Luronium; Luscinia; Lusciniola; Lutjanus;
Lutra; Lutraria; Lutreola; Lutrogale;
Luzula; Lybia; Lycaena; Lycaon; Lychnis;
Lycodon; Lycogale; Lycoperdon;
Lycopersicum; Lycopodiella;
Lycopodium; Lycopsis; Lycopus; Lycoris;
Lycosa; Lygus; Lymanopoda; Lymantria;
Lymnaea; Lymnocryptes;
Lymphocytozoon; Lynchia; Lynx;
Lyonetia; Lyrurus; Lysichiton;
Lysimachia; Lysmata; Lythrum; Lytta.

M

Macaca; Macacus; Machilis; Maclura;
Macoma; Macracanthorhynchus;
Macraspis; Macrobdella; Macrobiotus;
Macrocheira; Macrochelys; Macroclemys;
Macrocyclops; Macrodontia; Macrogastra;
Macroglossum; Macroglossus;

Macromonas; Macroperipatus;
Macropipus; Macroplectrum; Macropodia;
Macropus; Macroscelides;
Macrosternodesnus; Macrostoma;
Macrostomum; Macrostylis; Macrotermes;
Macrotus; Macrourus; Macrozoarces;
Mactra; Madhuca; Madoqua; Madrepora;
Madtsoia; Madurella; Maena; Magnolia;
Mahonia; Maianthemum; Majorana;
Makaira; Malachius; Malaclemys;
Malacocincla; Malacosarcus;
Malacosoma; Malania; Malapterurus;
Malassezia; Malaxis; Malcomia;
Malleomyces; Malletia; Mallomonas;
Mallotus; Malpighia; Malus; Malva;
Mamestra; Mammillaria; Mammuthus;
Manayunkia; Mandragora; Mandrillus;
Mangifera; Manica; Manihot; Maniola;
Manis; Manocirrhus; Mansonella;
Mansonia; Mansonoides; Manta;
Mantichora; Mantis; Maranta; Marasmius;
Marchantia; Margaritana; Margaritifera;
Margaropus; Marginaster; Marmosa;
Marmota; Marrubium; Marsdenia;
Marsilea; Martes; Marthasterias;
Marumba; Masticophus; Mastigamoeba;
Mastigoproctus; Mastodon; Mastomys;
Mastotermes; Matricaria; Matteucia;
Matthiola; Mauremys; Mayermys;
Mayetiola; Mazocraes; Meandrina;
Mecistocirrhus; Medicago; Medinilla;
Megaceryle; Megachile; Megaderma;
Megadyptes; Megalobatrachus;
Megaloceros; Megaloglossus;
Megalomma; Megaloprepus; Megalops;
Megalopyge; Megalorhina; Megalosaurus;
Megalosporon; Meganeura; Megaptera;
Megarhinus; Megarhyssa; Megascolides;
Megasea; Megaselia; Megasoma;
Megastoma; Megathura;
Megatrichophyton; Megicica; Megophrys;
Melaleuca; Melampyrum; Melanargia;
Melanerpes; Melanitis; Melanitta;
Melanocorypha; Melanogrammus;
Melanoides; Melanolestes; Melanoplus;
Melanosuchus; Meleagris; Meles; Melia;
Melica; Meligethes; Melilotus; Melinna;
Melipona; Melissa; Melitta; Melittangium;
Mellisuga; Mellivora; Meloe; Melogale;

Melolontha; Melophagus; Melopsittacus;
Melursus; Membranipora;
Membranoptera; Menacanthus; Menidia;
Meningococcus; Menispermum;
Menopon; Mentha; Menura; Menyanthes;
Meoma; Mephitis; Mercenaria;
Mercurialis; Mergus; Meriones;
Merluccius; Mermis; Merodon; Merops;
Merulius; Mesamphisopus;
Mesembryanthemum; Mesidotea;
Mesoacidalia; Mesocerus; Mesocricetus;
Mesocypris; Mesogonimus;
Mesonychoteuthis; Mesoplodon;
Mespilus; Messor; Metacrinus;
Metagonimus; Metasequoia;
Metastrongylus; Methanobacterium;
Methanococcus; Methanomonas;
Metridium; Metrosideros; Meum; Mibora;
Micraspides; Microbacterium;
Microbatrachella; Microbuthus;
Microcebus; Microchaetus; Micrococcus;
Microcoelum; Microcyclus; Microdon;
Microfilaria; Microgaster; Microglossus;
Microhierax; Microhydra; Microhyla;
Micromesistius; Micrometrus;
Micrommata; Micromonospora;
Micromyces; Micromys; Micropterus;
Micropteryx; Micropus; Micropyrum;
Microscilla; Microsorex; Microspira;
Microspironema; Microsporon;
Microsporum; Microstomus;
Microtrombidium; Microtus; Micrurus;
Midgardia; Miescheria; Mikiola; Milium;
Millepora; Milvus; Mima; Mimas;
Mimosa; Mimulus; Mimus; Miniopterus;
Minuartia; Mirabilis; Mirafra; Mirapinna;
Mirounga; Misgurnus; Misopates;
Mistichthys; Misumena; Mitra; Mitu;
Miyagawanella; Mnemiopsis; Mniotilta;
Mnium; Moehringia; Moenchia; Moho;
Mola; Molinia; Mollienesia; Moloch;
Molorchus; Molothrus; Molucella; Molva;
Momordica; Monacanthus; Monacha;
Monachus; Monadenium; Monas;
Monbretia; Moneses; Monias; Moniezia;
Monilia; Moniliformis; Monocelis;
Monocystis; Monodella; Monodon;
Monodonta; Monodontus; Monochus;
Monopsyllus; Monoraphis;

Monochotrema; Monosporium;
Monostoma; Monostomum; Monotropa;
Monsonia; Monstera; Montacuta;
Montastrea; Montbretia; Montia;
Monticola; Montifringilla; Moraxella;
Morchella; Morimus; Moringa;
Mormyrops; Morpho; Morus; Moschus;
Motacilla; Mucor; Mucuna; Muellerius;
Muggiaea; Mugil; Mullus; Multicalyx;
Multiceps; Multicotyle; Multidictus;
Mungos; Muntiacus; Muraena; Murex;
Murgantia; Murimyces; Mus; Musa;
Musca; Muscardinus; Muscari;
Muscicapa; Muscivora; Musculus;
Mustela; Mustelus; Mutilla; Mutinus;
Mutisia; Mya; Mycelis; Mycobacterium;
Mycocandida; Mycococcus; Mycoderma;
Myconostoc; Mycoplasma;
Mycotoruloides; Mycteria; Mycteroperca;
Myctophum; Mydas; Mygale; Myiarchus;
Myiatropa; Myiochanes; Myletes; Myleus;
Myocastor; Myopus; Myosciurus;
Myosotis; Myosoton; Myosurus; Myotis;
Myoxocephalus; Myrica; Myricaria;
Myriophyllum; Myriotrochus; Myristica;
Myrmecia; Myrmecobius;
Myrmecocystus; Myrmecodia;
Myrmecophaga; Myrmeleon; Myrmica;
Myroxylon; Myrrhis; Myrtophyllum;
Myrtus; Mysia; Mysis; Mystax;
Mytilicola; Mytilus; Myxilla; Myxine;
Myxobolus; Myxococcidium;
Myxococcus; Myxovirus; Myzomyia;
Myzorhynchus; Myzus.

N

Nabis; Naegleria; Naja; Najas; Nannizia;
Nannosciurus; Nanophyes; Nanophyetus;
Napaeozapus; Narcissus; Nardostachys;
Nardurus; Nardus; Narke; Narthecium;
Nasalis; Nasilio; Nasonella; Nasonia;
Nassarius; Nasturtium; Nasua;
Nasutitermes; Natica; Natrix; Naucrates;
Naumanniella; Naumburgia; Nausithoe;
Nautilius; Neanthe; Neanthes; Nebalia;
Nebaliopsis; Necator; Necrophorus;
Nectarinia; Nectonema; Necturus;

Negaprion; Neisseria; Nelumbium;
Nelumbo; Nemacheilus; Nematodirus;
Nematus; Nemertes; Nemobius;
Nemophora; Neobalaena; Neoceratodus;
Neodiprion; Neodrepanis; Neofiber;
Neomeris; Neomys; Neophilina;
Neophoca; Neophron; Neopilina;
Neorickettsia; Neoschongastia; Neotoma;
Neotragus; Neottia; Nepa; Nepenthes;
Nepeta; Nephila; Nephrolepis; Nephrops;
Nepthys; Nepticula; Neptunea; Neptunus;
Nereis; Nerine; Nerita; Neritina; Nerium;
Nerophis; Neslia; Nesoryctes; Nestor;
Netta; Neuroterus; Neurotrichus; Nevskia;
Nezera; Niadvena; Nicotiana;
Nicrophorus; Nigella; Nigritella;
Niphargus; Nipponia; Nippostrongylus;
Niptus; Nitrobacter; Nitrocystis;
Nitrosocystis; Nitrosogloea;
Nitrosomonas; Nitrosospira; Nocardia;
Nochtia; Noctiluca; Noctua;
Noemacheilus; Noguchia; Nomeus;
Nonnea; Nosema; Nosopsyllus; Notechis;
Notemigonus; Notharchus;
Nothobranchius; Nothocrax; Nothofagus;
Notiophilus; Notodelphys; Notodonta;
Notoedres; Notonecta; Notornis;
Notoryctes; Nototrema; Notropis;
Noturus; Nucella; Nucifraga; Nucula;
Nuculana; Numenius; Numida; Nuphar;
Nuttallia; Nyctalus; Nyctanassa; Nyctea;
Nycterutes; Nyctibius; Nycticebus;
Nycticorax; Nyctidromus; Nyctotherus;
Nymphaea; Nymphalis; Nymphicus;
Nymphoides; Nymphon; Nymphula;
Nyssa; Nyssorhynchus.

O

Obelia; Oberea; Oceanites; Oceanodroma;
Ocenebra; Ochlerotatus; Ochlodes;
Ochotona; Ochrobium; Ochromonas;
Ochromyia; Ocimum; Ocinebra; Ocotea;
Octolasium; Octomitus; Octomyces;
Octopus; Ocyphaps; Ocypode;
Odobaenus; Odobenus; Odocoileus;
Odonestis; Odontaspis; Odontaster;
Odontites; Odontogadus; Odontoglossum;

Odontomyia; Odynerus; Oecanthus; Oeciacus; Oecophylla; Oedemera; Oedipoda; Oenanthe; Oeneis; Oenothera; Oesophagostomum; Oestrus; Ogocephalus; Oiceoptoma; Oidium; Okapia; Olae; Olibrus; Oligomyrmex; Olindias; Oliva; Ollulanthus; Ommastrephes; Omphalodes; Onchidium; Onchocerca; Onchocercus; Onchorhyncus; Onciola; Oncocerca; Oncocercus; Oncomelania; Oncorhyncus; Ondatra; Oniscus; Onobrychis; Onoclea; Ononis; Onopordum; Onychogalea; Onychogales; Onychoteuthis; Oospora; Opalina; Opercularia; Operculina; Operophtera; Ophiacantha; Ophiactis; Ophiarachna; Ophidion; Ophidium; Ophiocephalus; Ophiodaphne; Ophioderma; Ophioglossum; Ophiomisidium; Ophion; Ophionotus; Ophiophagus; Ophiosaurus; Ophiosphalma; Ophiothrix; Ophisaurus; Ophiura; Ophrys; Opilio; Opisocrostis; Opisthocomus; Opisthognathus; Opisthonema; Opisthorcis; Opiura; Opsanus; Opuntia; Orcaella; Orchestia; Orchis; Orcinus; Orconectes; Oreamnos; Orectolobus; Oreochloa; Oreopteris; Oreotyx; Orester; Orgyia; Origanum; Oriolus; Orlaya; Orneodes; Ornithodoros; Ornithogalum; Ornithomyia; Ornithoptera; Ornithopus; Ornithorhynchus; Orobanche; Oropsylla; Orthezia; Orthilia; Orthochela; Orthotomus; Orya; Orycteropus; Oryctes; Oryctolagus; Oryx; Oryza; Oryzomys; Oryzoryctes; Oscanius; Oscarella; Oscillaria; Oscinella; Oscinis; Osmanthus; Osmerus; Osmia; Osmorhiza; Osmunda; Osmylus; Osphronemus; Osteodontornis; Osteoglossum; Osteolaemus; Ostericum; Ostertagia; Ostiolum; Ostracion; Ostrea; Ostrinia; Ostrya; Oswaldocruzia; Otaria; Otiorhynchus; Otis; Otobius; Otocentor; Otocolobus; Otodectes; Otomyces; Otus; Oulema; Ovibos; Ovis; Ovoplasma; Owenia; Oxalis; Oxeamnos; Oxybelis; Oxydendrum; Oxyporus; Oxyria; Osyruncus; Oxyspirura; Oxytropis; Oxyura; Oxyuranus; Oxyuris; Ozaena.

P

Pachymerium; Paecilomyces; Pagellus; Pagophilus; Paguma; Paguristes; Pagurus; Palaeopithecus; Palaeostenzia; Palaquium; Paleosuchus; Palomena; Pan; Panax; Pandaka; Pandinus; Pandion; Pangasianodon; Panicum; Panopea; Panorpa; Panstrongylus; Panthera; Panulirus; Panurgus; Panurus; Papaver; Papilio; Papio; Paracaudina; Parachordodes; Paracoccidioides; Paracolobactrum; Paradoxurus; Parafossularus; Paragnathia; Paragonimus; Paragordius; Paralichthys; Paralithodes; Parameba; Paramecium; Paramoeba; Paramphistomum; Pararge; Pararutilus; Parasaccharomyces; Parasitus; Paraspirillum; Parateuthis; Paravespula; Pardanthus; Parelephas; Parendomyces; Parietaria; Parnassius; Paroctopus; Parus; Passer; Passiflora; Pasteurella; Pasteuria; Patella; Patu; Paullina; Pausinystalia; Pecten; Pedaliodes; Pediculus; Pediococcus; Peganum; Pelagothuria; Pelamis; Pelargonium; Pelecanus; Pelecus; Pelmatohydra; Pelobates; Peloploca; Penicillium; Pennatula; Pentastoma; Pentatrichomonas; Pepsis; Peptococcus; Peptostreptococcus; Perameles; Perca; Percina; Perdix; Perforatella; Periophthalmus; Periplanea; Periplaneta; Periploca; Perla; Pernis; Petricola; Petrodomus; Petrognatha; Petromyzon; Petroselinum; Peucedanum; Peumus; Pfeifferella; Phacochoerus; Phalacrocorax; Phalaenoptilus; Phalangium; Phalaropus; Phalera; Pharbitis; Pharnacia; Phascolosoma; Phaseolus; Phasianus; Phialophora; Philanthus; Philomachus; Phlebotomus; Phobaeticus; Phobosuchus; Phoca; Phoenetria; Phoenicurus; Pholas; Pholcus; Phoneutria; Phormia; Phororhacos; Photobacterium; Phoxinus; Phragmidiothrix; Phreatoicopsis; Phronima; Phryganea; Phrynobatrachus; Phrynomerus; Phryxe; Phthirus; Phyllanthus; Phyllobates; Phyllobrotica;

Phyllodecta; Phyllomedusa; Phyllopertha; Phyllopora; Phylloscopus; Physalia; Physalis; Physaloptera; Physeter; Physocephalus; Physopsis; Phsostegia; Physostigma; Phytocoris; Phytolacca; Phytomonas; Pica; Picea; Picraena; Picrasma; Picromerus; Picrorhiza; Picrorrhiza; Picus; Piedraia; Pierus; Pilocarpus; Pimenta; Pimpinella; Pimpla; Pinguinus; Pinna; Pinnotheres; Pinoyella; Pinus; Piophila; Piper; Pipistrellus; Piptadenia; Piratinga; Pironella; Piroplasma; Pisaster; Pisaura; Piscidia; Pisidium; Pistacia; Pithecophaga; Pitymys; Pityophagus; Pityrosporon; Pityrosporum; Placentonema; Placobdella; Plagiodera; Plagiodontia; Plagionotus; Plagiostomum; Planaria; Planigale; Planorbarius; Planorbis; Plantago; Plasmodiophora; Plasmodium; Platanista; Platichthys; Platypoccilus; Plecotus; Plectrophenax; Pleodorina; Plethodon; Pleurobranchus; Pleurodeles; Pleuronectes; Pleuronichthys; Plexaura; Plocamium; Plumaria; Plumbago; Pluvialis; Pneumocystis; Pocadius; Podangium; Podiceps; Podilymbus; Podophyllum; Podura; Poecilia; Poephila; Polemaetus; Polemonium; Polistes; Pollachius; Polyangium; Polycelis; Polydesmus; Polydrusus; Polygala; Polygonatum; Polygonia; Polygonoporus; Polygonum; Polymnia; Polyodon; Polyommatus; Polyplax; Polyporus; Polypterus; Polystoma; Polystomum; Polytoma; Polytrichum; Polyxenus; Pomacentrus; Pomatoceros; Pomatomus; Pomatoshistus; Pomolobus; Pomoxis; Poncirus; Pongamia; Pongo; Pontederia; Pontoporia; Popillia; Populus; Porcellana; Porcellanaster; Porcellio; Porocephalus; Porthetria; Portunus; Porzana; Potamobius; Potamochoerus; Potamogale; Potentilla; Poterolebias; Potosia; Priapulus; Primula; Prinos; Prionace; Prionotus; Prionus; Pristinotherium; Pristis; Proactinomyces; Procambarus; Procavia; Procerodes; Profelis; Promachocrinus; Promicrops;

Propeomyces; Propionibacterium; Prosthagonimus; Prostherapis; Protaminobacter; Protea; Proteoglypha; Proteosoma; Proterorhinus; Proteus; Protocalliphora; Protoglossus; Protominobacter; Protospirura; Protostrongylus; Protula; Prowazekella; Prowazekia; Prunella; Prunus; Psammechinus; Psammodromus; Psammothuria; Psetta; Psettodes; Pseudacris; Pseudamphistomum; Pseudis; Pseudocarcinus; Pseudodiscus; Pseudogorilla; Pseudois; Pseudolynchia; Pseudomonas; Pseudomonilia; Pseudomydura; Pseudophyllanax; Pseudopleuronectes; Psidium; Psilocybe; Psithyrus; Psittacus; Psophus; Psoralea; Psoroptes; Psylla; Pteranodon; Pterocarpus; Pterocles; Pterois; Pteromys; Pteronura; Pterophyllum; Pteropus; Pterostichus; Ptinus; Ptiolcercus; Ptychocheilus; Ptychodera; Pudu; Pueraria; Pulex; Pullularia; Pulsatilla; Punica; Puntius; Pupilla; Pusa; Putorius; Pycnanthemum; Pycnopodia; Pygocentrus; Pygopristis; Pygoscelis; Pyracantha; Pyralis; Pyretophorus; Pyrgus; Pyrochroa; Pyroplasma; Pyrosoma; Pyrrhidium; Pyrrhocorax; Pyrrhocoris; Pyrrhula; Pyrrosoma; Pyrus; Python; Pyxicephalus.

Q

Quassia; Quelea; Quercus; Quillaja.

R

Raillietina; Raja; Rallus; Ramibacterium; Rana; Rangifer; Raphidia; Rattus; Rauwolfia; Recurvirostra; Regalecus; Regulus; Remijia; Remiz; Reticulitermes; Retortamonas; Rhabditis; Rhabdomonas; Rhabdonema; Rhacophorus; Rhagio; Rhagium; Rhagoletis; Rhagonycha; Rhamnus; Rhamphosuchus; Rhaphidophora; Rheum; Rhincodon;

Rhinencephalus; Rhineodon;
Rhinicephalus; Rhinichthys; Rhinobatos;
Rhinocephalus; Rhinocladium;
Rhinocricus; Rhinodrilus; Rhinoestrus;
Rhinolophus; Rhinoptera;
Rhinosporidium; Rhipicephalus;
Rhipsalidopsis; Rhizobium; Rhizoglyphus;
Rhizopus; Rhizostoma; Rhodactis;
Rhodeus; Rhodnius; Rhododendron;
Rhodomicrobium; Rhodopechys;
Rhodopseudomonas; Rhodospirillum;
Rhodospiza; Rhodotheca; Rhogogaster;
Rhopalopsyllus; Rhus; Rhyacophila;
Rhyncocyon; Rhyniella; Rhyparia;
Rhyssa; Ribes; Ricinus; Rickettsia;
Rickettsiella; Ricolesia; Riparia; Rissa;
Rivea; Rivulogammarus; Roccella;
Romericus; Rosa; Rosmarinus; Rossiella;
Rousettus; Rubia; Rubus; Rudbeckia;
Rupicapra; Ruta; Ruthenica; Rutilus.

S

Sabatinca; Sabella; Sabellastarte;
Sabouradites; Saccharomyces;
Saccoglossus; Saduria; Sagittaria; Saiga;
Saintpaulia; Salamandra; Salix; Salmo;
Salmonella; Salticus; Salvelinus; Salvia;
Sambucus; Sanguisuga; Sanninoidea;
Sansevieria; Santalum; Santolina;
Saperda; Saponaria; Saprolegnia;
Saprospira; Saraca; Sarcina; Sarcocystis;
Sarcophaga; Sarcophyton; Sarcoptes;
Sarda; Sardinia; Sardinops; Sardinus;
Sarothamnus; Sarracenia; Sassafras;
Saturnia; Saurophagus; Saussurea;
Saxicola; Saxidomus; Saxifraga; Scabiosa;
Scaeva; Scalaria; Scaphistostreptus;
Scardinius; Scarus; Scatophaga;
Schindleria; Schinopsis; Schinus;
Schistocerca; Schistosoma;
Schistosomatium; Schizoblastosporion;
Schizocardium; Schizophyllum;
Schizosaccharomyces; Schizosiphon;
Schizotrypanum; Schlumbergera;
Schoenocaulon; Scilla; Scindapsus;
Sciurus; Sclerostoma; Scolia; Scolopax;
Scolopendra; Scomber; Scomberomorus;

Scophthalmus; Scopolia; Scopula;
Scopulariopsis; Scopus; Scorpaena;
Scorpaenopsis; Scotoplanes;
Scrobicularia; Scutiger; Scyliorhinus;
Scytonema; Sebastes; Sebastodes; Secale;
Sedum; Segmentina; Selaginella;
Selasphorus; Selatosomus; Selenomonas;
Semnopithecus; Sempervivum; Senecio;
Sepia; Sepsis; Serenoa; Serica;
Sericopelma; Serinus; Serranelus;
Serrasalmus; Serratia; Sertularella;
Sertularia; Sesamum; Sesia; Sessinia;
Setaria; Shigella; Shorea; Sialis; Sicista;
Siderobacter; Siderocapsa; Siderococcus;
Sideromonas; Sideronema; Siderophacus;
Siderosphaera; Sigmodon; Silene; Silurus;
Simaruba; Simulium; Sinningia;
Sinodendron; Siphonochalina;
Siphonophanes; Siphonophora;
Siphunculina; Siren; Sirex; Sistrurus;
Sisyra; Sisyrinchium; Sitta; Smaragdina;
Smerinthus; Smilacina; Smilax;
Sminthillus; Sminthopsis; Soja; Solanum;
Solea; Solenoglypha; Solenopotes;
Solenostomus; Solidago; Somateria;
Somniosus; Sophora; Sophora;
Sorangium; Sorbus; Sorex; Spalax;
Spartium; Sparus; Spathiphyllum;
Spenceriella; Spermophilus; Sperosoma;
Sphaeranthus; Sphaeria; Sphaeridium;
Sphaerium; Sphaerodactylus;
Sphaerophorus; Sphaerophysa;
Sphaerotilus; Sphagnum; Sphenodon;
Spherophorus; Sphingonotus; Sphinx;
Sphyraena; Sphyrna; Spigelia; Spilanthes;
Spiraea; Spiranthes; Spirillum; Spirobolus;
Spirobranchus; Spirocerca; Spirochaeta;
Spironema; Spiroschaudinnia;
Spirostreptus; Spondylis; Spongia;
Spongilla; Sporothrix; Sporotrichon;
Sporotrichum; Sprattus; Sprekelia;
Squaliolus; Squalus; Squatina;
Stagmomantis; Staphylinus;
Staphylococcus; Stegomyia; Stegosaurus;
Stellaria; Stennama; Stenella;
Stenobothrus; Stenopus; Stenoteuthis;
Stenotomus; Stephanoberyx;
Stephanofilaria; Stephanurus; Stercorarius;
Sterculia; Sterigmatocystis;

Sterigmocystis; Sterna; Sternbergia; Sternotherus; Stichopus; Stigmella; Stillingia; Stipa; Stizostedion; Stomoxys; Storeria; Strangalia; Stratiomys; Strelitzia; Streptobacillus; Streptococcus; Streptomyces; Streptopelia; Streptosporangium; Streptothrix; Strigops; Strix; Strombus; Strongylocentrotus; Strongyloides; Strongylus; Strophanthus; Stropharia; Struthio; Strychnos; Sturnus; Stylaria; Stylosanthes; Styrax; Suberites; Succinea; Sufflamen; Sula; Suncus; Sus; Swertia; Syagrus; Sycon; Syctalidium; Sylvia; Symbiotes; Sympetrum; Symphoricarpus; Symphysodon; Symphytum; Symplocarpus; Synanceja; Synangium; Synapta; Syngamus; Syngnathus; Synodus; Synsepalum; Syphacia; Syringa; Syringospora; Syrinx; Syrnium; Systenocerus; Syzgium.

T

Tabanus; Tachina; Tadorna; Taenia; Taeniarhynchus; Taeniarhynus; Taeniorhynchus; Tagetes; Talauma; Talpa; Tamarindus; Tamias; Tamus; Tanacetum; Taniopygis; Takatogenos; Taraxacum; Tarbosaurus; Tarentola; Taricha; Tarpon; Tatusia; Taurotragus; Tautoga; Taxodium; Taxotes; Taxus; Tayassu; Tedania; Tegenaria; Teinopalpus; Telescopus; Tellina; Tenebrio; Tenodera; Tenthredo; Tenuibranchiurus; Teratornis; Teredo; Terminalia; Ternidens; Testudo; Tethys; Tetracanthagyna; Tetrachilomastix; Tetraclinis; Tetracoccus; Tetragnatha; Tetrameres; Tetramitus; Tetranychus; Tetrao; Tetraodon; Tetrapteris; Tetrapturus; Tetrasomus; Tetrastes; Tetrastoma; Tetratrichomonas; Tetrix; Tetropium; Tettigonia; Teucrium; Thalarctos; Thalassema; Thalassoma; Thalazia; Thalia; Thamnidium; Thamnophis; Thanasimus; Thapsia; Thaumatococcus; Thaumetopoea; Thea; Theileria; Thelazia; Thelohania; Thelyphonus; Theobaldia; Theobroma; Theragra; Theraphosa;

Therioplectes; Therizinosaurus; Thermoactinomyces; Thermopsis; Thevetia; Thiobacillus; Thiobacterium ; Thiocapsa; Thiocystis; Thioderma; Thiodictyon; Thiopedia; Thiopolycoccus; Thiosarcina; Thiospira; Thiospirillum; Thiothece; Thiovulum; Thracia; Threskiornis; Thuja; Thunbergia; Thunnus; Thyatira; Thylacinus; Thylacomys; Thymallus; Thymus; Thysania; Thysanosoma; Tibicen; Tibouchina; Tichodectes; Tichodroma; Tilapia; Tilia; Tilletia; Tinca; Tindaria; Tineola; Tinospora; Tipula; Titanus; Tityus; Toddalia; Tolypocladium; Tomicus; Torpedo; Tortrix; Torula; Torulopsis; Toxascaris; Toxocara; Toxoplasma; Toxopneustes; Toxostoma; Toxothrix; Trachelospermum; Tracheophilus; Trachinus; Trachurus; Trachybdella; Trachylobium; Trachys; Trachyspermum; Tradescantia; Tragelaphus; Tragia; Tragulus; Trechona; Treponema; Triactis; Triakis; Trianthema; Triatoma; Tribolium; Tricercomonas; Trichechus; Trichia; Trichina; Trichinella; Trichobilharzia; Trichocephalus; Trichodectes; Trichodina; Trichoglossus; Trichomastix; Trichomonas; Trichophaga; Trichophyton; Trichosoma; Trichosomoides; Trichosporon; Trichostrongylus; Trichothecium; Trichuris; Tridacna; Tridontophorus; Trigonella; Trillium; Trimastigamoeba; Tringa; Triops; Tripheustes; Triphleps; Tripleorospermum; Triteleia; Tritelia; Triticum; Tritoma; Tritonia; Triturus; Trochosa; Trogium; Troglodytella; Troglodytes; Troglotrema; Troides; Trollius; Trombicula; Trombidium; Tropaeolum; Tropinota; Truncatellina; Trutta; Trypanophis; Trypanoplasma; Trypanosoma; Trypanosomonas; Trypanozoon; Trypocastellanelleae; Tsuga; Tubifex; Tubipora; Tulipa; Tunga; Tunnus; Turbatrix; Turbinaria; Turdoides; Turdus; Turnera; Tursiops; Tussilago; Tydeus; Tylenchus; Tylophora; Tylosurus; Typhlocoelum; Typhlomolge; Typhlops;

Typhlotriton; Typhonium; Tyrannosaurus; Tyria; Tyroglyphus; Tyrothrix; Tyto.

U

Uca; Uleiota; Ulmus; Umbra; Umbrina; Uncaria; Uncia; Uncinaria; Uudulina; Unio; Upupa; Uragoga; Uranotaenia; Ureaplasma; Urechis; Urechites; Urginea; Uria; Urocerus; Urocyon; Urocystis; Uroglena; Uroglenopsis; Uromastyx; Uronema; Urophycis; Uroplatus; Urosalpinx; Ursus; Urtica; Usnea; Ustilago.

V

Vaccaria; Vaccinium; Vahlkampfia; Valeriana; Valerianella; Vallonia; Valvaster; Valvata; Vampyromorpha; Vampyrus; Vanda; Vandellia; Vanellus ; Vanessa; Vanilla; Varanus; Variimorda; Variocorhinus; Variola; Variolaria; Vaspa; Vateria; Veillonella; Velamen; Velella; Velia; Velifer; Veltheimia; Venus; Veratrum; Verbascum; Verbena; Vermicella; Vermivora; Veronia; Veronica; Verticillium; Vertigo; Vespa; Vespertilio ; Vespula; Vibrio; Vibrion; Viburnum; Vicia; Victoria; Villarsia; Vimba ; Vinca; Vincetoxicum; Viola; Vipera; Vireo; Viscaria; Viscum; Vitaliana; Vitex; Vitis; Vitrea; Vitreoscilla; Vitrina; Viverra; Viverricula; Viviparus; Volucella; Volvox; Vombatus; Vormela; Vorticella; Vriesia; Vulpes; Vulpia; Vultur.

W

Wahlenbergia; Wasielewskia; Weigela; Welwitschia; Willemetia; Wisteria; Withania; Wolffia; Wombatula; Woodsia; Wuchereria; Wulfenia.

X

Xanthium; Xanthoceras; Xanthochymus; Xanthomonas; Xanthoria; Xanthorrhiza; Xema; Xenicus; Xenophora; Xenopsylla; Xenopus; Xenosaurus Xenus; Xeranthemum; Xeris; Xerolycosa; Xestobium; Ximenia; Xiphias; Xiphophorus; Xiphorhamphus; Xiphorhynchus; Xiphosura; Xyleborus; Xyleutes; Xylocopa.

Y

Yersinia; Yoldia; Yponomeuta; Yucca.

Z

Zabrus; Zaglossus; Zalophus; Zanclus; Zannichellia; Zantedeschia; Zanthoxylum; Zapus; Zea; Zebrasoma; Zebrina; Zeledonia; Zelkova; Zenkerella; Zephyranthes; Zerynthia; Zeuglodon; Zeus; Zeuzera; Zingiber ; Zinnia; Ziphius; Zirfaea; Zoarces; Zonaginthus; Zoogloea; Zootermopsis; Zopfius; Zorotypus; Zostera; Zosterops; Zuberella; Zygadenus; Zygaena; Zygnema; Zygocotyle; Zygoena; Zymobacterium; Zymogenes; Zymomonas; Zymonema.

Index

Aardvarks/Earth-pigs → Orycteropus
Acanthocephalous parasites → Onciola
Acanthus/Hygrophylla plants → Acanthus, Asteracantha, Hygrophila, Jacobinia, Thunbergia
Adderstongue/Adder's Tongue plants → Botrychium, Ophioglossum
Agave plants → Agave
Algae → Oscillaria, Scytonema
    Blue-green → Aphanizomenon
    Brown → Ascophyllum, Chorda, Fucus, Halidrys, Himanthalia, Laminaria
    Flagellate → Eudorina
    Green → Acrosiphonia, Codium, Cosmarium, Enteromorpha, Volvox, Zygnema
    Red → Ceramium, Chondrus, Corallina, Delesseria, Digenea, Furcellaria, Gigartina, Halarachnion, Hildenbrandia, Laurencia, Lithophyllum, Membranoptera, Plocamium, Plumaria
Amaranth plants (family Amaranthaceae) → Achyranthes, Amaranthus, Celosio
Amaryllis plants (family Amaryllidaceae) → Sprekelia, Sternbergia
Amebas → Councilmania
    Ameboid bodies → Lymphocytozoon
    Coprozoic → Dimastigamoeba, Hartmanella, Naegleria, Vahlkampfia, Wasielewskia
    Flagellated → Histomonas, Trimastigamoeba
        (Rhizopod) → Mastigamoeba
    Non-pathogenic → Chaos
        (found in invertebrates) → Endamoeba
        (intestinal) → Endolimax, Iodamoeba, Karyamoebina
        (vegetable) → Blastocystis
    Pathogenic → Caudamoeba, Diendamoeba, Dientamoeba, Entameba, Entamoeba, Loeschia
    Testacean/Testate → Arcella, Difflugia
Anteaters/Numbats/Pangolins -
    Anteaters → Cyclopes
    Anteaters/Numbats → Myrmecobius
    Anteaters (toothless) → Myrmecophaga
    Pangolins (Scaly anteaters) → Manis
Antelopes (Antelope subfamily Antilopinae) -
    Addax Antelopes → Addax, Addox
    Antelopes → Tragelaphus
    Blackbucks → Antilope
    Blue Bull Antelopes → Boselaphus
    Bongos → Boocercus
    Chamois → Antilope
    Dwarf → Hylarnus
    Dwarf (Dik-dik) → Madoqua
    Elands → Taurotragus
    Gazelles → Ammodorcas, Gazella, Lithocranius, Litocranius
    Gnus/Wildebeest → Gorgon
    Goat-antelopes/Tahrs → Hemitragus
    Goats → Budorcas, Capra, Capricornis

      Hartebeest → Alcephalus
      Impala → Aepyceros
      Pronghorns → Antilocapra
      Pygmy (neotragine) → Neotragus
      Roan Antelopes → Hippotragus
      Saiga → Saiga
      Springboks → Antidorcas
Ant-lions - see under Lacewing insects
Ants → Anomma, Camponotus
      Army → Eciton
      Black → Lasius
      Bull/Bulldog → Myrmecia
      Driver → Dorylus
      Giant → Dinoponera
      Harvester → Messor
      Harvesting/Parasol → Atta
      Honeypot → Myrmecocystus
      Indian (Worker) → Oecophylla
      Long-lived (Westwoodi's) → Stennama
      Red → Myrmica
      Velvet → Mutilla
      Wood → Formica, Potosia
      Wood-driller → Megarhyssa
      Worker → Manica
      Worker (world's smallest ant) → Oligomyrmex
Aphids - see under Flies
Aphrodisiac plants → Sphaeranthus
Apocyanaceous plants - see under Trees - Apocyanaceous/Dogbane
Aquatic plants → Thalia
Aralia plants → Dizygotheca
Armadillos → Cabassous, Chaetophractus, Chlamyphorus, Dasypus
Aromatic and/or Medicinal plants -
      (Anthelmintic) → Veronia
      (Anti-asthma/diarrhea) → Tylophora
      (Antispasmodic) → Talauma
      (Basil/Mountain Mint) → Pycnanthemum
      (Benoil/Sajina) → Moringa
      (Boldin producing) → Peumus
      (Chaulmoogra Oil/Leprosy) → Hydnocarpus
      (Cocaine producing) → Erythroxylon, Erythroxylum
      (Damiana/aphrodisiac) → Turnera
      (False Spikenard) → Smilacina
      (Ginseng) → Panax
      (Gum-resin producing) → Garcinia
      (Hallucinogenic) → Piptadenia
      (Hydnocarpus Oil/Leprosy) → Taraktogenos
      (Hydrophyllaceous) → Eriodictyon
      (Mansa - "tonic") → Houttuynia
      (Nux Vomica) → Strychnos

(Passionflower) → Passiflora
(Poke/emetic) → Phytolacca
(Polychrest remedy) → Typhonium
(Purgative) → Xanthochymus
(Rosebay/Fireweed) → Chamaenerion
(Spigelin-producing) → Spigelia
(Talin producing/Sweetener) → Thaumatococcus
(Tannin producing Witch-hazel) → Hamamelis
(Trianthema producing) → Trianthema
(Willow-herb) → Epilobium
(Yellowroot) → Xanthorrhiza
Arrowroot plants → Maranta
Arthropods (Endoparasitic/Worm-like) → Pentastoma, Porocephalus
Arum plants → Acorus, Amorphophallus, Anthurium, Arisaema, Arisarum, Arum, Colocasia,
        Dieffenbachia, Dracunculus, Lysichiton, Monstera, Rhaphidophora, Scindapsus,
        Spathiphyllum
Aye-aye (Mammals) → Daubentonia

Badgers/Martens/Minks/Otters/Polecats/Skunks/Stoats/Weasels/etc. (family Mustelidae) -
        Badgers (Common) → Meles
        Badgers (Ferret) → Melogale
        Badgers (Hog) → Arctonyx
        Badgers (Honey/Ratel) → Mellivora
        Grisons → Grison
        Martens → Martes
        Minks → Lutreola
        North American Minks → Mustela
        Old World Otters → Lutra
        Otters → Lutrogale
        Otters (Giant Brazilian) → Pteronura
        Otters (Small-clawed) → Aonx
        Polecats → Ictonyx, Putorius
        Polecats/Weasels → Vormela
        Sea Otters → Enhydra
        Skunks (Striped) → Mephitis
        Wolverine/Glutton → Gulo
Balsam plants → Abies, Impatiens
Banana plants → Musa, Strelitzia
Bandicoots - see under Rodents - Rats
Barberry plants → Epimedium, Mahonia, Podophyllum
Basketstar Echinoderms (class Ophiuroidea) → Astroboa, Astrotoma, Gorgonocephalus
Bats → Barbastella
        Blood-sucking Vampire → Desmodus, Diphylla
        False Vampire → Megaderma
        Fruit/Fruit-sucking → Eidolon, Epomorphorus, Hypsignathus, Macroglossus, Megaloglossus,
                Pteropus, Rousettus
        Horseshoe → Rhinolophus
        Leaf-nosed → Macrotus, Vampyrus
        Long-eared (family Vespertilionidae) → Eptesicus, Myotis, Nyctalus, Pipistrellus, Plecotus,

       Vespertilio
     Long-winged → Miniopterus
     Mastiff → Eumops
Bearberry plants - see under Trees/Shrubs
Bears (family Ursidae) → Helarctos
     (Brown) → Ursus
     (Polar) → Thalarctos
     (Sloth) → Melursus
Bear's Breech plants → Acanthus
Beavers -
     (Dam-constructing) → Castor
     (Mountain) → Aplodontia
Bedstraw/Madder plants (family Rubiaceae) → Asperula, Myrmecodia
     (Catechu) → Uncaria
     (Coffee) → Coffea
     (Crosswort) → Cruciata
     (Cuprea bark) → Remijia
     (Gardenia) → Gardenia
     (Hedge) → Galium
     (Ipecacuanha) → Cephaelis, Uragoga
     (Madder) → Rubia
     (Quinine) → Cinchona
     (Yohimbe) → Corynanthe, Pausinystalia
Bees -
     Black → Panurgus
     Bumble → Bombus
     Carpenter → Xylocopa
     Cuckoo → Psithyrus
     Flowering clover → Melitta
     Hairy-legged Mining → Dasypoda
     Hive/Honey → Apis, Melipona
     Leaf-cutter → Megachile
     Mason (Poppy) → Osmia
     Parasitic → Leucopsis
     Solitary → Andrena, Anthophora, Anthrophora, Colletes, Halictus
Beetles → Bitoma, Chalcophora, Dictyoptera, Malachius, Oedemera, Uleiota
     (found in tree fungi) → Endomychus
     (larvae found in tree stumps) → Hylecoetus
     Bark (Fig insect) → Blastophaga, Cryphalus
     Bark (Spruce) → Ips
     Bark (Pine) → Tomicus
     Bark (Weevil-type) → Xyleborus
     Blister - see under Soldier Beetle
     Cardinal → Pyrochroa, Variimorda
     Carpet, Hide, Museum → Anthrenus, Attagenus, Dermestes
     Carrion → Aclypea, Necrophorus, Nicrophorus, Oiceoptoma
     Chafer/Dung/Scarab (Lamellicorn - Lamelliform) → Amphimallon, Amphimallus, Anisoplia,
         Anomala, Aphodius, Catharsius, Cetonia, Chalcosoma, Chiasognathus, Cladognathus,
         Copris, Dicranocephalus, Dynastes, Euchlora, Geotrupes, Goliathus, Golofa,

Megalorhina, Melolontha, Oryctes, Phyllopertha, Serica, Tropinota
Checkered → Thanasimus
Click → Agriotes, Agrypnus, Athous, Corymbites, Elater, Selatosomus
Coleopterous (Sheath-winged) → Dorcus, Lagria, Laria
Darkling/Mealworm → Tenebrio
Dung - see under Chafer
Dwarf → Nasonella
Elephant → Megasoma
Flour → Tribolium
Flower → Glischrochilus, Meligethes, Pityophagus, Pocadius
Flower (Shining) → Olibrus
Glow-worm → Lampyris
Ground → Agonum, Brachinus, Calosoma, Carabus, Cychrus, Harpalus, Notiophilus,
    Pterostichus, Zabrus
Histerid/Polyphagan → Hister, Hololepta, Hydrophilus
Japanese → Popillia
Ladybird → Adalia, Anatis, Chilocorus, Coccinella, Thea
Leaf → Agelastica, Cassida, Chrysomela, Cryptocephalus, Galeruca, Leptinotarsa, Lilioceris,
    Oulema, Phyllobrotica, Phyllodecta, Plagiodera, Smaragdina
Long-haired → Acanthocinus
Long-horned (Longicorn) → Acrocinus, Aegosoma, Aromia, Batocera, Callidium, Cerambyx,
    Clytus, Dorcadion, Harpium, Hylotrupes, Lamia, Leiopus, Leptura, Macrodontia,
    Molorchus, Morimus, Oberea, Petrognatha, Plagionotus, Prionus, Pyrrhidium,
    Rhagium, Saperda, Spondylis, Strangalia, Tetropium, Titanus
Metallic Green → Byctiscus
Metallic Wood-borer → Agrilus, Anthaxia, Buprestis, Burretis, Trachys
Oil - see under Soldier Beetle
Pill → Byrrhus
Raspberry → Byturus
Rose → Creophilus
Rove → Oxyporus, Staphylinus
Scarab - see under Chafer
Scavenging → Hygrobia
Soldier/Blister/Oil → Cantharis, Lytta, Meloe, Rhagonycha, Sessinia
Spider → Niptus, Ptinus
Stag → Lucanus, Sinodendron, Systenocerus
Swimming (Saucer Bug) → Hyocoris
Tiger → Cicindela, Mantichora
Water (Carnivorous) → Acilius, Agabus, Colymbetes, Cybister, Dytiscus, Graphoderus,
    Graptodytes, Hydraena, Hydroporus, Hyphydrus, Ilybius
Water-Scavenger → Hydrochara, Hydrous, Sphaeridium
Weevil → Anthonomus, Cionus, Curculio, Liparus, Otiorhynchus, Polydrusus
Weevil (Leaf Roller) → Apoderus, Attelabus, Deporaus
Whirligig (Scavenging) → Gyrinus
Wood → Xestobium
Woodworm (Furniture) → Anobium
Belladonna Leaf - see Nightshade plants
Bellflower plants (family Campanulaceae) → Adenophora, Campanula, Jasione, Legousia,
    Wahlenbergia

Berberine (an alkaloid) - see Trees/Shrubs (Barberry)
Betany plants → Betonica
Bindweed/Bineweed plants - see under Trees
Birds -
    Accentor → Prunella
    Albatrosses - see under Birds - Tubenose
    Auklets → Aethia
    Auks/Guillemots/Puffins/Razorbills (Shorebirds) → Alca, Cepphus, Eudromias, Fratercula,
        Phalaropus, Pinguinus, Uria
    Avocet/Stilt → Cladorhynchus, Himantopus, Recurvirostra
    Babblers - see under Birds - Nightingales
    Bee-eaters → Merops
    Bell-magpies/Butcherbirds/etc. → Cracticus
    Bitterns/Egrets/Herons → Ardea, Ardeola, Botaurus, Casmerodius, Egretta, Gorsachius,
        Ixobrychus, Leucophoyx, Nyctanassa, Nycticorax
    Blackbirds/Orioles → Agelaius, Dolichonyx, Euphagus, Icterus, Molothrus,
    Blackstarts → Cercomela
    Bluebirds → Irena
    Boatbills (Boatbill Herons) → Cochlearius
    Boobies/Gannets → Morus
    Broadbills → Calyptomena, Eurylaimus
    Bunting → Calcarius, Chloris, Emberiza, Plectrophenax
    Bustards → Ardeotis, Chlamydotis, Chorictis, Lissotis, Otis
    Butcherbirds - see under Birds - Shrikes
    Buzzards - see under Birds - Eagles
    Canary - see under Birds - Finches
    Cariama/Seriema → Cariama, Chunga
    Cassowary → Casuarius
    Cat/Mocking/Thrasher → Dumetella, Mimus, Toxostoma
    Chatterer/Cotinga → Cephalopterus, Chasmorhynchus
    Chough (Alpine) → Pyrrhocorax
    Cockatoos, Lories, Macaws, Parakeets, Parrots (Psittacidae) -
        Cockatoos → Cacatua, Kakatoe, Microglossus, Nymphicus
        Lories → Trichoglossus
        Macaws → Ara
        Parakeets, Parrots → Agapornis, Amazona, Conuropsis, Melopsittacus, Nestor
        Parrots → Psittacus, Strigops
    Condors - see under Birds - Vultures
    Coot/Gallinule/Moorhen/Rail → Aramides, Atlantisea, Atlantisia, Crex, Fulica, Gallinula,
        Gallirallus, Laterallus, Notornis, Porzana, Rallus
    Cormorants - see under Birds - Pelicans
    Courser/Pratnicole → Cursorius, Glareola
    Crab-plover → Dromas
    Cranes → Anthropoides, Anthyopoides, Balearica, Grus
    Creeper → Certhia
    Crows/Jays/Magpies -
        (Azure-winged Magpie) → Cyanopica
        (Blue Jay) → Cyanocitta
        (Crestless Jay) → Cyanocorax

(European Jay) → Garrulus

(House Crow) → Corvus

(Magpie) → Pica

(Nutcracker Crow) → Nucifraga

(Wattled Crow) → Callaeas

(Yelllow-billed Crow) → Coracia

Cuckoo → Centropus, Chalcites, Chrysococcyx, Clamator, Coccyzus, Coua, Crotophaga, Cuculus

Cuckoo (Road-runner) → Geococcyx

Cuckoo-shrike/Minivet → Campephaga, Coracina

Curassow → Crax, Mitu, Nothocrax

Curlew - see under Birds - Plovers

Darters → Anhinga

Dipper/Water-ouzel → Cinclus

Diver/Loon → Gavia

Doves/Pigeons → Caleonas, Columba, Columbigallina, Coura, Didunculus, Ectopistes, Goura, Leptotila, Ocyphaps, Streptopelia

Drongoes → Chaetorhynchus, Dicrurus

Ducks/Geese/Swans/etc. -

(Barbary/Muscovy Ducks) → Cairina

(Diving/Pochard Ducks) → Aythya

(Diving Ducks/Mergansers) → Mergus

(Ducks) → Aix, Anas, Bucephala, Oxyura

(Eider) → Somateria

(Geese) → Chlöephaga

(Harlequin Ducks) → Histrionicus

(Long-tailed Ducks) → Clangula

(Merganser Ducks) → Lophodytes

(Perching Ducks) → Dendronessa

(Pochard) → Netta

(Scoter/Sea Ducks) → Melanitta

(Sheldgeese) → Cereopsis

(Shelducks) → Casarca, Tadorna

(Swans) → Coscoroba

(Swan Geese) → Cygnopsis

(Tree/Whistling Ducks) → Dendrocygna

(Wild Swan) → Cygnus

Eagles/Buzzards/Harriers/Hawks/Kites/Old World Vultures (family Accipitridae) → Aguila, Aquila, Circaetus, Haliaeetus, Haliaetus, Harpia, Hieraetus, Pithecophaga

Buzzards → Buteo, Pernis

Harriers → Circus

Hawks (True) → Accipiter

Kites → Elanoides, Elanus, Haliastur, Ictinia, Milvus

Old World Vultures → Aegypius, Gypaetus, Gyps, Neophron

Elephant → Aepyornis

(Fairy) Bluebirds/Leafbirds/Iora - see under each

Falcons -

(Carrion Hawks) → Caracara

(Eagle) → Polemaetus

(Falconets) → Microhierax

(Gentle) → Accipiter

(Peregrine) → Falco

Finches → Carpodacus, Coccothraustes, Fringilla, Loxia, Montifringilla, Poephila, Pyrrhula

(Canary) → Serinus

(Darwin's) → Carduelis

(Desert) → Rhodospiza

(Mountain) → Rhodopechys

(Plush-capped) → Catamblyrhynchus

(Zebra) → Taniopygis

Flowerpeckers → Anaimos

Flycatchers (Tyrant/New World)/Kingbirds (family Tyrranidae) → Contopus, Ficedula, Muscivora, Myiarchus, Myiochanes, Oxyruncus

Frigate → Fregata

Fulmars - see under Birds - Tubenose

Gadfly Petrels → Bulweria

Geese → Alopochen, Anser, Branta

Goshawks → Accipiter

Grassfinches/Mannikins/Waxbills (family Estrildidae) → Estrilda, Lonchura, Zonaginthus

Grebes → Aechmophorus, Centropelma, Colymbus, Podiceps, Podilymbus

Grouse/Ptarmigan → Bonasa, Lagopus, Lyrurus, Tetrao, Tetrastes

(Sage) → Centrocercus

(Sand) → Pterocles

Guinea-fowls → Acryllium, Numida

Gulls - see under Terns

Harriers - see under Eagles

Hawks - see under Eagles

Hoatzin → Opisthocomus

Honey-eaters → Anthornis

Honey-guides → Indicator

Hoopoe → Upupa

Hornbills → Buceros, Lophoceros

Huia/Saddleback/Wattled Crows → Heteralocha

Hummingbirds → Acestrura, Archilochus, Archulochus, Heliactin, Loddigesia, Mellisuga, Scopus, Selasphorus

Ibises - see under Birds - Spoonbills

Jabirus - see under Birds - Storks

Jacanas → Actophilornis

Jagger/Skua → Catharacta

Jays - see under Birds - Crows

Kingfishers → Alcedo, Ceryle, Chloroceryle, Dacelo, Halcyon, Ispidina, Megaceryle

Kites - see under Birds - Eagles

Kiwis → Apterygiformes, Apteryx

Larks -

(Bush) → Mirafra

(Calandra) → Melanocorypha

(Crested) → Galerida

(Desert) → Ammomanes

(Magpie) → Corcorax

    (Magpie-Lark/Mudlark) → Grallina

    (Shore) → Eremophila

    (Short-toed) → Calandrella

    (Skylarks) → Alauda

    (Wood) → Lullula

Leafbirds → Chloropsis

Lilytrotters/Jacanas → Actophilornis, Jacana

Limpkins/Courlans → Aramus

Lory/Lories - see under Birds - Cockatoos, etc.

Lyrebirds → Menura

Macaws - see under Birds - Cockatoos, etc.

Magpies - see under Birds - Crows

Manakins → Chiroxiphia

Mannikins - see under Birds - Grassfinches

Martins - see under Birds - Swallows

Megapodes/Moundbuilders → Leiopa

Mesite/Monia/Roatelo → Monias

Mockingbird - see under Birds - Cat/Mocking/Thrasher

Nightingales/Babblers/Baldcrows/Old World Flycatchers/Robins/Thrushes/Warblers/etc.
    (families Muscicapidae; Sylviidae; etc.) → Acrocephalus, Alcippe, Cercotrichas,
    Cettia, Cyanosylvia, Dendroica, Enicurus, Erithacus, Eupetes, Hippolais, Hylocichla,
    Kittacincla, Locustella, Luscinia, Lusciniola, Malacocincla, Mniotilta, Monticola,
    Muscicapa, Oenanthe, Orthotomus, Phoenicurus, Phylloscopus, Regulus, Saxicola,
    Sylvia, Turdus, Turdoides, Vermivora, Zeledonia

Nightjars → Caprimulgus
    (Nighthawks) → Chordeiles, Nyctidromus

Nuthatches → Hypositta, Sitta

Old World Vultures - see under Birds - Eagles

O-o (Hawaiian) → Moho

Orioles → Oriolus

Ospreys → Pandion

Ostriches → Struthio

Ovenbird/Baker → Furnarius

Owls (Typical) → Aegolius, Asio, Athena, Athene, Bubo, Glaucidium, Nyctea, Otus, Strix,
    Syrnium, Tyto

Oxpeckers/Starlings (family Sturnidae) → Buphagus, Gracula, Sturnus

Oystercatchers/Sea-pies → Haematopus

Palm-chats → Dulus

Parakeets - see under Birds - Cockatoos, etc.

Parrots - see under Birds - Cockatoos, etc.

Partridges/Pheasants/Quails → Alectoris, Ammoperdix, Argusianus, Callipepla, Colinus,
    Coturnix, Crossoptilon, Crossoptilum, Cyrtonyx, Excalfactoria, Francolinus, Gallus,
    Lophophorus, Lophortyx, Lophura, Oreotyx, Perdix, Phasianus
    (Cock Pheasant) → Chrysolophus

Pelicans/Cormorants/Gannets → Pelecanus, Phalacrocorax, Sula

Penguins → Aptenodytes, Eudyptes, Eudyptula, Megadyptes, Pygoscelis

Petrels - see under Birds - Tubenose

Pheasants - see under Birds - Partridges

Pipits/Wagtails → Anthus, Motacilla

Plovers/Lapwings/Sandpipers/Waders/Snipes/Tattlers/Curlews -
    (family Charadriidae) → Anarhynchus, Aphriza, Arenaria, Charadrius, Haplopterus,
      Pluvialis, Vanellus
    (family Scolopacidae) → Actitis, Bartramia, Calidris, Capella, Catoptrophorus,
      Crocethia, Erolia, Gallinago, Heteroscelus, Limicola, Limnodromus, Limosa,
      Lymnocryptes, Numenius, Philomachus, Scolopax, Tringa, Xenus
Poor-will/Torpid birds (Nuttall's) → Phalaenoptilus
Potoos - see under Birds - Wood-nightjars
Puffbirds → Notharchus
Quails - see under Birds - Partridges
Quelea birds → Quelea
Razorbills → Alcaligenes
Robins - see under Birds - Nigtingales
Roller → Coracias, Eurystomus
Sandpipers - see under Birds - Plovers
Screamers → Anhima, Chauna
Scrub-birds → Atrichornis
Shearwater birds - see under Birds - Tubenose
Sheathbill → Chionis
Shoebills → Balaeniceps
Shorebirds - see under Birds - Auks; Plovers
Shrikes/Butcherbirds → Lanius
Skuas → Stercorarius
Snipes - see under Birds - Plovers
Sparrows/Weaver birds → Passer
Spoonbills/Ibises → Ajaia, Nipponia, Threskiornis
Starling - see under Birds - Oxpeckers
Storks/Jabirus → Anastomus, Ciconia, Ephippiorhynchus, Euxenura, Ibis, Jabiru, Leptoptilos,
       Leptoptilus, Mycteria
Sunbirds → Aethopyga, Anthreptes, Nectarinia
   (False) → Neodrepanis
Sunbittern → Eurypyga
Sun-grebes/Finfoot birds → Heliopais, Heliornis
Swallows/Martins → Acanthodactylus, Aeronautes, Cecropis, Delichon, Hirundo,
      Iridoprocne, Riparia
Swifts → Apus, Callocalia, Chaetura, Collocalia, Hemiprocne
Sylphs → Aglaiocercus
Tanager → Chlorophonia
Tattlers - see under Birds - Plovers
Terns/Gulls → Anous, Chlidonias, Gelochelidon, Hydropogne, Larosterna, Larus, Rissa,
      Sterna, Xema
Thick-knee → Burhinus
Thrasher - see under Birds - Cat/Mocking/Thrasher
Thrushes - see under Nightingales
Tinamous → Crypturellus, Eudromia
Tits → Aegithalos, Panurus, Parus, Remiz
Tree Creeper (Australian) → Climacteris
Trogons → Calures, Hapalodermes, Harpactes
Tubenose birds (Albatrosses/Fulmars/Petrels/ Shearwater) -

(Albatrosses) → Diomedea, Phoenetria
(Fulmars) → Fulmarus
(Petrels) → Hydrobates, Oceanites, Oceanodroma
Turkeys → Agriocharis, Meleagris
Vireo birds → Vireo
Vultures -
(Condors) → Gymnogyps, Vultur
(New World) (family Cathartidae) → Cathartes, Coragyps
(Old World) - see under Eagles
Waders - see under Birds - Plovers
Wagtails - see under Birds - Pipits
Walking Birds (Emu) → Dromaius, Dromiceius
Wall creeper → Tichodroma
Warblers - see under Birds - Nightingales
Waxbills - see under Birds - Grassfinches
Waxwings → Bombycilla, Hypocolius
Weaver birds - see under Sparrows
White-eyes/Silvereyes → Zosterops
Woodhewers → Xiphorhynchus
Wood-nightjars/Potoos → Nyctibius
Woodpeckers (True)/Wrynecks → Colaptes, Dendrocopos, Dryocopus, Jynx, Melanerpes,
Picus
Wood-swallows → Artamus
Wrens → Acanthisitta, Troglodytes, Xenicus
Wren-tits → Chamaea
Birdsnest plants → Monotropa
Birthwort plants → Aristolochia, Asarum
Bixaceous plants → Chaulmoogra, Gynocardia
Bogbean/Buckbean plants → Menyanthes, Nymphoides
Borage plants → Alkanna, Anchusa, Asperugo, Cerinthe, Cynoglossum, Echium, Eritrichium,
Heliotropium, Lappula, Lithospermum, Lycopsis, Myosotis, Nonnea, Omphalodes,
Symphytum
Brittlestar Echinoderms (class Ophiuroidea) → Amphiophiura, Amphipholis, Amphiura,
Ophiacantha, Ophiactis, Ophiarachna, Ophiodaphne, Ophioderma, Ophiomisidium,
Ophionotus, Ophiosphalma, Ophiothrix
(Sandstars) → Ophiura, Opiura
Broomrape plants → Lathraea, Orobanche
Bryony plants (Used in homeopathy) → Tamus
Brunfelsia (South American) plants → Brunfelsia, Franciscea
Buchu plants → Agathosma
Buckthorn plants → Frangula, Rhamnus
Bugs -
Bed → Acanthia, Cimex, Clinocoris, Clinophilus, Klinophilus
Capsid → Deraeocoris, Lygus, Phytocoris
Cone-nosed → Conorhinus, Entriatoma, Lamus, Panstrongylus, Triatoma
Cotton-stainer (Firebug) → Pyrrhocoris
Damsel → Nabis
Flower → Anthocoris, Triphleps
Green → Palomena

    Kissing/Corsair → Melanolestes
    Pill/Sow (Woodlice) → Armadillidium, Macrostylis, Porcellio
    Pond-skaters/Water-bugs/-striders → Aquarius, Gerris, Hydrometra, Ilyocoris
    Sow - see under Bugs - Pill/Woodlice
    Shield → Antestia, Murgantia, Nezera
    Spittle → Aphrophora, Cercopis
    Squash → Anasa, Mesocerus
    Swimming → Aphelocheirus
    Terrestrial → Blepharidopterus, Blissus
    True → Aelia, Eurydema, Graphosoma, Oeciacus, Picromerus
        (Plant) → Dolycoris
    Trypanosoma-transmitting → rhodnius
    Water Boatman (Corixid) → Corixa
Burseraceous plants -
    (Elemi) → Canarium
    (Myrrh) → Commiphora
Buttercup/Crowsfoot plants (family Ranunculaceae) -
    Aconite → Aconitum
    Aconite (winter) → Eranthis
    Adonidis → Adonis
    Baneberry → Actaea
    Berberine-producing → Coptis
    Bugbane, Black Snakeroot → Cimicifuga
    Calianthemum → Calianthemum
    Climbing → Clematis
    Columbine → Aquilegia
    Fennel-flower → Nigella
    Globe Flower → Trollius
    Hellebore → Helleborus
    Isopyrum → Isopyrum
    Liverleaf → Hepatica
    Marsh Marigold → Caltha
    Mousetail → Myosurus
    Paparraz/Stavesacre → Delphinium
    Pasque Flower/Pulsatilla → Anemone
    Pulsatilla/Anemone → Pulsatilla
Butterflies - see under Moths

Cactus plants → Aeonium, Ancistrocactus, Anhalonium, Aporocactus, Astrophytum, Cereus,
        Echinocactus, Epiphyllum, Ferocactus, Gymnocalycium, Leptocladodia,
        Leuchtenbergia, Lophophora, Mammillaria, Opuntia, Rhipsalidopsis, Schlumbergera,
        Sempervivum
    (Asafetida) → Ferula
Caecilian amphibians → Chthonerpeton
Camels or Dromedaries → Camelus, Lama, Llama
Candy Mustard plants → Aethionema
Carrot plants -
    Alpine Lovage → Ligusticum
    Ammoniacum → Dorema

Angelica → Angelica
Aniseed → Pimpinella
Caraway/Kummel/Ajowan → Carum, Trachyspermum
Cenolophium → Cenolophium
Chervil → Chaerophyllum
Cicely → Myrrhis, Osmorhiza
Cnidium → Cnidium
Conium/Hemlock → Conium
Coriander → Coriandrum
Cowbane/Water-hemlock → Cicuta
Cummin → Cuminum
Dill → Anethum, Peucedanum
Eryngo → Eryngium
Fennel → Foeniculum
Goutweed → Aegopodium
Hacquetia → Hacquetia
Hemlock-parsley → Conioselinum
Hogwood/Cow Parsnip → Heracleum
Laser → Laser
Laserwort → Laserpitium
Longleaf → Falcaria
Marsh Angelica → Ostericum
Masterwort → Astrantia
Orlaya → Orlaya
Parsley → Anthriscus, Carum, Petroselinum
    (Bur) → Caucalis
    (Celery) → Apium
    (Fool's) → Aethusa
Pennywort (Indian) → Centella, Hydrocotyle
Spignelmeu → Meum
Thapsia → Thapsia
Visnaga → Ammi
Wild → Daucus
Cathartic plants (family Polygonaceae) (Rhubarb) → Rheum
Cats (family Felidae) -
Cheetahs → Acinonyx
Domestic/Jaguars/Leopards/Lions/etc. → Felis
Golden (wild) cat → Profelis
Jaguars → Jaguarius
Jaguars/Leopards/Tigers/etc. → Panthera
Leopards/Ocelots → Leopardus, Uncia
Lions → Leo
Lynxes → Caracal, Lynx
Manul/Pallas's cat → Otocolobus
Marsupials → Dasyurops, Dasyurus
Servals → Leptailurus
Cattle (Horned Ungulates) (family Bovidae) -
Antelopes - see under Antelopes
Auroch, Bison, Wisent → Bison

Buffaloes -
- Dwarf/Pygmy → Anoa
- Ox (Wild) → Bibos
- Water → Bubalus

Chamois → Rupicapra
Cows → Bos
Goats → Oreamnos, Oxeamnos
Oryx → Oryx
Musk-ox → Ovibos
Sheep - see under Sheep

Cavy Mammals/Guinea Pigs → Dolichotis, Hydrochoerus
Centipedes (class Chilopoda) → Geophilus, Himantarum, Lithobius, Orya, Pachymerium, Scolopendra, Scutiger
(Earwigs) → Labia
Chestnuts (Horse) → Aesculus
Chickweed - see under Pink plants
Chinchillas - see under Rodents
Chlamydial microorganisms (family Chlamydiaceae) → Bedsonia, Chlamydia, Colesiota, Colettsia, Miyagawanella, Ricolesia
Chordate animals -
- (Acorn Worms) → Balanoglossus, Glandiceps, Glossobalanus, Harrimania, Protoglossus, Ptychodera, Saccoglossus, Schizocardium
- (Comb Jellies/Sea Walnuts) → Boeroë, Bolina, Bolinopsis, Callianira, Cestum, Euchlora, Leucothea, Mnemiopsis, Velamen
- (Comb Jellies, Tentaculate) → Coeloplana
- (Hydrozoan medusae) → Olindias
- (Jellyfishes) → Nausithoe, Rhizostoma
- (Jellyfishes - True) → Cassiopeia, Chrysaora, Chrysoara, Corymorpha, Cyanea, Leuckartiara
- (Jellyfishes - Sea wasp) → Charabdea, Chiropsalmus
- (Jellyfishes - Stalked) → Craterolophus, Haliclystus, Lucernaria
- (Jellyfishes - "Thimble") → Linuche
- (Lanceolets) → Amphioxus, Branchiostoma
- (Lampreys) → Eudontomyzon
- (Planktonic Salps) → Doliolum
- (Pterobranchids) → Cephalodiscus
- (Sea Squirts) → Amaroucium, Ascidiella, Botryllus, Clavelina, Cynthia, Dendrodoa
- (Sea Squirts/Sea Vases) → Ciona
- (Siphonophores) → Physalia

Cicadas - see under Locusts
Ciliate Protozoans → Chilodon, Chilodonella, Cothurnia, Ichthyophthirius, Trichodina, Troglodytella, Uronema, Vorticella
Cinnamon plants → Canella
Civet/Genet/Mongoose mammals → Cryptoprocta
- (Civets) → Arctictis, Civettictis, Fossa, Paguma, Paradoxurus, Viverra, Viverricula
- (Genets) → Genetta
- (Mongoose) → Herpestes, Ichneumia, Mungos
- (Otter Civets) → Cynogale
Climbing plants → Campsis, Plumbago, Tropaeolum
Cocoa plants - see under Soapberry plants

Coccobacilli → Achromobacter, Acinetobacter, Alcaligenes, Alkaligenes, Asterococcus,
    Astrococcus, Azotomonas, Brucella, Calymmatobacterium, Chlorobium,
    Chlorochromatium, Clathrocystis, Coccobacillus, Cylindrogloea, Dialister,
    Flavobacterium (group II), Francisella, Mima, Moraxella, Pasteurella, Photobacterium,
    Rhodopseudomonas, Sideromonas, Zygomonas
Cochineal producing Hemipterous Insects → Coccus, Dactylopius
Cockroaches/Mantids → Attaphila, Blatta, Cryptocercus, Ectobius, Periplanea, Perplaneta
    (Praying Mantids) → Stagmomantis, Tenodera
Cola/Kola plants → Cola
Comb Jellies - see Chordate animal
Copal plants -
    (Austrial copal) → Agathis
    (Brazilian copal) → Hymenoea
Corals -
    Astrangial → Astrangia
    Octocorals → Akabaria, Dendronephyta, Sarcophyton
        (Black/Thorny) → Antipataria, Antipathes
        (Horny) → Corallium, Eunicea, Gorgonaria, Gorgonia, Hymenogorgia, Isis
        (Horny/Sea Fans) → Eunicella, Plexaura
        (Pipe Organ) → Tubipora
        (Soft) → Helipora
    Sea Pens → Acanthoptilum, Pennatula
    Solitary → Balanophyllia, Caryophyllia
    Thorny → Cerianthus
    True (Stony) → Acropora, Favia, Fungia, Galaxea, Goniopora, Madrepora, Montastrea
        (Brain corals) → Meandrina
        (Yellow corals) → Turbinaria
Coypu - see under Rodents
Crabs - see under Custaceans - Decapod
Cranesbill plants - see under Geranium
Crassula plants → Echeveria, Jovibarba, Kalanchoe, Sedum
Crayfishes - see under Crustaceans - Decapod
Cress/Crucifer plants → Alliaria, Arabidobsis, Arabis, Cakile, Calepina, Camelina, Capsella
    (Alpencress) → Hutchinsia
    (Ball Mustard) → Neslia
    (Cabbage) → Conringia
    (Candytuft) → Iberis
    (Cuckoo Flower) → Cardamine
    (Flixweed) → Descurainia
    (Hairy Mustard) → Erucastrum
    (Hoary-cress) → Cardaria
    (Horseradish) → Armoracia
    (Horseradish, Scurvy-grass) → Cochlearia
    (Moonwort) → Lunaria
    (Night-scented Rocket) → Hesperis
    (Pepperwort) → Lepidium
    (Rock-cress) → Cardaminopsis
    (Rock Garden) → Alyssum, Aubrieta, Aubrietia, Aurinia
    (Rock Garden/Alyssum) → Lobularia

(Rock Hutchinsia) → Hornungia
(Rock Scurvy-grass) → Kernera
(Salt-cress) → Hymenobolus
(Sea-kale) → Crambe
(Stinkweed) → Diplotaxis
(Three-horned Stock) → Matthiola
(Treacle-mustard) → Erysimum
(Virginia Stock) → Malcomia
(Wallflower) → Cheiranthus
(Wart-cress) → Coronopus
(Watercress) → Nasturtium
(Whitlow-grass) → Draba, Erophila
(Woad) → Isatis
Crickets-
    Bush → Anabrus, Decticus, Leptophyes, Phyllopora, Tettigonia
    Camel → Ceuthophilus
    Large → Pseudophyllanax
    Mole → Gryllotalpa
    Primitive → Grylloblatta
    True → Acheta, Acneta, Gryllus, Oecanthus
        (Ground/Wood) → Nemobius
    Water → Velia
Crocodiles -
    (Alligator) → Alligator
    (Black Caiman) → Melanosuchus
    (Broad-fronted) → Osteolaemus
    (Caiman) → Paleosuchus
    (Caiman/Cayman) → Caiman
    (Indian Crocodile/Gavial) → Gavialis
    (Nile Crocodile, etc.) → Crocodylus
Croton plants → Codiaeum, Croton
Crowberry plants → Empetrum
Crowsfoot plants - see under Buttercup plants
Crucifer plants - see Cress plants
Crustaceans -
    Amphipod → Hyperia, Ingolfiella, Phronima
    Arthropod → Siphonophanes, Triops
    Branchiopod → Branchinecta, Lepidurus
    Branchiuran (Fish Lice) → Chonopeltis
    Barnacle/Cirripede → Balanus, Lepas
    Blind (Worm-like) → Derocheilocaris
        (Freshwater) → Micraspides
    Cephalopods (Squids) → Stenoteuthis
    Conchostracan → Cyclestheria
    Copepod → Balaenophilus, Calanus, Cecrops, Corycaeus, Cyclops, Demoixys, Diaptomus,
         Lepeophtheirus, Lernaea, Lernaeocera
        (Cyclops) → Macrocyclops
        (Cyclopod) → Mytilicola, Notodelphys
    Cumacean → Almyracuma

Decapod - Crabs → Callinectes, Carcinus, Lybia, Neptunus, Paguristes
    (Diminutive/Pea) → Pinnotheres
    (Edible) → Cancer
    (Fiddler) → Uca
    (Ghost/Racing) → Ocypode
    (Hermit) → Eupagurus, Pagurus
    (Horseshoe) → Limulus, Xiphosura
    (Land Hermit) → Birgus
    (Long-tailed) → Diogenes
    (Marine) → Eriocheir, Eriocheiris, Ethusina
    (Porcelain) → Porcellana
    (Porcelain Crabs) → Orthochela
    (Primitive Sponge) → Dromia
    (Short-tailed) → Dorippe, Ilia
    (Spider) → Hyas, Inachus, Macrocheira, Macropodia
    (Stone) → Lithodes, Paralithodes
    (Swimming) → Portunus
    (True/Land) → Geocarcoidea
    (Velvet) → Macropipus
    (Xanthid) → Pseudocarcinus
Decapod - Crayfishes → Astacopsis, Astacus, Astracopsis, Cambarellus, Cambarus,
      Orconectes, Potamobius, Procambarus, Tenuibranchiurus
Decapod - Lobsters/Prawns -
    (Common) → Homarus
    (Freshwater) → Euastacus
    (Marine) → Galathea
    (Prawns/Scampi) → Nephrops
    (Spiny) → Panulirus
Decapod - Others → Hoplophorus
Euphausiid → Bentheuphausia
Freshwater → Atyaephyra
    (Anthurans) → Cyathura
    (Harpacticoids) → Canthocamptus
    (Lakes/Conchostracans) → Limnadia
    (Thermosbaenacean) → Monodella
    (Water-fleas) → Daphnia
Isopod (Sowbugs/Woodlice) → Astacilla, Bathynomus, Mesamphisopus, Mesidotea, Oniscus,
      Paragnathia, Phreatoicopsis, Saduria
Leptostracine → Nebalia
    (Planktonic) → Nebaliopsis
Lobsters - see under Lobsters
Marine → Hutchinsoniella
Ostracod (seed Shrimps) → Gigantocypris, Mesocypris
Prawns (Decapods) → Hippolyte
Primitive → Chirocephalus
Sandhopper (Orchestiidae) → Orchestia
Sea-slater (Isopods) → Ligia
Shrimps (Decapods) → Caprella, Coronula, Corophium
    (Brine) → Artemia

        (Common) → Crangon
        (Freshwater) → Gammarus, Niphargus, Rivulogammarus
        (Marine) → Idotea, Leucifer, Lysmata
        (Marine/Cleaner) → Stenopus
        (Opossum - Freshwater) → Mysis
        (Seed/Ostracods) → Cypris, Cytherella
    Water-fleas → Leptidora
Cryptomonad Protozoans → Chilomonas
Cucumber/Gourd/Melon plants → Citrullus, Cucumis, Cucurbita, Ecballium, Momordica

Daffodil plants (family Amaryllidaceae) -
    Amaryllis/Barbados Lily → Hippeastrum
    Amazon Lily → Eucharis
    Belladonna Lily → Amaryllis
    Blood Lily → Haemanthus
    Cape Colony → Nerine
    Fairy/Rain/Zephyr Lily → Zephyranthes
    Garlic → Allium
    Narrow-leaved → Narcissus
    Sea Daffodil → Hymenocallis
    Snowdrop → Galanthus
    Spider Lily → Lycoris
Daisy plants (family Compositae) → Adenostyles, Anacyclus, Antennaria, Anthemis, Aposeris,
            Aracnospermum, Arctium, Arnica, Arnoseris, Aster, Bellis, Carpesium, Centaurea,
            Cosmos, Gazania, Gerbera, Gnaphalium, Ligularia, Mutisia, Xeranthemum, Zinnia
    (Absinthe) → Artemisia
    (African Daisy) → Dimorphotheca
    (Artichoke) → Cynara
    (Aster) → Callistephus
    (Bearsfoot/Leafcup) → Polymnia
    ("Blood of the Maccabeans") → Helichrysum
    (Cat's-ear) → Hypochoeris
    (Chamomile) → Chamaemelum, Tripleorospermum
    (Chicory) → Cichorium
    (Cone-flower) → Echinacea
    (Chrysanthemum, Tansy, etc.) → Chrysanthemum, Tanacetum
    (Cineraria) → Senecio
    (Cocklebur) → Xanthium
    (Coltsfoot) → Homogyne, Tussilago
    (Cone-flower) → Rudbeckia
    (Costus) → Saussurea
    (Cudweed) → Filago
    (Dahlia) → Dahlia
    (Dandelion) → Leonotodon, Taraxacum
    (Elecampane - Alant Starch) → Inula
    (Eupatorium) → Eupatorium, Hebeclinium
    (False Cudweed) → Micropus
    (Fleabane) → Conyza, Erigeron
    (Gaillardia) → Gaillardia

(Galinsoga) → Galinsoga
(Goldenrod) → Solidago
(Goldilocks) → Crinitaria
(Guar/Jaguar) → Cyanopsis
(Gum-succory) → Chondrilla
(Gumweed/Tarweed) → Grindelia
(Hawksbeard) → Crepis
(Hawkweed) → Hieracium
(Lavender Cotton) → Santolina
(Leopardsbane) → Doronicum
(Lion's foot) → Leonotopodium
(Marigold) → Calendula, Tagetes
(Milfoil) → Achillea
(Nipplewort) → Lapsana
(Para Cress) → Spilanthes
(Sand Jurinea) → Jurinea
(= Senecio) → Kleinia
(Sneezeweed/Sneeze Wort) → Helenium
(Spike Gayfeather) → Liatris
(Spring Snowflake) → Leucoium, Leucojum
(Sunflower/Artichoke/etc.) → Helianthus
(Thistle) → Carbenia, Carduus, Carlina, Cirsium, Cnicus, Echinops
(Thistle, Scotch -) → Onopordum
(Thistle, Sow-) → Cicerbita
(Tickseed) → Coreopsis
(Wild Camomile) → Matricaria
(Wild Lettuce) → Lactuca, Mycelis
(Willemetia) → Willemetia
Daphne/Mezereon plants → Daphne
Datura plants → Brugmansia
Deer (Mouse deer or Chevrotain) → Tragulus
Deer (True) (family Cervidae) -
    Axis/Chital/Spotted → Axis
    Elk/Moose → Alces, Cervus
    Fallow → Dama
    Milu/Père David → Elaphurus
    Muntjacs/Barking Deers → Muntiacus
    Musk → Moschus
    Pampas → Blastocerus
    Pudu → Pudu
    Reindeer → Rangifer
    Roe → Capreolus
    Water → Hydropotes
    White-tailed/Virginian → Odocoileus
Devil's Cotton → Abroma
Diapensia plants → Diapensia
Diatoms (plants) → Actinoptychus, Arachnoidiscus, Asterionella, Coscinodiscus, Diploneis
Dinoflagellate Protozoans (order Dinoflagellata) → Ceratium, Gonyaulaux, Gonyaulax,
        Gymnodinium, Noctiluca

Diplurid Insects → Anajapx, Campodea
Dock/Knottgrass plants -
    (Mountain Sorrel) → Oxyria
Dogbane plants - see under Trees - Apocyanaceous/Dogbane
Dogs/Dingoes/Coyotes/Foxes/Jackals/Wolves (family Canidae) -
    (Arctic Foxes) → Alopex, Canis, Cerdocyon
    (Fennecs) → Fennecus
    (Foxes) → Urocyon, Vulpes
    (Hunting Dogs) → Lycaon
    (Racoon-dogs) → Nyctereutes
    (Wild Dogs) → Cuon
    (Wolves) → Chrysocyon
Dolphins (family Delphinidae) -
    Bottlenosed → Tursiops
    Common → Delphinus
    Hose's Sarawak → Lagendelphis
    Right Whale → Lissodelphis
    River → Inia, Lipotes, Orcaella, Platanista, Pontoporia
    Spotted → Stenella
    White-beaked → Lagenorhynchus
Dolphin fishes - see under Fishes
Duckbills - see under Platypus
Duckweed plants → Lemna, Wolffia
Dugongs - see under Sea-cows
Dragonflies -
    Hawker → Aeschna, Aeshna

Earwigs (insects) → Arixenia, Forficula, Hemimerus
Elephants → Elephas, Loxodonta
Endopterygote insects → Climacia
Entoproctal/Kamptozoan animals → Loxosoma
Euglenoid Protozoans → Euglena
Euphorbiaceous plants - see under Spurge plants
Extinct Animals -
    Amphibians → Ishthyostega
    Apes (Anthropoid) → Dryopithecus, Palaeopithecus
    Arthropods (Crustaceans) → Karagassiema
        (Spiders) → Palaeostenzia
    Birds → Dinornis, Gigantornis
        (Condor-like) → Teratornis
        (Great Auk) → Pinguinus
        (Pelicans/Storks) → Osteodontornis
        (Ratite) → Phororhacos
    Carnosaurs → Deinocheirus, Saurophagus
    Crocodiles → Deinosuchus, Phobosuchus, Rhamphosuchus
    Deer/Elks → Megaloceros
    Dinosaurs (Anteaters, prehistoric) → Therizinosaurus
        (Carnosaurs) → Tarbosaurus, Tyrannosaurus
        (Icthyosaurs) → Leptophtergius

       (Large Lizards) → Megalosaurus

       (Pliosaurs) → Kronosaurus

       (Reptiles) → Apatosaurus, Diplodocus, Elasmosaurus, Hylaeosaurus, Hypselosaurus, Iguanodon

Fishes → Chondrosteus, Coelacanthus

Flies - (Dragon) → Meganeura

Hominids → Australopithecus

Hyaenas/Hyenas → Hyaenictis

Insects → Rhyniella

Land Mammals → Benaratherium, Pristinotherium

Lizards → Branchiosaurus

Mammoths/Elephants → Archidiskodon, Hesperoloxodon, Mammuthus, Mastodon, Parelephas

Pterosaurs (Flying Reptiles) → Pteranodon

Reptiles → Archerpeton, Hylonomus, Romericus, Stegosaurus

       (Snakes) → Gigantophis, Madtsoia

Rhinoceros → Baluchitherium, Indricotherium

Theropods (Meat Reptiles) → Brontosaurus

Whales → Basilosaurus, Zeuglodon

Feather-star Echinoderms - see under Sea Lilies

Ferns (Polypody plants) -

    Aspidium/Male → Dryopteris

    Bladder → Cystopteris

    Clover → Marsilea

    Club-moss → Selaginella

    Filmy → Hymenophyllum

    Lady → Athyrium, Matteucia, Onoclea

    Maidenhair → Adiantum

    Marsh → Oreopteris

    Oak → Gymnocarpium

    Parsley → Cryptogamma

    Primitive → Angiopteris

    Rock → Adiantum, Woodsia

    Rock Garden → Cetarach, Ceterach

    Royal → Osmunda

    Rusty-back (Spleenwort) → Asplenium

    Sword → Nephrolepis

    Tree → Cheilanthes, Cyathea

    Water → Azolla

Figwort (Snapdragon) plants (Scrophulariaceae) → Alectra, Anarrhinum, Antirrhinum, Calceolaria, Chaenorhinum, Cymbalaria, Digitalis, Erinus, Euphrasia, Gratiola, Kickxia, Limosella, Linaria, Lindernia, Mimulus, Misopates, Odontites, Picrorhiza, Picrorrhiza, Verbascum, Veronica, Wulfenia

Filarian parasites → Acuria, Tetrameres

Fishes -

    Alfonsino → Beryx

    Anchoveta → Cetengraulis

    Anchovy → Engraulis

Angel → Pomacanthus, Pterophyllum
Angle → Holacanthus
Angler/Goose → Antennarius, Borophyrne, Lophius
Arapaima/Bonytongue → Arapaima, Heterotis
Archer → Taxotes
Ateleopid → Ateleopus
Band → Lumpenus
Barracudas → Sphyraena
Bass - see under Fishes - Grouper
Bat-fishes → Halieutaea, Ogocephalus
Bird → Gomphosus
Bleak → Alburnus
Blenny → Aspidontus, Blennius, Chirolophis
Blind → Amblyopsis
Bluefish → Pomatomus
Bonito (Scombridae) → Katsuwonus
Bony → Albula
    (Flying) - see Flying Fish → Exocoetus
Bow/Grindle → Amia
Box → Tetrasomus
Breams (Freshwater) - see under Fishes - Carps
    (Marine) - see under Fishes - Sea Breams
Bristlemouth → Cyclothone
Bullhead/Sculpin → Cottus
Burr/Rabbit/Porcupine → Chilomycterus, Diodon
Butterfish → Centronotus
Butterflyfish → Chaetodon, Heniochus
Candiru → Vandellia
Capelin → Mallotus
Carp/Minnow/Barbel/Bream/Tench (family Cyprinidae) → Alburnoides, Alburnus, Aspius,
      Barbus, Idus, Labea, Labeo, Leucaspius
    (African Cyprinid) → Engraulicypris
    (Bitterling) → Rhodeus
    (Bream) → Abramis, Blicca, Chrysophris
    (Carps) → Ctenophyryngodon, Cyprinus, Variocorhinus
    (Carp/Gudgeon) → Gobio
    (Carp/Minnow) → Vimba
    (Catla) → Catla
    (Chubs) → Kyphosus
    (Colored) → Chrosomus
    (Common Shiner) → Notropis
    (Dace) → Leuciscus
    (Gardon) → Gardonus
    (Goldfish) → Carassius
    (Minnow) → Phoxinus
    (Mosquito-eating Minnows) → Gerardinus
    (Nase) → Chondrostoma
    (Roach) → Rutilus
    (Roach Carps) → Pararutilus

(Rudd) → Scardinius
(Silver Carps) → Hypophthalmichthys
(Stoneroller) → Campostoma
(Tench) → Tinca
(Tooth Carps) → Jordanella
(Ziege fish) → Pelecus
Carpsuckers - see under Fishes - Clingfishes
Cartilaginous → Callorhynchus, Chimaera
Cat → Ameiurus, Arius, Brachyplatystoma, Callichthys, Ictalurus, Kryptopterus, Loricaria,
     Malapterurus, Noturus, Pangasianodon, Piratinga, Silurus
Char - see under Fishes - Salmon
Characin → Copeina, Erythrinus, Gasteroplecus, Lucioharax, Myletes, Myleus,
     Xiphorhamphus
(Tigerfish) → Hydrocyon
Clingfishes/Skittlefishes/Suckers → Caularchus
     (Carpsuckers) → Carpiodes, Cycleptus
     (Clingfishes) → Gobiesox, Lepadogaster
     (Common Suckers) → Catastomus
     (Lumpsuckers) → Caraproctus, Cyclopterus
     (Rocksuckers) → Chorisochismus
     (Suckers - Buffalofishes) → Ictiobus
Clown → Amphiprion
Clupeiform (freshwater) → Osteoglossum
Cod/Haddock/Hake/Rockling/Whiting (family Gadidae) → Arctogadus, Boreogadus, Ciliata,
     Eliginus, Enchelyopus, Gadus, Gaidropsaurus, Lota, Melanogrammus, Merluccius,
     Micromesistius, Odontogadus, Theragra, Urophycis
     (Ling) → Molva
Coelacanth → Latimeria
Cornetfish/Snipefish → Centriscus, Fistularia
Crappie → Pomoxis
Croaker → Cheilotrema, Umbrina
Cusk → Brosme
Cusk-eel → Ophidion, Ophidium
Dace → Rhinichthys
Damsel → Amblyglyphiodon, Dascyllus, Pomacentrus
Darter - see under Fishes - Perch
Deep-sea → Bassogigas, Bathypterois, Haloporphyrus, Ijimaia
Dibranchoid → Dibranchus
Diretmid → Diretmus
Discus → Symphysodon
Dogfish - see under Fishes - Sharks
Dolphin → Coryphaena
Dragonet → Callionomys, Callionymus
Drum → Aplodinotus
Eelpouts - see under Fishes - Pouts
Eels → Anguilla
     (Conger) → Ariosoma, Conger
     (Deep, Marine, Venomous) → Histiobranchus
     (Electric) → Electrophorus

(Gulper) → Eupharynx
(Moray) → Echidna, Gymnothorax, Muraena
Flatfish/Flounder/Plaice/Sole → Achirus, Glyptocephalus, Hippoglossoides, Limanda,
      Liopsetta, Microstomus, Scophthalmus, Solea
Flounder - see under Fishes - Flatfish, etc.
Flounders (Left-eyed) - see under Fishes - Turbot
    (Right-eyed) - see under Fishes - Halibut
Flying (Winged)/Skippers → Camptopera, Cypselurus, Cypsilurus, Exocoetus
    (Gurnard) → Dactylopterus
Forehead Brooder/Kurtus → Kurtus
Frogfishes/Sargassum-fishes → Histrio
Frost/Cutlass/Scabbard fishes → Lepidopus
Four-eyed (viviparous) → Anableps
Gar → Belone, Lepisosteus
Goby → Aphia, Crystalogobius, Eviota, Gobionellus, Gobius, Mistichthys, Pandaka,
      Pomatoschistus, Proterorhinus
Golden-shiner → Notemigonus
Goosefishes - see under Fishes - Angler
Grayling/Salmoniform → Thymallus
Grenadier - see under Fishes - Rat-tail
Groupers/Sea Bass/Freshwater Bass/Sunfishes/Headfishes → Adioryx
    (Freshwater Bass/Sunfishes) → Lepomis, Micropterus
    (Groupers) → Epinephelus
    (Groupers/Sea Bass) → Grammistes, Serranelus
    (Headfishes) → Mola
    (Sea Bass) → Centropristes, Mycteroperca
Grunion → Leuresthes
Grunt → Anisotremus, Haemulon
Guitar-fish → Rhinobatos
Guppies/Four-eyed/Killifishes/Molly/Swordtails -
    (Guppy) → Acanthophacetus, Lebistes, Platypoccilus, Poecilia
    (Molly) → Mollienesia
Gyrinocheilid → Gyrinocheilus
Hag → Bdellostoma, Eptatretus, Heptatretus, Myxine
Hake - see under Fishes - Cod/etc.
Halibut/Right-eyed Flounders → Atherestes, Hippoglossus, Pleuronectes, Pseudopleuronectes
Hatchet → Argyropelecus
Head - see under Fishes - Groupers/Sunfishes
Herring → Alosa, Clupea, Opisthonema, Pomolobus, Sardinia, Sardinops, Sardinus, Sprattus
Horse-mackerel/Crevalle/Jack/Pompano/Saurel/Scad/etc. (family Carangidae) -
    (Jack) → Caranx
    (Pilot-fish) → Naucrates
      (Saurel) → Trachurus
    (Pompano) → Alectis
Jaw → Opisthognathus
Jelly → Atolla, Aurelia
Jenny → Eucinostomus
Jew → Promicrops
John Dory → Zeus

Killifishes → Austrofundilis, Fundulus, Nothobranchius, Poterolebias
Labyrinth → Osphronemus
Lampreys → Caspiomyzon, Lampetra, Petromyzon
Lancet → Alepisaurus
Lantern → Diaphus, Myctophum
    (Bunmallow fishes) → Harpodon
Leaf-fish → Manocirrhus
Left-eyed → Lepidorhombus
Lion-fish/Zebra → Pterois
Lizard → Synodus
Loaches → Botia, Cobitis, Misgurnus, Nemacheilus, Noemacheilus
Lumpsuckers - see under Fishes - Clingfishes
Lungfishes → Ceratodus, Lepidosiren, Malania, Neoceratodus
Mackerel → Sarda, Scomber, Scomberomorus
Marlin → Makaira, Tetrapturus
Meagre → Argyrosomus
Midshipmen - see under Fishes - Toadfishes
Miniature → Schindleria
Minnow - see under Fishes - Carp
Mirapinnid → Mirapinna
Moonfish → Lampris
Moorish Idol → Zanclus
Mormyrid → Gnathoremus, Mormyrops
Mosquito → Gambusia
Mosquito larva-eating fishes → Haplochilus, Puntius
Mouth-breeder → Tilapia
Mud Minnow → Umbra
Mud-skipper → Periophthalmus
Mullet → Cheylon, Mugil, Mullus
Nile Bichir → Polypterus
Oysterfishes - see under Fishes - Toadfishes
Paddle-fish → Polyodon
Parrot fish → Scarus
Pearlfishes → Carapus
Perch/Walleye/Darter → Acerina, Ammocrypta
    (Black Bass) → Grystes
    (Darter) → Etheostoma
    (Dwarf Perch) → Micrometrus
    (European Perch) → Perca
    (Labyrinth) → Anabas, Aphredoderus
    (Perch) → Embiotica, Lates
    (Pikeperch) → Lucioperca, Stizostedion
    (Ruffe/Pope) → Gymnocephalus
    (Shield Darter) → Percina
    (Sunfish) → Eupomotes
    (Zingel) → Aspro
Perciform (freshwater) → Cichlasoma
Pike/Pickerel/Mudminnow/etc. → Dallia, Esox, Essox
Pipefish - see under Fishes - Sea-horse

Piranha → Pygocentrus, Pygopristis, Serrasalmus
Plaice - see under Fishes - Flatfish, etc.
Pogge → Agenus, Agonus
Pollack → Pollachius
Pompano - see under Fishes - Horse-mackerel
Porgy/Sea Bream/Snapper → Dentex, Diplodus, Sparus, Stenotomus
Portuguese Man o'War → Nomeus
Pouts - Eel/Ocean → Macrozoarces, Zoarces
Predator → Tylosurus
Prickle → Acanthochaenus, Malacosarcus, Stephanoberyx
Psettodid → Psettodes
Puffer → Arothron, Tetraodon
Pupfish ("Thermal") → Cyprinodon
Rag → Acrotus, Icichthys, Icosteus
Rat-fish → Hydrolagus
Rat-tail/Grenadier → Krohnius, Macrourus
Ray/Skate → Dasyatis, Manta, Raja, Rhinoptera
     (Electric) → Narke, Torpedo
Razor → Aoliscus
Reed → Calamoichthys
Ribbon/Oar → Regalecus
Rock-fish → Sebastodes
Sailfish → Istiophorus
Salmon/Char/Trout → Brachymystax, Cristivomer, Hucho, Onchorhyncus, Oncorhyncus,
     Salmo, Salvelinus, Variola
Salt-water → Bax
Sand-eel → Ammodytes
Sandfish/Beaked Salmon → Gonorhynchus
Sawfish → Pristis
Sawtail → Idiacanthus
Schneider → Alburnoides
Scorpion → Agonus, Drascius, Sebastes, Synanceja
     (Bullhead) → Agenus
     (Humpbacked) → Scorpaena, Scorpaenopsis
     (Lump) → Inimicus
     (Sea-snail/Snail) → Liparis
Sculpin → Myoxocephalus
Sea Bass - see under Fishes - Grouper
Sea Breams (family Sparidae) → Maena, Pagellus
Sea Goldfish → Anthias
Sea-horse/Pipe/Snipefish → Doryicthys, Dunckerocampus, Dunkerocampus, Entelurus,
     Hippocampus, Nerophis, Solenostomus, Syngnathus
Sea-robin fish → Prionotus
Sergeant Major → Abudefduf
Sharks -
     (Angel) → Squatina
     (Angel/Dogfish) → Echinorhinus
     (Basking) → Cetorhinus
     (Carpet/Nurse) → Ginglymostoma, Ginglyostoma, Orectolobus

    (Cartilaginous) → Galeus
    (Cat) → Cephaloscyllium, Scyliorhinus
    (Cow/Six-gilled) → Heptranchias, Hexanchus
    (Dogfish) → Squalus
    (Dogfish/Smooth hound) → Mustelus
    (Hammerhead) → Sphyrna
    (Horn) → Heterodontus
    (Lemon) → Negaprion
    (Leopard) → Triakis
    (Mackerel) → Isurus, Lamna
    (Mackerel/Thresher) → Alopias, Alopius
    (Notidanid) → Chlamydoselachus
    (Requiem/Requin) → Carcharhinus, Galaeorhinus, Galeorhinus, Prionace
    (Sand) → Carcharias, Odontaspis, Zygoena
    (Sleeper) → Somniosus
    (Small) → Squaliolus
    (Sucker) → Echeneis
    (Tiger) → Galeocerdo
    (Whale) → Rhincodon, Rhineodon
    (White) → Carcharodon
Skate - see under Fishes - Ray
Skittlefishes - see under Fishes - Clingfishes
Siamese → Betta
Silverside → Atherina, Atherinopsis, Menidia
Smelt (Freshwater) → Osmerus
    (Surf) → Hypomesus
Snake Mackerel → Gempylus
Snakehead → Ophiocephalus
Snapper → Lutjanus
Soldier/Squirrel → Adioryx, Holocentrus
Sole - see under Fishes - Flatfish, etc.
Spade-fish → Chaetodipterus
Spiny Flounder → Acanthopsitta
Sqaw-fish → Ptychocheilus
Star-gazer → Astroscopus
Stickleback/Trumpetfish → Apeltes, Eucalia, Gasterosteus
Sturgeon/Beluga → Acipenser, Huso
Suckers - see under Fishes - Clingfishes
Sunfishes - see under Fishes - Groupers
Surgeon-fishes → Acantharus, Zebrasoma
Swallower → Chiasmodon
Swamp → Chologaster
Swamp-eel → Alabes, Amphipnous
Swordfishes → Xiphias
Swordtails → Xiphophorus
Tarpon → Megalops, Tarpon
Tautog → Tautoga
Tigerfish - see under Fishes - Characin
Tile → Lopholatilus

Toadfishes/Midshipmen/Oysterfishes → Opsanus

Trigger → Balistes, Balistipus, Monacanthus, Sufflamen

Trout - see under Salmon

Trumpetfishes - see under Stickleback

Trunkfishes → Lactophrys, Ostracion

Trutta → Trutta

Tuna/Tunny/Bluefin (Thunnidae) → Thunnus, Tunnus

Turbot/Left-eyed Flounders → Arnoglossus, Hypsopsetta, Paralichthys, Platichthys, Pleuronichthys

    (Turbot) → Psetta

Veliferid → Velifer

Viperfishes → Chauliodus

Wahoo → Acanthocybium

Weak → Cynoscion

Weever → Trachinus

Whalefishes → Cetomimus

Whitefishes → Coregonus

Whiting/Wall-eye - see under Fishes - Cod/Haddock/Hake/etc.

Wolf → Anarhicas, Annarrhicas

Wrasse → Anampses, Coris, Crenilabrus, Ctenolabrus, Halichoeres, Hemigymnus, Labroides, Labrus, Thalassoma

Zebra - see under Fishes - Lion-fish

Flagellates -

Bacterial (peritrichous) → Noguchia

Coprozoic → Bodo, Copromastix, Copromonas, Helkesimastix, Prowazekella, Prowazekia, Tetrachilomastix

Phytomast → Astasia

Protozoan → Craigia, Embadomonas, Enteromonas, Hexamita, Octomitus, Parameba, Paramoeba, Retortamonas, Tricercomonas, Uroglena, Uroglenopsis

    (Cryptomonads) → Chilomonas

    (Chrysomonads) → Chromulina, Gonium, Mallomonas, Ochromonas

    (Euglenoids) → Colacium

    (Leishmania) → Helcosoma, Leishmania, Ovoplasma

    (Leptomonads) → Phytomonas

    (Metamonads) → Diceromonas, Giardia, Lamblia, Megastoma

    (Phytomonads) → Pleodorina, Polytoma

    (Protomonads/Zoomast) → Crithidia

    (Trichomonads) → Pentatrichomonas, Tetratrichomonas

    (Trichomonad type) → Trichomastix

    (Zoomastigs) → Cercomonas, Cheilomastix, Chilomastix, Macrostoma, Tetramitus, Trichomonas

Trypanosome - see under Sporozoan parasites

Flax plants → Linum

Fleas -

Biting → Culicoides

Cavy → Rhopalopsyllus

Dog → Ctenocephalides, Ctenocephalus, Ctinocephalus

Man → Tunga

Man/Mouse/Rat/etc. → Ceratophyllus

Man/Rat/Dog/etc. (Pulicidae) → Pulex, Xenopsylla
Mouse/Rat → Ctenophthalmus, Ctenopsylla, Ctenopsyllus, Diamanus, Leptopsylla,
    Nosopsyllus
Rodent → Oropsylla
Siphonapterous → Echidnophaga
    (common rat) → Monopsyllus
    ("gigantic") → Hystricopsylla
    (ground squirrels/plague) → Hoplopsyllus
Squirrel → Opisocrostis
Water → Alonella, Alonopsis
Flies → Aphiochaeta, Atherix, Bibio, Carysomyia, Chrysomyia, Cordylobia, Cryptolucilia,
    Eusimulium, Hydrothaea, Hypobosca, Megaselia, Ochromyia
    (causes intestinal myiasis) → Piophila, Protocalliphora
    (in nasal passages of horses) → Rhinoestrus
    (transmitting yaws) → Oscinis
Alder - see under Flies - Snake
Ant-lions - see under Lacewing insects
Aphid → Myzus
    (Greenfly) → Lachnus
Bee → Bombylius
Blood-sucking → Haemophoructus, Lasiohelia, Leptoconops
Blood-sucking larval → Auchmeromyia
Blow → Calliphora, Lucilia, Phormia
Bot-fly/Warble fly → Cephenomyia, Dermatobia, Gasterophilus, Gastrophilus, Hypoderma,
    Oestrus
Cabbage root maggot → Hylemyia
Caddis → Arthripsodes, Limnephilus, Phryganea, Rhyacophila
Crane → Holorusia, Tipula
Damsel/Dragon → Agriocnemis, Agrion, Austrophlebia, Calopteryx, Coenagrion, Gomphus,
    Lestes, Libellula, Megaloprepus, Pyrrosoma, Sympetrum, Tetracanthagyna
    (Hawker) → Anax, Cordulegaster, Cordulia
Deerfly - see under Lousefly
Dobson → Archichauliodes, Corydalus
Dung → Scatophaga, Sepsis
"Eye-fly" → Siphunculina
"Eye-gnat" → Hippelates
Flesh-fly → Sarcophaga
Fruit → Rhagoletis
Fruit (Acalyptrate Insects) → Ceratitis, Dacus, Drosophila, Oscinella
Gall Gnat/Midge → Mayetiola, Mikiola
Horn (Two-winged) → Haematobium
Horsefly/Deerfly (Tabanid) → Chrysops, Chrysozona, Haematopota, Hematopota,
    Lepidoselaga, Tabanus, Therioplectes
House → Anthomyia, Fannia, Homalomyia, Lispa, Musca
Hover-flies (family Syrphidae) → Episyrphus, Eristalis, Helophilus, Merodon, Microdon,
    Myiatropa, Scaeva, Volucella
Ichneumon → Ophion, Pimpla, Rhyssa
Lacewings - see under Lacewing insects
Lantern → Laternaria

Leaf-miner → Agromyza
Lousefly (Deerfly) → Lipoptena, Ornithomyia
Maggot → Delia
May → Cloeon, Ephemera, Ephoron, Hexagenia
Parasitic on Ticks → Ixodiphagus
Pigeon → Lynchia, Pseudolynchia
Robber → Asilus, Laphria, Mydas
Sand (Biting/Blood-sucking) → Phlebotomus
Saw → Caliroa, Cimbex, Diprion, Lophyrus, Nematus, Neodiprion, Rhogogaster, Tenthredo
Scorpion → Boreus, Panorpa
Screw-worm → Callitroga, Cochliomyia, Compsomyia
Simulid → Simulium
Snake/Alder → Agulla, Ascalaphus, Raphidia, Sialis
Snipe (Biting) → Rhagio
Soldier → Hermetia, Odontomyia, Stratiomys
Spiny → Echinomyia
Stable/Leg-sticker → Stomoxys
Stone → Perla
Tabanid - see under Flies - Horsefly/Deerfly
Tachinid → Phryxe, Tachina
Tick → Hippobosca
Tsetse/Biting → Glossina
Warble - see under Flies - Bot-fly
Floss Flower plants → Ageratum
Flowering plants (Indoor) -
(Flying Gold Fish) → Columnea
(Ornamental) → Conophytum, Crossandra, Gloxinia, Sinningia
(Spider) → Cleome
Flukes - see under Worms (Trematode)
Foraminifers (Foraminiferous protozoan animals) → Elphidium, Globigerina
Foxes - see under Dogs/Foxes/Jackals/Dingoes/Wolves/etc.
Frogbit/Pondweed/Waterweed plants → Egeria, Elodea, Hydrilla, Hydrocharis
Frogs/Toads → Centrolenella, Crinia, Hyloxalus, Phrynomerus
Adhesive Toads → Microhyla
Aquatic (Toads) → Barbourula, Batrachuperus, Bombina, Prostherapis, Pseudis
Arrow-poison (Venomous) → Dendrobates, Phyllobates, Sminthillus
Bamboo → Afrixalus
Clawed (Frogs/Toads) → Xenopus
Common (Toads) → Bufo
Diminutive → Eleutherodactylus, Microbatrachella, Phrynobatrachus
Hairy → Astylosternus
Hoarned/Spadefoot Frogs → Megophrys
Horned Toads → Ceratophrys, Leptodactylus
Long-living (Bullfrogs) → Pyxicephalus
Microhylid → Gastrophyrne, Hypopachus
Painted → Discoglossus
Plant → Cornufer
Primitive → Ascaphus, Liopelma
Rattray's → Anhydrophryne

Spadefoot Toads → Pelobates

Tail-less → Breviceps

Tree → Acris, Atelopus, Brachycephalus, Gastrotheca, Hyla, Nototrema, Phyllomedusa, Pseudacris, Rhacophorus

True → Rana

Fumitory plants → Corydalis, Dicentra, Dielytra, Fumaria

Fungi/Molds/Mushrooms/Yeasts → Agaricus, Botrytis, Coprinus, Eurotium, Oospora, Paecilomyces

Achorion → Lophophyton

Ascomycetous (aflatoxins) → Aspergillus, Fusarium

Bacteriolytic → Cephalosporium, Clytocybe, Emericellopsis, Fusidium, Penicillium

Bracket → Daedalea, Fomes, Polyporus

Cantharell/Chanterelle → Craterellus

Cup → Morchella

Cup/Helvel → Gyromitra, Helvella

Earth-stars → Geaster

Edible mushrooms → Cantharellus

Fairy Club → Clavaria

Favus (Achorion) → Achorion

Gasteromycetous → Mutinus

Gill → Armillaria, Amanita, Hypholomoa, Inocybe, Lactarius, Lepiota, Marasmius, Psilocybe

Hydnum → Hydnum

Medicinal → Sphaeria

    (Immunosuppressive) → Tolypocladium

Parasitic/Pathogenic → Absidia, Allescheria, Arthroderma, Arthrographis, Basidiobolus, Blatella, Cercosporalla, Empusa, Enantiothamnus, Grubyella, Mycoderma, Octomyces, Otomyces, Pullularia, Rhinocladium

    (Actinomycosis/Mycetoma/etc.) → Actinomyces

    (Blastomycotic, etc.) → Scopulariopsis

    (Chromoblastomycosis) → Hormodendrum

    (Ergot) → Claviceps

    (Ergotism-producing) → Tilletia

    (Histoplasmosis) → Histoplasma

    (Human ear) → Trichothecium, Verticillium

    (Hyphomycetous) → Phialophora

    (in rats) → Haplosporangium

    (Maduromycosis) → Monosporium

    (Mycetoma - Granular f.) → Madurella

    (Mycetoma-causing) → Indiella, Sterigmatocystis, Sterigmocystis

    (Mycosis-causing) → Rhizopus

    (Mycosis in man) → Hemispora, Lichtheimia

    (Nail infections in man) → Hendersonula

    (Non-dermatophyte) → Syctalidium

    (Onychomycotic/Yeastlike) → Schizoblastosporion

    (Otomycosis) → Glenospora

    (Parasitic on hair) → Piedraia

    (Parasitic to man/animals) → Megatrichophyton

    (Phycomycetous/Pathogenic) → Rhinosporidium

    (Piedra/Hair) → Trichosporon

(Pityriasis versicolor) → Malassezia
(Ringworm) → Cercosphaera, Epidermophyton, Microsporon, Microsporum, Nannizia, Sabouraudites
(Skin/nails/hair) → Endodermophyton, Trichophyton
(Sporotrichous) → Sporothrix, Sporotrichon, Sporotrichum
(Trichophyton fungi) → Megalosporon, Pinoyella
(Vaginitis in women) → Leptomitus
(Yeast-like) → Blastomyces, Candida, Coccidioides, Cryptococcus, Endomyces, Monilia, Mycocandida, Mycotoruloides, Oidium, Paracoccidioides, Parasaccharomyces, Parendomyces, Pityrosporon, Pityrosporum, Propeomyces, Pseudomonilia, Saccharomyces, Schizosaccharomyces, Syringospora, Torula, Torulopsis, Zymonema
Phycomycete (non-pathogenic) → Mucor, Thamnidium
Plant-growth stimulators → Gibberella
Poisonous → Conocybe, Stropharia
Puffball/Earthball → Lycoperdon
Saprophytic → Cladosporium, Saprolegnia
Slime → Arcyria, Lycogala
Smut → Urocystis, Ustilago
Spore-bearing (Basidia) → Gomphidius
Tree → Grifolia
Truffle → Choiromyces
Wood → Merulius

Galactogogue plants → Chlorostigma
Geckos - see under Lizards
Geranium/Cranesbill/Storksbill plants (family Geraniaceae) → Erodium, Geranium, Monsonia, Pelargonium
Gentian plants → Lomatogonium, Villarsia
    Centaury → Centaurium, Erythraea
    Cicadia → Cicendia
    Chirat/Balmony → Swertia
    Rough → Gentianella
    Yellow → Gentiana
Gerrid insects (Marine) → Halobates
Ginger plants → Alpinia, Zingiber
    Cardamom → Amomum, Elletaria
    Ginger Lily → Hedychium
    Turmeric → Curcuma
Ginseng plants → Aralia
Giraffes/Okapis → Giraffa, Okapia
Globe-daisy plants → Globularia
Gnats -
    Fungus → Arachnocampa
Goats - see Antelopes
Goosefoot plants -
    Orache → Atriplex
    Sea Purslane → Halimione
    Summer Cypress → Kochia

Upright → Chenopodium
Gorillas - see under Monkeys
Grasses (family Gramineae) -
    Bamboo → Dendrocalamus
    Barley (Wood) → Hordelymus
    Barley (Common) → Hordeum
    Bent → Agrostis
    Bermuda → Cynodon
    Brush → Chrysopogon
    Bread-wheat → Aegilops, Triticum
    Citronella → Cymbopogon
    Cockspur → Echinochloa
    Corn/Maize → Zea
    Couch → Agropyron, Agropyrum
    Cow-wheat (Field) → Melampyrum
    Cut → Leersia
    Darnel (Common/Poisonous) → Lolium
    Dock/Knotgrass → Fagopyrum
    Dog's-tail (Crested) → Cynosurus
    Eel-grass → Zostera
    Eragrostis → Eragrostis
    Foxtail → Alopecurus, Alopercurus
    Grain (Moriyo Starch) → Panicum
    Gravel Fescue → Micropyrum
    Hair → Aira, Corynephorus
        (Finger) → Digitaria
        (Glaucous/Sand Hair) → Koeleria
        (Tufted) → Deschampsia
    Harestail → Lagurus
    Holy-grass → Hierochloe
    Knot-grass → Polygonum
    Loose Silky-bent → Apera
    Leyme → Elymus, Leymus
    Maize - see under Grasses - Corn
    Mat → Nardurus, Nardus
    Meadow → Festuca
    Moor → Molinia, Oreochloa
    Oat → Arrhenatherum, Avena
        (Meadow) → Helictotrichon
    Orchard → Dactylis
    Pampas → Cortaderia, Gynerium
    Rat's tail/Fescue → Vulpia
    Reeds (Small-) → Calamagrostis
    Reeds (Spanish) → Arundo
    Rice → Oryza
    Rough-grass → Achnatherum
    Rye → Secale
    Sand-grass → Mibora
    Sleepygrass → Stipa

Sorghum → Andropogon
Spikenard → Andropogon
Sweet-grass/Manna-grass → Glyceria
Valerian → Andropogon
Vernal → Anthoxanthum
Whorl-grass → Catabrosa
Wood Melick → Melica
Wood Millet → Milium
Yorkshire Fog → Holcus
Grasshoppers - see Locusts
Gum Arabic → Acacia
Gymnosperm plants → Welwitschia
Gymnostome Protozoans → Didinium, Dileptus, Holophrya

Hamsters - see under Rodents
Hares/Rabbits → Lepus, Oryctolagus, Pseudois
Heath/Heather - see under Trees/Shrubs
Hedgehogs → Echinosorex, Erinaceus, Hylomys
Heliozoan/Helizoan Protozoa → Actinophrys, Actinosphaerium, Clathrulina
Hemp & Hop plants → Cannabis, Humulus
Heteropterous insects -
    Water Boatman/ Backswimmer → Notonecta
Hippopotamus -
    (Great African) → Hippopotamus
    (Pigmy) → Choeropsis
Hogs - see under Pigs
Holotrichs -
    Suctorian → Acineta
Homopterous insects → Cryptococcus, Nothofagus
Honeysuckle plants (family Caprifoliaceae) → Abelia, Diervilla, Linnaea, Lonicera, Sambucus,
            Saponaria, Viburnum, Weigela
Horned-pondweed plants → Zannichellia
Horned Ungulates - see under Cattle
Horses/Asses/Zebras (family Equidae) → Equus
Horsetail plants → Equisetum
Hornwort plants → Ceratophyllum
Humans → Homo
Hyacinth (Water) plants → Eichhornia, Eichornia
Hyaenas/Hyenas → Crocuta, Hyaena
Hybrid plants → Fuchsia
Hydras/Hydroids/Sea Firs/Thecate Hydroids (class Hydrozoa) → Abietinaria, Aglaophenia,
            Chlorohydra, Chrysomitra, Clava, Clytia, Craspedacusta, Dyamena, Dynamena,
            Eudendrium, Gonionemus, Halecium, Halistemma, Hydra, Hydractinia, Hydrallmania,
            Microhydra, Lafoea, Millepora, Obelia, Pelmatohydra, Sertularella, Sertularia, Velella
Hydrozoan Protozoa → Coryne, Cunina, Distichopora
Hyenas - see Hyaenas
Hygrophylla plants - see under Acanthus plants
Hypotrich Protozoans → Euplotes
Hyrax Mammals → Dendrohyrax, Procavia

Iguanas - see under Lizards
Impala - see Antelopes
Incense → Boswellia
Indaceous plants (family Indaceae) → Sisyrinchium
Iris plants (family Iridaceae) → Crocosmia, Crocus, Freesia, Gladiolus, Hermodactylus, Iris, Ixia,
          Monbretia, Montbretia, Pardanthus, Tritonia
Ivy plants - see under Trees - Ivy

Jack-fruit - see under Trees - Mulberry
Jellyfishes - see under Chordate animals
Jerboas - see under Rodents
Joewood plants → Jacquinia

Kangaroos - see under Marsupials

Lacewing insects/Ant Lions (Lacewings) → Chrysopa, Formicaleon, Myrmeleon, Osmylus
Lampreys - see under Fishes - Lampreys
Leaf Insects - see under Stick Insects
Leeches - see under Worms - Annelids
Lemmings - see under Rodents
Lemurs/Lorises → Arctocebus, Bradicebus, Cheirogaleus
        (Bush-tailed Lorises) → Galago
        (Flying Lemurs) → Cynocephalus
        (Indris Lemurs) → Indri
        (Mouse Lemurs) → Microcebus
        (Ring-tailed Lemurs) → Lemur
        (Slender Lorises) → Loris
        (Slow Loris) → Nycticebus
Lice -
        (Bird/parasite) → Goniodes
        (Biting) → Haematomyzus
        (Biting/Bird) → Menacanthus, Menopon
        (Booklice) → Trogium
        (Cattle - Sucking) → Solenopotes
        (Cattle/Horses/Swine - Sucking) → Haematopinus
        (Dog) → Tichodectes
        (Dogs - Sucking) → Linognathus
        (Fish/parasite) → Argulus
        (Guinea pig/parasite) → Gliricola, Gyropus
        (Horse/Biting) → Trichodectes
        (Man/parasite/Sucking) → Pediculus, Phthirus
        (Mice/Rats/Sucking) → Polyplax
        (Plant Sucking/Jumping) → Psylla
        (Rabbits - Sucking) → Haemodipsus
        (Squirrel - Sucking) → Lignognathoides
        (Water/scavenger) → Asellus
Lichens/Liverworts/Moss plants → Alectoria, Alextoria, Cetraria, Cladonia
        Bog/Peat Moss → Sphagnum

Ceylon Moss → Gracilaria
Club Moss → Crassula, Hupertzia, Lycopodiella, Lycopodium
Common Liverwort → Marchantia
Fork-moss → Leucobryum
Haircap/Juniper Moss → Polytrichum
Lichens → Evernia, Lichen, Xanthoria
Lichens/Iceland Moss → Lobaria
Litmus → Roccella
Liverworts → Lophocolea
Medicinal → Variolaria
Primitive (land) → Hyloconium
Tree Lichens → Usnea
True/Palm-tree → Mnium
Willow → Fontinalis
Woodland Moss → Dicranum
Lily plants → Aculeatus, Agapanthus, Aspidistra, Canna, Convallaria, Crinum, Eucomis, Funkia,
Gagea, Hosta, Veltheimia
(Aloes) → Aloe
(Arum) → Calla, Zantedeschia
(Asparagus) → Asparagus
(Asphodel) → Asphodeline, Asphodelus
(Blackberry) → Belamcanda
(Belladonna) → Amaryllis
(Bogasphodel) → Narthecium
(Bourbon, Madonna, White Lily) → Lilium
(Bowstring Hemp) → Sansevieria
(Dragon) → Dracaena
(Fawn/Trout) → Erythronium
(Fritillary) → Fritillaria
(Glory Lily) → Gloriosa
(Grassnut) → Brodiaea, Triteleia, Tritelia
(Hellebore) → Veratrum
(Hyacinth) → Galtonia, Muscari
(Hyacinth/Spire Lily) → Hyacinthus
(Lemon Day Lily) → Hemerocallis
(May Lily) → Maianthemum
(Mound Lily) → Yucca
(Natal/Kaffir) → Clivia
(Ox-tongue/Aloes) → Gasteria
(Peruvian) → Alstroemeria
(Sabadilla) → Schoenocaulon
(Saffron) → Colchicum
(Sarsaparilla) → Smilax
(Snowdon) → Lloydia
(Solomon's seal) → Polygonatum
(Spanish Squill) → Endymion, Eremurus, Urginea
(Spiderwort) → Anthericum
(Squill) → Scilla
(Star-of-Bethlehem) → Ornithogalum

(Torch Lily) → Tritoma
(Torch Lily/Tritoma) → Kniphofia
(Tulip) → Tulipa
(Wake-robin) → Trillium
Liquorice - Indian → Abrus
Liverworts (& Mosses) - see under Lichens
Lizards (Reptiles) -
Agamid → Agama, Amphibolurus, Chlamydosaurus, Leiolepis, Moloch, Uromastyx
Anguid (Glass Snake/Slow-worm) → Ophiosaurus, Ophisaurus
Chameleon → Chamaeleo
Gecko → Coleonyx, Cyrtodactylus, Gekko, Gymnodactylus, Hemidactylis, Hemidactylus, Sphaerodactylus, Tarentola, Uroplatus
Girdle-tailed → Cordylus
Harlequin → Calotes
Helodermatids (venomous lizards) → Heloderma
Iguana → Amblyorhynchus
(Black) → Ctenosaura
(Collared Lizard/etc.) → Crotaphytus
(Common/Green) → Iguana
(Galapagos) → Conolophus
Lacertid → Psammodromus
Monitor → Varanus
Racer → Eremias
Salamander - see under Salamanders
Sand → Lacerta
Skink - see under Skinks
Slow-worm (Lateral fold) → Anguis
Snake → Chamaesaura
Very rare → Xenosaurus
Worm → Amphisbaena
Lobelia plants → Lobelia
Lobsters - see under Crustaceans - Decapod
Loco Weed - see under Pea plants
Locusts (Cicadas/Crickets/Grasshoppers) → Acrida, Anacridium, Calliptamus
(Cicadas) → Cicada, Cicadetta, Tibicen
(Crickets - see under Crickets)
(Grouse) → Tetrix
(Leaf-camouflage) → Cycloptera
(Leaf-hoppers) → Cicadella, Circulifer
(Migratory) → Locusta
(Short-horned) → Chorthippus, Megicica, Oedipoda, Psophus, Sphingonotus, Stenobothrus
(Short-horned - plague) → Chortoicetes
(Short-horned/Migratory) → Melanoplus
(Short-horned/Desert) → Schistocera
Loosestrife plants - see under Trees/Shrubs - Myrtle
Lorises - see under Lemurs/Lorises

Madder plants - see under Bedstraw plants (family Rubiaceae)
Malarial parasites → Laverania

Mallow plants (family Malvaceae) → Abutilon, Kitaibelia, Lavatera
    Althea/Rose of Sharon → Althea, Hibiscus
    Common → Malva
    Levant Cotton → Gossypium
    Marsh → Alcea, Althaea
Malpighiaceous plants (family Malpighiaceae) → Tetrapteris
Manatees - see under Sea-cows/Dugongs/etc.
Mantids - see under Cockroaches
Marestail/Mare's tail plants → Hippuris
Marine Echinoderm animals → Myriotrochus
Marmosets - see under Monkeys
Marmots - see under Squirrels
Marsupials -
    Carnivorous marsupials → Thylacinus
    Kangaroos → Dendrolagus, Macropus, Onychogalea, Onychogales
        (Hare-wallabies) → Lagorchestes
    Mice → Planigale
        (Dunnart) → Sminthopsis
    Opossums (American/Water) → Chironectes
    Opossums (American/Land) → Didelphis
    Opossums (American/Arboreal) → Marmosa
    Phalangers → Acrobates
    Primitive (Monotreme) → Zaglossus
Marvel plants → Mirabilis
Medusae (Colenterate Hydrozoan animals) → Geryonia, Liriope
Melia plants -
    (Neem) → Azadirachta, Melia
Mesembryanthemum plants (family Aizoaceae) → Dorotheanthus, Lithops, Mesembryanthemum
Mice - see under Rodents
Microorganisms (family Enterobacteriaceae) → Aerobacter, Alginobacter, Arizona, Edwardsiella, Enterobacter, Escherichia, Klebsiella, Palacolobactrum, Proteus
Microsporidian parasites → Encephalitozoon, Encephalocytozoon
    (Mosquito parasite) → Thelohania
Microzoa (parasitic) → Proteosoma
Midges (Non-biting, Nematoceran insects) → Chironomus
Midget insects → Forcipomyia
Milfoil - see under Daisy plants
Milkweed plants (family Asclepiadaceae) → Asclepias, Calotropis, Caralluma, Ceropegia, Fockea, Gonolobus, Hemidesmus, Heurnia, Hoya, Marsdenia, Periploca, Vincetoxicum
Milkwort plants → Chamaebuxus, Polygala
Millipedes (class Diplopoda) → Blaniulus, Glomeris, Graphidostreptus, Iulus, Julus, Macrosternodesnus, Polydesmus, Polyxenus, Rhinocricus, Scaphistostreptus, Schizophyllum, Siphonophora, Spirobolus, Spirostreptus
Mint plants - see under Thyme
Minute Arthropods/Animacules (phylum Tardigrada) → Macrobiotus
Mistletoe plants - see under Trees
Mites → Acarapis, Bdella, Dermatophilus, Leeuwenhoekia
    Ascarid/Acarine → Histiogaster, Otodectes, Parasitus
        (Barley/Itch) → Acarus, Sarcoptes

(Cheese/Flower/Copra itch) → Tyroglyphus
(Chigger/Harvest) → Leptotrombidium, Microtrombidium, Trombicula, Trombidium
(Chigger/Harvest - Larval form) → Leptus
(Chigger/Harvest/Red Bug) → Eutrombicula
(Chigger/Harvest/Spider) → Tetranychus
(Sugar-infesting) → Glyciphagus
Bird → Harpyrynchus
Bird/Chicken/Poultry → Dermanyssus
Cat → Notoedres
"Coolie-itch" → Rhizoglyphus
Diminutive → Tydeus
Feather/Hair → Analges
Horse → Psoroptes
House → Dermatophagoides
Mange → Demodex
Rat → Lelaps
Sarcoptid → Cnemidocoptes
Scabies/Itch → Acarus
Tyroglyphid → Carpoglyphus
Typhus-transmitting → Leiognathus, Liponyssus
Molds - see under Fungi
Moles -
Desman → Desmana, Galemys
Golden → Amblyosomus, Chrysochloris, Chrysospalax
Marsupial → Notoryctes
Shrew → Neurotrichus
Star-nosed → Condylura
Talpid → Talpa
Molluscs → Abra, Acanthinula, Cassidaria, Cerithium, Cuspidaria, Glossodoris, Neophilina, Neopilina, Perforatella
Ark Shell → Arca
Astarte-shell → Astarte
Bivalve (class Bivalvia) → Malletia
(Banded Cockle) → Cardium
(Basket Shell) → Aloides
(Carpet/Venus Shell) → Venus
(Clams) → Saxidomus, Tindaria
(Cockle) → Acanthocardia, Cerastoderma, Glossus, Glycymeris, Laevicardium, Montacuta
(Cupped Shell) → Crassostrea
(Cyprina) → Arctica
(Edible Oyster) → Ostrea
(False Angel Wings) → Petricola
(Fan-mussel) → Pinna
(Farrow Sunset Shell) → Gari
(File Shell) → Lima
(Flask Shell) → Gastrochaena
(Flower Shell) → Scrobicularia
(Freshwater Cockle) → Pisidium, Sphaerium

(Gaper) → Mya
(Giant Clams) → Tridacna
(Lantern Shell) → Thracia
(Marbled Mussel) → Musculus
(Mussel Shell) → Mytilus
(Otter/Gaper Shell) → Lutraria
(Pearl Mussel/Oyster) → Margaritana, Margaritifera
(Piddock) → Barnea, Pholas, Zirfaea
(Protobranchial) → Yoldia
(Razor Shell) → Cultellus, Ensis
(Rock Borer) → Panopea
(Saddle Oyster) → Anomia
(Ship Worm) → Teredo
(Swan Mussels) → Unio
(Tellin) → Tellina
(Tellin Shell) → Macoma
(Trough Shell) → Mactra
(Venus/Carpet Shell) → Callista, Dosinia, Mercenaria, Mysia
(Wedge Shell) → Donax
(Zebra Mussel) → Dreissena
Brachiopod -
(Lamp Shell) → Lingula
Cephalopod → Heteroteuthis
Cranchid molluscs → Mesonychoteuthis
Cuttle → Sepia
Octopuses → Argonauta, Eledone, Hapalochlaena, Octopus, Ozaena, Paroctopus
Squids → Alloteuthis, Architeuthis, Crynchia, Dosidicus, Loligo, Nautilius,
    Ommastrephes, Onychoteuthis, Parateuthis
Chiton → Acanthozostera, Chiton, Cryptochiton, Lepidochitona
Clams - see under Molluscs - Bivalve
Cockles - see under Molluscs - Bivalve
Date → Lithophaga
Flying Squids → Vampyromorpha
Gastropod → Achatina, Bielzia, Clathrus, Cochlodina, Cypraea, Epitonium, Monodonta,
    Scalaria, Vallonia
Clausiliid → Macrogastra, Ruthenica
Conch → Aporrhais, Fasciolaria
Cylindrical shell → Truncatellina
Doorshell → Clausilia
Earshell (Sea Ear) → Haliotis
Freshwater Snail → Planorbarius, Planorbis, Viviparus
Garden Snail → Arianta, Cepaea
Glass Snail (Pellucid) → Vitrina
Land Slug → Arion, Deroceras, Lehmannia
Land Snail → Eulota, Helicella, Helicigona, Helicodonta, Helix, Isognomostoma,
    Limax, Monacha, Oncomelania, Succinea, Trichia, Vertigo, Vitrea, Zebrina
Limpet → Acmaea, Ancylus, Atlanta, Axiolotus, Diodora, Lottia, Patella
Limpet (Keyhole) → Megathura
Marine (shell) → Mitra, Strombus

Naked Shell → Hexabranchia
Planorbid → Segmentina
Pond Snail → Lymnaea
Pupillid → Pupilla
Rounded Snail → Discus
Sea Slug/Hare → Actaeon, Aplysia, Bulla, Coryphella, Elysia, Oscanius,
        Pleurobranchus, Tethys
Sea Snail → Bittium, Cassis, Hydrobia, Murex, Neptunea, Syrinx
Sea Snail (Periwinkle) → Littorina
Snails → Isidora, Katayama, Limnaea, Parafossularus
Top Shell → Gibbula
Tusk Shell → Dentalium
Whelks → Buccinum, Busycon
Whelks (Dog) → Lora, Nassarius
Mesogastropod → Calyptraea
    (Limpet) → Capulus, Crepidula, Fissurella, Lamellaria, Natica
Neogastropod (Conches/Cone shells/Oysterdrills/Whelks) → Conus, Nucella, Ocenebra,
        Ocinebra, Oliva, Urosalpinx,
Nut Shell → Nucula, Nuculana
Primitive → Nerita, Neritina
Pteropod (Sea Butterflies) → Limacina
Scallop → Aequipecten, Chlamys, Pecten
Snails → Alocimua, Bulinus, Bythnia, Melanoides, Physopsis, Pironella
Swan Mussels → Anodonta
Spiral Shells → Xenophora
Top Shell (Archaeo-gastropod) → Astraea, Calliostoma, Emarginula
Valve Snails → Valvata
Mongooses -
    Marsh/Water → Atilax
Monkeys (Primates) → Cephus
    Apes - Anthropoid → Gorilla
        (Gibbons) → Hylobates
        (Chimpanzees) → Pan
        (Orangutans) → Pongo
    Baboons → Papio
    "Black Apes" (Old World Monkey) → Cynopithecus
    Capuchin → Cebus
    Catarrhine - Rhesus → Macaca, Macacus
    Douroucouli (Night Apes) → Aotes, Aotus
    Geonon (Old World) → Cercopithecus
    Gorillas → Pseudogorilla
    Guereza (Old World) → Colobus
    Howling → Alouatta
    Langur → Semnopithecus
    Mandrill (Old World) → Mandrillus
    Mangabey (Old World) → Cercocebus
    Marmoset/Tamarin → Callithrix
        (Pigmy) → Cebuella
        (Lion) → Leontocebus, Mystax

Patas/Hussar (Old World) → Erythrocebus
Proboscis (Old World) → Nasalis
Saki (New World) → Chiropotes
Spider → Ateles
Swamp → Allenopithecus
Tamarin - see under Marmoset
Titi → Callicebus
Uakari → Cacajao
Woolly (New World) → Lagothrix
Morning-glory plants (family Convolvulaceae) → Pharbitis, Rivea
Moschatel plants → Adoxa
Mosquitoes → Armigeres, Corethra, Desvoidea, Haemagogus, Taeniorhynchus, Theobaldia, Uranotaenia
(transmitting malaria/filariasis in Africa) → Pyretophorus
Anopheline → Anopheles, Myzomyia, Myzorhynchus
Culicine → Aedes, Culex, Janthinosoma, Mansonia, Mansonoides, Megarhinus, Nyssorynchus, Ochlerotatus, Stegomyia,
Moss animals (Polyzoa) - see under Sea Mats
Moss plants - see under Lichens
Moths (Butterflies) → Chimabacche, Cosmotriche, Johanssonia, Lemonia, Lithocolletis, Lyonetia, Nemophora, Nepticula, Nymphula, Teinopalpus, Troides, Yponomeuta
Alpine → Endrosa
Apollo → Parnassius
Bagworm → Apterona
Bell/Codling/Tortoise family (Tortricidae) -
    Bell → Adoxophyes, Choristoneura
    Codling → Carpocapsa, Cydia, Laspeyresia
Birdwing → Ornithoptera
Blue/Copper/Hairstreak → Albulina, Brephidium, Callophrys, Chrysophanus, Cyaniris, Heodes, Lycaena, Polyommatus
Browns (Satyridae) → Coenonympha, Erebia, Eumenis, Maniola, Melanargia, Oeneis, Pararge
Burnet/Forester → Agrumaenia, Zygaena
Cactus → Cactoblastis
Clearwing → Sesia
Clothes (Common) (Tineidae) → Tineola
Clothes (Tapestry) → Trichophaga
Cossid → Xyleutes
Eggar - see under Lappet
Emperor/Silkworm (Saturniidae) → Actias, Antheraea, Argema, Attacus, Coscinoscera, Endromis, Eudia, Graellsia, Hyalophora, Hyperchiria, Lonomia, Odonestis, Saturnia
Ermine → Hyponomeuta
Flour → Anagaster, Ephestia
Forester - see under Moths - Burnet
Fruit → Grapholitha
Geometer/Looper (Geometridae) → Abraxas, Alsophila, Aplocera, Eranis, Geometra, Gonodontis, Idaea, Lomaspilis, Operophtera, Scopula
Goat (Cossidae) → Cossus, Zeuzera
Hawk (Sphingidae) → Acherontia, Amorpha, Celerio, Coccytius, Daphnis, Deilephila,

Hemaris, Herse, Hyles, Hyloicus, Laothoe, Macroglossum, Marumba, Mimas, Smerinthus, Sphinx
Hooktip → Drepana
Io → Automeris
Lappet/Eggar → Gastropacha, Melacosoma
Lepidopterous → Lasiocampa, Sabatinca
    (Owlet) → Thysania
Meal → Pyralis
Migrating → Lymanopoda, Pedaliodes
Miniature → Stigmella
Monarch → Danaus
Noctuid/Owlet → Acronicta, Acronycta, Agrotis, Apamea, Apatele, Autographa, Diloba, Erebus, Mamestra, Noctua
Nymphalid → Limenitis, Polygonia
    (Fritillary) → Clossiana, Fabriciana
    (Satyrine) → Melanitis
Papillon (Papilionidae) → Zerynthia
Peach Blossom → Thyatira
Peach Tree Borer → Sanninoidea
Peacock → Inachis
Peppered → Biston
Pine - Lappet → Dendrolimus
Plant (Rustic) → Gortyna
Plume → Aciptilia, Alucita
Primitive → Micropteryx
Processionary → Thaumetopoea
Prominent/Puss → Cerura, Dicranura, Notodonta, Phalera
Puss (Flannel) → Megalopyge
Pyralid (Meal) → Asopia, Orneodes
Pyraustid → Acentropus, Ostrinia
Shark → Cuculia
Silkworm - see under Moths - Emperor
Silkworm (family Bombycidae) → Bombyx
Skipper → Carterocephalus, Erynnis, Euschemon, Ochlodes, Pyrgus
Swallow-tail/Apollo → Iphiclides, Papilio
Swift/Ghost/Hepialid → Hepialus, Leto
Tapestry - see under Moths - Clothes (family Tineidae)
Tiger (family Arctiidae) → Arctia, Callimorpha, Coscinia, Euplagia, Rhyparia, Tyria
Tortoiseshell (family Nymphalidae) → Aglais, Aglivus, Apatura, Araschnia, Argynnis, Cynthia, Euphydryas, Kallima, Libythea, Mesoacidalia, Morpho, Nymphalis, Vanessa
Tortrix/Oak → Tortrix
Triplasic → Abrostola
Tussock → Hemerocampa, Leucoma, Lymantria, Orgyia, Porthetria,
Tussock/Browntail → Dasychira, Euproctis
Underwing → Catocala
White → Bupalus
Whites & Yellows → Anthocharis, Aporia, Colias, Gonepteryx, Pierus
Muscarine (an alkaloid) → Agaricus
Mushrooms - see under Fungi/Mushrooms

Mustard plants -
    White → Brassica
Myrtle plants - see under Trees

Naiad plants → Najas
Neem - see under Melia plants
Nematode Worms - see under Worms
Nettle plants (family Urticaceae) → Parietaria, Urtica
Newts - see under Salamanders
Nightshade plants (family Solanaceae) -
    Aswagandha → Withania
    Belladonna → Atropa
    Bladder/Ground Cherry → Physalis
    Datura/Stramonium → Datura
    Duboisinine → Duboisia
    Henbane → Hyoscyamus
    Mandrake → Mandragora
    Pepper/Paprika → Capsicum
    Potato/Medicinal → Solanum
    Scopola → Scopolia
    Tobacco → Nicotiana
    Tomato → Lycopersicum
    Willow-herb → Circaea

Opossums - see under Marsupials
Orchids (family Orchidaceae) → Aceras, Anacaptis, Bletia, Catasetum, Cattleya, Cephalanthera,
       Chamaeorchis, Coelagyne, Coeloglossum, Corallorhiza, Cymbidium, Cypripedium,
       Dactylorchis, Dactylorhiza, Dendrobium, Dendrochilum, Epidendrum, Epipactis,
       Epipogium, Goodyera, Gymnadenia, Hammarbya, Herminium, Himantoglossum,
       Leucorchis, Limnodorum, Listera, Macroplectrum, Malaxis, Neottia, Nigritella,
       Odontoglossum, Onchidium, Ophrys, Orchis, Spiranthes, Vanda, Vanilla
Ornamental plants → Delospermum, Faucaria, Gloxinia, Lampranthus
Otters - see under Badgers
Oxalis/Wood-sorrel plants → Oxalis

Pandas → Ailuropoda, Ailurus
Parilla plants -
    (Moonseed) → Menispermum
Pansy plants - see under Violet/Viola plants
Pea plants/herbs/shrubs/trees (family Leguminosae) → Anagyris, Andira, Anthyllia, Anthyllis,
       Caesalpinia, Cercis, Mimosa
    (Alfalfa) → Medicago
    (Ashok/Asoka) → Saraca
    (Bean) → Phaseolus
    (Birdsfoot-trefoil) → Lotus
    (Black Pea) → Lathyrus
    (Bladder Senna) → Colutea
    (Broad Bean/ Vetch) → Vicia
    (Calabar Bean) → Physostigma

(Carob, Locust Bean, etc.) → Ceratonia
(Casca/Mancona/Sassy Bark) → Erythrophloeum
(Cockscomb/Coral Bean) → Erythrina
(Copaiba) → Copaifera
(Copal) → Trachylobium
(Cowage/Cowhage/Cowitch) → Mucuna
(Crown-vetch) → Coronilla
(Cube Root) → Lonchocarpus
(East Indian trees) → Pongamia
(Fenugreek) → Trigonella
(French Lilac) → Galega
(Golden Rain/Royal Broom/etc.) → Cytisus, Genista
(Greenweed) → Chamaespartium
(Hairy Dorycnium) → Dorycnium
(Heartwood/Logwood) → Haematoxylon
(Heartwood/Sandal Wood) → Pterocarpus
(Hedysarum) → Hedysarum
(Himalayan Indigo) → Indigofera
(Jaimaica Dogwood) → Piscidia
(Laburnum - see Cytisus) → Laburnocytisus
(Laburnum - Golden Rain) → Laburnum
(Lentil) → Lens
(Liquorice) → Glycyrrhiza
(Loco Weed) → Hosackia
(Loco Weed, Tragacanth, etc.) → Astragalus
(Lupin/Lupine) → Lupinus
(Medicinal/d-Sparteine) → Thermopsis
(Medicinal/l-Sparteine) → Sophora
(Melilot) → Melilotus
(Milk-vetch) → Oxytropis
(Pencil-flower) → Stylosanthes
(Peru Balsam) → Myroxylon
(Psoralea/medicinal) → Psoralea
(Pueraria) → Pueraria
(Sainfoin) → Onobrychis
(Scoparium) → Sarothamnus
(Senna) → Cassia
(Serradella) → Ornithopus
(Soya Bean) → Glycine, Soja
(Spanish/Weaver's Broom) → Spartium
(Sphaerophysine/Medicinal) → Sphaerophysa
(Spiny Rest-harrow) → Ononis
(Tamarind) → Tamarindus
(Vetch) → Hippocrepis
(Wild Tamarind) → Lucaena
(Wisteria) → Wisteria
Peccaries (Ungulate Mammals) (family Tayassuidae) → Tayassu
Pepper plants → Piper
Periwinkle plants - see under Trees - Apocynaceous/Dogbane

Phlox plants → Cobaea
Phytomonad Protozoans → Chlamydomonas
Picrotoxin → Anamirta
Pigs/Hogs -
    Wild → Hylochoerus, Phacochoerus, Potamochoerus
    Wild (Boars) → Sus
Pikas/Mouse-hair mammals → Ochotona
Pineapple plants/trees → Achmea, Ananas, Cryptanthus, Guzmania, Vriesia
Pink/Chickweed plants (family Caryophyllaceae) → Agrostemma, Cerastium, Corrigiola,
    Cucubalus, Dianthus, Gyposophila, Gypsophila, Heliosperma, Herniaria, Holosteum,
    Honkenya, Illecebrum, Lychnis, Minuartia, Moehringia, Moenchia, Myosoton, Silene,
    Stellaria, Vaccaria, Viscaria
Pinnipeds - see under Seals
Pitcher-plants -
    (New World) → Sarracenia
    (Old World) → Nepenthes
Plasmocytes → Haemapium
Platypus/Duckbills → Ornithorhynchus
Plumbago plants → Ceratostigma
Polecats - see under Badgers
Polygalaceous shrubs -
    (Rhatany) → Krameria
Pondweed plants → Groenlandia
Poppy plants → Chelidonium, Eschscholtzia, Eschscholzia, Glaucium, Papaver
Porcupines -
    New World → Coendou, Echinoprocta, Erethrizon
    Old World → Atherurus, Histrix, Hystrix
Porpoises → Neomeris
Prairie Dogs - see under Squirrels
Prawns - see under Crustaceans - Decapod
Praying Mantis (suborder Mantodea) → Hymenopus, Mantis
Predatory Insects → Empicoris
Prickly Thrift plants → Acantholienon, Fittonia
Primitive Insects → Machilis
Primrose plants (family Primulaceae) → Anagallis, Androsace, Cortusa, Cyclamen, Dodecatheon,
    Glaux, Hartmannia, Hottonia, Lysimachia, Naumburgia, Primula, Vitaliana
Protozoan animal (parasites) -
    Animal parasites → Myrtophyllum, Piroplasma, Pyroplasma, Pyrosoma, Sarcocystis
    Bee parasites → Nosema
    Cattle parasites → Babesia, Theileria
    Ciliate → Opercularia, Paramecium
    Dog/Horse parasites → Nuttallia
    Fish parasites → Myxobolus
    Mastigophorous → Opalina
    Minute/Solitary → Monas
    Opportunist parasites → Pneumocystis
    Rhabdophorinous → Lionotus
Psyll plants → Plantago
Punarnava (Punarnaba) plants → Boerhaavia

Purslane plants → Lewisia, Montia

Quassia plants - see under Trees - Simarubaceous
Quillwort plants → Isoetes

Rabbits - see under Hares
Racoons/Raccoons (family Procyonidae) → Bassariscus, Nasua
Rats - see under Rodents
Reptiles - see under Lizards, Salamanders, Snakes, or Turtles
Resinous plants -
    (Dammar) → Balanocarpus, Hopea, Shorea
Rhinoceros → Ceratotherium, Dicerorhinus, Diceros
Rhizopods (one-celled animals) → Plasmodiophora
Ricinuleid insects (Arachnids) → Cryptocellus
Rickettsial microorganisms (family Rickettsiaceae) → Cowdria, Coxiella, Dermacentroxenus,
      Ehrlichia, Grahamella, Haemobartonella, Neorickettsia, Rickettsia, Rickettsiella,
      Symbiotes
Rock garden plants → Capparis
Rockrose plants → Cistus, Fumana, Helianthemum
Rodents -
    Agouti → Dasyprocta
    Bandicoots -
        (Barred) → Perameles
        (Marsupial) → Choeropus
        (Rabbit) → Thylacomys
        (Short-nosed) → Isodon
    Beavers - see under Beavers
    Burrowing → Ammospermophilus
    Cavy/Paca → Cavia, Cuniculus
    Chinchilla → Chinchilla, Lagidium, Lagostomus
    Coypu → Capromys, Ctenomys, Geocapromys, Myocastor, Plagiodontia
    Crested Rats → Lophiomys
    Gophers (Pocket) → Geomys
    Hamsters → Cricetulus, Cricetus, Mesocricetus
    Hares - see under Hares
    Jerboa → Allactaga, Dipus, Jaculus
    Lemmings → Lemmus, Myopus
    Marmots - see under Squirrels
    Mice → Acomys, Apodemus
        (Banana/Tree) → Dendromus
        (Doormice/Dormice) → Dryomys, Eliomys, Glis, Graphiurus, Muscardinus
        (Field) → Microtus
        (Harvest/Diminutive) → Micromys
        (Jumping) → Eozapus, Napaeozapus, Sicista, Zapus
        (New Guinea) → Mayermys
    Muskrats/Musquashes → Ondatra
    Pacarana/False Paca → Dinomys
    Prairie Dogs - see under Squirrels
    Rabbits - see under Hares

Rats → Ctenodactylus
    (African Mole) → Bathyergus, Heterocephalus
    (Cotton) → Sigmodon
    (Kangaroo) → Dipodomys
    (Mole Rats) → Spalax
    (Old World) → Bandicota, Cricetomys, Mus
    (Old World/Black/Ind. Plague) → Epimys, Rattus
    (Pack/Trade/Wood) → Neotoma
    (Rice) → Oryzomys
    (Sand) → Gerbillus
    (Sand/Gerbils) → Meriones
    (Transmitting Lassa fever) → Mastomys
    (Water) → Crossomys, Neofiber
Shrews - see under Shrews
Squirrels - see under Squirrels
Voles (family Cricetidae) → Arvicola, Clethrionomys, Dolomys, Pitymys
Wombats → Lasiorhinus, Vombatus, Wombatula
Rose plants (family Rosaceae) → Agrimonia, Alchemilla, Amelanchier, Aphanes, Armeniaca,
    Dryas, Exochorda
(Agrimony) → Aremonia
(Almond, Peach, Plum, etc.) → Prunus
(Apple) → Pyrus
(Bitter Almond) → Amygdalus
(Cousso/Roe) → Brayera, Hagenia
(Crab Apple) → Malus
(Damask Rose) → Rosa
(Dropwort) → Filipendula
(Firethorn/Willowleaf/etc.) → Cotoneaster, Crataegus, Pyracantha
(Goatsbeard) → Aruncus, Spiraea
(Japanese Kerria) → Kerria
(Japanese Plum) → Eriobotrya
(Japanese Plum/Medlar) → Mespilus
(Mountain ash/Rowan) → Sorbus
(Quill) → Quillaja
(Quince) → Chaenomeles
(Quince Seed) → Cydonia
(Raspberry) → Rubus
(Scarlet Avens) → Geum
(Tormentil) → Potentilla
(Wild Strawberry) → Fragaria
Rotifer Metazoa → Callotheca, Conochilus, Euchlanis, Keratella
Rue plants (family Rutaceae) -
A medicinal → Toddalia
Armel/Harmel → Peganum
Bael/Citrus → Aegle
Buchu → Barosma
Burning Bush → Dictamnus
Caromy/Cusparia → Galipea
Citrus → Citrus, Poncirus

Jaborandi → Pilocarpus
Orange Blossom → Choisia
Prickly Ash → Zanthoxylum
Rue → Ruta
Rush plants → Hydrocleis, Juncus, Luzula

Salamanders (Lizards/Newts) (order Caudata) → Cryptobranchus, Cynops, Diemictylus,
        Euproctus, Hydromantes, Necturus, Taricha
    Blind Newts → Haideotriton
    Fire → Salamandra
    Giant → Dicamptodon, Ensatus, Megalobatrachus
    Lamper Eels → Amphiuma
    Mediterranean (western) → Pleurodeles
    Neotenic ("Larval") → Gyrinophilus
    Plethodontid → Bolitoglossa, Chiropterotriton, Desmognathus, Eurycea, Plethodon
        (Grotto/Ozark) → Typhlotriton
        (Texas Blind) → Typhlomolge
    Primitive → Hynobius
    Smooth Newts → Triturus
    Tiger → Amblystoma, Ambystoma
    Worm → Batrachoseps
Sapodilla Shrubs/Trees → Achras
Saxifrage plants (family Saxifragaceae) → Astilbe, Bergenia, Chrysosplenium, Hydrangea,
        Megasea, Saxifraga
Scabious/Teasel plants (family Dipsacaceae) → Dipsacus, Knautia, Scabiosa
Scale Insects - Coccidae (order Homoptera) → Laccifer, Lecanium, Orthezia
        (Armored) → Chionaspis
Schizomycetes (saprophytic microorganisms) → Podangium, Polyangium, Schizosiphon,
        Sorangium, Synangium, Vitreoscilla
Scorpions → Androctonus, Buthacus, Buthus, Centruroides, Hadogenes
        (Extremely venomous) → Leiurus, Tityus
        (False or Pseudo-scorpions) → Chelifer, Chthonius
        ("Huge") → Heterometrus, Pandinus
        (Micro-whip/Palpigrade) → Koenenia
        ("Midget") → Microbuthus
        (True) → Euscorpius
        (Water) → Nepa
        (Whip) → Hypoctonus, Mastigoproctus, Thelyphonus
Sea Anemones → Actinia, Actinodiscus, Adamsia, Aiptasia, Anemonia, Calliactis, Cereus,
        Corynactis, Dasyphyllia, Dendrogyra, Diadumene, Diploria, Discoma, Epizoanthus,
        Euptasia, Gyrostoma, Metridium, Rhodactis, Triactis
Sea Animals (Coelentrates) → Alcyonium
Sea-cows/Dugongs/Manatees/Sirens → Dugong, Hydrodamalis, Trichechus
Sea Cucumbers (Echinoderms, class Holothuroidea) → Cucumaria, Holothuria, Paracaudina,
        Pelagothuria, Psammothuria, Scotoplanes, Stichopus, Synapta
Sea Fans - see under Corals - Octocorals/Horny
Sea Firs - see under Hydras
Sea-Lavender plants → Armeria, Limonium
Sea Lilies/Feather-star Echinoderms (class Crinoidea) → Antedon, Bathycrinus, Heliometra,

Heterometra, Metacrinus, Promachocrinus

Sea Mats (Moss Animals → Bryozoa, Polyzoa) → Alcyonidium, Bugula, Crisia, Cristatella, Flustra, Membranipora

Sea Pens - see under Corals

Sea Slugs - see under Molluscs - gastropod

Sea Snails - see under Molluscs - gastropod

Sea Spiders → Nymphon

Sea Squirts - see under Chordate animals

Sea Urchins (Echinoderm animals) (class Echinoidea) → Asthenosoma, Centrostephanus, Colobocentrotus, Dendrogaster, Diadema, Dorocidaris, Echinocardium, Echinocyamus, Echinometra, Echinostrephus, Echinus, Encope, Heterocentrotus, Meoma, Sperosoma, Strongylocentrotus, Toxopneustes, Tripheustes

Sea Walnuts - see under Chordate animals

Seals/Pinnipeds -

    Eared (Fur) → Arctocephalus, Callorhinus, Eumetopias, Neophoca

        (Sea-lions) → Otaria, Zalophus

    Earless/True → Erignathus, Halichoerus, Hydrurga, Laptonychotes, Leptonychotes, Lobodon, Mirounga, Monachus, Phoca, Pusa

    Early (Hooded) → Cystophora

    Harp/Greenland → Pagophilus

    Walruses → Odobaenus, Odobenus

Sedge plants → Carex, Cyperus, Eleocharis, Eriophorum, Kobresia

Serendipity plants → Discoreophyllum

Sesame plants → Sesamum

Sharks - see under Fishes

Sheep → Bovidae

    Barbary → Ammotragus

    Domestic → Ovis

Shrews (primate Mammals) → Blarina, Crocidura, Cryptotis

    (Common) → Sorex

    (Elephant/Jumping) → Elephantulus, Macroscelides, Nasilio, Petrodomus, Rhyncocyon

    (Feather-tailed) → Ptiolcercus

    (Pigmy) → Microsorex, Suncus

    (Water) → Neomys, Potamogale

Shrimps - see under Crustaceans

Silverfish insects → Ctenolepisma

Siphonophore Coelenterates → Agalma, Muggiaea

Sipunculoid (worm-like marine) animals → Dendrostoma, Golfingia, Phascolosoma

Sirens (class Amphibia) → Siren

Skinks (Lizards) (Reptiles) → Abelpharus, Chalcides, Eumeces

Skunks - see under Badgers

Sloths -

    (Three-toed) → Bradypus

    (Two-toed) → Choleopus

Snails - see under Molluscs

Snakes (class Reptilia) -

    Adders/Vipers → Atractaspis, Bitis, Causus, Cerastes, Echis, Vipera

    Boa → Boa, Constrictor, Corallus, Epicrates, Eryx, Eunectes, Lichanura

    Cobras/Mambas/Kraits → Bungarus, Dendroaspis, Naja, Ophiophagus

Cobra-monil → Daboia
Colubrid → Boaedon, Boiga, Chrysopelea, Clelia, Coronella, Dasypeltis, Dispholidus,
      Heterodon, Lampropeltis, Licheteredon, Lycodon
    (Cat) → Telescopus
    (Coachwhip) → Masticophis
    (Corn) → Elephe
    (Grass) → Coluber
    (Grass/Ringed) → Natrix
    (Ladder) → Elaphe
    (Whip) → Dryophis
    (Water) → Herpeton
Copperhead Vipers → Agkistrodon, Ancistrodon
Death Adder → Acanthophis
DeKay's snake → Storeria
Elapid (Coral/Harlequin) → Elaps, Micrurus, Proteoglypha
Five-ringed snakes → Vermicella
Long-nosed → Ahaetulla
Ophid (Vine) → Oxybelis
Pit Viper → Bothops, Bothrops, Lachesis
    (Rattlesnake) → Crotalus
Pythons → Python
Rattlesnakes → Sistrurus, Solenoglypha
Ribbon → Thamnophis
Sea-snakes → Hydrophidus, Hydrophis, Laticauda, Pelamis
Shield-nose → Aspidelaps
Shovel-nose → Chionactis
Taipan → Oxyuranus
Thread/Worm/Blind burrowing snakes → Leptotyphlops, Typhlops
Tiger → Notechis
Snapdragon plants - see under Figwort plants
Soapberry/Cocoa plants → Paullinia, Xanthoceras
    (cacao) → Theobroma
Spiders -
    Communal/"Social" → Amaurobius
    Crab → Diaea, Misumena
    Funnel Web → Atrax, Trechona
    Daddy Longlegs → Pholcus
    Harvestmen → Opilio, Phalangium
    Highly colored → Gasterocanthus
    Jumping → Salticus
    Large → Heteropoda
        (Bird-eating) → Theraphosa
    Mite → Bryobia
    Orb Weaver → Nephila
    Orb Web → Araneaus, Araneus
    "Sun" → Galeodes
    Swamp → Dolomedes
    "Tarantula" → Avicularia, Dugesiella, Mygale
    Tarantula → Sericopelma

Tiny → Patu

Trapdoor → Liphistius

True → Argiope, Argyrodes, Eresus, Euctimsna, Garypus, Glyphesis, Glyptocranium,
  Grammostola, Hogna, Lactrodectus, Lasiodora, Loxosceles, Micrommata, Phoneutria,
  Pisaura, Tegenaria, Tetragnatha, Trochosa
  (American Tarantula) → Eurypelma

Water → Agyroneta, Argyroneta

Widow (Black) → Latrodectes, Latrodectus

Wolf → Alopecosa, Hysterocrates, Lycosa, Xerolycosa

Spiderwort plants (family Commelinaceae) → Tradescantia, Zebrina

Spiral microorganisms (family Spirillaceae) → Desulfovibrio, Paraspirillum, Selenomonas,
  Spirillum, Vibrio

Spiral-shaped microorganisms (family Athiorhodaceae) → Rhodospirillum

Spirally wound microorganisms (family Thiorhodaceae) → Thiospirillum

Spirochaetes → Borrelia, Entomospira, Leptospira, Microspira, Microspironema, Saprospira,
  Spirochaeta, Spironema, Spiroschaudinnia, Treponema

Spirotrich protozoans → Clevelandella, Epidinium

Sponges → Cordylophora

Ascon (Calcareous) → Clathrina, Leucosolenia

Calcareous → Cliona

  (Bath) → Euspongia, Spongia

  (Freshwater) → Euspongilla, Spongilla

  (Gelatinose) → Hymeniacidon

  (Gelatinose - siliceous) → Halichondria, Haliclona

  (Honeycomb) → Hippospongia

  (Marine) → Sycon

Freshwater → Ephydatia

Gelatinous → Myxilla, Oscarella

Glass → Hyalonema, Monoraphis

Horny → Cavochalina, Dactylochalina, Janthella

Marine → Aphrocalistes, Aphrocallistes, Calyx, Euplectella, Grantia, Hexactinella,
  Siphonochalina, Tedania

Siliceous → Desmacidon, Suberites

  (four-rayed) → Geodia

Sporozoan parasites → Lankesterella, Lankesteria, Leucocytozoon, Myxococcidium, Toxoplasma

Acephaline → Monocystis

Cephaline → Gregarina

Coccidian → Coccidium, Eimeria

Eucoccidian → Haemogregarina

Haemosporidian → Haemoproteus, Hemoproteus

Plasmodian → Laverania, Plasmodium

Sarcosporidian → Miescheria

Trypanosomes → Castellanella, Duttonella, Schizotrypanum, Tatusia, Trypanophis,
  Trypanoplasma, Trypanosoma, Trypanosomonas, Trypanozoon, Trypocastellanellae,
  Undulina

Springtail (Wingless) Insects → Anurida, Bourletiella, Holacanthella, Isotoma, Podura

Spurge plants (family Euphorbiaceae) → Dalechampia, Monadenium

  (Camala/Kamala) → Mallotus

  (Caoutchouc/Rubber) → Hevea

    (Cassava/Tapioca) → Manihot
    (Castor Oil) → Ricinus
    (Dwarf) → Euphorbia
    (Mercury) → Mercurialis
    (Ortiga) → Jatropha
    (Poisonous weed) → Tragia
Squirrels/Marmots/Prarie Dogs (family Sciuridae) -
    Flying → Glaucomys
    Ground (Chipmunk) → Eutamias, Tamias
    Ground (Prairie Dog) → Cynomys
    Marmots/Tree Squirrels → Arctomys, Marmota, Myosciurus
    Marmots (Fur-bearing) → Spermophilus
    Scaly-tailed (Flying) → Anomalurus
    Scaly-tailed (Flightless) → Zenkerella
    Tree → Sciurus
        (Flying) → Pteromys
        (Ground) → Citellus
    Tufted-ear → Nannosciurus
Starfishes (Echinoderms) (class Asteroidea) → Acanthaster, Asterias, Asterina, Astropecten,
        Coscinasterias, Echinaster, Eremicaster, Evasterias, Fromia, Gomophia, Heliaster,
        Linckia, Pycnopodia, Luidia, Marginaster, Marthasterias, Midgardia, Odontaster,
        Orester, Pisaster, Porcellanaster, Valvaster
Stick/Leaf Insects -
    (Indian/Oriental) → Carausius
    (Large) → Phobaeticus
    (Large/Tropical) → Pharnacia
    (Spiny) → Extastosoma, Heperopteryx
    (Walking - American) → Anisomorpha
    (Walking - Plasmid) → Diapheromera
Strawberry - see under Trees
Streptobacilli → Haverhillia
Suicide plants → Phyllanthus
Sundew plants → Aldrovanda, Drosera

Tamarins - see under Monkeys - Marmosets
Tenrecs - mammals related to the hedgehog and the shrew → Centetes, Dasogale, Hemicentetes
    (Hedgehog) → Ericulus
    (Rice) → Nesoryctes, Oryzorictes
Termites (order Isoptera) → Calotermes, Macrotermes, Nasutitermes, Zootermopsis
    (Diminutive) → Afrosubulitermes
    ("Magnetic") → Amitermes
    (Primitive) → Mastotermes
    (Subterranean) → Reticulitermes
Thecate Hydroids - see under Hydras
Thrips Insects → Heliothrips
Thyme/Mint/Sage plants (family Labiatae) → Cunila, Molucella, Prunella
    Basil-thyme → Acinos
    Bugle → Ajuga
    Calamint → Calamintha

(wild) → Clinopodium
Dead-nettle → Horminium, Lamium
Dragon-head → Dracocephalum
False Dragon-head → Physostegia
Germander → Teucrium
Gipsy-wort → Lycopus
Ground Ivy → Glechoma
Hemp-nettle → Galeopsis
Honey → Melissa
Horehound/Hoarhound → Marrubium
Hyssop → Hyssopus
Indoor → Coleus
Lavender → Lavandula
Lion's Ear → Leonotis
Majoram → Majorana
Motherwort → Leonurus
Peppermint → Mentha
Red/Golden Sage → Salvia
Rosemary → Rosmarinus
Sweet Basil → Ocimum
Sweet Marjoram → Origanum
Thyme → Thymus
Wild Cat-mint → Nepeta
Ticks -
Argasid → Ornithodoros
Argasid (Rabbits/Cattle) → Otobius
Bandicoot/Dog → Haemaphysalis
Bont, Amblyomma
Cattle → Boophilus, Hyalomma, Margaropus, Rhinencephalus, Rhinicephalus,
Rhinocephalus, Rhipicephalus
Cattle/Horses → Dermacentor, Dermatocentor, Otocentor
Chicken/Poultry → Argas, Neoschongastia
Geese → Holothyrus
Rat/Stable → Laelaps
Sheep → Ixodes, Melophagus
Toads -
Midwife (Bell) → Alytes
Tortoises - see under Turtles
Trees/Shrubs -
Acalypha → Acalypha
Acanthus → Daedolacanthus
Acerola - see under Cherry
Ailanthus - see under Trees - Simarubaceous
Alder → Alnus
Apocynaceous/Dogbane (family Apocynaceae) → Acocanthera, Acokanthera, Alstonia,
Catharanthus, Cerbera, Haemadictyon, Holarrhena, Lochnera, Nerium, Rauwolfia,
Strophanthus, Thevetia, Trachelospermum, Urechites, Vinca
Artocarpus (Upas) → Antiaris
Ash - see under Trees - Olive

Barberry → Berberis
Bayberry/Buckbean → Myrica
Bdellium (gum-resin) → Balsamodendron
Bean (Chinese) → Incarvillea
Bean (Indian) → Catalpa
Bean (Tonka) → Dipteryx
Beech → Castanea, Fagus, Lithocarpus, Nothofagus
Bindweed/Binewood/Convolvulus → Calystegia, Convolvulus, Cuscuta, Exogonium,
      Ipomoea, Operculina,
Birch/Ostrya → Alnus, Betula, Ostrya
Butterfly → Buddleia, Buddleja
Buttonweed → Cotula
Calumba/Tumeric → Coscinium, Jateorhiza
Caprifoliaceous → Symphoricarpus
Cashew (family Anacardiaceae) → Anacardium, Pistacia, Rhus, Schinus
      (Mango) → Mangifera
Cedar - see under Trees - Pine
Cherry/Acerola → Malpighia
Chinese Dove → Davidia
Cohosh/Papoose/Sqaw → Caulophyllum
Cornel/Dogwood → Aucuba, Benthamia, Chamaepericlymenum, Cornus, Cynoxylon,
      Helwingia
Crassula → Cotyledon
Currant (Grossulariaceae) → Ribes
Cycadaceous → Ceratozamia, Dioon
Cypress → Austrocedrus, Calocedrus
      (Incence Cedar) → Libocedrus
      (Juniper) → Juniperus
      (Sandarac) → Tetraclinis
      (Swamp) → Taxodium
Dammar/Copal → Dammara, Vateria
Dawn-redwood → Metasequoia
Dogbane - see under Trees - Apocynaceous
Elm → Ulmus, Zelkova
Euphorbicaceous → Anda, Stillingia
Ferns - see under Ferns (Polypody plants)
Fig - see under Trees - Mulberry
Gingko/Ginkgo/Maidenhair → Ginkgo
Gnetaceous shrubs → Ephedra
Grape - see under Vine plants
Guaiacum/Heartwood → Guaiacum
Guapi/Grape/Huapi Bark → Guaree
Gum (Gutta-percha) → Palaquium
Gum (Sour/Black) → Nyssa
Gymneman → Gymnema
Hazel → Carpinus, Corylus
Heartwood - see under Trees - Sandalwood
Heath shrubs (family Ericaceae) -
      (Andromeda) → Andromeda

(Azalea) → Loiseleuria
(Bearberry) → Arbutus, Arctostaphylos, Arctous
(Blueberry) → Vaccinium
(Calico/Laurel) → Kalmia
(Dwarf shrubs) → Chamaedaphne
(Fetterbush) → Leucothoe
(Heather) → Calluna, Erica
(Labrador Tea/Rosemary) → Ledum
(Moss Heather) → Cassiope
(Redvein) → Enkianthus
(Rhododendron) → Rhododendron
(Sorrel) → Oxydendrum
(Sweet Birch) → Gaulthoria
Hernanda → Hernanda
Honeysuckle - see under Honeysuckle plants
Ivy → Fatsia, Hedera, Kalopanax
Jasmine/Jessamine → Cestrum, Jasminium, Jasminum
Jute/Lime → Corchorus
Laurel → Cinnamomum, Laurus, Ocotea, Sassafras
Lime → Tilia
Machineel-yielding → Hippomane
Magnolia → Illicium, Liriodendron
    (Cucumber) → Magnolia
Maple/Sycamore → Acer
Medicinal → Protea
Melastomatous → Medinilla, Tibouchina
Mistletoe → Loranthus, Viscum
Mulberry → Dorstenia, Maclura, Morus
    (Jack-fruit) → Artocarpus
    (Fig) → Ficus
Myrobalan/Combretaceous → Terminalia
Myrtle (family Myrtaceae) → Callistemon, Calothamnus, Embelia, Eucalyptus, Feijoa,
    Metrosideros
    (Clove) → Caryophyllus, Eugenia, Jambosa
    (Common Myrtle) → Myrtus
    (Guava) → Psidium
    (Jamaica Pepper) → Pimenta
    (Jambul) → Syzygium
    (Loosestrife) → Lagerstroemia, Lythrum
    (Loosestrife - Henna) → Lawsonia
    (Tea) → Melaleuca
Nettle → Laportea
Nutmeg → Myristica
Oak → Quercus
Olacineous → Ximenia
Oleaster → Elaeagnus
    (Buckthorn) → Hippophae
Olive → Forsythia, Ligustrum, Olae, Osmanthus, Syringa
    (Ash) → Fraxinus

Palm → Carludovica, Licuala
    (Babassu/Pissaba Oil) → Attalea
    (Bdellium) → Borassus
    (Betel) → Areca
    (Cabbage) → Cordyline
    (Caranda/Carnauba) → Copernicia
    (Coconut) → Cocos, Lodoicea
    (Cocos/Weddel) → Microcoelum, Syagrus
    (Dwarf fan) → Chamaerops
    (Feathers) → Chamaedorea, Collinia
    (Feathers/Camedorea) → Neanthe
    (Indian) → Calamus
    (Oil) → Elaeis
    (Sago) - indoor → Cycas
    (Saw palmetto/Sabal) → Serenoa
Papaw/Papaya/Pawpaw → Asimina, Carica
Pea - see under Pea plants/trees
Philadelphian → Deutzia
Pine → Araucaria
    (Cedar) → Cedrus, Thuja
    (China-Fir) → Cunninghamia
    (Cypress) → Chamaecyparis, Cupressus
    (Larch) → Larix
    (Pine) → Pinus
    (Spruce) → Abies, Picea, Tsuga
Pineapple - see under Pineapple plants/trees
Pomegranate → Punica
Quassia - see under Trees - Simarubaceous
Quebracho → Aspidosperma, Schinopsis
Rose - see under Rose plants
Sandalwood/Heartwood → Eucarya, Santalum
Sandarac resin → Callitris
Sapodilla → Achras, Bassia, Madhuca
Sassafras → Atherosperma
She-oak → Casuarina
Silk-cotton → Ceiba
Silktree → Albizia
Simarubaceous/Quassia/Ailanthus → Ailanthus, Ailantus, Quassia, Simaruba
    (Jamaica) → Aeschrion, Picraena, Picrasma
Soapberry → Koelreuteria
Spindle → Euonymus
Spurge - see under Spurge plants
Sterculiaceous → Sterculia
Storax → Halesia, Styrax
Strawberry → Arbutus
Sweet-shrub → Calycanthus
Tamarisk → Myricaria
Tea → Camellia
    (African/Arabian) → Catha

      (Paraguay/Mate) → Ilex
      (Winterberry) → Prinos
    Tropical → Cananga, Dombeya
    Verbena/Vervain → Aloysia, Caryopteris, Clerodendrum, Lantana, Lippia, Verbena, Vitex
    Vine - see under Vine plants
    Walnut → Carya, Juglans
    Willow/Aspen/Poplar → Populus
    Winter Sweet → Chimonanthus
    Witch-hazel → Hamamelis, Liquidambar
    Yew → Taxus
    Zebra → Aphelandra
Trichostome Protozoans → Balantidium, Coelosomides, Colpoda
Trichostrongyline parasites → Oswaldocruzia
Trypanosomes - see under Sporozoan parasites
Tuatara Reptiles → Sphenodon
Tubocurarine plants → Chondrodendron
Turtles (Tortoises) -
    Freshwater (Terrapins) → Chrysemys, Clemmys, Emys, Kinosternon, Malaclemys,
        Mauremys
    Land → Gopherus, Kinyxis, Testudo
    Leathery/Leather-backed → Dermochelys
    Map (False) → Graptemys
    Marine → Caretta, Chelonia, Eretmochelys, Geochelone, Lepidochelys
    Short-necked → Pseudomydura
    Snake-necked → Chelodina, Chelys
    Snapping → Chelydra, Macrochelys, Macroclemys
    Stinkpot → Sternotherus

Valerian plants (family Valerianaceae) → Centranthius, Kentranthus, Nardostachys, Polemonium,
        Valeriana, Valerianella
Vampires - see under Bats
Verbena plants - see under Trees/Shrubs
Vine plants → Tinospora, Vitis
Violet/Pansy plants → Saintpaulia, Viola
Voles - see under Rodents

Walruses - see under Seals/Pinnipeds
Wasps -
    Braconid → Microgaster
    Common → Paravespula
    Cuckoo → Chrysis
    Digger → Ammophila, Cerceris, Crabro, Philanthus
    Eumenid → Odynerus
    Gall → Adleria, Andricus, Biorhiza, Cynips, Diplolepis, Neuroterus
    Hornets - see under Wasps - Vespid
    Horntail/Wood → Sirex
    Parasitic → Nasonia
    Potter → Eumenes
    Spider/Spider-hunting → Anoplius, Pepsis

True → Dolichovespula
Vespid (Hornet/Yellow Jacket) → Polistes, Vaspa, Vespa, Vespula
Wood → Urocerus, Xeris
Water-lily plants (family Nymphaeaceae) -
    (Cow-lily) → Nuphar
    (Lotus) → Nelumbium, Nelumbo
    (Royal) → Victoria
    (Spatterdock) → Niadvena
    (White) → Nymphaea
Water-milfoil plants → Myriophyllum
Water-plaintain plants → Alisma, Luronium, Sagittaria
Water-starwort plants → Callitriche
Waterwort plants → Elatine
Weasels - see under Badgers
Weeds -
    (Jimmey) → Aploppus
    (Pickerel) → Pontederia
    (Plantain) → Littorella
Weevil Insects → Balaninus, Hylobius
Whales -
    Beaked/Bottle-nosed/Toothed → Delphinapterus, Hyperoodon, Mesoplodon, Monodon,
        Ziphius
    Blue (Rorqual) → Balaenoptera
    Deep-diving → Berardius
    Grampus Whale/Dolphin Whale → Globicephala, Grampus, Orcinus
    Gray → Eschichtius
    Right → Balaena, Eubalaena, Neobalaena
    Rorqual (Humpback) → Megaptera
    Sperm → Physeter
Willow/Willow-herb plants → Ludwigia, Oenothera, Salix
Wingless Insects -
    (primitive) → Entomobrya, Lepisma, Lepismachilis
Wintergreen plants → Chimaphila, Moneses, Orthilia
Wombats - see under Rodents
Worms -
    Acorn - see under Chordate animals
    Annelids (Oligochaetes/Segmented Worms) -
        Ampharetid → Melinna
        Bamboo → Clymenella
        Chaetopterid → Chaetopterus
        Cirratulid → Cirratulus
        Earthworms → Aelosoma, Allolobophora, Aporrectodea, Digaster, Eisenia, Eiseniella,
            Glossoscolex, Lumbricus, Octolasium
        Earthworm (Giant) → Drawida, Megascolides, Microchaetus, Rhinodrilus, Spenceriella
        Enchytraeids → Enchytraeus
        Leeches (class Hirudinea) → Erpobdella, Glossiphonia, Haemadipsa, Haementaria,
            Haemopis, Helobdella, Hementaria, Hemiclepsis, Herpobdella, Heterodoxus, Hirudo,
            Limnatis, Macrobdella, Placobdella, Sanguisuga, Trachybdella
        Naidid → Stylaria

Oligochaete → Chaetogaster
Owenids → Owenia
Polychaete (Bristleworms) → Autolytus, Eulalia, Sabella
Polychaetes → Nepthys
Rag → Eunice, Hyalinoecia
Ragworms/Clam worms → Neanthes, Nereis
Scale → Arctonoe, Gattyana
Serpulid → Manayunkia, Pomatoceros, Protula
Syllid → Exogone
Tubificid → Tubifex
Velvet (Onychophorous) → Macroperipatus
Aphrodite → Aphrodite
Archoophorous → Macrostomum
Arthropod (Tongue worms) → Linguatula
Ascarid → Belascaris
Cestode → Amphilina, Axine, Gyrocotyle, Ligula, Mazocraes
Cooperid → Cooperia
Earthworms - see under Worms - Annelids
Echiuroid → Bonellia, Echiurus, Thalassema, Urechis
Feather-duster/Sabellid Fan → Bispira, Megalomma
Flat (Planarian/Tricladid) → Polycelis, Procerodes
Flat (Turbellarian) → Anonymus, Bothrioplana, Convoluta, Craspedella, Dendrocoelum,
      Dugesia, Monocelis, Planaria
Flukes - see under Worms - Trematode Worms
Gastrotrichs → Chaetonotus
Hair (Horsehair worms) (family Chordodidae) → Chordodes, Gordius, Parachordodes,
      Paragordius
Lobworms/Lugworms → Arenicola
Lungworms (sheep/goats) → Muellerius
Microscopic → Angiostrongylus
      (Rotifer) → Branchionus
Nematodes (Unsegmented Worms) (class Nematoda) → Aceraria, Agamomermis,
      Agamonematodum, Anguillula, Anguillulina, Aphelenchus, Ascocotyle, Eustrongylus,
      Gnathostoma, Gnathostomum, Habronema, Hepaticola, Heterakis, Lagochilascaris,
      Mecistocirrhus, Oesophagostomum
      (Capillary) → Capillaria
      (Eel) → Anguina
      (Filarial) → Mansonella, Setaria, Stephanofilaria
      (Filarial/Round) → Acanthocheilonema, Agamofilaria, Aprocta, Dipetalonema,
      Dirofilaria, Filaria, Litomosoides, Loa, Thalazia, Thelazia, Wuchereria
      (Filarial/Threadworms) → Gongylonema
      (Filiform) → Ostertagia
      (Hook) → Ancylostoma, Ankylostoma, Dochmius, Necator, Nippostrongylus, Uncinaria
      (In cats) → Protospirura
      (In cattle - Lung) → Protostrongylus
      (In dogs) → Spirocerca
      (In fowl & other birds) → Sclerostoma, Syngamus
      (Marine) → Nectonema
      (Marine/Round) → Placentonema

(Microfilaria) → Onchocerca, Onchocercus, Oncocerca, Oncocercus

(Microscopic) → Oxyspirura

(Minute/Rhabditoid) → Leptodera, Rhabditis, Rhabdonema

(Oxyurid rat parasites) → Symphacia

(Parasitic in swine) → Stephanurus

(Pin/Seat/Threadworms) → Enterobius, Oxyuris

(Prelarval stage) → Microfilaria

(Rhabditoid) → Turbatrix

(Round) → Amidostomum, Ascaris, Caenorhabditis, Cheyletiella, Dioctophyma, Dioctophyme, Hyostrongylus, Mermis, Metastrongylus, Moniliformis, Monochus, Nematodirus, Nochtia, Ollulanthus, Physaloptera, Strongyloides, Strongylus, Ternidens

(Round - Cats/dogs) → Toxocara

(Round - Cyst Eelworms) → Heterodera

(Round - Domestic fowls) → Trichosoma

(Round - Eelworms) → Ditylenchus

(Round - Guinea) → Dracunculus

(Round - Herb. animals) → Trichostrongylus

(Round - Large cats) → Toxascaris

(Round - Monkeys) → Tridontophorus

(Round - Plants) → Tylenchus

(Round - Pork-muscle) → Trichina, Trichinella

(Round - Rats) → Trichosomoides

(Sheep Wire) → Haemonchus

(Whipworm) → Trichocephalus, Trichuris

Nemertine (Ribbon) → Cephalothrix, Cerebratulus, Dichonemertes, Geonemertes, Lineus, Nemertes

Non-bursate → Physocephalus

Priapulid → Priapulus

Tapeworms (Merozoic Cestodes) (class Cestoda) → Amoebotaenia, Anoplocephala, Anthobothrium, Bertiella, Bothriocephalus, Calicotyle, Cenurus, Choanotaenia, Coenurus, Davainea, Dibothriocephalus, Diclodophora, Digramma, Diphyllobothrium, Diplogonoporus, Diplopylidium, Dipylidium, Disculiceps, Dracontium, Echinococcus, Fimbriaria, Haplobothrium, Hydatigera, Moniezia, Monodontus, Multiceps, Multidictus, Polygonoporus, Raillietina, Symplocarpus

(Beef) → Taenia, Taeniarhynchus, Taeniarhynus

(Bladder Worms) → Cysticercus

(Dwarf) → Diplacanthus, Hymenolepis

(Fringed) → Thysanosoma

Terebellid Worms → Amphitirte, Amphitrite, Lanice

Thorny-headed (Acanthocephalans) → Acanthocephalus, Echinorhynchus, Macracanthorhychus

Trematode Worms/Flukes/Flatworms (class Trematoda) → Agamodistomum, Alaria, Amphimerus, Amphistoma, Artyfechinostomum, Aspidogaster, Bilharzia, Bucephalus, Centrocestus, Cladorchis, Clonorchis, Coenogonimus, Cotylaspis, Cotylogaster, Cotylogasteroides, Dactylogyrus, Dicrocoelium, Diplozoon, Distoma, Distomum, Echinostoma, Fasciola, Fascioletta, Fascioloides, Fasciolopsis, Gastrodiscoides, Gastrodiscus, Gyrodactylus, Heterophyes, Loosia, Loxotrema, Macraspis, Mesogonimus, Metagonimus, Monochotrema, Monostoma, Monostomum, Multicalyx,

Multicotyle, Nanophyes, Nanophyetus, Opisthorcis, Ostiolum, Paragonimus, Paramphistomum, Plagiostomum, Polystoma, Polystomum, Prosthagonimus, Pseudamphistomum, Pseudodiscus, Schistosoma, Schistosomatium, Tetrastoma, Tracheophilus, Trichobilharzia, Troglotrema, Typhlocoelum, Zygocotyle

Tube worms/Tubicolous worms → Sabellastarte, Spirobranchus

Wort plants -

(St. John's) → Hypericum

Yams (Yam plants) → Dioscorea

Zebras → Hippotigris

Zorapterous Insects → Zorotypus

Zygadenous plants → Zygadenus